ANCIENT TEXTILES SERIES VOL. 8

TEXTILE TERMINOLOGIES

IN THE ANCIENT NEAR EAST AND MEDITERRANEAN FROM THE THIRD TO THE FIRST MILLENNIA BC

edited by

C. Michel and M.-L. Nosch

OXBOW | books
Oxford & Philadelphia

First published in 2010. Reprinted as a paperback in 2013 and 2021 by
OXBOW BOOKS
The Old Music Hall, 106–108 Cowley Road, Oxford, OX4 1JE

and in the United States by
OXBOW BOOKS
1950 Lawrence Road, Havertown, PA 19083

Paperback edition: ISBN 978-1-78297-391-1
Digital edition: ISBN 978-1-84217-753-2

A CIP record for this book is available from the British Library

The Library of Congress Cataloged the hardback edition as follows:

Textile terminologies in the ancient Near East and Mediterranean from the third to the first millennia BC / edited by C. Michel and M.L. Nosch.
p. cm. -- (Ancient textile series ; v. 8)
Includes bibliographical references.
ISBN 978-1-84217-975-8
1. Textile industry--Middle East--History--To 622--Terminology. 2. Textile industry--Mediterranean Region--History--To 1500--Terminology. 3. Bronze Age--Middle East . 4. Bronze Age--Mediterranean Region. 5. Middle East--Antiquities. 6. Mediterranean Region--Antiquities. I. Michel, Cécile. II. Nosch, Marie-Louise.
TS1413.5.T49 2010
677.00939'4--dc22
2010037610

For a complete list of Oxbow titles, please contact:

UNITED KINGDOM
Oxbow Books
Telephone (01865) 241249
Email: oxbow@oxbowbooks.com
www.oxbowbooks.com

UNITED STATES OF AMERICA
Oxbow Books
Telephone (610) 853-9131, Fax (610) 853-9146
Email: queries@casemateacademic.com
www.casemateacademic.com/oxbow

Oxbow Books is part of the Casemate Group

Front Cover: Artwork by Marianne Bloch Hansen, CTR, based on a shell mosaic from Mari (first half of the 3rd mill. BC) and various cloth logograms

---

Contents

1. Ole HERSLUND
2. Miguel Ángel ANDRÉS-TOLEDO
3. Benjamin FOSTER
4. Julia GALLIKER
5. Agnete WISTI LASSEN
6. Anna MICHAILIDOU
7. Michael WEIGL
8. Hartmut WAETZOLDT
9. Judy CRAIG
10. Jana JONES
11. Matteo VIGO
12. Lorenzo VERDERAME
13. Susanne LERVAD
14. Agnes GARCIA VENTURA
15. Cécile MICHEL
16. Klaas R. VEENHOF
17. Giulia BACCELLI
18. Marie-Louise NOSCH
19. Joanna CUTLER
20. Francis JOANNÈS
21. Stefan ZAWADZKI
22. Rachel FENTON
23. Sophie DESROSIERS
24. Eugenio R. LUJÁN
25. Juan-Pablo VITA
26. Eva ANDERSSON STRAND
27. Françoise ROUGEMONT
28. René LEBRUN
29. Agnès DEGRÈVE
30. Maria Giovanna BIGA
31. Margarita GLEBA

Absent from photo:
Anne-Claude BEAUGEARD
Catherine BRENIQUET
Mogens Trolle LARSEN
Maurizio DEL FREO
Pascaline DURY
Franco POMPONIO
Pierre VILLARD

Acknowledgements and research frameworks for the investigation of textile terminologies in the 3rd and 2nd millennia BC

The concept of the present volume and the exploratory workshop emerged in 2005 as a collaboration between us. That year, in Copenhagen, the Danish National Research Foundation's Centre for Textile Research launched its research programme with the aim to investigate Bronze Age textile production from an archaeological, experimental and linguistic point of view.[1] It encompasses two parts: *Tools and Textiles* (2005–2009) and *Texts and Contexts* (2007–2010). In the first part, *Tools and Textiles,* the mission was to gather information on textile tools from various types of Bronze Age contexts and sites in the Eastern Mediterranean area. This provided a new methodology for textile tool studies. Via tool studies, context studies, and experimental testing, the programme provided a clear picture of types and qualities of textiles, which derive from the tools. The second part, *Texts and Contexts* (2007–2010) focuses on the written records of the Eastern Mediterranean area in the 3rd and 2nd millennia BC, where we have references to a complex terminology of textiles, tools and techniques, decoration and specialised textile occupational titles. However, we often lack their precise meaning. The aim is thus to investigate textile terminology diachronically, and in a comparative approach. This stage also profits from the knowledge of textile quality and types gained from the typology research in the *Tools and Textiles* part.

In Nanterre, another research programme *The economy of wool in the ancient Near East* was also launched in 2005 by the team "Histoire et ARcheologie de l'Orient Cunéiforme" (HAROC), which belongs to a very large CNRS laboratory whose main topic is archaeology (Archéologies et Sciences de l'Antiquité).[2] The team is composed of both archaeologists and philologists who work on the different ancient Near Eastern cultures over a chronological time span defined by the use of cuneiform writing. One of the research themes is the Mesopotamian wool and textile economy. Within the framework of this research programme, several aspects are studied, such as, the manual treatment of wool, processing and manufacture of textiles, wool production, wool and textile trade, commercial structures, and the use and function of textiles. The research program is pluridisciplinary: the integrated collaborations and the association of data from different periods demonstrate several constant characteristics and allow isolating peculiar phenomena from the general developments.

Our aim was to interlink the French and Danish research programmes and to exchange knowledge. The scientific exchange was facilitated by the formal convention of collaboration between the Danish National Research Foundation (DNRF) and the Centre National de la Recherche

[1] http://ctr.hum.ku.dk/research/tools/

[2] http://www.mae.u-paris10.fr/arscan/-ArScAn-Histoire-et-Archeologie-de-.html.

Scientifique (CNRS), as well as a generous subvention from *La Mission de Coopération Scientifique et Universitaire*, Ambassade de France in Copenhagen, for which we are most grateful.

The European Science Foundation was our main sponsor of the exploratory workshop held in Copenhagen, March 2009. In addition, other institutions have graciously helped: The Italian Institute of Culture in Copenhagen generously sponsored the travels and stay in Copenhagen for Maria Giovanna Biga. Benjamin Foster was invited by the University of Copenhagen within the framework of the International Alliance of Research Universities, and he was also a guest of the Centre of excellence directed by Prof. Kim Ryholt, *Centre for Canon and Identity Formation in the Earliest Literate Societies,* University of Copenhagen. The travel and hotel costs of the French participants, as well as Marie-Louise Nosch's travels to Paris in autumn 2009 were defrayed by the convention between the DNRF and the French CNRS.

We are sorry that the expert on Mari textiles, J. M. Durand, did not attend the conference or contribute to this volume. Fortunately, his excellent monograph, *La nomenclature des habits et des textiles dans les textes de Mari*, appeared in spring 2009, and the authors therefore had the opportunity to integrate his results in their written contributions. Furthermore, we regret that the research on Hittite wool and textiles by Agnès Degraeve[3] and René Lebrun could not be published in the present volume.

The European Science Foundation is the main sponsor of the present volume, with additional support from The Danish National Research Foundation and the CNRS. It is our hope that this endeavour is a first step towards further collaboration and that it will open the way to even larger and longer projects.

June 2010
Cécile Michel (Nanterre)
Marie-Louise Nosch (Copenhagen)

[3] This research will be submitted as a PhD thesis in Louvain la Neuve.

Textile Terminologies

Cécile Michel and Marie-Louise Nosch

"Words survive better than cloth", writes textile scholar Elizabeth Barber in her monograph *Prehistoric Textiles*.[1] This is certainly true for the period under investigation, the 3rd to the 1st millennia BC, and for the ancient Near East and the Mediterranean where textiles rarely survive, with the notable exception of Egypt.[2] The richness and varieties of textual documentation, however, constitute a unique source of information of the ancient textiles, their production and consumption in these areas. Various scholars have over the years investigated this rich textile terminology data in comprehensive works on the role of textiles in ancient societies,[3] or in individual studies on single corpus terminologies;[4] here, for the first time, we attempt a comparative and diachronic study of ancient textile terminologies.

1. Chronological and geographical areas covered

The geographical and chronological framework for the present investigation in the Eastern Mediterranean and the Near East, focused on the period from the 3rd to the 1st millennia BC. During the 3rd millennium in Mesopotamia,[5] textile production developed from household production to standardised, industrialised, centralised production, on the basis of a division of labour. Sheep developed a white coat/wool through selective breeding,[6] wool was integrated into textile production as an alternative to plant fibres, which then provided the dynamics for the development of felting,[7] fulling, dye industries, colour extraction and intensive use of colour symbolism in dress and textiles.[8] Within this area we also have the development of palace economies and administrations, inscriptions with extensive records on production management, tools, glyptic, frescoes and relief iconography in which various types of dress are visible.

[1] Barber 1991, 260.
[2] Vogelsang-Eastwood 1999; Kemp & Vogelsang Eastwood 2001.
[3] Barber 1991; Gillis & Nosch (eds.) 2007; Breniquet 2008; Völling 2008; Burke 2010.
[4] Veenhof 1972; Waetzoldt 1972; 1981; Ribicini & Xella 1985; Van Soldt 1990; Archi 1999; Pasquali 1997; Barber 2001; Zawadzki 2006; Pomponio 2008; Durand 2009.
[5] Breniquet 2008; 2010.
[6] Ryder 1983.
[7] Burkett 1979.
[8] Cardon 2007; Alfaro & Karali 2008; Singer 2008.

The contributions analyse and discuss the parameters for the development of textile terminologies in these areas and periods. The textual analyses reveal how terms for tools, technology and textiles developed over the millennia to meet new demands. In the quasi-absence of Bronze Age archaeological textile remains, it is necessary to join forces and combine specialist knowledge, not only from the region itself, but also from elsewhere, such as in the Scandinavian experimental archaeological tradition,[9] textile expertise and tool studies from other areas,[10] and comparative linguistic explorations of how terminologies develop within a defined technical field.[11]

2. Sources, texts and language families

This volume contains studies of textile terminologies in the Semitic and Indo-European languages. In addition, the authors combine their analyses with data from other fields of research such as archaeology, which can yield information about textile remains,[12] imprints of textiles on clay,[13] mineralised textiles on metal objects, or textile tools.[14] Another rich source of information is iconography,[15] while other scholars include results from ethnographic studies[16] or experimental textile archaeology.[17]

The texts preserved from the Bronze Age and Early Iron Age in the eastern Mediterranean and ancient Near East are of a particular nature: each document has a specific function, and accordingly the data about textiles vary a great deal, depending on the category to which a document belongs. Some cuneiform documents are official texts, written for example for the king: accounts of royal victories, descriptions of the king as the builder of monuments and his cultic activities, and in such documents descriptions of luxurious textiles may occur; such types of textiles also occur in the cultic activities as gifts offered to the gods or to their statues.

Another category comprises texts describing economic and daily activities. This includes palace management of textile production (employees, production), accounts from large weaving workshops, rations for the textile workers, the administration and organisation of textile manufacture,[18] or the quantity and quality of wool needed or allocated.[19] The entire Linear B documentation belongs to this category.[20] However, despite the accuracy and details, such accounts rarely inform us about textile techniques or about the use of textiles.

A third category of texts, particularly well attested in the cuneiform corpus, consists of the diplomatic correspondence with its lists of gifts between royal courts among which are often textiles and clothes.[21]

---

9 Peacock 2001; Andersson & Nosch 2003; Andersson *et al.* 2008; Nosch forthcoming.

10 Hoffman 1964; Andersson 2003; 2010; Andersson & Nosch 2003; Gleba 2008.

11 Dury & Lervand 2010.

12 Recent archaeological textiles published in Frangipane *et al.* 2009; Andersson *et al.* 2010; Andersson Strand & Nosch (eds.) forthcoming.

13 Adovasio 1975/77; Möller-Wiering 2008.

14 Andersson *et al.* 2008; Mårtensson, Nosch, Andersson Strand 2009; Breniquet 2008.

15 Strommenger 1980/83; Barber 1991; Breniquet 2008; 2010; Foster 2010.

16 Hoffman 1964; Desrosiers 2010.

17 Andersson 2003; Andersson & Nosch 2003.

18 Waetzoldt 1972; Biga 2010; Pomponio 2010; Verderame 2008.

19 Waetzoldt 2010.

20 Del Freo, Nosch, Rougemont 2010; Luján 2010

21 Moran 1987; Lerouxel 2002; Biga 2008.

Finally, another rich category of textile related texts is the private archives documenting trade, daily use etc. The best example is the private correspondence between Assyrians trading textiles in Anatolia and their wives who wove at home in Aššur.[22]

Whereas in some languages, there is only one word to designate a type of fabric or material, other languages have developed – or preserved – a richer vocabulary. For example, for the primary textile plant fibre, modern English and German have two different words: "flax" (Engl.) and "Flachs" (Germ.) for the plant, and "linen" (Engl.) and "Leinen" (Germ.) for the cloth, whereas in French just "lin" is the term used for both the plant and the cloth. This recalls the situation in English in which there is a word for the living animal and another for its meat (e.g. "cow" / "beef"). Such parallel terms may reflect various situations, but we can only understand them if we combine linguistic, archaeological and technical knowledge. When the terminological enquiries, technical analyses of tools and archaeological textiles are woven together with the historical, ethnographical, anthropological knowledge and theoretical frameworks, the results yield not only stimulating perspectives but also new knowledge about textile production and its place in ancient societies.

3. Topology of textile terminologies

The textile terminology of the modern era testifies to trade routes, trends and traditions. We employ textile terms with multiple meanings. "Jeans" are garments from Gênes, Genoa; "denim" designates cloth "de Nîmes", from southern France, an area in which woad was cultivated, processed and used in the large scale dyeing manufacture of blue cloth. Generally speaking, such topographical indications are often employed to designate textiles. A 20th-century AD example of this is the artificial fibre "dederon" developed in the German Democratic Republic (DDR) as a copy of nylon and named after the acronym of DDR.[23] Likewise, the present volume reveals the crossing, development and exchange of textile terms between eras, areas, and cultures of the past.

Words change according to languages, but also to geography and chronology. In the cuneiform documentation, each dialect, each population has developed a specific vocabulary for textiles, which seems typically local. Despite geographical proximities or linguistic and etymological connections, communities in places such as Ebla, Mari and Aššur seem to have created their own textile vocabularies.[24] There are, nevertheless, terms which can be traced over wide geographical areas and through the millennia: The Greek word for a long shirt, *khiton*, *Ki-to* in Linear B, derives from a Semitic root, *ktn*. But the same root in Akkadian means linen, in Old Assyrian a garment made of wool, and perhaps cotton, in many modern languages. The Indo-Iranian and Indo-European linguistic reconstruction can contribute to identify the textile terminology which existed before Indo-Iranian was divided into the Indian and the Iranian language groups: some Old Indian and Old Iranian textile terms can be traced back to Indo-Iranian; Indo-Iranian words are furthermore connected to Indo-European textile terminology.[25]

These examples illustrate on the one hand how related some textiles terms are across time and space, but they also show how very carefully we must conduct the etymological and

[22] Veenhof 1972; Michel 2001; 2006; Michel & Veenhof 2010; Wisti Lassen 2010.

[23] Lehmann 1995.

[24] Ebla: Biga 2010; Mari: Durand 2009 and Beaugeard 2010; Aššur: Michel & Veenhof 2010.

[25] Andres-Toledo 2010.

terminological enquiry with constantly changing semantics as the common thread. Moreover, within a specific corpus such as the Neo-Babylonian, the same term was used for very different types of clothing.[26]

4. Textile terminologies and technologies: a methodology

In the field of textile terminology, classifications, concept systems and term collections usually include first the fibres, and then the yarns and the structures such as weaving or knitting.[27] As a large number of weave derivatives and variations can be created, it is almost impossible to find terms for each of them, and even more complicated to translate them from one language to another. Part of the solution to this problem resides in the use of non-verbal representations. The origin and use of a fabric cannot be represented easily by using graphic components, but the characteristics of form, structure and colour can conveniently be represented graphically. This solution is employed today in the modern textile industry and trade, and was also used in ancient societies, for example, in the form of logograms in Linear B.[28] Likewise, in Egyptian hieroglyphs, the "textile" category includes artefacts, verbs, adjectives and also expressions, which (today at least) seem foreign to the concept of textiles.[29]

Textile classification worldwide may use various criteria; one of them is the logic of the craft.[30] Another angle of approach is the functionality of textile tools, which outline and determine the technical possibilities of Aegean Bronze Age textile tools and thus the functional terminology.[31] The research on functionality is based on textile expertise, tool studies and the experimental testing of textile tools.[32] The tool studies, context studies, and experiments enable an assessment of the types and qualities of textiles, which derive from the tools.

Finally, the concept of *chaîne opératoire*, inspired from anthropology and archaeology, is a valid approach to textile production, and was also the theme of a workshop convened in Nanterre in 2007 on the topic of production systems of textiles.[33] Catherine Breniquet's recent monograph on weaving in Mesopotamia has introduced this concept in Mesopotamian iconography and a new reading of cylinder seal iconography along the processes of the textile production has been proposed.[34] This new reading of the proto-dynastic iconography seems to convey a much more realistic image than previously assumed; it is possible to see who weaves, and for what: the entire society is involved in weaving. These depictions may be used for their documentary significance but keeping in mind that they are not those of a hand weaver's manual. We are clearly within the symbolic world. The beginning and end of the weaving process as spinning and weaving, stretching and folding, which could be a metaphor for human life,[35] or two different activities

[26] Zawadzki 2010.
[27] Dury & Lervad 2010
[28] Del Freo, Nosch, Rougemont 2010.
[29] Herslund 2010.
[30] Desrosiers 2010.
[31] Andersson *et al.* 2008; Mårtensson, Nosch, Andersson Strand 2009; Frangipane *et al.* 2009.
[32] Andersson *et al.* 2008; Andersson Strand 2010.
[33] Breniquet ed. forthcoming. See also Lackenbacher 1982 for a text on textile finishing, and Joannès 1984 on the organisation of crafts.
[34] Breniquet 2008 and reviews by Michel 2008 and Biga 2009.
[35] On weaving as a metaphor for destiny, see Lyle ed. 2004 with a collection of papers dealing with the metaphorical

related to a cyclic perception of the year and time in which daily and gendered activities occur: churning and weaving, ploughing and weaving, etc. These scenes are parts of more complex systems, like series as they often appear to be combined in linear but not logical compositions. We can conclude that a quite different picture of weaving can be drawn and used for comparative perspectives, where sources are not in conflict and where iconography and archaeology can finally be linked with epigraphy.

The Akkadian period, with its closely dated works of art in which clothing plays a prominent role, as well as its rich administrative archives dealing with textiles and clothes, offers therefore a particularly rewarding opportunity to correlate visual and written evidence for continuity and change in fashion during this dynamic period of Mesopotamian history.[36] In a similar manner, the linen lists from the earliest Egyptian dynasties can be compared with the available archaeological textile data and this can shed new light on their interpretations.[37] Technology can also be used for the interpretation of the linguistic evidence, deriving from a practical knowledge from experimentations. Such practical knowledge is indeed a key for the understanding of the indications of the precise amounts and weight of warp and weft yarn as they are in some texts in the Ur III documentation.[38]

5. Specific methodological problems related to textile terminologies

In an investigation of textile terminologies, we encounter several fundamental difficulties when aiming at identifying a term with a tangible item or a technical reality.

The first difficulty is that textiles rarely survive in the archaeological context and thus we have no preserved tangible remains – in museums or in the hands of archaeologists – to target identification. More fortunate situations are when identifying terms for pottery such as the two-handled cup, *depas amphikypellon*, attested both in Homer's epics and in numerous specimens in Aegean museums; a similar situation is when we need to identify the names for plants. In these cases we may be able to verify an assumption by consulting an archaeologist or a palaeobotanist. The material culture sets up a defined range of possibilities and a framework in which we should search for correlations.

Another difficulty is to identify terms within a technology, which is completely foreign to us today. Basic textile knowledge, understanding of techniques, evaluations of possibilities and plausibilities, distinctions such as the fundamental difference between tabbies and twills, these no longer form part of acquired general knowledge among scholars. Furthermore, we hardly possess knowledge of textile terms in our modern languages, or master textile techniques.

An example of the difficulties in understanding and interpreting ancient textile terminology is the term *mazrum* attested at Mari. According to J.-M. Durand, "la laine *mazirtum napisṭum* doit être celle dont le fil a été tordu par simple cardage".[39] There are precise philological, lexical and etymological reasons for this translation. However, in terms of textile techniques, it remains

meaning of weaving in various cultures. This aspect is also discussed in Vogelsang 1986 and Pasquali 2010.

[36] Foster 2010.

[37] Jones 2010.

[38] Waetzoldt 2010.

[39] Durand 2009, 143, 600.

obscure: a thread cannot be twisted by carding; carding does not exist in the Bronze Age where wool fibres are instead combed or treated with a thistle.

Another difficulty is the fact that textile terms appear primarily in lists and inventories without pertinent data about the nature of the textiles. The aim of such lists is not to qualify the textiles (their quality could provably be verified by sight and touch in the storeroom). Instead they register the number, the recipient and sometimes the price of the textiles.[40]

6. Origins and textile categories of textile terminology

In some languages and cultures, textile terminology developed according to materials, in others, according to topography, techniques, colours, qualities, function and usage. The term "undergarment" indicates function and shape; "blue-collar" indicates colour, usage and social context; "lining" is not directly derived from "linen" but from Latin *linea* meaning a "linen thread, string, line"; French "soie de Chine" indicate fibre type and topography, just like the East German nylon type fibre "dederon". One of the most productive terminological Bronze Age categories for textiles seems indeed to be topology. However, this is perhaps also due to the fact that this topological category is the easiest for modern philologists to identify.

The exact meaning of the topographical indications connected with textile terms is not easily understood. Textiles are "from Akkad" or "Akkadian" in the Old Assyrian documentation, and this opens up the debate about whether the geographical designation indicates origin, place of production, or certain characteristics such as weave or decoration.[41] In Linear B, groups of female textile workers and their children are designated by Anatolian toponyms outside the Mycenaean palace area, and again we must ask whether they come from these places, were purchased or kidnapped at these places, or whether these women and children produce textiles of a quality which is typical for these places.[42]

The textile terminology thus develops and changes according to languages, but also to time and place; despite the overarching developments, textile terminologies are created locally and acquire their specific meanings within a limited area. In the Linear B documentation, we can furthermore investigate textile terminology on a personal level: palace scribe 103 at Knossos has a distinct handwriting and his records can be identified including his usage of the textile terminology.[43] It is, for example, his personal preference to classify textiles from previous years as *pa-ra-ja*, 'old', while his fellow scribes chose to designate such textiles with the term *pe-ru-si-nwa*, 'from the previous year'.[44] The two designations are employed as synonyms and depend entirely on personal style, and can therefore also form a defining feature for the identification of a scribal hand.

7. The nature and function of the items recorded in the texts: textiles or garments?

In recent years, several studies of ancient clothing have been published, in particular the clothing worn by rulers and the elite.[45] The majority of texts, however, do not clearly indicate the type or

---

40 Pomponio 2008; 2010; Vita 2010; Nosch 2006.

41 Michel & Veenhof 2010.

42 Chadwick 1988; Nosch 2003.

43 Luján 2010.

44 Killen 1972.

45 Biga 1992; Pasquali 2005; Sallaberger 2009.

quality or whether the item is a piece of textile or a piece of clothing. The issue of problematic generic translations such as "a garment", "a cloth" or "textile" and the nature – textile or garment – is addressed and discussed by several authors. Many of them reach the conclusion that the various Bronze Age archives record untailored fabrics rather than tailored ready-to-wear costumes. This again raises the question of how to define a garment, in a world of kilts, cloaks, capes, wrap-around garments, and a habit of using complex devices for attachments. Sewing often seems useless and tailoring a waste of resources.[46] A way to address the issue is to combine texts and iconography: we find types of wrap-around garments and togas in the Sargonic iconography and texts.[47] In Ur III, two different terms for textiles are used side by side mixing a piece of clothing with a type of weave.[48]

It should not be forgotten that textile is not only used for clothing:[49] In palace archives as Ebla and Mari, besides garments, the administrators also deal with large amounts of textiles for furnishing.[50] Furthermore, it must taken into account that a majority of the written documentation deals only with luxurious textiles and do not give a complete overview of the many types of textiles used in antiquity.[51] Or when they do, the data are very precise for luxurious garments but remain quite vague for the clothes of ordinary people.[52]

8. Colour indications: dyed textiles or the natural pigmentation of wool, or both?

This question is raised by several scholars, in particular inspired by the attestation of the term "multi-coloured" in various languages and cultures: In Linear B *po-ki-ro-nu-ka*, 'with multi-coloured fringes';[53] Numerous multi-coloured (Sum. gùn-a) textiles are mentioned in the texts from the royal estate of Garšana;[54] and in the Neo-Assyrian texts the standardised description of textiles as *lubulti birme u kitû*, 'multi-coloured textiles and linen textiles' occurs frequently.[55]

Furthermore, the recurrence of fabrics described as white, dark/black, and red/brown leads to the discussion of the available resources of both dyed and naturally pigmented wool. Several scholars come to the conclusion that the bulk of fabrics recorded with colour indications may possibly have been naturally pigmented.[56]

In the Ur III documentation, the natural colour of wool and clothing was light and white. Occasionally the wool of animals with various naturally pigmented wool hues was used to achieve colour effects. Generally, however, wool and textiles were only dyed in exceptional cases.

Colours are deliberately used to express status and symbolic meaning. Shining, yellow-dyed clothing was reserved for the king.[57] Colours of textiles bear a symbolic and ritual value, thus

---

[46] Wees 2005.
[47] Foster 2010.
[48] Vogelsang 1986; Waetzoldt 2010.
[49] Waetzoldt 2007.
[50] Durand 2009; Beaugeard 2010; Pasquali 2010.
[51] Vigo 2010.
[52] Joannès 2010.
[53] Del Freo, Nosch & Rougemont 2010.
[54] Waetzoldt 2010.
[55] Villard 2010.
[56] Nosch 2004; Waetzoldt 2010.
[57] Waetzoldt 2010.

in Ebla we find black textiles for purification rituals after death.[58] In the Hittite documentation many luxurious linen textiles are blue,[59] which can only be obtained through dyes containing indigotin, probably from plants, or, alternatively, purpurin from murex.[60] In the Neo-Assyrian corpus the red colour dominates; but here again, it is primarily valuable textiles that are quoted in the documentation.[61]

* * *

These overarching themes and classification frameworks for terminologies are relevant for most languages and cultures of the 3rd to the 1st millennia BC and even beyond. Textile terms indicate origin, material, techniques, at least in their first stage. With time, and over longer distances, these meanings then become blurred or fade, or the terms acquire a new meaning appropriate to a new context. Furthermore, textile terminology seems closely linked to expressions for destiny, cosmology and myths. The Indo-European root **u̯es-* "to dress" was also used in Indo-European poetic formulas, for example **u̯esº u̯es-* "to dress a dress", and applied to gods who dressed the sky.[62]

There is no doubt that textiles generate a comprehensive vocabulary via the development of technologies and the emergence of specialised occupations and division of labour. The costume development and experimenting with wrapped clothing, fibulae, fixation devices, and tailored garments generate yet new terms for the clothing elements, and for the ensemble and combination of such elements.

The present survey includes textile terminologies in various languages and cultures but it also demonstrates the need to carry this investigation further. Diachronic studies and interdisciplinary approaches are the only viable way to continue this endeavour. In a future perspective, it would be interesting to review the relationship between textile terminology, textile production and labour, in continuation of the 1987 publication *Labor in the Ancient Near East.*[63] Furthermore, gender in production and costume use should be further explored. The interaction and cross-craft aspects between textile terminologies and terminologies in other crafts would also be a stimulating approach in a future study.

Bibliography

Adovasio, J. (1975–1977) The Textiles and Basket Impressions from Jarmo. *Paléorient* 3, 223–230.

Alfaro, C. & Karali, L. (eds.) (2008) *Purpurae vestes II: Vestidos, Textiles y Tintes: Estudios sober la produccion de bienes de consumo en la antiguidad*, Valencia.

Andersson, E. (2003) *Tools for Textile Production from Birka and Hedeby*. Birka Studies 8.

Andersson, E. & Nosch, M.-L. B. (2003) With a Little Help from my Friends: Investigating Mycenaean Textiles with the help from Scandinavian Experimental Archaeology. In Foster, K. P. & Laffineur, R. (eds.) *METRON. Measuring*

[58] Biga 2010.
[59] Vigo 2010.
[60] Cardon 2007.
[61] Villard 2010.
[62] Andres-Toledo 2010.
[63] Powell (ed.) 1987.

the Aegean Bronze Age. Proceedings of the 9th International Aegean Conference / 9ᵉ Rencontre égéenne internationale, Yale University, 18–21 April 2002. Aegaeum 24, 197–205 and table XLV.

Andersson, E. Mårtensson, L., Nosch, M.-L., Rahmstorf, L. (2008) New Research on Bronze Age Textile Production, *Bulletin of the Institute of Classical Studies* 51, 171–174.

Andersson, E., Felucca, E., Nosch, M.-L., Peyronel, L. (2010) New Perspectives on the Bronze Age Textile Production in the Eastern Mediterranean. The First Results with Ebla as a Pilot Study. In P. Matthiae, F. Pinnock, L. Nigro, N. Marchetti (eds.) *Proceedings of the 6th International Congress on the Archaeology of the Ancient Near East, Rome 5–10 May 2008*, 159–176.

Andersson Strand, E. (2010) The Basics of Textile Tools and Textile Technology: From fibre to fabric. In Michel & Nosch 2010, 10–22.

Andersson Strand, E. & Nosch, M.-L. (eds.), forthcoming. *Tools, Textiles, and Contexts. Investigations of Textile Production in the Bronze Age Eastern Mediterranean.* Ancient Textiles Series. Oxford.

Andrés-Toledo, M. A. (2010) Some Considerations about Vedic, Avestan and Indo-Iranian Textile Terminology. In Michel & Nosch 2010, 430–444.

Archi, A. (1999) Clothes in Ebla. In Y. Avishur & R. Deutsch R. (eds.) *Michael. Historical, Epigraphical and Biblical Studies in Honor of Prof. Michael Heltzer*, Tel Aviv-Jaffa, 45–53.

Barber, E. J. W. (1991) *Prehistoric Textiles. The Development of Cloth in the Neolithic and Bronze Ages with Special Reference to the Aegean.* Princeton.

Barber, E. J. W. (2001) The Clues in the Clothes. Some Independent Evidence for the Movement of Families. In R. Drews (ed.) *Greater Anatolia and the Indo-Hittite Language Family. Papers presented at a Colloquium hosted by the University of Richmond, March 18–19, 2000.* Journal of Indo-European Studies, Monograph Series 38, 1–14.

Beaugeard, A.-C. (2010) Les textiles du Moyen-Euphrate à l'époque paléo-babylonienne d'après un ouvrage récent. In Michel & Nosch 2010, 283–289.

Biga, M. G. (1992) Les vêtements neufs de l'Empereur. *Nouvelles Assyriologiques Brèves et Utilitaires*, 19.

Biga, M. G. (2008) *Au-delà des frontières: guerre et diplomatie à Ébla.* Orientalia, 77, 289–334.

Biga, M. G. (2009) Review of Breniquet 2008. *Mesopotamia* XLIV, 177–179.

Biga, M. G. (2010) Textiles in the Administrative Texts of the Royal Archives of Ebla (Syria, 24th Century BC) with Particular Emphasis on Coloured Textiles. In Michel & Nosch 2010, 146–172.

Breniquet, C. (2008) *Essai sur le tissage en Mésopotamie, des premières communautés sédentaires au milieu du IIIᵉ millénaire avant J.-C.* Travaux de la Maison René-Ginouvès 5, Paris.

Breniquet, C. (2010) Weaving in Mesopotamia during the Bronze Age: Archaeology, techniques, iconography. In Michel & Nosch 2010, 52–67.

Breniquet, C. (ed.) (forthcoming), *Systèmes de production et circulation des textiles*, Cahiers des thèmes transversaux d'ArScAn IX, 2007–2008, Nanterre.

Burke, B. (2010) *From Minos to Midas. Cloth Production in the Aegean and Anatolia.* Ancient Textiles Series 7. Oxford.

Burkett, M. (1979) *The Art of the Felt Maker.* Abbot Hall Art Gallery, Kendal.

Cardon, D. (2007) *Natural Dyes, Sources, Tradition, Technology and Science.* London.

Chadwick, J. (1988) The Women of Pylos, *Minos supplement* 19, 43–95.

Del Freo, M., Nosch & M.-L., Rougemont, F. (2010) The Terminology of Textiles in the Linear B Tablets, with some Considerations on Linear A Logograms and Abbreviations. In Michel & Nosch 2010, 338–373.

Desrosiers, S. (2010) Textile Terminologies and Classifications: Some methodological and chronological aspects. In Michel & Nosch 2010, 23–51.

Durand, J.-M. (2009) *La nomenclature des habits et des textiles dans les textes de Mari.* Archives Royales de Mari 30. Paris.

Dury, P. & Lervad, S. (2010) Synonymic Variation in the Field of Textile Terminology: A study in diachrony and synchrony. In Michel & Nosch 2010, 1–9.

Foster, B. (2010) Clothing in Sargonic Period: Visual and written evidence. In Michel & Nosch 2010, 110–145.

Frangipane, M., Andersson, E., Nosch, M.-L., Laurito, R., Rast-Eicher, A., Möller-Wiering, S., Wisti Lassen, A. (2009) Arslantepe (Turkey): Textiles, Tools and Imprints of Fabrics from the 4th to the 2nd millennium BC, *Paléorient* 35:1, 5–30.

Gillis, C. & Nosch, M.-L. B. (eds.) (2007) *Ancient Textiles – Production, Crafts and Society. Proceedings of the First*

International Conference on Ancient Textiles, held in Lund, Sweden and Copenhagen, Denmark, March 19–23, 2003, Ancient Textiles Series 1. Oxford.

Gleba, M. (2008) *Textile Production in Pre-Roman Italy.* Ancient Textiles Series 4. Oxford.

Herslund, O. (2010) Cloths – Garments – and Keeping Secrets. Textile classification and cognitive chaining in the ancient Egyptian writing system. In Michel & Nosch 2010, 68–80.

Hoffman, M. (1964) *The Warp-Weighted Loom. Studies in the History and Technology of an Ancient Implement.* Studia Norvegica 14.

Joannès, F. (1984) Produits pour le travail du bois, du cuir, et du tissu. In J.-M. Durand & D. Charpin (eds.), *Archives Administratives de Mari* 1, ARMT 23, Paris, 133–191.

Joannès, F. (2010) Textile Terminology in the Neo-Babylonian Documentation. In Michel & Nosch 2010, 400–408.

Jones, J. (2010) The 'linen list' in Early Dynastic and Old Kingdom Egypt: Text and textile reconciled. In Michel & Nosch 2010, 81–109.

Kemp, B. & Vogelsang-Eastwood, G. (2001) *The ancient textile industry at Amarna.* With A. Boyce, H. G. Farbrother, G. Owen, P. Rose. London.

Killen, J. T. (1972) Two Notes on the Knossos Ak Tablets. In M. S. Ruipérez (ed.) *Acta Mycenaea. Actes du cinquième Colloque international des études mycéniennes, tenu à Salamanque, 30 mars–3 avril 1970. Vol.* II. Salamanca, 425–440.

Lehmann, W. P. (1995) The Spindle or the Distaff. In B. B. Kachru & H. Kahane (eds.), *Cultures, Ideologies, and the Dictionary. Studies in Honor of Ladislav Zgusta.* Lexicographica series Maior 64. Tübingen, 53–60.

Lerouxel, F. (2002) Les échanges de présents entre souverains amorrites au XVIIIe siècle av. n. è. D'après les archives royales de Mari. In D. Charpin & J.-M. Durand (eds.), *Recueil d'études à la mémoire d'André Parrot.* Florilegium marianum VI, Paris, 473–463.

Luján, E. (2010) Mycenaean Textile Terminology at Work: The KN Lc(1)-tablets and the occupational nouns of the textile industry. In Michel & Nosch 2010, 374–387.

Lyle, E. ed. (2004) *Cosmos. The Journal of the Traditional Cosmological Society* 20.

Lackenbacher, S. (1982) Un texte vieux-babylonien sur la finition des textiles, *Syria* 59, 129–149.

Mårtensson L., Nosch, M.-L., Andersson Strand, E. (2009) Shape of Things: Understanding a Loom Weight, *Oxford Journal of Archaeology* 28(4), 373–398.

Michel, C. (2001) *Correspondance des marchands de Kaniš au début du IIe millénaire avant J.-C.*, Littératures anciennes du Proche-Orient 19. Paris.

Michel, C. (2006) Femmes et production textile à Aššur au début du IIe millénaire avant J.-C., *Techniques et culture* 46, 281–297.

Michel, C. (2008) Review of Breniquet 2008. Les comptes rendus HISTARA, publié en ligne le 2008–12–30: http://histara.sorbonne.fr/cr.php?cr=588.

Michel, C. & Nosch, M.-L. (eds.) (2010) *Textile Terminologies in the ancient Near East and Mediterranean from the Third to the First Millennia BC.* Ancient Textiles Series 8. Oxford.

Michel, C. & Veenhof, K. R. (2010) The Textiles Traded by the Assyrians in Anatolia (19th–18th Centuries BC). In Michel & Nosch 2010, 210–271.

Möller-Wiering, S. (2008) A Textile Impression from Abu-al-Kharaz. In P. M., Fischer (ed.), *Tell Abu al-Kharaz in the Jordan Valley, Vol. I: The Early Bronze Age,* Austrian Academy of Sciences Press, Vienna, 399–400.

Moran, W. L. (1987). *Les lettres d'El-Amarna*, Littératures anciennes du Proche-Orient 13. Paris.

Nosch, M.-L. B. (2003) Centre and Periphery in the Linear B Archives. In N. Kyparissi-Apostolika & M. Papakonstantinou (eds.), *The Periphery of the Mycenaean World. 2nd international interdisciplinary Colloquium, 26–20 September, Lamia 1999. Proceedings*, Ministry of Culture, 14th Ephorate of Prehistoric and Classical Antiquities, 63–70. Athens.

Nosch, M.-L. B. (2004) Red Coloured Textiles in the Linear B Inscriptions. In L. Cleland & K. Staers (eds.) *Colour in the Ancient Mediterranean World.* BAR International Series 1267, 32–39.

Nosch, M.-L. (2006) More Thoughts on the Mycenaean *ta-ra-si-ja* system. In M. Perna (ed.), *Fiscality in Mycenaean and Near Eastern Archives. Proceedings of the Conference held at Soprintendenza Archivistica per la Campania, Naples 21–23 October 2004*, 161–182.

Nosch, M.-L. B. forthcoming. L'archéologie expérimentale et les tests systématiques d'outils de la production textile. In F. Blondé & A. Müller (eds.), *L'artisanat en Grèce, table ronde, École française d'Athènes, octobre 2007.*

Peacock, E. (2001) The contribution of experimental archaeology to the research of ancient textiles. In Walton Rogers, P., Bender Jorgensen, L., and Rast-Eicher, A. (eds.), *The Roman Textile Industry and its Influence. A Birthday Tribute to John Peter Wild,* Exeter, 181–192.

Pasquali, J. (1997) La terminologia semitica dei tessili nei testi di Ebla. In P. Fronzaroli (ed.), *Miscellanea Eblaitica* 4, *Quaderni di Semitistica* 19, 217–270.

Pasquali, J. (2005) Remarques comparatives sur la symbolique du vêtement à Ébla. In L. Kogan, N. Koslova, S. Loesov and S. Tishchenko (eds.), *Memoriae Igor M. Diakonoff, Babel und Bibel* 2, Winona Lake, 165–184.

Pasquali, J. (2010) Les noms sémitiques des tissus dans les textes d'Ebla. In Michel & Nosch 2010, 173–185.

Pomponio, F. (2008) *Testi amministrativi: assegnazioni mensili di tessuti. Periodo di Arrugum, ARET* 15/1, Roma.

Pomponio, F. (2010) New Texts Regarding the Neo-Sumerian Textiles. In Michel & Nosch 2010, 186–200.

Powell, M. A. (ed.) (1987) *Labor in the Ancient Near East,* American Oriental Series 68.

Ribichini, S. & Xella, P. (1985) *La terminologia dei tessili nei testi di Ugarit.* Roma.

Ryder, M. L. (1983) *Sheep and Man.* London.

Sallaberger, W. (2009) Von der Wollration zum Ehrenkleid. Textilien als Prestigegüter am Hof von Ebla. In B. Hildebrandt & C. Veit (ed.), *Der Wert der Dinge – Güter im Prestigediskurs. Münchner Studien zur Alten Welt* 6. München, 241–278.

Singer, I. (2008) Purple-dyers in Lazpa. In B. J. Collins, M. R. Bachvarova & I. C. Rutheford (eds.), *Anatolian Interfaces. Hittites, Greeks and Their Neighbours,* 21–43. Oxford.

Strommenger, E. (1980/83) Kleidung B. Archaeologisch. *Reallexikon der Assyriologie* 6, 31–33. Berlin.

Van Soldt, W. H. (1990) Fabric and Dyes at Ugarit, *Ugarit-Forschungen* 22 (1990), 321–357.

Veenhof, K. R. (1972) *Aspects of Old Assyrian Trade and its Terminology.* Studia et Documenta ad Iura Orientis Antiqui Pertinentia, vol. 10, Leiden.

Verderame L. (2008) Il controllo dei manufatti tessili a Umma. In M. Perna & F. Pomponio (eds.) *The Management of Agricultural Land and the Production of Textiles in the Mycenaean and Near Eastern Economy,* Napoli, 101–133.

Vigo, M. (2010) Linen in Hittite Inventory Texts. In Michel & Nosch 2010, 290–322.

Villard, P. (2010) Les textiles néo-assyriens et leurs couleurs. In Michel & Nosch 2010, 388–399.

Vita, J.-P. (2010) Textile Terminology in the Ugaritic Texts. In Michel & Nosch 2010, 323–337.

Vogelsang, M.E. (1986) Meaning and Symbolism of Clothing in Ancient Near Eastern Texts. In H. Vanstiphout (ed.) *Scripta signa vocis,* 265–284.

Vogelsang-Eastwood, G. M. (1999) *Tutankhamuns Wardrobe,* Rotterdam.

Völling, E. (2008) *Textiltechnik im Alten Orient. Rohstoffe und Herstellung,* Würzburg.

Waetzoldt, H. (1972) *Untersuchungen zur neusumerischen Textilindustrie.* Studi economici e technologici 1, Rome.

Waetzoldt, H. (1981) Kleidung, *Reallexikon der Assyriologie* VI/1–2, 18–31. Berlin – New York.

Waetzoldt H. (2007) The Use of Wool for the Production of Strings, Ropes, Branded Mats, and Similar Fabrics. In C. Gillis and M.-L. B. Nosch (eds.), *Ancient Textiles. Production, Craft and Society,* Ancient Textiles Series 1, Oxford, 112–124.

Waetzoldt, H. (2010) The Colours of Textiles and Variety of Fabrics from Mesopotamia around 2050 BC. In Michel & Nosch 2010, 201–209.

Wees, H. van (2005) Clothes, Class and Gender in Homer. In D. Cairns (ed.) *Body Language in the Greek and Roman Worlds,* Swansea.

Wisti Lassen, A. (2010) Tools, Procedures and Professions: A review of the Akkadian textile terminology. In Michel & Nosch 2010, 272–282.

Zawadski, S. (2006) *Garments of the Gods. Studies on the Textile Industry and the Pantheon of Sippar according to the Texts from the Ebabbar Archive.* Orbis Biblicus et Orientalis 218. Freibourg-Göttingen.

Zawadzki, S. (2010) Garments in non-Cultic Context (Neo-Babylonian Period). In Michel & Nosch 2010, 409–429.

1. Synonymic Variation in the Field of Textile Terminology: A study in diachrony and synchrony

Pascaline Dury and Susanne Lervad

This chapter sets out to examine the terminology of textiles from a linguistic point of view, and endeavours to show that studies in the field of terminology may prove very useful to archaeological studies. In the first part of the chapter, we will present the basis of terminology work and give the main founding principles of terminology regarding concepts, concept structures and synonymic variation. The second part of this chapter will give examples of synonymic variation and conceptual analysis in the field of textiles.

Though our main interest here is textiles, we also give some examples from other domains of activity, in order to illustrate the main terminology principles.

1. What is terminology and what do terminologists do?

Many dictionaries, such as the *Longman Dictionary of Contemporary English*, define the word *terminology* as: "The technical words or expressions that are used in a particular subject: Computer terminology, textile terminology, etc." Or, as the *Webster's Online Dictionary*, as: "A system of words used in a particular discipline: "legal terminology"; "the language of sociology".

Terminologists would rather define their field of work as the study of terms, i.e. words and compound words that are used in specific contexts, or "as a number of practices that have evolved around the creation of terms, their collection and explication and finally their presentation in various printed and electronic media".[1]

Terminology work therefore consists in:

- Analyzing the concepts and concept structures used in a specialized field or domain of activity,
- Identifying the terms assigned to the concepts and making definitions,
- In the case of bilingual or multilingual terminology, translating terms in the various languages,
- Compiling glossaries in databases,
- Managing these databases,
- And creating new terms, as required.

[1] Sager 1990, 1.

1.1. *Terminology – the study of concepts*

"Concepts are mental constructs, abstractions which can be used in classifying the individual objects of the inner and outer world".[2] One of the founding principles of terminology is that the study of concepts and concept structures or concept systems is essential. Terminology work is based on concepts and their delimitation.

Concepts are not independent phenomena. They are always related to other concepts in one way or another, and form concept systems, which can vary from fairly simple to extremely complex. In terminology work, an analysis of the relations between concepts and an arrangement of the concepts into concept systems, are prerequisites for the successful drafting of definitions[3].

Moreover, concepts are made up of what are called *notional elements*, also called *notional* or *conceptual characteristics*. In terminology theory, conceptual characteristics are regarded as the smallest elements of concepts which serve to identify these concepts and to distinguish them from each other. Conceptual characteristics, which can be considered concepts themselves, can be used for describing, classifying and defining concepts.

There are common and delimiting characteristics that correspond to the objects they describe.

1.2. *Delimiting characteristics*

There are usually a great number of characteristics in any concept. Many of these characteristics are so *common* or so *atypical* that they alone are not adequate for identifying a concept or differentiating it from other concepts (for example **TREES** and **GARDEN BENCHES** can be both hard and green).

Delimiting characteristics are those *typical*, or *relevant* characteristics which alone determine a concept, and differentiate it from other concepts (for example *having a hard self-supporting trunk* is a delimiting characteristic of **TREES** in relation to **CLIMBING PLANTS).** Therefore, only delimiting characteristics should be used in definitions, but it is not possible to select and name all of them in a term, because the term would then be too long to be written or spoken. Therefore, only a small number of conceptual characteristics are usually selected and named in terms. Which characteristics are selected in a term changes from one culture to another and from one language to another, and one concept existing in one linguistic community may not exist at all or only partially in another linguistic community, as shown below:

> "A point worth mentioning is that the concepts existing in one linguistic community may not exist at all or only in part in other linguistic communities. [....] A well-known example of culturally dependent concept formations is the concept systems of colours of certain Indian tribes in the Amazon which distinguish among 300 different sorts of green"[4]

---

[2] *British Standard Recommendation for the Selection, Formation and Definition of terms*, BS 3669, 1963.

[3] As also shown in part 3 below, non verbal elements like drawings or formulas are also considered vital elements for the successful drafting of definitions.

[4] Weissenhofer 1995, 2.

1.3. *The panlinguistic definition scheme*

Boisson (1996) analyzed the denomination of the slide rule in 41 languages in order to construct an extensive panlinguistic 'definition scheme'. He showed that each language typically selects from this scheme a couple of notional elements, which can vary from one language to the other.

> "The history of the slide rule is retold in a systematic way, and its denomination is analysed in 41 languages. A comparison of these terms allows us to postulate for the object an extensive panlinguistic "definition scheme", which might provide an empirical approximation to an analysis of the concept /slide rule/. Each language typically selects from this scheme a couple of notional elements, so that these terms look like elliptical definitions of the object".[5]

For example *slide rule* in English contains the notional element "slide", which does not exist in the French *règle à calcul* (which uses instead the notional element "something that is used to calculate" as in Italian and Spanish). The term is translated below in Danish, Italian, German and Spanish.

> → Regnestok/regolo calcolatore/Rechenschieber/regal de cálculo

2. Synonymic variation

The kind of conceptual characteristics which are named in terms also changes from one period of time to another, which may give rise to what is called 'synonymic variation', especially in specialized lexicons.

One specialized field of knowledge in which synonymic variation is known to be prevalent is medicine, but it is also the case in natural sciences and in textiles.

We will show here that, when they are examined diachronically, synonyms do have a role to play in the shaping of specialized lexicons. We will illustrate this point with the main results obtained from a corpus-based investigation into the semantic development of some synonyms of the term *petroleum* in 19th-century English. The corpus used was specially designed for the study and contains specialized texts[6] (mostly book chapters and scientific articles) relating to the field of mineralogy, chemistry and petroleum geology published in the 19th century and the first half of the 20th century.

The word count is at present a little over 257,000 words (=257,864 words). The corpus was investigated using a lexical analysis computer software called *Wordsmith Tools* and was divided into three sub-periods. The delimitation of these sub-periods was mainly governed by extralinguistic criteria. These periods roughly correspond to major events relating to the field of petroleum geology:

- Period one contains texts published between roughly 1800 and 1860 and corresponds to the period just before the first successful drilling of an oil well by "Colonel" Drake in Pennsylvania in 1859.
- Period two contains texts published between 1860 and 1900 and corresponds to the oil boom in Pennsylvania and subsequently in Texas.
- Period three contains texts published between 1900 and 1960, corresponding to the development of the internal combustion engine car and therefore to the large-scale industrial exploitation of petroleum.

The decision to base this study on the 19th century and the first half of the 20th century was made for extralinguistic reasons, too. People have used petroleum since ancient times (first as a

[5] Boisson 1996, 525.

[6] Also see Dury 2008 and 2009.

medicine and later to light their homes) and occurrences of oil throughout the world have been the objects of study by geologists since the 17th century.

In the first half of the 19th century, most geologists, mineralogists and chemists knew that petroleum was basically made of hydrogen and carbon and that it reached the surface of the earth at many places in circumstances which were in some ways peculiar. However at the time, the importance of oil was not fully appreciated and the scientists' understanding was also circumscribed by limited knowledge of stratigraphy, structure, geological and chemical processes. Until the first half of the 20th century, scientists had no detailed nor extensive knowledge of what exactly petroleum was, where it came from, and what could be produced from it.

2.1. *Semantic flexibility*

The 'limitations' imposed by the contemporary state of chemical knowledge had two consequences for the language: Firstly, it led to what may be called the 'semantic flexibility' of some terms which were sometimes used with exactly the same meaning as the term *petroleum*, but which were also sometimes used with a different meaning and secondly, it also led to a wide range of what may be called 'occasional' or 'temporary' synonymy, or the simultaneous co-existence of several terms to express the particular notion of *petroleum.*

Indeed, the information extracted from the corpus shows that a variety of terms (16) can be considered as synonyms of the term *petroleum*, and were used as such by authors.

Some of these synonymous terms, like *rock oil* or *earth oil* refer to where petroleum is found in nature; some other synonyms, like *empyreumatic oil* rather describe the smell of petroleum, while other terms, like *dark pitch, carbon oil, mineral oil,* put the emphasis on what petroleum is made of or what it looks like. And a last group of synonyms, which is also the largest, is made up of toponyms or terms based on place names which describe the regions, or the countries where petroleum was known to be found. This group includes for instance: *Barbadoes tar, Gabian oil, Sicilian oil, Trinidad bitumen, Persian rock oil, Genesee oil, Seneca oil, Seneca rock oil, Rangoon petroleum*, etc.

The data extracted from the corpus indicate that the vocabulary of petroleum went through a second stage after 1860, as the terms which proved to be semantically 'flexible' in the previous period became fixed in the vocabulary, and as most of the synonymous terms progressively disappeared from the language. This is the reason why they may be called 'occasional synonyms' or 'temporary synonyms', since they were only used in the vocabulary of petroleum at a time when petroleum geology was still in its early stages. However does this synonymic variation also apply to other fields of knowledge, like textiles?

3. From fibers to structures

In the field of textile terminology, classifications, concept systems and term collections usually include the fibers first and then the yarns and the structures such as weaving, knitting etc., as shown below in the definition of the term *man-made fiber*, which details the most essential conceptual characteristics: "Staple fiber are filament of polymers produced by manufacturing processes."[7]

[7] USTC-01-Nomenclature.

In this case, the definition does not mention the term *man-made*, but it uses the term *manufacturing*. A number of other synonyms also express the conceptual characteristics given in the definition above. This is the case for *manufactured fiber* which can also be directly derived from the definition and which is often understood as a short version of the definition. The terms *synthetic* and *artificial fiber* are often used as synonyms as well, which can sometimes prove problematic. The variants used to describe the different fibers, like *textile fiber* and *natural fiber* are much less ambiguous.

The next phase in textile production corresponds to the *construction* or the *structures*.

The examples chosen to illustrate synonymy in this case are weavings. Susanne Lervad inherited a background in weaving from her parents and grandparents who produced looms for hand-weaving for a century. Furthermore, she studied silk fabric, notably in Lyon, France, where the collections of these textiles and documentation are very rich. The patterns of these silk fabrics and the terminology was described in her Ph.D. thesis, and the experience in the trilingual terminology of fibers, threads and fabrics acquired while researching has shown how non-verbal aspects can be used to describe concepts in fields such as textiles.

3.1. *Non-verbal aspects*

This Ph.D. in textile terminology is both traditional, using primarily verbal definitions, and innovative as it attempts, in specific cases, to unify the definition and the designation. The innovative aspect of this work is that it shows that representing a concept using an illustration can unify the designation and description of this concept. Differently put, what is traditionally identified as a designation (most commonly a term), and the concept descriptions (definition) disappear in some of the examples studied. Representing textile concepts in a "multimodal" manner therefore seems to be a constructive and useful approach.

There are several types of illustrations used to represent a concept:

- Symbols
- Pictograms
- Diagrams
- Line drawings / sketches

In the field of textiles, representing a concept using an illustration is more universal than using a given language, but the effectiveness of these signs is dependent on a common understanding. Both the party who transmits and the party who receives the sign must share this understanding.

The diagrams below illustrate how concepts in this field are represented in the terminology. The diagram also works as a step-by-step guide to producing fabric. We will also try to show the limits of illustrations and non-verbal signs: it is clear that the image and text are complementary and the text dictates our conception of the image.

In the field of textiles, texts are particularly useful in explaining the characteristics which cannot be easily conveyed by means of an image – for example the softness of the fabric or other aspects requiring a verbal explanation. The examples below deal with the micro-structure of the fabric i.e. the weave. The macro-structures (for example the design) are not dealt with here.

> "**Weave**: System of interlacing the threads of warp and weft according to defined rules".[8]
> "**Weave unit**: The smallest cycle of interlacement of warp and weft that is constantly repeated in a weave or a binding system".[9]
> "**Binding system**: System in accordance with which ends and picks are bound".[10]

The illustrations used may be representative images (photos, paintings, drawings) or abstract images (diagrams, line drawings, etc.). The degree of abstraction determines the function of the graphic components. The graphic components can explain or clarify verbal definitions, or function independently providing a full representation of the concept in question. In this case, the verbal component serves only to provide a complementary explanation.

The examples below illustrate how the graphic components replace the verbal definitions to a greater or lesser extent in each case. The diagrams represent the structure of the fabric.

In order to describe a fabric as a concept and its characteristics, one should always start at the most basic level, i.e. the point at which two threads meet – the weave. The combination of basic weaves creates a wide variety of textures perfected in fabric production in French silk factories in Lyon. The weave can be represented using pictograms or diagrams of varying degrees of abstraction (see Fig. 1.1).

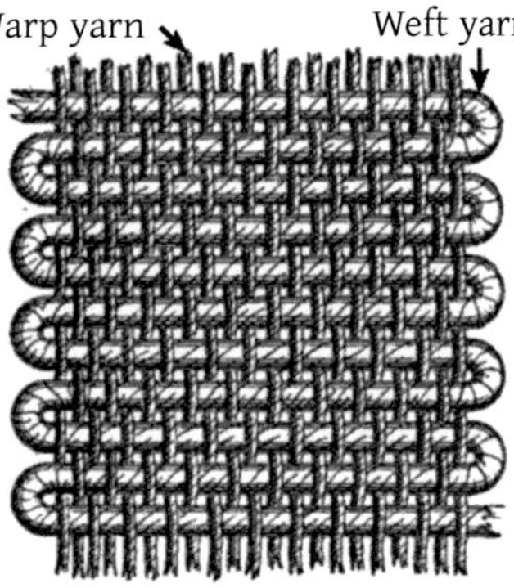

Fig. 1.1. Plain weave.

Figure 1.1 shows how a fabric is made up of vertical threads – the warp – and horizontal threads which cross over – the weft. There are an infinite number of ways of combining different types of crossovers. Figure 1.1 shows the simplest of these crossovers / weaves – plain weave. Another example is a diagram in binary form, the language of computers. Each thread has a numerical value of 0 or 1, i.e. one thread over or one thread under, which easily translates into the binary system.

In his book (1982), Hugues deals with the common ground occupied by one of the most ancient crafts, weaving, and the modern world of computers:

> "Indeed, a piece of fabric is constructed from combinations based on a binary code from the structure of the weave (one thread over, one thread under) and computers function using combinations translated by a code consisting of a series of 1 or 0".[11]

This binary system was used very early on in the French textile industry in the Jacquard weaving mechanism created in the *Croix Rousse* district of Lyon two hundred years ago, and which could be considered as one of the world's first computers.

The diagrams representing concepts mainly describe the weave – the smallest unit which is used to multiply and repeat structures in order to create the surface of the fabric.

The examples below show the three basic weaves: *plain, twill and satin.*

This weave is different from any other as the horizontal and vertical threads cross over

[8] Burnham 1980, 179.
[9] Burnham 1980, 179.
[10] Burnham 1980, 6.
[11] Hugues 1982, 449–50.

alternately. This basic unit is made up of 2 × 2 threads.

The concept is designated by two terms *plain weave* and *tabby* in the literature and this does not cause any problems.

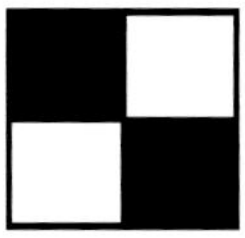

Fig. 1.2. Tabby/ Plain weave.

"**Tabby**: The binding system or weave based on a unit of two ends and two picks, in which each end passes over one and under one pick. The binding points are set over one end on successive picks".[12]

There are a large number of plain weave derivations such as *rib weave / rep weave and panama /hopsack weave*. These are easy to show in a diagram but difficult to designate using terms. In this case it is easier to show just the diagrams of the basic weaves and the derived weaves plus a code.

3.2. *Formulation of a code*

The international standard ISO 9354 establishes a code for the systematic numerical notation for basic weaves and their simple derivatives.

The code for any basic weave or one of its simple derivatives is made up from digit number elements that are separated from one another by hyphens. These elements indicate, in sequence, the following characteristics of the weave:

> First element: the kind of weave,
> Second element: the sequence of interlacing of the yarns, i.e. warp up or down,
> Third element: the warp thread grouping, i.e. the warp yarns weaving singly or in groups,
> Fourth element: the step or move number. The step number indicates the number of threads by which the point of intersection is offset each time.

For plain weave, the code is 10-010101 00 (ISO 9354 standard).

The second basic weave is *twill.* Basic twill weave consists of 3 × 3 threads with four possible combinations, one of which is shown above: *2/1 twill*, in which each time a weft thread passes over a warp thread, it then passes below the next two warp threads. In addition, there are four possible 2/1–1/2 twill weaves Z or S spun. Both weft twills or warp twills exist, and the points at which the threads cross over create a diagonal pattern.

Fig. 1.3. Twill weave.

There are a large number of variations / derivatives of basic twill weave such as the *5-end stitched twill, "Z" direction.* These figures can be written separated by a point 3.1.1.1 or a slash 3 1/1 1 representing the point of intersection.

The derivatives are easy to represent in diagrams and codes but almost impossible to translate from one language to another using verbal components. The code for 3/1 twill, "Z" direction is: 20–01 03–01–01 (ISO 9354)

The third basic weave is satin and here we also show below the diagrams, the definition and the code:

---

[12] Burnham 1980, 139.

> **"Satin:** Binding system or weave based on a unit of five or more ends, and a number of picks equal to, or a multiple of, the number of ends. Each end either passes over four or more adjacent picks and under the next one, or passes under four or more adjacent picks and over the next one. The binding points are set over two or more ends on successive picks and are distributed in an unobtrusive manner to give a smooth appearance".[13]

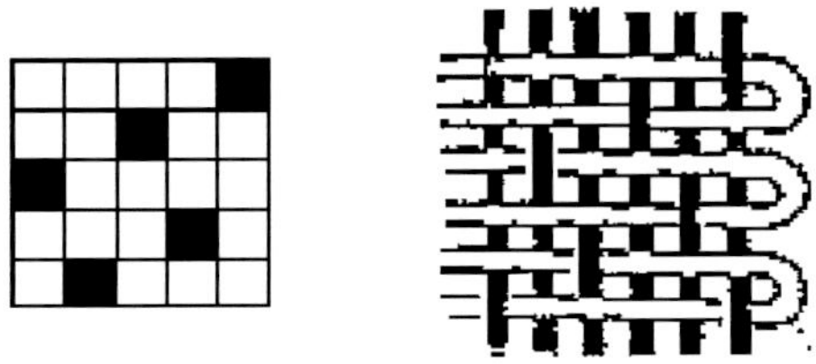

Fig. 1.4. Satin.

The step number indicates the number of threads by which the point of intersection is offset each time, and regular satins are produced by consistently using the same step number, while irregular satins are produced using several different step numbers in succession.

> Example: 6-end cross warp satin, steps 3,4,4,3,2.
> The code is: 30–05 01–01–02 04 04 03 02

The number of possible combinations of the basic plain, twill and satin weaves is infinite.

The use of graphic components to represent the weaves, as recommended in the ISO 9354 standard, bypasses the need to use long and complicated terms, which are of little use in conveying the concept. The image of the weave can be combined with a code, thus minimizing the need to produce a definition and a term.

The characteristics reflecting the origins and use of a fabric cannot be represented easily by using graphic components, but the characteristics of form, structure and colour can conveniently be represented graphically.

The work to create international standards within the framework of the ISO 9354 also shows, that in this field, definitions are often being replaced by diagrams, and terms by codes. When work on the standard started, a verbal explanation of the code was included, but this verbal element only served to explain the code itself. There is therefore a global consensus that the representation of derived weaves in the form of diagrams will greatly facilitate work on the terminology, as shown above.

As a large number of weave derivatives and variations can be created, it is almost impossible to find terms for each of them, and even more complicated to translate them from one language to another. The examples above show this clearly and part of the solution to this problem resides in the use of non-verbal representations and codes.

* * *

The first aim of this chapter was to explain what the field of terminology is and what its main founding principles are. The objective was also to demonstrate what kind of terminology work can be done in specialized lexicons and how concepts and concept structures are used to draft definitions, analyze and translate languages, whatever the cultural context, and whatever the time period.

A second conclusion to this chapter is that corpus evidence shows that 'temporary' or

[13] Burnham 1980, 113.

'occasional' synonymy as well as the 'semantic flexibility' of terms, far from being something to avoid in terminology, may well be considered as a phenomenon which crops up naturally at some point in the history of a scientific lexicon and which occurs as part of a conceptual formation stage, as is the case here with petroleum geology, and as is also the case in textiles.

Abbreviations

ISO 9354, (1989) *Weaves, Coding systems and examples*, case postale, 56, CH-211 Geneva, Switzerland.
USITC-01: *Norme internationale concernant le commerce du textile, US International Trade Commission*

Bibliography

Boisson, C. P. (1996) Les dénominations de la « règle à calcul », *Meta*, vol. 41, N. 4, 525–566.

British Standard Recommandation for the Selection, Formation and Definition of Terms (1963) BS 3669.

Burnham, D. (1980) *A textile terminology, Warp and Weft*. Routledge & Kegan Paul, London and Henley.

Budin, G. & Wright, S. E. (1997) *Handbook of Terminology Management, Volume 1.* John Benjamins Publishing Company, Amsterdam, Philadelphia.

CIETA (1997) *Vocabulaire français, allemand, anglais, espagnol, italien, portugais et suédois.* Centre International d'Etude des Textiles Anciens, Lyon.

Dury, P. (2009) Synonymic Variation in the 19th-Century Lexicon of Petroleum. In *Proceedings of the Second Symposium on New Approaches in English Historical Lexis, Hel-lex 2, Helsinki, 25–27 April 2008*, Cascadilla Press, 49–59.

Dury, P. (2008) Les noms du pétrole : une approche diachronique de la métonymie onomastique, *Lexis, E-Journal in English Lexicology*, http://screcherche.univ-lyon3.fr/lexis/.

Dury, P. & Lervad, S. (2008) La variation synonymique dans la terminologie de l'énergie : approches synchronique diachronique, deux études de cas, *LSP and Professional Communication*, Vol. 8, N.2, 66–79.

Hugues, P. (1982) *Le Langage du Tissu.* Editions Textile/Art/Industrie, Paris.

Lervad, S. (1999) Les éléments graphiques dans la terminologie des textiles, *LSP and Professional Communication*, Vol. 22, N.2 (48), 38–47.

Lervad, S. (1991) *En analyse af den fagsproglige kommunikation i tekstilområdet*, Ph.D. thesis, Handelshøjskole Syd, Kolding

Lervad, S. (1998) Analyse comparative de trois ouvrages de lexicographie spécialisée dans le domaine de textiles concernant les définitions comme rerésentation de connaissances. In *La banque des mots*, N.8 spécial, *Qualité et terminologie.*

Longman Dictionary of Contemporary English (http://www.ldoceonline.com/)

Sager, J.C. (1990) *A Practical Course in Terminology Processing.* John Benjamins Publishing Co.

Webster's Online Dictionary (http://www.websters-online-dictionary.org/).

Weissenhofer, P. (1995) *Conceptology in Terminology Theory, Semantics and Word-formation*, Termnet.

Wüster, E. (1984) *Einführung in die Allgemeine Terminologielehre*, *LSP Centre*, Copenhagen Business School, Denmark

2. The Basics of Textile Tools and Textile Technology: From fibre to fabric

Eva Andersson Strand

The production of a textile from fibre to finished product is complex and includes many stages. The general stages in textile production are fibre procurement, fibre preparation, spinning, weaving and finishing and each stage includes several processes.

The focus of this article is to briefly explain the different stages and processes in order to provide a better understanding of the complexity of textile production.[1] Furthermore, specific textile techniques will be related to textile archaeological experiments that have been designed and conducted in the Tools and Textiles – Texts and Contexts (TTTC) research programme of the Danish National Research Foundation's Centre for Textile Research.

1. Fibres for producing textiles

Several different fibres, both plant- and animal, can and have been used for producing textiles. The fibres in use in the Bronze Age Eastern Mediterranean according to written sources were primarily sheep wool and flax.[2] However, other fibres were also used, for example goat hair as has been shown via a textile analysis carried out on a textile from Arslantepe, Turkey dated to 3000–2750 BC (Fig. 2.1).[3] Another textile dated to the Late Bronze Age that includes one nettle thread has been found in Khania, Crete.[4] Furthermore, fibres such as tree bast and hemp but also horse and camel wool could have been used.

The stage from flax stems to processed fibres includes many steps. When flax is ripe it is pulled up by the roots and the seeds are rippled. The flax then has to be retted. In this process the stems can either be placed in water or spread on the ground. The moisture assists in the process of dissolving the pectin between the bunches of fibre in the bark and the stem. The next step is breaking, when a wooden club is used to break up the stem and the bark which have to be separated from the fibres. Thereafter, the flax has to be scutched with a broad wooden knife, which scrapes away the remainder of stem and bark. Finally the fibres are hackled or combed.

[1] For more detailed information, see for example Barber 1991; Breniquet 2008.

[2] Barber 1991.

[3] Frangipane *et al.* 2009.

[4] Moulherat & Spantidaki 2006; Andersson Strand & Nosch forthcoming.

Fig. 2.1. Sample 086/2002; microscopic photograph. The sample shows very fine s-spun threads of an exceptionally thin diameter of 0.1 mm. Photo by Antoinette Rast-Eicher.

The tools used in these production processes are primarily made of wood and it may be surmised that this is why these tools are rarely found in archaeological contexts.[5]

There is no information in ancient sources on the amount of fibres one would obtain from a field of flax, but today, the calculations of the outcome based on a 100 m^2 field are:[6]

- 100 m^2 = 1 working day to pull the flax stems by hand
- 100 m^2 = *c.*25 kg yarn in different qualities + 14 kg of fibres for rope etc.
- 25 kg yarn = 287 500 m thread[7]
- 11 threads per cm in warp and weft = 2200 m thread/ m^2 = 130 m^2 fabric

Different types of sheep breeds existed in ancient times.[8] There is a large variation in wool fibres: the quality varies between various breeds, but there is also a difference between individuals within the same breed, and depending on if the wool is from a lamb, a ewe, a ram or a wether. Furthermore, a great variety is found in the coarseness of wool fibres depending on from which part of the sheep the wool is obtained. Wool from the thighs, for example, is coarser and longer than the wool from the side and shoulders. The wool is obtained by shearing or cutting, but it has been considered that the oldest method is pulling.[9] Shears may have been used, but it is likely that knives were the oldest tools used. A sheep can be sheared twice a year while it can be plucked only once a year.

The wool can be spun immediately after it has been shorn or plucked from the sheep but usually it is first teased by hand or combed with the aid of wool combs. In this way, the long hair is also separated from the wool.

Sheep in ancient times were smaller than today's sheep and it is difficult to determine the amount of wool that could be obtained from one animal. Icelandic sources from the early 19th

[5] Andersson 2003.

[6] Fröier & Zienkiewicz 1991.

[7] The numbers of meters of thread and the calculations are based on a spinning test with flax fibres in the research programme. Mårtensson *et al.* forthcoming.

[8] Barber 1991; Breniquet 2008.

[9] Barber 1991.

century AD state that an ewe could yield 1–1.25 kg of washed wool and a wether between 1.75–2.5 kg. Please note that this is the weight of the wool for spinning and not the total weight of the fleece.[10]

2. Spindles and spinning

A spindle consists of a spindle shaft that is generally made of wood and often also of a spindle whorl. The spindle whorls vary as regards material, shape and size (Fig. 2.2).[11]

After the preparation, the wool is twisted by hand into a short thread which is attached to the shaft. The spindle can then hang freely, a so-called suspended spindle, but one can also spin with a supported spindle. The shaft is rotated while the spinner simultaneously draws out the fibres, and it is the twisting of the fibres around their own axis that forms the thread.

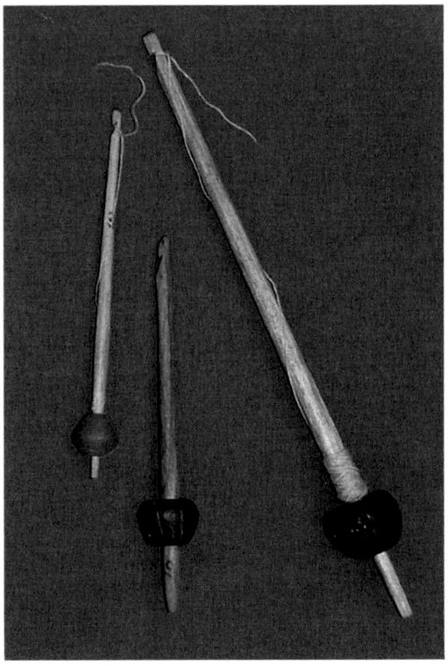

Fig. 2.2 (above). Spindle shafts and spindle whorls. Photo by Linda Olofsson.

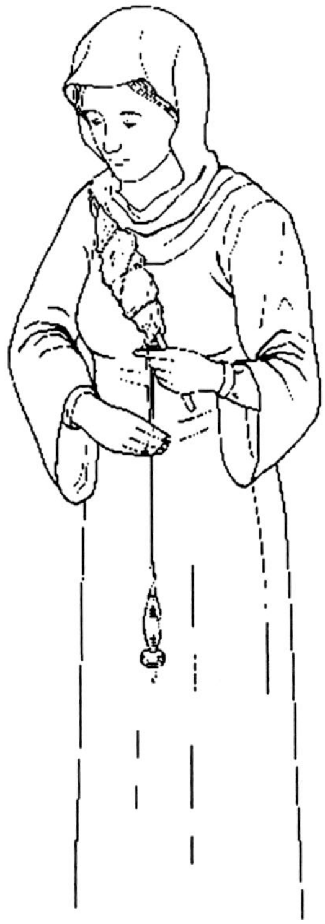

Fig. 2.3 (right). Spinning with a spindle and using a short distaff. Drawing by Tina Borstam.

When spinning, it is easier to hold the raw material on a distaff, so that the prepared fibres are not mixed up again (Fig. 2.3). When one has spun a certain length, the thread is wound up on the spindle and it is possible to continue spinning. This is repeated until the spindle shaft has been filled with thread. The spun yarn is then wound onto a reel.

In the TTTC research programme, several experiments with different spindle whorls of various sizes have been undertaken.[12] The tools used in the spinning tests were reconstructed copies[13] of biconical and conical ceramic whorls from the Bronze Age site Nichoria, Greece, one weighing 18 g, another weighing 8 g, and a third ceramic whorl weighing 4 g.[14]

A white fleece, weighing 2.7 kg was selected. After sorting, removing felted parts, dirt and the most irregular parts, 1.1 kg of rather homogeneous wool was left for the experiments. Furthermore, as much as 22 percent of the cleaned wool was also discarded in the combing process. The wool was then formed into bands of fibres and rolled into balls, ready for spinning. Altogether it took about 6 hours (h) for two technicians to prepare 170 g of wool.[15]

---

10 Adalsteinsson 1991, 286.

11 See for example: Barber 1991; Carington Smith 1992; Gleba 2008; Frangipane *et al.* 2009.

12 Andersson *et al.* 2008; Olofsson *et al.* forthcoming.

13 The tools were reconstructed by ceramist Inger Hildebrandt at the Lejre Experimental Centre, Denmark, and we are grateful for this collaboration.

14 Carington Smith 1992.

15 The written sources record *pe-ki-ti-ra*$_2$ translated as "wool-carders" (Ventris & Chadwick 1973, 570), but carding is a

The test spinning of the 4 g whorl demonstrated that fine, soft and washed wool was necessary.[16] In this case it took approximately 9 h for the technicians to prepare 66 g wool after washing and drying.[17]

The spinners employed the suspended spinning method during the experiment and a total of 56 tests were conducted. The spinning was carried out until the spindle was filled, and only up to the point that the weight of the yarn affected the rotation negatively. It required more concentration to spin with the 4 g and 8 g whorl than with the 18 g whorl. In the tests with the 4 g whorl, the spinners also added more twist with the hand than they did when spinning with the 8 g and 18 g whorls, and it was more time-consuming to spin a thread with a light whorl than it was with a heavy whorl. Furthermore, there is an obvious difference in how many fibres the thread contains when using different whorls, and the thread spun with the 4 g whorl contains remarkably less fibres than the others.[18]

The time estimate only takes the actual spinning time into consideration, and time for the cleaning, sorting and preparation of the wool has to be added. The average output of thread/h is:

c.50 m of yarn/h (18 g whorl)
c.40 m of yarn/h (8 g whorl)
c.35 m of yarn/h (4 g whorl)

This denotes that if 2 km of thread is needed for the weaving of a piece of textile, it would take:

40 h to spin 2 km of yarn (18 g whorl)
50 h to spin 2 km of yarn (8 g whorl)
57 h to spin 2 km of yarn (4 g whorl)

Contrary to what one might imagine, 2 km of thread is by no means sufficient for the production of a garment, but enough for 1 square m of cloth with a thread count of 10 threads/cm^2.

The thread length/100 g of prepared wool was measured and as can be seen in Figure 2.4 the tests clearly demonstrated that the outcome of metre yarn per 100 g wool is larger when using a light spindle. It should also be noted that the impact of the tools is greater than that of the individual spinners. The difference in yarn length can be explained by the fact that the lighter the spindle

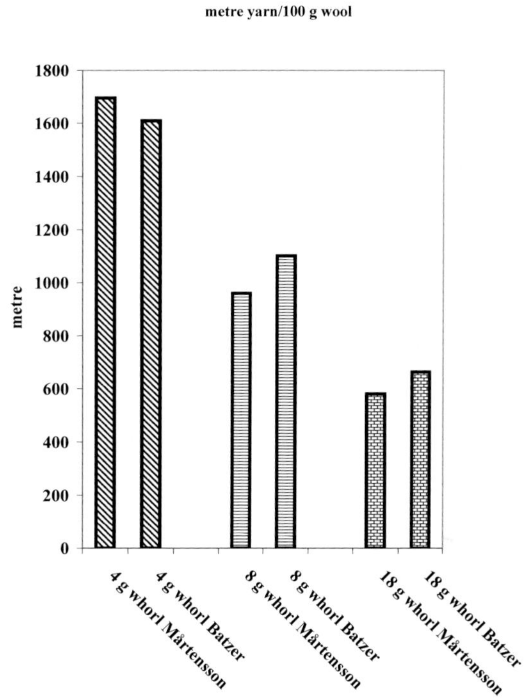

Fig. 2.4. Calculation of metre yarn per 100 g spun wool, 4 g, 8 g and 18 g whorls.

textile process of the Iron Age. In the Bronze Age, it is surmised that combs were in use, and finds from the Caspian Sea area (Shishlina *et al.* 2000) suggest that combs were made of wood.

16 It is known from Bronze Age written sources that wool was occasionally washed before spinning (Waetzoldt 1972, 109–119). The washing was conducive to the spinning. In total, 254 g wool was washed in 40–60 centigrade hot water in three stages for about 6 minutes. The wool lost 17 g in weight in this process.

17 Andersson *et al.* 2008; Olofsson *et al.* forthcoming.

18 Andersson *et al.* 2008; Mårtensson *et al.* 2006a; Olofsson *et al.* forthcoming.

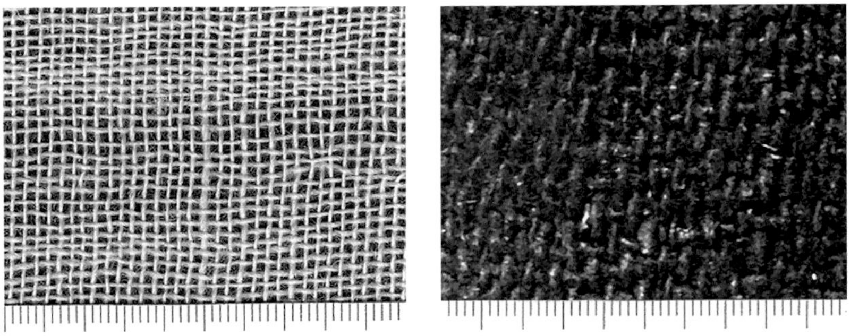

(i) (ii)

Fig. 2.5. Two fabrics, both with 6 warp threads per cm, but totally different, one open and with a thread diameter of c.0.5 mm (i) and one dense and with a thread diameter of c.0.9 mm (ii).

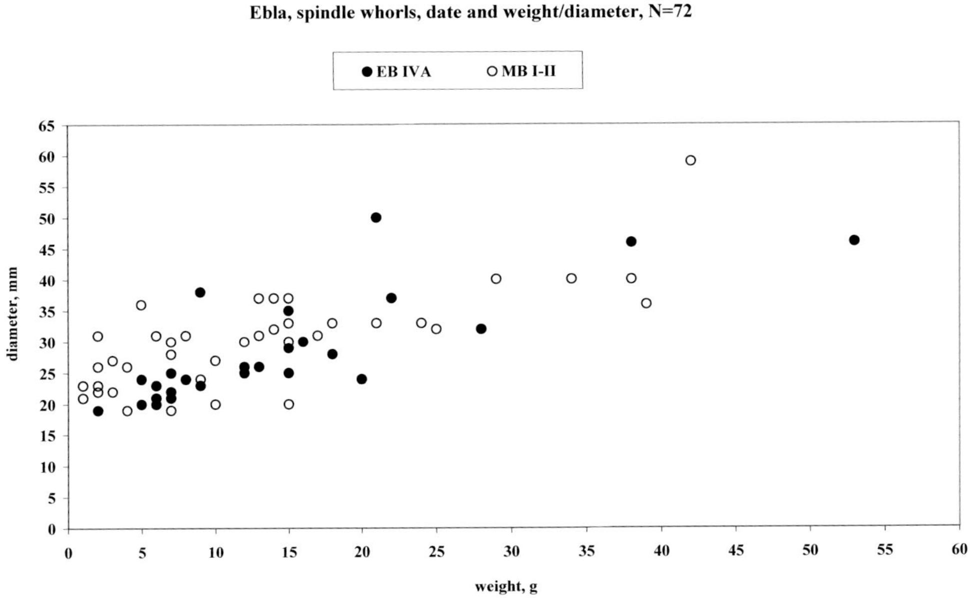

Fig. 2.6. Ebla, chronological distribution of the recorded spindle whorls and the relationship between spindle whorl weight/diameter. By courtesy of Luca Peyronel and the Italian Archaeological Expedition at Ebla.

whorl, the less fibre is used per metre – the lighter the spindle, the thinner the thread; the heavier the spindle whorl, the thicker the thread. The difference in thin and thick yarn can be expressed in terms of the diameter of the thread. A thin yarn has a small thread diameter, while a thicker yarn has a larger thread diameter; in general, a yarn with a larger thread diameter contains more fibres than a yarn with a smaller diameter.[19]

Even a small difference in thread diameter, sometimes not even visible to the eye, will affect the final product, the fabric (Fig. 2.5). However, it is important to note that also the quality of the fibres affect the result: if a coarser fibre were used, a coarser thread would result.

[19] Andersson *et al.* 2008.

The results of the spinning tests help us to interpret the type of yarn that may have been spun on a certain site on the basis of the finds of spindle whorls from an archaeological excavation. For example, in Tell Mardikh – Ebla, Syria, the graph demonstrates a varied production of different types of thread with a focus on thinner threads (Fig. 2.6).

3. Looms, loom weights and weaving

The most common archaeological evidence for weaving in the Eastern Mediterranean is the presence of loom weights, which indicates the use of the warp-weighted loom (Fig. 2.7). In addition to the warp-weighted loom, other forms of looms were also used in the Mediterranean area during the Bronze Age; principally, the vertical two-beam loom and the horizontal ground loom. These two loom types are made of perishable material and do not require loom weights. The additional use of alternative types of loom cannot be excluded.[20]

A fabric is the result of weaving two thread systems, which cross each other at right angles. One of the systems, the warp, runs parallel to the side of the loom and is kept stretched during weaving. On a warp-weighted loom, the vertically hanging warp threads are kept taut by the weight of the attached loom weights. The other system, the weft, runs alternately over and under the warp threads (Fig. 2.8).[21]

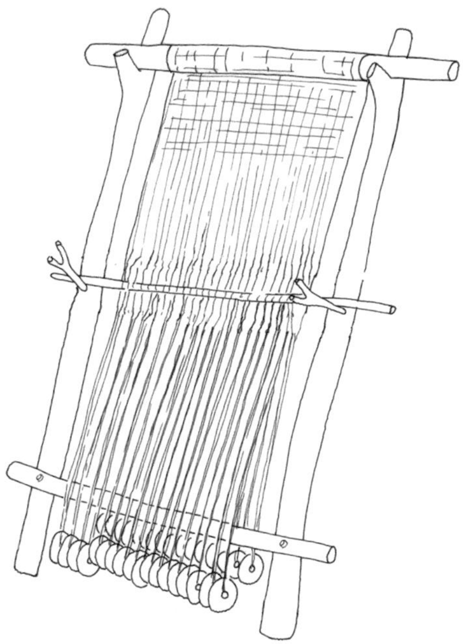

Fig. 2.7. The warp-weighted loom with a tabby setup with two rows of loom weights. Drawing by Annika Jeppsson, after Andersson 2003.

A loom can operate in a number of ways, depending on the type of weaving technique used. Tabby is considered to be the most common technique used in the Eastern Mediterranean during the Bronze Age. However, finds of Bronze Age textiles are extremely rare in the Mediterranean region generally, and this conclusion is based on only a handful of preserved textile fragments, which indicates that the existence of other weaving techniques therefore cannot be discounted.[22]

In a tabby weave, the weft runs under one warp thread, over one warp thread, under one, over one, and so on (Fig. 2.8). Another technique is twill weaving (Fig. 2.9). This entails the use of more than two layers of warp threads, in order to create different sheds, one behind the other. There are many variations of twill; for example, in a 2/2 twill the weft runs over two warp threads, under two warp threads, over two, under two, and so on. In this

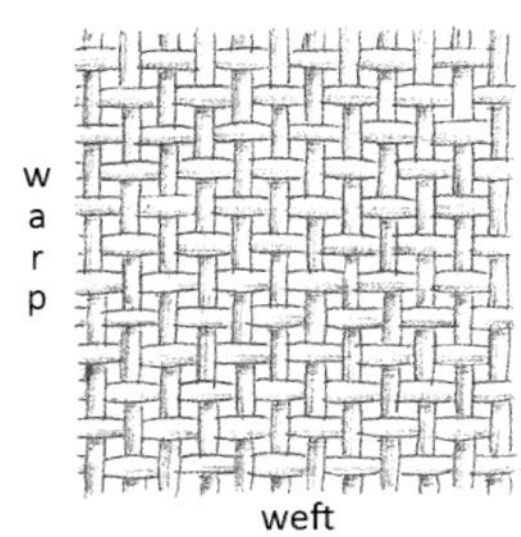

Fig. 2.8. Warp and weft in a tabby weave. After Stærmose Nielsen 1999.

[20] See also Breniquet this volume.
[21] Andersson 2003.
[22] Barber 1991; Spantidaki & Moulherat forthcoming; Andersson Strand & Nosch forthcoming.

(i)

(ii)

Figure 2.9. A 2/2 twill weave woven on a warp-weighted loom with four rows of loom weights. The fabric has 15 warp threads and 15 weft threads per cm. 15 warp threads are attached to each loom weight. (i) The warp-weighted loom with the setup, (ii) the four rows of loom weights. Please note that the warp threads are fastened to a loop of yarn that is attached to the loom weight; i.e. the warp threads are not directly attached to the loom weight. By courtesy of Ulla Lund Hansen and the Vorbasse project, photo by Linda Olofsson.

(i) (ii)

Fig. 2.10. 2/2 (i) and 2/1 twill (ii). After Stærmose Nielsen 1999.

technique, there are four layers of warp threads (as opposed to the two layers/sheds needed for a tabby weave). An alternative twill technique is 2/1 twill, in which the weft thread passes over two warp threads and under one, over two and under one, and so on (Figs. 2.9 and 2.10).

A fabric can be balanced with an equal number of weft and warp threads per square centimetre (cm), or the number of warp and weft threads may differ. The weave can be open, with a few threads per cm, or dense, with the threads packed closely together (Fig. 2.5). Different types of yarn may also be used in the warp and in the weft respectively. The thread count refers to the number

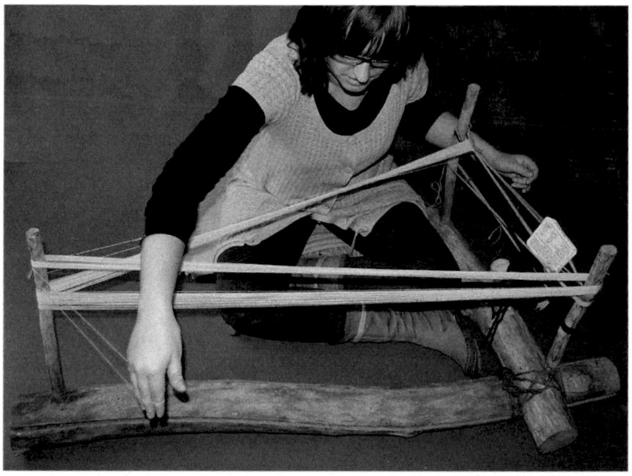

Fig. 2.11 (above left). Warping. By courtesy of Ulla Lund Hansen and the Vorbasse project, photo by Mette Bruun.

Fig. 2.12 (left). The warp threads are tied to the starting border rod and to the loom weights. By courtesy of Ulla Lund Hansen and the Vorbasse project, photo by Linda Olofsson.

Fig. 2.13 (above right). Heddling. By courtesy of Ulla Lund Hansen and the Vorbasse project, photo by Linda Olofsson.

of warp and weft threads per square cm and is often used when describing an archaeological textile.[23]

Prior to weaving, the type of fabric to be produced has to be decided and the amount of yarn needed has to be calculated. The choice of weaving technique can be due to several factors, cultural, social, economic but foremost the functional purpose of the fabric, for example, clothing, wall hanging, or sailcloth. Thus, an open tabby weave will not function as a sailcloth but can be used as a veil. A dense and coarse twill fabric can be excellent for an outer garment and will protect the wearer from rain, wind and cold. A finer twill and tabby would be very functional for different types of inner garments, but perhaps also to protect the wearer from sun and heat.

The first step is to warp the warp threads (Fig. 2.11). Depending on the weaving technique, on the length of the fabric and on the number of threads per cm, the number of metres of yarn has to be calculated. For example, if the fabric is one metre wide and 2 metres long and has 20 threads per cm, it requires approximately 4000 m of warp yarn. After warping, the starting

[23] For more information see for example Barber 1991; Breniquet 2008.

border/warp threads is/are tied to the starting border rod, and the loom weights must be tied to the warp threads (Fig. 2.12). In order to be able to change the shed, the next step is to heddle the warp threads to the heddle rods and thereby divide the warp threads in the different layers (Fig. 2.13). When the set up is completed, can one start weaving.

Different types of yarn need different tension when they are used in the warp on a loom. On the warp-weighted loom, the tension is provided by the attached loom weights. To a large extent it is the thread diameter that defines the appropriate warp tension, but the amount of tension required is also affected by how hard the thread is spun, the fibre quality, and the degree of fibre preparation. If too much tension is applied, the thread will break, whereas if the tension is not sufficient, the thread will not be held taut enough which will affect the weaving process, since it will be more difficult and time-consuming to change the shed. In order to be strong enough to be able to support the tension, the warp yarn is generally spun hard; the weft yarn can be more loosely spun.[24]

The results from the TTTC weaving experiments clearly demonstrate that: the *weight* of a loom weight dictates how many threads of a particular type can be fastened to it while the *thickness* of a loom weight controls how closely threads of a particular diameter will be spaced in the fabric. The tests have confirmed that if the weaver wants to produce an open fabric with thick yarn, (s)he should choose heavy, thick loom weights; if a coarse, dense fabric is desired, s(he) should chose heavy but thin loom weights. On the other hand, if an open fabric with thin threads is required, light, thick loom weights should be chosen. Finally, if a dense fabric with thin yarn and many threads per cm^2 is needed, thin loom weights should be selected.[25]

In Arslantepe, Turkey we have studied a total number of 86 loom weights.[26]

The loom weights are generally dated to 3800–3350 BC (period VII), 3350–2750 BC (period VIA–VIB1) and 2000–1750 BC (period VA). As can be seen in Figure 2.14, there is a difference in shape and material between the periods.

However, when comparing the loom weights and thickness another result is obtained (Fig. 2.15). No large variation in weight and thickness can be observed between period VII and VIA–VIB1, while the loom weights during period V are lighter and thinner.

		fired clay	unfired clay
VII	hemispherical		18
VI A–VI B1	conical	17	
	discoid elliptical	1	
	hemispherical	1	
	spherical ovoid	2	
VI B2–VI D	discoid elliptical	2	
	spherical ovoid	1	
V	discoid elliptical	28	8
	flat rectangular	2	2
	flat trapezoidal	2	
	hemispherical	1	1

Fig. 2.14. Arslantepe, chronological distribution of loom weights according to shape and material. No. 86.

24 Mårtensson *et al.* 2009; Andersson Strand & Nosch forthcoming; Cutler *et al.* forthcoming.
25 Mårtensson *et al.* 2009; Andersson Strand & Nosch forthcoming; Cutler *et al.* forthcoming.
26 Mårtensson *et al.* 2009; Frangipane *et al.* 2009.

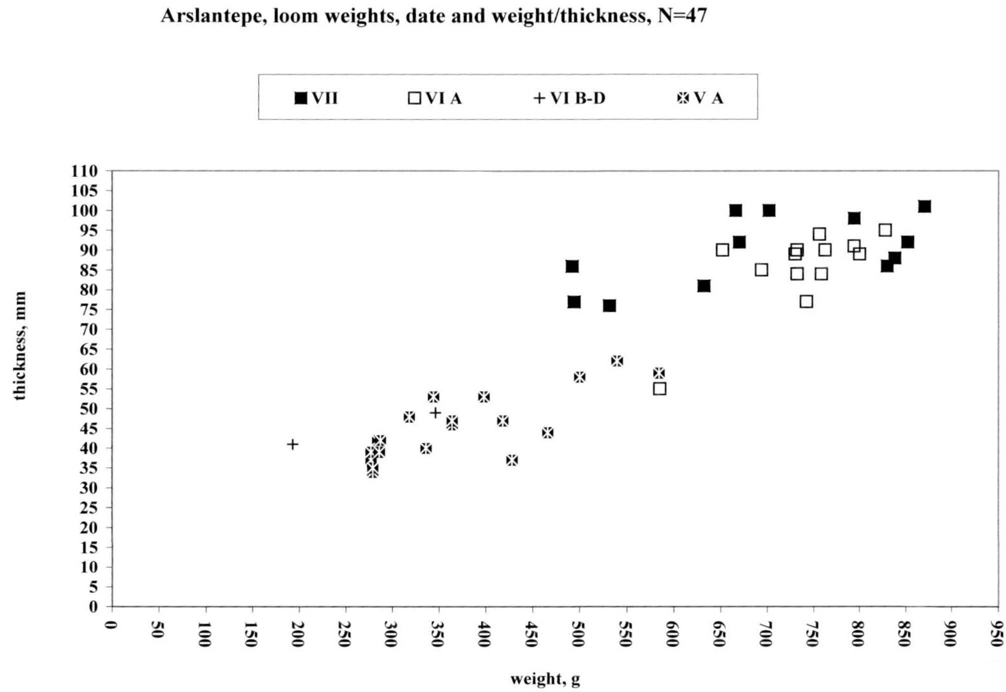

Fig. 2.15. Arslantepe, chronological distribution of loom weights according to weight and thickness, No. 47. Please note that only loom weights with preserved thickness and weight are included in this figure.

It may be concluded that the variation in the production of different qualities of fabrics is larger during the later period. The change in the type of loom weight, at least between period VII and VIA–VIB1, cannot be explained by a technological change; the answer has to be found elsewhere, perhaps in cultural or economic changes.

The results of the loom weight analyses enable us to interpret the types of fabrics that could have been produced with specific types of loom weights, not only in Arslantepe but at every site, region and period where loom weights have been recovered.[27]

4. Dyeing[28]

While plant fibres result in uniformly coloured shades of grey-white, wool comes in a variety of natural colours which was utilized in textile manufacture. Different sheep have various shades of brown and grey and the same sheep can have several natural shades. The different colours can be sorted and spun separately, taking advantage of the shades in the weave. Flax fibres can be bleached by various methods.[29]

---

[27] Mårtensson *et al.* 2009; Andersson Strand and Nosch forthcoming.

[28] For more information on dyes and dyeing techniques see Barber 1991; Cardon 2007.

[29] Fröier & Zienkiewicz 1991.

Furthermore, according to iconography and written sources, it is clear that some textiles were dyed.[30] The earliest evidence for dyeing comes from late 3rd millennium BC Mesopotamia, but there are indications that dyes were already being used in earlier periods.[31]

The way a textile is processed enables the dyer to colour textile material while at the fibre, yarn or cloth stage. In practice, the stage at which dyeing takes place varies mostly according to the textile fibres used and the effects desired.[32] There are different dyeing techniques, for example:[33]

- *Direct dyeing*: which involves soaking or boiling certain plants in water, and the fibres are then immersed in the resulting dye bath. The majority of the dyeing molecules, extracted from plants and animals, do not bind strongly with the textiles fibres.[34] In order to make them colourfast the extracts can be combined with various metallic salts, the so-called mordant dyeing.[35]
- *Vat dyeing*: which is the technique used to dye with indigo and molluscs. As Cardon excellently describes: "In their coloured form, indigo and shellfish purple are insoluble, therefore the various methods used to dye with these pigments can be explained by the necessity to render them soluble and allow them to impregnate the textile fibres. The methods all involve a reduction process in alkaline conditions. The fibres absorb the dyes in their soluble reduced forms that are barely coloured, being a greenish-yellow. When the fibres are taken out of the vat and exposed to oxygen in the air, indigo and purple precipitate again both inside the textile fibres and at the surface of the textile, which gradually take on blue or violet shades respectively."[36]

Furthermore, there is an endless amount of combinations to obtain different nuances of colours. For example, if a grey yarn was dyed in a yellow dye bath, the yarn will become greenish; if one dyes an indigo coloured yarn in a red dye bath, the yarn will be purple.

Numerous plants can have been used, for example, a blue colour could have been obtained from woad (*Isatis tinctoria L*), a red colour from dyer's madder (*Rubia tinctorum L.*) and yellow from dyer's weed (*Reseda luteola L.)* and saffron (*Crocus sativus L.)*.[37] However there are very few plants that give a red and blue colour, while for example yellow, brown, and green colours can be obtained from several plants.

The archaeological evidence for dyeing is scarce but it is nevertheless important to study:

- Water installations; when dyeing, large quantities of water are needed and the fibres/yarn/textiles generally also have to be washed before and after dyeing.
- Different tools, for example grinders, pestles and mortars, that could have been used order to prepare the dye material and mordants.
- Raw material debris such as for example, pollen from dye plants and murex shells.

5. Finishing

After the fabric has been taken off the loom, the fabric can be fulled or smoothed. Fulling is done in order to make the textiles denser. If a textile is fulled, it will become more waterproof which

[30] Nosch 2004.
[31] Barber 1991, 223–225.
[32] Cardon 2007, 6.
[33] Cardon 2007, 4–6.
[34] Cardon 2007, 4.
[35] Gleba 2008, 76.
[36] Cardon 2007, 4.
[37] Gleba 2008.

is preferable when producing an outer garment or a woollen sail cloth for example. The fabric is kneaded, stomped and pounded in wet and preferably warm conditions until the surface is matted to the degree desired.[38] Smoothing can be done in order to give the fabrics, especially linen, a shining and smooth plain surface. A simple plain stone or a stone of glass can be used for this process.

5.1. *Sewing*

Since there are no preserved costumes or other complete textiles from this region and period, there is little information on sewing and sewing techniques. However, finds of sewing needles demonstrate that, at least, some textiles were sewn and could also have been embroidered. When sewing a fine fabric, it is best to use a thin needle, while it is easier to sew with a larger bone needle in coarser and fulled fabrics.

5.2. *Other Textile Techniques*[39]

The knowledge of other textile techniques is scarce, which is partly due to the fact that these techniques do not require any tools at all and/or the tools are made of perishable material. However, from the iconography it can be seen that textile craftspeople have had knowledge of producing different types of bands and different braiding techniques and probably many others that cannot be seen in the archaeological record.[40]

* * *

In the TTTC programme, we are investigating textile tools from different sites and periods of the Bronze Age, and considering their possible function. Although very few textiles are preserved from this area, but by combining the textile tool data with contexts and function, together with the information provided by the available texts and also iconography, more detailed knowledge of what may have been produced is emerging. The results clearly demonstrate that textile production in the Eastern Mediterranean area during the Bronze Age was very varied, and that a wide range of fabrics, from plain to elaborate, were being manufactured.

Acknowledgments

The Danish National Research Foundation's Centre for Textile Research, University of Copenhagen. I warmly thank Marie-Louise Nosch, Ulla Mannering, Margarita Gleba, Joanne Cutler and Linda Olofsson for providing constructive criticism and wise viewpoints, for which I am always grateful. I would also like to thank Ulla Lund Hansen and Karen-Hanne Stærmose Nielsen for their support and for kindly letting me use their illustrations. This article is written with the support of the Culture Programme of the European Union in connection with the project DressID: Clothing and Identities in the Roman World.

[38] Barber 1991, 216.
[39] For more information see for example Barber 1991; Breniquet 2008.
[40] Barber 1991.

Bibliography

Adalsteinsson, S. (1991) Importance of sheep in early Icelandic agriculture. *The Norse of the North Atlantic, Acta Archaeologica 61*, 285–291.

Andersson, E. (2003) *Tools for Textile Production from Birka and Hedeby.* Birka Studies volume 8. Stockholm.

Andersson, E. B., Mårtensson, L., Nosch, M.-L. & Rahmstorf, L. (2008) New Research on Bronze Age Textile Production, *BICS 51*, 171–174.

Andersson Strand, E. & Nosch, M.-L. (eds.) (forthcoming). *Tools, Textiles, and Contexts. Investigations of Textile Production in the Bronze Age Eastern Mediterranean.* Ancient Textiles Series. Oxford.

Barber, E. J. W. (1991) *Prehistoric Textiles. The Development of Cloth in the Neolithic and Bronze Ages with Special Reference to the Aegean.* Princeton.

Breniquet, C. (2008) *Essai sur le tissage en Mésopotamie, des premières communautés sédentaires au milieu du IIIe millénaire avant J.-C.* Paris.

Cardon, D. (2007) *Natural Dyes, Sources, Tradition, Technology and Science.* London.

Carington Smith, J. (1992) Spinning and weaving equipment. In W. A. MacDonald & N. C. Wilkie (eds.), *Excavations at Nichoria I*, Minneapolis, 674–711.

Cutler, J., Andersson Strand, E., Nosch, M.-L., Firth, R., Poursat, J.-C. & Rougemont, F. (forthcoming) Textile production in Quartier Mu. In J.-Cl. Poursat (ed.), *Le Quartier Mu V* Études Crétoises. Athens.

Frangipane, M., Andersson Strand, E., Laurito, R., Möller-Wiering, S., Nosch, M.-L., Rast-Eicher, A. & Wisti Lassen, A. (2009) Arslantepe Malatya (Turkey): Textiles, Tools and Imprints of Fabrics from the 4th to the 2nd Millennium BC. *Palēorient No. 35.1*, 5–29.

Fröier, K. & Zienkiewicz, H. (1991) (1st ed. 1979) *Linboken, hemodling och hemberedning.* Stockholm.

Gleba, M. (2008) *Textile Production in Pre-Roman Italy.* Ancient Textiles Series 4. Oxford.

Moulherat, C. & Spantidaki, Y. forthcoming. Textiles from the Bronze Age to the Roman Period Preserved in Greece. In M. Gleba & U. Mannering (eds.) *Textiles and Contexts.* Ancient Textiles Series. Oxford.

Mårtensson L., Nosch, M.-L. & Andersson Strand, E. (2009) Shape of Things: Understanding a Loom Weight, *Oxford Journal of Archaeology 28(4)*, 373–398.

Nosch, M.-L. B. (2004) Red Coloured Textiles in the Linear B Inscriptions. In L. Cleland and K. Staers (eds.) *Colour in the Ancient Mediterranean World.* BAR International Series 1267, 32–39.

Olofsson, L., Andersson Strand E. & Nosch, M.-L. (forthcoming) Experimental testing of Bromze Age textile tools. In E. Andersson Strand & M.-L. Nosch (eds.) *Tools, textiles and Contexts. Textile Production in the Bronze Age eastern meidterranean.* Ancient Textiles Series. Oxford.

Shishlina, N. I., Orfinskaya, O.V. & Golikov, V. P. (2000) Bronze Age textiles of the Caspian Sea Maritime Steppes. In J. Davis-Kimball (ed.), *Kurgans, Ritual Sites, and Settlements: Eurasian Bronze Age and Iron Age.* BAR International series 890, 109–117.

Spantidaki, Y. & Moulherat, C. (2006) Rapport Identification des fibres. Copenhagen. http://ctr.hum.ku.dk/upload/application/pdf/f51d6748/Identification%20des%20fibres.pdf (090915)

Stærmose Nielsen, K.-H. (1999) *Kirkes væv. Opstadvævens historie og nutidige brug.* Forsøg med Fortiden 6, Historisk-Arkæologisk Forsøgscenter. Lejre.

Waetzoldt, H. (1972) *Untersuchungen zur neusumerischen Textilindustrie.* Studi economici e tecnologici 1, Roma.

Ventris, M. & Chadwick, J. (1973) (1st ed. 1956). *Documents in Mycenaean Greek.* Cambridge.

3. Textile Terminologies and Classifications: Some methodological and chronological aspects

Sophie Desrosiers

The field of textiles is very diverse and through the examination of iconography and a variety of archaeological artifacts, it is possible to observe that this diversity existed very early. Nevertheless, in the ancient Near East and the Mediterranean area from the 3rd to the 1st millennium BC, most research has focused on loom-weaving with flax and sheep's wool, leaving aside other materials and techniques. This paper will provide a tool to help interpret textile terms found in ancient documents, through an examination of the technical classifications undertaken within the last century. It will include the broad span of textile aspects and, without attempting to be exhaustive, it will give major examples of early textile archaeological finds. This discussion will extend the definition of the field of textiles while underlining the most significant categories that may have existed and were given specific names in the past. Last but not least, these classifications have been used as a basis to create modern descriptive terminologies, which are accepted by a majority of scholars and might prove to be useful to define the historical terms found in ancient documents.

1. The variety of the field of textiles

When, on the first class of the year, I asked my students to tell me what they consider as belonging to the world of textiles, invariably they look at their garments and often touch the cloth they are made of. Instantly, they consider that textiles consist in a soft material suitable for clothing. Then they extend their observations to interior furnishings with towels, curtains, and carpets, for example; and sometimes to public spaces with textiles used to protect, to bear messages, or to carry decorations. Later on, they recognize also the importance of textiles in technical activities such as transport with sails, carrying nets, and cordage, and many other activities including health care and art (Fig. 3.1). Such an exercise is aimed at showing to students the diversity of aspects, qualities, uses and functions of textiles. At present, this diversity can be epitomized by the juxtaposition of denim with a carrying net (Fig. 3.2).

Long ago at a period when paper and plastic did not exist, one can imagine how useful and varied textiles must have been, as they were needed for clothes, home, and public spaces, for technical uses such as fishing and hunting, animal harnessing, transport, and protection and for

a

b

c

d

Fig. 3.1. Variety of the field of textiles: a) Embroidered towel; b) Crown of flowers made of textiles for a funerary context (Paris, February 26th 2009; c) Knotted net made of animal hide strips with grass inside used to carry oranges (Oruro market, Bolivia, 1979); d) Cordage used on boats.

a

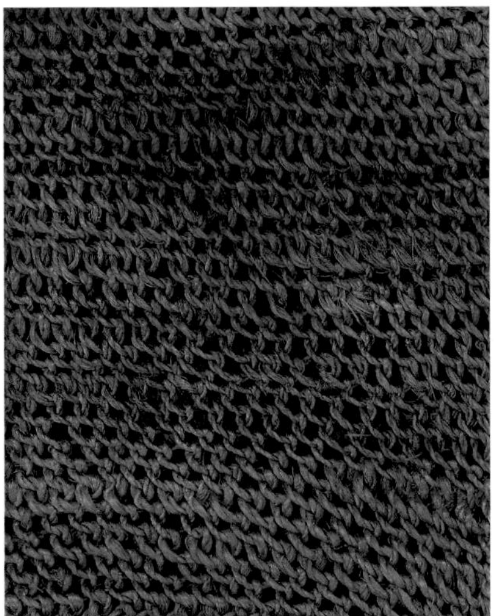

b

Fig. 3.2. a) Denim b) detail of a looped carrying net.

art. The fact that the brittleness of some textiles has seldom allowed them to survive must not interfere with our capacity to imagine the diversity not only of the linen and woolen textiles woven on the loom, but also of the other types of textiles executed for various uses, from other materials and with other techniques. Such a variety of textile types and forms has already been incorporated into numerous classifications.

2. Textile classifications

Besides an abundance of works classifying specific items,[1] two general and systematic classifications have been published during the second half of the 20th century. They offer complementary rather than contrary viewpoints.[2] The earliest one, *The Primary Structures of Fabrics* written by Irene Emery has undergone several editions at the Textile Museum of Washington DC since 1966.[3] A discussion of some parts of it and a Spanish translation of many of the terms have been proposed by Ann P. Rowe in 1984 and 2006. In 1973 appeared the *Systematik der Textilen Techniken* elaborated at the *Museum der Kulturen* in Basel (Switzerland) by Annemarie Seiler-Baldinger. This work is now known through a revised and expanded edition published in 1991, and by its English translation issued by the Smithsonian Institution Press in 1994 under the title *Textiles: A Classification of Techniques.*

These two classifications are the result of long-term research conducted in museums by the authors and conceived with different orientations. Emery's system has been built on the observation of archaeological and ethnographic textiles preserved in the collections of various institutions. It classifies "structures", a term used by Emery to designate the relationships between elements in finished fabrics.[4] Therefore it is better adapted to the understanding of preserved textiles whose methods of production are unknown and cannot be always reconstructed. Illustrations show enlarged models made with thick generic threads that represent idealized textile structures and tend to separate them from associations with specific fabrics and fabric qualities.[5] The Swiss classification reflects the anthropological background of Seiler-Baldinger and of the "Bühlers' school" she belonged to. It is based primarily on the classification of techniques, or methods of production, that anthropologists can observe during fieldwork, and secondarily on the interworked elements themselves.[6] These two levels are illustrated by two types of photos – one type showing people at work with their instruments and, another type showing textiles themselves – as well as by numerous drafts of the various structures obtained with each technique.[7] Both Emery and Seiler-Baldinger's classifications encompass simple techniques – i.e. without complex machinery – and they follow the same principle – from the simplest form to the

[1] For example: interlaced fabrics (Larsen and Freudenheim 1986); silks (CIETA 1964 and 1973); tapestry (Viallet, 1971); basketry (see classifications cited by Wendrich 1999, 24–56).

[2] A comparison between these two systems has been proposed by Balfet and Desrosiers, 1987.

[3] Emery 1980.

[4] *ibid.*, xi.

[5] *ibid.*, xii–xvi.

[6] Seiler-Baldinger 1994, xv–xvi.

[7] Additional photos matching the chapters of the classification have been published in the catalogue of an exhibition held at the Basel museum in 2000 (*Textil* 2000). One regrets that there are too few illustrations of people at work and that the drafts included in the classification do not show the trajectory of the elements as it is often the case in another well illustrated book dealing with a similar variety of fabrics: Collingwood 1987.

Part One: Components of Fabric Structures
I. Material Content
II. Structural Make-up

Part Two: Classification of the Structures of Fabrics
I. Felted Fibres
II. Interworked Elements
A. Single Element
B. Two Single Elements
C. One Set of Elements
D. Two or More Sets of Elements

Part Three: Structures Accessory to Fabrics
I. Accessory Stitches
II. Accessory Fabrics and Fabric Complexes
III. Accessory Objects

Fig. 3.3. Main divisions of The Primary Structures of Fabrics by Emery (1980).

most complicated – but, apart from their classification basis, they differ noticeably in two more aspects: their main internal divisions and their scope.

Part Two of Emery's classification considers the various ways fibers are organized, dividing "felted fibers" from "structures made of separate interworked elements" (or threads). She orders the later according to the number of elements (and sets of elements) involved (Fig. 3.3). Emery's Parts One and Three respectively present the variety of the materials used to make them, and some additional functional or decorative structures. Strictly speaking, she does not include specific aspects of basketry, but her terminology can be used for the numerous forms found in common with those used in clothes.[8] By contrast, the Basel

The Techniques of Element Production

The Techniques of Fabric Production
Primary Textile Techniques (made by hand or with the aid of very simple implements)
- Fabric Production with a Single Continuous Element : Production of Mesh Fabrics
- Fabric Production with Two or More sets of Elements (thread systems)
- Transitional Forms to Plaiting with Active Systems and Advanced Techniques of Fabric Manufacture

Advanced Textile Techniques (invariably require equipment)
- Warp Methods
- Half-weaving
- Weaving

The Technique of Fabric Ornamentation
Ornamentation by Additional Elements During Fabric Production
- Formation of Pile or Tuft Fabrics
- Beadwork
- Making of Borders and Fringes

Ornamentation after Production of the Fabric
- Ornamentation with Solid Materials
- Ornamentation with Liquid Materials

The Techniques of Fabric Processing (joining of fabrics)

Fig. 3.4. Main divisions of Textiles. A Classification of Techniques by Seiler-Baldinger (1994).

[8] Emery 1980, 208.

system considers techniques through the implements potentially used in production processes. Between two short presentations of the techniques dealing with element production and fabric processing, are the two main parts presenting fabric production and ornamentation (Fig. 3.4). Among the techniques of fabric production are distinguished "*Primary Textile Techniques*" made by hand or with the aid of very simple implements, and "*Advanced Textile Techniques*" which invariably require equipment and include "weaving".[9] In principle, only fabrics made by the interworking of elements are included. Excluded are those items made directly from fibers such as barkcloth and felt, but it does include matting and basketry which fit perfectly within the discussion of primary techniques using no or very simple implements. As for ornamentation, Seiler-Baldinger creates a division of two groups, defined according to when this is added: whether during or after fabric production. The volume includes long discussions on beadwork and on the making of borders and fringes which are not considered by Emery. On the whole, with the exception of felted textiles, the textile field delimited by the Swiss classification is more extensive.

3. Descriptive terminologies

Considering the question of terminology, both systems put forward descriptive terms applicable to the subject they cover. These terms are intentionally free, as much as possible, from specific contextual associations, in order to escape the confusions often made by the use of terms which had different meanings through time and space. For instance, Emery uses "textile patterned with supplementary wefts" instead of "brocade" which has acquired specific but diverse structural connotations through time.[10] Nevertheless, it is not possible to escape the use of some contextualized terms, for instance the substantive "textile" which is either a generic or a specific term according to the language, and to the environment where it is used.

In the English terminology established by Emery:[11]

> *"The terms* fabric *and* textile *are differentiated on the basis of their literal meanings and derivations:* fabric *(from the Latin* fabricare, *to make, to build, to 'fabricate') as the generic term for all fibrous constructions;* textile *(from the Latin* texere, *to weave) to refer specifically to* woven *(i.e.* interlaced *warp-weft)* fabrics."

But in German and French, at least, "*Textilien*" and "*textile*" are usually considered as generic terms, while "*Gewebe*" and "*tissu*" may be restricted to woven fabrics.[12] As pointed out by the

[9] Seiler-Baldinger 1994, xv–xvi.

[10] Emery 1980, 171–172; The CIETA vocabularies (1964; 1973) – and Burnham's *Warp and Weft* (1980), to a great extent inspired by it – must be used with care because their terms have been borrowed from the terminology current in the industry by the middle of the 20th century. As some terms have changed meaning through time, mistakes might occur when such changes are not taken into account. For instance, according to Diderot's Encyclopedia descriptions, many figured silks considered as damask during the 18th century do not fit with the meaning of damask in CIETA vocabulary and Burnham's book (Desrosiers 1988, 106–107). Therefore, the careless reading of 18th century documents might be misleading.

[11] Emery 1980, xvi; Barber accepts the same meaning in her book *Prehistoric Textiles*, even if she has to confront the presence of sprang and felt among the archaeological artifacts she takes into account (Barber 1991, 122–124; 215–222). Nevertheless, the *Webster's Dictionary* (1986) gives a generic meaning to the noun "textile", and a broader meaning to "fabric" considered as any product of building (see the entries "textile" on p. 2366; "cloth" on p. 428; "fabric" on p. 811)

[12] "restricted" because when used in the industrial field, the French "*tissu*" has a wider meaning. The *Petit Robert* (1993,

above quotation, this situation depends on the respective position of the available terms in a scale varying from generic to specific. In French for example, on one hand there is no equivalent to "fabric".[13] And on the other hand, "*tissu*" is in balance with the verb "*tisser*", while "*textile*" whose early meaning refers to "textile fibers" has a more generic connotation because of its standard usage not only in relation with fibers, but also with techniques, industry and art – four domains which refer to a field much larger than weaving.[14] Despite Emery's effort to standardize a scientific terminology, "textile" is also a generic term for many English speaking textile specialists, for example the "Textile Society of America" (TSA) and the "Center for Textile Research" (CTR).[15] To match with the terminology used at CTR and the workshop title on "textile terminology", the present article uses "textile" in its broad sense. And as it will follow the Swiss classification for techniques (see further), it will consider that textiles in a broad sense – or fiberwork – includes both products needing no tension to interwork the elements – i.e. matting and basketry – and those needing some type of tension or a frame to interwork the elements – i.e. cloth or textiles in a narrow sense.[16]

Coming back to classifications, they will be examined now in two steps in an attempt to underline first the diversity of the material, and second the diversity of the techniques of production.

4. Main categories of materials

The selected use of specific fibers constitute a fundamental issue as they are evidence of the relationship of societies to their environment and because they influence the type of textiles that can subsequently be manufactured with them.[17] If, for example, the use of flax and good quality wool have fostered the development of loom-weaving in many places of the ancient Near East and the Mediterranean area, then the presence of other raw materials has likely encouraged other developments. It would be interesting to have a list of the animal and plant species with a textile potential that were available in the regions to know more about other types of textile productions. This is somewhat of a side issue, however, as our primary point of discussion rather will be to define the main categories of material employed in textile constructions, then to give a few examples demonstrating their distinctive uses.

Materials are usually classified according to their origins – animal, plant or mineral. Further they are classified by their location within the source – such as the animal parts – external or internal fibers, or secreted filaments –, plant parts – seed and fruit hairs, leaf, bast, bark and

2259) gives even a generic meaning to "*tissu*". This difficulty comes from the fact that mechanically produced items do not correspond to those considered by the two general classifications.

[13] In the English issue of CIETA vocabulary (1964, 17), « Fabric » is translated by « *étoffe* », a fabric which, according to the *Petit Robert* (1993, 832), is a surface used for garments and furniture, so emphasizing only its shape and function.

[14] In CIETA French vocabulary (1973, 48), « *textile* » is defined either as « *matière propre au tissage* » (material for weaving), or as « *tissu, ou objet de toute sorte, créé par entrelacement de fils* » (woven textile, or object of any kind, created by the interworking of threads).

[15] This is the case for Margrethe Hald (1980, 124) for example.

[16] About the distinction between matting/basketry and textiles in a narrow sense, see Leroi-Gourhan 1954, 269. For a broader discussion, see Emery 1980, 208, and Rast-Eicher & Dietrich forthcoming.

[17] A greater attention has been given to them recently, in particular for early periods: Médard 2006; Beugnier & Crombé 2007; *Archéologie* 2008.

1. Animal
A. External fibers: wool and hair of sheep, goat, camel, rabbit, llama, alpaca, vicuña, buffalo, ovibos, cow, horse, pig, dog, beaver ... as well as human hair, feathers and so on. Strips of fur, rawhide, partially tanned skin, leather and so on, sometimes guilded.
B. Internal fibers: Sinew, strips of guilded guts.
C. Secreted filaments: caterpillars including the silk from the cocoons of the cultivated silkworm (*Bombyx Mori*), the wild silk from a number of silk moth caterpillars which cannot be domesticated and from communal larval nests, spider silk, byssus or pinna silk from the beard by which certain marine molluscs (*Pinna nobilis* and related varieties) attached themselves to the rock or sand floor of the sea.

2. Plant
A. Seed and fruit hairs: cotton, kapok, "vegetable silk" from milkweed floss.
B. Leaf fibers: from the leaves of monocotyledonous plants as sisal, yucca, abaca, pineapple, banana, esparto, agave, phormium, aloe, yucca, from palm-leaf (raffia)
C. Bast fibers: from the stem of dicotyledonous plants as flax (linen), jute, hemp, ramie, apocynum, nettle, and from the inner bark of certain trees and shrubs as lime tree ...
D. Bark and root fibers: some bark as cedar that can be shredded, and fibers found in some root structures (e.g. broom)
E. Miscellaneous: from various plant sources as palm-leaf segments (raffia), nuthusk fibers (coir), stem-fibres from monocotyledonous plants (Spanish or Southern moss), pappus down and bristles, reeds, grasses, et cetera.

3. Mineral
A. Asbestos: from several minerals having a fibrous structure.
B. Worked metals: gold, silver, and other metals are used in the form of wire or flat metal ribbon frequently wound around a core of another material

Fig. 3.5. Main categories of materials as classified by Emery (1980: 4–5).

root fibers, and miscellaneous –, or mineral types – asbestos or worked metals. Emery lists them in this way (Fig. 3.5).[18] Four examples concerning the area and period under examination will be considered now with more details: goat hair and low quality sheep wool, horse hair, wild silk, and gold.[19]

On the basis of Sumerian documents and their specific textile terminology, Hartmut Waetzoldt noted that, around 2000 BC in Mesopotamia, goat hair and some 3rd and 5th quality sheep wool were used to make strings, ropes and cords, and the webbing of beds, stools and chairs as shown by terracotta models and some archaeological artifacts (Fig. 3.6).[20] Boat sails were probably also made with goat hair. Concerning sheep wool, it is interesting to note the differentiation between

[18] Gordon Cook's classification (1993, vol. 2, 678–680) is slightly different, in particular regarding metallic threads which are considered as man-made fibers and not natural ones. Their number greatly exceeds the 15 main natural fibers considered by "FAO International Year of Natural Fibres 2009" presented on http://www.naturalfibres2009.org.

[19] The case of other fibers as cotton – known in Jordan by the 5th millennium BC and in Pakistan one millennium before –, hemp, nettle, and other plant fibers, as well as Byssus or marine silk have been recently examined elsewhere (Betts *et al.* 1994; Moulherat *et al.* 2002; Breniquet 2008, 101; for details on the production of marine silk, see Meader *et al.* 2004). In Neolithic Europe, beside flax, Médard has demonstrated the use of bast from several trees and insisted on the likely use of many other plants in textile productions (Altorfer & Médard 2000, 54–58; Martial & Médard 2007).

[20] Waetzoldt 2007.

Fig. 3.6. *Reconstruction of a bed in an Old Babylonian grave from Baguz (Syria) (After Waetzoldt 2007, Fig. 18.3).*

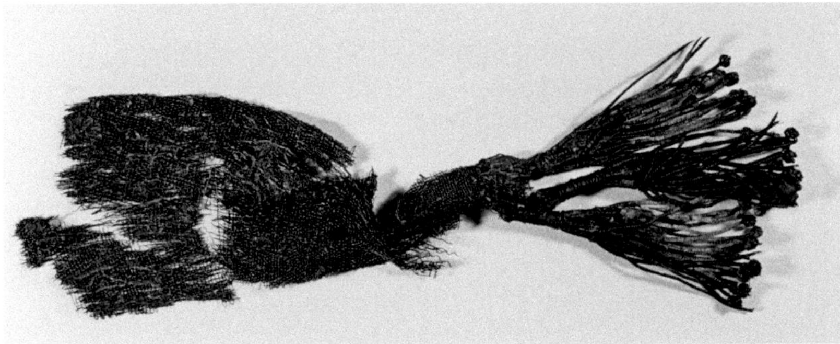

Fig. 3.7. One fragment of a horsehair woven band found with the Cromaghs hoard, Armoy, Ireland (Courtesy of the National Museum of Ireland, Inv. No. IA:1906.13).

various qualities that existed and also, the practice of using the lower quality materials for ordinary utilitarian objects. The same rules are present in the Mari texts investigated by Jean-Marie Durand and this practice has been regularly observed by specialists of archaeological textiles when comparing within one specific textile tradition the qualities of the wool in relation to the qualities of the various types of textiles woven with them.[21]

More unusual is horsehair. Found in very few Bronze Age and later sites in Northern Europe – for example the hairnet from Skrydstrup, Denmark in the Middle Bronze Age – and more often in Southern Siberia and Xinjiang sites of the second half of the first millennium BC, horsehair is stiff and favors the production of three-dimensional objects (For example, ethnographic collections from Tibet include snow glasses made from horsehair).[22] One particular piece – a band woven in broken twill with elaborate ornamental tassels (Fig. 3.7) – discovered with the Cromaghs Hoard in Ireland (9th–8th century BC and later) has been compared by Elisabeth W. Heckett with horse trappings represented on bas-reliefs from Nimrud (*c.*875–860 BC).[23] Whatever the place of production of such a fine object, this comparison provides evidence of the probable early use of horsehair for textile construction in the Near East.

Regarding wild silk, several indicators show that Lepidoptera producing filaments were present as early as the Bronze Age in the Aegean area.[24] The discovery of what is probably a cocoon of *Pachypasa otus* (L.) on the island of Thera (Santorin) in the middle of the 2nd millennium BC is one clue. This evidence can be seen in the context of an existing rich iconography of "butterflies" on late Minoan I and Mycenaean seals and wall paintings, with round markings on their wings, that recall those of the moth *Saturnia Pyri* Den., another silk producing insect present in the area (according to the research of zoologists). The "butterfly" motif appears not only on gems and seals, but also on a gold balance from Mycenae interpreted either "for ritual purposes, perhaps the weighting of the soul of the departed", for the weighting of gold for currency, or as now possibly

[21] Durand 2009, 15; Waetzold (1972, 43) found a similar situation in Sumerian texts.

[22] Heckett 1998, 34–36; Desrosiers 2001, 146 Fig. 3, 203 No. 89; Ronge (n.d.) photo p. 159 right.

[23] Heckett 1998, 35–36.

[24] Panagiotakopulu *et al.* 1997.

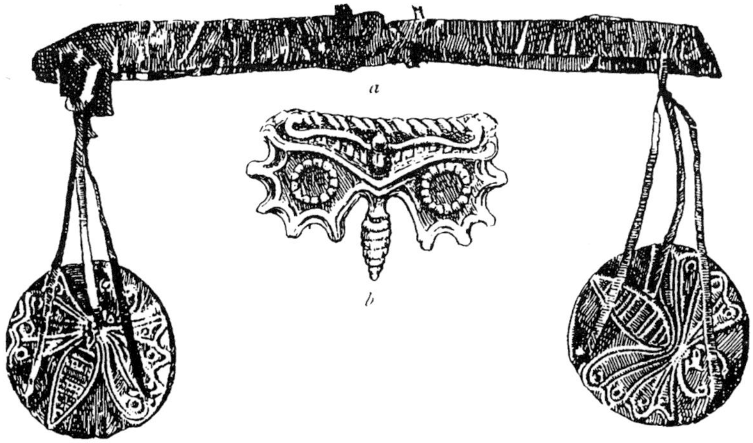

Fig. 3.8. Gold balance with moths on its scales and gold moth pendant from the third Shaft Grave, Mycenae (After Evans 1930, 150, Fig. 100).

used to weight silk thread (Fig. 3.8).[25] It is also represented on the "ship flotilla" wall painting at Thera.[26] While the literary evidence suggests several terms for luxurious materials, no candidate has proved to refer inevitably to wild silk so far.[27] But its use in the Aegean area prior to classical times remains possible. The recent identification of wild silk from at least two different species in two important Indus sites – Harappa and Chanhu-Daro – proves that by *c.*2450–2000 BC the use of wild silk for textile construction existed outside of China.[28]

Finally, as shown by Maria Giovanna Biga during the workshop, fine gold strips interworked to make small objects looking like miniature mats have been found in Ebla in a 3rd millennium BC context (Fig. 3.9). They might fit with the term *zimidatum* – possibly a band in gold thread used as ornamentation for ceremonial garments – mentioned in a text found nearby.[29] These were made well before the earliest interworking of gold in a fabric identified to date as a tapestry woven with gold strip and shellfish purple dyed wool which had been buried with Philip II of Macedonia (372–336 BC) in Vergina.[30] The presence of these gold constructions cause us to pose the interesting question whether the gold lamella were interwoven by a weaver, or whether the textile technique was imitated by a jeweller.

5. Main categories of techniques

From the two general textile classifications presented above, the election of one or the other depends more on the context of the research subject and on the questions posed than on a

---

[25] *ibid.*, 424.
[26] *ibid.*, 425–426, Figs. 7–8.
[27] *ibid.*, 426–428; Talon 1986; Breniquet 2008, 100.
[28] Good *et al.* 2009.
[29] See Biga this volume; Matthiae 2006, 454–455 and Fig. 8; Ramazotti forthcoming; Pomponio (2008, 101, 107) comments the association of textiles with metallic decorative objects in various texts from Ebla, and Durand (2009, 23) in those from Mari.
[30] Flury-Lemberg 1988, 234–237. An earlier piece with metallic thread (gold and silver wrapped around a fiber core) could be the embroidered linen fragment from Koropi (Attica), end of the 5th century BC (Schuette & Müller-Christensen 1963, 25, pl. 2, No. 2 ; Barber 1991, 206)

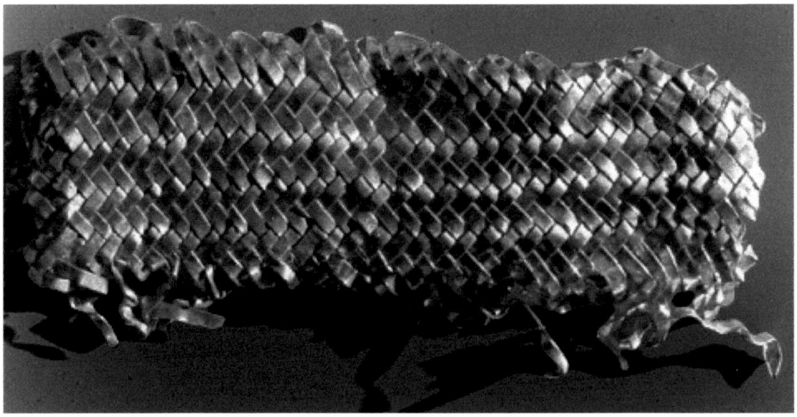

Fig. 3.9. One out of six gold bands from the Royal Palace G in Ebla (Syria), 2400–2300 BC. Gold lamella, diagonal plaiting in 2.2 twill, c.4.5 × 1.2 cm (Idlib, Ebla archaeological museum, inv. No. TM.04.G.170a. Copyright Missione Archeologica Italiana in Siria, Roma, 2005).

preliminary decision about the superiority of one system over the other. In the present case dealing with 3rd to 1st millennium BC documentation that is characterized by a scarcity of material remains and the fact that written documents exists that describe materials, crafts and craftsmen, there is no doubt that the Basel classification approach to the diversity of techniques will prove more adaptable to this subject.[31] As will be shown further, however, it does not allow the easy classification of archaeological artifacts which may have been made with multiple techniques. But it provides an interesting approach to help to understand the relationship between extant fiber remains and the tools found with them.

The classification will be presented under a very simplified and slightly enlarged form where I have excluded some examples and the numerous variants which can be found in the original book, but have made some additions. One, as a category in itself, I include the direct transformation of fibers into long elements and flat products (though, as specified above, Seiler-Baldinger considers as « *Textilien* » only those items made by the interworking of elements and not the elements themselves, nor those made directly from fibers). As another addition, I have extended the definition of the techniques of ornamentation to include dyeing and finishing processes which aim at embellishing textiles. Finally, here and there, I have included some techniques identified in archaeological remains and not listed in the Swiss classification either because they are not practiced any more, or because they had not been identified yet. When it is necessary, each category will be illustrated with one example showing one process (among several potential ones) or else one product resulting from it, and by a schematic drawing of the way elements are interworked.

The direct transformation of fibers can be considered as a preliminary group of primary textile techniques composed of two sub-groups.

The first sub-group encompasses the production of flexible elements such as threads, strings, ropes, and stiffer elements more often used for matting and basketry. Some need almost no processing. Others need specific techniques, the main ones being: reeling of long lengths of threads (Chinese silk), knotting of short elements (horsehair, raffia...), twisting between the hands or on a surface (such as the thigh) for relatively long and stiff fibers (tree and leaf bast...), splicing of flax

[31] An inventory of early archaeological textile remains found in the Near East has been published recently by Breniquet 2008, 55–58.

Fig. 3.10. Preliminary group of Primary Textile Techniques I. Production of flexible elements: Three types - and three ages - of ropes: sewn and twisted strips of leather (background), braided threads of spun black and white lama wool (middle), long thick strip of tyre (cut continuously like an orange) (Oruro, Bolivia, 1979).

as in ancient Egypt, spinning with a hand spindle for softer fibers as flax, hemp, cotton, wool, wild silk, and so on, as well as plying by twisting several threads together in order to make them stronger (Figs. 3.1d, 3.10).[32]

Although they cannot be considered as fiberwork in the strict sense of the word, the cutting of strips in flat flexible materials, sometimes twisted (as leaves, leather, skin, woven textiles, gold leaf and many others) and the drawing of metal threads that can be used for textile constructions are added here (Figs. 3.1c, 3.9, 3.10). This long list shows the diversity of techniques used to obtain elementary textiles and the need for no or few specific implements besides hand spindles whose shape may vary from a simple stem to a hook.[33] If the presence of spindle whorls in archaeological contexts supports the local practice of hand spinning, their absence cannot be used to prove the contrary – that no elementary textiles were produced – since the various implements potentially used for that, including spindle whorls, may have been made with a perishable material.[34] In regards to textile terminology, the making of threads, strings, ropes and other elements can be traced through early written documents of the ancient Near East and the Mediterranean area.[35]

The second sub-group includes the production of flat fabrics such as beaten barkcloth and papyrus obtained from natural plant forms, and felt and paper resulting from the agglomerating of massed disconnected fibers (Fig. 3.11).[36] On the basis of geographical and chronological reasons,[37] felting is the only technique directly relevant to the textile terminologies under examination. As explained by Barber,[38] it "is the process of matting wool or hair into a stable fabric by a combination of pressure, warmth, and dampness. Heat and moisture cause the tiny scales on the surface of the hairs to stick out; and prolonged kneading when they are in this condition makes them catch on each other until they are inextricably interlocked."[39] Being the only natural fibers

---

[32] Seiler-Baldinger 1994, 2. See also Bellinger 1959; Barber 1991, 39–78; Altorfer & Médard 2000, 47–54 and Médard 2006 for examples in Neolithic Europe.

[33] For instance see Barber 1991, 41–78; Breniquet 2008, 110–126; and for valuable details Andersson Strand in this volume.

[34] Karen Hardy concurs that the discovery of perforated objects among very old remains suggests the construction of elementary textiles as strings at a very early date: 300,000 for a few objects, and 200,000 for beads (Hardy 2008, 272). Fragments of cordage are known from the Upper Paleolithic in the Levant (Ohalo II, Israel, *c.*17,000 BC) and Europe (Lascaux cave, France, *c.*15,000 BC (Nadel *et al.* 1994; Glory 1959). Imprints of fine complex woven textiles dating to around 27,000 BC were also discovered in Moravia (Pavlov I, Czech Republic) (Adovasio *et al.* 1996 ; Hardy 2008, 273).

[35] See for example: Waetzoldt 1972, 120–123, and 2007; Ribichini & Xella 1985, 19–20.

[36] Following Emery 1980, 20–24.

[37] Beaten barkcloth is considered to be limited to tropical and sub-tropical regions, *Cyperus papyrus* is native to the Upper Nile, and paper was invented in China during the last centuries BC.

[38] Barber 1991, 215.

[39] Fulling, practised on woven textiles and later on knitted ones, is a process very close to felting but it aims at changing the touch (hand) and the appearance of textiles, not at making them. It will be considered among the techniques of textile ornamentation.

a b

Fig. 3.11. Preliminary group of Primary Textile Techniques I. Production of flat felted fabrics: a) Beaten barkcloth (inner bark of paper mulberry tree) on a stone anvil with wooden beater (experiment, Hawaii 2008); b) Close view of felt.

with scales, animal hairs are the only ones to felt. In the areas where it is still practiced, felting requires warm water and a mat (to apply the uniform pressure) whose specific function in and of himself is not identifiable would it be found among archaeological remains.[40]

Felt is usually considered to have been developed by nomads of Eurasia whose life is, to the present day, still highly dependent upon it. In the Altai region *c.*300 BC, the grave mounds of Pazyryk have preserved an enormous quantity of felt fabrics used for clothing, rugs, tents, horse blankets and saddles, and decorative objects.[41] Towards the west, in Anatolia, large quantities of felt have been found from 700 BC at Gordion – the capital created by the Phrygians recently arrived from the steppes.[42] From the point of view of textile terminology, the words for felted or felt-like goat-hair bed-pads tentatively identified in Sumerian by Steinkeller have been contradicted by Weatzold who showed "that these craftpeople made strings, ropes, different types of plaited straps, mats and similar products".[43]

The first group of primary textile techniques distinguished by the Basel classification consists in the production of mesh fabrics "with a single continuous element".[44]

Some mesh fabrics are made "with a continuous element of limited length" allowing the maker to introduce the thread into the meshes of the preceding row as in the case of linking, looping, and knotting (Figs. 3.1c, 3.2b, 3.12). These techniques require either no implement, or simple ones – poles, gauges or sticks very difficult to identify within an archaeological context,

[40] Laufer 1930, Gervers & Gervers 1974.

[41] Rudenko 1970, 35, 285; Barber 1991, 219–220.

[42] Bellinger 1962; Barber (1991, 216–221) judges the presence of felt in 6th millenium BC Çatal Hüyük unlikely as sheep were then "predominantly kempy rather than woolly". She considers that the presence of felt during the 3rd millenium BC in Beycesultan could be possible... But Lloyd & Mellaart (1962, 45) give a too vague description of "a thick deposit of some hairy substance which resemble partially burnt felt" to consider the Beycesultan mention as an evidence.

[43] Steinkeller 1980; Weatzold 2007.

[44] Seiler-Baldinger 1994, 7–25.

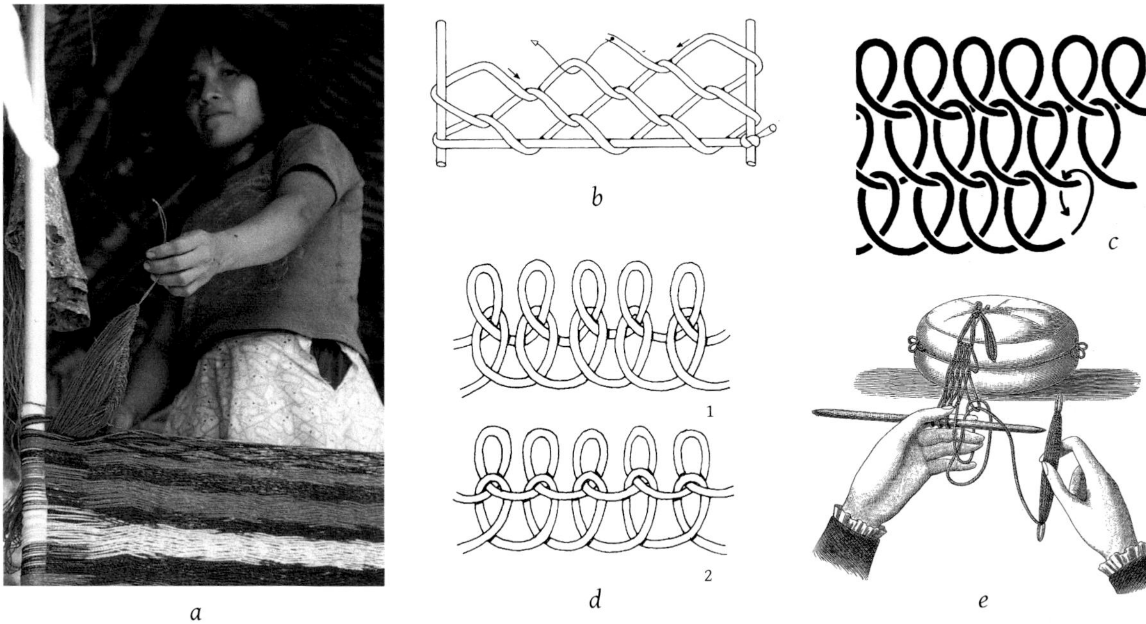

Fig. 3.12. Primary Textile Techniques I. Production of mesh fabrics with a continuous element of limited length: a) Yagua woman from the Atacuari river making a hammock: complex linking (linking with skipping of rows) Loreto, Peru, 1981 (Copyright A. Seiler-Baldinger); b) Simple linking (After Collingwood 1987, 8); c) Simple looping; d) Cross-knitt looping: 1. obverse and 2. reverse (obverse looking like knitting) from Nahal Hemar (After Schick 1988, fig. 9ab); e) Knotting (After Dillmont 1886, Fig. 638).*

**According to conventions established with present indigeneous populations, any use of the graphic motives reproduced in the following figures showing people at work or ethnographic pieces is forbidden without previous agreement with the group directly concerned.*

or needles stronger than those used for sewing.[45] Nevertheless, looping and knotting have been used very early in various areas of the world as shown by archaeological remains. In the Near East, early evidence comes from Nahal Hemar cave (Israel, mid-7th millenium BC), Jarmo (Iran, 7th millenium BC) and in Europe from Friesack 4 (Germany, 8th millenium BC).[46] Their usually opened structure and their flexibility make them appropriate for carrying and fishing nets, hunting nets, hair nets, and so on. An outstanding example is the net-like headgear from Nahal Hemar (Fig. 3.13) which combines dense knotting alternating with open areas and a wide band produced with the help of another technique.[47] The interlinked threads of the band does not seem to have been produced with sprang because there is no observable reversal at the center (a key element of sprang: Fig. 3.17a, c). However, it is not clear whether this structure is the result of working with "an element of limited length" (as on Fig. 3.12ab) or whether it comes from another "warp method" close to sprang (as on Fig. 3.17a) but worked with free ends.

Other mesh fabrics are made "with a continuous element of unlimited length" allowing the maker to introduce only loops of thread into the preceding meshes. In this category are two well-

[45] Various types of needles have been illustrated by Hald 1980, 278–279, Figs. 281–283.
[46] Adovasio 1975–1977; Kernchen & Gramsch 1989; Gramsch 1992; Schick 1988, 34–36.
[47] Schick 1988, 35–36, Fig. 12 and pl. XVII–XVIII.

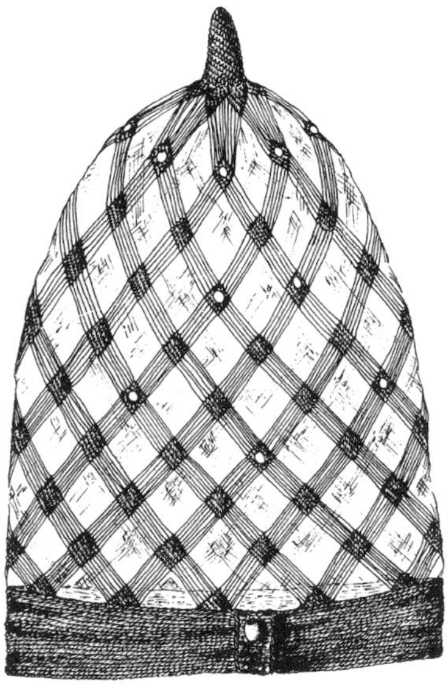

Fig. 3.13. Net-like headgear from Nahal Hemar (After Schick 1988, Fig. 12).

known techniques: crocheting and knitting made again with no implement, or simple ones – a hook, one or several sticks (Fig. 3.14). The earliest knitted pieces identified so far date from the Abbasid.[48] Knitting, and also crocheting, appear as late techniques which fall out of the range of ancient textile terminology.

The second group of primary techniques consists of systems for "fabric production with two or more sets of elements (thread systems)" and it is also divided into two sub-groups.[49]

The first sub-group includes "plaiting with a passive and an active system" : among others, binding where the crossing of two or more passive sets of elements "are fixed with the help of a running active thread, a set of elements or short thread pieces", twining where "Two or more active elements are twined together in such a way that with every twist they fix one or more elements of the passive system", or coiling where "The elements of the passive systems, sometimes a bundle of threads, are fixed by an active element (...)" (Fig. 3.15). These techniques require either no implement or a kind of hook or needle used when coiling with a flexible

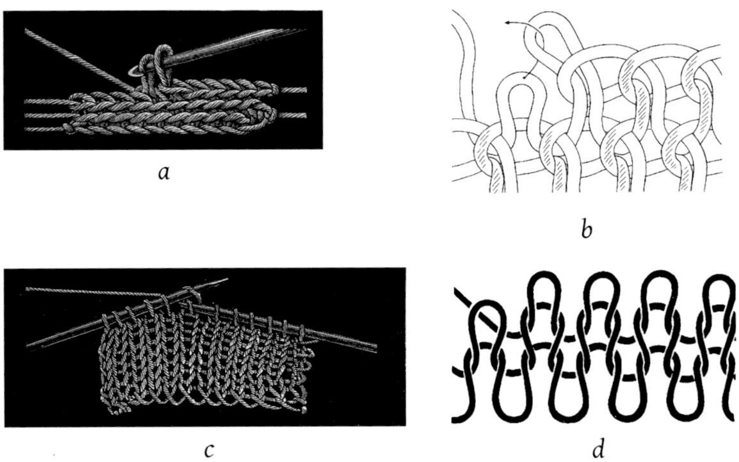

Fig. 3.14. Primary Textile Techniques I. Production of mesh fabrics with a continuous element of unlimited length: a) Crocheting; b) Simple crochet stitch (After Collingwood 1987, 35); c) Knitting; d) Plain knitting stitch (a and c: after Dillmont 1886, Figs. 445, 392).

---

[48] Cornu *et al.* 1993, 260–261, No. 159. The piece from Dura-Europos (Syria, mid-3rd century AD) considered by Forbes (1956, 179) as knitted does not seem so from observing the illustration published in Pfister & Bellinger 1945, 54–56: Nos 265–266, pl. I and XXVI. It might be the result of cross-knit looping, a variation of looping (looking like a knitted fabric) which has been observed by Schick among the Nahal Hemar fiber remains (Fig. 3.12d. See Schick 1988, 34, Fig. 9 and pl. XVI$_2$). See also Barber 1991, 122.

[49] Seiler-Baldinger 1994, 26–47.

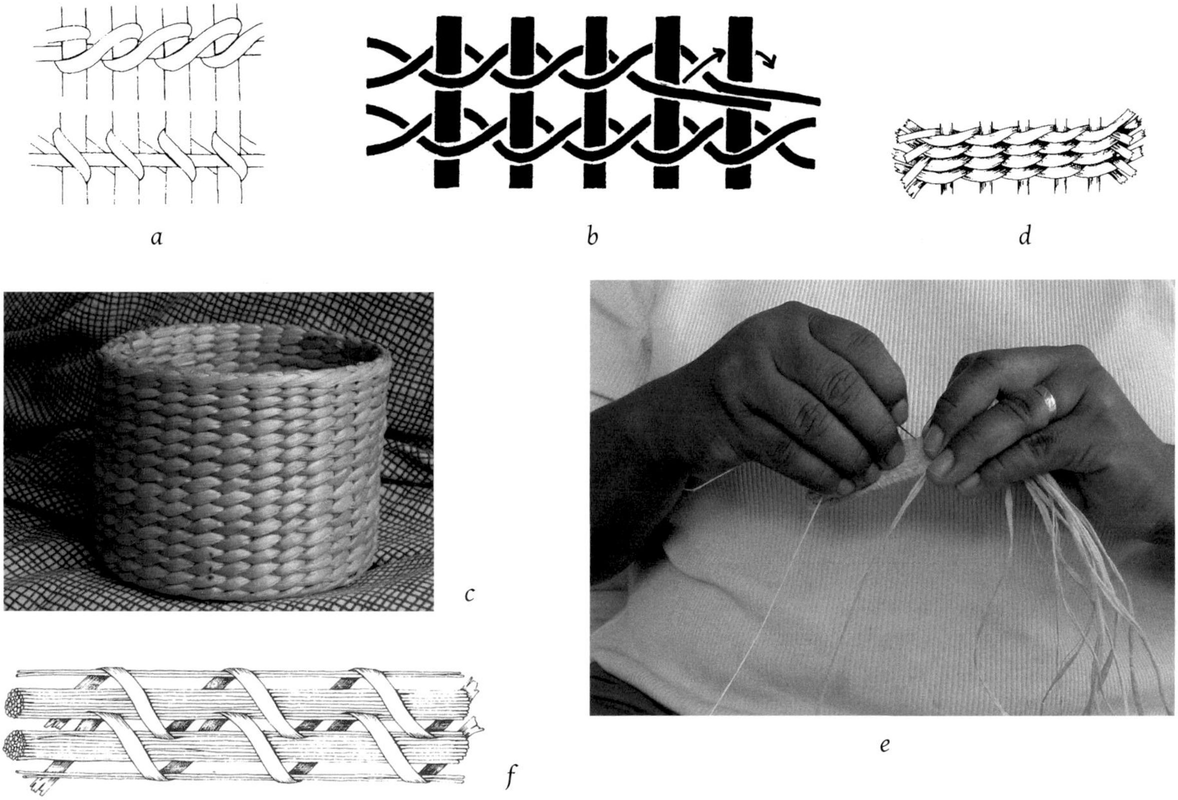

Fig. 3.15. Primary Textile Techniques II. Production with two or more sets of elements. Plaiting with a passive and an active system: a) Binding: example found in Nahal Hemar (the obverse looks like twining, the reverse is different); b) Two-strand twining; c) Two-strand twining: box with internal and external faces identical, Ikea, c.2000; d) Close twining found in Nahal Hemar; e) Coiling: Arawak-Lokono woman making a flat element, Balaté, Saint Laurent du Maroni, French Guyana, 2003 (Copyright D. Davy); f) Split coiling: example found in Nahal Hemar (a, c and e: after Schick 1988, Figs. 3, 13, 4)

active thread or in a tight manner. They are usually considered as the speciality of basket weavers because there is no set of elements under tension. Archaeological remains are also known from early sites: for example, Nahal Hemar, Jarmo, Çatal Hüyük, and in Europe Noyen-sur-Seine, Zurich and Hornstaad ... .[50]

The second sub-group includes "plaiting with active systems" in two, three and more directions: right-angled plaiting, diagonal plaiting (as illustrated by the Ebla gold piece on Fig. 3.9), braiding (for instance the camelid braid on Fig. 3.10), oblique intertwining, and ply-splitting (Fig. 3.16). Ply-splitting is composed of plyed elements i.e. threads or cords made of at least two elementary threads twisted together, and the process consists of introducing the threads of one direction through those of the other direction. Although ply-splitting has been identified only quite recently

[50] Adovasio 1975–1977; Schick 1988, 33–34, Fig. 3, pl. XV; Schick 1989, 42–47; Mordant & Mordant 1992; Rast-Eicher 1992; Müller 1994.

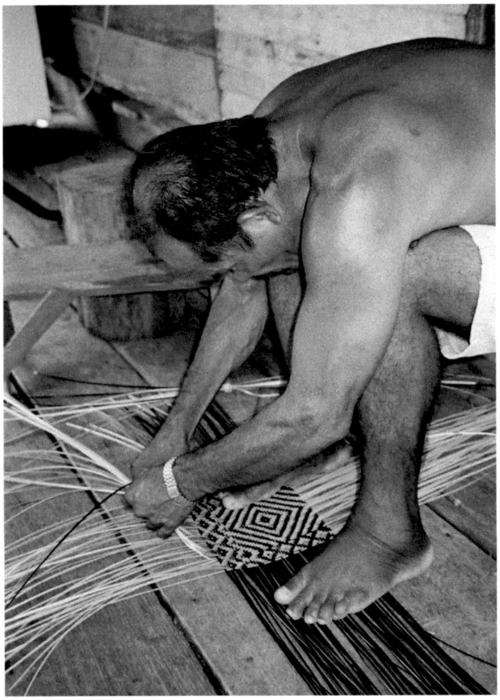

a

b

g

c

d

e

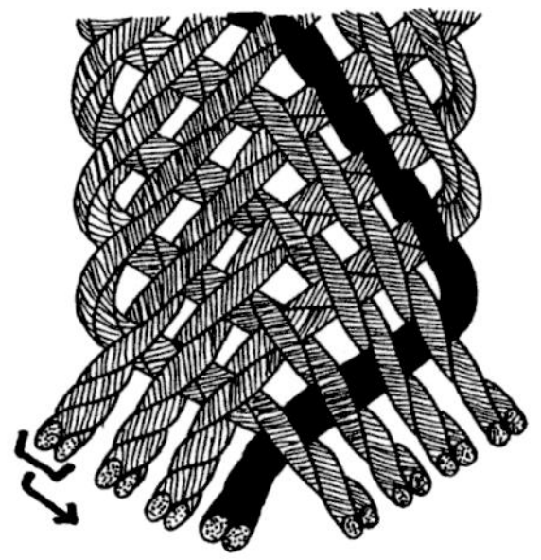

f

Fig. 3.16. Primary Textile Techniques II. Production with two or more sets of elements. Plaiting with active systems in two directions: a) Right-angled plaiting: Palikur man making a sieve, Saint Georges de l'Oyapock, French Guyana, 2003 (Copyright D. Davy); b) Oblique plaiting: Basket making by Luiz da Silva Baniwa, Itacoatiaramirim, São Gabriel da Cachoeira, Amazonas, Brasil, 2006; c) Decorative braid on a woolen tunic from Djoumboulak Koum, Keriya River, Xinjiang, China, 5th to 3rd century BC (After Debaine-Francfort & Idriss 2001, 185; copyright Mission Archéologique franco-chinoise au Xinjiang/X. Renaud/ Fondation EDF); d) oblique intertwining; e) right-angled ply-splitting; f) oblique ply-splitting; g) Plaiting in three directions: first step in the making of a carrying basket by Elídio Isidoro Coelho, Espírito Santo, Santa Isabel do Rio Negro, Amazonas, Brasil, 2006; (b and g: photos L. Emperaire, IRD, projet Pacta IRD-CNPq/Unicamp. D to f: after Desrosiers 2001, 148, Figs. 5b, 2–3).

as a process – in 1976 according to Seiler-Baldinger's bibliography, and is still very frequently used in India for making camel harnesses, it has a long history. Two bands made with this technique have been identified among 5th to 3rd century BC archaeological finds in Xinjiang together with pieces in oblique intertwining.[51] These techniques require either no implement or simple ones: such as a needle or hook, and potentially small bobbins use to wind the threads.[52]

The second main category of "advanced textile techniques" requires the "use of a warp, i.e. a tensioned and fixed set of elements". Therefore there is a need for one or two posts or a frame to hold the tension of the threads. This category is divided into three groups: "warp methods" without other implements, "half-weaving" when one shed is formed automatically, and "weaving" when at least two separate sheds can be formed by mechanical or automatic means called looms.[53]

"Warp methods" using no implement other than the one giving tension to the threads are divided between those productions with an active warp and those with a passive warp.

In the first case, warps are fixed at one end or at both ends and are manipulated with the hands through warp-twining or sprang techniques. Sprang consists in plaiting with threads stretched between two parallel beams in such a way that work progresses towards both ends until the two areas of interworked threads meet in the middle and are fixed to keep them from unravelling (Fig. 3.17a, b, c). Examples have survived in Europe from the Early Bronze Age and later.[54] They appear much later in the Mediterranean, but certainly from the 5th century AD as frames painted on Greek vases and often considered as "tapestry or embroidery frames" may be mis-identified as they in fact show all the characteristics of a sprang process (Fig. 3.17d).

In the second case, the passive warp is usually fixed by the weft in various ways, producing weft-faced or open-work according to the density of the wefts. Three types are illustrated by Seiler-Baldinger: weft-wrapping used in "soumak" carpets, weft-twining, and weft interlacing – tapestry being an important variation allowing the composition of elaborate designs dependent on the localized use of wefts of various colors (Fig. 3.18).

"Half-weaving" – with one shed formed automatically – was probably a step towards weaving with a loom (Fig. 3.19). It is difficult to deduce from ancient textiles that they were produced this way as there are few details which may prove it. But webbing such as found used for the bed from Baguz (Fig. 3.6) was probably made by using the upper beam of the wooden bedframe to form one shed. Therefore, it can be considered as half-weaving.

Regarding "weaving", the classification shows a progression according to "the fully automatic shedding achieved by implements specially designed for the purpose". It separates again weaving with an "active" warp – finger weaving, turning weights weaving and tablet weaving (Fig. 3.20) – from weaving with a "passive" warp – with a rigid heddle and with the aid of individual heddles (where I would distinguish the case with heddle and shed rod from the other with only heddles) (Fig. 3.21). These issues will not be examined in detail here as there is an abundant bibliography on these techniques, and as heddle weaving has been extensively presented by Andersson Strand and Breniquet in this volume.[55]

---

51 Seiler-Baldinger 1994, 43; Collingwood 1998; Desrosiers 2001, 148–149 Figs. 2–3, 5b and 7, 166 No. 51, and 182 No. 73.

52 Bobbins used for braiding are illustrated by Hald 1980, 242–245.

53 Seiler-Baldinger 1994, respectively 49–67, 68–70, 71–102.

54 Hald 1980, 274 and 251–275 for a general view on sprang technique and history; Barber 1991, 122–124.

55 See Barber 1991, 118–122.

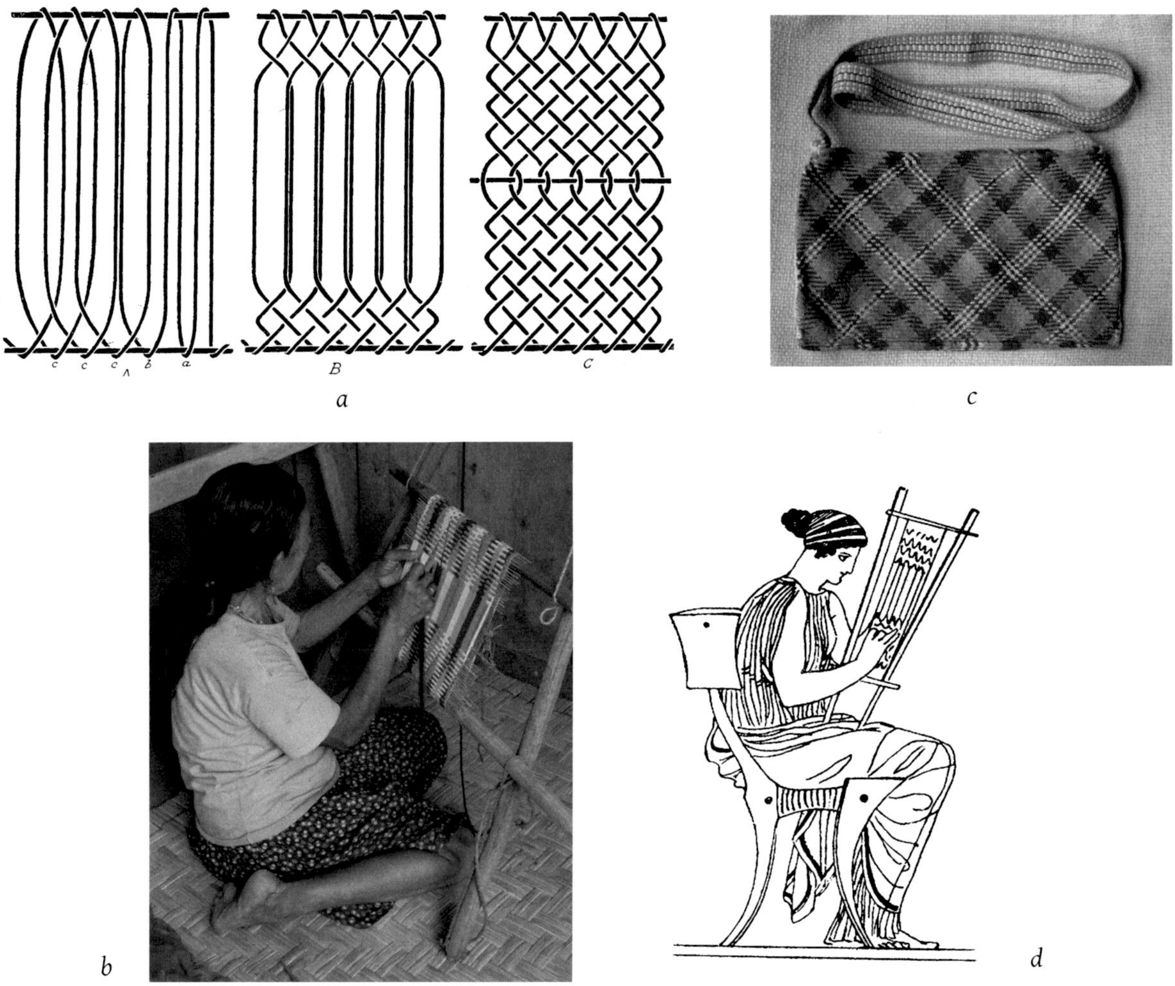

Fig. 3.17. Advanced Textile Techniques I. Warp methods with an active warp: a) Interlaced sprang in plain weave: three main steps (After d'Harcourt 2008, 80, Fig. 49); b) Mosetene Doña María making a bag interlaced in 2.2 twill on a sprang frame in San José [Covendo], Beni, Bolivia, 2009 (Copyright I. Daillant); c) Mosetene bag in sprang with the loops securing the meeting line positioned at its bottom after folding the sprang fabric (I. Daillant's collection); d) Greek woman working on a frame showing some specific features of a sprang frame, including the absence of heddles, while showing a work executed at the center area of the warp and the symmetrical effects at its top and bottom. The sticks used to work the sprang are missing (After Roth 1978, Fig. 29b).

With the examples having two or more sets of elements (Figs. 3.15 to 3.21), the limits of the Basel classification become more obvious when applied outside of ethnographic observations.[56] Although in many cases a detail identified under the close observation of an archaeological textile or the co-existence of implements in the same archaeological context might gives the clue to help identify the technique used to make it, it is not always possible to distinguish between

[56] This limit is perfectly described in Seiler-Baldinger's book foreword (1994, xv).

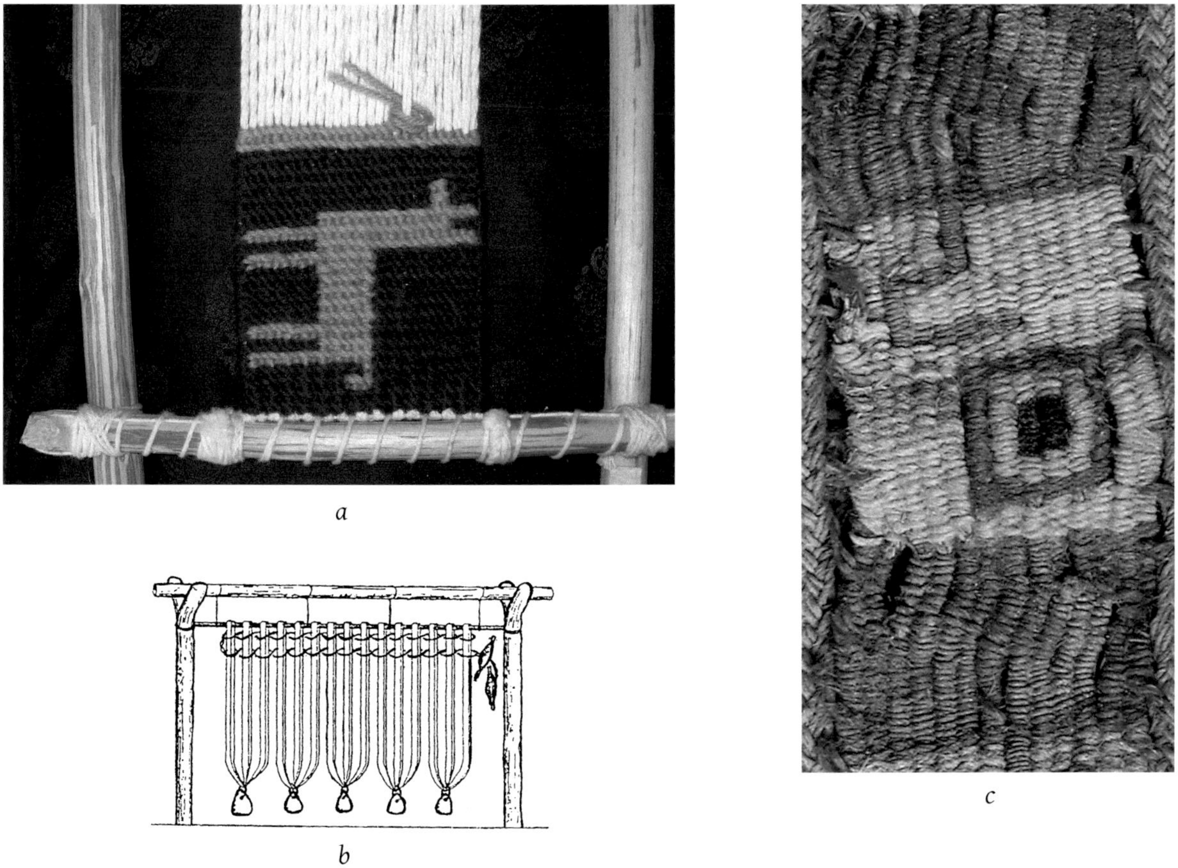

Fig. 3.18. Advanced Textile Techniques I. Warp methods with a passive warp: a) weft-wrapping: reconstruction of a Bolivian frame for belts; b) Warp-weighted frame for weft-twining probably used in Switzerland during the Neolithic period (After Winiger 1995, Fig. 32a1); c) Decorative band in tapestry on a woolen tunic from Djoumboulak Koum (same legend as Fig. 3.16c).

techniques that can produce the same end product. It may be possible to differentiate twining and right-angled plaiting as primary techniques (Figs. 3.15b, c, d & 3.16a) from twining and warp and weft interlacing with warp threads under tension, therefore as an advanced technique (Figs. 3.18b, c–3.21), according to the stiffness or flexibility of the material. However, when confronted with small flexible fragments of textiles with warps and wefts interlaced at right angle or twined, it is not always easy to understand which kind of advanced technique has been used : a frame - without shedding device or with only one shed formed automatically – (Figs. 3.18–3.19), or a loom – with mechanical shedding – (Figs. 3.20–3.21).[57] Specific woven borders and weaving mistakes may help to answer this question as well as fine comparisons with better preserved fragments

[57] Seiler-Baldinger (1994, 63) considers tapestry among warp methods even if she recognizes that "in practice this is often achieved by automatic shedding" – i.e. true weaving (Figs. 3.18c, 3.21d, e). The difficulty to identify the kind of advanced technique used to make such textiles appears clearly in Alfaro's study of the Tell Halula (Syria, 8th millenium BC) small imprints (Alfaro 2002).

a

b

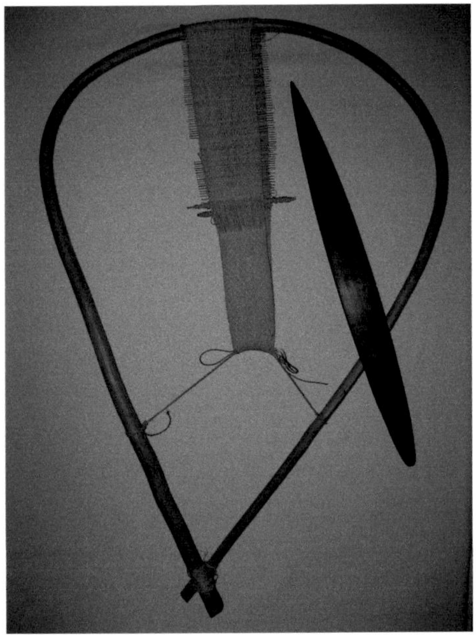

c

Fig. 3.19. Advanced Textile Techniques II. Half-weaving: a) Formation of the first shed by moving down the shed rod; b) Formation of the countershed by hand by picking up alternate threads (a and b: after Seiler-Baldinger 1994, Fig. 129a, c); c) Matsiguenga frame with sheds reserved on fine sticks later pulled out to be replaced by the weft, Puerto Huallana, Cuzco department, Peru, 1980 (F.-M. Casevitz's collection).

from the same area and period or with ethnographic observations. Recent finds on Swiss sites tend to shed more light on this complex problem already examined with perspicacity by Marta Hoffmann and Margrethe Hald.[58] They show that the presence of loom-weights in an archaeological context proves only that warp-threads were under tension on a frame. It does not mean that true weaving on a loom was practiced unless the position of the weights in lines (each corresponding to a shed) has been preserved.

Techniques of textile ornamentation are examined in an additional chapter. They are grouped according to the time when the decoration is added: during or after fabric production, and the type of material – solid or liquid – used to make it. Embellishing techniques such as dyeing and finishing can be considered among them.

During fabric production, ornamentation may occur with the formation of pile or tufts, and the making of beadwork, borders and fringes.[59] The formation of pile or tuft textiles appears historically in association with meshwork, plaiting, or an advanced textile technique. The carpet found in Pazyryk (3rd century BC) is usually given as the earliest evidence of knotted pile technique but a few fragments of plain weave with tufts knotted with symmetrical knots have been discovered in the Neolithic site of Charavines in France as early as the beginning of the 3rd millennium BC,[60] thus showing the antiquity of such a technique (Fig. 3.22a, b). Earlier evidence is the pile or tuft fixed in meshwork or weft-twining from various Neolithic sites in Europe.[61]

Beadwork is also associated with various textile techniques: mainly meshing, plaiting, and

[58] Hoffmann 1974; Hald 1980, 207–218; Rast-Eicher 1994; Altorfer & Médard 2000, 65–68; Rast-Eicher & Dietrich (forthcoming)

[59] Seiler-Baldinger 1994, 104–135.

[60] Barber 1991, 200–203; Desrosiers 1989; Cardon 1998.

[61] For instance Rast-Eicher 1992.

a

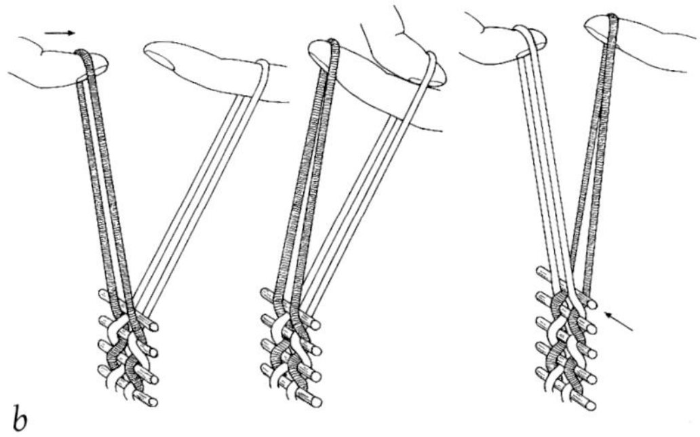

b

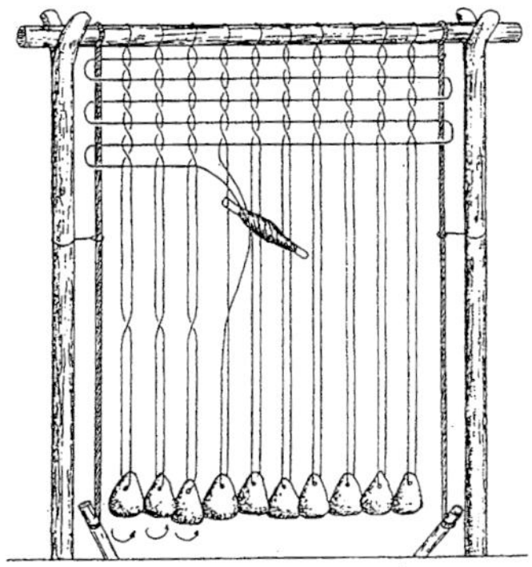

c

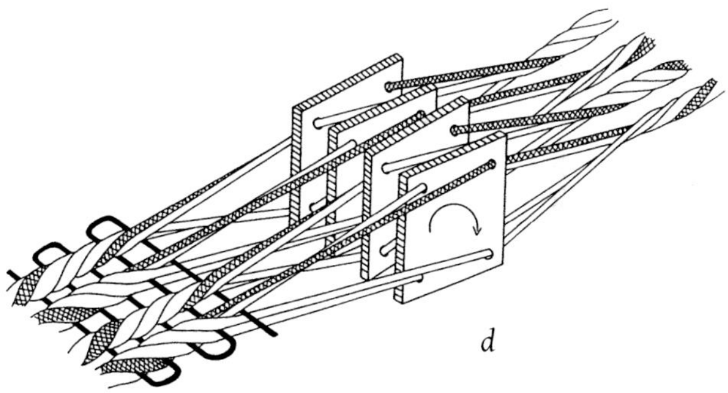

d

Fig. 3.20. Advanced Textile Techniques III. Weaving on a loom with an "active" warp: a) Finger weaving by a burnous maker in Bou-Saâda, Algeria (After Nacib 1986, 52. Copyright J. Evrard); b) Finger weaving (After Collingwood 1987, 39); c) Turning weights weaving: hypothetical loom reconstructed by Reinhard (1992) (After Winiger 1995, Fig. 32b1); d) Tablet weaving (After Collingwood 1982, Fig.20).

weaving (Fig. 3.22c, d). Pharaonic Egypt certainly highly valued beadwork as shown by the various examples found in Tutankhamun's wardrobe including at least a tunic, an apron, and a kilt.[62] In the Mediterranean area, Philip P. Betancourt demonstrated that Minoan art provides evidence of net-like fabrics which are considered as knotted nets. However, some of them, which clearly show beads intact where the elements meet, were likely produced as beadwork (Fig. 3.22e).[63]

Borders and fringes demonstrate the use of a wide range of techniques both during and after

[62] Vogelsang-Eastwood 1999, Figs. 1:8, 4:3 and 4:7.

[63] Betancourt 2007, at least Figs. 30.2 and 30.3.

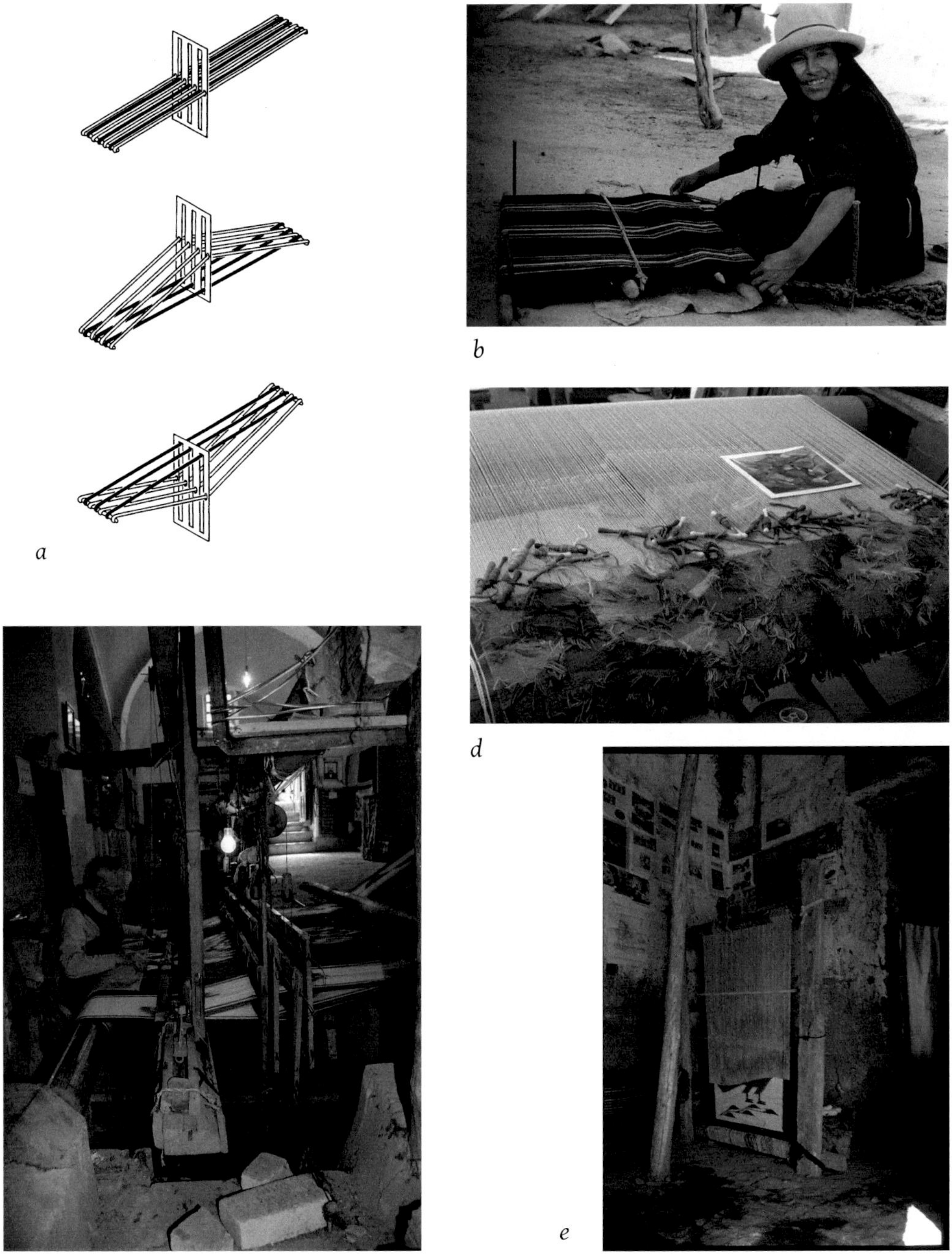

Fig. 3.21. Advanced Textile Techniques III. Weaving on a loom with a "passive" warp: a) with a rigid heddle (After Seiler-Baldinger 1994, Fig. 141); b) with heddle and shed rod on a ground loom in Bolivar, Cochabamba, Bolivia, 1983; c) with four heddles on a treadle loom with a resist-dyed (ikat) warp. Yazd, Iran, 2006 (Copyright J. Burkel); d) Tapestry woven on a two heddles horizontal loom, M. Veauvy's workshop, Crest, Drôme, France, 2007 (Copyright M. Veauvy); e) Tapestry woven on a vertical loom with heddle and shed rod, Don Corsino's workshop, Villa Ribero, Cochabamba, Bolivia, 1988.

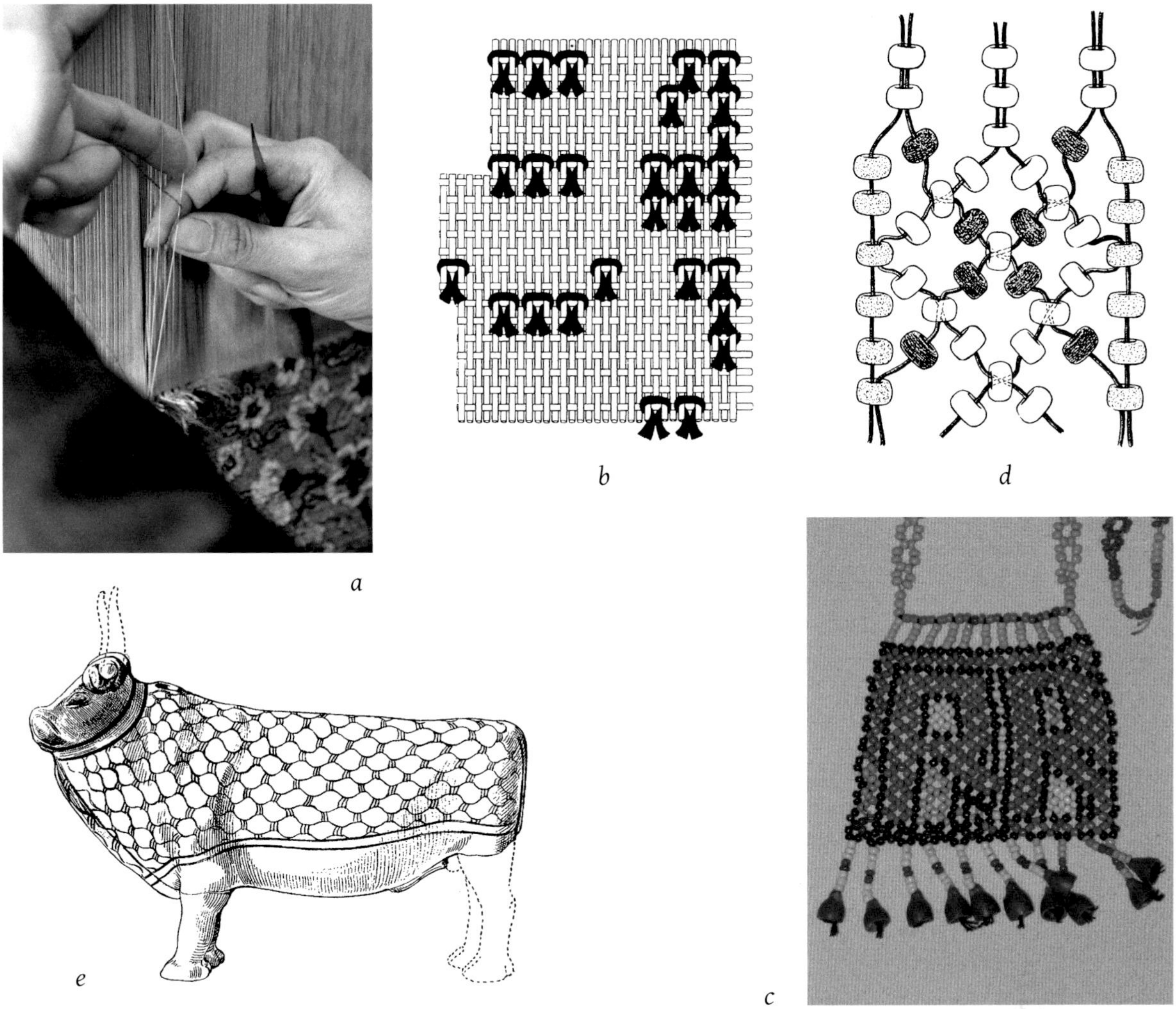

Fig. 3.22. Techniques of Textile Ornamentation I. During fabric production: a) Mrs Miraym Hashemi Isfahani-Mahdie introducing pile during weaving, Ispahan, Iran, 2004 (After Burkel & Burkel 2007, Fig. 181. Copyright R. Ghilini); b) Plain weave with knotted piles from the Neolithic site of Charavines (France) (After Desrosiers 1989); c) Necklace in beadwork from the Matsiguenga, Kirigueti, Cuzco department, Peru, 1980–90 (F.-M. Casevitz's collection); d) Similar work from Rio Napo in Ecuador (After Orchard 1975, Fig. 125); e) Minoan figurine of a bull covered with beadwork (?) (After Evans 1930, Fig. 139b).

fabric production. Thanks to their specificity and diversity, they are important characteristics that enable us to reconstruct the techniques used to produce a textile and to identify some textile traditions (Figs. 3.7, 3.22c, 3.23c).

After fabric production, various techniques may be applied to make motives by quilting, by sewing patches of textiles in layers, or by embroidery with various materials (threads, beads, pieces of textiles, hair, ...) (Figs. 3.1a, 3.23a). Or they can be achieved by painting, printing, and

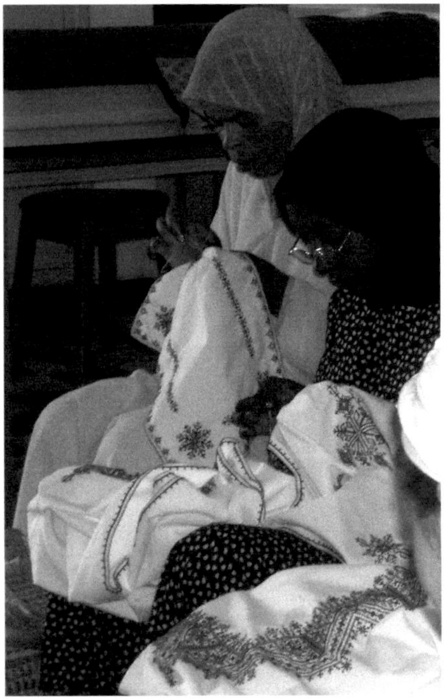

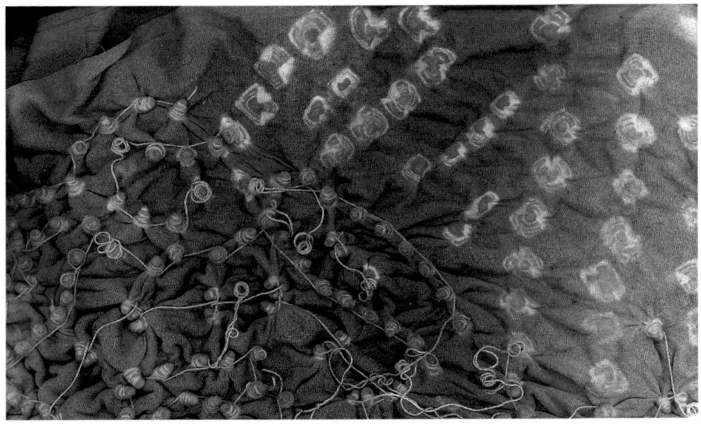

b

a

c

Fig. 3.23. Techniques of Textile Ornamentation II. After fabric production: a) with solid material: embroidery in the workshop of the Franciscan sisters of Mary, Midelt, Morocco, 2006 (Copyright R. Varrault); b) with liquid material: tie-dyed textile (plangi) with parts of the fabric pulled into a knob-like form tied with threads. On the right, undone knots let the characteristic undyed circles appear; c) Fringes with pearls of knotted graminae and pompoms on woolen fragment with griffins from Djoumboulak Koum (Keriya River, Xinjiang, China, 5th to 3rd century BC) (After Debaine-Francfort & Idriss 2001, 211. Copyright Mission Archéologique franco-chinoise au Xinjiang/X. Renaud/ Fondation EDF).

a range of resist dyeing techniques (with certain parts of the fabric protected by folds, stitches, ties, stencils or paste prior to dyeing) (Figs. 3.21c, 3.23b).

Dyeing can be performed on elementary textiles as well as those with interworked elements - and also directly on fibers. The same applies to bleaching aimed at whitening fibers. The finishing of woolen textiles by fulling, napping and shearing, or the glazing of linen cloths, and pleating observed at least on Egyptian tunics[64] should be added also to the list established by Seiler-Baldinger (Fig. 3.24). They were probably all known in the ancient Near East and the Mediterranean area at an early date.[65]

* * *

[64] For instance Barber 1991, Fig. 5.2; Rutchowskaia 1998.

[65] Among many references, see Waetzold 1972, 151–166; Lackenbacher 1982; Barber 1991, 223–243; Cardon 2003; and Andersson Strand in this volume.

a

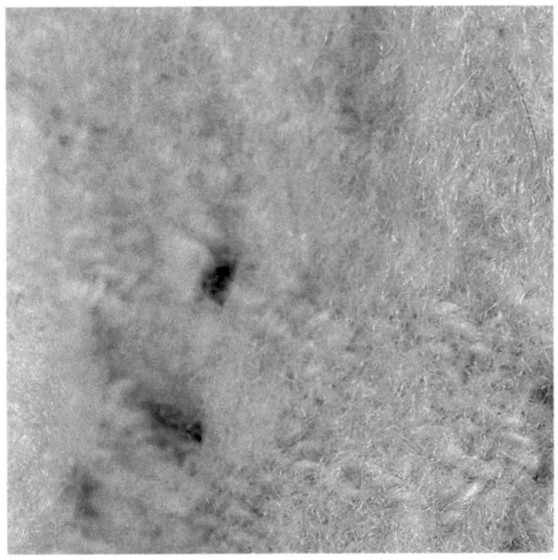

b

c

Fig. 3.24. Techniques of Textile Ornamentation III. For embellishing: a) Dyeing a skein of thread in Parham Sayahi's workshop, Ispahan, Iran, 2006 (After Burkel & Burkel 2007, 91. Copyright J. Burkel); b) Fulling; close view of a heavily fulled woolen cloth whose twill weave appears under the nap eaten by moths; c) Pleated linen tunic, Egypt, Middle Empire (Musée des Tissus de Lyon, Inv. No. MT 46841 – photo: Stephan Guillermond).

This examination of the Basel classification presents only some fabric categories without entering into their numerous variants because it seemed important to make the logic of the classification understandable – examples of the variants being available in the original book. It has the advantage of showing that a high number of techniques – so-called "primary" – requires no or very simple implements that are difficult to identify in archaeological contexts. Therefore, the lack of textile implements in an excavation does not mean that there was no textile construction nor that the constructions were not elaborate. The knotted headgear from Nahal Hemar (Fig. 3.13), the plaited gold objects from Ebla (Fig. 3.9), and the knotted nets and/or beadworks identified on Minoan art (Fig. 3.22e) are three examples of complex and luxurious textiles made with "primary" techniques that deserve our attention. Even if the word "primary" refers to a basic grade, it is used by the Swiss classification in a value system limited to the tool's complexity. It does not take into account the elaborate "savoir-faire" often necessary to operate the tools – "savoir-faire" which, in the case of textiles, may call for complicated abstract mathematical concepts as shown by the Yagua and Ticuna hammock weavers (as the one illustrated on Fig. 3.12a) investigated by Seiler-Baldinger.[66]

In order to make techniques more easily understandable, this article has been illustrated as much as possible by craftsmen in action, showing this way that many techniques are still in practice in some areas of the world. Keeping in mind Marta Hoffmann's outstanding investigation that helped her, in the middle of the 20th century, to find people who knew how to weave on the warp-weighted loom that had already been in use during the Neolithic era, and a more recent work of Ian Thompson and Hero Granger-Taylor on the vertical drawloom presently used at Meybod (Iran) and probably created at the beginning of our era,[67] one believes that other techniques described in ancient documents have been preserved as well. This is the case of bed webbing identified by Waetzoldt in Mesopotamia during the 2nd millennium BC and still performed in the Nile valley as Willeke Wendrich has demonstrated.[68] More observations of present textile techniques which can find parallels with those described in ancient documents would likely help to understand the complexity of the "savoir-faire" and to solve some other particular questions of ancient terminology.

Acknowledgements

I must thank my colleagues, friends and photographs, for their help when collecting the images or for correcting my English expression: Anath Ariel de Vidas, Tatiana Benfoughal, Maria Giovanna Biga, Jean Burkel, France-Marie Casevitz, Catherine Choron-Baix, Isabelle Daillant, Damien Davy, Corinne Debaine-Francfort, Laure Emperaire, Jacques Evrard, Marie-Hélène Guelton, Aline Hemond, Patricia de Largentay, Elena Phipps, Maria-Anne Privat-Savigny, Marco Ramazotti, Antoinette Rast-Eicher, Annemarie Seiler-Baldinger, Frieda Sorber, Richard Varrault, Mireille Veauvy, Elayne Zorn.

Experiments and reconstructions appearing in the figures are from the author. When not specified, photos are also from the author, with copyright reserved.

Bibliography

Adovasio, J. M. (1975–1977) The textile and basketry impressions from Jarmo. *Paleorient* 3, 223–230.

Adovasio, J. M. , Soffer, O. & Klíma, B. (1996) Upper Palaeolithic fibre technology: interlaced woven finds from Pavlov I, Czech Republic, *c.* 26,000 years ago. *Antiquity* 70, 526–534.

Alfaro, C. (2002) Étoffes cordées du site néolithique de Tell-Halula (Syrie – VIIIe millénaire avant J.-C.). *Bulletin du CIETA* 79, 17–25.

Altorfer, K. & Médard, F. (2000) Nouvelles découvertes textiles sur le site de Wetzikon-Robenhausen (Zürich, Suisse). Sondages 1999. In D. Cardon and M. Feugère (eds.) *Archéologie des textiles des origines au Ve siècle.* Montagnac. 35–75.

Andersson Strand, E. (this volume) The basics of textile tools and textile technology – from fibre to fabric.

Archéologie (2008) *Archéologie des textiles et teintures végétales. Les Nouvelles de l'archéologie* 114. Paris.

Balfet, H. & Desrosiers, S. (1987) Où en sont les classifications textiles? *Techniques & culture* 10, 207–212.

Barber, E. J. W. (1991) *Prehistoric Textiles: The Development of Cloth in the Neolithic and Bronze Ages.* Princeton.

[66] Seiler-Baldinger 1980. The relation between Mathematics and Textiles was probably already obvious to ancient people as shown by an Old Babylonian mathematical text dedicated to the calculation of the number of workdays needed to achieve various lengths of woven textiles (Waetzold 1972, 139).

[67] Hoffmann 1974 ; Thompson & Granger-Taylor 1995–1996.

[68] Wendrich 1999, 374–378.

Bellinger, L. (1959) Craft Habits. Part II: Spinning and Fibers in Warp Yarns. *The Textile Museum Workshop Notes*, no. 20.

Bellinger, L. (1962) Textiles from Gordion. *The Bulletin of the Needle and Bobbin Club* 46, 4–33.

Betancourt, P. P. (2007) Textile Production at Pseira: the Knotted Net. In C. Gillis & M.-L. Nosch (eds.) (2007), 185–189.

Betts, A., van der Borg, K., de Jong, A., McClintock, C. & van Strydonck, M. (1994) Early Cotton in North Arabia. *Journal of Archaeological Science* 21, 489–499.

Beugnier, V. & Crombé, P. (eds.) (2007) *Plant Processing from a Prehistoric and Ethnographic Perspective*. BAR International Series 1718. Oxford.

Biga, M. G. (this volume) Textiles in the Administrative Texts of the Royal Archives of Ebla (Syria, 24th Century BC) with Particular Emphsis on Coloured Textiles, pp. 146–172.

Breniquet, C. (2008) *Essai sur le tissage en Mésopotamie des premières communautés sédentaires au milieu du IIIe millénaire avant J.-C.* Paris.

Breniquet, C. (this volume) Weaving in Mesopotamia during the Bronze Age: Archaeology, techniques, iconography, pp. 52–67.

Burkel, J. & Burkel, D. (2007) *Tapis d'Iran. Tissage & techniques d'aujourd'hui*. Paris.

Burnham, D. (1980) *Warp and Weft. A Textile Terminology*. Toronto.

Cardon, D. (1998) Neolithic Textiles, Matting and Cordage from Charavines, Lake of Paladru, France. In *NESAT* 6, 3–22. Göteborg.

Cardon, D. (2003) *Le monde des teintures naturelles*. Paris (revised edition in English. *Natural Dyes. Sources, Tradition, Technology and Science*. London. 2007)

CIETA (1964) *Vocabulary of Technical Terms. English, French, Italian, Spanish*. Lyon.

CIETA (1973) *Vocabulaire français*. Lyon.

Collingwood, P. (1982) *The techniques of tablet weaving*. New York.

Collingwood, P. (1987) *The Maker's Hand*. Ashville/Loveland.

Collingwood, P. (1998) *The Techniques of Ply-Split Braiding*. London.

Cornu, G. & Martiniani-Reber, M. (1993) *Tissus d'Egypte. Mémoires du monde arabe VIIIe-XVe siècles. Collection Bouvier*. Genève.

Debaine-Francfort, C. & Idriss, A. (eds.) *Keriya, mémoires d'un fleuve. Archéologie et civilisations dans les oasis du Taklamakan*. 144–155. Paris.

Desrosiers, S. (1988) Comment la Fabrique lyonnaise a-t-elle exécuté les commandes royales au XVIIIe siècle? In P. Arizzoli-Clémentel & Ch. Gastinel-Coural (eds.), *Soieries de Lyon. Les commandes royales au XVIIIe siècle (1730–1800)*, 103–109. Lyon.

Desrosiers, S. (1989) Des fragments de textiles du Néolithique Final provenant du lac de Paladru (Isère). In *Tissage, Corderie, Vannerie. IXe Rencontre Internationale d'Archéologie d'Antibes, octobre 1988*, 129–139. Juan-les-Pins.

Desrosiers, S. (2001) Une culture textile raffinée. In Debaine-Francfort, C. & Idriss, A. (eds.)

Dillmont, T. de (1886) *Encyclopédie des ouvrages de Dames*. Dornach.

Durand, J.-M. (2009) *La nomenclature des habits et des textiles dans les textes de Mari*. T. 1. Paris.

Emery, I. (1980) *The Primary Structures of Fabrics* (1st edition 1966). Washington D.C.

Evans, Sir A. J. (1930) *The Palace of Minos*. Vol. 3. London.

FAO http://www.naturalfibres2009.org.

Flury-Lemberg, M. (1988) *Textile Conservation and Research*. Bern.

Forbes, R. J. (1956) *Studies in Ancient Technology IV*. Leiden.

Gervers, M. & Gervers, V. (1974) Felt-making Craftsmen of the Anatolian and Iranian Plateaux. *Textile Museum Journal* IV (1). 15–29.

Gillis, C. & Nosch M.-L. (eds.) (2007) *Ancient Textiles. Production, Craft and Society*. Ancient Textiles Series 1. Oxford.

Glory, A. (1959) Débris de corde paléolithique à la grotte de Lascaux (Dordogne). *Mémoires de la Société Préhistorique Française* 5, 135–169.

Good, I. L., Kenoyer, J. M. & Meadow, R. H. (2009) New evidence for early silk in the Indus civilization. *Archaeometry* 51 (3), 457–466.

Gordon Cook, J. (1993) *Handbook of Textile Fibres*. 2 vol. Durham.

Gramsch, B. (1992) Friesack Mesolithic Wetlands. In *The Wetland revolution in prehistory*. B. Coles (ed). 65–72. Exeter/London.
Hald, M. (1980) *Ancient Danish Textiles from bogs and burials*. Copenhaguen.
Harcourt, Raoul d' (2008) Les textiles anciens du Pérou et leurs techniques. Paris.
Hardy, K. (2008) Prehistoric string theory. How twisted fibres helped to shape the world. *Antiquity* 82, 271–280.
Heckett, E. W. (1998) A Late Bronze Age Horsehair Ornament from Cromaghs, Armoy in Ireland. In *NESAT* 6, 29–37. Göteborg.
Hoffmann, M. (1974) *The Warp-Weighted Loom*. Oslo.
Kernchen, I. & Gramsch, B. (1989) Mesolitische Netz- und Seilreste von Friesack, Bezirk Postdam, und ihre Konservierung. *Veröffentlichungen des Museums für Ur- und Frühgeschichte Postdam* 23, 23–28.
Lackenbacher, S. (1982) Un texte vieux-babylonien sur la finition des textiles. *Syria* LIX, 129–149.
Larsen, J. L. & Freudenheim, B. (1986) *Interlacing, the Elemental Fabric*. Tokyo.
Laufer, B. (1930) The Early History of Felt. *American Anthropologist* 32, 1–18.
Leroi-Gourhan, A. (1954) *L'Homme et la matière*. Paris
Lloyd, S. & Mellaart, J. (1962) *Beycesultan*. London.
Maeder, F., Hänggi, A. & Winderlin, D. (2004) *Bisso marino: Fili d'oro dal fondo del mare. Muschelseide: Goldene Fäden von Meeresgrunde*. Basel.
Martial, E. & Médard, F. (2007) Acquisition et traitement des matières textiles d'origine végétale en Préhistoire: l'exemple du lin. In Beugnier & Crombé (2007), 67–82.
Matthiae, P. (2006) Un grand temple de l'époque des archives dans l'Ébla protosyrienne: fouilles à Tell Mardikh 2004–2005. *Comptes rendus de l'Académie des Inscriptions et Belles-Lettres* janvier-mars, 447–493.
Médard, F. (2006) *Les activités de filage au Néolithique sur le Plateau Suisse. Analyse technique, économique et sociale*. Paris.
Mordant, D. & Mordant, C. (1992) Noyen-sur-Seine: a Mesolithic Waterside Settlement. In *The Wetland revolution in prehistory*. B. Coles (ed.), 55–64. Exeter/London.
Moulherat C., Tengberg, M., Haquet, J.-F. & Mille, B. (2002) First Evidence of Cotton at Neolithic Mehrgarh, Pakistan: Analysis of Mineralized Fibres from a Copper Bead. *Journal of Archaeological Science* 29, 1393–1401.
Müller, A. (1994) Geflechte und Gewebe aus Hornstaad/Hörnle I. In *NESAT* V, 27–33.
Nacib, Y. (1986) *L'oasis de Bou-Saâda*. Liège/Bruxelles.
Nadel, D., Danin, A., Werker, E., Schick, T., Kislev, M. E. & Stewart, K. (1994) 19,000-Year-Old Twisted Fibers from Ohalo II. *Current Anthropology* 35 (4), 451–457.
Orchard, W. C. (1975) *Beads and beadwork of the American Indians*. New York.
Papagiotakopulu, E., Buckland, P. C., Day, P. M., Doumas, C., Sarpaki, A. & Skidmore, P. (1997) A Lepidopterous cocoon from Thera and evidence for silk in the Aegean Bronze Age. *Antiquity* 71, 420–429.
Petit Robert (1993) *Le nouveau Petit Robert. Dictionnaire alphabétique et analogique de la langue française*. Paris.
Pfister, R. & Bellinger, L. (1945) *The Excavations at Dura-Europos. Final Report IV. Part II: The Textiles*. New Haven.
Pomponio, F. (2008) Note sui tessili della documentazione di Ebla. Il periodo di Arrugum. In M. Perna & F. Pomponio (eds.). *The Management of Agricultural Land and the Production of Textiles in the Mycenaean and Near Eastern Economies*. 101–109. Studi egei e vicinorientali 4.
Ramazotti, M. (forthcoming) Royal Administration During the Conquest. New Archaeological and Epigraphic Discoveries in the Royal Palace G at Ebla/Tell Mardikh.
Rast-Eicher, A. (1992) Neolitische Textilien im Raum Zürich. In *NESAT* 4, 9–19.
Rast-Eicher, A. (1994) Gewebe in Neolithicum. In *NESAT* V, 18–26.
Rast-Eicher, A. & Dietrich, A. (Forthcoming) *Neolitische und bronzezeitliche Gewebe und Geflechte aus dem Kanton Zürich*.
Reinhard, J. (1992) Étoffes cordées et métiers à pierres, *Helvetica archaeologica* 23, 51–55.
Ribichini, S. & Xella, P. (1985) *La terminologia dei tessili nei testi di Ugarit*. Roma.
Ronge, V. (n.d.) Das Handwerkertum, In C. C. Müller & W. Raunig (eds.), *Das Welt zum Dach der Welt*, 153–201. Innsbruck/Franckfurt am Main.
Roth, H. Ling (1978) *Ancient Egyptian and Greek Looms*. Mcminnville (first edition 1913).

Rowe, A. P. (1984) After Emery: Further Considerations of Fabric Classification and Terminology. *Textile Museum Journal* 23, 53–71.

Rowe, A. P. (2006) Términos textiles en Castellano. In V. Solanilla D. (ed.) *III Jornadas Internacionales sobre Textiles Precolombinos*, 443–469. Barcelona.

Rudenko, S. I. (1970) *Frozen Tombs of Siberia*. Berkeley.

Rutchowskaia, M.-H. (1998) Tunique plissée. In *Guide des collections du Musée des Tissus, Lyon*, 12–13. Lyon.

Schick, T. (1988) Nahal Hemar Cave - cordage, basketry and fabrics. *Atiqot* (English series) 18, 31–43.

Schick, T. (1989) Early Neolithic twined basketry and fabrics from the Nahal Hemar cave, Israel. In *Tissage, Corderie, Vannerie. IXe Rencontre Internationale d'Archéologie d'Antibes, octobre 1988*, 41–52. Juan-les-Pins.

Schuette & Müller-Christensen (1963) *La Broderie*. Paris.

Seiler-Baldinger, A. (1980) Hängematten – Kunst: Textile Ausdrucksformen bei Yagua- und Ticuna- Indianern Nordwest – Amazoniens. *Verhandlungen der Naturforschenden Gesellschaft* 90, 59–130.

Seiler-Baldinger, A. (1991) *Systematik der Textilen Techniken*. Basel (revised and expanded from the first 1973 edition).

Seiler-Baldinger, A. (1994) *Textiles. A Classification of Techniques*. Basel (English translation of Seiler-Baldinger, A. [1991]).

Steinkeller, P. (1980) Matresses and Felt in Early Mesopotamia. *Oriens Antiquus* 19, 79–100.

Talon, P. (1986) Le coton et la soie en Mésopotamie. *Akkadica* 47, 75–78.

Textil (2000) *Textil, Technik, Design, Funktion*. Catalogue of the exhibition at the Museum der Kulturen. Basel.

Thompson, J. & Granger-Taylor, H. (1995–1996) The Persian Zilu Loom of Meybod. *Bulletin du CIETA* 73, 27–53.

Viallet, N. (1971) *Tapisserie. Principes d'analyse scientifique*. Paris.

Vogelsang-Eastwood, G. M. (1999) *Tutankhamun's Wardrobe*. Rotterdam.

Waetzoldt, H. (1972) *Untersuchungen zur neusumerischen Textilindustrie*. Studi Economici e Tecnologici 1. Rome.

Waetzoldt, H. (2007) The Use of Wool for the Production of Strings, Ropes, Braided Mats, and Similar Fabrics. In C. Gillis & M.-L. Nosch (eds.) (2007), 112–121.

Webster's Dictionary (1986) *Webster's Third New International Dictionary of the English Language unabridged*. Springfield.

Wendrich, W. (1999) *The World According to Basketry*. Leiden. CNWS 83.

Winiger, J. (1995) Die Bekleidung des Eismannes und die Anfänge der Weberei nördlich der Alpen. 119–187. In *Der Mann im Eis. Neue Funde und Ergebnisse*, vol. 2. Vienne.

4. Weaving in Mesopotamia during the Bronze Age: Archaeology, techniques, iconography

Catherine Breniquet

Mesopotamia is the Greek name for the valley of the Tigris and the Euphrates. Its history begins with the first Neolithic villages and ends with Alexander the Great's conquest. Mesopotamia is known to be the "the land of wool" as Hammurabi said in the 18th century BC. Texts also refer to industrial textile workshops from the 3rd millennium BC, where women and their children work. Most of these workshops transformed wool into clothes and were part of the economic organization of the Sumerian country.

Questioning cuneiform sources about textile terminology is the aim of the symposium and of the present volume. Before that, it can be useful to have an idea of what we can do with the documentation on weaving techniques. Different sources exist but appear to be in conflict. The aim of this contribution is to merge these sources to renew the approach. We shall deal mainly with the early Bronze Age, the first half of the 3rd millennium, but we shall not limit our remarks to these periods or areas, as evidence is scarce. Five elements will be discussed: sources, raw materials, techniques, methodological approach, and iconography which is the newest area of attention. New perspectives will appear at the end of this contribution.

1. Sources

Five sources can be used to study weaving:

– *Pieces of preserved fabrics or imprints* on materials (clay, bitumen, etc.).[1] Most of the time, the first have not survived: the most famous example is the discovery of a *kaunakes* among other fabrics in several tombs at the Royal Cemetery of Ur.[2] None was photographed, nor drawn, or sent to a museum. They probably disappeared because of the effects of light and air. This is the most famous example but often pieces of fabric are not fully studied, and sometimes never even collected at all. In the opposite case, when they are collected or when imprints are studied, scholars have to deal with such small pieces on which it is often hard to observe any technical aspect, and

[1] Stordeur 1989.

[2] Barber 1991, 164.

especially the structural ones. The way that the warp and the weft are interlaced is the main (and often the only) observation made.[3]

– *The natural sciences* have helped in recognizing plants and animals, from recent excavations at least, and specific studies with electronic microscopes have led to the identification of fibres. This doesn't mean that it is an easy task: for example, textiles from Çatal Höyük were first identified as wool, then flax, then both![4]

– *Cuneiform texts* document production because they are economic records.[5] From the Uruk period, pictograms refer to clothes[6] involved in local exchanges (for the ration system), and devoted to external exchanges. During protodynastic times, the same phenomenon is observed,[7] but techniques are not clearly described and fabrics received various and often unclear names.

– *Archaeology* brings technical information to the table with its study of spindle whorls, weights, parts of structures, etc. Each of them needs to be analysed and interpreted[8] but the task is not easy because they are often out of context or badly damaged by erosion: in Mersin, it is hard to know if the loom was found in context during the excavations or if we were dealing with stored weights. The loom found at Troy, which is perpendicular to the wall, could be a double warp weighted loom or a single four heddle rods' loom! The Nahal Mishmar loom is said to be a horizontal loom but its size and the shapes of the small beams suggests that it – probably – belongs to a backstrap loom. The same uncertainty exists around the most ancient El Kowm loom.[9] Buildings like the Northern Palace of Tell Asmar were interpreted as industrial textile workshops but this hypothesis is only supported by the discovery of economic tablets in the upper layers of the tell. They do not seem to have any direct connexion with the building itself.[10] Nothing inside the building can be seen as a weaving device or structure.[11] As already pointed out by architectural analysis, this building (which is part of a series including other large "palaces") is more likely a residential unit rather than an economic one.[12] Workshops did exist but should be thought of as small structures,[13] as seen in the Egyptian models, and not as the large factories developed since the Industrial Revolution. These workshops, with few people working inside, were probably dispatched in the cities.

– *Iconography* at least, shows the fabrics (when worn) on sculptures or reliefs, and exceptionally the loom itself is portrayed (for example on the famous bulla from Susa: Fig. 4.7a). In both cases, rules of composition and iconography alter the final reading of these representations and are usually passed over in silence.

[3] Breniquet 2008, 55–58.

[4] Burnham 1965; Helbaek 1963; Ryder 1965.

[5] Waetzoldt 1972.

[6] Szarzinska 1988.

[7] Crawford 1973.

[8] Breniquet 2008, 135–194; See also Frangipane, Andersson Strand, Laurito *et al.*, 2009.

[9] Stordeur 2000, 50; Breniquet 2008, 142.

[10] Delougaz, Hill & Lloyd, 1967, 197–198.

[11] Margueron 2008, 178–182.

[12] Henrickson 180.

[13] Grégoire 1992.

Separately, each of these sources brings considerable information and precious details. But this complex set of data is hard to gather and synthesise.[14] This often leads to conclusions that need to be discussed. Among the most popular misinterpretations, we can mention those suggesting that:

- the first fibre to be used is wool,
- flax is used for oil and medicine,
- felt could be the first wool fabric and basketry, the first plant one.
- Mesopotamia knows only the horizontal loom and fabrics are thought to be poor in quality

It is easy to refute all these conclusions one by one. All of them are much too old and were produced by using inadequate models, especially the evolutionary or economic ones.

2. Raw materials

Plant fibres were the first materials to be used for weaving and this continued everywhere in the ancient world until the end of the Palaeolithic era.[15] Not all of them are identified,[16] but flax is among the first cultivated plants in the Near East.[17] This is attested to have been around 9000 BC at Jerf el'Ahmar.[18] As seen above, the common explanation is that the plant was used for oil or medicine, as ensuing historical texts suggest.[19] But nothing in archaeobotany authorizes such a conclusion. The first known fabrics from Nahal Hemar or Eridu are made of flax,[20] and flax is most probably a textile (or multipurpose) plant, cultivated in gardens as it needs plenty of water. The use of flax as textile plant is not limited to the early periods in the ancient Near East.[21] It doesn't disappear when wool is introduced. A kind of self-implemented balance between the two fibres, emerged based on their prestige and degree of generalization. For example, the use of flax is restricted to ritual fabrics in the 3rd millennium (curtains for sanctuaries, clothes for priests or divine bodies, etc.), and wool could be used for daily life objects.[22] The most famous example is the linen bed-cover woven for the Sacred Marriage in the production of which Inanna herself is involved. Other plant fibres suitable for weaving are not documented in near eastern prehistoric archaeology.

The use of wool probably first appeared during the Neolithic Age, but direct evidence is scarce as wool does not resist well. The evidence comes from sheep butchery curves and their comparison with theoretical models from actual breeding practices showing the age for slaughtering.[23] However, data are often biased by the small size of the samples and the use of different models. Alive, the animal offers the opportunity to be used for "secondary products", milk and wool, rather than just meat. The secondary products revolution was supposed to have appeared in the

[14] See for example, Ellis 1972.
[15] Hardy 2008.
[16] Good 2007, 182–183.
[17] Helbaek 1959; Van Zeist & Bakker-Heeres 1974.
[18] D. Stordeur, pers. com.
[19] Van Zeist 1985. See also Reculeau, 2009, for an up-to-date conclusion on the controversy flax/sesame.
[20] Schick 1989; Gillet 1979.
[21] Waetzoldt 1980.
[22] Waetzoldt 2007.
[23] Helmer 1992.

4th millennium as A. Sherratt suggested,[24] but it seems that it appeared earlier. Some special selections of sheep are recorded during the Halafian period, but wool was probably collected as soon as domestication occurred. Human selection of animals affects their morphology and biology as well that of wool in the case of sheep. On the basis of iconography and for a long time, zoologists thought that three or five kinds of sheep existed in Sumer, as could be determined by the shape of their horns, their decreasing size, the presence of a fleece, the existence of a fat tail.[25] At least till the middle of the 3rd millennium BC, domesticated flocks present a wide range of characteristics.[26] But we need to consider that the Neolithic "wool" was very different from our wool.

The ancestor to the sheep, the wild sheep, the still extant *Ovis orientalis*, has no wool. Its coat is made of two kinds of hairs: long coarse hairs and a fine under wool which are both shed annually in spring.[27] As domesticated animals they are plucked or combed for their wool, and sheep tend to produce "naturally" more and more under wool, which becomes also finer. The model for the appearance of wool in ancient times has been suggested by M. Ryder based on zoology, iconography and comparisons with the Soay sheep which is supposed to be a primitive race. Wool is supposed to have appeared in five evolutionary stages (wild coat, hairy medium, generalised medium wool, woolly type, fine wool).[28] Even if this model needs improvement, it is generally accepted by specialists. One of the major problems in Antiquity is the date for the genetic mutation which brought about the continuous growth of the fleece. Indeed, the archaeological remains, the iconography or the written mentions are often ambiguous. For example the so-called sheep from Tepe Sarab, which is supposed to be an intermediate example from coat to wool with "staples" is probably nothing else than a depiction of a goat![29] Wool with continuous growth appears later, not before the 2nd millennium or the end of the Bronze Age. Also, the colour of the coat changes with domestication. Originally brown for the wild sheep, a domesticated sheep's coat is either black, grey, spotted or white which is the rarest. It is also the most desirable colour, restricted to royal purposes.[30] Also, the variety of sheep breeds, with or without fleece, needs to be investigated for the later periods.

It is hard to evaluate the impact of the environment. Sheep are among the domesticated animals of early Mesopotamia and were part of a diversified subsistence system.[31] But at least since the middle of the Ubaid period, c. 5000 BC, animals were probably a part of the political economy of the country, suggesting that wool was not commonly distributed but acted instead as a form of primitive money.[32] More than thirty variations of the sheep ideogram are known, mixing the properties of the raw material and its uses. The geographic origin of wool, the kind of sheep (fat tailed sheep, mountain, from Drehem, etc.), the animal feed are always mixed up.[33]

---

24 Sherratt 1982.
25 Davis 1993.
26 Steinkeller 1995, 55.
27 Ryder 2007, 16.
28 Ryder 2007, 46; Breniquet 2008, 93–94 with other references.
29 Breniquet 2008, 94, with references.
30 Breniquet 2008, 96, with references.
31 Davis 1993.
32 Breniquet 2008, 200–219.
33 Waetzoldt 1972, 5; Steinkeller 1995, 57.

From my point of view, very little can be known from direct ethnoarchaeological surveys,[34] as actual herders and herding practices are quite different from those of Mesopotamia when the country was at the stage of a primary state formation. Environmental adaptation is a parameter among others, although it is true that herding practices became more and more nomadic with the growth of urban space, especially from the end of the Uruk period[35] It is true also that fat tailed sheep seem to have appeared at that time when breeding was colonizing the arid edges of the plain. But the growth of textile factories during the protodynastic period is not only a matter of evolution and complexity.[36] It needs to be understood within a specific and complex historical frame (rather than an economic or competitive one): the impoverishment of people who had lost access to agricultural land and had to work, as well as wars between cities and states, which brought prisoners of war who were employed for manual tasks.[37]

Other fibres are not present in the 3rd millennium.[38] Cotton may have appeared as a wild product around 4000 BC at Dhuweila, and as a cultivated plant around 700 BC in the royal park of Sennacherib in Niniveh. We have no clear evidence of silk in Mesopotamia until the end of the 1st millennium BC. Goat hairs need further investigations.[39]

3. Techniques

For a comprehensive study of ancient weaving techniques, one must have a look at the properties of raw materials and the evidence that fabrics furnish. But this study is difficult as the origins and uses of techniques are unclear as seen above.

It is no longer possible to assume that felt is the primary technique for making wool fabrics, and that basketry is the primary one for flax. Recent discoveries from salvage excavations in the Levant show that weaving and basketry developed alongside each other, around 9000 BC.[40] These two techniques are closely connected because parts of their *chaîne opératoire* look alike, but are not linked to an evolutionary process. Basketry doesn't need a loom and doesn't need warping. There is no evidence of early felt, except stylistic comparisons done with the Çatal Höyük paintings and actual felt carpets,[41] and maybe some ambiguous terms found in cuneiform texts. These cannot be considered as the proofs we expect/need. Felt is a technique which has nothing to do with weaving. Fibres are compressed together with hot water, yarn is not required and felt probably required large quantities and a special quality of wool. The first true example is the carpet from the Pazyryk's tombs, c. 3rd century BC although it could have existed earlier (Beycesultan).

Flax and wool do not share the same properties.[42] Flax is often pulled out in order to keep the maximum length but the plant is never used directly. It needs to stay in water in order to dissolve the pectose, which keeps the fibres together. Very little is known in the Near East on the original making of the yarn. In Egypt, some fabrics show that the yarns are not spun as is

[34] Ochenschlager 1993.
[35] Green 1980.
[36] Breniquet 2008, 213–219.
[37] Gelb 1972, 1973.
[38] Breniquet 2008, 97–101.
[39] Ryder 1993. For new evidence, see Frangipane, Andersson Strand, Laurito *et al.*, 2009, 27.
[40] D. Stordeur, pers. com.
[41] Burkett 1977 and 1979, 7–9.
[42] McCorriston 1997, 521–524.

the case with wool, but joined together with a kind of splice.[43] This technique could have been the original way of obtaining a thread.

Wool is covered by microscopic scales which allow the fibres to hang together. It means probably that the historical use of wool is probably at the origins of new technical practices such as:

- the spinning with spindle, and the spindle whorl and distaff with which it is possible to stretch and to twist the fibres in a single gesture,
- the use of the warp weighted loom on which people weave against gravity,
- the use of an original loom which is probably not the horizontal loom.

All of them appear between the Neolithic and the Uruk period but it is hard to assign a precise date to their implementation. They are present when we have direct evidence, but not necessarily at the moment of their introduction.

Archaeological remains offer a wide range of early techniques. The first fabrics are twined, which is obtained by hand or with a rudimentary loom.[44] On the basis of the El Kowm structure, we made the hypothesis that one of the earliest looms is the backstrap loom. But other looms exist. The Andean loom is made of 4 wooden sticks, one on each side, and it has the particularity of having no upper and no lower beam as the weaver turns it upside down to finish the weaving.[45] From my point of view, the Sumerian pictogram for a loom, the so-called wooden frame,[46] could be a picture of this loom. Lexical terms for technological details only appear later in history.[47]

From 7000 BC, imprints on clay from Jarmo[48] suggest that tabby fabrics are obtained on a loom with a "mechanized" shed as they are orthogonal. Twined fabrics had not disappeared by that time for they remain in use until the Iron Age, probably for specific uses. Around 3000 BC, twill may already exist as suggested by the discoveries from Alişar.[49] Other archaeological artefacts can be interpreted as tablets for weaving and from 5000 BC,[50] with the great number of weights, unfortunately out of context, suggesting s the possible existence of the warp weighted loom (Fig. 4.1). So it is easy to conclude that weaving is a much more complex activity than previously assumed, and involves many techniques.

Fig. 4.1. Ubaidian loom (?) weights found out of context, tell el'Oueili. Courtesy of Prof. J.-L. Huot.

4. Methodological approach

In order to investigate the matter, we have included the iconographic sources. According to the traditional interpretation, Mesopotamian iconography does not depict handicraft activities,

[43] Barber 1997, 47, fig. 2.9.
[44] Schick 1989.
[45] Desrosiers 2000, 118–119.
[46] Joannès 2001, 856.
[47] Ellis 1976; Ochenschlager 1993, 55.
[48] Adovasio 1975–1977.
[49] Osten 1937a, 44–51; 1937b, 334–335.
[50] Breniquet 2008, 186–192, with references ; for an early example, see Özdöğan 1995, 96.

but focuses on religious scenes.[51] Probably, iconographic studies used or are influenced by the old city-temple model, and this element is not linked with the sources we use, but with historiography. The images, especially the most numerous ones hang on cylinder-seals, are never seen as "documents". We will not deal with the matter in this paper which affects the conception of history of arts and the way we think the ancient world, as it needs further developments. However, they can be used to complete historical information in an illustrative perspective.

Cylinder-seals appear in the Uruk period and have to be seen as administrative devices. Their iconography shows geometric pictures or realistic ones. But during the 3rd millennium, two engraving processes coexist, one with a point for making detailed pictures, the second with a drill which gives more schematic designs. As the cylinder-seals are small in size, this last process makes the shapes hard to recognize. The result is that we interpret the pictures immediately as we do with our modern photographs without trying to understand how the Sumerian engravers perceived their own world.

The first step of this research was to rediscover technical gestures and attitudes (in front of the loom, standing, sitting, squatting, gestures of the arms, etc.) and tools. Ethnography helps as it shows what may appear as the most ancient techniques. Ancient iconography from other eras such as Greece, Etruscan Italy, and Egypt depicts also spinning or weaving in various ways. The clearest and the most helpful objects for us are without contest the Greek *lecythus* attributed to the Amasis painter and kept in the Metropolitan Museum of New York, and a bronze pendant from pre-Roman Bologna[52] (Fig. 4.2). From their study, we tried to reconstruct the entire *chaîne opératoire* from the gathering of fibres to the finished fabrics. Surprisingly, all the steps can be documented by iconographic analysis of archaic cylinder-seals,[53] representing between 11 and 14% of the entire corpus.[54]

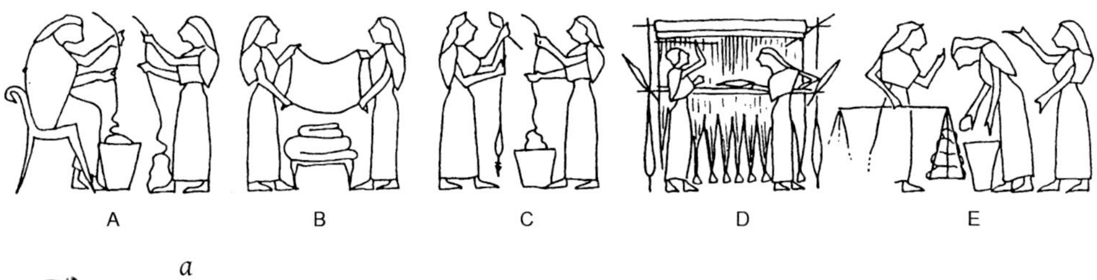

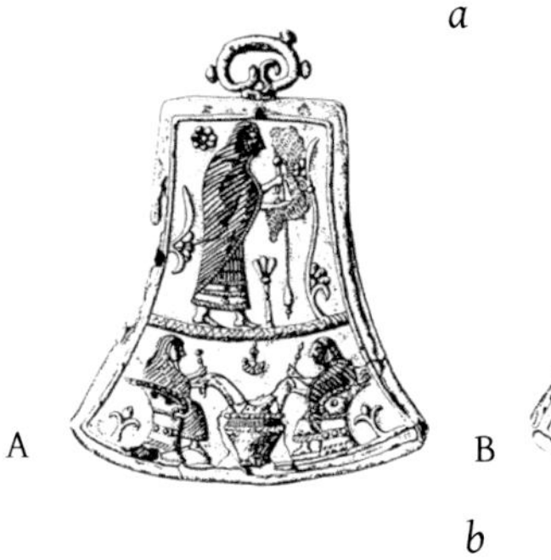

Fig. 4.2. Main comparisons with antique objects: a) Greek lecythus attributed to the Amasis Painter, Metropolitan Museum of New York (schematic drawing), Fletcher Fund, 1931 (31.11.10), 6th century BC. H = 5 cm (A: stretching, B: folding, C: spinning and stretching, D: weaving, E: weighing); b) Tintinnabulo, Tomba degli Ori, Bologna, 7th century BC. H = 11.5 cm (A: spinning and stretching, B: weaving and warping).

[51] Ellis 1976, 77.

[52] Breniquet & Mintsi 2000.

[53] Most of the following examples are from Amiet 1981 and Rova 1993.

[54] Breniquet 2008, 322.

5. The *chaîne opératoire*

At least eight main stages can be easily recognized.[55] We could expect the *chaîne opératoire* to begin with the collection of the fibres. But, hard to reproduce by engraving, this stage is not shown or is difficult to identify. However, the working of the wool is the most illustrated process on cylinder-seals and this fact testifies to the economic impact of this raw material.

Actually, the traditional way to work wool which will be hung on the distaff is to comb or card the material. During Antiquity, people used to make a kind of ribbon by stretching it by hand.[56] This also helps to finish the cleaning of the raw wool (Fig. 4.3a). This ribbon will then be rolled on a short distaff which is quite different from those we know. The distaff is held by hand. Wool to be stretched is stored usually in a basket (the Greek *kalathos*) or in a small pot into which distaffs waiting for use are stuck (for example, Fig. 4.2a: A and Fig. 4.2b: A). People work standing or sitting, probably inside the house or very close to it. The traditional interpretation

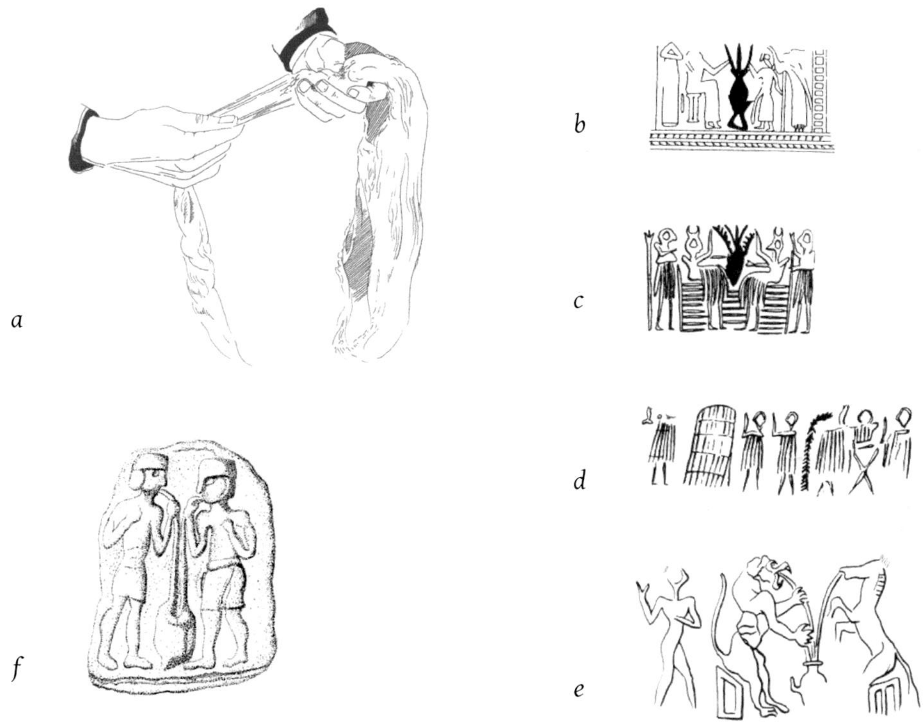

Fig. 4.3. Stretching the wool: a) Stretching the wool by hand (after Crokett 1979, 129); b) Drawing of a cylinder seal from Brak (after Matthews 1997, n° 511); c) Drawing of a cylinder seal from Kish (after Amiet 1981, n° 1359); d) Drawing of a cylinder seal, unknown provenance (after Amiet 1981, n° 1455); e) Akkadian banquet scene on a cylinder seal (after Amiet 1981, n° 1313); f) Old Babylonian terra cotta from Haradum showing a banquet scene H = c.10 cm. Courtesy of Mme C. Kepinski, CNRS.

[55] For a complete overview, see Breniquet 2008, 269–324.

[56] Crockett 1979, 129.

of such depictions is the banquet, suggesting that we are dealing with straws used for drinking beer. During the protodynastic period, most of the so-called banquet scenes depict in fact this handicraft activity as other ancient images did (Fig. 4.3b, 3c and 3d where the raw material is depicted as a bristling stalk). However, it is true that few details remain unclear. Banquet scenes become numerous from the Akkadian period (Fig. 4.3e) to Old Babylonian times (Fig. 4.3f), and are composed on the same iconographic basis.

After stretching the wool, workers make yarns. A lot of handy techniques exist all over the world as probably too in archaic Mesopotamia. Archaeology suggests that people used a spindle with a spindle whorl (the first records appear during the Neolithic time with the use of baked clay). Lines of pig-tailed ladies in the Uruk iconography show probably this activity (Fig. 4.4a). They are holding a stick with two rounded ends, one for the wool on the distaff, the other for the spindle whorl at the bottom of its spindle. They are standing up, shown as walking. Sitting women spin also with a belt shaped whorl (Fig. 4.4b).

Yarns were sometimes (or always?) twisted again or plied. The gesture was difficult to perform as the worker was standing up, holding two spindles together while the yarn passed over the shoulder as depicted in Egyptian iconography (Fig. 4.4c).[57]

The making of skeins is only shown on a standard from Mari. Women work two by two, one offering her arms to the other who unwinds the spun yarn (Fig. 4.4d).

It is hard to know at which step of the *chaîne opératoire* wool was weighed. One can assumed at the beginning of the process, when the wool was collected from the animal, but iconography suggests small quantities (balls of yarn?). Scales are easy to recognize with their beam held by hand (Fig. 4.5a). Three workers are usually involved in the process, one holding the balance, the second weighing, the third recording (Fig. 4.5b, compare with Fig. 4.2a: E).

a b c

d

Fig. 4.4. Making the yarn: spinning, plying, making the skeins: a) Spin and drop: drawing of a cylinder seal from Djemdet Nasr (after Rova 1993, n° 830); b) Spinning with a belt shaped whorl: drawing of a cylinder seal from Choga Mish (after Rova 1993, n° 80); c) Plying: drawing of a cylinder seal from Susa (after Rova 1994, n° 356); d) Makin the skeins: detail of a standard from Mari (drawing after Barber 1991, 57, fig. 2.18).

[57] Broudy 1979, 40, fig. 3–2.

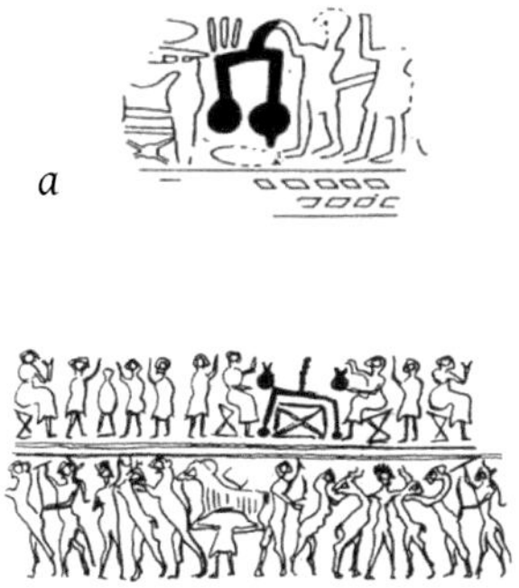

Fig. 4.5. *Weighing; a) Drawing of a cylinder seal from Brak (after Matthews 1997, n° 218); b) Drawing of a cylinder seal from Mari (after Amiet 1981, n° 1788).*

Coloured fabrics (especially yellow and green ones) are recorded since the Uruk times.[58] Dyeing was probably made with plant rather than with animal products. We don't know at which stage dyeing was performed: on the raw material, the yarns or the fabrics.

Warping depends on the loom. Two sticks sunk into the soil or on the wall, as well as the beams themselves, are suitable for the horizontal and two beams looms (Fig. 4.6a). With the warp weighted loom itself still in use in Norway,[59] a special device called a warping loom is used (Fig. 4.6b). A narrow band (Fig. 4.6b: A) is woven between two poles (Fig. 4.6b: B). The weft of this band is stretched (Fig. 4.6b: C) to a third stick located behind it (Fig. 4.6b: B). The strings of the weft are joined together in small balls (Fig. 4.6b: D). The narrow band will then be sewn on the single upper beam of the loom, the balls dropped down and the original weft will be automatically changed into warp for the future fabric. The warping of the warp-weighted loom is without doubt depicted on this print from Khafadjeh where band, balls, weft and sticks are clearly drawn in a kind of perspective (Fig. 4.6c).

Thus, at least two kinds of loom did exist in Mesopotamia during the Bronze Age, the horizontal loom where people work squatting (Fig. 4.7a), and the vertical warp-weighted one where workers

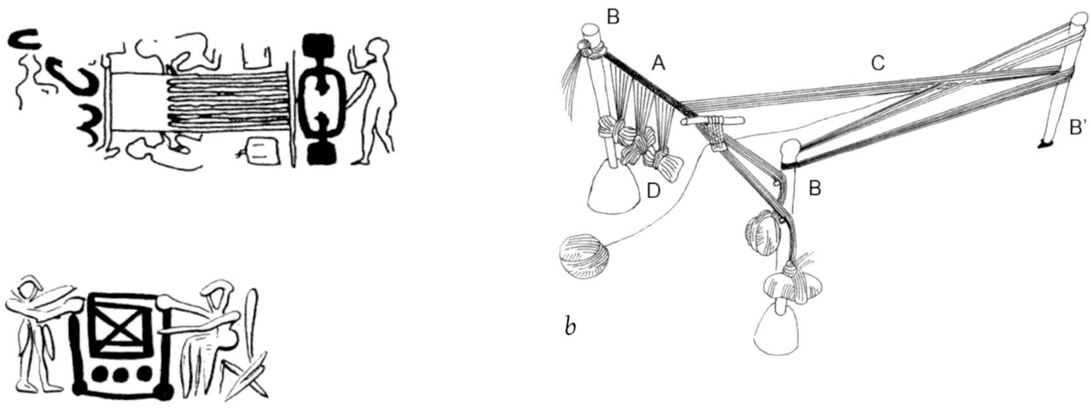

Fig. 4.6. Warping; a) Warping the horizontal loom: drawing of a cylinder seal from Susa (after Rova 1993, n° 364); b) Warping loom; c) Warping the warp weighted loom: drawing of a cylinder seal from Khafadjeh (after Amiet 1981, n° 1340).

[58] Englund 1998, 98.

[59] Hoffman 1974.

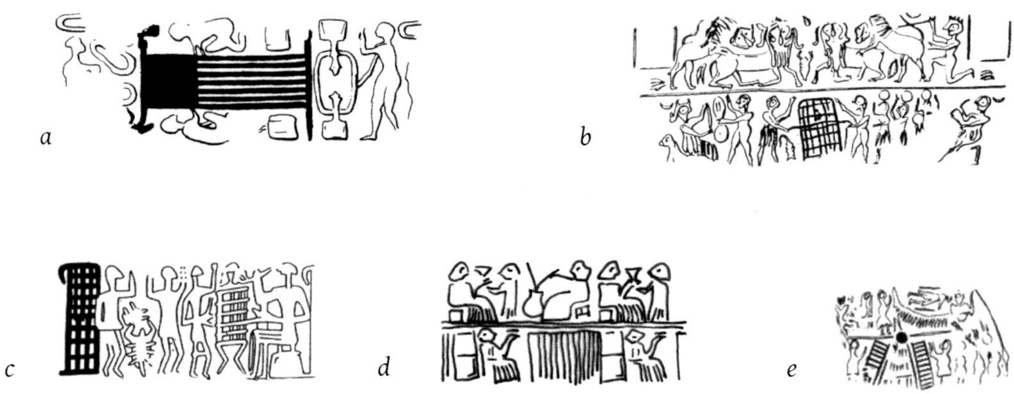

Fig. 4.7. Weaving; a) The horizontal loom: drawing of a cylinder seal from Susa (after Rova 1993, n° 364); b) The warp weighted loom: drawing of a cylinder seal from Tell Asmar, (after Amiet 1981, n° 1452); c) The warp weighted loom: drawing of a cylinder seal from Brak (after Matthews 1997, n° 508); d) The two beams loom: cylinder seal from Mari (drawing from Breniquet 2008, fig. 89–1, p. 301); e) A double loom ? cylinder seal, unknown provenance (after Amiet 1981, 1447).

stand up, in front of the shed beam (Fig. 4.7b). This loom was probably not placed over the wall as the Scandinavian one but vertical, with two vertical wooden feet or braces. This kind of loom is made for long rectangular pieces of cloth and probably had a winding beam (Fig. 4.7c). These two pieces of mosaic from Uruk (which are always mixed up with some of the so-called Inanna's poles) are probably a depiction of it with a fabric displayed on it.[60] The fabric has a tablet woven border as fishbone pattern suggests. The existence of the warp weighted loom is also proved by this small pompom made with strings from the warp and weft in the upper corners of the fabric depicted on the attendant's skirt on the Uruk vase.[61]

Moreover, a third kind of loom could be the two beams vertical loom where people work sitting (Fig. 4.7d). Due to the size of the cylinder seals, all the depictions are schematic but realistic too. The looms always appear as hatched squares, door-like devices. With time, some of the structures may change. They appear first parallel to the wall and then probably perpendicular to it, connected to it in its upper part. They could be "double", as in the case of the Troy structure, [62] perhaps with two opposite groups of weavers standing opposite each other (Fig. 4.7e). The two vertical looms were probably not common, but were factory instruments.

When finished, the fabrics can be used as garments. Their length increased with time. They were used without being tailored, as draped skirts or dress in the Uruk times for the "Priest King", in Akkadian sculpture for Maništušu for example, in the neo-Sumerian period for Gudea's dress. We must clarify here a difficult point. Clothes became more and more common with time but were probably part of a communication system which distinguished the elites from the others

[60] Breniquet 2008, 307–308.
[61] Breniquet 2008, 162–163.
[62] Barber 1991, 93, fig. 3.14.

Fig. 4.8. Folding the fabrics; a) Drawing of a cylinder seal from Khafadjeh (after Amiet 1981, n° 1454); b) Drawing of a cylinder seal from Kish (after Amiet 1981, n° 1444).

as suggested in the depictions on the Uruk vase. But other uses of cloth exist.[63] Fabrics could be stored for a later use and then had to be folded. We are sure that we can recognize this operation in the so-called ziggurat's building scenes (Fig. 4.8a and 8b, compare with Fig. 4.2a: B).

If we accept this reading of the proto-dynastic iconography, we can go further in our interpretation. It is possible to understand who weaves, and for what:

- Women first who take care of the domestic production (Fig. 4.9a).
- Men appear during the 3rd millennium as weavers, professional craftsmen, connected with palace economy (Fig. 4.9b).
- Among both of these two categories, some may be *arua* who are the only workers to be fully specialized (Fig. 4.9c). The *Arua* institution[64] is difficult to understand and involves many aspects such as labour, religion, ethnic origins of the female workers, etc.
- Others could be prisoners of war but are difficult to identify.
- Undifferentiated workers which give a general and generic significance to the depictions (Fig. 4.9d).
- Gods, who are easy to identify with their horned head dress (Fig. 4.9e).

Fig. 4.9. Who weaves? a) Women in domestic context? cylinder seal from Choga Mish (after Rova 1993, n° 319); b) Men in workshop? cylinder seal from Tell Asmar (after Amiet 1981, n° 1452); c) The Arua? standard from Mari (drawing from Barber 1991, 57, fig. 2.18); d) Undifferentiated people. Cylinder seal from Susa (after Rova 1993, n° 364); e) The gods. Cylinder seal from Khafadjeh (after Amiet 1981, n° 1454).

[63] Gelb 1965.

[64] Gelb 1972.

* * *

We can conclude that the entire society is involved in weaving. Our iconography is much more realistic than previously assumed. It is possible to use these depictions for their documentary significance. But we can also conclude that these depictions are not those of a hand weaver's manual. We are here far from historical explanations. What is depicted is not the economic world of textile workshops or domestic units, although wool is the most important raw material. We are clearly within the symbolic world. All these scenes are parts of more complex system-like series, as they appear to be often combined in linear, but not logic, compositions (Fig. 4.10a). The protodynastic iconography probably shows primary symbolic aspects of the world. It acts as a graphic metaphor to express human existence as well as the cycle of the time, both linked in a society organized by agriculture and references to the ancestors.[65]

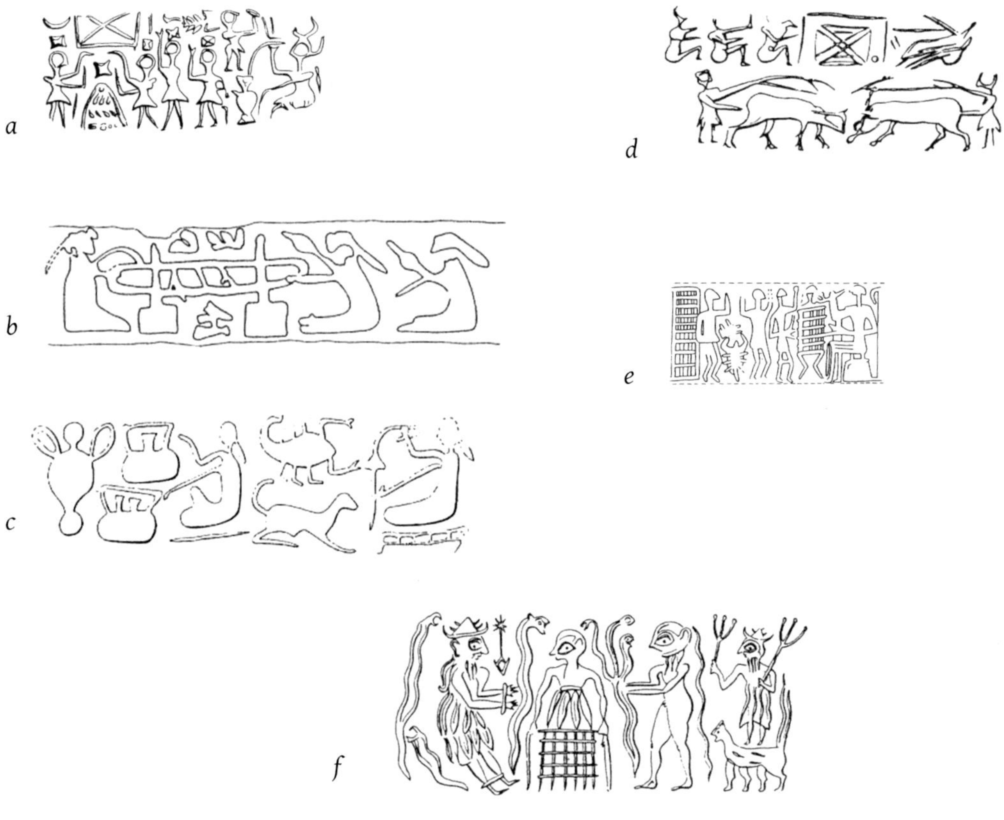

Fig. 4.10. What is depicted? a) Composite scene. Cylinder seal from Tell Asmar (after Amiet 1981, n° 1463); b) Two steps: spinning and weaving. Cylinder seal, unkown provenance (after Amiet 1981, n° 319); c) Two (spring?) activities: churning and spinning. Cylinder seal from Choga Mish (after Rova 1993, n° 80); d) Gender activities? weaving and ploughing. Cylinder seal from Ischali (after Amiet 1981, n° 1469; e) Metaphor for human life: birth and dead? Cylinder seal from Tell Brak (after Matthews 1997, n° 508); f) Birth from a loom. Cylinder seal, unknown provenance (after Amiet 1981, n° 1389).

[65] Breniquet 2008, 359–380.

As an example, two stages form the main articulation of the depiction: the beginning and end of the weaving process as spinning and weaving (Fig. 4.10b), stretching and folding, which could be a metaphor for human life (Fig. 4.10e), or two different activities related to a cyclic perception of the year and time in which daily or gender activities occur: churning and weaving (Fig. 4.10c), ploughing and weaving (Fig. 4.10d), etc. Another aspect is to depict the birth from a loom (Fig. 4.10f), a birth which is not biological but symbolic, for which we have no direct reference to the related myth. Lastly, weaving has also to be understood as a primary attribute for the gods. As said previously, this iconography represents between 11 and 14 % of the protodynastic repertoire which is probably part of a more complex perception of the world.

We can conclude that a quite different history of weaving can be drawn, used for comparative perspectives, where sources are not in conflict, and where iconography and archaeology can finally be linked with epigraphy.

Bibliography

Adovasio, J. (1975–1977) The Textiles and Basket Impressions from Jarmo. *Paléorient* 3, 223–230.

Amiet, P. (1981) *La glyptique mésopotamienne archaïque,* deuxième édition revue et corrigée avec un supplément, éd. du CNRS, Paris.

Barber, E. J. (1991) *Prehistoric Textiles. The Development of Cloth in the Neolithic and Bronze Ages, with Special Reference to the Aegean,* Princeton University Press, Princeton.

Breniquet, C. (2008) *Essai sur le tissage en Mésopotamie, des premières communautés sédentaires au milieu du 3e millénaire avant J.-C.* De Boccard, Paris. Travaux de la Maison René-Ginouvès, 5.

Breniquet, C. & Mintsi, E. (2000) Le peintre d'Amasis et la glyptique mésopotamienne pré- et protodynastique. Réflexions sur le tissage et sur quelques prototypes orientaux méconnus. *REA* 102 3/4, 333–360.

Broudy, E. (1979) *The Book of Looms. A History of the Handloom from Ancient Times to the Present.* University Press of New England. Hanover and London.

Burkett, M. (1977) An Early Date for the Origins of Felt. *Anatolian Studies* 27, 111–115.

Burkett, M. (1979) *The Art of the Felt Maker.* Abbot Hall Art Gallery, Kendal.

Burnham, H. R. (1965) Çatal Hüyük. The Textiles and Twined Fabrics. *Anatolian Studies* 15, 169–174.

Crawford, H. (1973) Mesopotamia's Invisible Exports in the Third Millenium. *World Archaeology* 5, 231–241.

Crockett, C. (1977) *The Complete Spinning Book.* Watson-Guptill Publications, New York.

Davis, S. (1984) The Advent of Milk and Wool Production in Western Iran. Some Speculations. *In* J. Clutton-Brock & C. Grigson (eds.) *Animals and Archaeology 3. Early Herders and their Flocks*, Oxford, 265–278.

Davis, S. (1993) The Zooarchaeology of Sheep and Goat in Mesopotamia. *Bulletin on Sumerian Agriculture* 7–1, 1–7.

Delougaz, P, Hill, H.D. & Lloyd, S. (1967) *Private Houses and Graves in the Diyala Region.* University of Chicago Press, Chicago.

Desrosiers, S. (2000) Le tissu comme un être vivant ? A propos du tissage à quatre lisières dans les Andes. *In* F. Cousin, S. Desrosiers, D. Geimaert & N. Pellegrin (eds.) *Lisières & Bordures,* Actes des premières journées d'études de l'Association Française pour l'Etude du Textile, Paris, 13–14 juin 1996, 117–125, éd. Les Gorgones, Bonnes.

Ellis, R. S. (1976) Mesopotamian Crafts in Modern and Ancient Times: Ancient Near Eastern Weaving. *AJA* 80, 76–77.

Englund, R. K. (1998) The Texts from the Late Uruk Period. *In* J. Bauer, R.K. Englund & M. Krebernik *Mesopotamien Späturuk-Zeit und Frühdynastiscghe Zeit,* Universitätsverlag, Freiburg Schweiz, Vanddenhoevk & Ruprecht, Göttingen, 15–233.

Frangipane, M., Andersson Strand, E., Laurito, R., Möller-Wiering, S., Nosch, M.-L., Rast-Eicher, A. & Wisti Lassen, A. (2009) Arslantepe (Turkey): Textiles, Tools and Imprints of Fabrics from the 4th to the 2nd Millennium BC. *Paléorient No. 35.1*, 5–29

Geijer, A. (1979) *A History of Textile Art*, Pasold Research Fund in Association with Sotheby Parke Bernet Publications, Philip Wilson Publishers Ltd, London.
Gelb, I. (1965) The Ancient Mesopotamian Ration System. *JNES* 24, 230–243.
Gelb, I. (1972) The *Arua* Institution. *RAAO* 66–1, 1–32.
Gelb, I. (1973) Prisoners of War in Early Mesopotamia. *JNES* 32, 70–98.
Gillet, J. (1979) Botanical Samples. *In* F. Safar, M. Ali Mustafa & S. Lloyd *Eridu*, Baghdad, 317–318.
Good, I. (2007) Invisible Exports in Aratta: Enmerkar and the Three Tasks. *In* C. Gillis & M.-L. Nosch (ed.) *Ancient Textiles. Production, Craft and Society*, Ancient Textiles Series 1, Oxbow Books, Oxford, 179–184.
Grégoire, J.-P. (1992) Les grandes unités de transformation des céréales: l'exemple des minoteries de la Mésopotamie du sud à la fin du IIIe millénaire avant notre ère. *In* P. Anderson (dir.) *Préhistoire de l'agriculture : nouvelles approches expérimentales et ethnographiques*, Paris, CNRS, 321–339.
Green, M. (1980) Animal Husbandry at Uruk in the Archaic Period. *JNES* 39–4, 1–35.
Hardy, K. (2008) Prehistoric string theory. How twisted fibres helped to shape the world. *Antiquity* 82, 271–280.
Helbæk, H. (1959) Note on the Evolution and History of Linseed. *Kuml*, 103–129.
Helbæk, H. (1963) Textiles from Çatal Hüyük. *Archaeology* 16, 39–46.
Helmer, D. (1992) *La domestication des animaux par les hommes préhistoriques,* coll. Préhistoire, Masson, Paris – Milan – Barcelone – Bonn.
Helmer, D. (2000) Etude de la faune mammalienne d'El Kowm 2. In Stordeur, D. (dir.) *El Kowm 2. Une île dans le désert. La fin du Néolithique précéramique dans la steppe syrienne*, CNRS-Editions, Paris, 233–257.
Henrickson, E. F. (1980) Functional Analysis of Elite Residencies in the Late Early Dynastic of the Diyala Region. *Mesopotamia* 17, 5–33.
Hoffmann, M. (1974) *The Warp-Weighted Loom: Studies in History and Technology of an Ancient Implement*, Universitetsforlaget, Oslo (reprinting of the 1964 edition).
Joannès, F. (2001) Tissage. *In* F. Joannès (ed.) *Dictionnaire de la civilisation mésopotamienne*, Ed. Robert Laffont, Paris, 854–856.
Margueron, J.-Cl. (2008) Note d'archéologie et d'architecture orientale 15. Installations hygiéniques ou artisanales ? *Syria* 85, 175–222.
Matthews, D. M. (1997) *The Early Glyptic of Tell Brak. Cylinder Seals of the Third Millenium Syria*. University Press, Fribourg, Göttingen, Vandenhoeck & Ruprecht.
McCorriston, J. (1997) The Fiber Revolution. Textile Extensification, Alienation and Social Stratification in Ancient Mesopotamia. *Current Anthropology* 38–4, 517–549.
Ochsenschlager, E. (1993) Village Weavers: Ethnoarchaeology at Al-Hiba. *Bulletin on Sumerian Agriculture VII: Domestic Animals of Mesopotamia, Part I*, Cambridge, 43–62.
Osten, H. von der (1937 a) *The Alishar Hüyük. Seasons of 1930–32. Part I*, University of Chicago Press, Chicago.
Osten, H. von der (1937 b) *The Alishar Hüyük. Seasons of 1930–32. Part III*, University of Chicago Press, Chicago.
Özdoğan, A. (1993) Life at Çayönü during the Pre-Pottery Neolithic Period (according to the artifactual assemblage). *Readings in Prehistory. Studies presented to Halet Çambel*, Yazarlar, Istambul, 79–100.
Reculeau, H. (2009) Le point sur la "plante à huile": réflexions sur la culture du sésame en Syrie-Mésopotamie à l'âge du Bronze. *Le Journal des Médecines Cuneiformes* 13, 13–37.
Rova, E. (1993) *Ricerce sui sigilli a cylindro vicino-orientali del periodo di Uruk/Jemdet Nasr*, Istituto per l'Oriente C. A. Nallino, Roma.
Ryder, M. (1965) Report of Textiles from Çatal Hüyük. *Anatolian Studies* 15, 175–176.
Ryder, M. (1993) The Use of Goat Hair. An Introductory Historical Review. *Anthropozoologica* 17, 37–46.
Ryder, M. (2007) *Sheep and Man*, Duckworth, London (reprinted from the 1983 edition).
Schick, T. (1989) Early Neolithic Twined Basketry and Fabrics from the Nahal Hemar Cave, Israel. *In Tissage, corderie, vannerie. Approches archéologiques, ethnographiques, technologiques*, éd. APDCA, Juan-les Pins, 41–52.
Sherrat, A. (1982) Plough and Pastoralism: Aspects of the Secondary Products Revolution. *In* I. Hodder, G. Isaac et N. Hammond (eds.) *Pattern of the Past. Studies in Honour of David Clarke*, Cambridge University Press, Cambridge, London and New York, 261–305.
Steinkeller, P. (1995) Sheep and Goat Terminology in Ur III Sources from Drehem. *Bulletin on Sumerian Agriculture* 8, 49–70.

Stordeur, D. (1989) Vannerie et tissage au Proche-Orient néolithique: IXe–V^{e} millénaires. *In Tissage, corderie, vannerie. Approches archéologiques, ethnographiques, technologiques*, éd. APDCA, Juan-les Pins, 19–39.

Stordeur, D. (ed.) (2000) *El Kowm 2. Une île dans le désert. La fin du Néolithique précéramique dans la steppe syrienne*, CNRS-Editions, Paris.

Szarzinska, K. (1988) Records of Garments and Clothes in Archaic Uruk/Warka. *Altorientalische Forschungen* 15, 220–230.

Van Zeist, W. (1985) Pulses and Oil Crop Plants. *Bulletin on Sumerian Agriculture* 2, 33–37.

Van Zeist, W. & Bakker-Heeres, J. (1974) Evidence for Linseed Cultivation before 6000 BC, *Journal of Science* 2, 215–219.

Waetzoldt, H. (1972) *Untersuchungen zur neusumerischen Textilindustrie*, Centro per la Antichita e la Storia dell'Arte del Vincino Oriente, Roma.

Waetzoldt, H. (1980) Leinen. *Reallexikon der Assyriologie und vorderasiatischen Archäologie*, 583–594.

Waetzoldt, H. (2007) The Use of Wool for the Production of Strings, Ropes, Braided Mats and Similar Fabrics. In C. Gillis & M.-L. Nosch (ed.) *Ancient Textiles. Production, Craft and Society*, Ancient Textiles Series 1, Oxbow Books, Oxford, 112–121.

5. Cloths – Garments – and Keeping Secrets. Textile classification and cognitive chaining in the ancient Egyptian writing system

Ole Herslund

In recent years Egyptologists have shown a growing interest in the study of classification as a way of gaining insight into the ancient Egyptians' conception of the world.[1] Egyptology, with written evidence that extends over more than 3500 years, is particularly well suited for studying classification. The ancient Egyptian hieroglyphic writing system and its cursive counterpart hieratic incorporate a multi-levelled classificatory system, combining signs that give the phonetic value of a word with an added sign that indicates membership in a category. For example the phonetic signs for words that relate to [time] must be followed by a sign depicting the sun and the writing of words that relate to [action] can end with an extended forearm arm and a hand holding a stick.

[time]	*ꜣt*, moment	[action]	*bꜣk*, to work
	hrw, day		*nḫ*, to protect

This added classifying sign is called, using Egyptological terminology, a determinative.[2] It can be used to categorise broad groups such as the occupations and personal names of men () and women (). It can also be very specific, conveying the exact meaning of words such as *miw* "cat" and *ibꜣ* "dance". There are also signs which categorise abstract categories. Neutral or positive abstracts are categorised with the papyrus roll sign, in words such as *rḫ* "to know" and *nfr* "good". Negative phenomena are identified with the addition of the sparrow found in words such as *bin* "bad" and *ḥḳr* "hunger".[3] Determinatives are also used to differentiate words that are graphic homonyms. The consonant combination *hb* can mean "plough" when followed by the plough sign or "to traverse" if followed by a pair of walking legs.

Traditionally, Egyptologists have regarded the determinative as a reading aid that assists in

[1] See e.g. David 2000; Frandsen 1997; Goldwasser 2002; Lincke (forthcoming), Shalomi-Hen 2000; 2006.

[2] Gardiner 1957, 31–33; Allen 2000, 3; 28–29.

[3] See now David 2000.

identifying the exact meaning of a word. More recently it has been demonstrated that these signs, as "graphemic classifiers", represent a system of classification and that they can be used to understand the structure of this system.[4] The graphemic classifiers (i.e. determinatives) make it possible to differentiate between a more overarching and inclusive superordinate category and the specific basic level members of that category. This approach also treats the signs that essentially only "repeat" the information found in the phonetic writing as functionally distinctive, as "repeaters" that reiterate the meaning of the word, rather than classifying it. This distinction is found in the examples given above. The sun sign 𓇳, found in the word *hrw* (𓉔𓂋𓅱𓇳) "day" indicates that the word belongs to the superordinate category of [time] and thus 𓇳 functions as a genuine classifier. The sign 𓃠, however, that follows the word *miw* (𓏇𓇋𓅱𓃠) "cat" is a repeater; it reiterates the meaning of the phonetic sign combination.

The sign 𓉐 is often used to illustrate the way in which determinatives function as graphemic classifiers. It is the simplified image of the floor plan of a house with an entry. 𓉐 can also be used phonetically as *pr*, and written with the stroke that indicates that there is an equation between sign and meaning, the sign means "house". When employed as a graphemic classifier written in the end of other words, the house sign shows membership of the category [habitat].

The sign 𓉐 as classifier for the category [habitat]*		
	*pr*	house
	*mnnw*	fortress
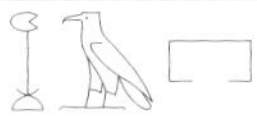	*ḫ3*	archive
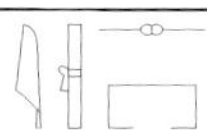	*is*	tomb
	*iḥw*	stable
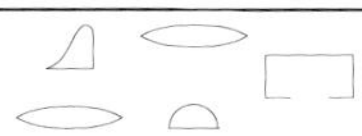	*ḳrrt*	cave
	*šš*	birds nest
	*imw*	tent
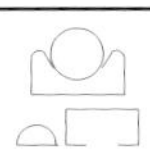	*3ḫt*	horizon

[1] After Goldwasser 2005, 97

The sign 𓉐 as classifier for the category [habitat]. After O. Goldwasser, "Where is Metaphor?: Conceptual Metaphor and Alternative Classification in the Hieroglyphic Script", *Metaphor and Symbol* 20/2 (2005), 97.

---

[4] Goldwasser 2002, 7–38.

The Egyptian vocabulary (lexicon) contains hundreds of words (lexemes) that stand for textiles and garments. But when consulting the dictionaries of Egyptian for basic terms signifying textiles and garments, the translations provided are often rather generic, like 'a kind of textile' or 'a kind of garment'. That the lexemes in question do in fact signify basic kinds of textiles or garments is clear from either textual context or from the cases where they can appear as small labels or comments to pictorial representations. However, given the nature of the highly canonised and rule bound Egyptian art and iconography, it is rarely easy to actually relate a given pictorial form in art to a specific kind of garment, let alone a specific kind of textile. A third way of detecting that a given lexeme stands for kinds of [textiles] or [garments] is through graphemic classification in writing.

1. Cognitive chaining

In graphemic classification, the sign 𓋳 stands for the category [textile] during the majority of the 3rd and 2nd millennium. It was employed as a classifier for a variety of lexemes including nouns signifying kinds of textiles, kinds of garments and a variety of objects made fully or partially of textile. The sign 𓋳 was also used in the writing of names of certain divinities and even verbs denoting actions.

This raises the question of what is the semantic relation between nouns for manmade objects, names for mythological beings and kinds of actions. Moreover, what was the motivation for the ancient Egyptian scribe for writing terms for textiles, garments, and objects, mythological names and verbs with the 𓋳 [textile] classifier? In the classic theory of classification, stretching back to antiquity, category membership is only based on a fixed set of shared attributes by the category members, which in turn defines the category. But how can a wide range of different objects, mythological beings and actions share a fixed set of attributes?

Following his critique of the classic model of classification, linguist George Lakoff has furnished evidence for how category members are linked or chained together on the level of cognition. Most categories are cognitively arranged around core members, so called prototypes. From the prototype(s), the categories chain out to include other members of the category.[5] An often cited example of such cognitive chaining comes from the Dyirbal lingual classifier *Balan*, which is placed before any noun concerned with women, fire and dangerous things.

The Dyirbal noun classifers		
I.	*Bayi*	(human) males; animals
II.	*Balan*	(human) females, fire, fighting, danger
III.	*Balam*	Non-flesh food
IV.	*Bala*	everything not in other classes

The prototype of the category II. *Balan* are *women*, which through Dyirbal mythology become related to *fire*, because the sun is female (i.e. a goddess). Somewhat unexpected the plant "*the hairy mary grub*" appear in the category II. *Balan*, where one would expect it to appear in the category IV. *Bala* with objects that are not animate or edible. It appears in the category *Balan* with women and fire because its sting is felt like a burning pain and it is a dangerous thing like fire can be.[6]

[5] Lakoff 1987, 91–144.
[6] Lakoff 1987, 92–104.

Cognitive chaining in the Dyirbal category II. *Balan*
Category prototype: [women]
[women] – [the sun] – [fire] – [dangerous things] – [hairy mary grub]

The cognitive chaining from the prototype women, to the sun (goddess), to fire, to the "the hairy mary grub" plant reveals some important features of human classification. The grouping of the members marked by *Balan* is not motivated by a fixed set of shared attributes. In this case it is mythological belief systems and bodily experiences that motivate the chaining from the prototype [women] to [fire], and on to other [dangerous things].[7]

According to Lakoff, cognitive category chaining is always motivated so as to make sense, but the structure of category chaining is not predictable, as emphasised by both the bodily experience as well as myth-as-motivator principle. This circumstance makes the category 𓋳 [textiles] a highly interesting case of ancient Egyptian knowledge organisation, when it is explored in terms of motivated cognitive chaining.

2. Textiles

According to Alan H. Gardiner's sign list, number S28 𓋳 depicts a "strip of cloth with fringe combined with the folded cloth 𓋴 S29".[8] That this is indeed the iconic signified by 𓋳 is clear when actual pieces of ancient Egyptian cloth are related to artistic forms like highly detailed hieroglyphs and sculptures grasping folded pieces of cloth.[9] The upper part of the sign 𓋳 depicts the most visually striking and prototypical gestalt form of an Egyptian textile, being a square flat sheet form, ending in a fringe. When combined with the folded piece of cloth (𓋴), it simultaneously encapsulates the embodied notion of the flexible and bendable materiality of cloth. The sign exists in the earliest writing system of the Archaic Period (*c*.3200–2700 BC)[10] and throughout the history of the hieroglyphic writing system ending in the first centuries AD.

Kinds of cloths stand as core members of the category marked by 𓋳 [textile]. Not only is the iconic signified of the sign 𓋳 a rendering of cloth, but words for kinds of cloth can only appear with the textile sign as graphemic classifier. Cloth terms cannot appear in any of the categories marked by other signs. Kinds of cloth are simply prototypical of the category marked by the sign 𓋳. Only later, towards the end of the 2nd millennium BC, textile and garment terms became normatively written with the 𓍱 classifier that signifies the combined category 𓍱 [textile + rope].

The shift in normative classifier from 𓋳 [textile] to 𓍱 [textile + rope] is primarily attested to in texts written in hieratic of the Ramesside Period (*c*.1295–1069 BC).[11] In both the 3rd and the earlier 2nd millennium the categories [textile] and [rope] were differentiated by the signs 𓋳 for [textile] and 𓍢 for [rope]. The motivation for grouping 𓋳 [textile] and 𓍢 [rope] together in the later Ramesside Period was simply a matter of reaching a step back in the *chaine d'opératoire* to "strings", the "material" that both [textile] and [rope] are "made of". Such schematic "made of" classification is a central motivator in ancient Egyptian classification in writing.[12]

---

[7] Lakoff 1987, 102–104.

[8] Entry to GSL S28, Gardiner 1957, 507.

[9] Gardiner 1957, 507, S28 n. 1; S29 n. 1.

[10] Kahl 1994, 698.

[11] Entry to GSL V6, 4, Gardiner 1957, 522.

[12] Goldwasser 2002, 33–35; for an overview of the material classifiers see Gardiner 1957, 32–33.

Most ancient Egyptian words for cloths refer to either quality or form. As mentioned above the specific meaning of most ancient Egyptian textile terms are now lost, but some of the terms in question do have relatively clear etymologies. Yet the overall problem of relating them to actual and specific textiles from archaeological contexts persists.

A group of basic terms that stand for kinds of coloured cloth are founded on nisbe-adjectives for colours. While the pictorial record often show ancient Egyptians clad in pure white garments, coloured textiles do occasionally show up in imagery and are likewise attested in a broad spectrum of the archaeological record.[13]

 **[textile] as classifier for kinds of cloth qualities**		
	*insy*	bright red linen
	*irtiw*	blue linen
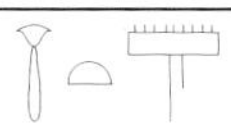	*w3dt*	green linen
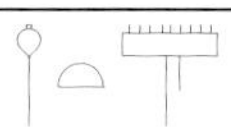	*ḥḏt*	white linen
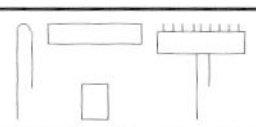	*sšp*	light coloured linen

Culturally significant textile terms for cloth qualities and forms are numerous. The examples include words that refer to differences in the cloth types' form and materiality, or economic and cultural value. The relative differentiation in economic and cultural value can often be discerned when a number of cloth terms appear as entries in lists of economic transactions and commodities or lists of temple and mortuary offerings.[14]

 **[textile] as classifier for kinds of cloth qualities and forms**		
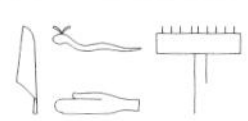	*ifd*	linen sheet
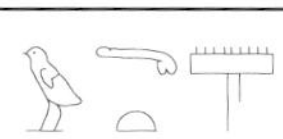	*wmt*	thick linen
	*wnḫw*	linen strips
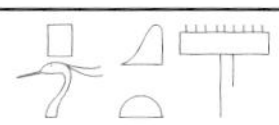	*p3ḳt*	fine linen
	*si3t*	fringed linen
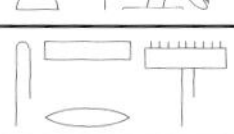	*sšr*	high quality linen

[13] Vogelsang-Eastwood 1995, 31–33.

[14] See e.g. Janssen 1975, 249–298; 436–438; 443–444; Kemp & Vogelsang-Eastwood 2001, 427–449.

Another way of lexicalising certain cloth qualities are through terms founded on nisbe-adjectives signifying relative size or position, from *ꜥꜣ* “great”, *ḥꜣt* “front” and *tpy* “first”. The derived cloth terms translates into something like “great linen”, “front class linen” and “first class linen”. As is the case with the textile terminology in general, the specific meaning of these quality designations remains unclear, although they clearly signify positive qualification and high value given their embodied semantics.

[textile] as classifier for kinds of cloth qualities		
	ꜥꜣt	great linen
	ḥꜣtiw	first (lit. front) class linen
	tpw	first class linen

3. Garments

From the prototypical core constituted by textiles, the [textile] category chain out to include garments. The relationship between the [textile] sign and the garment terms it can appear with is a simple meronymic relationship, i.e. a “made of” relationship. Given the prototypical relation between textiles and garments, garment terms can only in the rarest examples appear outside the [textile] category in writing. In some examples the word *mss* “bag tunic (?)” can be written with a [leather] or [metal] classifier when signifying “armour”, in *mss n ꜥḥꜥ*.[15] The vast majority of garment types are, like the cloth types above, prototypical core members of the category [textile].

[textile] as classifier for garments		
	ꜥfnt	head cloth
	ḫꜣt	royal head wear
	swḥ	kilt/skirt
	šndwt	kilt

As an alternative to writing garment terms with the [textile] classifier, some garment terms can appear with a repeater sign that visually reiterates the meaning of the phonetic sign combination. The repeater signs do not only visually specify the signified of the phonetic sign combination but simultaneously offer a pictorial rendering of what a given piece of ancient Egyptian clothes looked like.[16]

[15] WB II, 149, 7 = Urk. IV, 664; 732.

[16] Goldwasser 2002, 15.

kinds of garments written with a repeater sign			
		*ꜥfnt*	head cloth
		*ḫꜣt*	royal head wear
		*swḥ*	kilt/skirt
		*šnḏwt*	kilt

4. Objects made entirely or partially of textile

From cloths to garments, the category marked by [textile] chains further out to include a host of terms that stand for other manmade objects. The objects appearing in the [textile] category are highly varied in terms of materiality, function and social area of usages. They can roughly be divided into kinds of furniture, containers, ship's parts, architectural features and objects used in medical and cultic practices. Some of the objects are made of textiles whereas others are only partially made of textiles.

The chaining is motivated here by two kinds of meronymic relationships, a "made of" relationship or a "stuff-object" relationship. The "made of" relationship only covers those members that are made of the single stuff "textiles". More composite objects that are only partially made of textile activate a meronymic "stuff-object" relation. A "stuff-object" relation (e.g. hydrogen – water) is activated when a substance only constitutes a portion of the total material of which the objects are made. Since it is not always clear whether an object is made of a single material or from several, the border between "made of" relations and "stuff-object" relations can be rather elusive in ancient Egyptian graphemic classification.[17]

[textile] as classifier for kinds of objects		
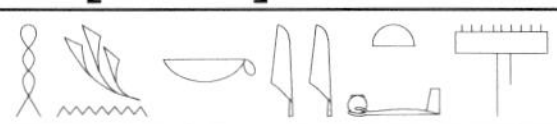	*ḥnkyt*	bed/bedding
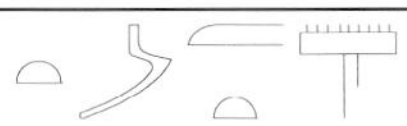	*tmꜣyt*	sitting mat
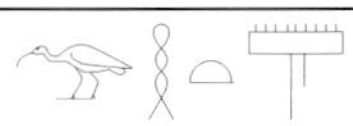	*gmḥt*	lamp wick
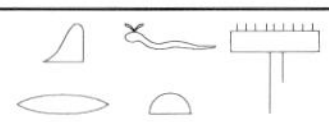	*ḳrft*	(medicine) bag
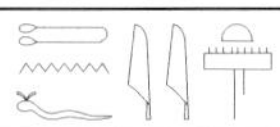	*ṯnfyt*	(tool) bag
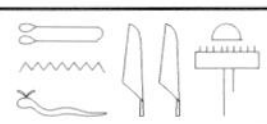	*ṯnfyt*	sail
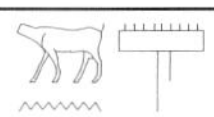	*ḫn*	tent

[17] See discussion by Goldwasser 2002, 33–35.

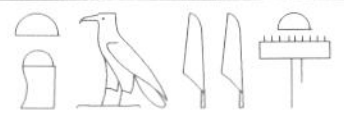	*tꜣyt*	curtain
	*wryt*	filtering cloth
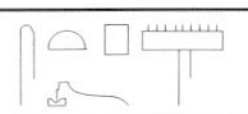	*stp*	patch
	*dꜣiw*	linen ball
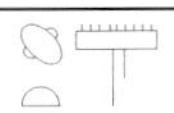	*wt*	mummy bandage

5. Mythological beings related to textiles

A special group of nouns to be written with the [textile] classifier are names of certain divinities and lesser mythological beings. The most prominent of these divinities is the goddess of weaving Tait. In different mythological and ritualised contexts, Tait produces and supplies the gods and the mummified dead with textiles and garments.[18] Other names of mythological entities to appear in the category [textile] stand for beings of the underworld that protect and aid the living dead, including the *wtꜣw* - "they who wrap (the mummy)"[19] and *dbꜣ* - "the one who adorns (the mummy)".[20] On the level of mythology, these beings stand in a contiguous relationship with textiles and garments, given their roles as e.g. "dressers of gods" or "wrappers of mummy bandages". Thus, the motivation for writing these mythological names with the textile-classifier is to be found in the belief system, which relates them metonymically to [textiles].

[textile] as classifier for mythological beings			
	*tꜣit*	Tait	weaving goddess
	*wtꜣw*	Wetaw	"They who wrap (the mummy)"
	*dbꜣ*	Deba	"He who adorns (the mummy)"

6. Actions

The experience of dressing and wearing clothes along with an embodied conception of garments as prototypical "covers" motivates the writing of a number of verbs with the [textiles] classifier as an alternative to the / [action] classifier. The verbs can in some examples be written with both a textile and an action classifier [textile + action] in a form of double classification.[21]

A number of the verbs denote, in their core meaning, acts of dressing or undressing. Writing them with the [textile] classifier was simply motivated by the part-whole, metonymic relationship

[18] WB V, 231; 232, 1–11.
[19] WB I, 380, 6.
[20] WB V, 435, 9–10.
[21] Goldwasser 2002, 16–17.

between acts of dressing and clothes. The part “clothes” is prototypical of [textile] and integral to the whole “dressing”, as well as integral to the antonymic notion of “undressing”.

[textile] as classifier for verbs		
	wnḫ	to clothe
	ḥꜣi	to bare
	ḥbs	to clothe
	sfḫ	to loosen (a band/garment)
	ḏbꜣ	to clothe

In extended meanings a term like *ḥbs* “to clothe” can carry the semantic value “to cover”. For example, in the coronation inscription of Hatshepsut (*c.*1480–1460 BC), the verb is used about the king’s domain – *ḥbs nbt pt* “all that heaven covers”.[22] In the Poem of the Battle of Kadesh from the reign of Ramesses II (*c.*1280–1215), the enemies facing the Egyptians are so numerous that *ḥbs=sn dww intw* “they covered mountains and valleys”.[23]

The semantic extension from “to dress” to the more generic “to cover” is paralleled by the writing system when the category [textile] chains out to include verbs for “covering”, and from there on to verbs for “hiding”. The intimate experience and daily practice of covering the body with textiles, created the cognitive link, and following the motivation for writing verbs for “covering” with the [textile] classifier. Thereby, the writing system identifies [textile] as kinds of prototypical “covers” in ancient Egyptian world organisation.

Examples include *skꜣp* ”to cover”, which can be used about covering a building with a roof or the covering of a clouded sky.[24] The verb *ḥꜣp* is used in the sense “to cover with something”, like a piece of cloth, a pot or with the hands.[25] In extended meaning *ḥꜣp* signify “to hide” and “to keep secret”. For example, in a letter from a haunted husband to his deceased wife, he makes it explicit that he did not hide (i.e. *ḥꜣp*) anything from her during her life.[26] In metaphorical use, *ḥꜣp* “to hide” can be used passively about *mdt* “words” - *mdt ḥꜣpt* “hidden words” (i.e. secrets).[27]

22 Deir el Bahari, Temple of Hatshepsut = Urk. IV, 244, 12.
23 Kadesh Poem = KRI II, §51.
24 WB IV, 316, 17; 19.
25 WB III, 30, 6–8.
26 WB III, 30, 16; P. Leiden 371, rt. 17 = Gardiner 1928, VII, 17.
27 WB III, 30, 17; Stele of Sehetepibre, Cairo CG 20538, rt. 10 = Lange & Schäfer 1908, 147; Stele of Kares, 7–8 = Urk. IV, 46, 15.

[textile] as classifier for verbs		
	ḥ3p	to cover / to hide
	sk3p	to cover

Similarly, the [textile] classifier can be used in the writing of a group of antonymic verbs for "to cover" like "to bare" and "to expose". The writing follows the same bodily-experience-motivation as for the verbs "dressing" and "covering" they are antonymic to. The verb *sḥ3i* "to strip" can be used about the removal of peoples' clothes, the baring of a body, but also about the unveiling of mysteries.[28] Another example is *kfi* "to uncover", from clothes, a wound from bandage, teeth when laughing, a secret, etc.[29]

[textile] as classifier for verbs		
	sḥ3i	to strip / to bare
	kf	to uncover

* * *

The survey of the category [textile] has shown how the category chained out from the prototypical core to include a whole range of nouns and verbs. Category membership was not based on a set of shared attributes, which in turn defined the category. Rather, the category was structured according to a selection of cognitively motivated relationships to the category prototype [textile], which enabled the chaining from the icon prototype to textiles, garments, objects, mythological beings and actions.

The iconic signified of the sign and the fact that textile terms must appear with as graphemic classifier for the major part of the 3rd and 2nd millennium securely identifies [textiles] as the category core prototype. The chaining from the category prototype ( ) to [textile] to the category's other lower ranking members is based on a variety of identifiable motivated relationships. The survey presented here identified three overall semantic motivations for writing a given term or name with the [textile] classifier.

The manmade cloths, garments and miscellaneous objects, are all related meronymically to [textile]. The schematic relationships come in two kinds, "made of" relation or "stuff-object" relationships that covers objects made fully of textiles or composite objects made only partially of textiles:

[28] WB IV, 10; 11–13.

[29] WB V, 119, 4; 6–8.

Cognitive chaining in the category [textile]
Category prototype: [textiles] - [textiles] - [textiles] – [garments] - [textiles] – [objects]

A second cognitive chain comprises the names of divinities and mythological beings with the [textile] classifier. Mythology and iconography furnish the evidence for these mythological beings' contiguous relationship to textiles and garments. They are metonymically related to textiles through their roles in different cultic practices and mythological places.

Cognitive chaining in the category [textile]
- [textiles] – [garments] – [mythological beings]

Bodily experiences of dressing and covering with textiles motivate the category to chain out to a range of actions, along with antonymic senses of these actions. The metonymic relation between textiles and dressing creates the first link in the chain of verbs for "dressing". To these should be added the verbs that can mean "to dress" in a more extended sense, like the wrapping of a mummy or the adorning of a cult statue.

Cognitive chaining in the category [textile]
- [textiles] – [garments] – [dressing]

A number of the verbs polysemiously signify "to dress with clothes", "to cover", and "to hide" thereby identifying a context bound semantic transition between [to dress], [to cover] and/or [to hide]; a semantic transition that arose from shared embodiment and bodily experiences of "dressing" as a prototypical "covering". The semantic move from "textiles", to "dress", "cover" and on to "hide" and "keep secret" is of course well known for the modern English and Danish speakers alike. Our own more or less corresponding terms "to cover" (Danish: at dække) and "to veil" (Danish: at sløre) likewise arise from a shared embodiment and bodily experiences of textiles as prototypical [covers]. The examples include "to cover" in generic senses as well as metaphorical senses when it is abstract phenomena like "secrets", which are hidden. This is paralleled in graphemic classification when verbs for [to cover] in both general and metaphorical sense take the [textiles] classifier.

Cognitive chaining in the category  **[textile]**
 - [textiles] – [garments] – [dressing] – [covering] – [hiding] – [keeping secrets]

This paper has demonstrated how the study of graphemic classification in ancient Egyptian writing can contribute to semantic studies of the ancient Egyptian lexicon and world organisation. The case of the  [textile] category presented here demonstrates how the writing system provides additional layers of semantic information to a given word, and how this semantic information can be used to rediscover some basic features of ancient Egyptian knowledge organisation. Without visual classification in writing, this crucial phenomenon would have evaded the Egyptologists. Secondly the case of the  [textile] category provides an example of how human categorisation is not necessarily based on fixed sets of shared attributes among members. Instead, the category  [textile] was chained together following a number of identifiable motivations: meronymy, metonymy, mythological belief systems, bodily experience of the real world and material culture.

More extensive studies of ancient Egyptian object categorisation and classification are awaited. Classification theories, including graphemic classification as part of the ancient Egyptian writing system will be important tools in future analyses. The focus on cognitive chaining is just one aspect of the application of an Egyptological perspective to the study of cognitive and anthropological issues, which should result in a greater understanding of the language and writing systems, and above all, ancient Egyptian world organisation and material culture.

Acknowledgements

I am grateful to Paul John Frandsen, Rune Nyord and Kim Ryholt for their insightful comments and corrections to my English.

Abbreviations

GSL Gardiner Sign List, Gardiner, A. H. (1957) *Egyptian Grammar* (3rd ed. rev), 438–548.
KRI Kitchen, K. A. (1975–1989) *Ramesside Inscriptions,* Vol. I–VII. Oxford.
URK IV Sethe, K. (1906–1909) *Historisch-biographische Urkunden der 18. Dynastie*, Bd. I–IV. Leipzig.
WB Erman, A. & Grapow, H. (1971) *Wörterbuch der Aegyptischen Sprache*, Bd. I–V. Berlin.

Bibliography

Allen, J. P. (2000) *Middle Egyptian. An Introduction to the Language and Culture of Hieroglyphs.* Cambridge – New York.

David, A. (2000) *De l'inferiorité à la pertubation. L'oiseau du "mal" et la catégorisation en Egypte ancienne.* Wiesbaden.

Frandsen, P. J. (1997) "On Categorization and Metaphorical Structuring: Some Remarks on Egyptian Art and Language", *Cambridge Archaeological Journal* 7, 71–104.

Gardiner, A. H. (1928) *Egyptian Letters to the Dead, mainly from the Old and Middle Kingdoms.* London.

Gardiner, A. H. (1957) *Egyptian Grammar* (3rd ed. rev).

Goldwasser, O. (2002) *Prophets, Lovers and Giraffes: Wor(l)d Classification in Ancient Egypt.* Wiesbaden.

Goldwasser, O. (2005) "Where is Metaphor?: Conceptual Metaphor and Alternative Classification in the Hieroglyphic Script", *Metaphor and Symbol* 20/2, 95–113.
Janssen, J. J. (1975) *Commodity Prices from the Ramessid Period.* Leiden.
Kahl, J. (1994) *Das System der Ägyptischen Hieroglyphenschrift in der 0.-3. Dynastie.* Wiesbaden.
Kemp, B. J. & Vogelsang-Eastwood, G. (2001) *The Ancient Textile Industry at Amarna.* London.
Lakoff, G. (1987) *Women, Fire, and Dangerous Things. What Categories Reveal about the Mind.* Chigaco – London.
Lange, H. O. & Schäfer, H. (1908) *Catalogue Général des Antiquités du Musée du Caire Nos 20001–20780, Grab und Denkstein des Mittleren Reichs.* Berlin.
Lincke, E. S. (forthcoming), *Das Schriftsystem der Pyramidentexte, Zu den Prinzipien der Klassifizierung im Altägyptischen.*
Shalomi-Hen, R. (2000) *Classifying the Divine, Determinatives and Categorisation in CT 335 and BD 17.* Wiesbaden.
Shalomi-Hen, R. (2006) *The Writing of Gods, The Evolution of Divine Classifiers in the Old Kingdom.* Wiesbaden.
Vogelsang-Eastwood, G. (1995) *Fra Faraos Klædeskab* (transl. Nielsen & Nielsen). København.

6. The 'linen list' in Early Dynastic and Old Kingdom Egypt: Text and textile reconciled

Jana Jones

The favourable conditions for organic preservation in ancient Egypt have yielded significant quantities of surviving textiles, providing an archaeological record spanning some 5,000 years from the Neolithic to the Roman periods and beyond. The earliest extant example of an Egyptian woven linen textile dates to the early 5th millennium BC, from a Neolithic settlement in the Fayum oasis.[1] The majority of textiles from ancient Egypt originate in funerary contexts, and as such would not necessarily reflect the full spectrum of textile technologies. However, the re-use of household textiles in the mummification process and the inclusion of offerings of textiles and clothing in burials result in a wide range of archaeological evidence.

The archaeological material is supported by written evidence from as early as the Protodynastic period (*c.*3250 BC) on commodity labels from Tomb U-j at Umm el-Qa'ab at Abydos.[2] Textual and iconographic evidence on private funerary stelae begins in the Early Dynastic period[3] (*c.*2900–2545^{+25} BC) and continues into the Old Kingdom, specifically into the reign of Khufu in the Fourth Dynasty (*c.*2509–2483^{+25} BC).[4] From the Old Kingdom onwards, iconographic and textual evidence for organised flax cultivation and processing, textile production and distribution

---

1 Caton Thompson & Gardner 1934, 46, Pl. 28.3; Jones & Oldfield 2006, 33–35. The textile was examined by Jana Jones and Ron Oldfield at the School of Biological Sciences, Macquarie University, Sydney Australia. Microscopic examination confirmed that the ultimate fibres exhibit the characteristic features of *L. usitatissimum*, the principle fibre used in ancient Egyptian textiles. The thorough preparation suggests an already well-developed knowledge of the complex procedures involved in flax processing in the Neolithic period. Fragments of the textile are in the collections of the Petrie Museum of Egyptian Archaeology (UC 2943), the British Museum (EA 58761), and Bolton Museum and Art Gallery (56.26.109/1 and 2). Permission to sample the Bolton textile for further analysis was granted by Angela Thomas, formerly Keeper of Egyptology.

2 Dreyer 1998.

3 The Early Dynastic period is the earliest historical period, when the Egyptian 'dynastic' state was formed, and a unified, complex society emerged.

4 Dynastic period chronologies are based on Hornung, Krauss & Warburton 2006; Predynastic chronologies are based on Hendrickx 1999 and Hendrickx 2006. Absolute dates for the early stages of Egyptian chronology remain tentative, and earlier dates for the beginning of the Early Dynastic period have been proposed.

complement the archaeological material. The processes are illustrated in wall paintings in tombs, wooden models of weaving workshops and by administrative titles.[5]

The private funerary stelae with offering scenes originate in the necropolises of the ancient capital city, Memphis, namely Helwan, Saqqara and Giza.[6] The inventories of textiles (the so-called 'linen lists') on funerary stelae first appear in the late First Dynasty at Helwan *c.*2800 BC, and are the source of the greatest body of textile and textile-related terminology in the early 3rd millennium BC. The linen lists reach their full development on the Giza slab stelae of the Old Kingdom, specifically dated to the Fourth Dynasty during the reign of Khufu, the builder of the Great Pyramid. After this time they occur very rarely.[7]

In 2003, Peter Der Manuelian noted: " ... there is still a need to reconcile more of the textual evidence with the material remains. Our understanding of the linen lists of the Old Kingdom would greatly benefit from such an interdisciplinary approach."[8]

There are possibilities but also problems inherent in this approach. The linen lists specify particular qualities, dimensions and quantities of textiles or clothing to be offered for the use of the deceased in the afterlife. The meaning of most of these signs remains largely enigmatic, despite the attempts of philologists to translate them. Although ideograms representing items of clothing are mostly recognisable from comparison with three-dimensional iconographic evidence, the textiles pose a different problem. In order to fully reconcile 'text' and 'textile', that elusive body of evidence is needed: intact lengths of textile inscribed with signs specifying size or quality.

The writer has examined large quantities of textiles on site in Egypt and in museums, but they have either lacked inscriptions or the textile has been cut, as in the case of the inscribed fragment from the pyramid of Pepi I, discussed below. In some cases the ground textile has been degraded by the corrosive chemical action of the pigments so that the signs are illegible. In the past, inscriptions have been recorded but textile analyses were not carried out and the material has been lost.[9] Furthermore, many textiles in museum collections are undated or without provenance, so even if they were inscribed it would be difficult to relate the technological evidence to a specific period. Textiles from new excavations often are not studied systematically, and modern protocols governing the unwrapping of bodies limit the availability of material for examination.

This contribution will attempt to relate textile terminologies to archaeological and iconographic evidence within the parameters of available material, correcting some long-standing

[5] For interpretations based on tomb scenes and Middle Kingdom models see Winlock 1955; Bellinger 1959; Baines 1989, 20–21; Vogelsang-Eastwood 2000, 268–298. For a discussion of spinning techniques based on tools and ethnographic models see Crowfoot 1931; spindles and whorls are discussed in Kemp & Vogelsang-Eastwood 2001, 265–277. For looms and weaving see Roth 1978 and Kemp & Vogelsang-Eastwood 2001, 307–404. Jones 2000 lists a number of Old Kingdom titles related to textile production, e.g., 170 no. 649 "overseer of every royal bundle of flax"; 512 no. 129 "overseer of the houses of weaving women of ... of the *itiwy*-cloth of the king (?)".

[6] These sites are located south of modern day Cairo. Giza and Saqqara were positioned on the west bank of the Nile in close proximity to the king who resided at Memphis, and Helwan on the east bank. Saqqara and Helwan were the two main necropolises of the Early Dynastic period, but Saqqara was the prime burial site for the highest-ranking elite and aristocracy. Minor royalty, middle and lower ranking courtiers, lesser members of the priesthood and artisans were buried at Helwan.

[7] The variations of the linen lists that date beyond the Fourth Dynasty will not be discussed in this paper.

[8] Manuelian 2003, 155.

[9] For example, linen wrappings from the Fourth Dynasty mummy from tomb G 2220 B I inscribed in hieratic script included the quality *šmꜥ.t nfr.t* that appears as a 'heading' in contemporary Old Kingdom linen lists. See Manuelian 2003, 156, Figs. 232, 233.

misconceptions and proposing alternate translations of textile terms in the process. Further, it will be shown that there is a clear development from the Early Dynastic textile inventories to the 'fully developed' linen lists of the Fourth Dynasty Giza slab stelae.

1. A brief overview of the principles of Egyptian writing systems

The ancient Egyptian writing system is based on the principles of rebus, alphabet and complement.[10] In order to facilitate understanding of the terminology considered in this paper, the following definitions summarising the main elements of the developed system are offered.[11]

Ideograms (or logograms) are words represented pictorially and are etymologically or semantically related to the sign's own meaning.

Phonograms are used in any word for notating mono or bi-consonantal phonemes, generally according to the rebus principle.

Determinatives classify a word according to its semantic sphere, and appear at the end of the word.

Phonetic complements are uniliteral signs added to clarify the reading of multi-literal signs that may have more than one phonetic value.

The first evidence for writing in Egypt (and the earliest source of references to textiles) is from the elite, Protodynastic Tomb U-j at Umm el-Q'aab, Abydos, dated NIIIA1 (*c.*3250 BC).[12] Ideograms, phonograms and determinatives were already in use at the date of Tomb U-j, but phonetic complements were not, and the stock of hieroglyphic signs was still quite limited. A more developed form and an almost complete syllabary are not apparent until the First Dynasty, probably by the time of King Den *c.*2800 BC.[13] The fully developed form is evident only at the beginning of the Third Dynasty.[14] In developed hieroglyphic writing, the signs can be used to fulfil several functions, either or both as ideograms, determinatives, phonograms and phonetic complements.[15] Some of the characteristics of archaic writing that make it quite difficult to translate and often ambiguous, are that spelling, orientation and sequence of the signs is completely variable.[16]

2. Textile terminology: the written record

2.1. *Commodity labels*

Almost 200 bone commodity labels were discovered in Tomb U-j at Abydos in Middle Egypt, mostly in chamber U-j 11.[17] The incised signs have been interpreted as designating the quantity and origin of a variety of commodities. It was deduced that a number of these labels had been attached to textiles. The chamber also yielded fragments of wooden boxes in which the textiles may have been housed.[18] Two labels are incised with ideograms that represent garments – a sleeved dress and a cloak with ties.[19] Three other labels have a sign that may be a garment, but that identification is

[10] Kahl 2001, 105.

[11] Kahl 2001, 116; Ockinga 1998, 1–6.

[12] Dreyer 1998. Chronology according to Hendrickx 1999, 31, 76.

[13] Kahl 2001, 124. Kahl proposes an earlier date of *c.*2900 BC.

[14] Kahl 2001, 116; *c.*2592–2544^{+25} BC

[15] Kahl 2001, 116. See also Kahl 1994, 52 for methods of creating hieroglyphic signs.

[16] Kahl 2001, 113.

[17] Dreyer 1998.

[18] Dreyer 1998, 139.

[19] Dreyer 1998, 132–133, Fig. 81.161–162, Pl. 34. Both are referred to as *Gewand*.

uncertain.[20] The excavator suggested that the ideograms are generic signs for clothing and textiles rather than designations of the actual garments, and refer to the origin of the garments in the royal stores.[21] A number of other labels are incised with short strokes arranged either vertically or horizontally in groups of six to 12.[22] The vertical signs[23] are understood to be numerals,[24] but it was suggested that the horizontal signs[25] might be interpreted as designations of the surface area of the textile, where one sign represents 10 square cubits, as on the linen lists of the Old Kingdom slab stelae. [26] The sign 𓍢 *šn.t* [27] that represents the number '100' occurs on five labels. [28] The latter sign also appears in the linen lists of the Third and Fourth Dynasty stelae inside a square or rectangular ideogram of a textile (sometimes fringed) and represents 100 square cubits. These signs will be discussed below in the context of the linen lists.

By analogy with the information derived from other commodity labels from Tomb U-j, it is highly probable that textile production and distribution was under some form of centralised control.

2.2. *Private funerary stelae*

2.2.1. *The Helwan relief slabs: archaeological background*

The Helwan corpus of 46 known relief slabs[29] (or stelae as they are commonly described) is the largest group of securely provenanced, early, inscribed material of its kind and is the largest, single source of early textile terminology. Thirty-two of these relief slabs were excavated by the Egyptian archaeologist Zaki Saad between 1942 and 1954.[30] During that period, approximately 10,000 graves of the 'middle class' elite were uncovered. Although similar stelae were known from other sites such as Saqqara and from the antiquities market in the early 20th century, these relief slabs were especially exciting because of their supposed origin in a primary archaeological context of the Second Dynasty. Saad observed that the slabs were located in the substructure of the tomb, specifically in the ceiling above the deceased; hence they were referred to as 'ceiling stelae'. In 1971 this premise was disputed. It was established that the slabs were found in a secondary context, originally having been in the superstructure of the tomb but later were reused as fill in robbers' shafts.[31]

The Australian mission to Helwan, under the direction of E. C. Köhler, began recording the few Helwan artefacts on display in the Egyptian Museum, Cairo in early 1997. The location of the

[20] Dreyer 1998, 59, 132, Fig. 88.160, 163, Pl. 34.
[21] Dreyer 1998, 144.
[22] Dreyer 1998, Figs. 74–75, Pls. 27–28.
[23] Gardiner 1988, sign list no. Z 1.
[24] Dreyer 1998, 113–117, 139–140.
[25] Gardiner 1988, sign list no. Aa 12.
[26] Dreyer 1998, 140. After Posener-Kriéger 1977, to be discussed below.
[27] Gardiner 1988, sign list no. V 1. Gardiner 1988, 521 noted that the sign could be transliterated either as *š.t* or *šn.t*. The latter transliteration is used in this paper.
[28] Dreyer 1998, 113–118, 139–140, Pls. 27–28. Only one sign for 100 originates from (above) U-j.
[29] Köhler & Jones 2009. Forty-one of the slabs are included in the catalogue, and the remainder will be published subsequently. The authors have chosen to follow the terminology of Müller 1933, 166–167. He differentiated between tectonic monuments (stand alone, such as stelae), and architectonic monuments (dependent, such as false doors). The Helwan slabs were integrated within the superstructure of the tomb, so they should not be considered as tectonic monuments. Therefore the term 'stela' is technically not correct.
[30] Saad 1951.
[31] Haeny 1971, 143–164.

main corpus of the Helwan reliefs was unknown until the discovery of the 'lost' Helwan artefacts in the basement of the Museum. In 1999 the relief slabs were found inside ten boxes, and were photographed and drawn. Additional complete slabs and fragments continue to be found in each season's excavations at Helwan.

2.2.2. Description

The relief slabs are of white, fine, dense limestone probably originating in the limestone quarries near the site at Tura or Ma'asara. They are of different sizes and volume, but the average is approximately 50–60 cm long, 20–25 cm high and 6–10 cm thick.[32] Although the form of each is different, generally the slabs have an undressed back and side flanges, with a dressed central front panel that is carved either in raised relief and/or incised into the surface. Traces of paint remain on some slabs. Although none of the slabs was found in situ, the evidence suggests that they were incorporated into the mud-brick superstructure of the tomb above the false door.

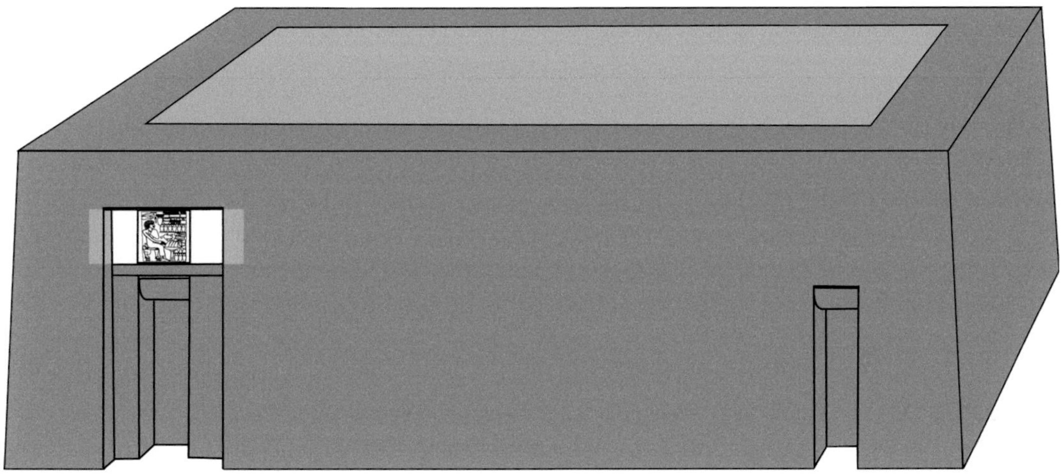

Fig. 6.1. Mastaba reconstruction with possible position of relief slab[33]

The tomb owner is depicted seated on a stool or chair before an offering table laden with loaves of bread. Inventories of offerings comprise items of food such as bread and baked goods, meat and fowl, fruit and agricultural produce, beverages, cosmetics, oils and fats, incense and textiles,[34] and can include utilitarian objects such as tools, boxes, head rests and beds. The tomb owner generally grasps the shoulder knot of his tight-fitting cloak with the left hand and with his right reaches towards the bread. The female tomb owners do not hold the knot, placing the open palm or fist of the left hand on the chest. His or her name and titles appear above the head. The inventory of textiles and clothing, or linen list, generally constitutes a large proportion of the total iconography. Signs designating different qualities, sizes and quantities of textiles and items

[32] Köhler & Jones 2009, 17.
[33] Köhler & Jones 2009, Fig. 30.
[34] Thirty-one of the total of 41 Helwan relief slabs studied include textiles.

Fig. 6.2. Relief slab of Bat or Khenmet (?) EM99-18 (mid- to late Second Dynasty).

of clothing occur early in the corpus, and gradually increase in number and variety. The textiles are prominently positioned, often at the head of the lists of offerings and finally in an ordered, compartmentalised section, generally on the right hand side of the relief slab.

The reliefs play a two-fold part: the tomb owner is identified at the cult place of the tomb through his or her name and titles, and the offerings, which are specified and generally quantified, are a symbolic form of sustenance and provisioning. These are spiritually activated in the mortuary ritual for the eternal use of the deceased in the afterlife. The textiles represent an important part of the owner's estate and tomb assemblage. Later royal and non-royal funerary texts allude to the significance of specific types of textiles and clothing that are essential equipment for the deceased in the passage into the afterlife.[35]

2.2.3. *Seriation and relative dating*

A relative chronology of the relief slabs was established by using multivariate correspondence analysis and seriation, based on certain diagnostic iconographic elements. Briefly, where the same iconographic variables occur among different units, they show chronological proximity; the fewer features they have in common, the further distant they are. Known parallels from other sites were included in the seriation matrix. As a result, the Helwan reliefs were randomly divided into three date groups: Early ('A', First Dynasty, *c.*2900–2730^{+25}), Middle ('B', Second Dynasty *c.*2730–2590^{+25}BC) and Late ('C', Third and early Fourth Dynasties, *c.*2592–2543^{+25} BC).[36]

[35] The "utterances" of the Pyramid Texts (inscribed on the internal walls of pyramids, dated to the end of the Fifth and the Sixth Dynasties *c.*2321–2153 BC) and the later "spells" of the Middle Kingdom Coffin Texts (inscribed on the inside of wooden coffins *c.*2080–1760 BC) contain numerous references to clothing or wrapping and binding of the body. For example, Faulkner 1969, 151, in P.T. Utt. 453, the king is attired in the textile offerings so that he may appear before the gods; Faulkner 1977 Vol. II, 197, C.T. Spell 608 evokes the textiles given by Horus to Osiris as being "*idmy* cloth, *mnḫ.t* ('cloth') *ʿ3.t* cloth and *ḥḏ.t* ('bright one')".

[36] Köhler & Jones 2009, 23–25.

2.2.4. *Previous studies and translations of the Early Dynastic and Old Kingdom linen lists*

Studies of offering scenes and lists appeared from the early 20th century. These focused principally on the religious and symbolic significance in the context of the funerary ritual of the Early Dynastic and Old Kingdom periods.[37] The textile terminology of the linen list particularly attracted the attention of scholars, because the list is such a prominent element in the reliefs.[38] Winfried Barta published a range of offering lists from various periods, including the Helwan stelae, which had originally been published by Saad in 1957.[39] He suggested a possible origin and development and attempted to establish a chronological sequence.[40] The work on Early Dynastic inscriptions by Peter Kaplony was published in the same year.[41] He collated all the known sources of Early Dynastic offering scenes, including the original examples from Helwan, and subsequently published another eight to ten that had been in the site museum.[42] Both scholars proposed translations of the textile terminology and noted that many of the items in the Old Kingdom offering lists also appeared earlier, providing a point of reference to the translation of the often-problematical archaic Egyptian hieroglyphs.[43]

The study of the Old Kingdom linen list by William Stevenson Smith in 1935 has been the most comprehensive and most quoted interpretation of the elements of the linen list and the meaning of ancient Egyptian textile and clothing terms.[44] Smith listed 43 occurrences of linen lists on the then known monuments, including Early Dynastic stelae, Old Kingdom stelae and false door panels, and offerings on the walls of burial chambers and chapels. More recently, Peter Der Manuelian has re-examined the Giza group of slab stelae, provided translations and detailed commentary, and included all the known Early Dynastic stelae in an appendix.[45]

The latest comprehensive study of all archaic inscribed material, including the Helwan relief slabs, is by Jochem Kahl, who assembled all known inscriptions dating from Dynasty 0 to the Third Dynasty.[46] Although the work is invaluable as a compilation of the inscriptions and an analysis of the early writing systems, Kahl has relied heavily on previous scholars' interpretations of the textile terminology, much of which can be questioned.

2.2.5. *The Early Dynastic linen list*

The recent examination of the Early Dynastic and early Old Kingdom funerary relief slabs from Helwan confirms that there is a clear progression from the textile inventories that made their first appearance in the late First Dynasty, to the fully developed Old Kingdom linen list.[47] Until

---

[37] Most notable were Junker 1925–1955; Hassan 1932–1960; Reisner 1932, 324–331; 1936; Reisner & Smith 1955.

[38] For example, Lacau 1904, 91; Griffith 1905, 33; Weill 1908, 228–229; Jéquier 1921, 34; Sethe 1923, 216; Blackman 1924, 47–48; Junker 1929, 177–178; Smith 1935, 134–149; Edel 1975, 13–30; Barguet 1986, 85–86.

[39] Saad 1957.

[40] Barta 1963.

[41] Kaplony 1963, Vols. I–III.

[42] Kaplony 1966.

[43] See discussion on early writing systems above, and Kahl 2001, 101–134.

[44] Smith 1935, 134–149.

[45] Manuelian 2003.

[46] Kahl 1994. Kahl devised a system of numbering the early signs based on Gardiner's sign list (Gardiner 1988, 438–548). He followed this work with translations of Third Dynasty inscriptions (Kahl, Kloth and Zimmerman 1997) and began a dictionary of archaic Egyptian terminology (Kahl 2002–).

[47] Köhler & Jones 2009.

the Third Dynasty, the arrangement of the linen list was somewhat arbitrary. In certain examples the offerings were unstructured (Fig. 6.14), organised in horizontal rows (Fig. 6.5), shown on 'shelves' (Fig. 6.2), or within defined compartments (Figs. 6.3, 6.4). Yet the textiles were always positioned prominently, generally at the head of the offerings. The first occurrence of textiles in the Helwan corpus is on a late First Dynasty, well-preserved relief slab that still retains traces of red and black paint (Fig. 6.3). A rectangular textile with a short fringe heads the list of offerings. As yet there are no indicators of quality or quantity.

Fig. 6.3. Relief slab of Ini-sw-skit (or -skr ?) SO1-38 (late First Dynasty).

A slightly later relief slab dated late First to early Second Dynasty,[48] shows three textiles in the uppermost register, each with a different number of long fringes, but includes the quantity of two below each item (Fig. 6.4). To the left of the vertical separator, a narrower textile with only two strands of fringe is associated with vessels of calcite. The significance of this juxtaposition is discussed below (under 'Clothing'). The variation in the number of strands of fringe suggests that the textiles were graded according to specific criteria, which to date are not fully determined. There are no designations of quality.

Fig. 6.4. Relief slab of Wa(t) EM99-14 (late First to early Second Dynasty).

From the mid-Second Dynasty there is a noticeable increase in the type, variety and quantity of textiles depicted. Items of clothing now appear as offerings.[49] Furthermore, the three main qualities characteristic of the Third and Fourth Dynasty stelae occur together on one of the late Second Dynasty Helwan slabs (Fig. 6.5). In the uppermost register of the relief, signs designating specific qualities appear as headings above textile signs showing different lengths of fringe.[50] From right to left, the Horus falcon on a standard signifies the finest, or 'royal' quality *iti.wy/idmy*;[51] next is the lesser quality

[48] Kaplony 1966, 266, No. 1096.

[49] For example, EM99-4, -5, -18. Signs designating items of clothing do not continue in the Old Kingdom linen lists.

[50] See discussion of fringe type as indicator of size, below.

[51] Gardiner 1988, sign list no. G 7, a combination of G 5 on R 12, with the addition of a folded (or pleated) cloth projecting from the 'perch' of the standard. The transliteration of *iti.wy* has been adopted for the Early Dynastic corpus, and *idmy*,

šsr,[52] represented by the arrow, then another quality *ꜥ3*,[53] represented by the tent pole or column. This is followed by *m3ṯ* ('dress') written phonetically, and finally *šsr* is repeated. The numerals specifying quantities to be offered appear below each textile sign. These include quantities of two, three, four and 20.

Four relief slabs dated to the Third Dynasty have been found at Helwan.[54] Although each has offerings in a compartmentalised section on the right hand side, EM99-19 is the most 'developed', with a linen list that takes up almost half of the entire slab and is devoted entirely to textiles (Fig. 6.6). Although it anticipates the Fourth Dynasty linen lists in all its components, a garment *šnḏw.t* (kilt) is included amongst the textiles. Items of clothing do not appear in the linen lists of the Giza stelae.

Fig. 6.5. Relief slab of Heken EM99-4 (Late Second Dynasty).

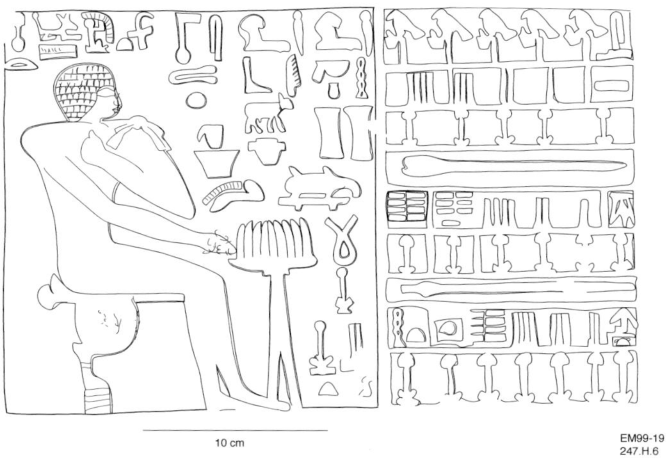

Fig. 6.6. Relief slab of Nefer EM99-19 (Third Dynasty).

Before proceeding to a detailed study of the textile-related signs in the Early Dynastic corpus and a discussion of surviving textile evidence, an overview of the structure of the Old Kingdom linen lists and meaning of the individual elements is warranted. This will be followed by a brief discussion of the construction of fringes, on which calculations of linear dimensions of textiles are based.

which appears to be a later form, for the Old Kingdom corpus. See Edel 1975, 24–27. For a discussion of *idmy* v. *iti.wy* see also Manuelian 2003, 153. The traditional translation of *idmy* as 'red linen' has been refuted by Edel 1975, 24, and independently by the writer in an unpublished M.A. thesis at Macquarie University, 1998. Colour in early Egyptian textiles will not be addressed in this paper, as there is limited reliable textual or archaeological evidence for coloured textiles in this period. There seems to have been a preference for pure white linen in the Early Dynastic period (Jones 2008, 122–123) but there is some archaeological evidence for the use of mineral-based dyes in the Predynastic period (Jones 2002, 327–328). See also Germer 1992.

[52] Gardiner 1988, sign list no. T 11.

[53] Gardiner 1988, sign list no. O 29.

[54] Köhler & Jones 2009 EM99-13, 19, 32 and EM97-43. The latter still retains a haphazard arrangement of items of linen.

2.2.6. The arrangement of the Old Kingdom linen list

The linen list shows textile quality, dimensions of individual pieces, and quantity to be offered. It is arranged into three (or sometimes four) groups of three horizontal registers, generally divided further into a number of vertical registers. The uppermost register group consists of the Horus falcon on a standard in the top register, signifying *idmy*[55] 'royal' linen, the finest quality of textile; the second register usually consists of designative signs that indicate specific sizes of textiles, and the third contains numerals specifying quantities. This configuration is repeated, with different 'headings', most commonly *šsr* (represented by the arrow) and *ʿꜣ* (tent pole or column) in the first register of each new group. These denote lesser qualities of linen. Other less frequently occurring headings in the Old Kingdom lists include *šmʿ.t nfr.t* (which is first attested in the Third Dynasty) and *idmy*, written phonetically.[56]

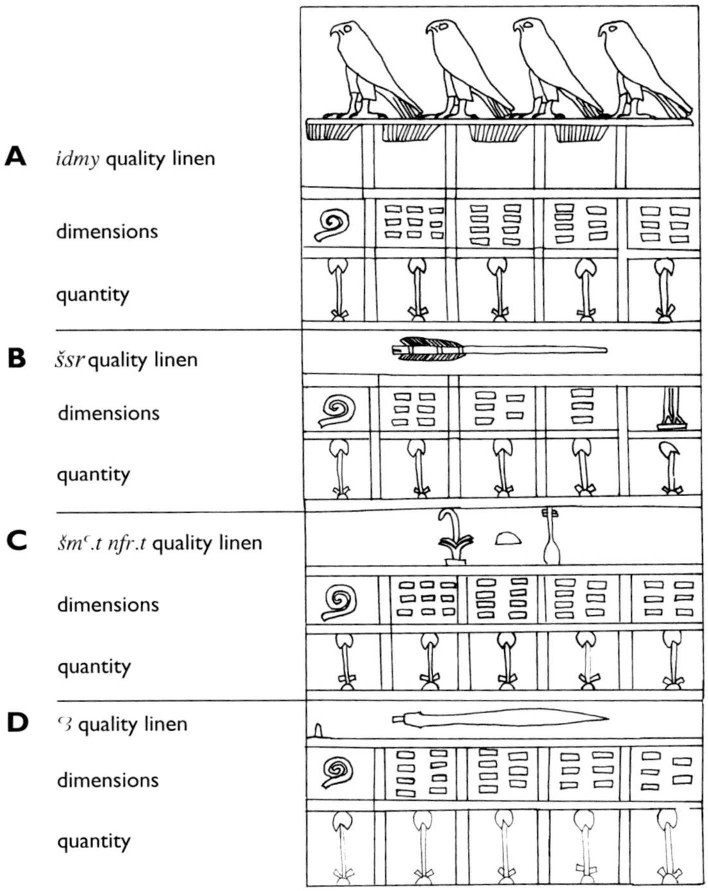

Fig. 6.7. Elements of the fully developed Old Kingdom linen list. After Manuelian 2003, Pl. 16, slab stela of Ini. Drawn by Mary Hartley, Macquarie University, Sydney Australia. Here 'dimensions' describe the function of the signs in the second register of each group, contra Manuelian, who used 'width'. Note that otherwise unrealistic measurements of width are obtained when the horizontal sign, which in all probability denotes area, is used as a linear measurement.

The following reading of the linen list is proposed, based on the criteria put forward by Posener-Kriéger.[57] Briefly, she suggested that each of the horizontal signs in the second register of each group represents an area of 10 square cubits multiplied by the number of times the sign is written, and that the upright 'fringes' represent linear measurements of 1 cubit each (width) × 10 cubits (length).[58] The sign 𓍶 *šn.t* represents 100 square cubits.[59] These formulae for calculation of dimensions, and the inherent problems, will be discussed in detail presently.

[55] See n. 53 on the use of *idmy* v. *iti.wy* in the Old Kingdom lists.

[56] No two examples follow exactly the same arrangement. Cf. Smith 1935, 147.

[57] Posener-Kriéger 1977, 86–96.

[58] Posener-Kriéger 1977, 92. One cubit (52.3 cm) is equivalent to 7 palms and 28 digits.

[59] Posener-Kriéger 1977, 93.

The groups of signs would be interpreted as:

A *idmy quality linen ('royal' linen)*
(Right to left) 1,000 pieces × 60 square cubits (each) in area; 1,000 × 70 square cubits; 1,000 × 80 square cubits; 1,000 × 90 square cubits; 1,000 × 100 square cubits

B *šsr quality linen*
(Right to left) 1,000 pieces × 2 cubits wide × 10 cubits long;[60] 1,000 × 30 square cubits; 1,000 × 40 square cubits; 1,000 × 60 square cubits; 1,000 × 100 square cubits.

C *šmꜥ.t nfr.t quality linen*
(Right to left) 1,000 pieces × 60 square cubits (each) in area; 1,000 × 70 square cubits; 1,000 × 80 square cubits; 1,000 × 90 square cubits; 1,000 × 100 square cubits.

D *ꜥꜣ quality linen*
(Right to left) 1,000 pieces × 50 square cubits (each) in area; 1,000 × 60 square cubits; 1,000 × 70 square cubits; 1,000 × 80 square cubits; 1,000 × 100 square cubits.

The signs representing quality in all four 'headings' are oriented left to right, *i.e.*, facing the tomb owner, but the writer has chosen to read the second register containing textile dimensions from right to left, contra Manuelian.[61] This reading is based on the orientation of *šn.t* and the arrangement of the dimensions in ascending, rather than descending, order. It should be conceded that the orientation of signs appears to be variable and inversions occur. In the Giza slab stelae the offerings face largely toward the seated owner, whereas in the Helwan slabs the majority face right, i.e., away from the owner.

2.2.7. *Construction of the fringes*

The construction of the long, twisted 'forked' or 'inverted V' vertical fringe that expresses linear measurements was first identified and interpreted by Jéquier,[62] based on the representation of fringed cloths on the coffin of Djehutihetep in the Egyptian Museum in Cairo. He postulated that the number of fringes represents the width of the cloth, but did not specify a unit of measure. This fringe type is formed when an equal number of warp yarns are twisted together at regular intervals, forming a decorative fringe at the finished end of the textile (Fig. 6.8). The 'raw' warp fringe occurs when the yarns are cut from the warping beam on completion of weaving and are not worked further. The short weft fringe is inserted in bundles of yarn into the selvedge during the weaving process, and in ancient Egyptian linens is generally found on the left hand side of the textile.[63] (See Fig. 6.9 for a reconstruction of the elements of a woven textile).

The long, twisted upright fringe and the short weft fringe are graphically illustrated in the linen list of Kanefer (Fig. 6.10). The upright 'inverted V' fringe in the second register of the first group, under the heading *idmy*, indicates linear measurements. In the second group signs denoting measurement of surface area are enclosed in short-fringed cloths.

[60] Expressed in linear measurements, indicated by the two vertical fringes. Discussed in detail below.
[61] Manuelian 2003, 70.
[62] Jéquier 1921, 31.
[63] See Kemp & Vogelsang-Eastwood 2001, 123–144 for fringe construction.

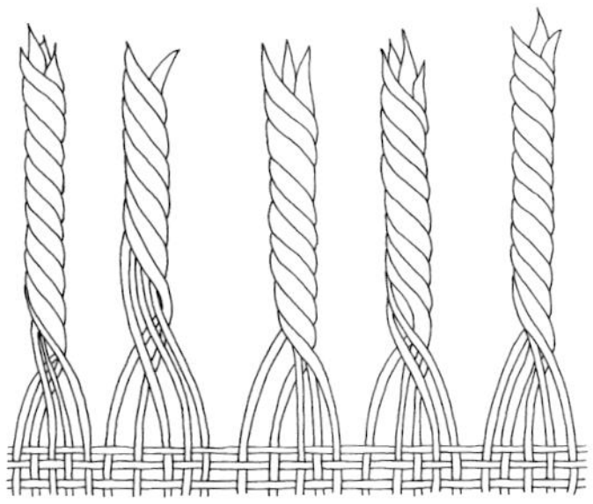

Fig. 6.8. (left) Construction of the vertical 'forked' or 'inverted V' fringe. Drawn by Mary Hartley, Macquarie University.

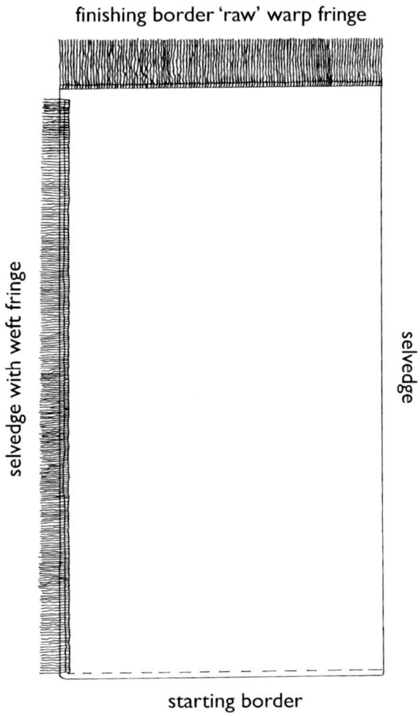

Fig. 6.9. Reconstruction of a fringed textile. Drawn by Mary Hartley, Macquarie University, adapted from Kemp and Vogelsang-Eastwood 2001, 90, fig. 4.1(b). Note that the orientation of the textile has been altered to conform to known ancient Egyptian weaving techniques, i.e. the weft fringe is on the left hand edge and the starting border at the lower end. This is from the perspective of the weaver, who worked from the starting border.

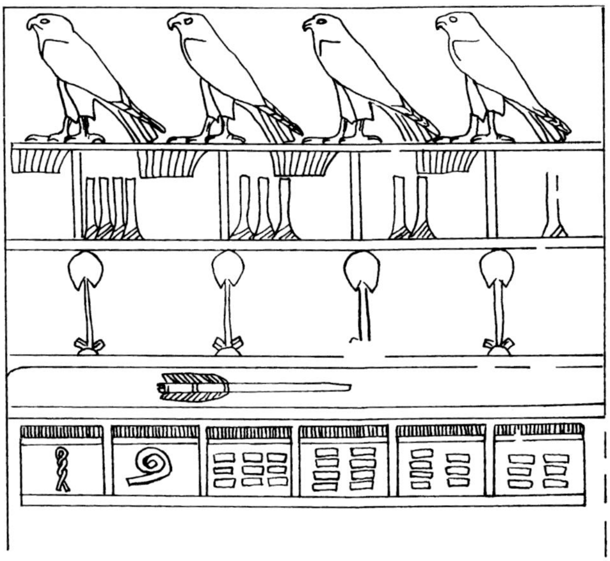

Fig. 6.10. Upper section of linen list of Kanefer (G 1203) showing two fringe types. After Manuelian 2003, Pl. 4. Drawing by Mary Hartley, Macquarie University. Although damaged, all remaining registers except the first show measurements in surface area. The first register shows the 'inverted V' fringe indicating linear measurements.

2.2.8. *Calculation of dimensions in the Old Kingdom linen lists*

Smith considered a number of interpretations (including some made by previous scholars) regarding the meaning expressed by the varying number of fringes. For example, that they could represent the number of yarns in a specific length of cloth or the quality of the yarn, estimated by the number of fibres per thread. He quite correctly dismissed this as being technically impossible, and proposed that the number of vertical fringes depicted on the textile suggest a measurement of width, based on the number of "hand-breadths". He also suggested that "special terms" such as *ssf, tmꜣ* may also indicate "unusual" widths, either extra wide or narrow, or plaited bands for girdles.[64] His observations have been further expanded, refined, and corrected.

64 Smith 1935, 142; 148–149. His interpretation of the meaning of "special terms" have been accepted by some but

Paule Posener-Kriéger attempted more exact calculation of dimensions on the Old Kingdom stelae, based on analysis of a Fourth or Fifth Dynasty papyrus from the Gebelein archives, which was devoted to the accounting of cloth deliveries.[65] She applied the principles therein to the calculation of size of the textiles on the Giza slab stelae. As discussed briefly above, she proposed that each of the vertical signs (i.e. fringes ⊥) is equivalent to one cubit, so that the width of a textile can be calculated by the number of vertical fringes, up to a total of four or five in the Old Kingdom linen lists. When placed over the horizontal sign[66] ═ that she postulated to be equivalent to 10 units, the length of the textile would be 10 cubits. For example, ⊥⊥ = 2 cubits wide × 10 cubits long.[67] When the number of vertical signs was greater than four or five, Posener-Kriéger noted that generally the size was written as a series of ═ each of which was equivalent to 10 square cubits, *i.e.*, giving a measurement of surface area.[68]

The use of horizontal signs as well as the vertical fringes to represent the smaller dimensions e.g. 30 square cubits, implies that there is a clear distinction in meaning between the two types of signs.[69] The sign ⊥⊥⊥ would designate a textile with a specific width of 3 cubits × 30 cubits in length, but ≡ (30 square cubits) would provide a number of possibilities: 1 cubit × 30 cubits, 2 cubits × 15 cubits, 3 cubits × 10 cubits, 5 cubits × 6 cubits. The first measurement is unlikely, although technically possible, nor is the last probable. This leaves the second measurement as the most likely, as distinct from the third (3 cubits × 10 cubits) that would have been written with the upright fringes if a width of 3 cubits had been intended. Similar observations can be made regarding the larger sizes: 60 square cubits has six possible combinations of linear dimensions of width and length, 70 square cubits has four possibilities, 80 square cubits has five possibilities, and 90 square cubits has six possibilities.[70]

Posener-Kriéger applied the formula to the Heb-Sed cloak on the reliefs at Abu Gurob, made of *idmy* quality linen, 30 or 40 square cubits in size.[71] The calculation based on 1 cubit as the base unit yields measurements of 1.60 m. × 5.25 m., or 2.10 m. × 5.35 m. This result appears to be unrealistic for the construction of a cloak that in representations finishes above the knees, except perhaps if the material were doubled. By calculating the surface area on the basis of 1 cubit × 1 palm, a relatively reasonable measurement of 0.75 m. × 2.1 m. was reached.[72]

---

questioned by others, including the writer.

[65] Posener-Kriéger 1977, 86–96.

[66] Gardiner 1988, sign list no. Aa 12.

[67] Posener-Kriéger 1977, 88–89.

[68] Posener-Kriéger 1977, 92. This interpretation has been followed by Kahl, Kloth & Zimmerman 1995, 175–179 and Köhler & Jones 2009, throughout. Manuelian in his translation of the linen lists used the formula to calculate width, which gives unrealistic measurements of e.g. 90 cubits (wide), 80 cubits (wide) etc. Posener-Kriéger 1977, 91 discussed the nuances inherent in the vertical and horizontal signs and concluded that the horizontal sign indicates area.

[69] As in the second register of the second group in the linen list of Ini, Fig. 6.7 above.

[70] I wish to thank John Croucher, Professor of Management, Macquarie Graduate School of Management, Sydney Australia for many insightful comments regarding the calculations.

[71] Bissing & Kees 1923, 6–7, Pls. 16, 22. However, the reason that Posener-Kriéger chose 30 or 40 square cubits is unclear, because the inscription very clearly shows four upright fringes i.e. 4 cubits x 10 cubits = 40 square cubits. The Heb Sed cloak was worn by the pharaoh during the ceremony to celebrate his continued rule, the first generally held after 30 years of reign.

[72] Posener-Kriéger 1977, 94–95.

3. Textile and textile-related signs in the Helwan corpus[73]

3.1. *Designations of quality*

It has been noted previously that three designations of quality appear in the Helwan linen inventories.

iti.wy: 'royal' linen, the finest quality (Fig. 6.11: 1)
This designation of quality is the second to occur in the Helwan corpus, towards the end of Group B in the mid-Second Dynasty, on relief EM99-23. Although there are eight occurrences of the ultra-fine 'royal' linen in the lists, only three of the individuals had royal associations according to their titles. Hence the significance of the inclusion of that superior quality in a private individual's tomb requires further investigation. The question arises how, and in fact whether, in reality these non-royal persons were able to access a commodity that probably should have been reserved for the consumption of the elite.[74]

šsr: a lesser quality (Fig. 6.11: 2)
This indicator of quality is the first to occur in the Helwan corpus (EM99-15), also in the mid-Second Dynasty, but earlier in Group B than *iti.wy*.

ꜥꜣ: another quality (Fig. 6.11: 3)
This is the last quality that appears in the corpus, dated mid- to late Second Dynasty (see Fig. 6.2, EM99-18).

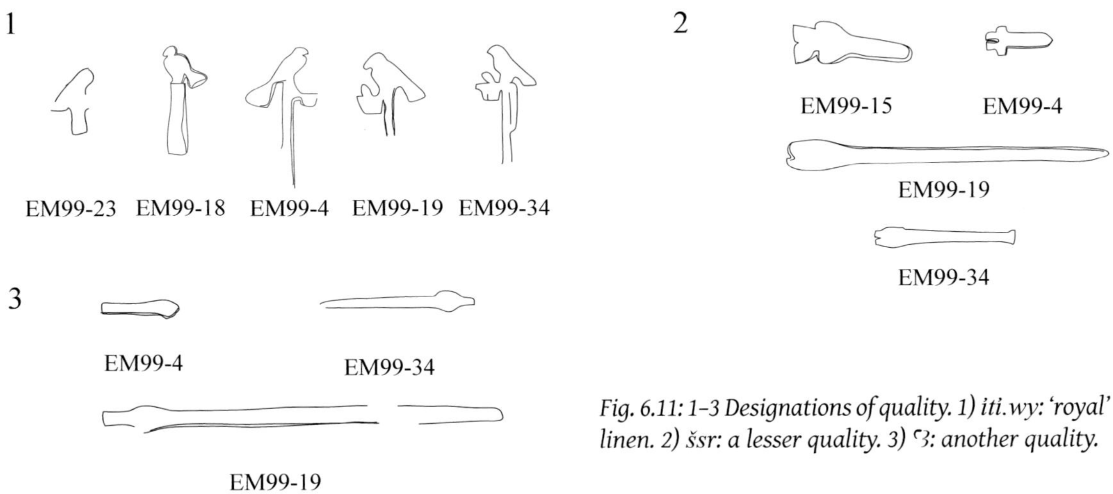

Fig. 6.11: 1–3 Designations of quality. 1) iti.wy: 'royal' linen. 2) šsr: a lesser quality. 3) ꜥꜣ: another quality.

The signs provide implicit evidence that linen was graded according to quality and assessed by specific criteria, which would have included at least the density of the weave and the fineness of the yarn. However, there is an absence of evidence to determine the standard for grading of quality, and the specific meaning of the signs remains unknown. Textiles inscribed with designations of

[73] Figs. 6.11, 6.12, 6.16, 6.17, 6.19 are reproduced from Köhler & Jones 2009, 48–50.
[74] Köhler & Jones 2009, 95.

quality would require analysis of thread count, yarn diameter, preparation of the fibre, etc., and comparison with other examples bearing similar signs in order to collect any significant data. (The problems encountered with this approach have been discussed in the introduction).

Nor is it possible to quantify textile output on the basis of the textual or archaeological evidence, except to note that large quantities of fine quality textiles appeared in burials in this period, including articles of clothing. The greatest quantity of extant textiles is from the large mastabas at Tarkhan that are contemporary with the First Dynasty stelae.[75] Mastaba 2050, despite plundering, still contained 17 different qualities of linen textiles when excavated (Petrie 1914; Midgley 1915) as well as the oldest surviving dress (Hall 1982).[76] Furthermore, very fine linen was being produced. When analysing the textiles, Midgley commented on the excellent preparation of the flax fibre and the quality of the woven textiles, which he stated were finer than the finest contemporary Irish linen (Midgley 1915, 48). The finest quality textile had a count of 200 warp and 48 weft yarns per inch (approx. 80 × 20 per cm).[77]

More than 600 years earlier in the Predynastic period, evidence shows that specialist workshops produced textiles of uniform quality for funerary purposes. At Hierakonpolis in the 'working class' cemetery HK43 (*c.*3500 BC), only two qualities of textile were used as shrouds and wrappings throughout the cemetery – a fine and a medium.[78] At the contemporary Cemetery 7000 at Naga ed-Dêr three qualities – fine, medium, coarse – are reported.[79]

3.2. *Clothing*

mꜣṯ: long-sleeved, V-necked dress, or garment (Fig. 6.12: 1)

Perhaps the most interesting of the signs denoting clothing is *mꜣṯ*.[80] It represents ideographically one of the few surviving items of clothing: the long-sleeved, V-necked dress with horizontal pleats. The identification is supported further by iconographic evidence from the Helwan corpus. Smith was the first to relate the form of the Sixth Dynasty Old Kingdom dresses from the Reisner excavations at Naga ed-Dêr[81] and the Petrie excavations at Deshasheh[82] to the 'dress' ideograms in the linen list on the Second Dynasty stela of Sehefner from Saqqara.[83] Since that time, the earliest extant dress dated First Dynasty from Mastaba 2050 at Tarkhan has been found in the storerooms of the Petrie Museum and conserved,[84] together with two of the Deshasheh dresses.[85]

Rosalind Hall, Elizabeth Riefstahl and Gay Robins noted that the horizontally pleated dress with sleeves and bodice was not depicted on the monuments, and all stated that depictions in

[75] Textiles from Tarkhan Mastabas 1060, 2038 and 2050 are in the Petrie Museum of Egyptian Archaeology in London and the Bolton Museum and Art Gallery, Bolton, U.K.

[76] Petrie Museum of Egyptian Archaeology UC 28614B1.

[77] Mastaba 2050, textile 9 (Midgley 1915, 50). The most common weave in ancient Egypt is in the ratio of two warp yarns to one weft.

[78] Jones 2008, 116–119.

[79] Lythgoe & Dunham 1965, 33.

[80] All occurrences in the Helwan corpus date to the Second Dynasty.

[81] Smith 1935, 139, Fig. 1. Museum of Fine Arts, Boston, u 34.56 from tomb N 94. This is one of nine such garments.

[82] Smith 1935, 138; Petrie 1897, 31, Pl. 35.

[83] Quibell 1923, Pl. 26.

[84] Hall 1982, 27–30. Petrie Museum of Egyptian Archaeology UC 28614B1.

[85] Hall 1981, 168–171; 1982, 27–45. Petrie Museum of Egyptian Archaeology, London, UC 31182–3.

Fig. 6.12: 1–6. Clothing signs. 1) mꜣṯ: long sleeved V-necked dress or garment. 2) tmꜣ.t (EM99–10), mš.ṯ (EM99–13): cloak. 3) bꜣhy.t: penis sheath (?). 4) šnḏw.t: kilt. 5) šsr: linen (generic). 6) mnẖ.t: clothing (generic).

art differ from surviving clothing.[86] Both Hall and Riefstahl suggested that the sleeved dress was a logical development out of the Old Kingdom V-necked dress with shoulder straps, as depicted on the monuments.[87] It is surprising that Hall shared this view, as she knew of the existence of the First Dynasty dress, which predates depictions of the sleeveless Old Kingdom dress by some 300 years.[88]

However, the recent study of the Helwan material has shown that two female tomb owners

[86] Hall 1981, 170; Riefstahl 1970, 244; Robins 1990, 45–46.

[87] Hall, 1984, 139; Riefstahl 1970, 251.

[88] Despite the great quantities of textiles in Old Kingdom burials, the only reported evidence of the typical sleeveless, V-necked dress depicted on the monuments was a 'cut out' placed over the Fourth Dynasty mummy from Giza G 2220 B. It was inscribed with the quality *šmʿ.t nfrt*. An excavation photograph is all that remains. See Vogelsang-Eastwood 1993, Pl. 25; Ikram and Dodson 1998, Pl. 174.

Fig. 6.13 (left). Dress with pleated short sleeves and pleated bodice. Relief slab of N(i)t-mah EM99-5 (mid-Second Dynasty)
Fig. 6.14 (right). Dress with short, tightly pleated sleeve protruding from cloak. Relief slab of Sep EM99-12 (early to mid-Second Dynasty).

on Second Dynasty reliefs are depicted wearing V-necked dresses with finely pleated short sleeves (Figs. 6.13, 6.14). Apart from the Louvre figurine[89] of a female wrapped in a fringed cloak revealing one short, pleated sleeve in the style of EM99-12, the Helwan reliefs represent the only two-dimensional iconographic evidence for pleated, short-sleeved dresses known to us.

An incorrect transliteration based on a misreading of the phonetic writing of *mꜣṯ* on the textile label from the Third Dynasty unfinished pyramid of Sekhemhket[90] has entered the literature and should be reconsidered. This sign group was read by Wolfgang Helck as *mꜣṯ.t*,[91] and followed by Kahl[92] but not Kaplony.[93] Inspection of the label in the Egyptian Museum, Cairo[94] has shown that the sign group does not contain the sign ⌓ '*t*' that is shown in the Helck drawing. The sign group is comparable to that on EM99-4 dated late Second Dynasty (Fig. 6.5), which contains the first occurrence of the phonetic writing of the sleeved garment as *mꜣṯ*. The sign does not occur in the linen lists beyond the late Second Dynasty,[95] and that appearing on the Third Dynasty label

[89] Louvre E11888. Ziegler 1997, 19.
[90] Goneim 1957, Pl. 65b.
[91] Helck 1957, 72–76.
[92] Kahl 1994, 715, n. 2161; 2002–, 175. (Kahl no. s 7).
[93] Kaplony 1963, Vol. I, 328.
[94] Egyptian Museum Cairo, JE 92679. Permission to examine and publish the label was granted by the former director of the Egyptian Museum, Mohammed Saleh.
[95] The use of the hieroglyph representing a sleeved garment or dress does not occur on the Old Kingdom Giza slab stelae, contra Hall, who stated that the sleeved garment occurs as a determinative to *ssf* in the Old Kingdom linen

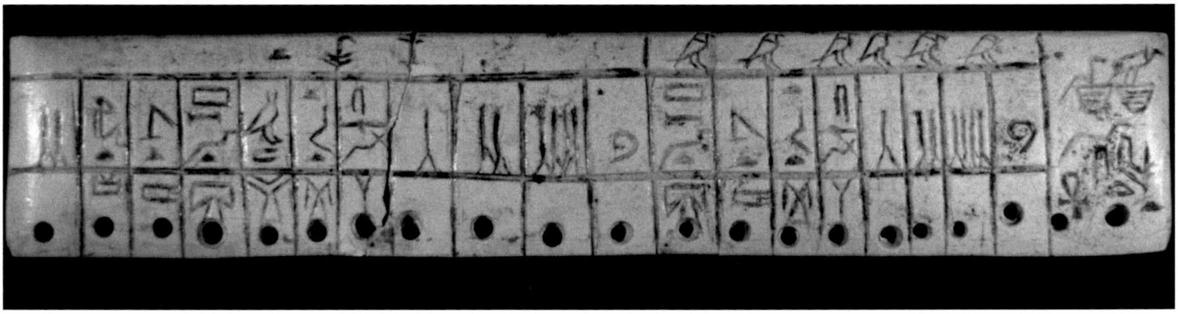

Fig. 6.15 Textile label from the unfinished pyramid of Sekhemkhet. m3ṯ is the 5th sign group from the left. Photograph © J. Jones

is the last occurrence known to the writer. Consequently there appears to be no evidence for the transliteration of the sign group as *m3ṯ.t*.

tm3.t , *mš.ṯ*: cloak (Fig. 6.12: 2)

The identification of cloaks appears to be the most challenging. There are numerous signs that have invited interpretation,[96] and the cloak ideogram often occurs in names and titles.[97] There are two apparent occurrences of cloaks with ties in the Helwan group (*tm3.t* Kahl sign s 19# on EM99-10; *mš.ṯ* Kahl sign s 28 on EM99-13). It is proposed that a third sign, on EM99-10, identified as a fringed textile by Kaplony and Kahl,[98] may be interpreted as the rectangular short-fringed cloak which is depicted wrapped around the shoulders of archaic sculptures such as the Louvre figurine discussed above. The internal lines on both the garments in EM99-10 may indicate creases, suggesting that they have been folded inside storage chests.[99] The latter could be compared to the textile label of a cloak with ties from Tomb U-j and one from earlier excavations at Abydos.[100]

b3hy.t: penis sheath (?) (Fig. 6.12: 3)

The sign that has been read as *b3hy.t* occurs only once in the Helwan corpus (EM99-23) and appears to be an isolated occurrence amongst Early Dynastic reliefs. The standard translation as 'penis sheath' may be colourful, but the form of the sign does not allow further interpretation. Nor is there any archaeological evidence to clarify the meaning.

šnḏw.t: kilt (Fig. 6.12: 4)

The ideogram read as *šnḏw.t*, a 'kilt' occurs only on EM99-19 (Fig. 6.6). This translation is based on the phonetic writing on the bone textile label from Sekhemkhet, where the ideogram appears

lists. Hall 1981, 169; Ogdon 1983, 81–83 follows Hall.

96 Kaplony 1963, Vol. I, 322–325; Kaplony 1963, Vol. II, 1026–1028; Kaplony 1966, 27–28; Edel 1975, 102, 28–30. See Staehelin 1966, 36 for a discussion of cloaks made from spotted animal skins.

97 For example, Kaplony 1963, Vol. I, 326–327; Kaplony 1963, Vol. II, 1031 (n. 1705).

98 Kaplony 1963, Vol. I, 324–325; Kahl 1994, 719.

99 Bochi 1996, 221–253 discusses the social significance of creases on garments depicted in Middle Kingdom statuary.

100 Dreyer 1998, Fig. 81.162; Pl. 35 X190 (Brussels 6143) and Index 190. Also Kaplony 1966, Pl. 22.

as a determinative (Fig. 6.15).[101] Again, there are no occurrences in the Old Kingdom linen lists, but the form of the garment is clearly identifiable from two- and three-dimensional art.

šsr: linen (generic) (Fig. 6.12: 5)
The sign *šsr*, a generic term for 'linen', or articles made of linen, appears only once in the Helwan corpus, in EM99-12 (Fig. 6.14). It occurs in isolation beside the offering table, and no other textiles are designated.

mnḫ.t: clothing (generic) (Fig. 6.12: 6)
A very early occurrence of *mnḫ.t*[102] appears in EM99-14 (Fig. 6.4). This is a generic term for 'clothing' and is an early example of a grouping of linen and *šs*, 'vessels of calcite', commonly found in later periods.[103] They appear together above the offering table, in a register separated from the rest of the textiles by a vertical division. In EM99-6, the two signs occur in the lower right hand corner of the panel, beyond the offering table. This combination also occurs in EM99-19 (Fig. 6.6), where they appear beside the offering table. The latter is a standard arrangement on the slab stelae of the Giza group.

3.3. *Textiles*

Fringed cloths. The rectangular piece of textile with a varying number of fringes is the most commonly occurring sign on the Early Dynastic Helwan relief slabs, where it appears on 30 out of the 31 reliefs with textiles.[104] Three distinct types of fringe have been noted, and have been arranged into three categories (see Fig. 6.16: 1–3).

Scholars to date have not distinguished between the short and long fringe types. When Kahl studied the signs, he saw no difference between the fringes, noting that they are interchangeable, and their use is based on available space. He suggested that Type 3, the short, upright fringe is a variant of Type 2, but represents large measurements.[105] Kaplony noted the distinction, and suggested further investigation.[106] Examination of the fringe types in context shows that when both long and short-fringed textiles occur on the same slab, often in the same register, their writing does not appear to be governed by space restrictions (as is clear in Fig. 6.5).[107]

By relating the three different representations to actual textiles, the following observations can be made.

[101] Goneim 1957, Pl. 65b; Helck 1957, 76; Kaplony 1963, Vol. I, 332.
[102] Gardiner 1988, sign list no. S 27.
[103] Gardiner 1931, 161–183.
[104] Fringes are an intrinsic decorative component of woven linen textiles throughout ancient Egyptian textile history. The earliest archaeological evidence for the presence of a weft fringe is from the prehistoric 'painted linen' from Gebelein (Naqada IC-IIA, *c.*3700 BC), Museo Egizio, Turin. Suppl. 17158.
[105] Kahl 1994, 712, n. 2138. A new number s 37 has been allocated to the short-fringed textile to distinguish it from Kahl number s 2, which encompasses all fringe types. Köhler & Jones 2009, 47–48.
[106] Kaplony 1966, Vol. 1, 321.
[107] For example, Köhler & Jones 2009, EM99-4, -5, -7, -15, -17 -23, S99-5, S05-135. The difference is also clear on the Second Dynasty stela of Sehefner at Saqqara, Quibell 1923, Pl. 36. The First Dynasty wooden label from Saqqara Tomb X (Neska) unmistakably depicts both long and short-fringed textiles amongst other commodities, Emery 1958, 115, Fig. 65.

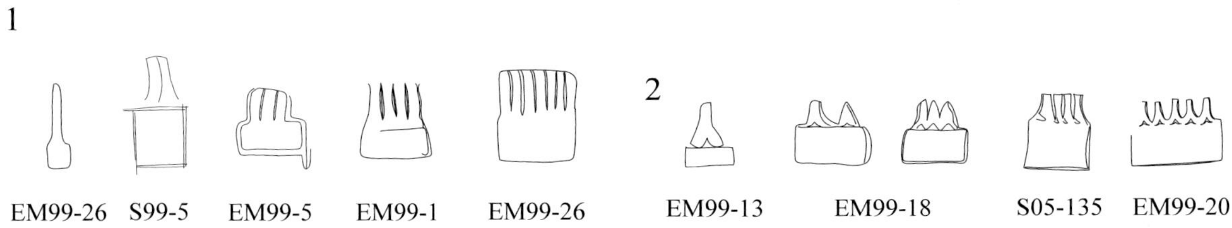

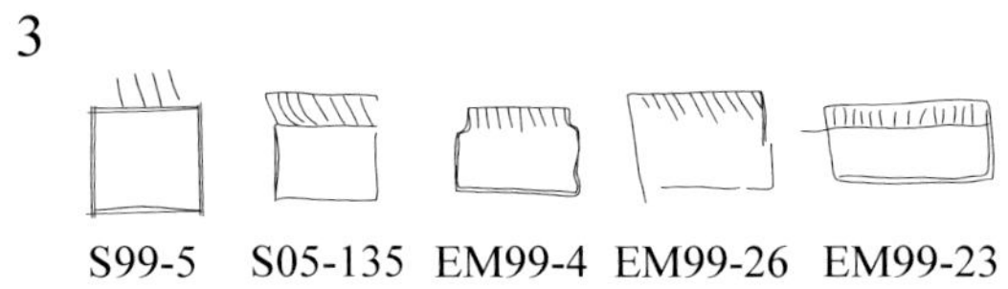

Fig. 6.16: 1-3 Fringed cloths. 1) Type 1: the long, vertical fringe. 2) Type 2: the long, twisted 'forked' or 'inverted V' fringe. 3) Type 3: the short fringe, upright or sloping to the left.

Type 1: the long, vertical fringe (Fig. 6.16: 1)

The fringe type may be interpreted as the 'raw' warp fringe (Fig. 6.9). This fringe type does not occur with Type 2 on any of the reliefs. The fringes generally range in number from one to eight.

Type 2: the vertical 'forked' or 'inverted V' fringe (Fig. 6.16: 2; see also Figs. 6.8, 6.10)

The number varies between one and five in the Helwan lists. The first occurrence at Helwan is in the early Second Dynasty,[108] with one occurrence in the mid- to late Second Dynasty.[109] In the Third and Fourth Dynasties it is the standard long fringe type, and Type 1 no longer occurs. Consequently it appears that Types 1 and 2 are interchangeable, until Type 2 supersedes Type 1.[110]

Type 3: the short fringe (Fig. 6.16: 3)

This is indicated by multiple strokes, up to a maximum of 15. It is proposed that the short fringe represents the inlaid weft fringe inserted into the left hand selvedge (Fig. 6.9). The orientation to the left in some of the examples may graphically illustrate the weft fringe. When fringed textiles are used in garment construction, the weft fringe is often positioned vertically on the left side, so that the component threads tend to slant downwards.

After taking into consideration the above observations that the two types of long fringe are probably interchangeable, and that there is a clear distinction between the long and short fringes shown on the textiles, the function of the latter in the Early Dynastic inventories still remains a mystery. Furthermore, the horizontal sign indicating area does not occur in the Early Dynastic linen lists. The question of how surface area in the Early Dynastic lists was measured remains in doubt, and is the subject of ongoing investigation by the writer.

[108] EM99-20.

[109] EM99-18.

[110] The writer is not aware of extant textiles with Type 2 fringes from Early Dynastic contexts.

3.4. *Designations of area in the Early Dynastic and Old Kingdom Linen Lists*

The hypothesis put forward by Posener-Kriéger that each of the horizontal signs represents 10 square cubits, and the area of the textile is calculated by multiplying that unit by the number of times the sign is written, has been accepted as the most credible based on currently available evidence.[111] In the discussion of the fringed textiles in the Early Dynastic lists it has been noted that the horizontal signs that denote area and which are so common on the Old Kingdom linen lists are completely absent from any known Early Dynastic linen inventories. They do not occur until the Third Dynasty on EM99-19 (Fig. 6.6), when the list begins to acquire the structure of the developed Fourth Dynasty lists.[112] The hiatus of some 650 years between the occurrence of the horizontal signs on the bone labels from Tomb U-j at Abydos and the Third and Fourth Dynasties is difficult to interpret. One possible explanation is that the so-called 'horizontal' signs were in fact numerals, and appeared to be oriented horizontally as a result of the random placement of the hole that was drilled into one corner of each label.

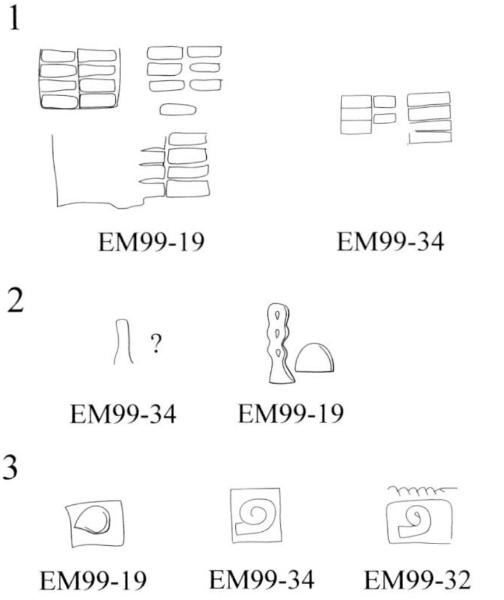

Fig. 6.17: 1-3 Signs designating surface measurement in cubits. 1) Horizontal signs designating square cubits. 2) ḥ; ḥ.t a value greater than 100 (square?) cubits and less than 1,000. 3) šn.t 100 square cubits.

ḥ; ḥ.t: an area greater than 100 (square?) cubits, less than 1,000 (square?) cubits (Fig. 6.17: 2)

A less frequently occurring sign *ḥ*[113] or *ḥ.t* (sometimes written inside a fringed rectangle) represents a known value probably greater than 100 or 200, but less than 1,000 cubits. Posener-Kriéger tentatively favoured a linear measurement of 300 cubits.[114] However, the frequent occurrence of the sign in the same register as signs expressly denoting area would suggest that the notion that it represents a linear measurement may require further investigation.[115] The sign occurs in the same register as *šn.t* that designates area in the two Helwan stelae where it appears (see below). Where the sign is seen in the lists of the Giza group, it is also together with *šn.t.*[116] The sign does not appear in the Helwan corpus until the Third and early Fourth Dynasties. However, *ḥ* occurs in the linen list of the Second Dynasty stela of Sehefner from Saqqara.[117]

[111] Posener-Kriéger 1977, 86–96.

[112] Köhler & Jones 2009, EM99-19; Kahl *et al.* 1995, 172–175; 176–177). Also on the false door of Saqqara S 3073 Murray 1905, pls. 1 and 2; Kahl *et al.* 1995: 188–189, 194–195.

[113] Gardiner 1988 sign list no. V 28. The writing *ḥ.t* appears to be an isolated occurrence. It is always written *ḥ* in the Giza slab stelae.

[114] Posener-Kriéger 1977, 93–94, n. 27.

[115] Exceptions are Manuelian 2003, Pl. 6 G1205; Pl. 8 G1207; Pl. 14 G1227; Pl. 30 G4860, each of which contains a sign denoting linear measurements of 2 × 10 cubits in a register devoted to measurements of area.

[116] Manuelian 2003, Pl. 4 G 1203; Pl. 8 G 1207; Pl. 14 G 1207; Pl. 20 G 2135; Pl. 26 G 4150.

[117] Quibell 1923, Pl. 26.

šn.t: an area of 100 square cubits (Fig. 6.17: 3)

The number '100' written inside a square or rectangle, in all probability represents an area of 100 square cubits. As has been noted, the horizontal sign designating area occurs only up to a total of nine i.e. 90 square cubits, never ten.[118] In EM99-32, the sign is written inside a fringed rectangle. It is possible that the fringes were applied in pigment on EM99-19 (Fig. 6.6), where missing vertical lines suggest that other sections of the list may originally have been painted.

The short fringe on the Giza slabs appears only on the fringed rectangles enclosing signs representing area. (For example, Fig. 10 Kanefer).[119] In the complete absence of the horizontal signs in the Early Dynastic textile inventories, how was surface area indicated? Is it possible that the Early Dynastic sign of the rectangular, short-fringed textile is the precursor of the fringed cloths that generally contain signs designating area on the Giza slab stelae? Consequently, could the short-fringed textile in the Early Dynastic stelae denote a measurement of area? This proposal is a working hypothesis pending future evidence.

Calculations of size applied to archaeological textiles

The measurements of five intact Middle Kingdom textiles from excavations at the temple of Mentuhotep at Deir el-Bahari by the German Archaeological Institute were made available to Posener-Kriéger.[120] Yet, when these measurements are converted into both linear and square cubits, they involve fractions of a cubit and not whole numbers. For example, one of the textiles measures 1.20 m. × 18.40 m. = 2.29 × 35.18 cubits = 80.56 square cubits.[121]

Clearly a complete textile inscribed with one of the signs designating size would give irrefutable evidence for the units of measure. Posener-Kriéger cited the only inscribed example known to her, but stated that she was unaware of the dimensions.[122] The linen, from the Sixth Dynasty pyramid of Pepi I, was inscribed in ink headed by the textile sign denoting width 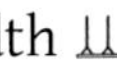, followed by the inscription  *šps.t iri.t n nsw bi.ty Ppy ʿnḫ(.w) ḏ.t*, "costly/noble (cloth) which was made for the King of Upper and Lower Egypt Pepy, may he live eternally".[123] Based on Posener-Kriéger's reckoning, the textile in its original form should have measured the same as the larger Heb-Sed cloak mentioned above.

The writer examined the textile in the Egyptian Museum, Cairo (Fig. 6.18).[124] It had been

118 Posener-Kriéger 1977, 93.

119 See also Manuelian 2003 G 1201 Pl. 1, the stela of Wepemnefret, where it occurs in the last register with other signs of area. The fringed textile occurs twice on the stela from G 4860 (Manuelian 2003, Pl. 30) enclosing *ḫ* and *šn.t*, and on the Fifth Dynasty linen list on the east wall in the chapel of Seshemnefer I (Manuelian 2003, 156, Fig. 234 and Appendix 3, Figs. 310–312.) The squares of the latter list appear to be empty, but close inspection of the original expedition photographs (from 1931) shows traces of relief inside each, and possibly a short fringe surmounting the squares.

120 Posener-Kriéger 1977, 92 n. 22.

121 The measurement of textiles in the linen lists is currently under investigation with John Croucher.

122 Posener-Kriéger 1977, 94 n. 29.

123 Grammatically, it would also be possible to read the inscription as " ... which Pepy the king of Upper and Lower Egypt made ...". That would indeed make it an interesting artefact! I am grateful to Boyo Ockinga, Macquarie University, Sydney, Australia for clarifying the alternate readings of the inscription.

124 Temporary No. 21/3/33/1. Located in Upper Corridor 44. Permission to examine and publish was granted by the former director of the Egyptian Museum, Mohammed Saleh. Yarn is single, s-spun; the weave is a medium quality tabby with a thread count of 24 × 16. Inscription published by Wiedemann 1884, 211; Maspero 1903, 472, n. 57; Sethe 1932, 97.16; Brugsch 1968, Vol. V, 1212; Porter & Moss 1981, 424; Labrousse 1994, 160, 161 n. 33. I am grateful to Audran Labrousse for generously providing references and information on the precise provenance of the textile, which was

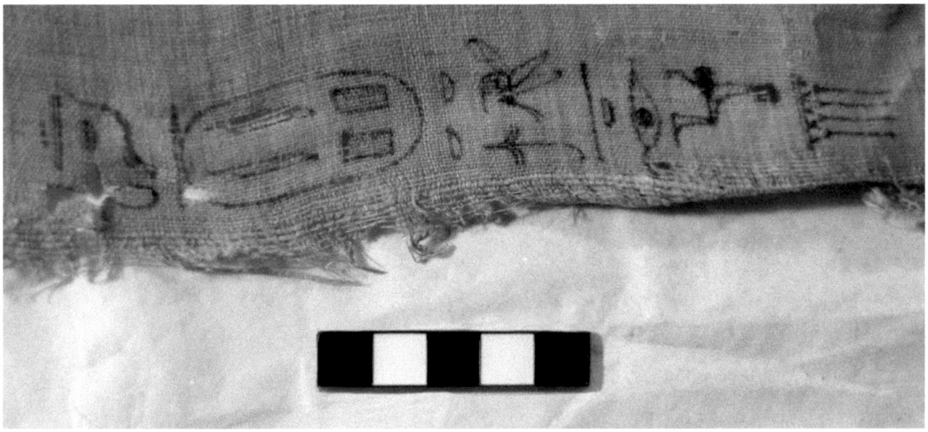

Fig. 6.18. Pepi I inscribed textile. Photograph © J. Jones

cut from the complete textile, and folded to display the inscription. Doubled over, it measured approximately 29 cm; the total width was stated as 45 cm in the *Journal d'Entrée*. The textile was too fragile to unroll to measure the length, but the size of the fragment would not have been relevant to any calculation of dimensions. The inscription was located at the finished end near the raw warp fringe that originally would have measured approximately 8 cm. Additional signs not noted in publications of the inscription were observed using a stereo-microscope at 10× magnification. They were located 2 cm above and 2.5 cm to the left of the beginning of the inscription. They comprised the two horizontal signs designating area arranged one above the other, approximately 1 cm in length, with one sign to the right of these, approximately half that length. Unfortunately we have lost a most important diagnostic specimen.

The calculations proposed by Posener-Kriéger still remain problematical. The papyrus was incomplete, and the exact value of the units of measure remains to be confirmed. Often some quite fantastic measurements of size are obtained if the cubit is used as the basic unit in the linen lists. Posener-Kriéger's study is a valuable point of departure for further research. The possibility that the unit of measure in the linen lists, like the 'ideal' quantity of 1,000 has become purely symbolic and no longer bears any relation to reality, cannot be discounted without further inquiry.

3.5. *sf*, a less common textile sign

A less frequently occurring textile sign and largely misunderstood term is *sf* [hieroglyphs]. There are five confirmed occurrences of the phonetic writing of the term *sf* in the Helwan linen lists[125] and four on the Giza slab stelae.[126] In the course of investigation into the term, it became clear that there is confusion and disagreement surrounding the meaning amongst various scholars, who have proposed the following translations.

not noted in the *Journal d'Entrée*.

[125] In addition to the occurrences reproduced in Fig. 6.19, *sf*, written with no apparent determinative, occurs on EM97-43 dated Third Dynasty.

[126] Manuelian 2003, Pl. 14, G 1227; Pl. 18, G 2120 (four occurrences); Pl. 22, G 2115(?); Pl. 30, G 4860.

3.5.1. *sf* as a 'sleeved garment'

The misconception that *sf* or *ssf*[127] refers to a sleeved garment appears to have its origins in the flawed interpretation of the sign by Kaplony. This translation has continued to be perpetuated in the literature.[128] In the linen list of EM99-1, dated to the late First to early Second Dynasties, the first of the three textile signs (reading right to left) in the top register was read as *sf* by Kaplony,[129] with the determinative of a sleeved garment below (Fig. 6.20). Kahl followed this interpretation and allocated the number s 3 (*Hemd*) to the sign group.[130] Although the phonemes appear to be clear, albeit roughly executed, the reading of the determinative is problematical. If it were interpreted as a sleeved garment, it would have only one 'sleeve'. There is insufficient space on the panel for the other 'sleeve' to be depicted, even if fused with *s*. Kaplony refers to this writing as a 'lesser known' form of *sf*[131] but there do not appear to be comparable occurrences of the sign.

It should be noted that parts of the relief are heavily eroded, with salt and sinter incrustation, which contribute to the difficulty in deciphering the signs. Examination of the relief has not convinced the writer that Kaplony's reading of the sign is possible. A second reading of *sf* as a garment by Kaplony in the textile inventory of EM99-11[132] is even less plausible than that of EM99-1, because of the poor execution of the relief in roughly incised lines.

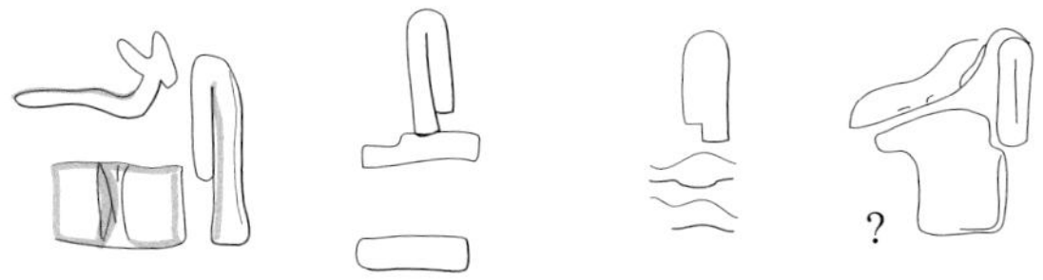

EM99-24 EM99-19 EM99-34 EM99-1

Fig. 6.19. sf

Fig. 6.20. Relief slab of Khui-itef EM99-1 (late First to early Second Dynasty).

[127] *ssf* is written with the redundant '*s*'. This writing appears in EM99-19, -34. For discussion of the redundant '*s*' see Kahl 1994, 66–70. *sf* is the first textile sign in the Early Dynastic corpus to be written phonetically.

[128] As noted above, Hall 1981, 169; Ogdon 1983, 81–83; Kahl 1994, 67, n. 78; 714.

[129] Kaplony 1963, Vol. I, 324, 331; 1966, Pl. 146, Fig. 850.

[130] Kahl 1994, 67, n. 78; 714.

[131] Kaplony 1963, Vol. I, 331.

[132] Kaplony 1963, Vol. I, 324.

3.5.2. *sf as a 'narrow cloth'*

The sign has been interpreted as a narrow cloth of a specific size by Posener-Kriéger[133] and followed by Kahl, who described it as "*ein schmaler Stoff mit bekannten Abmessungen*".[134] Smith had also proposed that *sf* represents a narrow width of cloth, erroneously basing that premise on the perceived narrowness of the Y-shaped determinative[135] on the stela of Sehefner at Saqqara.[136] In fact, this sign represents the tie binding bolts of linen together, as suggested below.

3.5.3. *sf as a 'bolt of linen tied in the centre'*

Henry G. Fischer argued for the interpretation of the sign as a bolt of cloth tied in the centre, represented by the hieroglyph adapted from N 18 in Gardiner's sign list.[137] He based this reading on a comparison of the determinatives of *sf*/*ssf* occurring in offering lists, as well as representations in Old Kingdom burial chambers of bales of cloth bound together.[138] He conceded that the sign might well represent an unknown, specified dimension, but not a narrow cloth. His interpretation is graphically illustrated by the determinative of *sf* on the early Second Dynasty relief slab EM99-24[139] as well as on a newly discovered, unpublished relief slab from Helwan. Although the tied ends are not shown protruding from the top of these determinatives, this feature does not occur in every case, as can be seen in the examples collected by Fischer.[140] Furthermore, Fischer noted that early examples show the tie in the centre.[141] Where *sf* appears in the lists, it is always in the register or section designating size. This would corroborate the premise that it represents a specific, pre-determined dimension. The writer has chosen to follow Fischer in the interpretation of this sign as a bolt of linen, possibly of specific dimensions, tied in the centre.

* * *

The inscriptions on the private funerary stelae of the Early Dynastic period have been an important source of study for philologists, because they provide some of the earliest evidence for hieroglyphic writing. By their very prominence on the stelae, the Early Dynastic and Old Kingdom linen lists have attracted many attempts at 'deciphering' the textile terminology. The archaeological evidence has been very rarely, if ever, systematically examined and related to the lexicographic evidence. Consequently many hypotheses with no basis in technical realities of production have been advanced and perpetuated.

133 Posener-Kriéger 1977, 94.

134 Kahl, Kloth & Zimmerman 1995,175 n. 5; see Kahl 1994, 65–70 for an analysis of the various forms.

135 Gardiner 1988, sign list no. O 30.

136 Smith 1935, 149; Quibell 1923, Pl. 26 (tomb 4164E); Smith 1978, Pl. 32a. The determinative also appears on the Sekhemkhet textile label, Fig. 15.

137 Fischer 1977, 148–155. The sign follows S 26 (*šnḏw.t* a 'kilt') in the sign list in Gardiner 1988, as referenced by Fischer 1977, 148 n. 41.

138 Fischer 1977, 150–154.

139 Fischer 1977, 151 n. 50 discusses the occurrence of the sign on Helwan EM99-24, citing it as the earliest attested example. However, he dates the relief to the Third Dynasty based on the research of earlier scholars, and reproduces the determinative incorrectly (Fischer 1977, 153, Fig. 15a).

140 e.g. Fischer 1997, 153, Fig. 12 b, e, f, g, h, i, j.

141 Fischer 1997, 151.

Close examination of the inscriptions *in situ* has enabled a number of misunderstood or flawed interpretations of textile terminology to be rectified. The observations on calculation of dimensions of the Old Kingdom textiles made by the late Posener-Kriéger have been by far the most 'scientific' because they are based on papyri dealing with textile deliveries. However, she cautioned that there are many lacunae and that the base unit of measure still remains to be established. Moreover, the criteria for measurement of surface area in the Early Dynastic textile inventories remains unclear because of the absence of the horizontal signs that represent area in the Old Kingdom linen lists. Another element of the linen list that needs further investigation is the determination of the quality of the textiles. Meaningful solutions to these problems are largely dependent upon the availability of inscribed textiles for analysis and examination of new, comprehensive textual evidence.

The Helwan corpus has also clarified the origins and development of the linen list, confirming that the Early Dynastic textile inventories are indeed antecedents of the Old Kingdom linen lists. In addition to the evolution of an ordered arrangement of the textiles, it can be observed that the type and variety depicted gradually increased and signs representing items of clothing were included. The quantities of offerings also increased and signs designating quality of the textiles were introduced. The archaeological evidence from contemporary burials containing large amounts of textiles and garments corroborates that there was a real demand for a wide variety of textile qualities and clothing as part of the funerary ritual, and that the large-scale production of fine, high quality linen textiles was already a significant factor in the early Egyptian economy.

Yet the textile inventories of the early 3rd millennium BC offer insights into a broader range of issues in relation to the society of the time. In the early 'lists' we see a rather random and possibly individual choice of offerings that increasingly become more ritualised and formalised, reflecting changing protocols in funerary religion and beliefs. The social ranking of the owners of the relief slabs was probably the highest in the necropolises. The deceased, who was identified and named, joined the community of the afterlife for eternity, and required that his or her status be upheld through the size and provisioning of the tomb. In addition to the symbolic function of the offerings inscribed on the slabs, the deceased was equipped with perishable and non-perishable provisions that reflected the ability or the wishes of the tomb owner and the family to indulge in mortuary expenditure. We are indeed fortunate that textiles and textile inventories reflecting the final 'wish-list' of Egyptian individuals have survived five millennia, affording us a glimpse of their desires and aspirations for the afterlife.

Acknowledgements

I am grateful to E. Christiana Köhler, Macquarie University, Sydney Australia, for giving me access to the Helwan stelae and the privilege of working with her on the publication of *Helwan II. The Early Dynastic and Old Kingdom Funerary Relief Slabs* (2009). Artwork (unless otherwise specified) is by E. C. Köhler and appears in the publication. The hieroglyphs in this paper have been reproduced using JSesh: an open source hieroglyphic editor.

Abbreviations

ÄA cf. Äg.Abh.
Äg.Abh. Ägyptologische Abhandlungen
Beitr.Bf Beiträge zur ägyptischen Bauforschung and Altertumskunde herausgegeben vom Schweizerischen Institut für ägyptische Bauforschung and Altertumskunde in Kairo
BIFAO Bulletin de l'Institut français d'archéologie orientale
CdE Chronique d'Égypte
DE Discussions in Egyptology
EA Egyptian Archaeology
GM Göttinger Miszellen
JEA Journal of Egyptian Archaeology
LÄ Lexikon der Ägyptologie
MÄS Münchner Ägyptologische Studien
MDAIK Mitteilungen des deutschen archäologischen Instituts Abteilung Kairo
NAWG Nachrichten von der Akad. der Wissenschaften zu Göttingen, phil-hist. Kl.
RdE Revue d'Égyptologie
SASAE Annales du Service des Antiquités de l'Égypte
WZKM Wiener Zeitschrift für die Kunde des Morgenlandes
ZÄS Zeitschrift für ägyptische Sprache und Altertumskunde

Bibliography

Baines, P. (1989) *Linen; Handspinning and Weaving*. London.
Barguet, P. (1986) *Les textes des sarcophages égyptiens du Moyen Empire*. Paris.
Barta, W. (1963) *Die Altägyptische Opferliste von der Frühzeit bis zur griechisch-römischen Epoche*. MÄS 3. Berlin.
Barta, W. (1982) s.v. Opferliste. *LÄ* IV, 586–589.
Bellinger, L. (1959) Craft Habits, Part II: Spinning and Fibers in Warp Yarns. *The Textile Museum Workshop Notes* 20. Washington.
Bissing von, F. W. & Kees, H. (1923) *Das Re-Heiligtum des Königs Ne-Woser-Re (Rathures)*. Vol. 2. Leipzig.
Blackman, A. M. (1924) *The Rock Tombs of Meir*. Part IV. London.
Bochi, P. (1996) Of Lines, Linen, and Language: A Study of a Patterned Textile and its Interweaving with Egyptian Beliefs. *CdE* LXXXI, Fasc. 142, 221–253.
Brugsch, H. K. (1968) *Thesaurus Inscriptionum Aegyptiacarum: Altaegyptische Inschrifen* Vols. V-VI. Graz. (First published 1883–1891, Leipzig).
Caton Thompson, G. & Gardner, E. W. (1934) *The Desert Fayum*. London.
Crowfoot, G. M. (1931) *Methods of Handspinning in Egypt and the Sudan*. Bankfield Museum Notes, Series 2 no. 12. Halifax.
Dreyer, G. (1998) *Umm el-Qaab I. Das Prädynastische Königsgrab U-j und seine frühen Schriftzeugnisse*. Archäologische Veröffentlichungen 86. Mainz.
Edel, E. (1975) Beiträge zum ägyptischen Lexikon VI. *ZÄS* 102, 13–30.
Edel, E. (1987) Eine althieratische Liste von Grabbeigaben aus einem Grab des späten Alten Reiches der Qubbet el-Hawa bei Assuan. *NAWG* 6, 104–105.
Faulkner, R. O. (1969) *The Ancient Egyptian Pyramid Texts*. Oxford.
Fischer, H. G. (1977) An Elusive Shape within the Fisted Hands of Egyptian Statues, in C. Aldred (ed.), *Ancient Egypt in the Metropolitan Museum Journal* Vols. I–II (1968–1976), 143–155. New York.
Gardiner, A.H. (1988) *Egyptian Grammar*, 3rd ed. revised. Oxford.
Germer, R. (1992) *Die Textilfärberei und die Verwendung gefärbter Textilien im Alten Ägypten*. *ÄA* 53. Wiesbaden.
Goneim, Z. (1957) *Horus-Sekhemkhet. The Unfinished Step Pyramid at Saqqara*. Cairo.
Griffith, F. Ll. (1905) in Murray, M. A. *Saqqara Mastabas Part I*. Egypt Research Account 10. London.

Haeny, G. (1971) Zu den Platten mit Opfertischszene in Helwan und Giseh. In *Geburtstag von Herbert Ricke. Beitr. Bf* 12, 143–164. Wiesbaden.

Hall, R. M. (1982) Garments in the Petrie Museum of Egyptian Archaeology. *Textile History* 13 (1), 27–45.

Hall, R. M. (1981) Two linen dresses from the Fifth Dynasty site of Deshasheh now in the Petrie Museum of Egyptian Archaeology, University College London. *JEA* 67, 168–171.

Hall, R. M. & Pedrini, L. (1984) A Pleated Linen Dress from a Sixth Dynasty Tomb at Gebelein now in the Museo Egizio, Turin. *JEA* 70, 136–139.

Hassan, S. (1932-1960) *Excavations at Giza*. Vols. I–VI. Oxford.

Helck, W. (1957) Das Kleidertäfelchen aus der Pyramide des *sḫm-ẖt*. *WZKM* 54, 72–76.

Hendrickx, S. (1999) La chronologie de la préhistoire tardive et des débuts de l'histoire de l'Egypte. *Archéo-Nil* 9, 13–81.

Hendrickx, S. (2006) Predynastic to Early Dynastic Chronology. In E. Hornung, R. Krauss & D. Warburton (eds.), *Ancient Egyptian Chronology*, 52-93. Leiden/Boston.

Hornung, E., Krauss R. & Warburton, D. (2006) (eds.) *Ancient Egyptian Chronology*. Leiden/Boston.

Ikram, S. & Dodson, A. (1998) *The Mummy in Ancient Egypt. Equipping the Dead for Eternity*. London.

Jéquier G. (1921) *Les Frises d'Objets*. Cairo.

Jones, D. (2000) *An Index of Ancient Egyptian Titles, Epithets and Phrases of the Old Kingdom* Vol. I. British Archaeological Reports, Int. ser. 866 (I) Oxford.

Jones, J. (2002) The Textiles from Abydos: New Evidence. *MDAIK* 58, 323–340.

Jones J. & Oldfield. R. (2006) Egypt's Earliest Linen. *EA* 29, 33–35.

Jones, J. (2008) Pre- and Early Dynastic Textiles. Technology, Specialisation and Administration during the Process of State Formation, in: B. Midant-Reynes, Y. Tristant, J. Rowland, S. Hendrickx (eds.), *Egypt at its Origins 2. Proceedings of the International Conference 'Origin of the State. Predynastic and Early Dynastic Egypt'*, Toulouse (France), 5th–8th September 2005, 99–132. Leuven.

Junker, H. (1929) *Giza. Bericht über die von der Akademie der Wissenschaften in Wien auf gemeinsame Kosten mit Wilhelm Pelizaeus unternommenen Grabungen auf dem Friedhof des Alten Reiches bei den Pyramiden von Giza. Band 1, Die Mastabas der IV. Dynastie auf dem Friedhof in Verbindungen mit K. Holey.* Wien & Leipzig.

Kahl J. (1994) *Das System der ägyptischen Hieroglyphenschrift in der 0.–3 Dynastie*. Wiesbaden.

Kahl, J., Kloth N. & Zimmerman U. (1995) *Die Inschriften der 3. Dynastie. Eine Bestandsaufnahme.* Äg. Abh. 56. Wiesbaden.

Kahl, J. (2001) Hieroglyphic Writing During the Fourth Millennium BC: an Analysis of Systems. *Archéo-Nil* 11.

Kahl, J. (2002-) *Frühägyptisches Wörterbuch*. Wiesbaden.

Kaplony, P. (1963) *Die Inschriften der ägyptischen Frühzeit*, Vols I-III. Wiesbaden.

Kaplony, P. (1966) *Kleine Beiträge zu den Inschriften der ägyptischen Frühzeit*. Wiesbaden.

Kemp, B. J. & Vogelsang-Eastwood, G. (2001) *The Ancient Textile Industry at Amarna*. Egypt Exploration Society 68. London.

Köhler, E. C. & Jones, J. (2009) *Helwan II. The Early Dynastic and Old Kingdom Funerary Relief Slabs*. Rahden.

Labrousse, A. (1996). *L'architecture des pyramides à textes, I - Saqqara Nord. BIFAO* 114/1. Cairo.

Lacau, P. (1904) *Sarcophages antérieurs au Nouvel Empire*. Vol. II. Cairo.

Lythgoe, A. M. & Dunham, D. (1965) *The Predynastic Cemetery N7000. Naqa-ed-Dêr*. Part IV. Berkeley/Los Angeles.

Manuelian, P. D. (2003) *Slab Stelae of the Giza Necropolis*. New Haven & Philadelphia.

Maspero, G. (1903) *Guide to Cairo Museum: 1st Edition*. Cairo.

Midgley, W. W. (1915) Reports on Early Linen. In W. M. F. Petrie & E. MacKay *Heliopolis Kafr Amar and Shurafa*. British School of Archaeology in Egypt & Egyptian Research Account 18th Year, 1912. London, 48–50, Pls. 57–58.

Ockinga, B. G. (1998) *A Concise Grammar of Middle Egyptian. An Outline of Middle Egyptian Grammar by Hellmut Brunner Revised and Expanded*. Mainz.

Ogdon, J. R. (1983) Studies in Archaic Epigraphy VII. On the long-sleeved dress logogram and its phonetic values. *GM* 68, 81–83.

Petrie, W. M. F. (1897) *Deshasheh*. Fifteenth Memoir of The Egypt Exploration Fund. London.

Petrie, W. M. F. (1914) *Tarkhan II*. British School of Archaeology in Egypt & Egypt Research Account 26. London.

Porter, B. & Moss, R. (c. 1981) *Topographical Bibliography of Ancient Egyptian Hieroglyphic Texts, Reliefs and Paintings.* Vol. 3, pt. 2 no. 1 (Second edn). Oxford.
Posener-Kriéger (1977) Les mésures des étoffes à l'Ancien Empire. *RdE* 29, 86–96.
Quibell, J. E. (1923) *Archaic Mastabas, Excav. Saqq. 1912-14.* Cairo.
Reisner, G. A. (1932) *A Provincial Cemetery of the Pyramid Age, Naqa ed-Der* III. Oxford.
Reisner, G. A. (1936) *The Development of the Egyptian Tomb down to the Accession of Cheops.* Oxford.
Reisner, G. A. & Smith, W. S. (1955) *A History of the Giza Necropolis*, Vol. II. Cambridge.
Riefstahl, E. (1970) A Note on Ancient Fashions. Four Early Egyptian Dresses in the Museum of Fine Arts, Boston. *Boston Museum Bulletin* 354, 244–259. Boston.
Robins, G. (1990) Problems in Interpreting Egyptian Art. *DE* 17, 45–46.
Saad, Z. Y. (1951) *Royal Excavations at Helwan (1945-1947).* SASAE 14. Cairo.
Saad, Z. Y. (1957) *Ceiling Stelae in Second Dynasty Tombs from the Excavations at Helwan.* SASAE 21. Cairo.
Saad, Z. Y. (1969) *The Excavations at Helwan. Art and Civilization in the First and Second Egyptian Dynasties.* Oklahoma.
Sethe, K. (1923) *Dramatische Texte zu altaegyptischen Mysterienspielen.* Leipzig.
Sethe, K. (1932) *Urkunden des Alten Reichs.* Vol. I. Leipzig.
Smith, W. S. (1935) The Old Kingdom Linen List. ZÄS 71, 134–139.
Smith, W. S. (1978) *A History of Egyptian Sculpture and Painting in the Old Kingdom.* New York.
Staehelin, E. (1966) *Untersuchungen zur ägyptischen Tracht im Alten Reich.* MÄS 8. Munich
Vogelsang-Eastwood, G. (1993) *Pharaonic Egyptian Clothing.* Leiden.
Vogelsang-Eastwood, G. (2000) Textiles, in: Nicholson, P. T. & Shaw, I. (eds.), *Ancient Egyptian Materials and Technology.* Cambridge.
Wiedemann, A. (1884) *Ägyptische Geschichte.* Gotha.
Winlock, H. E. (1955) *Models of Daily Life in Ancient Egypt from the Tomb of Meket-re at Thebes.* Cambridge, Massachusetts.
Weill, R. (1908) *Des monuments et de l'histoire des IIe et IIIe Dynasties égyptiennes.* Paris.
Ziegler, C. (1997) *Egyptian Antiquities.* Paris.

7. Clothing in Sargonic Mesopotamia: Visual and written evidence

Benjamin R. Foster

I am the loom,
I weave threads,
I clothe common folk,
I make the king splendid.

The Tamarisk and the Date Palm

Mesopotamian art and documents attest to a great variety of textiles and textile terminologies, as well as to aspects of the chain of production leading from raw materials to finished goods. The artistic evidence for Mesopotamian clothing during the late fourth and third millennia BC has been studied by several scholars, who reached different conclusions.[1] The most intensively examined written evidence has been the administrative records for state-sponsored textile production under the Third Dynasty of Ur (2112–2004 BC), though philological data for textiles from earlier and later periods have been collected and summarized.[2] This chapter focuses on the Sargonic, or Akkadian, period (2334–2154 BC), considering both art and written records. It will argue that events of this era caused important changes in elite dress, some of which became normative, especially in portrayals of the ruler, for centuries to come. After a brief survey of the political, social, and economic history of the time, this chapter documents the rapid changes in clothing styles resulting from increased international contacts, influx of wealth, and the growth of a new class of notables and administrators directly subservient to the king, rather than to traditional local entities such as city-states or temples.

1. Historical background

The Sargonic or Akkadian period is defined principally by the reigns of five kings, covering a period of 150 years: Sargon, Rimuš, Maništušu, Narām-Sîn, and Šarkališarri. Six succeeding kings are counted with the dynasty, but are generally considered post-Sargonic or post-Akkadian.

[1] Strommenger 1971. I thank Karen Polinger Foster for drawing Figures 1–14 for this article, for assisting with collation of the monuments in the Louvre, and for numerous suggestions adopted here.
[2] Waetzoldt 1972, 1980/83.

"Sargonic" refers to the dynasty founded by Sargon of Akkad, while "Akkadian" refers both to the region of Akkad, originally the area around the confluence of the Diyala and Tigris rivers, and to the Akkadians, the easternmost branch of the broad arc of Semitic-speaking peoples who lived in northern Mesopotamia, the mid- and upper Euphrates Valley, and northern Syria in the third millennium. There was a city of the same name, called Agade for convenience in modern books, to distinguish it from the land. Its location is unknown, but it was probably near the confluence of the Diyala and Tigris, hence not far from modern Baghdad.[3]

The history and culture of the Akkadians are closely associated with the Sargonic dynasty's political, military, economic, and cultural unification of Mesopotamia, Syria, and southwestern Iran, which is often considered the first of the great empires or territorial states of antiquity.[4] In Mesopotamian historical memory, the rise, expansion, prosperity, grandeur, and ignominious collapse of this empire were new and dramatic phenomena in history, worthy of study and reflection.[5] The site of Agade was explored by later kings for traces of its ancient royal palace and its famous temple to the war goddess, Ištar; Akkadian inscriptions were copied and studied in school, and a rich body of later literature in both Akkadian and Sumerian focused on events and personalities of the Akkadian period.[6] Would-be conquerors of later ages named themselves after their two most famous Akkadian predecessors, Sargon and Narām-Sîn, and imitated their monuments, including their dress. In short, in second- and first-millennium Mesopotamia, the Akkadian period was considered a pivotal age, with the Sargonic kings the first rulers of history for whom authentic sources and historical tradition were available.

The founder of the dynasty, Sargon of Akkad, was remembered as a warrior-king and empire-builder. Long after his death, stories were told about his birth and youth: he had been born in secret to a high priestess, who set him adrift in a basket on the Euphrates, from which unpromising beginning he rose to become the greatest king of all time;[7] he had been in the service of the king of Kiš, the major power in northern Babylonia, the city where monarchy began, when the goddess had singled him out for her favor: "Holy Inanna was there, all the time beside him!"[8] These stories were the easier to tell because his inscriptions do not name his father, as became customary in later times, and say nothing of his rise to power. A list of Mesopotamian kings, compiled in Sumer long after his death, considered Sargon one of eight remarkable rulers of the past who were not of royal birth.[9] The importance of these stories is not their historicity but that they show the singular glamour that attached to this ruling family and its fortunes.

In his own inscriptions, Sargon boasts that he was victorious in 34 campaigns,[10] an achievement that sets him apart from even the greatest conquerors of history, such as Alexander the Great and Suleiman the Magnificent. His triumphs began in Sumer and culminated there in the defeat of Umma, Lagaš, Ur, and Uruk, the major city of the region, enabling him to extend his domains to the headwaters of the Gulf.

---

[3] Foster 1997.
[4] A. Westenholz 1999.
[5] J. Westenholz 1983; Glassner 1986.
[6] Kraus 1963; Charpin 1986, 425–428; J. Westenholz 1997.
[7] Lewis 1980.
[8] Cooper & Heimpel 1983, 76.
[9] Glassner 2004, 122–125.
[10] Frayne 1993, 28.

To the south and east, Sargon invaded Elam and Susa, their neighboring territories of Awan and Sabum, and even routed forces from Marhaši, probably the region around Kerman, famed in Mesopotamia as a source of precious stones, calcite vessels, and other exotic luxury goods.[11] Few future Mesopotamian dynasties would rule at Susa, protected as it was by the desert that wore down even the most formidable fighting forces. Sargon's Elamite campaign may be commemorated in a massive stone victory stele, showing Sargon himself and his retinue (Fig. 7.7).[12] Someone later made a determined effort to destroy it, battering it with hammers, breaking it, and attempting to saw it into smaller pieces.

To the north and west, Sargon enjoyed the submission of Mari, which controlled the mid-Euphrates, and Ebla, south of Aleppo, one of Mari's major rivals. By Sargon's own account, the weather god Dagan bestowed upon him the Upper Lands, the territories and cities of the upper Euphrates region. His armies pushed into Anatolia as well, known to the Akkadians as a land of cedar trees and mountains where silver was mined.[13]

Besides conquests, Sargon wished to be remembered for a new style of administration, in which he placed Akkadians in governorships in the conquered lands; for bringing international trade to his city, Agade; and for having sufficient resources at his disposal to feed daily 5400 able-bodied men in his service.[14] His daughter, Enheduanna, became high priestess of the moon-god at Ur, and is credited with the first literature of any civilization that can be identified with a specific author.[15] The land of Akkad, hitherto a sparsely populated and marginal place,[16] though with considerable agricultural potential, had entered world history.

Sargon was succeeded by two of his sons, Rimuš, who ruled for nine years, and Maništušu, who ruled for fifteen. Rimuš faced a revolt in Sumer, which had clearly resented his father's rule. In a series of brutal campaigns, Rimuš re-established his father's dominion there, boasting that he killed in battle over 23,000 men and taking twice that number captive, presumably killing them too.[17] He also expelled thousands of men of fighting age from the defeated cities and subjected them to cruel punishment, presumably mass execution or forced labor, or both. Among other punitive measures, Rimuš apparently expropriated some 134,000 hectares of prime agricultural land near Lagaš to create a new royal domain to distribute to his retainers, thereby endowing a new landed class in Sumer which had, in principle, no allegiance to the old city-states or to the great temples of the region. This is the largest single transaction ever recorded in a third-millennium Mesopotamian contractual or administrative document.[18]

Rimuš next renewed Sargon's invasion of Elam. A coalition had been formed, led by the ruler of Marhaši, who had extended his authority west into Elam, creating a strategic alliance that could counterbalance any Akkadian influence in the Gulf. Rimuš's forces routed the coalition and captured its commanding officers, bringing Elam and Susiana once again under Akkadian influence or direct rule. The booty from this campaign was enormous. At the sanctuary of Nippur alone,

[11] Steinkeller 1982a, 2006; A. Westenholz 1999, 91–92.

[12] Strommenger 1962, 115; Moortgat 1969, 125; Amiet 1976, 8, 72; Börker-Klähn 1982b, 18d–i.

[13] Frayne 1993, 28–29.

[14] Frayne 1993, 29.

[15] Hallo & Van Dijk 1968; J. Westenholz 1989; Zgoll 1997.

[16] Adams 1965, 42–43.

[17] Frayne 1993, 41–51.

[18] Foster 1985; figures revised in accordance with suggestion of J. Friberg.

Rimuš dedicated over 15 kilograms of gold, 1800 kilograms of copper, and 300 slaves.[19] Stone bowls, vases, sea shells, mace heads and other objects were dedicated at sanctuaries throughout the land, examples of which have turned up at Kiš, Aššur, Nippur, Sippar, Ur, Šuruppak, and Tuttub, even as far away as the Khabur Valley. Rimuš proudly recorded that his craftsmen produced the first known statue of tin, the rarest industrial metal in the Mesopotamian world.[20] Rimuš's eventful reign was brought to an end by his assassination.[21]

Maništušu, his brother, inherited a firmly-founded state in southern Mesopotamia and southwestern Iran. His only major historical inscription tells that he continued to campaign in Iran and pushed down the Gulf, conquering 32 cities "across the sea," either on the Iranian coast or in Oman. There he quarried diorite, the hard black stone of the mountains, and had it hauled back to Akkad in ships.[22] This stone was challenging to work but took a smooth, high polish. Used only rarely in Mesopotamia before the Akkadian conquests, diorite became more common during and after Maništušu's reign, especially for statues apparently intended to memorialize the ruler himself, with softer stones used for reliefs that commemorated his military victories.[23] Maništušu, like his brother, expanded the domains of the new class called "Akkadians." This was no longer an ethnic term but referred to notables who served the king. Maništušu purchased several large tracts of land in northern Babylonia to distribute to 49 of these men, including administrators and military officers.[24] Also like his brother, he was assassinated.[25]

The thirty-seven-year reign of Maništušu's son and successor, Narām-Sîn, was the apogee of the Akkadian Empire. No other ruler of the third millennium, save Sargon, made such an impression on Mesopotamian historical tradition. The surviving art and commemorative prose of his reign allow us to glimpse extraordinary military achievements rivaling or eclipsing those of his grandfather, Sargon, as well as the creation of proud imperial culture.

Narām-Sîn summarized his own conquests as stretching from Marhaši in southeastern Iran, as far as the Cedar Forest, meaning the slopes of the Amanus or Lebanon, and from the Mediterranean to the "lands beyond the sea," perhaps Oman. Within this vast territory, he campaigned in the central and northern Zagros and the Iranian plateau. He marched to the sources of the Tigris and Euphrates and smashed through the massive walls of the great cities of Syria, such as Armanum (Tell Bazi on the mid-Euphrates), the citadel fortifications of which stood 44 cubits high. He claimed to be the first king to conquer Ebla as well.[26]

The most dramatic event of Narām-Sîn's reign was a revolt that began in Mesopotamia itself, where pretenders appeared at Kiš and Uruk, the ancient northern and southern rivals for power in Sumer and Akkad. In due course, Kiš was joined by a coalition of Babylonian cities, such as Cutha, Borsippa, and Sippar, as well as by bands of Amorite tribesmen from the Jebel Bishri region. Narām-Sîn routed the Kišite army, taking many important captives. The rebels fell back on Kiš itself, where, in a second battle, Narām-Sîn was again victorious. Fighting continued in

---

[19] Frayne 1993, 55.

[20] Frayne 1993, 68 (considers the metal "meteoritic iron").

[21] Goetze 1947, 256.

[22] Frayne 1993, 76.

[23] Amiet 1972.

[24] Foster 2000.

[25] Goetze 1947, 157.

[26] Frayne 1993, 135; Otto 2006.

the city itself, leaving the Euphrates choked with bodies of the dead. Resolved to obliterate the city, Narām-Sîn dismantled its walls and diverted the river to flood it.[27]

In the meantime, Amar-girid, a rebel king at Uruk, had rallied most of the cities of Sumer, including Adab, Isin, Lagaš, Nippur, Šuruppak, and Umma, to his cause, as well as the conquered cities in the Gulf. From this, and from the desperate resistance of the northern allies, one can gauge the deep unpopularity of Sargonic rule. Amar-girid sent messages to the rulers of the cities in the Upper Euphrates and Assyria, begging them to support him, but they hesitated. Narām-Sîn, flush from his victory over Kiš, moved first against the Amorite tribesmen, who had supported the cause of Uruk after the defeat of Kiš. He then crushed the Sumerian coalition and followed up with an invasion of the Gulf, booty from which he dedicated in various sanctuaries.[28]

From the Akkadian standpoint, beleaguered Agade had been delivered from enemies on all sides in a brilliant series of nine major victories in a single year, for which the king had been obliged to call up his Akkadian forces nine times. According to Narām-Sîn's own account, the people of Agade asked of the great gods that they might henceforth worship him as the god of their city, and they built a temple to him there.[29] With the deification of the king, a new emphasis appears in art on the king's godlike qualities.

In addition to being a warrior, Narām-Sîn was a builder, paying particular attention to the temples of Ištar throughout Mesopotamia, as befit the patron deity of his dynasty. He ordered work on her temples at Nineveh in Assyria, at Zabala and Adab in Sumer, and possibly at Babylon, as well as in Agade itself. He also ordered work on the temple of the moon-god at Ur and installed his daughter, Enmenanna, as high priestess, successor to his long-lived aunt, Enheduanna.[30]

Narām-Sîn, like his predecessors, claimed special support from the god Enlil, principal god of Sumer, with his sanctuary at Nippur. There he presented captured and defeated rulers to the god, as Sargon had before him. There too he began a vast project to rebuild Enlil's temple, Ekur, in a lavish imperial style, no doubt as a showpiece of the glory of his reign. His son and crown prince, Šarkališarri, was in charge of the work, as well as his major-domo, Su'aš-takal. Hundreds of kilograms of silver, gold, and bronze, and tons of copper, were used in the work; one contemporaneous record of the project lists 77 woodworkers, 86 goldsmiths, 10 sculptors, 54 carpenters, plus engravers and other craftsmen, at work at the same time. They were drawn from all over the realm and quartered on the local citizenry and notables. This great undertaking was incomplete at Narām-Sîn's death, but resumed by his son and successor, Šarkališarri.[31]

Šarkališarri's twenty-five-year reign is poorly known. To judge from his surviving inscriptions, he too was an effective field commander, but some of the battles he fought were close to home, so one has the impression of increasing instability in the realm, for which many explanations have been offered.[32] The empire collapsed and Agade itself was overrun by barbarous mountain peoples, the Gutians, whom the Akkadians held in contempt. A period of anarchy followed, which the Sumerian King List summarized by noting "Who was king? Who was not king?"[33]

[27] Frayne 1993, 104–107.
[28] Frayne 1993, 107; Sommerfeld 2000.
[29] Frayne 1993, 113–114.
[30] Frayne 1993, 145–146.
[31] A. Westenholz 1989.
[32] Glassner 1986; Weiss & Courty 1993.
[33] Glassner 2004, 122–123.

Mesopotamia was thereafter broken into a small kingdoms and principalities: Gutian rulers were in the Diyala region, Umma, and Adab; a vigorous, prosperous state at Lagaš built on its special position as administrative capital of a large Sargonic province; an ambitious dynasty was founded at Ur and Uruk, now known as the Third Dynasty of Ur;[34] and a new dynasty established itself at Agade for several generations of long-lived kings. The splendid example for them all was the Akkadian state, and each appropriated Sargonic tradition in its own way.

At Mari and Elam, independent rulers were proud to adopt Sargonic military titles, *šakkanakku* (general) and *sukkal-mah* (high commissioner). The peoples of the Zagros, including the Gutians, celebrated themselves in Akkadian-style reliefs and inscriptions. The Amorites of Upper Mesopotamia and the Diyala region considered themselves the direct heirs to the Akkadian tradition; rulers of Aššur and kings of Ešnunna, not far from Agade, called themselves Sargon and Narām-Sîn and cultivated literature and costume in the Akkadian style; the early second millennium warrior-king Šamšī-Adad spent time at Agade and may have let it be thought that he was somehow a descendant of the Sargonic royal family. The rulers of Lagaš maintained the Akkadian practice of making portrait statues of the ruler in diorite and wore the special garments typical of the Akkadian elite. They also inherited the great domains that Rimuš and his successors had carved out in the area, there developing and refining Akkadian agrarian policies. The rulers of Ur were destined to rebuild in a more detailed, Sumerian style the Akkadian administrative apparatus, with its refinements of documentation, use of mathematical models, unified metrology, and special handwriting, while developing to an extreme the royal personality cult that made Narām-Sîn such a subject of fascination to later writers. What in modern scholarship are called the "reforms of Šulgi" are in fact Narām-Sîn's administrative measures.

The rulers of Ur, however, associated themselves with the ancient primacy of Uruk, rather than with Agade, though they honored the memory of the Akkadian kings. Uruk was a city so old that no one knew its origins, and the kings of Ur promoted legendary, pre-Akkadian, rulers of Uruk, such as Enmerkar, Lugalbanda, and Gilgameš, rather than Sargon and Narām-Sîn of more recent memory, as fit subjects for epic poems, more witty and entertaining than heroic in the Akkadian style.[35] At Ur and Lagaš, the vibrant brilliance of Akkadian art dimmed to a staid, hierarchical formalism, leaving Akkadian sculpture and seals the "Classical" art of early Mesopotamia, its innovations palely imitated later.[36]

2. Economy and society

Although Narām-Sîn and his predecessors had commemorated acquisition of wealth by conquest, other cornerstones of the empire's prosperity and dynamism were overlapping royal and local administration, agrarian management, urban governance, and control of temples, for all of which important documentation is available.[37] In the present consideration of clothing, five aspects of Mesopotamian society in the Sargonic period should be emphasized: imperial patronage of a new class of royal servants and notables, internationalization of styles through broad programs of conquest, a massive influx of new materials and technologies in a short period of time,

[34] Sallaberger 1999, 132–134.
[35] Vanstiphout 2003.
[36] Moortgat 1969.
[37] Foster 1982c, 1985.

patronage of both traditional and experimental industries, and development of techniques for their management.

One of the most effective means the Akkadian government had at its disposal for managing its subjects and resources was its power of patronage, exercised over a wide network of people linked by common economic interests and their dependency on the ruler and his notables. A distinctively Akkadian form of this was the practice of notables to surround themselves with an entourage of personal followers. On the lowest level, one group of these was called "the select" (*nisqu*), who were clients, perhaps armed, ready to do a notable's bidding. To judge from later sources, there was a second group of such people, called "boys" (*ṣuhārū*), who were menials or servants. Such groups were new to Sumer, so were referred to with Akkadian terms. In addition, Akkadian documents often refer to subordination in a more general way, "the one of," meaning a client, dependent, or family member of somebody of superior status. Thus Akkadian Mesopotamia was a society in which gradations of status were carefully noted and a rhetoric of subservience could be part of official and legal discourse and self-identification.[38]

By expropriating, purchasing, or developing extensive areas of agricultural land outside the major cities, the Akkadian king had immense wealth to bestow on whom he wished, so a tripartite structure of dependency was established. First were the king's governors, whom he appointed and who served at his pleasure, being dignitaries of political, military, and economic importance; second were the ancient cult centers, which, in Sumer especially, disposed of considerable agricultural land; and third were notables, men of preferment who received fields for their support from the king's land office, portions of which they, in their turn, could distribute or lease to followers and tenants. A Sumerian poet considered this tripartite hierarchy the foundation of Akkadian prosperity:

> The governors of cities, the managers of temples,
> The scribes who parceled out the farmland in the steppe.[39]

The governor, or *ensi*, held land, buildings, labor, livestock, and wealth which it was his agreeable task to administer, with the support of a large staff. Some governors were school graduates, thus educated men.[40] To get and keep their important and lucrative posts, they must have availed themselves of many opportunities to display their loyalty, such as attending court festivals in the capital, staging lavish entertainment for the royal family and key notables when they were in the area,[41] ensuring that the king's revenues flowed freely from their cities and domains, supporting local and national cults and praying for the royal family, reporting regularly matters of intelligence value, and honoring the king in inscriptions and dedications. As will be suggested below, accession of a new king may well have occasioned applications to receive or retain governorships, in the form of expressions of fealty and lavish presents, including clothing, to the new king. Governors were experienced administrators, with detailed understanding of management, no doubt a good sense of how to deal with local Akkadian notables, and perhaps belonged to a prominent local family through birth and marriage, so were able to rely on kinship ties among their subordinates, as is well known from later periods.[42]

---

[38] Gelb 1979; Abrahami 2008, 5–7.

[39] Cooper 1983, 52:51–52.

[40] Schileiko 1914.

[41] Foster 1980; Visicato 2001.

[42] Hallo 1972; Zettler 1984; Dahl 2007.

Their literacy meant that the governors belonged to a small, proud elite who could read and write their own letters and documents in the elegant new Akkadian hand, quote aptly literature they had studied in class, and understand the technical details of measuring land, estimating crops, and apportioning income.[43] One of Narām-Sîn's own sons served as governor in Akkad itself.[44] City governors were, as the Sumerian poet recorded, the first of the king's subjects in the maintenance of his state and its prosperity.

These governors and their entourages were among the leaders of the new class Sargon and Maništušu called "Akkadians," men who owed their primary allegiance to the king rather than to local or family interests, who were prepared to leave their place of birth to serve the king where he willed. Although they were called "Akkadians," many of them were not born in Akkad or Agade or even had Akkadian names. In return for their loyalty, they expected the king to give them land and preferment, just as Maništušu did in his purchase of land and Rimuš in his appropriation of it.[45] This class no doubt enjoyed locally the esteem that collaborators generally do who profit from the exploitation of conquered territories, as evidenced by the determined rebellions in Sumer and northern Babylonia against Akkadian rule.

Religious office was, like a governorship, a calling worthy of a member of the royal family. Sargon and Narām-Sîn appointed their daughters to be high priestesses, so, as with governorships, family connections may well have been important for temple officials too.[46] The upper echelons of the temple hierarchy, like the upper echelons of provincial administration, were educated people, sharing interests and background, if not a family tree, with their counterparts among the new Akkadians around them.

Temples were traditionally the largest buildings in Sumerian cities, organized like manors or households, with an enclosed central complex for religious rites and ancillary structures, such as staff residences, barns, and storage buildings.[47] The leading cultic figure was a high priest or priestess, the leading administrator the *sanga*. It was the *sanga* who administered the temple lands, gardens, herds, flocks, and other resources, acting, in Sumer at least, as the counterpart to the local governor and the representatives of the king's household. There were temple stewards, herdsmen, building attendants, weavers, cultivators, and boatmen, as well as laborers, some of whom were dedicated to temple service by their families, while others were indigent, foundlings, handicapped, or mentally deficient people put to work in temple gardens and orchards to be usefully employed and to have a way to live. There were also temple women who were not married nor under male authority, forming a class by themselves in a male-dominated society. Certain temples specialized in specific industries or activities, such as the temple of Gula at Isin, which kept dogs for healing the sick. Temples had, therefore, large vested interests in a conquered dominion, whose divine householders had entrusted their subjects to the king.[48]

The relationships among the royal and temple centers of concentrated resources, with their different regional and local traditions, must at times have been delicate to work out, but, in favor of developing mutual interests, royalty and the religious establishment shared a hierarchical

[43] Kienast & Volk 1995, 53–54.
[44] Frayne 1993, 54.
[45] Foster 2000.
[46] Hallo 1972, Zettler 1984.
[47] Oppenheim 1961.
[48] Renger 1967, 1969; Gelb 1972.

view of the world: gods were like rulers and rulers were like gods, exercising their powers in the same ways over the same countryside. Rulers paid homage to the gods and credited them with their successes, lavishing their households with gifts, endowments, and projects. A strong and effective ruler enjoyed divine favor and requited that by sharing the fruits of his success with the gods who had singled him out to rule, as Narām-Sîn did with Enlil, chief god of Sumer, by commemorating his triumphs in monuments placed in the temple courtyard. Therefore, on the local horizon, priests and temple managers enjoyed and expected the king's favor, even as land-rich temples no doubt needed the resources of labor, tools, transport, and draught animals the king's patronage could bring them.

In addition to their reliance on the governor's and temple households, the king and royal family had their own local interests and their own agents to look after them. The Sumerian poet who mentioned the "scribes who parceled out land in the steppe" was thinking of these agents, whose work can sometimes be followed in detail through their lists of lands and cultivators, as well as their surveys, calculations of areas, reckoning of expenses of cultivation, and balances of harvests due.[49] The scribes who carried out these tasks were subject to an exalted personage called by a Sumerian title, *šassukku*, "the registrar." This royal official had thousands of hectares of good land at his disposal, divided into districts and subdivisions, to be distributed among members of the administration, ruling elite, professionals and their followers, families, and lessees. The registrar, working with mathematical models, recorded blocks of territory, systematically cut up into schematic sub-sections, from which actual or schematic parcels were measured off by survey to those privileged to receive them.[50] One registrar in Sumer controlled over 6200 hectares of arable land in three large districts near Lagaš, which had some of Sumer's most fertile fields, fifteen times the size, for example, of a parcel held locally by the high priestess of a patron deity of the Akkadian dynasty.[51]

The Akkadian notables, moreover, controlled areas that sometimes dwarfed those of the local administrators and temple staff. Yetib-Mer, the *šāpiru*, or chief steward, of Narām-Sîn early in his reign, held at Umma and Lagaš alone close to a thousand hectares straddling the two provinces, surveyed by royal scribes and recorded by the registrar. Governors, priests, and administrators must have approached this man with the respect due his exalted status and were careful not to encroach on his domains and clients.[52]

Into this new society flowed the fruits of the Akkadian conquests — specie and exotic materials, products, animals, and people. Foreign names were scrupulously recorded by scribes in conquest narratives; foreign hairstyles, clothing, and objects, such as vessels and daggers, were carefully rendered by artists.[53] New titles appeared in the royal titulary in response to unexampled conquests.[54] Small wonder, then, that new styles of art rapidly developed among the rising social and managerial elite and notables, along with new styles of verbal expression, and new fashions in dress. New technologies included using the lost-wax technique for large-scale metal sculpture, first attested in Akkadian Mesopotamia. New materials included exotic stones, such

[49] Foster 1982b, 21.
[50] Foster 1982b, 21.
[51] Foster 1982b, 63.
[52] Foster 1982b, 36.
[53] Mellink 1963.
[54] Hallo 1980.

as diorite, known previously in very limited quantities, and serpentine, which enjoyed a vogue in the production of cylinder seals.[55]

As for textiles, on the basis of documents from the Third Dynasty of Ur, one might posit Sargonic beginnings for their industrial production, but little evidence has turned up so far in the thousands of administrative documents from the Akkadian period. The various references to workshops and teams of workers may refer to special projects, such as the rebuilding of the temple of Enlil at Nippur, or to royal workshops for furniture and art objects, as abundantly documented in later periods. While mass production of some commodities, such as pottery, is well attested, one seldom finds specific reference to groups of weavers.[56] The only Akkadian structure sometimes identified with industrial production, in fact, industrial weaving, is the Northern Palace at Tell Asmar, ancient Ešnunna.[57] There is, however, little archaeological basis for this proposal, such as numerous loom weights or other evidence for weaving one would expect to find. A document from the archives of this building records rations to 105 male and 585 female workers, some teams of whom are identified as túg-ni, perhaps something to do with textiles, but the meaning of the word is unclear.[58]

We may for the present consider state-sponsored textile production an innovation of the succeeding Ur III period, in which the king's government attempted to take over and quantify as much of the chain of production as feasible. This was part of a broader trend, in which the kings of Ur hoped to dominate and regulate important aspects of their economy using large numbers of dependent workers, unlike the Akkadian rulers, who had relied on a corporate management style combining patronage, force, and community involvement.

3. Art of the Akkadian period

Few works of Akkadian art survive, compared to preceding and later periods, and, in the case of sculpture and relief, many pieces were vengefully defaced, broken, and mutilated. What remains, however, includes some of the most intensively studied and most often reproduced objects from ancient Mesopotamia. In considering the visual evidence for Sargonic clothing, the following points should be kept in mind. First, Akkadian monumental art emerged from prior tradition, as known in both Sumer and Akkad, and, in the course of two or three generations, rapidly developed a new, imperial style. Royal patronage of sculpture attracted the most skilled artists and resulted in a striking uniformity in production for a given reign, as if statues and reliefs using new artistic conventions were produced centrally and distributed throughout the realm, with a growing emphasis on the overwhelming power of the sovereign. It was a time of much innovation: the first life-size or larger free-standing statues of human beings; the first naturalistic depictions of landscape; the first convincing renderings of musculature, folds of flesh, contours of limbs, and hints of sexual allure. By the middle of the Akkadian period, the registers traditionally used in pictorial narrative had been dissolved into one unified scene, and various essays in hierarchical size (representing important figures as taller or larger than other figures in the same scene) had

[55] Moorey 1994, 26–27; Collon 1982, 26.
[56] A. Westenholz 1999, 101–102. See below p. 145.
[57] Breniquet 2008, 64–65.
[58] Gelb 1962 (=MAD 1) No. 163; A. Westenholz 1999, 65.

been tried. Artists even experimented with the orientation of writing on monuments to convey victory and defeat.[59]

The portrayal of Sargonic clothing in sculpture develops in this artistic context, reflecting the elite status of a small ruling group with access to enormous resources. Akkadian stone carvers were the first to show garments draping over the body, as well as their thinness or transparency. Of particular importance to the historian of clothing is a growing tendency to place commemorative, triumphal, even legal inscriptions on carved stone monuments, allowing precise dating of a significant percentage of the surviving fragments.[60]

Akkadian seal carving likewise shows a high degree of experimentation.[61] New materials were used, and new styles and subject matter appear, gradually replacing or modifying earlier ones.[62] Despite the seals' small scale, close attention was given to the garments worn by deities and human beings, and new fashions were continually incorporated into the repertory.[63] As with sculpture and relief, the growing practice of inscribing seals with the names of owners and their patrons helps to date them precisely, and, as will be suggested below, to correlate visual and written evidence.[64]

This dynamism in art suggests that textile production, even if not directly patronized by the ruling elite, also experienced significant changes, if for no other reason than because of the empire's increasing contacts with Iran, Anatolia, Palestine, and Syria, even possibly Egypt. This being so, one is surprised that dyeing technology was not imported, along with new fabrics and designs, as there is strikingly little evidence for the practice of this very ancient and near-universal art in late third millennium Mesopotamia, even in the textile mills of the kings of Ur.[65]

4. Styles of dress in Akkadian art

The sources for Sargonic clothing are of two types: visual, including commemorative statues, reliefs, and seals; and written: consisting of administrative and legal records mentioning textiles of various kinds, including clothing. Although the garments pictured in reliefs and seals have been discussed in detail, and modern efforts made to reproduce them, none of these studies has attempted to incorporate archival evidence.[66] Most philologists, for their part, are unwilling to make the leap of faith that proposing identifications of words with artifacts or ancient images requires. Yet, for the Akkadian period, the written evidence does offer suggestive correlations,

[59] Lambert 1965, 181.

[60] Strommenger 1959.

[61] Boehmer 1965; Nagel & Strommenger 1968; Collon 1982; Collon 1987.

[62] Frankfort 1970, 83–91, who writes of a "grim world of cruel conflict, of danger and uncertainty, a world in which man is subjected without appeal to the incomprehensible acts of distant and fearful divinities whom he must serve but cannot love"; Spycket 1981: 143–174; Moortgat 1967, 45–54, who writes of "heroic spirit and turbulent energy," rapid change, even rejection of the static. Though Moortgat's interest in the "bearing and physiognomy" of Semites and what he saw as racial influences in Mesopotamian art of the Akkadian period invokes ideas no longer fashionable in modern scholarship, his presentation of Akkadian art was, despite this, deeply insightful and influential on later studies, such as those of Börker-Klähn 1982a and 1982b and Bänder 1995.

[63] Collon 1982, 27–32.

[64] Nagel & Strommenger 1968; Edzard 1968/69.

[65] Breniquet 2008, 126–129; Waetzoldt in this volume.

[66] Houston & Hornblower 1920, 48–49 (study and modern reproduction of the toga-garment); Heuzey & Heuzey 1935; Strommenger 1971, 1980/83; Collon 1982, 27–29.

and clothing had such special significance in elite life that artists showed it in great detail. For the first time in the history of Mesopotamian art, garment folds were shown in sculpture, not to suggest the body in motion, as in some artistic traditions, but to reveal the contours of a powerful body in repose, under the soft clothing.[67]

The core imagery of Sargonic clothing comes from sixteen mostly fragmentary monuments, the majority of a military and commemorative nature. Seven of these can be dated to specific reigns by their inscriptions, while the others may be dated on stylistic grounds, with a high level of agreement among different studies. The monuments are briefly summarized in the following table, with reference to the most informative photos and drawings of them, a classification of each garment shown, and in some cases discussion of aspects of the imagery not readily visible in published photographs, based on study of the original monuments.

Sargonic Statues and Reliefs showing Clothing (* = dated by inscription, otherwise dated by style)

Reign	*Photo*	*Drawing*
SARGON		
1. diorite stele* from Susa	Amiet 1976 No. 1; Strommenger 1962, 115; Moortgat 1969, 125; Börker-Klähn 1982b, 18b, c	Amiet 1976, 8; Börker-Klähn 1982b, 18d–i; Bänder 1995, XXXIV; Fig. 7.7

Sargon wears shaggy, wrap-around garment [“Wickelgewand,” “Flammengewand,” “robe en toison laineuse à mèches ondulées qui dégage exceptionellement l’épaule droite” (Spycket 1981, 146–147)], twisted sash, and plain necklet, perhaps with pendant hidden by beard; *attendant* wears shoulder sash; *soldiers* wear shaggy, wrap-around garment, long fringed sash.

Sargon’s extended right hand grasps one or two human heads by the neck, possibly associated with a battle net, whose shape and diagonal patterns are just discernible (see Amiet 1976, 72); compare Strommenger 1960, 55. Although some drawings show Sargon carrying a staff or mace, rulers hold these at the bottom of the shaft and they do not have human heads.

2. diorite stele from Susa	Amiet 1976 No. 5; Moortgat 1969, 138; Strommenger 1962, 114; Hansen 2003, 199	Amiet 1976, 11; Börker-Klähn 1982b, 19, 20

Soldier wears short shoulder sash with no fringe.

3. limestone disk* of Enheduanna from Ur	Moortgat 1969, 130; Hansen 2003, 200	Amiet 1976, 15, Bänder 1995, LXI Fig. 7.1

Enheduanna wears tiered, ruched, or ruffled garment.

[67] Strommenger 1962, 26.

RIMUŠ

1. limestone stele from Telloh	Amiet 1976 No. 25; Moortgat 1969, 134–5; Strommenger 1962, 117; Hansen 2003, 200–201	Strommenger 1971, 42 Bänder 1995, XLIII Fig. 7.13

Rimuš? wears long shoulder sash, fringe on one side.

MANIŠTUŠU (see Strommenger 1988 for more details)

1. diorite statue* from Susa	Amiet 1976 No. 13, Moortgat 1969, 141	Strommenger 1960, pl. 9 Heuzey 1935, 57 Fig. 7.9

Maništušu wears closed, wrap-around garment with tasseled fringe ("Reichsakkadisches Wickelgewand").

2. limestone statue from Susa	Amiet 1976 No. 15, Moortgat 1969, 142	

Maništušu? wears closed, wrap-around garment with tasseled fringe.

3. diorite statue from Aššur	Moortgat 1969, 139–140, Strommenger 1971, 116	Amiet 1976, 22, Strommenger 1971, 48 Fig. 7.5

Maništušu? wears necklace of large beads and a long counterpoise, and, according to most analyses, a thin garment similar to Nos 1 and 2 above, through which his shoulder blades and spine show. Some see the garment as secured by a strap passing over the back, others that the "strap" is the garment edge passing across the back (Amiet 1976, 21; Spycket 1981, 155).

4. seated diorite statue a*	Amiet 1976, No. 11	
seated diorite statue b	Amiet 1976, No. 12	
both from Susa	Moortgat 1969, 147–9 = a+b	

Maništušu wears closed, wrap-around garment with tasseled fringe.

5. green alabaster stele from Nasiriyya	Strommenger 1971, 118–19 Moortgat 1969, 136–7 (2 pieces)	Amiet 1976, 27 (3 pieces); Strommenger 1971, 43 (1 piece); Börker-Klähn 1982b, (22a 3 p.); Bänder 1995, XLVII (3 pieces); Hansen 2003, 204, Fig. 7.13, left

Soldier wears long, fringed shoulder sash, as in Sargon 1.

6. stele fragments from Susa	Amiet 1976, No. 26 (green alabaster)	
	Amiet 1976, No. 21 (diorite) = Amiet 1966, 160	
	Amiet 1976, No. 20 (diorite) = Amiet, 1966, 162	
		Börker-Klähn 1982b, 23

Court officials? some wearing shaggy, wrap-around garments, necklaces.

7. alabaster stele from Eridu	Hansen 2003, 205	

Woman? wears multiple-strand choker.

8. granite throne base from Susa — Amiet 1976 No. 22 = Amiet 1966, 161

   Court official? wears shaggy, wrap-around garment and necklace.

NARĀM-SÎN (see Strommenger 1998/2001 for details)

1. limestone stele* from Susa — Amiet 1976, No. 27; Moortgat 1969, 155–6; Strommenger 1971, 122–3; Börker-Klähn 1982b, 26d–j; Hansen, 2003, 196; Reimpell 1921, Abb. 35; Heuzey 1935, 56; Börker-Klähn 1982b, 26a, k; Bänder 1995, LXIII, LXIV; Fig. 7.12

   Narām-Sîn wears fringed cape or shawl knotted at the hip, also thin body shirt and kilt, elaborate necklace and 2 wristlets; see Vigneau 1935, 1:215. Descriptions and analyses of the clothing shown in this relief vary widely, as do drawings of it; Bänder 1995 corresponds most closely to my own observations of the original monument.

2. basalt stele* from Pir Husseyn — Moortgat 1969, 153; Amiet 1976, No. 21; Hansen 2003, 203; Börker-Klähn 1982b, 25; Fig. 7.2

   Narām-Sîn wears tiered, ruched, or ruffled garment with banded edges.

3. rock relief from Darband-i-Gaur — Strommenger 1963, pls. 15–18; Moortgat 1969, 157

   Narām-Sîn? wears closed, wrap-around garment, with body shirt, necklace, wristlets. Some scholars date this piece much later (Moortgat 1967, 52) but consider it a close imitation, while Strommenger adduces strong arguments for an Akkadian date.

4. diorite statue of Su'aš-takal from Susa — Moortgat 1969, 150; Strommenger 1960, pl. 13; Fig. 7.10

   Su'aš-takal, šāpiru, or head of the royal household late in the reign of Narām-Sîn, wears toga-garment. Collation of this monument in artificial and natural light strongly suggests that there are two pins where the garment is tucked in, one perhaps to hold the fabric, the other to hold a seal (for wearing of seals, Collon 2001).

5. limestone mould provenance unknown — Hansen 2000, 206–7

   Narām-Sîn? sits, bare-chested, in an intimate moment with the goddess Ištar, who puts the lead ropes of defeated peoples into his hand.

Based on the visual evidence, we may distinguish four main types of garments shown in Sargonic art.[68] First is the garment variously called "flounced," "fleecy," "tiered," "ruched," "ruffled," "striped," "plissée" or "tuyautée," covering much or all of the body.[69] This is attested throughout the third millennium. There is no agreement as to what it was made of or why it looked the way it did. Men and women wore it, and there were various styles of it. In the Akkadian period, Sargon's daughter Enheduanna and Narām-Sîn wear this garment (Figs. 7.1 and 7.2), as do gods and goddesses in cylinder seals. One may suggest that a person wearing the flounced garment in Akkadian art was to be understood as performing some act of religious significance. Enheduanna,

[68] Strommenger 1971.

[69] Strommenger 1960, 55, 88–89 (with bibliography of earlier studies); Strommenger 1971, 52–3; Rashid 1967, 28.

Fig. 7.1 (left). Flounced garment: Enheduanna, from Disk of Enheduanna, from Ur (University Museum of Archaeology and Anthropology, Philadelphia).

Fig. 7.2 (right). Flounced garment: Narām-Sîn, from Pir Husseyn Stele of Narām-Sîn (Eski Şark Museum, Istanbul).

high priestess to the moon god at Ur, is shown participating in a ritual (Fig. 7.1). Narām-Sîn tells us, on the very relief that depicts him wearing this garment, that he heaped up a burial mound over his slain enemies, so he too may wear it in connection with some ceremony (Fig. 7.2). It is conceivable that this stele was actually set up at the burial mound, to remind people of the fate in store for the king's enemies.[70] Some historians of art and of clothing consider the flounced or shaggy garment an old, conservative form of dress, which would be appropriate for religious garb, since this usually lags well behind other forms of fashion. In modern scholarship it is sometimes referred to by a term used in Classical Greek, *kaunakes*, often said to be a non-Greek word, and first associated with Babylonia by the Greco-Egyptian rhetorician Julius Pollux, active in the second half of the second century AD, who wrote on exotic textile terminology among many other subjects.[71]

The second major item of third-millennium Mesopotamian dress was the "Wickelrock," a skirt for women and a kilt for men (Fig. 7.3).[72] This was made of a rectangular piece of cloth, the exposed seam of which normally hung straight down on the right side, at least when worn by men. It was

[70] A. Westenholz 1970.

[71] L. Heuzey 1887. Heuzey attempted to reconstruct the manufacture of the garment, and wrote, "J'ai consulté, sur la question de fabrication, l'expérience technique de mon ami M. Alfred Darcel, lorsqu'il était encore directeur de la manufacture nationale des Gobelins. Il m'a montré, sur les métiers mêmes, que ces combinaisons devaient se produire assez facilement. L'ouvrier nouait pour cela, sur plusieurs lignes parallèles, les fils de la trame à ceux de la chaîne, et il les laissait retomber sur l'une des faces du tissu, en longues boucles pendantes. L'espèce de nœud qui assujettit ainsi chaque boucle entre deux fils de la chaîne s'appelle encore, dans la tradition des ouvriers des Gobelins, *nœud sarrazin*, ce qui prouve bien l'origine orientale du procédé ... Pour fabriquer le *kaunakès*, on laissait ... intactes les boucles tombantes ... Parfois, dans une autre variété du même tissu, on se contentait de couper ces longues boucles à leur extrémité, de manière à former des étages plus réguliers; souvent, aussi, on les ondulait, afin de leur donner l'aspect de la laine vivante" (Heuzey 1887, 264). For a proposed modern reconstruction, see Rashid 1967, 28.

[72] Strommenger 1971, 37–41; modern reconstructions in Rashid 1967, 5, 9–13.

Fig. 7.3 (left). Kilt: attendant, from Uruk Vase (Iraq Museum, Baghdad).

Fig. 7.4 (right). Girdle: votive figure, from Tuttub (University Museum of Archaeology and Anthropology, Philadelphia).

evidently secured in three ways: (1) by rolling the upper edge, leaving the upper corner of the rolled part of the vertical seam exposed; (2) by a girdle at the waist, which need not have been tied but mounted such that the two end pieces hung down in back, from under the roll of the skirt (Fig. 7.4); and tied by a separate cord passed several times around the waist. According to Strommenger, the cord style is not attested in art until the very end of the third millennium. If this is correct, the Sargonic skirt was secured by a simple roll or held with a separate girdle or waist piece. There are many variants of the skirt itself, which was suitable for work and physical exertion. It could be worn with an outer garment covering the upper body, such as a tunic or cape. There is no indication in art as to whether anything was worn under the kilt, a question often asked of Scotchmen. I propose below that textual evidence suggests that a sash or undergarment tied around the waist could be worn with or under the kilt or skirt.

The third important style of dress was the "Wickelgewand," or wrap-around garment, of which Strommenger argues that there were two main sub-types, closed and open. The former, as in Fig. 7.5, is characteristic of the Akkadian period. The open type, shown in Fig. 7.6, appears in art only later, beginning at Mari. Since there was an earlier closed type made of shaggy cloth, such as Sargon wears on his stele (Fig. 7.7), the closed wrap-around garment had a longer history of use than the open one.[73]

A new closed sub-type, made of smooth cloth, perhaps linen, appears in the reign of Maništušu, seen first in his statues (Figs. 7.5 and 7.9) and in the seal of Ubil-Ištar (Fig. 7.8).[74] The weft ends

[73] Strommenger 1971, 48–49. For discussion of Egyptian wrap-around garments, and estimates of the cloth required to make them, see Hallmann 2007, 17 with note 24.

[74] Strommenger 1971, 48–49; Spycket 1981, 152 describes the tasseled garment as follows: "Le tissu, probablement en fine laine tissée, est bordé d'un galon auquel sont fixées plusieurs sorts de franges, formées par l'aboutissement de la trame et de la chaîne: la frange est courte et fine dans le bas, plus longue, et faite de sept brins de laine liés à l'extrémité pour constituer un gland, sur le côté." Moortgat 1969, 49, on the other hand, considers it to be of wool.

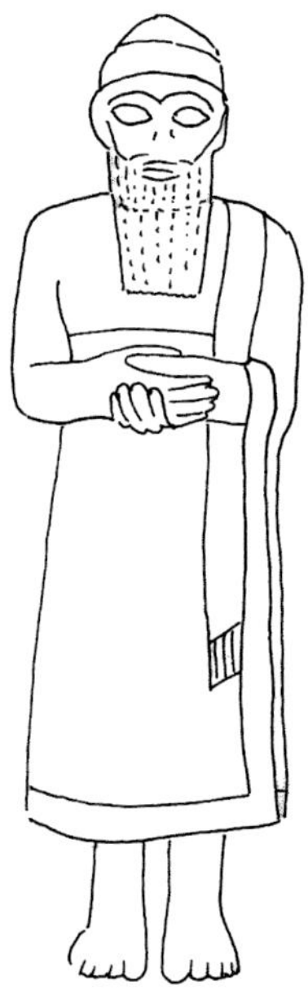

Fig. 7.5 (top left). Closed, wrap-around, tasseled garment: Maništušu(?) statue, from Assur (Staatliche Museen, Berlin). Fig. 7.6 (bottom left). Open, wrap-around garment: statue of Ištup-Ilum of Mari (Aleppo Museum). Fig. 7.7 (top right). Closed, shaggy, wrap-around garment: Sargon and attendant, from Sargon Stele, from Susa (Musée du Louvre, Paris). Fig. 7.8 (bottom right). Closed, wrap-around, tasseled, knee-length garment: Kalaki the scribe, Ubil-Ištar seal (British Museum, London).

are finished with what looks like a band or fringe, while the warp ends are finished in elaborately tasseled fringes, which hang vertically to the wearer's left. In the seal, however, the tassels hang to the wearer's right, conceivably a mistake of an artist who had little experience depicting this new style. This seal is discussed further in section 5 below. The closed garment was worn by laying a fairly long, rectangular piece of cloth from the left hip up over the left breast and shoulder, then drawing it crossways over the back to the right hip. It was secured by rolling down the upper edge such that, at the second wrapping, the outer and inner segments could be rolled together. Next,

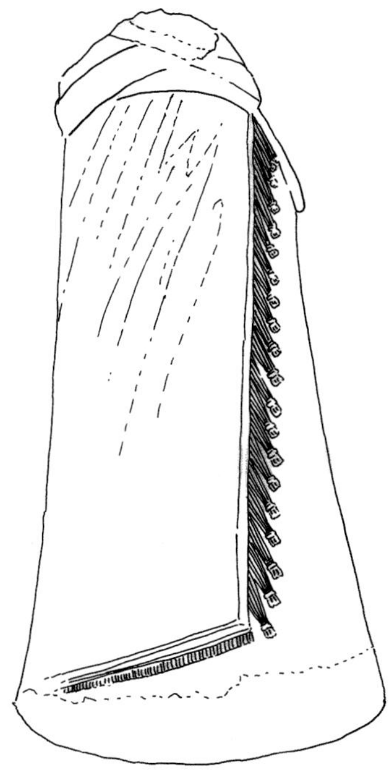

Fig. 7.9. Closed, wrap-around, tasseled garment: Maništušu statue, from Susa (Musée du Louvre, Paris).

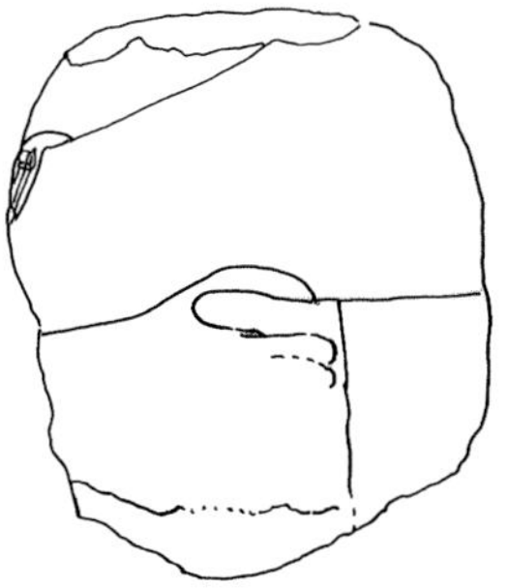

Fig. 7.10. Toga-garment: Su'aš-takal statue, from Susa (Musée du Louvre, Paris).

the cloth left hanging over the left upper front of the body was wrapped inside, leaving the right upper half of the body free. The rolled material could be visible as a horizontal band across the waist. This garment varied from knee- to full-length.[75]

The open wrap-around garment left more of the body exposed than the closed (Fig. 7.6).[76] A fringed, rectangular piece of cloth was wrapped around the lower body, then drawn over the right hip, over the back, and down in front over the left shoulder, covering the left arm. There was a more elaborate version of this with a piece that wrapped around the neck, ornamented with tassels. The open wrap-around garment appears first in the north and west, rather than in Sumer, during the post-Akkadian period, so may have originated in Anatolia or Syria.

The fourth major Sargonic clothing type was the toga-garment, which is known in art first in the reign of Narām-Sîn and remained in royal fashion until the early Old Babylonian period.[77] This was a long, wide, rectangular piece of cloth whose warp threads were extended into a fringe, like the Akkadian wrap-around. The width of the cloth covered the height of the body. The wearer drew one end over the left shoulder, so that the fringe hung vertically from shoulder to foot. The garment was then passed across the back and under the right armpit, and secured by tucking the upper right corner into the cloth covering the right breast. We may see in the Su'aš-takal statue (Fig. 7.10) that a long pin could be inserted at this point. Modern replication suggests that a piece of cloth 1.4 × 3 meters would achieve the look of the toga-garment worn by Gudea in his statuary.[78]

The toga-garment appears first in sculpture in a damaged figure of Su'aš-takal, the *šāpiru*, or head of the royal household of the deified Narām-Sîn (Fig. 7.10). The title and office were Akkadian in origin and may have designated the highest-ranking member of the civil administration. In lists of gifts to dignitaries and members of the royal family, this official is sometimes listed immediately after the king, queen, and royal family.[79] At one point, Su'aš-takal was entrusted with supervision of the work on the temple of Enlil at Nippur. A settlement named for him near Gasur (later Nuzi) may well have been his home estate.[80]

---

[75] Strommenger 1971, 48.
[76] Strommenger 1971, 49–50.
[77] Strommenger 1971, 46–47.
[78] Rashid 1967, 26; Houston & Hornblower 1920, 48–49; Strommenger 1971, 46; Spycket 1981, 156.
[79] Foster 1980, 29.
[80] Strommenger 1959, 42–43; 1960: 61; A. Westenholz 1989, 55; Frayne 1993, 164.

Fig. 7.11. Toga-garment: Lugal-ušumgal seal impression, from Telloh (Musée du Louvre, Paris).

The Akkadian toga-garment also appears on a seal impression of Lugal-ušumgal, governor of Lagaš, in which he states that he was a servant of the deified Narām-Sîn (Fig. 7.11), so he was Su'aš-takal's contemporary. He was a graduate of the scribal academy and could write beautifully the imperial Akkadian script. Lugal-ušumgal may have been related by marriage to the governor of Uruk, Dudu. Thus he was perfectly representative of the Akkadian governing elite in Sumer during the glory days of the reign of Narām-Sîn.[81] As with the toga-garment worn by Su'aš-takal, it appears that this one was similarly secured by a large pin at the point it was tucked in.

The toga-garment was a stylistic innovation, presumably among the elite of society. What the two types of full-body wrap-around garments and the toga have in common, as opposed to the skirt or kilt, according to Strommenger and Spycket, is that the wearer could not do much besides stand or move in a stately fashion, so these may be assigned to the ceremonial or court category of dress, especially since the wearers are usually shown barefoot.[82]

5. Styles of dress in Akkadian administrative documents

Administrative documents of the Akkadian period are a rich source of contemporaneous evidence for the structure and functioning of record-keeping institutions and groups, but have thus far received little consideration by modern historians of the period, who tend to rely on inscriptions, historical memory, later Akkadian literature, and the major works of art noted above. Administrative documents imply accountability, the obligation to keep records for property that is not one's own. Nearly all known Akkadian administrative archives were found at sites out of their original contexts, or were dug up by looters and sold on the antiquities trade, so they must be regrouped today into dossiers, files, and archives. The most significant of these belong to the archives of important regional administrative centers, such as Adab and the Northern Palace at Ešnunna, or of prominent administrators, such as Lugal-ušumgal, governor of Lagaš, and Mesag, governor of Umma. There are also archives of local administrative activities, such as those from Gasur (later Nuzi), or from the rural estate of a certain Mesag, who may be identical with the governor of that name. Most of these documents date to the reign of Narām-Sîn and Šarkališarri.[83]

During the Akkadian period, the practice of drawing up witnessed, written contracts between private parties to a transaction, known already in the later Early Dynastic period, became normative in both Sumer and Akkad.[84] Family archives, sometimes extending over several generations, attest to the spread of writing and record-keeping to private business affairs, especially in instances

[81] Delaporte 1920, T. 105 (Planche 9, 3–4); Strommenger 1959, 45; A. Westenholz 1975a, 435; Bauer 1987/90.

[82] Strommenger 1971, 48; Spycket 1981, 156 calls the toga-garment "engonçant."

[83] Survey in Foster 1982c, 1986.

[84] Neumann 2003; Wilcke 2007.

where one person was managing resources in which multiple family members or investors had interests.[85]

Thanks to these documents, therefore, the historian of clothing may identify possible references to specific items of clothing, pinpoint changes in clothing styles, and mark occasions or circumstances for their use. For the first major Akkadian type of garment identified above, the flounced or tiered garment (Figs. 7.1 and 7.2), one expects a word preferably attested throughout the third millennium, referring to garments worn by both men and women, associated with religious ritual, and of special value. The evidence for religious garments is too rich and varied to make a definitive proposal. One attractive identification, however, is the garment called *pala*, with a "kingly" or "priestly" variety. When Elamites made a raid and took booty in pre-Sargonic Lagaš, for example, five of these items were recovered in a reprisal attack and deposited in a local sanctuary for safekeeping, so they were presumably of special value. In the Sargonic period, a *pala* was provided with a protective garment bag or case.[86]

For the skirt or kilt (Fig. 7.3), the linguistic situation seems better. One expects a word with a long history of usage in documents about clothes in different social contexts, referring to both men and women. One reasonable candidate is the Sumerian ibbadu (íb-(ba)-dù, perhaps also ib-ba-ru), meaning "worn at the middle/waist." An advantage of this word is that it occurs from the Fara through the Ur III periods; a disadvantage is that it is less common than one would expect if it referred to a common item of apparel. What may be the same or a related garment, called simply ib "middle/waist," has likewise a long history of usage, and may be a kind of skirt or girdle, but of leather, wool, even metal.[87] According to late Early Dynastic documents, the ibbadu was worn by both men and women, in different social contexts. In the Akkadian period, one such garment weighed about 1.5 kilograms, heavy enough to be a kilt rather than a belt, as is sometimes proposed.[88]

The skirt or kilt was part of the clothing worn for religious festivals. A woman of high status at late pre-Sargonic Lagaš, first-ranked of several dressing for the festival of the goddess Ba'u, one of the most important in the cultic calendar, was issued seven garments and a piece of jewelry: two ibbadu, that is, according to this analysis, the basic skirt; two niglal "things that tie," one kudda "cut-off," one gada-bar-dul_5 "linen outer wrap," one iblal "waist tie," and one dub-šir "pendant."[89] No other women in the same group received the pendant, so it was presumably a marker of special status, as is discussed below, section 6 Suit No. 1. Enheduanna apparently wears a large pendant on her necklace, perhaps a moon (Fig. 7.1). Other women who were given garments for the festival of Ba'u received one called salla "thin" or "narrow," instead of the "cut-off," apparently an upper-body garment, perhaps of the sub-type in which one breast was exposed.[90]

A tie, called niglal (níg-lál), "thing that ties," came with each skirt given to the head woman. The niglal is a well-attested item, down to the Old Babylonian period, when it was worn by soldiers, and could be fairly heavy, though in other contexts it can weigh only a few grams.

[85] A. Westenholz 1989, 59–86.

[86] Waetzoldt 1980/83, 26–28; Kienast & Volk 1995, 25–29; (ITT 2, 4522); for the bag, see p. 139.

[87] Calmeyer 1957/71, 721–722; Waetzoldt 1980/83, 23.

[88] Gelb, Steinkeller & Whiting 1991, 294 ("belt", "loin band"); A. Westenholz 1975b, 60 (= ECTJ 108).

[89] Bauer 1972, 471–474 (= No. 167).

[90] This garment is attested at Sargonic Lagaš as well (ITT 1, 1096: bar sal-la).

Therefore it seems to have been outer, upper-body wear, sometimes of a protective nature.[91] For the nigla l, therefore, the fixed girdle shown in third-millennium art is a tempting identification. It encircles the waist, with its ends hanging down in back, as in Figure 7.4.

There is no evidence for what the "cut-off" and the "linen outer wrap" might be, but perhaps they are a dress and shawl or shoulder garment, worn by women on ceremonial occasions, attested in art from Uruk times down to the second millennium, of which there were many variations.[92]

As for the iblal, or "waist tie," by etymology it could be a sash or belt worn with the skirt, although Strommenger argues that belts were not worn with skirts until after the Akkadian period. Some historians, however, consider the waistband depicted on the statue from Aššur (Fig. 7.5), for example, to be a separate piece of cloth, rather than rolls of the closed wrap-around garment, so that too could be a "waist tie," like a modern cummerbund.[93] Another possibility for the iblal, if one accepts that separate, outer belts were only post-Sargonic, would be an undergarment, wrapped around the waist, worn under the skirt. When the head woman was given her ceremonial clothes for the festival of Ba'u, however, she received only one iblal, but two skirts and two "things that tie," so an outer sash or belt may, after all, be a preferable identification.[94] In any case, the iblal could be made of wool, linen, or leather.

Comparing these garments to male costume, we find that the ibbadu, or kilt, and bar were worn by men in the same period and place. Thus, the inventory of personal belongings in the estate of the queen's deceased brother at Lagaš, who was a contemporary of the women dressing for the festival, mentions three garments, an ibbadu, a bar, and a bardul, or outer cloak. One suspects that these garments were for his interment.[95] The Sumerian signs bar-dùl, or "outside covering," were later used to write the Akkadian word *kusītu* "cloak," but in Early Dynastic documents it seems best to assume that this was a native Sumerian piece of clothing with a native Sumerian word for it. Waetzoldt shows that bardul was a full-body garment of linen and could be heavy, weighing anywhere from 1.5 to 4 kilograms (in the Akkadian period nearer 1.5); its manufacture was time-consuming, and it could be up to four meters long.[96]

As for the smooth, tasseled, Akkadian closed, wrap-around garment, first seen in the Maništušu statues (Figs. 7.5 and 7.9) and the Ubil-Ištar seal (Fig. 7.8), one expects a word first attested in early Old Akkadian. An obvious candidate presents itself, the garment called šusega (šu sè-ga), according to etymology "set on" or "put over" the hand or arm, first mentioned in the Maništušu Obelisk so precisely contemporaneous with the statues. The Obelisk records Maništušu's purchase of large tracts of land from a group of families. The payment includes numerous gifts of valuable commodities and pieces of adornment that Maništušu distributes among the sellers and their male relatives, but these garments were given to only a few, who were leading members of the selling families.[97] In another major land sale of the Sargonic period, known as the Sippar Stone,

---

91 Akkadian-period references in Gelb 1959, 191; comparative material cited by Waetzoldt 1980/83, 25.

92 Strommenger 1971, 50–53.

93 Spycket 1981, 155.

94 Waetzoldt 1980/83, 23.

95 Selz 1993, 628–633 (BIN 8, 388).

96 A. Westenholz 1975b, 59 (= ECTJ 104); Waetzoldt 1980/83, 21.

97 For the word in the Maništušu Obelisk, see Gelb, Steinkeller & Whiting 1991, 295 ("a type of cloth"). It occurs also in A. Westenholz 1975b, 62 (= ECTJ 112), an administrative document older than Narām-Sîn's reign, so a good fit for Maništušu.

the same garment is mentioned, but spelled differently (šu zag-ga), which suggests that there was no standard Sumerian word sign for this garment. Both the Maništušu Obelisk and the Sippar Stone were written in Akkad in the Akkadian language, so the Sumerian signs may well have been intended to convey an Akkadian word (perhaps resembling the *sāgu* of Middle Assyrian sources?) by giving it an imaginary but apt Sumerian etymology, a common scribal practice.[98] "Set on" or "put over the hand" well describes the closed, wrap-around garment, in which the left arm is concealed: "couvre l'omoplate gauche en biais, l'épaule et le haut du bras, englobant le coude."[99]

This style seems to have gone out of fashion during Narām-Sîn's reign, as it does not appear in art after that, except in a few seals of uncertain date, and the word šusega disappears at the same time. This was, then, a new, short-lived word for a new, short-lived style.[100] Furthermore, so far as I am aware, the šusega does not occur in pre-Sargonic or Sargonic documents from the time of Narām-Sîn and later, so it seems the ideal candidate for Maništušu's new-style, Akkadian, smooth garment, as it shares a comparable limited distribution in time and place.

In the seal of Ubil-Ištar, Kalaki the scribe, holding his tablet, wears a knee-length version of the smooth, tasseled Akkadian garment (Fig. 7.8); the longer form, as seen in Figure 7.9, may have been reserved for men of higher status. The man in the center of the seal image, with the bun typical of Akkadian images of royalty, is Ubil-Ištar himself, the "king's brother" (šeš lugal), whom Kalaki the scribe serves. Kalaki may be an example of the new Akkadian elite drawn from non-Akkadian circles, as his name is both unique and exotic. Of the two men depicted in the more old-fashioned shaggy or fleecy garments of the type Sargon wears in his stele (Fig. 7.7), one is a soldier carrying an axe, and the other evidently an administrator, carrying a staff of office. An archer wears a smoother, shorter shoulder garment and special shoes turned up at the toes, ending in balls. The soldier, archer, and administrator face the two central figures, Ubil-Ištar and Kalaki, eyes right and left, apparently undergoing a review or inspection, as is well attested in Sargonic administrative records. Two servants appear at the right beneath the inscription, bearing a chair and other items.[101]

Boehmer, Nagel, Strommenger date the Ubil-Ištar seal to the time of Sargon, believing that Ubil-Ištar was Sargon's brother, though Collon found this dating problematic.[102] But the designation

---

98 For etymologizing word signs, see Hallo 1973.

99 Spycket 1981, 155.

100 Dury & Lervad, this volume.

101 Strommenger 1962 No. 113 = Collon 1982 No. 141 = Collon 1987 No. 641 = Hansen 2003 No. 150; the best photo of this seal is Reade 1991, 63. The chair-bearer, or gu-za-lá, is well attested in third-millennium sources as a court official. Boehmer 1965, 128, 43 considers this a "hunting scene" and the figures to be "heroes," whereas Frankfort 1939, 140 considers this purely a "secular" scene, the central figure being "some official guided by tribesmen of the mountains and followed by a shaven scribe or priest ... two servants are carrying their master's requisites for the wild districts which he is visiting, namely a bed and some provisions." Frankfort's idea that the official was visiting mountains was presumably suggested by the pointed shoes of the archer, called, variously, "Schnabelschuhe," "tsarouchi," "à la poulaine," which some writers, including Collon, describe as mountaineer's shoes, for reasons unknown to me. Salonen 1969 Plate 4,2 reproduces the Ubil-Ištar seal, but does not discuss these shoes specifically. He considers the Victory Stele of Narām-Sîn to be the first illustration of a sandal in Mesopotamian art and suggests that the shoe or half-shoe (Akkadian *mešēnu*), first mentioned in the Akkadian period, and etymologically related to the word for sandal, may have been of northern or Anatolian origin because of the colder climate there, requiring more coverage of the foot. For inspection in the Sargonic period, see Foster 1986, 49–50.

102 Boehmer 1965, 128 (= No. 1686); Nagel & Strommenger 1968, 156; Collon 1982, 26.

"king's brother" occurs elsewhere only in the Maništušu Obelisk, referring to a certain Yabarim, the father of one of the Akkadian witnesses. There is thus good reason to date Ubil-Ištar and his seal to Maništušu's reign, just when the tasseled garment appeared in sculpture. In any case, the "king's brother" likely meant "the king's kinsman," not necessarily brother.[103] In the Obelisk, the witness-son of Yabarim was Maništušu's contemporary. He bears a name formed from a rhetorical question: Ali-ahi, "Where is my brother?" (meaning that he has none), comparable to Maništušu's own rhetorical question name, "Who is with him?" (meaning no one). Kinship was evidently an issue of royal importance at the time, as Maništušu was later said to been the older brother of Rimuš, so was somehow passed over in the immediate succession to Sargon.[104]

In sum, the smooth tasseled garment was a fashion that appeared among the ruling elite in the reign of Maništušu, worn by the king and other members of the elite. Whereas the basic form of the garment has Mesopotamian precedents, its special features do not, so could well be of foreign origin. The central political event recorded in Maništušu's inscriptions, as noted above, was a spectacular victory in Iran over an alliance of Elam and Marhaši, and the exotic appearance of the enemy troops was carefully conveyed in the fragments of art that may commemorate this or other victories in the region, such as the limestone statue from Susa often attributed to Maništušu.[105] Sargon had campaigned extensively in Iran, and the young Maništušu may well have served there. Thus one may suggest that the new court dress might have been inspired by fabrics and styles observed or captured during Maništušu's campaigns in Iran. In fact, in the Victory Stele of Narām-Sîn, the defeated ruler of the Lullubi, a people of the Zagros, wears a comparable tasseled garment, as he stands facing Narām-Sîn himself. The tassels are difficult to see in photographs, but collation of the stele revealed traces of three of them towards the bottom of his garment. Indeed, Calmeyer has suggested that tassels were particularly characteristic of dress in southwestern Iran.[106] Maništušu also commemorated the mining of hard black stone somewhere along the Gulf, the very diorite imported for many of his statues, so stone and style may have had the same origin.

Besides the šusega-garment, the Akkadians of the Sargonic period knew a garment they called *kusītu*, but in Akkad and at Sippar, they spelled it out syllabically, rather than use Sumerian signs, so they did not recognize a Sumerian equivalent. This Akkadian practice thereafter disappeared, and the Akkadian *kusītu* was presumably written using the Sumerian signs, bar-dùl, as the word does not occur as such in Classical Sargonic administrative documents. The Old Assyrians, on the other hand, continued to write *kusītu* syllabically. Reimpell drew attention to a later version of the Akkadian wrap-around garment, depicted on Old Assyrian seals from Kanesh, calling it the "Kappadokischer Mantel," which is more likely to be the northern Mesopotamian cloak, or *kusītu*.[107] All of this goes to suggest that there were very definite northern styles of the wrap-around, outer

[103] Gelb, Steinkeller & Whiting 1991, 123 (= MO A xi 1).

[104] The title šeš ensi "brother of the governor" also appears in an inscription of Rimuš, prior to the time of Narām-Sîn: Frayne 1993, 44.

[105] Bänder 1995, LV Nr. 14. Other Akkadian notables are depicted wearing tasseled garments on their seals. Šu-ilišu, the Meluhha-interpreter (Collon 1987 No. 637 = Boehmer 1965, No. 1299, fig. 557), is depicted wearing an ankle-length tasseled wrap-around garment (Meluhha is usually thought to be the Indus Valley), as is a certain Ikrub-Il (Boehmer 1965 No. 1060, fig. 560; compare also Boehmer 1965 No. 1301).

[106] Calmeyer 1957/71, 692.

[107] Reimpell 1921, 33–34.

cloak, but either the style became generalized throughout Mesopotamia, or the administrative scribes soon chose to write all varieties with the same Sumerian word signs as they had used to write the Sumerian-style outer cloak, another instance of simplifying terminology over time.

Finally, a word is needed for the toga-garment (Figs. 7.10 and 7.11), first known in art from Narām-Sîn's reign, though he himself is not shown wearing it. Accordingly, this should be a word that appears first in sources from the time of Narām-Sîn. Here a candidate presents itself immediately: šagida (šà-gi-da$_5$), a word conceivably of Akkadian origin (*šaqītu*).[108] This too denotes a garment worn "over the heart," or so the scribes spelled it, using the Sumerian word for heart, even though that too may have been a fanciful etymology.

If we look for its social context in documents, we find that a well-dressed notable from Umma, a certain Ur-Sîn, who probably lived in the reign of Narām-Sîn, left one šagida among his personal effects when he died.[109] By my analysis, this would be the new-style Akkadian toga-garment. He also left a bardul, by my analysis, the more traditional Sumerian or Akkadian wrap-around cloak, what the Akkadians called a *kusītu*. In fact, Ur-Sîn was a well-fitted man, as his estate included eighteen other assorted pieces of cloth or clothing, plus twenty bolts of woolen cloth and thirty bolts of linen cloth, far more than the amount normally accounted for in inventories of this kind. Ur-Sîn also owned four mirrors, enough for the vainest of men.

The word šagida seems restricted to Classical Sargonic documents, from Lagaš, Umma, Nippur, Adab, and in Akkad, all from the time of Narām-Sîn, and in small numbers, usually only one, and then the word disappears from the sources. Here we may have a situation similar to that with *kusītu*. The toga, a new type of garment that appeared in the time of Narām-Sîn, was first designated with a specialized, conceivably Akkadian, word, but the word disappeared after a generation or two, after the style became generalized, and some other word was used for it, perhaps bardul, which meant both the *kusītu* and the Sumerian cloak. Who could guess, after a few millennia, that "le smoking" in no way resembles in design, material, or use the English "smoking jacket," or that "hose" in English of the 17th and 19th centuries are quite different garments?

It is important to note that the šagida, or some form of it, could evidently be worn by upper-class women as well, in Akkad during the reign of Narām-Sîn. At Ešnunna, for example, two of these garments were part of a lavish bridal trousseau that included silver, livestock, wool, food, bedding(?), and other commodities, handed over in the presence of several witnesses.[110] There is another, even more lavish list of commodities from Tuttub, which could well be another bride gift. It includes two šagida-garments, along with such items as a gold ring, a wagon and draught beast, livestock, and other clothing.[111] As to whether or not women wore togas in Akkad in the same style as those worn by men in Sumer, one should always allow for a regional difference in usage, as with the English "jumper," which in the United States is a one-piece, sleeveless garment worn by girls and women over a blouse or other top, but in England is a pull-over, or upper body garment, worn by both sexes, called "sweater" in the USA.

---

[108] Steinkeller & Postgate 1992, 48, with, however, evidence that *šaqītum* could be a pre-Sargonic word. The tablet and related tablets Pomponio, Visicato & Westenholz 2006 No. 148, dated by the editors "Medio-Sargonico," in which the word occurs, on epigraphic grounds could well be contemporaneous with the tablet cited in note 109, which belongs to a group I would date to the first decades of the reign of Narām-Sîn (Foster 1982a, 53).

[109] Steinkeller & Postgate 1992, 47 (= TLATIM No. 21).

[110] Milano 1987 (= MAD 1 129).

[111] Sommerfeld 1999, 104–105 No. 47 (= MAD 1 266).

As for the origins of the toga-garment, perhaps it was a novelty that spread among the Sumero-Akkadian elite in Sumer as a result of one of Narām-Sîn's conquests, enjoying a vogue at court and making its way into the wardrobes of local notables, among them Ur-Sîn at Umma and the handsomely provided for bride at Ešnunna. Was it an import from North Syria, perhaps a trend-setting region for Akkadian culture?[112] Syria may also have been the home of the open wrap-around garment (Fig. 7.6). In any case, in the reign of Narām-Sîn, the toga-garment was a new style of dress, which in a few generations would become the standard dress of the rulers of Lagaš, such as Gudea, and the later kings of Ur. Clearly, it was seen as marking a person of highest status.

One other significant sartorial innovation of this period, the Narām-Sîn shawl (Fig. 7.12), may belong in the realm of military accoutrements, for combat or parades. This was hung over the left shoulder, then knotted rakishly at the right hip. It may be a dress form of the "shoulder-hung garment" of the pre-Sargonic and early Sargonic military (Fig. 7.13), the battle form of which protected the body in combat while leaving the arms free.[113] Thus the safest course is to identify it with the word gu'e (gú-è) ("that which comes down at the neck"), a word attested first in Sargonic documents, though forms of the shoulder-hung garment appear earlier in art.[114] There were as many styles of the gu'e as there were styles of military uniforms over time and place. Elaborate versions occur in the monuments of the Sargonophile early Old Babylonian rulers in Akkad, Assyria, and Upper Mesopotamia, such as the kings of Ešnunna, one of whom was named Narām-Sîn. The triumphant king in the Mardin relief, sometimes thought to be Šamšī-Adad I, wears one of these as he smites his enemies (Fig. 7.14), though the knot in the garment has apparently been misunderstood as a decorative patch. The inscription informs us that he installed governors in the cities he conquered, just as Sargon himself tells us he did in vanquished Sumer.[115] Therefore this garment had rich imperial Akkadian associations centuries later, in the same region. Historians of clothing doubt that the Narām-Sîn shawl was ever practical battle gear, but suggest it was shown in combat or victory scenes to demonstrate divinely sanctioned triumph in warfare.[116]

One may also consider the possibility that an undershirt might be worn beneath cloaks, mantles, or festival garments. Strommenger mentions this only in passing, as it is seldom visible in art.[117] The best candidate for this is the garment called šagadu (šà-ga-dù), a Sumerian word meaning, appropriately, "worn at the heart/abdomen," as opposed to the ibbadu, "worn at the hips/loins/waist."[118] For this garment, the earliest attestations are late Early Dynastic sources. Šagadu, borrowed into Akkadian as *šakattu*, may even have been a general term for any sort of garment worn over the abdomen. Later šagadu's could be made of hide or linen. The style and the word seem gradually to go out of general usage after the Old Babylonian period but it

[112] Börker-Klähn 1982a, 76; Bänder 1995 both stressed Egyptian parallels to Akkadian triumphal art, for which Syria or Palestine could be likely places of contact. For bibliography on Egyptian clothing, see Hallmann 2007.

[113] Strommenger 1971, 42–44.

[114] Waetzoldt 1980/83, 22.

[115] Strommenger 1971, 43 Figure 18 (considers to be Narām-Sîn of Ešnunna); Grayson 1987, 63–65 (considers to be Šamšī-Adad I). See further Moortgat 1969, 72, 84.

[116] Nagel 1957/58, 98–99.

[117] Strommenger 1971, 43.

[118] Waetzoldt 1980/83, 22 (though he considers it worn about the loins); Steinkeller 1982b, 362: "loin band."

Fig. 7.12 (top left). Narām-Sîn shawl: Narām-Sîn, from Victory Stele of Narām-Sîn, from Susa (Musée du Louvre, Paris). Fig. 7.13 (top right). Military uniforms: left: Soldier, from Nasiriyya Stele (Iraq Museum, Baghdad), right: Rimuŝ(?), from Victory Stele, from Telloh (Musée du Louvre, Paris). Fig. 7.14 (right). Narām-Sîn shawl: Old Babylonian king, stele fragment, from Mardin or Sinjar (Musée du Louvre, Paris).

was very much at home in the Amorite milieu. At Mari it was mostly made of linen, and Durand considers it a kind of belt.[119] Waetzoldt notes that the šagadu was fairly light, in Ur III weighing from 125 to 150 grams.[120]

It seems reasonable, on the etymological basis alone, to move the šagadu away from the waist or loins, where most Assyriologists locate it, to become a short-sleeved undershirt or mid-body wrapping (Strommenger and Bänder's "Oberkörperkleidung" or Waetzoldt's "Rock, auch Untergewand"). A stele fragment found at Susa depicts a figure with a large bun in the manner of Akkadian royalty, though he is sometimes interpreted as a captured enemy, wearing a fleecy outer garment and an undershirt. It extends above the fleecy garment to the neck and

[119] Durand 1983, 458 (of linen).
[120] Waetzoldt 1980/83, 22.

is knotted behind the neck.[121] Narām-Sîn wears a cap-sleeved undershirt in his stele (Fig. 7.12), its apparently light material hugging his body. Bänder suggests that it was nearly transparent, so it was presumably of very fine linen.[122]

Thinness of textiles was, perhaps, another Akkadian innovation, beginning with the Akkadian wrap-around garment of the time of Maništušu. Moortgat, writing of the Aššur statue of Maništušu (Fig. 7.5) considers that the wrap-around too was very light: "the limbs of the body look as if they were showing through"; similarly he finds Narām-Sîn's garment in the Pir-Husseyn stele (Fig. 7.2) "thin and clinging" so the curvature of his chest "can be noticed through it."[123] Spycket too was struck by the thinness of the garment on the Aššur statue ("tissu mince"), through which features of the back were visible (Fig. 7.5).[124] Thus production of very thin textiles that revealed body contours, or at least their representation in art, may be deemed an Akkadian innovation.

According to Hansen, the Rosen mold shows Narām-Sîn not wearing a revealing shirt but bare-chested, the only instance of royal nakedness in Mesopotamian art; furthermore, he is seated, rather than standing respectfully in the presence of the goddess Ištar, for which again there is no parallel in Mesopotamian art.[125] The scene is no doubt calculated to depict Narām-Sîn as the goddess's lover, just as he suggests in his inscriptions. When Narām-Sîn stripped down for battle, as shown in his Victory Stele, he removed his outer mantle, leaving on his šagadu-undershirt and ibbadu-kilt, then donned his gu'e-shawl and knotted it. But alone with the goddess Ištar, as shown on the Rosen mould, he took off his undershirt as well.

Another aspect of the šagadu may also be mentioned. Strommenger has drawn attention to the fact that the toga-garment should reveal part of the wearer's leg in the open triangular space where the garment crosses itself in front.[126] If the šagadu was sometimes tunic-length, it would cover the leg and this could explain the absence of this anatomical detail in art in which the body is otherwise depicted attentively. Another explanation for the absence of visible leg in a toga-clad figure might be purely practical: the artist ignored the leg to have enough stone to hold up the figure, a problem solved less adroitly by the mass of stone left around the feet in the Maništušu statues.[127]

6. Suits of clothes in Akkadian documents

Akkadian administrative documents sometimes list suits of clothes given to people for some official purpose. One may consider first a group of related documents from the early Sargonic period, that is, from before the time of Narām-Sîn; I would date them to the reign of Rimuš.[128] These record the ceremonial presentation of suits of clothes to men in positions of responsibility or honor.

[121] Amiet 1976, Pl. 22.
[122] Bänder 1995 LXIV.
[123] Moortgat 1969, 49–50, 51.
[124] Spycket 1981, 155.
[125] Hansen 2002, 2003, 197.
[126] Strommenger 1971, 47.
[127] Amiet 1976, 20.
[128] Foster 1982a, 38–39, 46–47.

Suit No. 1 (RBC 3015)

1. 1 ur$_5$ ùl-la x [ ]
2. ki-lá-bi 1/3–ša ma-[na]
3. 1 túg šà-ga-dù
4. 1 túg níg-lá
5. 1 túg lamahuš
6. 1 túg <erasure?> dùl-x [ ]
7. ugula Uš-àm
8. *Ì-lu-ra-bí* é-mah ensí Dilmun-dar-ra-na^{ki}-ka-ke$_4$
9. é-mah ensí-ra mu-na-DU-a e-ba
10. 1 túg šà-ga-dù
11. [1 túg] lamahuš
12. [ ]-*bí-ì-lum* gudá
    dingir-kala-ke$_4$ e-ba
13. uš-bar-àm
14. [*Ì*]-*lí-a-ḫi*
15. maškim-bi
16. ensí-ke$_4$ A-ga-dè^{ki}-a
17. Pù-ma-lí E.UR-a
18. Umma^{ki}-šè du-ni
19. Dilmun-dar-ra-na^{ki} e-ne-ba
20. túg Ur-lú dub-sar-da gi-gi-[da$^?$] e-da-g[ál]
    6 mu [ ]

"One medallion of [silver?], weighing 1/3 mina, one undershirt, 1 cloth tie-girdle, one red festival-garment, one head covering(?)."
(*Occasion*): "Under the supervision of Uš, Ilu-rabi, in the governor's headquarters at Dilmun-darra, having brought it to the governor, gave it to him in the governor's headquarters."

Suit No. 2 (RBC 3015)
"1 undershirt, 1 red festival-garment,"
(*Occasion*): "Dingir-kalag gave to [x] the priest. It is woven piece work(?). Ili-ahi oversaw the transaction."
(*Occasion of both transactions*): "When the governor was in Agade and ... came to Umma, he gave these to them in Dilmun-darra. The textiles were with Ur-lu the scribe, as he returned(?)."

Suit No. 3 (BIN 8 331)
"1 undershirt, 1 cloth tie-girdle, 1 red festival-garment, one head covering(?)."
(*Occasion*): "To Ib'um, the man of Lugal-ra. The governor gave them to him in the headquarters of Umma. Il-uda the courier was supervisor of this transaction."

Suit No. 4 (Nik DV II 49)
"1 undershirt, one tie-girdle, 1 red festival-garment."
(*Occasion*): "To Išda'um, foreman to the cupbearers. When Lugal-ra came from the south, the

governor gave them to him in Zabala when he sent the work troops to Sabum. Di-Utu the courier was supervisor of this transaction."

Suits No. 5–9 (ECTJ 108)
"1 red festival-garment? to Urununuz-zi, land registrar. 1 red festival-garment to [ ], of Umma. (both received by third parties). 1 NI.TUG, 1 aba: PN. 1 ibbadu [ ]; 1 ibbadu: PN."
(*Occasion*): "Being garments Lugal-ra brought from Umma."

The record for Suit No. 1 tells more about the transaction than is normal in Sargonic accounts, so we must assume that it was exceptionally important to the governor of Umma, perhaps the presentation of a special suit of clothes for his investiture. The precious metal object of adornment listed first seems too heavy, at about 250 grams, for a ring, as was given to visiting dignitaries under the Third Dynasty of Ur.[129] One may suppose it was silver and circular on the basis of the object called "silver ulla-ring" (ha-ar ùl-la kù-babbar) in another, later Sargonic administrative text (weight of that object not given)[130]. Therefore I suggest it was a pendant or medallion that high officials wore around the neck, following royal example. Sargon seems to initiate the practice of kings wearing necklets or necklaces, as Sumerian rulers are not shown with them. The figure from Aššur (Fig. 7.5) wears heavy beads with a large counterpoise in back, but if there was anything hanging in front, it is concealed by the beard. Narām-Sîn wears a complicated necklace in his Victory Stele, with a large central bead (Fig. 7.12), and Old Babylonian kings, such as the victorious ruler on the Mardin stele (Fig. 7.14), wear circular medallions hanging from a necklace. Inventories of necklaces from Narām-Sîn's time mention gold, silver, and stones, but not medallions.[131]

The festival-garment, presented here and with the other suits, was red outer wear, presumably made from red wool. It was fairly light weight in the Akkadian period, weighing about 1.5 kilograms, even though it could use 3.5 meters of fabric, both in length and width.[132] There seems also to have been some kind of headgear, of which there are numerous types in Sargonic art, or a parasol, such as is carried behind Sargon (Fig. 7.7).[133] Note that no shoes are provided for any of these men, although in later times sandals were standard gifts from the kings of Ur to visiting dignitaries. Perhaps at Sargonic court events one went barefoot, for the series of Maništušu statues always show a barefoot man wearing the full-length, ceremonial body garment. Narām-Sîn is the first Akkadian king depicted wearing footgear.[134] Whereas we might expect a toga-garment to be part of this presentation, we recall that this article would not appear until later in the Sargonic period, during the reign of Narām-Sîn; as suggested above, the record of Suit 1 is earlier.

Suits Nos. 2, 3, and 4 seem to have been on the same general order, but are less elaborate. Since

[129] Michalowski 1978.
[130] (Charpin & Durand 1981, no. 29; also MCS 9 260).
[131] Limet 1972, 5, 7. Round silver pendants were found in a late(?) Akkadian-period horde at Brak (Maxwell-Hyslop 1971, 29–30), but the excavation report does not give their weights to compare with the textual record (Mallowan 1947, 176).
[132] Waetzoldt 1980/83, 22; (ITT 5, 9996). For red wool, a particularly fine and valuable variety, see Waetzoldt 1972, 51.
[133] Boehmer 1980/83, 204; Bänder 1995, LXVII, LXVIII (tabulations of hair styles and modes of securing them, see in general Börker-Klähn 1972/75).
[134] Lambert 1973, 175.

there is no tie-girdle in Suit No. 2, this may not have been an essential part of the outfit. The term ušbar in Suit No. 2, translated above as "piece work?," often refers in Ur III sources to inferior, rapidly produced weaving, but this seems inappropriate for a garment given ceremonially, so I have chosen another word ušbar that, in Sargonic texts, refers to valuable metal objects produced under supervision, though its precise significance is not clear. In the document published in the Appendix, only a few of these are sent to the capital, in contrast to many toga- and festival-garments.[135] Lugal-ra, mentioned in connection with Suits Nos. 3 and 4, was apparently a dignitary of some importance, acting at both Umma and Nippur, but I cannot document him further. Suits Nos. 5–9 continue in descending order of simplicity, listing mostly festival-garments and kilts.

7. Gifts of clothing to the royal family

Administrative records also document gifts of clothing from people of lower rank to higher. When the royal family (I believe Šarkališarri and his court) came to Sumer, perhaps for his coronation at Nippur, they were presented with fine clothing and many other gifts by a man (Mesag?) possibly hoping to receive or retain the governorship of Umma:

> [ ] kilograms of gold; 20 silver cups weighing x+2 kilograms; 1 bronze platter(?), 40 bronze cups(?), 2 garment bags (*našpārum*); 1 red festival-garment; 1 undershirt; 1 toga-garment; 10 shawls; 2 wagons; [ ] draught animals ... oblation to the king; 500 grams of gold; 1 calf, 1 lamb ...; 10 baskets of ...; oblation to the queen; 1 kilogram of silver ... ; 1 red ... festival-garment ... oblation to [the king's son].[136]

First among the textiles are two items called *našpārum*, literally, "sending containers," here translated "garment bags." It seems very likely that these were to keep fine clothes clean when they were transported, as they were also part of the Akkadian bride gifts mentioned above (see note 110), and one came with a tiered garment (see note 86). Next come a red festival-garment, the light undershirt to go with it, a toga-garment, and ten shoulder-garments or Narām-Sîn shawls. Here the undershirt was the basic foundation, over which could be worn a choice of garments, depending on the occasion, the festival-garment, the shoulder shawl, or the toga.

The queen receives no textiles in this list of gifts, only food. We may thus suggest that it was inappropriate to give clothes to Akkadian royal women in the way one did to male dignitaries or royalty. The crown prince receives no food or objects, but he does get a special type of red festival-garment, perhaps a child's version of it.

* * *

In late third millennium Akkad and northern Mesopotamia, a wrap-around garment was worn, different styles of which may have had different words in Akkadian, but which were eventually subsumed under the Sumerian signs bardul "body covering garment." In the reign of Sargon, this was shaggy in appearance, as in earlier periods, but under Maništušu, probably as a result of his conquests in Iran, a new version appeared that was thin, smooth, and elaborately fringed. This garment was called šusega "put over the arm" and was worn by royalty and people of high status. In the time of Narām-Sîn, the toga (šagida), appeared, likewise perhaps a fashion inspired

[135] Waetzoldt 1980/83, 22; Foster 1982a, 132.

[136] Foster 1982a, 133–134.

by foreign campaigns. It spread among male and female members of the elite in Akkad and Sumer, but in successive generations became the garment worn, perhaps exclusively, by royalty in Sumer, such as Gudea of Lagaš and the kings of Ur, and the Old Akkadian word for it soon disappeared.

I have also suggested terms for the short skirt or kilt worn by men and women (ibbadu), for the tie-girdle sometimes worn over it (niglal), and for an undergarment tied around the waist (iblal). In addition, I have proposed that some outer body clothing, such as the red festival-garment, shawl, and body-wrapping garment, might be worn over a sleeveless undershirt (šagadu). A specifically military item, a long, shoulder-hung garment in various styles (gu'e), may well underlie the Narām-Sîn shawl, seen first on his Victory Stele, which, in Akkad itself, was the garment later kings were depicted wearing on their own victory monuments.

The Akkadian period, with its closely dated works of art in which clothing is important, as well as its rich administrative archives dealing with textiles and clothes, offers therefore a particularly rewarding opportunity to correlate visual and written evidence for continuity and change in fashion during this dynamic period of Mesopotamian history.

Appendix

Fine Garments to the Capital

The tablet published herewith, NBC 11441 (5.7 × 4.2 cm), written in a fine Akkadian scribal hand (Sommerfeld 1999, 14, Duktus II or III), lists commodities shipped to Agade by a royal scribe, including an enormous number of toga- and festival-garments, an example of the wealth extracted from local producers and sent to the royal capital. This was a shipment fit for a king.

A cart is the only object in the bill of lading that is not a textile. Hand carts were vehicles drawn by a team of mules to convey Akkadian notables and were no doubt a sign of rank and prestige. Maništušu gave a few of these to important sellers in the Obelisk of Maništušu (Gelb, Steinkeller, & Whiting 1991, 116–140) and one is listed among the valuable goods in the document from Tuttub cited above, note 111. An ambitious underling pressed his Akkadian patron in a letter to give him a two-wheeled cart, perhaps for the same purpose (Kienast & Volk 1995, 48–50).

The find spot of this tablet is unknown. I know of no other reference to this scribe. Old Akkadian names with *apsu* are very rare and can be associated with the prominence of Ea in the Sargonic onomasticon of Babylonia (A. Westenholz 1999, 79). The "royal scribe" is likewise not otherwise attested (Visicato 2000).

1.	2 túg *na-áš-pár*-[ ]	2 garment bags
2.	120 túg lamahuš	120 festival garments
3.	120 túg šà-ga-dù	120 undershirts
4.	120 túg šà-gi-da_5	120 toga-garments
5.	7 túg uš-bar	7 piece-work(?) garments
6.	2 ^{giš}gigír níg-šu	2 hand carts:
7.	*Be-lí*-ABZU	Beli-apsu
8.	dub-sar lugal-ke_4	the royal scribe
9.	A-ga-dè^{ki}-šè	loaded on the boat
10.	má-a ba-gar	for Agade.

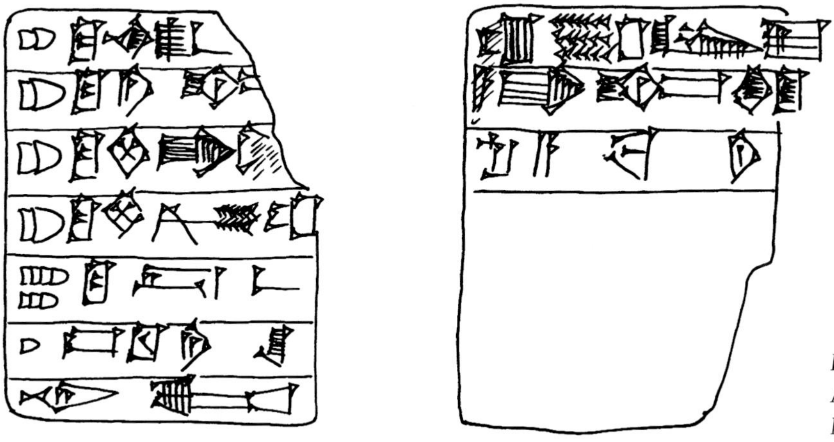

Fig. 7.15. List of garments sent to Agade (Yale Babylonian Collection, New Haven).

Afterword

A document indicating industrial production of garments in the Akkadian period has been published by M. Maiocchi, *Classical Sargonic Tablets Chiefly from Adab in the Cornell University Collections, Cornell University Studies in Assyriology* 13 (2009), no. 147. This records a delivery of over 500 garments "in the third installment," including 103 toga-garments, subscribed túg šà sila-ka-kam. I interpret this enigmatic phrase to mean garments through "(regular supply) channels" (sila-a gál-la); compare (MVN 3 125), an Ur III contract in which a man undertakes to supply lard for rations if it is not available through regular supply channels (sila-ta è-è ... tukumbi nu-ù-um-ta-è). Maiocchi no. 144 is an interesting school exercise giving a fantastic balanced account for woolen garments (see writer, *Archiv Orientalni* 50 [1982], 238–241), in which, for example, two turbans weigh 8 kg each, and a huge arrears is humorously charged to a tavern singer.

Abbreviations

BIN — Babylonian Inscriptions in the Collection of J. B. Nies. New Haven.

ECTJ — Westenholz, A. (1975b) *Early Cuneiform Texts in Jena: Pre-Sargonic and Sargonic Documents from Nippur and Fara in the Hilprecht-Sammlung vorderasiatischer Altertümler, Institut für Altertumswissenschaften der Friedrich-Schiller-Universität, Jena. Det Kongelige Danske Videnskabernes Selskab, Historisk-Filosofiske Skrifter* 7, 3. Copenhagen.

ITT — F. Thureau-Dangin, H. De Genouillac & L. Delaporte, *Inventaire des tablettes de Tello, conservées au Musée Impérial Ottoman*, tomes I–V. Paris, 1910–1921.

MAD — Gelb, I. (1962) *Sargonic Texts from the Diyala Region. Materials for the Assyrian Dictionary* 1, second edition. Chicago.

MCS — *Manchester Cuneiform Studies*. Manchester, UK.

NBC — Nies Babylonian Collection. Yale University, New Haven. Museum number.

Nik DV II — M. V. Nikol'skij, *Dokumenty chozâjstvennoj otčetnosti drevnej Chaldei iz sobraniâ N. P. Lichačeva II: Epocha dinastii Agade i epocha dinastii Ura*. Drevnosti Vostočnyâ V. Moscow, 1915.

TLATIM — Steinkeller, P. & Postgate, N. (1992) *Third-Millennium Legal and Administrative Texts in the Iraq Museum, Baghdad. Mesopotamian Civilizations* 4. Winona Lake, Ind.

Bibliography

Abrahami, P. (2008) L'armée d'Akkad. In P. Abrahami & L. Battini (eds.), *Les armées du Proche-Orient ancien (IIIe-Ier mill. av. J.-C.)*, 1–22. *Actes du colloque international organisé à Lyon les 1er et 2 décembre 2006, Maison de l'Orient et de la Méditerranée.* BAR International Series 1855. Oxford.

Adams, R. Mc. (1965) *Land Behind Baghdad: A History of Settlement on the Diyala Plains*. Chicago.

Amiet, P. (1966) *Elam*. Paris.

Amiet, P. (1972) Les statues de Maništušu, roi d'Agadé. *Revue d'Assyriologie* 66, 97–109.

Amiet, P. (1976) *L'art d'Agadé au Musée du Louvre*. Paris.

Bänder, D. (1995) *Die Siegesstele des Naramsîn und ihre Stellung in Kunst- und Kulturgeschichte*. Idstein.

Bauer, J. (1972) *Altsumerische Wirtschaftstexte aus Lagasch. Studia Pohl* 9. Rome.

Bauer, J. (1987/90) Lugal-ušumgal. *Reallexikon der Assyriologie* 7, 155.

Boehmer, R. (1965) *Die Entwicklung der Glyptik während der Akkad-Zeit*. Berlin.

Boehmer, R. (1980/3) Kopfbedeckung. *Reallexikon der Assyriologie* 6, 31–33. Berlin.

Börker-Klähn, J. (1972/75) Haartrachten, Archäologisch. *Reallexikon der Assyriologie* 4, 1–12. Berlin.

Börker-Klähn, J. (1982a) Die Reichsakkadische Kunst und Ägypten. *Wiener Zeitschrift für die Kunde des Morgenlandes* 74, 57–94.

Börker-Klähn, J. (1982b) *Altvorderasiatische Bildstelen und vergleichbare Felsreliefs. Baghdader Forschungen* 4. Mainz am Rhein.

Breniquet, C. (2008) *Essai sur le tissage en Mésopotamie des premières communautés au milieu du IIIe millénaire avant J.-C.* Paris.

Calmeyer, P. (1957/71) Gürtel. *Reallexikon der Assyriologie* 3, 689–693. Berlin.

Charpin, D. (1986) *Le Clergé d'Ur au siècle d'Hammurabi (xixe-xviiie siècles av. J.-C.)*. Geneva-Paris.

Charpin, D. & Durand, J.-M. (1981) *Documents cunéiformes de Strasbourg. Recherche sur les grandes civilisations Cahier* n° 4. Paris.

Collon, D. (1982) *Catalogue of the Western Asiatic Seals in the British Museum, Cylinder Seals II, Akkadian-Post Akkadian Ur III-Periods*. London.

Collon, D. (1987) *First Impressions, Cylinder Seals in the Ancient Near East*. Chicago.

Collon, D. (2001) How Seals Were Carried, The Archaeological and Iconographic Evidence. In T. Abusch, P.-A. Beaulieu, J. Huehnergard, P. Machinist, & P. Steinkeller (eds.), *Historiography in the Cuneiform World, Proceedings of the XLVe Rencontre Assyriologique Internationale* (1998), 2, 15–30. Bethesda.

Cooper, J. (1983) *The Curse of Agade*. Baltimore.

Cooper, J. & Heimpel, W. (1983) The Sumerian Sargon Legend. *Journal of the American Oriental Society* 103, 67–82.

Dahl, J. 2007 *The Ruling Family of Ur III Umma: A Prosopographical Analysis of an Elite Family in Southern Iraq 4000 Years Ago*. Leiden.

Delaporte, L. (1920) *Musée du Louvre, catalogue des cylinders orientaux, cachets et pierres gravées de style oriental.* I *Fouilles et missions*. Paris.

Durand, J.-M. (1983) *Textes administratifs des Salles 134 et 160 du Palais de Mari. Archives Royales de Mari* 21. Paris.

Edzard, D. (1968/1969) Die Inschriften der altakkadischen Rollsiegel, *Archiv für Orientforschung* 22, 12–20.

Foster, B. (1980) Notes on Sargonic Royal Progress. *Journal of the Ancient Near Eastern Society of Columbia University* 12, 29–42.

Foster, B. (1982a) *Umma in the Sargonic Period. Memoirs of the Connecticut Academy of Arts & Sciences* 20. Hamden, Ct.

Foster, B. (1982b) *Administration and Use of Institutional Land in Sargonic Sumer. Mesopotamia. Copenhagen Studies in Assyriology* 9.

Foster, B. (1982c) Archives and Record-keeping in Sargonic Mesopotamia. *Zeitschrift für Assyriologie* 72, 1–27.

Foster, B. (1985) The Sargonic Victory Stele from Telloh. *Iraq* 47, 15–30.

Foster, B. (1986) Archives and Empire in Sargonic Mesopotamia. In K. Veenhof (ed.), *Cuneiform Archives and Libraries, Compte rendu de la trentième Rencontre Assyriologique Internationale*, 46–52. Leiden.

Foster, B. (1997) Akkadians. In *The Oxford Encyclopedia of Archaeology in the Near East* 1, 49–54.

Foster, B. (2000) The Forty-nine Sons of Agade. In S. Graziani (ed.), *Studi sul Vicino Oriente Antico dedicati all memoria di Luigi Cagni*. Istituto Universitario Orientale, Dipartimento di Studi Asiatici, *Series Minor* 61, 309–318. Naples.

Frankfort, H. (1939) *Cylinder Seals: A Documentary Essay on the Art and Religion of the Ancient Near East*. London.
Frankfort, H. (1970) *The Art and Architecture of the Ancient Orient. The Pelican History of Art*. Baltimore.
Frayne, D. (1993) *Sargonic and Gutian Periods (2334–2113 BC). The Royal Inscriptions of Mesopotamia, Early Periods* 2. Toronto.
Gelb, I. (1959) Hurrians at Nippur in the Sargonic Period. In R. Kienle (ed.), *Festschrift Johannes Friedrich zum 65. Geburtstag am 27. August 1958 gewidmet*, 183–194. Heidelberg.
Gelb, I. (1962) *Sargonic Texts from the Diyala Region. Materials for the Assyrian Dictionary* 1, second edition. Chicago.
Gelb, I. (1972) The Arua Institution. *Revue d'Assyriologie* 66, 1–32.
Gelb, I. (1979) Household and Family in Early Mesopotamia. In E. Lipiński (ed.), *State and Temple Economy in the Ancient Near East* 1, 1–97. *Orientalia Lovaniensia Analecta* 5. Leuven.
Gelb, I., Steinkeller, P. & Whiting, R. (1991) *Earliest Land Tenure Systems in the Near East: Ancient kudurrus. Oriental Institute Publications* 104. Chicago.
Glassner, J.-J. (1986) *La Chute d'Akkadé: l'événement et sa mémoire. Berliner Beiträge zum Vorderen Orient* 5. Berlin.
Glassner, J.-J. (2004) *Mesopotamian Chronicles. Writings from the Ancient World* 19. Atlanta.
Goetze, A. (1947) Historical Allusions in Old Babylonian Omen Texts. *Journal of Cuneiform Studies* 1, 253–265.
Grayson, A. (1987) *Assyrian Rulers of the Third and Second Millennia BC (to 1115 BC). Royal Inscriptions of Mesopotamia, Assyrian Periods* 1. Toronto.
Hallmann, A. (2007) The "Kushite Cloak" of Perkartor and Iriketakana: Novelty or Tradition? *Journal of the American Research Center in Egypt* 43, 15–27.
Hallo, W. (1972) The House of Ur-Meme. *Journal of Near Eastern Studies* 31, 87–95.
Hallo, W. (1973) Choice in Sumerian. *Journal of the Ancient Near Eastern Society of Columbia University* 5, 165–172.
Hallo, W. (1980) Royal Titles from the Mesopotamian Periphery. *Anatolian Studies* 30, 189–195.
Hallo, W. & Van Dijk, J. (1968) *The Exaltation of Inanna. Yale Near Eastern Researches* 1. New Haven.
Hansen, D. (2002) Through the Love of Ištar. In L. Werr (ed.), *Of Pots and Plans: Papers on the Archaeology and History of Mesopotamia and Syria Presented to David Oates in Honour of His 75th Birthday*, 91–112. London.
Hansen, D. (2003) Art of the Akkadian Dynasty. In J. Aruz (ed.), *Art of the First Cities: The Third Millennium BC from the Mediterranean to the Indus*, 189–209. New York.
Heuzey, L. (1887) Une étoffe chaldéenne (le kaunakès). *Revue Archéologique* 9, 257–272.
Heuzey, L. & Heuzey, J. (1935) *Histoire du Costume dans l'Antiquité Classique, l'Orient, Egypte - Mésopotamie - Syrie - Phénicie*. Paris.
Houston, M. & Hornblower, F. (1920) *Ancient Egyptian Assyrian and Persian Costumes and Decorations*. London.
Kienast, B. & Volk, K. (1995) *Die sumerischen und akkadischen Briefe des III. Jahrtausends aus der Zeit vor der III. Dynastie von Ur. Freiburger Altorientalische Studien* 19. Stuttgart.
Kraus, F. (1963) Altbabylonische Quellensammlungen zur Mesopotamischen Geschichte. *Archiv für Orientforschung* 20, 153–155.
Lambert, M. (1965) Inscription royale de l'époque d'Agadé. *Revue d'Assyriologie* 59, 177–182.
Lambert, M. (1973) Comptes rendus. *Revue d'Assyriologie* 67, 174–176.
Lewis, B. (1980) *The Sargon Legend: A Study of the Akkadian Text and the Tale of the Hero Who Was Exposed at Birth*. ASOR Dissertation Series. Cambridge, Mass.
Limet, H. (1960) *Le Travail du métal au pays de Sumer au temps de la IIIe Dynastie d'Ur*. Paris.
Limet, H. (1972) Les Métaux à l'époque d'Agadé (2370–2250 av. J.-C.). *Journal of the Economic and Social History of the Orient* 15, 3–34.
Mallowan, Max (1947) *Excavations at Brak and Chagar Bazar. Iraq* 9.
Maxwell-Hyslop, K. (1971) *Western Asiatic Jewellery c.3000–612 BC*. London.
Mellink. M. (1963) An Akkadian Illustration of a Campaign in Cilicia? *Anadolu* 7, 101–115.
Michalowski, P. (1978) The Neo-Sumerian Ring Texts. *Syro-Mesopotamian Studies* 2,3. Malibu.
Milano, L. (1987) OAkk. BAN-*ḫa-tum* = *tirḫatum* "bridal price". *Orientalia* 56, 85–86.
Moorey, P. (1994) *Ancient Mesopotamian Materials and Industries, The Archaeological Evidence*. Oxford.
Moortgat, A. (1969) *The Art of Ancient Mesopotamia: The Classical Art of the Near East*. Translated by Judith Filson. New York.
Nagel, W. (1957/58) Ein altassyrisches Königssiegel. *Archiv für Orientforschung* 18, 97–103.

Nagel, W. & Strommenger, E. (1968) Reichsakkadische Glyptik und Plastik im Rahmen der mesopotamisch-elamischen Geschichte. *Berliner Jahrbuch für Vor- und Frühgeschichte* 8, 137–206.

Neumann, H. (2003) *Recht im antiken Mesopotamien.* In U. Manthe (ed.), *Die Rechtskulturen der Antike, Vom Alten Orient bis zum Römischen Reich*, 55–122. Munich.

Oppenheim, A. (1961) The Mesopotamian Temple. In G. Wright & D. Freedman (eds.), *The Biblical Archaeologist Reader*, 1: 158–69. Garden City, NY.

Otto, A. (2006) Archaeological Perspectives on the Localization of Narām-Sîn's Armanum. *Journal of Cuneiform Studies* 58, 1–26.

Pomponio, F., Visicato, G., & Westenholz, A. (2006) *Tavolette cuneiformi di varia provenienza delle collezioni della Banca d'Italia.* Rome.

Rashid, S. A., ed. (1967) *Costumes of Iraq: Sumerian Costume.* Baghdad.

Reade, J. (1991) *Mesopotamia.* London.

Reimpell, W. (1921) *Geschichte der babylonischen und assyrischen Kleidung*, ed. by E. Meyer. Berlin.

Renger, J. (1967) Untersuchungen zum Priestertum der altbabylonischen Zeit. I. *Zeitschrift für Assyriologie* 58, 110–188.

Renger, J. (1969) Untersuchungen zum Priestertum der altbabylonischen Zeit. II. *Zeitschrift für Assyriologie* 59, 104–230.

Sallaberger, W. (1999) Ur III-Zeit. In W. Sallaberger & A. Westenholz, *Mesopotamien: Akkade-Zeit und Ur III-Zeit*, 121–390. *Orbis Biblicus et Orientalis* 160/3. Göttingen.

Salonen, A. (1969) *Die Fussbekleidung der alten Mesopotamier nach sumerisch-akkadischen Quellen. Annales Academiae Scientiarum Fennicae* Series B 157. Helsinki.

Schileiko, W. (1914) Die sechsseitige Tonprisma Lugal-ušumgal's aus der Sammlung Lichatschew. *Zeitschrift für Assyriologie* 29, 78–84.

Selz, G. (1993) *Altsumerische Wirtschaftsurkunden aus amerikanischen Sammlungen. Freiburger Altorientalische Studien* 15. Stuttgart.

Sommerfeld, W. (1999) *Die Texte der Akkade-Zeit, 1. Das Dijala-Gebiet: Tutub. Imgula* 3/1. Münster.

Sommerfeld, W. (2000) Narām-Sîn, die "Große Revolte" und MAR.TU^{ki}. In J. Marzahn & H. Neumann (eds.), *Assyriologica et Semitica: Festschrift für Joachim Oelsner. Alter Orient und Altes Testament* 252, 420–436.

Spycket, A. (1981) *La Statuaire du Proche-Orient Ancien.* Leiden.

Steinkeller, P. (1982a) The Question of Marḫaši: A Contribution to the Historical Geography of Iran in the Third Millennium B.C. *Zeitschrift für Assyriologie* 72, 237–265.

Steinkeller, P. (1982b) Two Sargonic Sale Documents Concerning Women. *Orientalia* 51, 355–368.

Steinkeller, P. (2006) New Light on Marhaši and Its Contact with Makkan and Babylonia. *Journal of Magan Studies* 1, 1–17.

Steinkeller, P. & Postgate, N. (1992) *Third-Millennium Legal and Administrative Texts in the Iraq Museum, Baghdad. Mesopotamian Civilizations* 4. Winona Lake, Ind.

Strommenger, E. (1959) Statueinschriften und ihre Datierungswert. *Zeitschrift für Assyriologie* 53, 27–50.

Strommenger, E. (1960) Das Menschenbild in der altmesopotamischen Rundplastik von Mesilim bis Hammurapi. *Baghdader Mitteilungen* 1, 1–103.

Strommenger, E. (1962) *Fünfjahrtausende Mesopotamien: Die Kunst von den Anfängen um 5000 v. Chr. bis zu Alexander dem Großen.* Munich.

Strommenger, E. (1963) Das Felsrelief von Darband-i-Gaur. *Baghdader Mitteilungen* 2, 83–88.

Strommenger, E. (1971) Mesopotamische Gewandtypen von der frühsumerischen bis zur Larsa-Zeit. *Acta Praehistorica et Archaeologica* 2, 37–55.

Strommenger, E. (1980/83) Kleidung B. Archaeologisch. *Reallexikon der Assyriologie* 6, 31–33. Berlin.

Strommenger, E. (1998/2001) Narām-Sîn, König von Agade B. *Reallexikon der Assyriologie* 9, 174–177.

Vanstiphout, H. (2003) *Epics of Sumerian Kings: The Matter of Aratta. Writings from the Ancient World Series* 20. Atlanta.

Vigneau, A. (1935) *Encyclopédie photographique de l'Art: Le Musée du Louvre.* Paris.

Visicato, G. (2000) *The Power and the Writing: The Early Scribes of Mesopotamia.* Bethesda.

Visicato, G. (2001) The Journey of the Sargonic King to Aššur and Gasur. In T. Abusch, P.-A. Beaulieu, J. Huehnergard,

P. Machinist, & P. Steinkeller (eds.), *Historiography in the Cuneiform World, Proceedings of the XLVe Rencontre Assyriologique Internationale* (1998), 1, 467–472. Bethesda.

Waetzoldt, H. (1972) *Untersuchungen zur neusumerischen Textilindustrie.* Centro per le Antichità e la Storia dell'Arte del Vicino Oriente, *Studi Economici e Tecnologici* 1. Rome.

Waetzoldt, H. (1980/83) Kleidung A. Philologisch. *Reallexikon der Assyriologie* 6, 18–31. Berlin.

Weiss, H. & Courty, M. (1993) The Genesis and Collapse of the Akkadian Empire: The Accidental Refraction of Historical Law. In M. Liverani (ed.), *Akkad, the First World Empire: Structure, Ideology, Traditions*, 131–155. *History of the Ancient Near East/Studies* 5. Padua.

Westenholz, A. (1970) barūtum, damtum, and Old Akkadian KI.GAL: Burial of Dead Enemies in Ancient Mesopotamia. *Archiv für Orientforschung* 23, 27–31.

Westenholz, A. (1975a) Old Sumerian Administrative Documents in the Pontifical Biblical Institute in Rome. *Orientalia* 44, 434–438.

Westenholz, A. (1975b) *Early Cuneiform Texts in Jena: Pre-Sargonic and Sargonic Documents from Nippur and Fara in the Hilprecht-Sammlung vorderasiatischer Altertümler, Institut für Altertumswissenschaften der Friedrich-Schiller-Universität, Jena. Det Kongelige Danske Videnskabernes Selskab, Historisk-Filosofiske Skrifter* 7, 3. Copenhagen.

Westenholz, A. (1989) *Old Sumerian and Old Akkadian Texts in Philadelphia Part Two: The 'Akkadian' Texts, the Enlilemaba Texts, and the Onion Archive. Carsen Niebuhr Institute Publications* 3. Copenhagen.

Westenholz, A. (1999) The Old Akkadian Period: History and Culture. In W. Sallaberger & A. Westenholz, *Mesopotamien: Akkade-Zeit und Ur III-Zeit*, 17–117. *Orbis Biblicus et Orientalis* 160/3. Göttingen.

Westenholz, J. (1983) Heroes of Akkad. *Journal of the American Oriental Society* 103, 327–336.

Westenholz, J. (1989) Enḫeduanna, En-Priestess, Hen of Nanna, Spouse of Nanna. In H. Behrens, D. Loding, & M. Roth (eds.), *DUMU-E$_{2}$-DUB-BA-A, Studies in Honor of Åke W. Sjöberg*, 539–556. *Occasional Publications of the Samuel Noah Kramer Fund* 11. Philadelphia.

Westenholz, J. (1997) *Legends of the Kings of Akkade. The Texts. Mesopotamian Civilizations* 7. Winona Lake, Ind.

Wilcke, C. (2007) *Early Near Eastern Law. A History of Its Beginnings*, revised edition. Winona Lake, Ind.

Zervos, C. (1935) *L'Art de la Mésopotamie de la fin du quatrième millénaire au XVe siècle avant notre ère, Elam, Sumer, Akkad.* Paris.

Zettler, R. (1984) The Genealogy of the House of Ur-Me-me: A Second Look. *Archiv für Orientforschung* 31, 1–9.

Zgoll, A. (1997) *Der Rechtsfall der En-hedu-Ana im Lied nin-me-šara. Alter Orient und Altes Testament* 246. Münster.

8. Textiles in the Administrative Texts of the Royal Archives of Ebla (Syria, 24th Century BC) with Particular Emphasis on Coloured Textiles

Maria Giovanna Biga

1. The Eblaite written sources on textiles

The city of Ebla was the capital of a Syrian kingdom which, around the middle of the 3rd millennium played a notable role in Syria's history as can be reconstructed by the texts found in its archives. The history of the city dates to the first half of the 3rd millennium BC. The Ebla texts enable us to conclude that, at the onset of the 3rd millennium BC, Syria witnessed the creation of the first urban centres by Semitic tribes who settled there, thus giving rise to small states. Syria at that time was divided into a series of kingdoms. Some had a king with the same title of *en*, "king", as Ebla's king, and others had titles such as *badalum,* apparently also indicating an independent ruler. It is possible to conclude from the Ebla documents that, this city-state, from its very start, had created a vast trade network and maintained friendly relations with many kingdoms. Thus, the exchange of gifts and the possibly of trade are attested, although this latter factor emerges less frequently from the texts.

The texts from the royal archives of Ebla provide a wealth of information on the textiles produced, used and exchanged in Ebla itself and in the Syrian region and Upper and Central Mesopotamia of the time. A frequently encountered type of text is the monthly accounts of textile deliveries, registering the deliveries of fabrics to various people on several occasions in any one month. Most of these accounts are of ceremonial gifts sent to the courts of Ebla's closest allies, who, in turn, made gifts of wine, animals and various goods to Ebla.

Among the first 42 tablets discovered in 1974 were several quoting different types of textiles, wool, and woollen articles used as equipment for donkeys and equids.

In the great archive L.2769, *c.*650 inventory numbers of tablets recorded monthly accounts of textile deliveries. They register the outgoing of precious cloth, undoubtedly Ebla's most important product.

In 1976, a buried wooden plank was discovered in the Audience court; on this plank were tablets, some intact and some broken. They were later published by E. Sollberger in the volume ARET VIII (and then by G. Pettinato in MEE V) as monthly accounts of textiles and annual accounts of metals. Joins to some of these broken tablets were found in the principal archive. This indicates that the tablets were seized after the destruction of the archives. It is possible that these may constitute parts of tablets that the conquerors of the town carried off as war booty. This hypothesis can explain the lack of texts concerning some major events at the Ebla court, such as the funeral and the funerary gifts for King Irkab-damu of Ebla, which we know about from some tablets. However, it is not as extensive as we would wish to have in a monthly account of textiles with the complete list of textiles and other objects of the funerary array!

In the texts in archive L.2712 (roughly 1,000 inventoried tablets and fragments of the same), a few quote some deliveries of textiles but they deal, *inter alia*, with rations of food, wheat, bread, objects of wood, some silver, one of gold, or oil.[1] No rations of wool are referred to in these texts. The wool rations are instead normally registered in the monthly accounts of textiles. Usually, the last columns of the reverse of a monthly account of textiles are devoted to the registering of wool for different purposes, including buying some products at fairs.

In a list of precious objects probably brought as booty to the royal palace, ARET X 55, and its parallel ARET X 2, several textiles are quoted.

Thus far only 25% of the monthly accounts of textile deliveries have been published.[2]

Textiles are also quoted in other types of texts: the ritual texts, first of all in the texts of the great ritual of royalty[3] and in other rituals; in the "mu-DU" texts of incoming goods to the Eblaite administration; in the annual accounts of metals where some silver is registered as being used to purchase various amounts of different textiles.

Work is obviously still in progress on these texts. The monthly accounts of textile deliveries are a mine of information, not only for the textiles of Ebla and of the Syrian region, but also for the reconstruction of the political, social and economic history of Syria and Upper Mesopotamia during the 24th century BC. Many of these aspects have already been studied. In the Ebla archives, there are several very important texts, such as the treaty with Abarsal, the letter of King Enna-Dagan of Mari, or the letter of a functionary of the king of Ḫamazi. However, in order to write the history of Ebla in the period of the archives, the texts of textile deliveries are the most useful, as they, above all, are full of news of many events quoted only in this type of text.

Therefore, to trace the history of the kingdom of Ebla and other kingdoms of the region, it was necessary first of all to arrange the texts in chronological order.[4] Thus, in 1993, I began a programme of systematic study of all the texts of textile deliveries in the large Ebla archive, L.2769. I studied the prosopography (so important in chronological reconstruction), the events and all the elements useful to reconstruct the chronology of the texts, to then write the history of Syria and Upper Mesopotamia in the mid 3rd millennium BC.

Given that many texts of textile deliveries are fragmentary, I began a systematic search for joins. This task, which still continues at the time of writing, has been possible because all the fragments found by archaeologists in room L.2769 are preserved.

Having finally established a reasonably relative chronological order for the annual accounts of metals,[5] through the crucial help provided by the monthly documents of textiles, we correlated several of these monthly accounts of textiles to the corresponding annual account of metals. However, this task is not yet complete. Then the joins of the previous years' result in certain texts

[1] See ARET IX.

[2] See ARET I, II, IV, VII, VIII, XV; MEE II, VII, X, XII.

[3] See ARET XI, 1,2,3.

[4] See Biga 2003a, 345–367.

[5] Archi & Biga 2003, 8–9.

requiring further prosopographic study in order for them to be inserted into the chronological table and sometimes this requires certain texts to be moved within the relative chronology.

As we will see below, I have correlated many of the textile texts to the corresponding annual account of metals for the first three years of the vizier Ibbi-zikir[6], but I am still searching for other texts pertaining to these three years. The remaining textile texts for all the other periods have yet to be correlated to their corresponding annual metal account.

Due to the large array of topics, in many articles not only the study but even the quotation of the types of textiles have been neglected. In several articles devoted to other topics, quoting passages of monthly accounts of textiles we wrote: textiles (1,1,1 or 2,2,1,1,1) etc. and then the personal names of those receiving the textiles and then the occasion of the delivery.

Today, after thirty years of prosopographical studies, we can state that the large archive, which is very extensive since it must have held about three to four thousand tablets, covered a period of roughly 50 years of the city and kingdom's history. Four kings ruled during this 50 year span, the first of these, King Kun-damu, being documented in a very limited number of texts. More texts, including some regarding textiles, can be attributed to the third from last king, Igriš-Ḫalab, even more to the penultimate king, Irkab-damu, who ruled for about ten years and a very large number to the last king, Išar-damu, son of King Irkab-damu. It is the reign of King Išar-damu, that is known best of all. This is because all, or almost all, of the large documents recording goods entering and leaving the kingdom are available to us, from the moment Išar-damu ascended to the throne right up to the destruction of the city, a period of approximately 36 years. This last king of Ebla, who reigned so long having ascended to the throne when still a child, continued his father's work and extended Ebla's kingdom before witnessing the destruction of the capital city, which was sacked.

The kings of Ebla were assisted in their rule by a varying group of elders, primarily constituting certain principal members of important Eblaite families. These were powerful people who channelled their great wealth towards the royal palace, perhaps by collecting tributes. They acted as chiefs of the administration and were called "lugal", "lord". Their precise role remains difficult to define. In the texts from King Igriš-Ḫalab's reign, the most important of the "lords"were Darmia and Tir. In King Irkab-damu's reign the most important "lord" was Arrugum. In King Išar-damu's reign, the most important "lord" who acted as a vizier and was also at the head of the Eblaite army, was first, Ibrium and then, after his death, his son Ibbi-zikir.[7]

As mentioned at the outset, among the documents of these kings, around 650 monthly documents register outgoing quantities of precious cloth, undoubtedly Ebla's most important product. These textiles were given to members of the court on many different occasions, to people from other cities in the kingdom and sent to other countries with which Ebla had diplomatic, political and commercial relations.

In fact, the designation of the monthly account of textiles is not arbitrary and was not made by the epigraphers of Ebla or by the archaeologists. It was made by the Eblaite scribes who counted, (at the very end of every text), the partial total (AN.ŠÈ.GÚ) or the general total (šu-nígin) of the delivered textiles leaving Ebla's storerooms. Thus, they were called monthly accounts of textiles because almost all have at the end the totals (partial and general) and are dated with the month's name. Some, but very few, are *pluri-mensuel*, covering some successive months.

---

[6] Biga 2010, 39–57.

[7] For the "lords", see Archi 2000, 19–58.

Fortuitously, the Ebla scribes also registered, in the monthly accounts of textiles, the precious objects, mostly of silver and gold, which were delivered with the textiles to the same people. They sometimes also wrote the weight of these objects in an abbreviated form.[8] The occasion of the delivery too is sometimes mentioned: always an event of the Ebla court and of other Syrian kingdoms commemorating a birth, a marriage, a death, a ceremony, a ritual, a military campaign, or news brought to the Ebla court. These events are fundamental for reconstructing the history of Ebla and the other kingdoms of the time.

Of course, in the case of some important events such as marriages, death of the vizier or a princess of the court, or in the case of a procedure (that despite several efforts by some scholars, remains difficult to understand) of the transfer of a bracelet or a dagger from one member of the court to another, the list of precious objects is longer with many objects of metals listed even in the monthly accounts of textiles. However, the Eblaite scribes always considered the texts as accounts of textiles and in the final total, they counted only the delivered textiles. Therefore, it is impossible to consider some texts as anomalous. Further proof that the scribes consider those texts as normal monthly accounts of textiles is that they put the tablets in chronological order on the same shelf where other monthly account of textiles of the same year were placed.

Several monthly accounts of textiles are heavily damaged and need to be completed. There are, in fact, several thousands of fragments, some large. Two volumes of fragments have been published (ARET III and XII), but more than two thousand fragments remain. We hope to be able to join them to their appropriate tablet and not to have to publish another volume of fragments.

Other typologies of texts are also important for the study of the textiles and to evaluate the number of movements of textiles and wool in the period covered by the archives. The so-called "mu-DU" texts apparently register the goods entering the palace storerooms. The "lords" are able to contribute great quantities of textiles and different metals to the Eblaite storerooms. Especially during the reign of King Išar-damu, his vizier, Ibrium and later his son, Ibbi-zikir are responsible for bringing vast quantities of textiles to Ebla. The origin of these goods registered in the "mu-DU" texts remains unclear. There are four "mu-DU" documents from the time of Vizier Arrugum, nineteen documents from the time of Vizier Ibrium, (including the document of the year in which King Irkab-damu died), and twelve "mu-DU" documents from Vizier Ibbi-zikir's time. In all of these texts, but especially in those of Ibrium and Ibbi-zikir, the list of goods brought by the vizier begins with a long list of hundreds of different types of textiles entering the Eblaite storerooms. Some "mu-DU" texts are already published, as, for example, MEE II 1, ARET II 13, MEE VII 38, and MEE XII 3. From these texts, it is evident that the production of textiles in Ebla itself (which we are not able to quantify) was integrated with large amounts of textiles coming from other towns of the Eblaite kingdom and from other countries in some way subordinate to Ebla.[9]

There are then 36 large annual accounts of metals registering the deliveries of different metals for many purposes. All these texts record (some) deliveries of silver to buy different types of textiles (even though where these textiles are bought remains un-specified). Large quantities of wool were bought especially from the city of Mari on the Middle-Euphrates and from the city of

[8] Biga & Pomponio 1993, 12–19

[9] See also Archi 1993, 43–58.

Armi, possibly Tell Bazi-Banat on the Euphrates too, between Karkemiš and Emar.[10]

Some of these annual accounts of metals are already published in MEE VII 34, MEE 10 20 and 29, and MEE XII 35, 36, 37.

Textiles then are of significance in some texts describing rituals. The use of particular textiles and their colours are important factors for the understanding of the type of ritual in question. The texts concerning the ritual of the Eblaite royalty – which was performed, at least, by King Irkab-damu and his queen and then by King Išar-damu and his queen Tabur-damu- quote several textiles used during the ceremonies.[11]

The identification of the type of fabrics quoted in the texts is not an easy task. It is still extremely difficult to understand the purposes and kinds of cloth, for which the different types of fabrics were used. This is a common problem with every type of text from all periods in the Ancient Near East.[12] It is uncertain if we are dealing with whole fabrics, or fabrics already cut and ready to be worn, wrapped around the body, or with clothes presumably cut and sewn. Etymology is often not helpful here.

From the Mari letters, we can sometimes ascertain that we are dealing with clothes. However, in the case of Ebla, it seems more likely that we are dealing with pieces of fabric of unknown size.

In Egyptian tombs, Egyptologists found complete arrays of textiles of different sizes, some as large as a sheet, some smaller and cut in bands of various size and length, normally kept in wooden boxes.[13] For all these reasons I would prefer not to use the word "clothes" for Eblaite textiles.

Almost all the names of textiles in the Ebla administrative texts were recognized and listed by G. Pettinato in his pioneer volume MEE II (Napoli 1980), and several items were discussed in the commentary to the monthly account of textiles published in that volume. In 1985, A. Archi briefly offered an excursus on some Eblaite textiles in ARET I, 227–229, considering them to be clothes. Then for many years scholars devoted their attention to other problems of interpretation of these texts, primarily to the history of the kingdom, although, several comments were made on the different types of textiles in the editions of monthly accounts of textiles in ARET and in the MEE series.

However, it is only in recent years that some excellent articles and lengthy studies have been devoted to the different types of textiles and their lexicon. In the *Thesaurus Inscriptionum Eblaicarum, Volume A*/1,2, Roma 1996, it is possible to find long lists of quotations and passages of several texts where names of textiles, expecially *ʾà-da-um*-TÚG and aktum-TÚG, are attested. A long, detailed and important study of the Semitic textile terms is given by J. Pasquali (1997). G. Conti (1997) produced an important study on the numerous objects made of wool which form part of the equipment for donkeys, equids and wagons. In A. Archi (1999), there is a brief excursus

[10] See Otto 2006, 1–26.

[11] For these texts see ARET XI, 1,2,3; for another ritual where coloured textiles have great importance, see Biga 2003b, 54–67; Pasquali 2005a, 165–184.

[12] See the important remarks of Veenhof 1972, 89–97 and Durand 2009, 9–20.

[13] See, for example, the description of the rich tomb of Gebelein during the Vth Dynasty, excavated by Schiaparelli in 1911, in which three wooden boxes full of textiles of various sizes were found. One of these boxes is now at the Egyptian Museum of Turin. The cover of the box has in its inner side a very interesting text with a list of textiles, one of the few examples of lists of textiles not for ritual use. The names of the textiles written in the list are difficult to interpret and translate. Egyptologists, even if they have many textiles still preserved in the tombs, are often not able to identify these textiles with the names they have in the texts. See Roccati 1970, 1–10. I thank A. Roccati for the bibliographical references.

on some of the most common types of fabrics (considered cloth) and a list of their prices. In his detailed commentary on single terms concerning the textiles mentioned in the texts, H. Waetzoldt (2001), quotes all the interpretations and translations for every textile term in the previous bibliography. The symbolism related to the colours of the fabrics used in different rituals is studied by J. Pasquali (2005a). An important study on the social significance of the gifts of different fabrics and the attempt to identify some textiles with the help of the iconography of small statues and other pieces of sculpture found in the Ebla palace was done by W. Sallaberger (2009).

For the most recent paper on the Semitic lexicon of the textiles, see Pasquali in this volume.

It has been clear for some time that the study of monthly accounts of textiles is more productive in conjunction with the relative chronology – as chronology has been important to trace the story of the kingdom of Ebla.[14] Hence, it was decided to publish the volumes dedicated to textile texts in chronological order. F. Pomponio, having published in the volume ARET XV,1 the first part of the monthly accounts of textiles written under Vizier Arrugum, will soon publish the second volume. Knowing all the texts of the period of around 5–6 years during which Arrugum was vizier, provided Pomponio with a better understanding of the distribution of different types of textiles as can be seen in his important article (2008). For some textiles quoted in lexical lists at Ebla see Civil 2008, 92–98.

In the following, I would like to offer first of all some general remarks on some questions that remain open and then examine some particular deliveries of textiles, in particular coloured textiles, to different groups of people in the service of the Palace.

2. The production of the textiles, their conservation in storerooms, and their dyeing

The state of Ebla controlled many flocks of sheep. In certain small texts, an extraordinary large number of sheep is quoted.[15] Hence the palace of Ebla had large quantities of wool available for the production of fabrics. Apparently, Eblaite wool was insufficient as the texts quote wool bought by merchants in the kingdom of Mari on the Middle Euphrates, in the kingdom of Armi and in other kingdoms.

At the end of each monthly account of textiles, in the last columns of the verso, the deliveries of wool (sometimes specified as white or black) are registered.

The wool was given to women, including the queen and the queen mother to make *inter alia* fabrics, blankets, pillows, cushions, carpets, or ropes[16]; to workers to make fabrics; it was used to purchase goods at fairs[17], (especially gall-nuts and other natural dye materials for fabrics and wool[18]); it was given as payment to several categories of workers; and it was given to functionaries responsible for the equipment of donkeys and equids and chariots of the members of the royal family and the vizier.[19]

---

[14] See also Rositani 2001, 261–270.

[15] See Archi 1980, 1–33.

[16] For ropes, see Waetzoldt 2007, 112–124.

[17] See Biga 2002, 277–288; Biga 2003c, 55–68.

[18] See Archi 2003, 27–30.

[19] The names of the functionaries charged with receiving quantities of wool for wagons (and related products) belonging to the king and the most important persons of the court were an important element in the dating of several texts, see Biga Pomponio 1990, 199–201.

Often, the quantities of delivered wool are not counted in the totals at the end of the text. Only when large quantities of wool are delivered are they counted and summed up at the end of the text.

Many textiles were probably produced in Ebla's royal palace, where several women worked under the supervision of some of the ladies of the court.[20] Apparently, some textile ateliers were situated elsewhere in Ebla, but we do not know very much about them from the texts. Then there was domestic production. Wool was often delivered to the queen mother, the queen and many other court ladies, probably for their personal use.

Many textiles came to the Eblaite storerooms as "mu-DU" of the lords, and from some countries allied with Ebla, or under its political control, and from Ebla's commercial and trading partners.

The annual accounts of metals illustrate, for instance, when the silver necessary to buy some fabrics is noted,[21] that some textiles were purchased abroad. Wool is also quoted as arriving from Mari or from Armi, both large production centres as is clear from the name of some wool qualified as "wool of Mari" and "wool of Armi". The wool was bought by Eblaite merchants in these places.

Different types of textiles were produced in Mari, Armi and Ebla itself. The toponym which sometimes follows the names of textiles probably indicates a particular type of weaving typical of that city.[22] Usually the texts, especially the annual account of metals, do not register where the textiles were bought, but Eblaite merchants purchased textiles in other countries as is clear, for example, from this text:

> 75.G.1741 (King Išar-damu, Vizier Ibrium, month še-gur$_x$-mìn) obv. II 14–III 4:
> 1 gu-mug-TÚG 1 íb+III-gùn-TÚG dumu-nita *A-ba-da-an* dam-gàr du-du *si-in* Kiš^{ki} níg-sa$_{10}$ túg-túg,"
> 1 gu-mug-textile, 1 íb-textile to the son of Abadan, the merchant, who goes to the town of Kiš to buy textiles".

There were several workers involved in the preparation and dyeing of fabrics, male and female weavers (túg-nu-tag, dam túg-nu-tag), sometimes quoted with a personal name but more often anonymous, the producers of felts (lú túg-du^2), and the dyers and apprentice dyers (gùn and dumu-nita tur gùn).

Several dyers, some anonymous, some well known by their personal name, are documented in many texts as receiving food rations[23] and also textiles and wool as remuneration. It is clear that their work was done both in Ebla itself and also often outside Ebla (uru-bar). In fact, a large area and large quantities of water are necessary to dye the wool.

> 75.G.1263 (King Išar-damu, Vizier Ibrium, month lost) obv. VIII' 7–10:
> 13 gu-dùl-TÚG 13 sal-TÚG 13 íb+III-gùn-TÚG dumu-nita-dumu-nita ur$_4$ gùn 12 gu-dùl-TÚG 12 íb+III-gùn-

[20] See Biga 1988, 159–170.

[21] See, for example, 75.G.2462 (King Išar-damu, Vizier Ibbi-zikir second year) rev. II 31–III 8:
10 lá-1 ma-na ša-pi 1 bar$_6$:kù níg-sa$_{10}$ 61 *ʾà-da-um*-TÚG-I níg-sa$_{10}$ 1 zara$_6$-TÚG níg-sa$_{10}$ 4 aktum-TÚG 30 túg-NI.NI 15 gíd-TÚG 73 níg-lá-sag 3 ma-na 10 bar$_6$:kù níg-sa$_{10}$ 2 *mi-at* 90 lá-3 na$_4$ siki *Ma-rí*^{ki},
"9 mina and 40 sicles of silver, price of 61 ʿà.-textiles, price of 1 zara-textile, 4 aktum-textiles, 30 NI.NI-textiles, 15 gid-textiles, 73 headbands, 3 mina, 10 sicles of silver, price of 287 measures-na$_4$ of wool of the city of Mari". See also, for example, MEE 12, 37 rev. XX 8–XXI 2.

[22] As already noted by Dercksen 2004, 16–17, the toponym which follows the name of a textile primarily refers to the type and not to the place of manufacture.

[23] See ARET IX, s.v. dar.

TÚG dumu-nita-dumu-nita gùn,
"13 gu-dùl-textiles, 13 sal-textiles, 13 íb-textiles for the apprentice-dyers "ur"? and 12 gu-dùl-textiles, 12 íb-textiles to the apprentice-dyers".
75.G.1298 (King Išar-damu, Vizier Ibrium, month *i-ba₄-sa*) obv. III 14–IV 2:
11 gu-dùl-TÚG 11 sal-TÚG 11 íb+IV-gùn-TÚG dumu-nita-dumu-nita gùn ur_4 12 gu-dùl-TÚG 12 íb+III-gùn-TÚG dumu-nita tur gùn.
75.G.1318 (King Išar-damu, Vizier Ibrium, month lost) obv. XV 1–2:
11 sal-TÚG 11 gu-dùl-TÚG 11 íb+III-gùn-TÚG dumu-nita ⌜ur_4⌝ gùn
75.G.1524 (King Išar-damu, Vizier Ibrium, month MAxGÁNA*tenû*-sag) obv. IX 14–17:
1 gu-dùl-TÚG *Da-i-bù* gùn *Mu-rí-gú*[ki]
75.G.1890+11556 (King Išar-damu, Vizier Ibbi-zikiz, month *i-rí-sa*) rev. V 2–10:
12 gu-dùl-TÚG 12 sal-TÚG 12 íb+IV-TÚG-gùn gùn-gùn "ur $_4$" 9 gu-dùl-TÚG 9 íb+II-TÚG-gùn gùn-gùn uru-bar,
"12 gu-dùl-textiles 12 sal-textiles 12 multi coloured íb-textiles for the dyers of the city "ur"?, 9 gu-dùl-textiles, 9 multi coloured íb-textiles for the dyers (working) outside (Ebla, in the neighbourhood)".

Syrians were depicted in the Egyptian tombs with multicoloured clothes and they were easily identified among other foreigners, due to these clothes.[24]

The textiles were kept in the storerooms of the royal palace where mostly textiles but also other items were kept. It is clear that metal objects were kept in the é-siki, "house of wool", which at the beginning probably was a storeroom for wool where later other types of goods too were stored. Several monthly accounts of textiles quote textiles from the storeroom é-ti-TÚG and others from the é-siki, and some texts give the huge number of textiles present in these storerooms. It is impossible to identify these storerooms in the preserved and excavated part of the great palace building complex that archaeologists have called Royal Palace G, of which so far roughly 2700 square meters have been revealed.

The scribes are sometimes very careful to register even the delivery of one textile from a storeroom, but more often, they do not register the storeroom from where the delivered textiles are taken. See for example:

75.G.1881(King Išar-damu, Vizier Ibrium, month lost) rev. VI 14–VII 7:
1 *ʾà-da-um*-TÚG-II lú é-ti-TÚG 1 aktum-TÚG 1 íb+IV-TÚG-sa₆-gùn 1 íb+III-ú-ḫáb *Ib-rí-um in* ud ì-ti *mi-nu* níg-kas₄ *Kak-mi-um*[ki],
"1 *à*.-textile of the house of the textiles-ti, 1 aktum-textile, 1 íb-textile of good quality and multicoloured, 1 íb-textile red to Ibrium when he came back from the military campaign against the kingdom of Kakmium".

Moreover, we know that in the é-ti-TÚG storeroom, all types of textiles were kept. The term is easy to translate as "house of the ti-TÚG", dedicated to this type of textiles; if the translation of ti-TÚG is "pleated textile" it is a little strange to admit an entire storeroom completely devoted to this type of textiles. In fact, in the é-ti-TÚG, all types of textiles and other goods were kept.

75.G.1744 (King Išar-damu, Vizier Ibbi-zikir, month lost, text type ARET I, 1–8) rev.V 17–25–VI 1: 2 gada-TÚG 1 aktum-TÚG 2 *la-da-lum* 1 IGI.NITA *I-bí-zi-kir* 1 íb+III-TÚG-sa₆-sag AN.ŠÈ túg-túg é-ti-TÚG *Iš₁₁-gi-bar-zú*," 2 linen textiles, 1 aktum-textile, 2 *ladalum*, for an equid of Ibbi-zikir, 1 íb-textile, of good quality, the best quality, to sum up to the textiles (already taken?) from the storeroom-ti to Išgibarzu".

---

[24] See, for example, one of the rock tombs (Dyn. XII) in Beni Hasan in which, in a procession of foreigners, an Asian family wearing multicoloured clothes is depicted arriving in Egypt, see Porter & Moss 1934, 146.

3. The most frequently attested types of textiles

3.1. ’*à-da-um*-TÚG

Among the most frequently attested textiles is the ʾ*à-da-um*-TÚG. Most often, it is quoted followed by the number I or II, although this is not always the case. What the numbers I or II in this context refer to is still uncertain. Archi (1985, 227) translated I as “scempio” (single weave) and II “doppio, di doppia tessitura” (double weave). Pasquali (1997, 218) considered them “una indicazione di qualità” (indications of quality). See for example the text:

> 75.G.1522 (King Išar-damu, Vizier Ibrium, month *i-ba$_4$-sa*) obv. III 8–IV 1:
> 1 ʾ*à-da-um*-TÚG 1 aktum-TÚG 1 íb+III-sa$_6$-gùn-TÚG *Iš$_{11}$-lu-mu Ra-*ʾ*à-ak*^{ki} giš-dug-DU *Ib-rí-um si-in* ḫúl na-rú,
> “1 ʾà.-textile, 1 aktum-textile, 1 multicoloured and of good quality íb-textile to Išlumu Ra’ak ... to Ibrium for the festival of his stele”;
> rev. I 3–9:
> 1 ʾ*à-da-um*-TÚG 1 aktum-TÚG 1 íb+III-sa$_6$-gùn-TÚG *I-ti-Aš-dar* dumu-nita *Dur-du-lum in* ud šu mu-nígin-*sù* šu mu-tak$_4$

The ʾ*à-da-um*-TÚG-I is not frequently found alone. This type of textile is usually distributed with other types, mostly with aktum-textile and íb-textile, but sometimes it is delivered alone, as for example in:

> 75.G.1523 (King Išar-damu, Vizier Ibrium, month *ig-za*) obv. II 15–III 7:
> 1 ʾ*à-da-um*-TÚG-I *Ib-du-*^{d}*Aš-dar* dumu-nita *A-su-mu* 2 ʾ*à-da-um*-TÚG-I 1 *gú-li-lum* a-gar$_5$-gar$_5$ kù-gi *A-si-ma-lik wa Iš-má-da-mu* 2 dumu-nita *Rí-dam-ma-lik* ugula *Šu-ti-gú*^{ki}.
> 75.G.1887 +ARET III 255 (King Išar-damu, Vizier Ibrium year 13, month lost) obv. XI 8–14:
> 1 ʾ*à-da-um*-TÚG-I *Ib-rí-um in* ud níg-kas al-kur$_6$ še *Kak-mi-um*^{ki},
> “1 ʾà.-textile to Ibrium when he went to cut the barley of Kakmium”.[25]

The ʾ*à-da-um*-TÚG-I can be of good quality (sa$_6$)

> 75.G.1707 (King Išar-damu, Vizier Ibrium, month *ga-sum*) obv. II 19–III 2: 1 ʾ*à-da-um*-TÚG-I-sa$_6$ *Ib-rí-um*

The ʾ*à-da-um*-TÚG-II is not given alone frequently either.

> 75.G.1884 (King Išar-damu, Vizier Ibrium, month *i-rí-sá*) obv. II 3–9: 2 ʾ*à-da-um*-TÚG-II *Bíl-za-ì wa En-na-Il* dumu-nita-dumu-nita šeš-šeš *Ib-rí-um*

The ʾ*à-da-um*-TÚG-I/II can be of first quality (sag)

> 75.G.1221 (King Išar-damu, Vizier Ibbi-zikir, month *ḫa-li*) obv. IX 12–17: 1 ʾ*à-da-um*-TÚG-II sag *Ḫa-ba* dumu-nita en *in* ‘*À-za-an*^{ki} šu ba$_4$-ti

In a few cases, this textile is mentioned as multicoloured (gùn).

> 75.G.1748 (King Išar-damu, Vizier Ibrium, month *za-*[ʾ*à*]*-na-*⌜*at*⌝) rev. II 6–14: 1 ʾ*à-da-um*-TÚG 1 gada-TÚG mu$_4$.mu 1 gada-TÚG SIG$_4$.KI en 1 ʾ*à-da-um*-TÚG-gùn 1 aktum-TÚG *Ib-rí-um* du-du *si-in* ʾ*À-da-*NI^{ki}

During the time of Vizier Ibbi-zikir, this is usually the first textile quoted in the groups of textiles given together, but in the time of Vizier Ibrium, other textiles, such as *guzitum* or túg- gùn often headed the lists. It is attested as either coloured or white.

---

[25] In this text, many deliveries are devoted to people of different countries participating in a military campaign against Kakmium conducted by Ibrium. He (at least according to this text) does not receive particular fabrics to go to war. In obv. X 12–13, he receives some linen textiles, but it is not specified if they are given for the military campaign. Only the ʾ*à-da-um*-textile is specified as given on the occasion of the campaign, but it is not known if he wore the textile during the campaign.

The translation of the term is uncertain. A similarity with the Semitic root **ḥtl* "to wrap" is quite possible, but so far a specific translation of the word has not been possible.[26]

3.2. *aktum-TÚG*

This type of textile, well known from the Mesopotamian texts, is given to important and less important people, often as part of three pieces of fabric, registered in second position after *ʾà-da-um* or another textile; sometimes it is also given alone. It is used, cut, for the equipment of 2 donkeys. This use does not allow a translation as a "tunic".[27] The translation of the term is uncertain.

> 75.G.1728 (King Išar-damu, Vizier Ibrium, month lost) rev. III 6–10:
> 1 aktum-TÚG maš-maš *ma-da-ra* 2 IGI-NITA *Ib-rí-um*,
> "1 aktum-textile to cut to make bands for two donkeys of Ibrium".
> 75.G.1830+ARET III 139+ARET 457 (King Išar-damu, Vizier Ibbi-zikir year 9, month MAxGÁNA*tenû*-úgur) rev. X' 8–11: 1 aktum-TÚG *ma-da-lum* 2 IGI.NITA *I-bí-zi-kir.*

It is attested as coloured, reddish-brown (ú-ḫáb) and white (babbar)

> 75.G.1881 (King Išar-damu, Vizier Ibrium, month lost) rev. III 9–10: 1 aktum-TÚG ú-ḫáb 4 aktum-TÚG babbar *Ti-*[...]

The aktum-TÚG mu_4.mu is a type of aktum often given but difficult to translate. It is translated aktum-textile "to dress"[28] or "for the dressing ceremony".

> 75.G.1328 obv. VII 3–8: 2 aktum-TÚG ti-TÚG 1 aktum-TÚG mu_4- mu *Du-bù-ḫu-ma-lik* lú *Ib-gi-da-mu* šu-ra *in Ḫal-sum* [ki]

The aktum-TÚG ti-TÚG is translated by some scholars as "ribbed, pleated cloth"[29] and the storeroom é-ti-TÚG is translated as "drapery warehouse". Textiles of this type are often given to the king and the vizier in large numbers.

> 75.G.1794+ARET III 469 (King Išar-damu, Vizier Ibrium, month lost) obv. III 18–23: 12 aktum-TÚG ti-TÚG 1 gada-TÚG šu-DAG 1 gada-TÚG pad-*sù* en *si-in ʿÀ-za-an*[ki].
> 75.G.1763 (King Išar-damu, Vizier Ibbi-zikir, month lost) obv. III' 25–IV 1: 7 aktum-TÚG ti-TÚG *Íl-ʿà-ak-da-mu* lú é-nun-*sù;*
> obv. IV' 9–13: 4 aktum-TÚG ti-TÚG *Gi-ša-du* ga-du_8 lú tu-ra-*sù.*
> 75.G.1756 (King Išar-damu, Vizier Ibrium, month lost) rev. VI 3–5: 10 aktum-TÚG ti-TÚG *Ib-rí-um.*

ti-TÚG is a qualification also of other types of textiles as for example:

> 75.G.1785 (King Išar-damu, Vizier Ibrium, month lost) obv. VII 1–6: 1 íb+III-gùn-TÚG *Iš-má-zi-kir Kak-mi-um*[ki] *ma-za-lum* maškim *Ib-rí-um* 1 gada-TÚG pad-ti-TÚG *Ib-rí-um.*

Aktum-TÚG túg-ZI.ZI

It seems quite certain that it refers to the turban, one of the symbols of Eblaite royalty, even if

[26] Archi 1999, 45–53 translates the term as "cloak". Pasquali 1997, 219–220 does not propose a translation.
[27] For a translation of "tunic", see Archi 1999, 45–53; there is no evidence that the aktum was "the basic article of clothing". Civil 2008, 97: "a heavy garment".
[28] See Archi 1999, 45–53.
[29] See Archi 1999, 47; Civil 2008, 94: "a type of cloth or garment". "A meaning 'ribbed' derived from ti 'ib' seems unlikely".

we do not know when the king wore it.[30] The number of occasions this turban was given to the king requires further study.

75.G.1894 (King Išar-damu, Vizier Ibbi-zikir, month lost) obv. II 6–9: 1 aktum-TÚG túg-ZI.ZI sag en.
75.G.1729 (King Išar-damu, Vizier Ibbi-zikir, month MAxGÁNA*tenû*-úgur) obv. I 1–4: 1 aktum-TÚG túg-ZI.ZI sag en.

3.3. *dùl-TÚG*

It is not certain that this is an abbreviation of gu-dùl-TÚG. Often, it is qualified as produced in Mari.

75.G.1262 (King Išar-damu, Vizier Ibrium, month *ga-sum*) obv. IV 29–V 4: 1 dùl-TÚG *Ma-rí*[ki] 1 aktum-TÚG 1 íb-TÚG-sa$_6$-gùn *En-na-ni-il* lú *I-bí-zi-im* lú-kar.
75.G.1704+ARET III 738+ARET XII 501 (King Išar-damu – Vizier Ibrium, month *i-ba$_4$-sa*) obv. V 6–9: 1 dùl-TÚG *Ma-rí*[ki] 1 gada-TÚG mu$_4$.mu *Ib-rí-um.*
75.G.1888+11723 (King Išar-damu, Vizier Ibrium, month *ga-sum)* obv. III 13–IV 5: 1 dùl-TÚG *Ma-rí*[ki] 1 aktum-TÚG 1 íb+III-TÚG-sa$_6$-gùn *I-bí-zi-kir* lú *Zi-da in* ud ì-ti *mi-nu* Kiš[ki].[31]

3.4. *du-za-mu*[32]

On several occasions, this term qualifies other textiles, probably indicating a particularly fine quality of the textiles, as for example in:

75.G.1898+2447 (King Išar-damu, Vizier Ibbi-zikir, month *i-si)* obv. II 12–III 4: 5 *du-za-mu* ᾿*à-da-um*-TÚG-I 6 *du-za-mu* dùl-TÚG 1 *du-za-mu* ᾿*à-da-um*-TÚG-I 5 íb+III-TÚG ú-ḫáb 4 íb+III-TÚG sa$_6$-gùn 2 íb+III-TÚG-sal mu-DU *I-bí-zi-kir ma-lik-tum* šu ba$_4$-ti.

3.5. *gada-TÚG*

Flax was cultivated in Syria and Mesopotamia in the upper valleys of the Tigris and Euphrates from Neolithic times as archaeological evidence proves. It was used to produce linen textiles that are already quoted in texts from 3rd millennium Mesopotamia (of the Third Dynasty of Ur) and Syria.[33] Linen fabrics are frequently attested in all monthly accounts of textiles, although not in large quantities. They are given to all levels of society, but most frequently to the king, the queen, members of the royal family and to the vizier. Several times, linen textiles are given to equip equids and chariots of the king or other members of the court.

75.G.1363 (King Išar-damu, Vizier Ibrium, month ì-nun) obv. IX 16–X 1: 1 gada-TÚG mu$_4$.mu *Ib-rí-um.*
75.G.1704+ARET III 738+ARET XII 501 (King Išar-damu – Vizier Ibrium, month *i-ba$_4$-sa*) obv. V 6–9: 1 dùl-TÚG-*Ma-rí*[ki] 1 gada-TÚG mu$_4$.mu *Ib-rí-um* 2 gada-TÚG 2 IGI.NITA-*sù*
75.G.1894 (King Išar-damu, Vizier Ibbi-zikir, month lost) obv. V 20–24: 1 íb+III-TÚG-sag 1 gada-TÚG mu$_4$-mu *Du-bù-ḫu-*[d᾿]*À-da* en níg-dug-DU.
75.G.1249+ARET III 937+10082 (King Išar-damu, Vizier Ibbi-zikir last years, month *za-*᾿*à-tum*) rev. XV 3–8: 1 ᾿*à-da*-II 1 aktum-TÚG 1 íb-sa$_6$ 1 gada-TÚG 1 níg-lá-sag *Bù-gu-a-nu A-da-áš*[ki] du-du *si-in Šè-ti-lum*[ki].

[30] See Biga 1992, 19; Pinnock 1992, 18.
[31] It should be noted that Ibbizikir returned from (a trip to) Kiš; when someone returns from a military campaign the scribe writes this (*mi-nu* níg-kas$_4$ GN).
[32] See Pasquali 1997, 231–233.
[33] For an excellent recent study on flax and linen in Mesopotamia, see Breniquet 2008, 85–90.

Several mu-DU texts register the mu-DU of the kingdom of Dulu, including among various types of textiles, several linen textiles and threads of different materials. Moreover, some monthly accounts of textiles register mu-DU from Dulu when they consist of textiles:

> 75.G.1900 (King Išar-damu, Vizier Ibbi-zikir, month *i-rí-sá*) obv. IV 8–VI 8: 10 gada-TÚG mu_4.mu 3 gada-TÚG 2 *kir-na-nu* ... 5 gu si_4 5 gu gi 2 *mi-at* gu-li en 2 *kir-na-nu* 2 gu si_4 2 gu gi *ma-lik-tum* en *Du-lu*[ki] mu-DU 10 gada-TÚG mu_4.mu 5 gada-TÚG *kir-na-nu* 1 íb+III-TÚG gada-TÚG 1 šu-kešda gada-TÚG 5 KA-LAGABxA 3 gu si_4 3 gu gi mu-DU *Ar-ra-ti-lu.*

Sometimes, the linen textiles are distributed to workers involved in the fabrication of important objects as in the case of text 75.G.1781, where carpenters in the town of Abarru receive linen textiles to cut in order to make bowls.

> 75.G.1781 (King Išar-damu, Vizier Ibbi-zikir, month MAxGÁNA*tenû*-sag) obv. III 3–8: 3 gada-TÚG nagar-nagar TAR.TAR GIŠ-ban GIŠ-ban *in A-bar-*ru_{12}[ki].

3.6. *gàr-su*

The term, following Semitic etymology, can be translated as "band", as proposed by Pasquali (1997, 234).[34] It can often be of different colours, more often black (gi_6). The black band is also delivered in non-funerary contexts.

> 75.G.1888+11723 (King Išar-damu, Vizier Ibrium, month *ga-sum*) obv. II 1–5: 1 *ʾà-da-um*-TÚG-II 1 aktum-TÚG 1 íb+III-TÚG-sa_6-gùn 1 dib 50 kù-gi 1 *gàr-su* gi_6 *ba-da-lum Ḫa-ra-an*[ki35]; obv. XII 1–8: 1 *ʾà-da-um*-TÚG-II 1 aktum-TÚG 1 íb+III-TÚG-sa_6-gùn 1 *gàr-su* gi_6 *En-na-ni-il* dumu-nita *Ik-rí-iš* TUŠ.LÚxTIL *in Ḫal-sum*[ki].

3.7. *gíd-TÚG*

It is quoted in LL857: gíd-TÚG= *gu-da-núm/nu-um*, Akk. *qutānum.* AHw s.v. "das Dünne".

Sometimes it is quoted in the first position before the other textiles. It can be coloured. The meaning "long" of the Sumerian "gíd" suggested a translation as "long garment".[36]

> 75.G.1221 (King Išar-damu, Vizier Ibbi-zikir, month *ḫa-li*) obv. VI 2–8: 1 $zara_6$-TÚG 1 gíd-TÚG 2 *bu-di* 10 kù-gi 1 dumu-mí en *Kak-mi-um*[ki] *si-in* ÉxPAP Gaba-*da-mu* šu mu-tak_4.
> 75.G.1324 obv. IV 4–13 : 1 $zara_6$-TÚG babbar 1 $zara_6$-TÚG ú-ḫáb 2 gíd-TÚG ú-ḫáb GIŠ-*ir-zú* 2 níg-lá-ZI.ZI ú-ḫáb babbar 2 *bu-di* GÁxLÁ šú+ša gín DILMUN kù-gi *ma-lik-tum Ma-nu-wa-at*[ki] ÉxPAP *Ìr-am*$_6$*-da-ar* di-ku_5 *A-bu* šu mu-tak_4, "1,1,2 textiles, 2 brooches weighing 20 sicles of gold for the queen of Manuwat, for her tomb, Iram-dar, the judge and Abu delivered".

3.8. *gu-dùl-TÚG*

This fabric is given to both important and less important functionaries. Often it is mentioned first when a group of textiles is distributed and is followed by sal-TÚG.[37]

> 75.G. 1243 rev. III 8–9: 20 gu-dùl-TÚG simug-simug, "20 gu-dùl-textiles to the smiths".
> 75.G.1524 (King Išar-damu, Vizier Ibrium, month MAxGÁNA*tenû*-sag) obv. V 1–5: 2 $zara_6$-TÚG 2 gu-dùl-TÚG 2 *bu-di* šú+ša gín-DILMUN bar_6:kù *Da-um* dam en *A-*ru_{12}*-ga-du*[ki], "2 zara-textiles 2 gu-dùl-textiles 2 brooches weighing 20 sicles of silver for Daum, woman of the king (in the palace) of Arugadu".

---

[34] For the translation as "cinghia", see Pasquali 1997, 233–235. A translation as "sacca" is proposed by Pomponio 2008, 107 following Archi 2003, 34.

[35] In this case, the black textile is sent to the badalum of Harran, but not on the occasion of a funeral.

[36] See Archi 1999, 48.

[37] Due to its position in the groups of textiles, Archi supposed that "it seems to be a kind of second quality cloak" (1999, 50).

3.9. *gu-mug-TÚG*

The gu-mug textile is often given alone to people of very different status, and therefore, it is considered a textile of low quality. The textile is considered a garment by Archi (1999, 47), a "very simple article of clothing originally made of shoddy wool". Nothing in the texts proves that we are dealing with a garment and not with a textile. It is often given to workers but also to important people and it is sometimes in first place in some groups of textiles. The group can be composed of a gu-mug-textile, sal-textile and íb-textile.

It is not counted in the totals of textiles at the end of the texts, but counted separately.

> 75.G.1524 (King Išar-damu, Vizier Ibrium, month MAxGÁNA*tenû*-sag) obv. X 17–XI 7: 1 gu-mug-TÚG 1 sal-TÚG 1 íb+III-gùn-TÚG 3 níg-lá-DU *I-sar wa* maškim-*sù* du-du *Ša-nam-ti-um*^{ki}, " 1 gu-mug-textile, 1 sal-textile, 1 multicoloured íb-textile, 3 bands for the feet, to Išar and his maškim-functionary who are going to the town of Šanamtium".
> 75.G.1249+ARET III 937+10082 (King Išar-damu, Vizier Ibbi-zikir last years, month *za-'à-tum*) obv. XIV 6–9: 2 gu-mug-TÚG 2 ugula bìr-BAR.AN en *Ù-ti-gú*^{ki}.
> 75.G.1351 (King Išar-damu, Vizier Ibrium, month MAxGÁNA*tenû*-sag) rev. IX 1–5: 6 gu-mug-TÚG ir$_{11}$-ir$_{11}$ en *En-na-il* šu be-ti;
> rev. X 11–15: 20 lá-2 gu-mug-TÚG ir$_{11}$-ir$_{11}$ en *Ù-rí* šu ba$_4$-ti.
> 75.G.1728 (King Išar-damu, Vizier Ibrium, month lost) rev. I 7–13: 2 gu-mug-TÚG *Ti-ti-nu* maškim *Ib-rí-um* šu ba$_4$-ti *in A-te-na-at*^{ki};
> obv. II 13–III 5: 10 gu-mug-TÚG *Ti-ti-nu* maškim *Ib-rí-um* šu ba$_4$-ti túg-nu-tag gá-gá tu.mušen é *Ib-rí-um*, "10 gu-mug-textiles Titinu, maškim-functionary of Ibrium, has received for the weavers of the dovecotes of the palace of Ibrium".[38]

gu-mug-TÚG-I

Occasionally, the textile is specified by a number I, whose significance is uncertain.

> 75.G.1888+11723 (King Išar-damu, Vizier Ibrium, month *ga-sum*) obv. V 8–15: 1 *'à-da-um*-TÚG 1 aktum-TÚG 1 íb+III-TÚG-sa$_6$-gùn *Bur-ḫa-áš* 1 gu-mug-TÚG-I 1 sal-TÚG 1 íb+II-TÚG-gùn *A-ma-lik* maškim-*sù* šu-du$_8$ in *Ḫal-sum*^{ki}; obv. VIII 9–11: 1 gu-mug-TÚG-I *Bù-ma-ù* NE.DI; obv. IX 13–16: 1 gu-mug-TÚG-I dumu-nita NE.DI *A-bar-sal*$_4$; obv. X 6–10: 1 gu-mug-TÚG-I 1 sal-TÚG 1 íb+III-TÚG-gùn *Ìr-a-um* ur$_4$ é *Ib-rí-um* lú *Dur-bí-du*^{ki}.[39]

3.10. *gu-zi-tum*-TÚG

cf. Akk. *kusītum, AHw* s.v. "Gewand".

In the texts of the time of Vizier Ibrium, this textile is often quoted in the first position where in the texts of Vizier Ibbi-zikir *'à-da-um*-I/II is mentioned.

> 75.G.1707 (King Išar-damu, Vizier Ibrium, month *ga-sum*) obv. IV 8–15: 1 *gu-zi-tum*-TÚG 1 aktum-TÚG 1 íb+III-TÚG-sa$_6$-gùn 1 *gú-li-lum* a-gar$_5$-gar$_5$ kù-gi TAR 2 1 íb-lá 1 gír-kun bar$_6$:kù ša-pi gín DILMUN 1 gír mar-tu ti kù-gi *Bù-da-il* TUŠ.LÚxTIL *Za-ḫi-ra-an*^{ki} *Maš-bar-rá*^{ki} [(...)] [*x*]-⌜*la?*⌝-*ga*^{ki}

3.11. *íb+I/II/III/IV/V-TÚG*

It is certainly the most commonly given textile in Ebla, either alone or with other textiles. It seems to be almost always multicoloured (gùn), and often of good quality (sa$_6$), sometimes of first quality (sag), white (babbar), white and woven in the city of Mari (babbar *Ma-rí*^{ki}), in the city of Armi (*Ar-mi*^{ki}), long (gíd), sal (uncertain translation), red (ú-ḫáb and ú-ḫáb-sa). It is

[38] For the presence of dovecotes at Ebla, see Biga 2009a, 11–16.

[39] In the totals at the end of the text in rev. IX 3: 7 gu-mug-TÚG without the specification I.

given to the king, the vizier and to all kinds of people of different status.

Despite various proposals by scholars, the significance of the numbers I,II,III,IV,V written in the sign íb or immediately after the sign is uncertain.

In the list of Egyptian textiles written on a box found in a tomb in Gebelein, some textiles are qualified with the number of their threads, from 1 to 9.[40] Very often it is given alone, but is often part (normally in third position but also in second position) of a group of textiles.

Given the meaning "hip" of the Sumerian word "íb", all scholars agree that we are dealing with a piece of fabric to wrap around the hips and to tie in some way. Different translations are given: Archi translates "waistband"; Pasquali "gonnellino", Waetzoldt "Hüfttüch"; Sallaberger "Hüfttuch" and Civil "sash". Many examples of people receiving íb-textiles are already published.

An íb-textile made with gold and silver thread for particularly precious clothing is possibly attested in some texts.[41] Some archaeological remains found in 2004 in the Throne room of Palace G of Ebla seem to support the hypothesis of such textiles very well documented in the Bronze Age Mediterranean area and in Mycenae.[42]

3.12. *níg-lá-sag*

A headband. Sometimes it is given alone, but more often with other textiles. It seems in some way more commonly given to people leaving on a journey, to protect the head.

> 75.G.1704+ARET III 738+ARET XII 501 (King Išar-damu, Vizier Ibrium, month *i-ba$_4$-sa*) obv. I 1–4: 1 *ʾà-da-um*-TÚG-II 1 aktum-TÚG-II 1 íb+III-TÚG-sa$_6$-gùn 1 níg-lá-sag 2 níg-lá-DU *Mu-ne-a-du* ugula *Ar-ḫa-du*[ki].
> 75.G.1934+10022 (King Išar-damu. Vizier Ibbi-zikir, month lost) obv. VI 21–VII 4: 1 níg-lá-sag *Iš-má-ga-lu* NE.DI *in Ì-ab* (without ki) šu ba$_4$-ti.
> 75.G.1249+ARET III 937+10082 (King Išar-damu, Vizier Ibbi-zikir final years, month *za-ʿà-tum*) rev. XV 3–8: 1 *ʾà-da*-II 1 aktum-TÚG 1 íb-sa$_6$ 1 gada-TÚG 1 níg-lá-sag *Bù-gu-a-nu A-da-áš*[ki] du-du *si-in Šè-ti-lum*[ki],
> "1 ʾà.-textile 1 aktum-textile 1 íb-textile of good quality 1 linen-textile 1 headband for Buguanu of the town of Adaš who is going to the town of Šetilum".

3.13. *níg-lá-gaba*

A chest-band. Sometimes these textiles are given alone and sometimes with other textiles.

> 75.G.1797 (King Išar-damu, Vizier Ibrium, month lost) obv. XI 16–XII 5: 1 níg-lá-gaba 1 íb+III-TÚG-sa$_6$-gùn 1 íb-lá gùn[43] *Ib-rí-um in* ud é-dingir-dingir-dingir du-du.
> 75.G.1917 (King Išar-damu, Vizier Ibbi-zikir, month lost) rev. I 14–II 11: 15 níg-lá-gaba *Ìr-ni-ba Ḫu-ba-an* Puzur-*ra-ma-lik Gur-da-lum*[ki] *Ba-du-lum A-za$_x$-bar-zú Še-bal Ib-ga-ì* ugula bìr-[BAR.AN] en *En-*⸢*na*⸣-BE *Bù-*⸢*gu*⸣ *Ìr-ḫuš-zé Dur-rí*[ki] *Mi-ga-ì Gú-šu Ib-dur-i-šar* ugula bìr-BAR.AN *I-bí-zi*-kir,
> "15 chest-bands to 15 Personal Names of the superintendants of the charioteers of the king and of Vizier Ibbizikir".

3.14. *níg-lá-DU*

A band for the feet, sometimes given to people travelling.

> 75.G.1221 (King Išar-damu, Vizier Ibbi-zikir, month *ḫa-li)* obv. II 7–15: 4 *ʾà-da-um*-TÚG-II 4 aktum-TÚG 4 íb+IV-TÚG-sa$_6$ 4 níg-lá-DU 2 gada-TÚG 1 gír mar-tu TI 10 bar$_6$kù níg-kaskal *En-ḫar-li-im En-na-ì* du-du

[40] See Roccati 1970, 7. According to Civil 2008, 93 "The number n is an indication of the length of the sash".
[41] See also Pasquali 2005b, 290.
[42] See Gleba 2006, 61–77. See also the contribution of Sophie Desrosiers in this volume.
[43] It should be noted the multicoloured belt (íb-lá gùn).

Du-gú-ra-su^{ki}, " 4 ʾà.-textiles 4 aktum-textiles 4 íb-textiles of good quality 4 bands for the feet 2 linen-textiles 1 martu-dagger weighing 10 sicles of silver, provision for the trip of Enhar-lim and Ennai who are going to Dugurasu".

3.15. *pad-TÚG*

A veil, well documented also in the ritual for royalty.

3.16. *sal-TÚG*

Textile given to any type of person; it can be cut and used to protect both feet. It is translated as "light", but there is no evidence to show if it was a "light textile".

> 75.G.1761 (King Išar-damu, Vizier Ibrium, month lost) obv. X 18–20: 1 sal-TÚG 2 DU *Ib-rí-um,* "1 sal-textile for both feet of Vizier Ibrium".
> 75.G.1777+ARET III 35 obv. III 15–IV 3: 1 sal-TÚG 2 DU *I-bí-zi-kir* Puzur-*rí* šu ba_4-ti, "1 sal-textile for both feet of Vizier Ibbizikir Puzurri received".

Sometimes qualified as ti-TÚG

> 75.G.1411 (King Išar-damu, Vizier Ibrium year 14, month lost) obv. X 8–14: 2 sal-TÚG ti-TÚG *Ga-da-ba-an* lú *A-gi* 1 sal-TÚG ti-TÚG *Ar-si-a-ḫa* šu-i *Ib-rí-um.*

3.17. *túg-gùn*

A multicoloured, precious fabric often in the first position in the group of textiles, given instead of ʾ*à-da-um*-textile.

> 75.G.1328 (King Išar-damu, Vizier Ibrium, month *i-rí-sá*) obv. I 1–9: 1 túg-gùn 1 aktum-TÚG 1 íb+III-sa_6-gùn 1 íb-lá 1 *si-ti-tum* 1 gír-kun 1 ma-na kù-gi lú mu-DU *Ib-rí-um* in-na-sum *I-bí-zi-kir in* ud TUŠ.LÚxTIL *Ḫal-sum*^{ki}.
> 75.G.1708 (King Išar-damu, Vizier Ibrium, month *ga-sum*) rev. VII 10–14: 1 túg-gùn *ù-ra* 1 aktum-TÚG 1 íb+III-sa_6-gùn-TÚG 1 íb-lá bar_6:kù TAR.TAR kù-gi 1 dib GÁxLÁ šú+ša kù-gi ⌜*x*⌝-⌜*x*⌝-*ša* ugula *Ù-si-gú*^{ki};
> rev. X 5–8: 1 túg-gùn *ù-ra* 1 sal-TÚG 1 íb+III-gùn-TÚG *Du-lu*^{ki} 1 túg-NI.NI 1 íb+III-gùn-TÚG maškim-*sù.*
> 75.G.1741 (King Išar-damu, Vizier Ibrium, month še-gur_x-mìn) obv. XII 1–9: 2 túg-gùn 2 aktum-TÚG 2 íb+IV-TÚG-sa_6-gùn *Zi-ba-da wa A-šum.*

3.18. *túg-NI.NI*

This is one of the most quoted textiles given to men, women and animals, although the less important women of the court received them more often than the more important women. A reddish-brown coloured túg-NI.NI was given to animals to be sacrificed during important festivals.

> 75.G.1221 (King Išar-damu, Vizier Ibbi-zikir, month *ḫa-li*) obv. IV 9–13: 1 túg-NI.NI ú-ḫáb 1 am ^{d}*Ra-sa-ap gú-nu* en ì-na-sum, "1 reddish-brown textile-NI.NI for a bull (to be sacrificed) to the god Rasap gunu the king has given".

This textile is sometimes qualified as being of good quality (sa_6)

> 75.G.1522 (King Išar-damu, Vizier Ibrium, month *i-*ba_4*-sa*) obv. XII 11–14: 1 túg-NI.NI-sa_6 *A-da-da* ama-gal *Du-si-gú.*
> 75.G.1263 (King Išar-damu. Vizier Ibrium, month lost) obv. VI' 16–VII' 18: 20 sal-TÚG to 20 Personal Names KÍD-sag,
> "20 sal-textiles to 20 people, gatekeepers".

3.19. *zara$_6$-TÚG*

This fabric is the most frequently given textile to the important women of the court. However, it is given to men and gods, too, several times on various occasions. It is rarely coloured, although it can also be in different colours, including white.

> 75.G.1522 obv. VII 7–13: 1 zara$_6$-TÚG babbar 1 zara$_6$-TÚG ú-ḫáb 2 *bu-di* šú+ša gín DILMUN bar$_6$:kù *Ma-ù-ut* dumu-mí en *gal-tum Dur-du-lum.*
> 75.G.1791(King Išar-damu, Vizier Ibrium, month lost) rev. VI 4–10: 1 zara$_6$-TÚG ú-ḫáb níg-ba ^{d}*Ra-sa-ap* lú ʾ*À-da*-NI^{ki} *Du-si-gú* in-na-sum.
> 75.G.1917 (King Išar-damu, Vizier Ibbi-zikir, month lost) obv. I' 8'-11': 1 zara$_6$-TÚG ú-ḫáb-sa níg-ba *ma-lik-tum* ^{d}*Ra-sa-ap.*

Brooches (*bu-di*) were often given with zara$_6$-TÚG and also with túg-NI-NI, gíd-TÚG and gu-dùl-TÚG, probably to fix these wrapped textiles when worn by women.

4. Individual Textiles and groups of textiles given to various palace employees.

4.1. *Groups of textiles*

It is useful to look at the textiles given to various people in the service of the Eblaite court, such as merchants, acrobats, singers, dancers and not least, soldiers. While the king and the queen and members of the Eblaite and other courts receive textiles as precious gifts, those employed in the palace receive the fabrics as payment for services rendered. Apart from the aktum-TÚG túg-ZI.ZI that is always only given for the head of the king, almost all the other textiles are given to any type of person.

There are no fabrics restricted to men or women. Hence, these pieces of fabrics could be used to make clothes for either gender.

It may be deduced that the textiles they receive vary in type according to the importance of the work done. Several groups of textiles given together are frequently attested. Among the more frequently attested groups of textiles are: ʾ*à-da-um*, aktum and íb; gudùl, sal and íb; gumug, sal and íb; ʾ*à-da-um* and íb (without aktum); ʾ*à-da-um*, sal and íb; *guzitum* and íb; túg-NI.NI and íb; aktum and sal; gudùl and íb.

As already observed by several scholars, the group of textiles composed by ʾ*à-da-um*, aktum and íb is the most attested in the MAT during Vizier Ibbi-zikir's time.

Already in the time of Vizier Ibrium, this group of textiles is often attested even if other groups of textiles are even more attested. In Vizier Ibrium's time, the textiles-aktum and then íb are always in second position but the first fabric varies and can be a *guzitum*-textile or a gùn-textile; while several other groups of textiles are attested as well. It is improbable that the group composed of three textiles is a group of fabrics necessary for a full garment.

A distinction in the type of deliveries to a king and to his elders or maškim-functionaries was surely made very often, most probably according to the status of the recipient, as is evident in many published and unpublished texts.

> 75.G.1524 (King Išar-damu, Vizier Ibrium, month MAxGÁNA*tenû*-sag) obv. I 1–10: 1 ʾ*à-da-um*-TÚG 1 aktum-TÚG 1 íb+III-TÚG-sa$_6$-gùn en *Gú-da-da-núm*^{ki} 3 gu-dùl-TÚG 3 sal-TÚG 3 íb+III-TÚG-gùn maškim-*sù in* ud du-du *si-in* SA.ZA$_x$^{ki},

"1,1,1 textiles to the king of Gudadanum, 3,3,3 textiles to his maškim-functionaries on the day when they went to the Saza (i.e. Ebla)."[44]
75.G.1225 obv. VI 1–11 : 1 *'à-da-um*-TÚG-II 1 aktum-TÚG 1 íb+III-TÚG-sa$_6$ en *I-bu-íb*^{ki} 1 zara$_6$-TÚG *ma-lik-tum* 1 aktum-TÚG 1 sal-TÚG 2 dumu-mí-*sù* 2 aktum-TÚG 2 íb+III-TÚG-sa$_6$-gùn 2 dumu-nita-*sù* 2 *'à-da-um*-TÚG-II 2 *'à-da-um*-TÚG-I 2 aktum-TÚG 2 sal-TÚG 2 íb+IV-TÚG-gùn abba$_x$-*sù*,
"1,1,1 textiles for the king of Ibuib, 1 textile for the queen, 1,1 textiles for two of their daughters, 2,2 textiles for 2 of their sons, 2,2,2,2 textiles for their elders";
75.G.1225 obv. VII 22–VIII 8: 1 *'à-da-um*-TÚG-II 1 aktum-TÚG 1 íb+III-TÚG-sa$_6$-gùn en *Bur-ma-an*^{ki} 1 zara$_6$-TÚG *ma-lik-tum* 2 sal-TÚG nin-ni en 1 aktum-TÚG šeš-mu-*sù* 2 *'à-da-um*-TÚG-II 2 *ʾà-da-um*-TÚG-I 2 aktum-TÚG 2 sal-TÚG 2 íb+III-TÚG-sa$_6$-gùn 2 íb+III-TÚG-gùn ABxÁŠ-*sù*,
"1,1,1 textiles for the king of Burman, 1 textile for the queen, 2 textiles for the sisters of the king, 1 textile for his brother, 2,2,2,2,2,2 textiles for his elders".

However, even anonymous messengers receive the most delivered group of textiles.

75.G.1225 (King Išar-damu - Vizier Ibbi-zikir) rev. V 4–11 : 2 *'à-da-um*-TÚG-II 2 aktum-TÚG 2 íb+IV-TÚG-sa$_6$-gùn 2 kas$_4$ *Sa-nap-zu-gúm*^{ki} 2 aktum-TÚG 2 íb+III-TÚG-gùn 2 maškim-*sù* 4 *'à-da-um*-TÚG-I 4 aktum-TÚG 4 íb+III-TÚG-gùn 4 kas$_4$ *Ḫa-ra-an*^{ki},
"2, 2,2 fabrics to 2 messengers of the kingdom of Sanapzugum, 2,2,2 fabrics to two maškim-functionaries, 4,4,4 fabrics to 4 messengers of Harran".

Several different groups of fabrics are already attested and well documented in many of the published texts. See also, for example,

75.G.1888+11723 (King Išar-damu, Vizier Ibrium, month *ga-sum*). In the text, several people receive the group composed of *ʾàdaum*, aktum, íb, on different occasions, such as, news brought to the palace or a purification ritual. Yet, others receive different groups, some composed of three textiles, but also two. Many, of course, receive only one type of textile, mostly íb:

obv. III 13–V 10: 1 dùl-TÚG *Ma-ri*^{ki} 1 aktum-TÚG 1 íb+III-TÚG-sa$_6$-gùn *I-bí-zi-kir* lú *Zi-da in* ud ì-ti *mi-nu* Kiš^{ki} 1 *ʾà-da-um*-TÚG-I 1 sal-TÚG 1 íb+III-TÚG-gùn *I-na-bal*! (KUL) *I-ni-bu*^{ki} 1 sal-TÚG 1 íb+III-TÚG-gùn maškim-*sù* giš-dug-DU 1 gu-dùl-TÚG 1 aktum-TÚG 1 íb+III-TÚG ú-ḫáb-sal *zú-ba-lum* ⌜*Gi?*⌝-⌜*rí*⌝ dumu-nita *Ib-rí-um* šu mu-tak$_4$ *in* ud ḫúl 2 na-rú-*sù* 1 *ʾà-da-um*-TÚG-II 1 aktum-TÚG 1 íb+III-TÚG-sa$_6$-gùn *Bur-ḫa-áš* 1 gu-mug-TÚG-I 1 sal-TÚG 1 íb+II-TÚG-gùn *A-ma-lik* maškim-*sù* šu-du$_8$ *in Ḫal-sum*^{ki}.
75.G.1761 (King Išar-damu, Vizier Ibrium, month lost) obv. VI' 7–VII' 4: 9 *ʾà-da-um*-TÚG-I 9 sal-TÚG 9 íb-[...] *La-bù-du I-rí-ik-da-mu Ib-za-ḫa-lam Ir*-NE *I-dè-ni-ki-mu Ìr-am$_6$-da-mu Du-bù-uš*-KA-*su-gú Zi-íb-da-mu Du-bù-uš-da-mu* dumu-nita-dumu-nita en *Ma-nu-wa-at*^{ki}.
75.G.1891+ARET III 31 (King Išar-damu, Vizier Ibrium, month lost) rev. IV 3–4: 1 aktum-TÚG 1 íb+III-babbar-TÚG 1 níg-lá-sag 1 níg-lá-gaba *I-ti-*^{d}*' À-da* (a prince of Ebla).

Sometimes, the chief of a kingdom receives the same textiles as his maškim-functionaries, as the case of Kuntisu, a well-known king of the kingdom of Ursaum, mentioned in several texts and more often qualified as en (king).

75.G.1375 (King Išar-damu, Vizier Ibrium, month ì-nun) obv. VIII 9–IX 2: 3 *ʾà-da-um*-TÚG-II 3 aktum-TÚG 3 íb+III-TÚG-sa$_6$-gùn *Kùn-ti-su* ugula *Ur-sa-um*^{ki} *Su-mi*-NE *Su-ru$_{12}$-si* maškim-*sù* 1 sal-TÚG 1 íb+II-TÚG-gùn *ma-za-lum-sù*.[45]

[44] See also Biga 2008, 293 for further examples.

[45] The story of the kingdom of Ursaum and its relationship with Ebla's kingdom is very interesting and requires further in-depth study. In this text, Kuntisu is qualified as ugula of Ursaum, later in several texts he is called king of Ursaum, sending always (as tribute?) several gifts to the Eblaite court and leaving his son as *valet de chambre* to the king of Ebla, see Biga 2008, 304.

75.G.1754 (King Išar-damu, Vizier Ibbi-zikir, month lost) obv. II' 14–III 8: 2 *ʾà-da-um*-TÚG-I 4 aktum-TÚG 2 gada-TÚG mu_4.mu 2 aktum-TÚG ti-TÚG *Ik-su-ub-da-mu Zé-da-mu* 2 dumu-nita máḫ 2 *ʾà-da-um*-TÚG-I 2 aktum-TÚG 2 gada-TÚG mu_4. mu *Zi-íb-da-mu Ga-du-um* dumu-nita tur en,
"2,4 textiles, 2 linen textiles, 2 textiles for Iksub-damu and Ze-damu 2 older sons (of the king) 2,2 textiles, 2 linen textiles for Zib-damu and Gadum small sons of the king".

The study of the totals at the end of every monthly account of textiles can be useful to understand the type of fabrics we are dealing with. They also serve to verify the types of textiles that are normally counted together, and in the order in which they are quoted. A textiles hierarchy cannot be discerned in these totals, only a list of the different types, and the various types of occasions in which they were used.

The scribes, who inscribed some types of textiles together, perhaps considered them as being similar.

75.G.2499 (King Išar-damu, Vizier Ibbi-zikir, first three-four years, month *gi*-NI) rev. XI 1–12: AN.ŠÈ.GÚ 36 *ʾà-da-um*-TÚG-II [20]+24 aktum-TÚG 12 gu-dùl-TÚG gada-TÚG *ʾà-da*-I 2 $zara_6$-TÚG 2 níg-lá-gaba 3 túg-NI.NI 8 sal-TÚG 1 íb+III-TÚG-sag-sa_6-gùn 44 íb+III-TÚG-sa_6-gùn 21 íb+III-TÚG-gùn and after the general total 1 gu-mug-TÚG is counted.
75.G.1888+11723 (King Išar-damu, Vizier Ibrium, month *ga-sum*) rev. X 1–8: [AN]. [ŠÈ].GÚ 20 lá-2 *ʾà-da-um*-TÚG-II *gu-zi*-TÚG dùl-TÚG *Ma-rí*[ki] 30 lá-2 aktum-TÚG 3 $zara_6$-TÚG 3 túg-NI.NI 20 lá-1 *ʾà-da-um*-TÚG-I gu-dùl-TÚG gada-TÚG 13 sal-TÚG 20 lá-1 ⌜íb⌝+III-TÚG-sa_6-gùn 1 íb+III-TÚG ú-ḫáb-sal 20 lá-1 íb+III-TÚG [...
75.G.1760+10130 (King Išar-damu, Vizier Ibrium, month MAxGÁNA*tenû*-úgur) rev. VIII 1–11: AN.ŠÈ.GÚ 11 *gu-zi-tum*-TÚG *ʾà-da-um*-TÚG-II 22 aktum-TÚG 3 túg-NI.NI 10 lá-2 gu-dùl-TÚG *ʾà-da-um*-TÚG-I gada-TÚG/ 16 sal TÚG 1 lá-gaba 13 íb+IV-sa_6-gùn-TÚG 30 lá-3 íb+III-TÚG-gùn šu-nígin 120 túg-túg è iti MAxGÁNA*tenû*-úgur; and, in another column: 17 gu-mug-TÚG.

It is often difficult to know if the textiles were distributed in Ebla itself, or taken by merchants to the different towns. Sometimes, the scribes registered that textiles were received in a certain town. At other times, it is clear that there is a sort of itinerary followed by the merchants or by the army. When the army was on the march, textiles were distributed for different purposes. In some cases, it seems quite probable that the quoted textiles were given at Ebla, when people visited for different reasons.

4.2. Textiles given to merchants (lú-kar) employed by the Palace

75.G.1322 (King Išar-damu, Vizier Ibrium, month lost) obv. XII 5–18:12 *ʾà-da-um*-TÚG 12 aktum-TÚG 12 íb+III-sa_6-gùn-TÚG *Du-bí-zi-kir Íl-ba-šum A-ma-ma-lik I-rí-ik-il Bù-da-ma-lik A-ti Bù-da-ma-lik-II En-zi-ma-lik En-na-ma-lik Wa-na A-du-bù* ⌜*x*⌝-⌜*x*⌝ [lú-ka]r.
75.G.1324 (King Išar-damu, Vizier Ibrium, month *za-ʾà-tum*) obv. VII 4–VIII 5: 6 *gu-zi-tum*-TÚG 6 íb+III-TÚG-sa_6-gùn *Ib-dur-iš-lu Zi-la-il Ḫu-la Ìr-am₆-ma-lik* ⌜*x*⌝-*[x] [.....]* lú-kar máḫ 4 gu-dùl-TÚG 4 íb+III-gùn-TÚG NI-*zi-ma wa Iš*$_{11}$*-gi-ma-lik Ib-gi Gú-li-im* lú-kar tur gibil.[46]
See also 75.G.1427 (King Išar-damu, Vizier Ibrium, month lost) obv. XI 12–XII 10: 11 íb+III-TÚG-sa_6-gùn to 11 personal names máḫ lú-kar; obv. XIII 4–XIV 5: 11 íb+III-TÚG-gùn to 11 personal names tur lú-kar.

However, it must be noted that in the same text, another merchant receives a different textile, see rev. VI 4–9: 1 sal-TÚG *Ar-si-a-ḫa* lú-kar *A-ru*$_{12}$*-ga*-du[ki] maškim *Bíl-ma-lik*

75.G.1298 (King Išar-damu, Vizier Ibrium, month *i-ba₄-sa*) obv. IX 6–8: 1 níg-lá-gaba 1 níg-lá-sag *I-za-iš-lu* lú-kar.

[46] It must be noted that there are some small differences between the types of textiles given to the principal merchants (lú-kar máḫ) and to those given to the other merchants (lú-kar tur).

There were surely standard items with which the Eblaite merchants were equipped when travelling, but what we know so far is that this consisted of silver that they received for the expedition, as attested by the metal texts.[47] As the textiles they received were of different types, we do not know of any standard types issued to the merchants for their journeys.

The texts do not register large amounts of textiles given to merchants to sell. Rather, the textiles and wool (siki) distributed to the merchants are only as payment for services rendered.

> 75.G.1262 (King Išar-damu, Vizier Ibrium, month *ga-sum*) obv. X 14–XIII 10: 36 íb+III-gùn-TÚG 56 KIN siki to 36 personal names [lú]-kar ⌜*A*⌝-⌜*da*⌝-*bí-ik*^{ki}.

4.3. *Textiles for brides and for the great ritual of the renewal of royalty*

On the occasion of the interdynastic marriages of some daughters of the kings of Ebla, the trousseaux given to these girls included many textiles of different types. Some of these lists have already been published[48] and some conclusions have recently been reached on the types and colours of textiles used during the marriage ceremony.[49] However, the trousseaux have not all been studied yet and compared with those that some of the newly deceased princesses of Ebla were accorded at their funerals.

After their wedding, King Išar-damu and his queen, Tabur-damu, celebrated a complex ritual of royalty, fertility and renewal.[50]

It is still hard to determine the type of ritual we are dealing with, and whether or not it resembles the ritual of renewal of the Egyptian *Sed* festival.

In the ritual texts, several textiles of different colours are used by the king, the queen and other participants in the ritual.

The marriage of the royal couple and the ritual performed immediately afterwards are also quoted in some administrative texts. The annual account of metals 75.G.1730 (=MEE VII 34) and two monthly accounts of textiles I identified some years ago as mentioning the same ritual, enable us to further our knoweldge of the types of textiles worn by the participants in the ritual.[51]

4.4. *Textiles for a newborn grandchild of the king of Ebla or for births in general*

As previously mentioned, some Ebla texts mention weddings of Eblaite princesses to foreign royalty. Other later texts often mention the birth of a child in a foreign court, a grandchild of the king of Ebla. This event was the occasion for the Eblaite court to send gifts of textiles and jewels to the mother, her newborn child, and its relatives.

The birth of the first son of the royal couple, King Išar-damu and Queen Tabur-damu, was particularly celebrated. As already noted,[52] two monthly accounts of deliveries of textiles (ARET I 15 and ARET IV 7) were necessary to register all the deliveries to members of the court and for the ceremonies on the occasion of this important birth.

---

[47] See Biga 1996, 34. For similar issue for the merchants of Ugarit, see Liverani 1979, 495–503.

[48] Biga 1996, 63–72; Biga 1998, 17–22; Archi 2002b, 161–199.

[49] Pasquali 2005a, 165–184.

[50] See ARET XI, 1,2,3.

[51] See Biga 2009b.

[52] Biga & Pomponio 1993, 108–111.

75.G.1704+ARET III 738+ARET XII 501 (King Išar-damu, Vizier Ibrium, month *i-ba$_4$-sa*) obv. VIII 9' – IX 1 : 1 zara$_6$-TÚG 2 *bu-di* šú+ša gín DILMUN bar$_6$-kù *Zi-mi-ni*-bar$_6$-kù *ma-lik-tum Bur-ma-an*[ki] *in* ud dumu-mí-*sù* tu-da 1 íb+III-TÚG *Ma-rí*[ki] dumu-mí-*sù* *I-[lul]-⌜za$_x$⌝-[ma-lik]* [šu mu-tak$_4$], "1 textile, 2 brooches weighing 20 sicles of silver to Zimini-barku, queen of Burman, when she gave birth to a daughter, 1 textile for her daughter, Ilulza-malik delivered".

It is possible that some of the clothes for a newborn baby were made using the íb-textile, often quoted as delivered to newborn children.

4.5. *Textiles for dancers (NE.DI)*

The male and female dancers working in the Eblaite palace often received gifts of textiles of all types, even if the íb-textiles are the most frequently given. However, it is impossible to know from the texts how they were dressed when performing their dances. We can suppose that they used the fabrics received to also make clothes for their performances.

75.G.1524 (King Išar-damu, Vizier Ibrium, month MAxGÁNA*tenû*-sag) obv. VIII 6–8: 1 gu-mug-TÚG 1 íb+III-gùn-TÚG *I-ti-a-gú* NE.DI.
75.G.1748 (King Išar-damu, Vizier Ibrium, month *za*-[*'à*]-*na*-⌜*at*⌝) rev. IV 15–VI 10: 29 íb+III-gùn-TÚG 1 íb+III-TÚG-sa$_6$-gùn *Iš-má-ga-lu Ìr-am$_6$-da-ar Ik-bù-ul-ma-lik Ib-ḫu-úr A-du-mu Ra-ba En-na-ma-lik Ìr-ni-ba Dùl-da-ma-ḫu* NE-*zi-ma-ì* Puzur$_4$-*ra-ma-lik U$_9$-bù I-šar Ab-ra-ḫu I-ti-a-gú Kum-zé Ku-bar-ru$_{12}$ Iš-la-da-du* NI-*zi-ma-aḫ Ìr-an-šar Gur-a Iš*-NE-*du Iš-ra-il En-na-ì I-ti-a-gú*-II *Za-a-rúm Iš$_{11}$-a-ma-lik Téš*-NI *A-bù-ma-lik Ib-ga-il* NE.DI-NE.DI.
75.G.1787 rev. IV 8–10: 2 túg-NI.NI [(1)]+1 dam NE.DI *Ša-nu-gú*[ki].
75.G.1776 (King Išar-damu, Vizier Ibrium, month lost) obv. X 5–8: 1 túg-NI.NI dam ugula *Ša-dab$_6$*[ki] NE.DI.
75.G.1781 (King Išar-damu, Vizier Ibbi-zikir, month MAxGÁNA*tenû*-sag) obv. IX 3–7: 3 gu-mug-TÚG 3 íb+III-TÚG-gùn *A-da-ša Bù-ma-ù* NI-*bù-ul-ì* NE.DI.
75.G.1748 (King Išar-damu, Vizier Ibrium, month *za-['à]-na-⌜at⌝*) rev. VII 15–18: 1 íb+III-gùn-TÚG "*Bu*"-*ma-il Gú-ra-ra-ab*[ki] NE.DI.

4.6. *Textiles for singers (nar)*

Several groups of singers in the service of the Palace are well known. They receive various textiles, but mostly íb-textiles for their service. In the case of the singers too, we do not know what they wore when performing.

75.G.1748 (King Išar-damu, Vizier Ibrium, month *za-['à]-na-⌜at⌝*) rev. VI 17–VII 14: 13 íb+III-gùn-TÚG *Á-da-mi-gú I-ti*-[d]En-ki *Ti-ga-lum Ìr-az$_x$-il I-ku-'à-bù En-na-ì Da-šè Tu-an Ša-za-iš Gú-la-gú En-da-za Zi-rí-gu Wa-at-ra-im* nar-nar;
rev. XII 1–3: 1 KIN siki íb-lá nar-nar, "1 KIN measure of wool for belts for singers".

4.7. *Textiles for acrobats (ḪÚB.KI)*

Acrobats, especially coming from Nagar, often performed at the Eblaite court.[53] In exchange for their services, they received different types of textiles. The most important acrobats normally received the group of textiles composed of *ʾà-da-um*, aktum and íb, while the young acrobats received other types of textiles. We do not know in which clothes they performed.

75.G.1754 rev. II 7–9: 1 *ʾà-da-um*-TÚG-I 1 aktum-TÚG *Gi-da-na-im* ḪÚB.KI.
75.G.1781 (King Išar-damu, Vizier Ibbi-zikir, month MAxGÁNA*tenû*-sag) obv. X 10–17: 5 íb+III-gùn-TÚG *Dur*-NI-NE-*zu Bar-i Wa-da-ʾà Bù-šu Ba-du-lum* dumu-nita-dumu-nita ḪÚB SA.ZA$_x$[ki].

[53] See Archi 1998, 10–11.

75.G.1833+ARET III 107+ARET XII 939 (King Išar-damu, Vizier Ibbi-zikir, month *za-lul*) obv. II 1–III 1: 5 *ʾà-da-um*-TÚG-I 5 gu-mug-TÚG 10 aktum-TÚG 10 íb+II-TÚG-sa$_6$-gùn *Dur-ʾà*-NE-*zu Ba-du-lum Wa-ʾà*-NE-*su Bar-i Bù-šu Gú-ba U$_9$*-NE-*lum En-na-ì, En-na-ì [x]-zi* ḪÚB máḫ 9 gu-mug-TÚG 9 sal-TÚG 9 íb+II-TÚG-gùn *Ga-da-na Zu-lum-šu A-wa-i-šar Bù-da-ì A-bù-*^{d}KU-*ra* NI-*ba-ì Zu-za-ga-mu Zi-ru$_{12}$-šu Zi-zi-gú* ḪÚB-ḪÚB tur.

4.8. *Textiles for* pašišu-*priests*

The priests[54] of the principal gods received the usual textiles given to important people at the court. However, it is not known if they wore particular types of clothes when performing their religious rites.

75.G.1221 rev. I 18–II 1: 2 gada-TÚG mu$_4$. mu *Kéš-ma-lik I-ti-ga-da-mu* pa$_4$:šeš ^{d}*ʾA$_5$-da-"bal" A-ru$_{12}$-ga-du*^{ki}.
75.G.1252 rev. XI 1–5: 1 *ʾà-da-um*-TÚG 1 aktum-TÚG 1 íb+II-TÚG-sa$_6$-gùn *An-da*-NI pa$_4$:šeš ^{d}*ʾ À-da* lú *Ḫa-lab$_x$*^{ki}.
75.G.1524 (King Išar-damu, Vizier Ibrium, month MAxGÁNA*tenû*-sag) obv. VII 7–15: 1 túg-NI.NI 1 íb+III-TÚG-gùn *A-du-lu* fašeš ^{d}KU-*ra in* ud ⌜mu$_4$⌝-mu ^{d}Tu lú é ^{d}KU-*ra*, " 1,1 textiles to Adulu, priest of the god Kura, on the day of the ceremony of dressing the goddess Nintu of the temple of Kura".

4.9. *Textiles to the šeš-II-ib*

The men and the boys of the most important Eblaite families served for some periods as šeš-II-ib, performing rituals and participating in ceremonies in certain towns where important sanctuaries, especially of the god Adabal, were located.[55] On these cult occasions, they all received the same type of textiles. However, the groups of textiles differ greatly from one occasion to another. Often, the most classic group of textiles including *ʾadaum*, aktum and íb is delivered, although, also different groups, for example consisting only of *ʾadaum* and íb too are encountered. The difference in the textiles does not seem to be connected to the different rituals performed. It may be surmised that, the fabrics they received on the occasion of the performed ceremony was a gift of the Eblaite court, but which were not used during the ceremony. The fabrics were in some way a payment for their services.

Several examples of lists of šeš-II-ib have already been published in the volumes of ARET and MEE. A study collecting almost all the names of šeš-II-ib quoted in the Ebla documentation has been made by Archi. To quote only some examples:

75.G.1783 (King Išar-damu, Vizier Ibrium, month MAxGÁNA*tenû*-úgur) obv. XI 4–rev. I 13: 8 *ʾà-da-um*-TÚG-II 8 íb+III-TÚG-sa$_6$-gùn Puzur-*ra-ma-lik wa Šu-ma-lik* lú *Iš$_{11}$-da-mu En-na-ni-il* lú *Ša-ù-um Iš-má-da-ba-an* lú *Iš-má-da-mu Bíl-ma-lik* lú *Ba-ḫa-ga I-si-rúm* lú *Ib-gú-nu* EN-*ga-da-ba-an* lú *Iš$_{11}$-gi-da-ar Iš$_{11}$-gi-bar-zú* lú *Sá-gú-šum* šeš-II-ib in šu mu-nígin ^{dʾ}*A$_5$-da-"bal" Lu-ba-an*^{ki}.
75.G.1760+10130 (King Išar-damu, Vizier Ibrium, month MAxGÁNA*tenû*-úgur) obv. I 1–II 5: [1]+4 *ʾà-da-um*-TÚG-II 5 aktum-TÚG 5 íb+III-sa$_6$-gùn-TÚG *I-bí-zi-kir wa Ù-ti* dumu-nita *Ib-rí-um En-na-Il* lú *En-mar Iš$_{11}$-gi-bar-zú* lú *Sá-gú-sum In-gàr* lú *A-a-da-mu* šeš-II-ib *in* šu mu-nígin ^{dʾ}*A$_5$-da-"bal" Lu-ba-an*^{ki}.
75.G.1417 (King Išar-damu, Vizier Ibrium first years, month *gi*-NI) obv. III 6–11: 2 *ʾà-da-um*-TÚG 2 íb+III-sa$_6$-gùn-TÚG *En-zi-li-im wa Du-bí* šeš-II-ib ^{d}*Ga-mi-iš*.
75.G.1356 (King Išar-damu, Vizier Ibbi-zikir, month *ig-za*) obv. II 5–13: 2 íb+x-TÚG-sa$_6$-gùn *Du-bí* lú *Zé-kam Du-bù-ḫu-ma-lik* lú NI-*a*-BAD šeš-II-ib kéš-da nídba ^{d}*Ti-mu-du*.
75.G.1772 (King Išar-damu, Vizier Ibrium, month lost) obv. I 1–9: 2 *ʾà-da-um*-TÚG-I 2 íb+III-TÚG-sa$_6$-gùn *I-si-rúm* lú *Ib-gú-nu wa Íl-da-kas* lú *En-zú-wa-rúm* šeš-II-ib kéš-da SA.ZA$_x$^{ki}.
75.G.1772 (King Išar-damu, Vizier Ibrium, month lost) obv. VI 3–10: 2 íb+III-TÚG sa$_6$-gùn *I-si-rúm* lú *En-na-il Du-bí-šum* lú NI-*gul* šeš-II-ib kéš-da SA.ZA$_x$^{ki}.

54 Biga 2006, 17–37.

55 Archi 2002a, 23–55; Biga 2006, 31–34.

75.G.1775 (King Išar-damu, Vizier Ibbi-zikir year 4 because parallel to 75.G.1918, month *za-'à-tum*) obv. V 1–9: 2 íb+III-TÚG-sa_6-gùn <*I*>-*dè-ni-ki-mu* lú *Bíl-ma-lik In-ma-lik* lú *Zu-ma-na-an* šeš-II-ib kéš-da *in Úr-lu*^{ki}.

Thus, we can conclude that the Ebla texts do not seem to reveal how the šeš-II-ib were dressed when performing their rituals.

4.10. *Textiles to soldiers* (*guruš*)

Several military campaigns are well documented in the Ebla texts. Many of those who accompany the vizier on military campaigns are often quoted by their personal names. Sometimes, a large number of anonymous soldiers are referred to by the term guruš.

The word guruš was used by Eblaite scribes to indicate a "valid man who can work and eventually go to war". In several cases, the texts quote soldiers going to war under the vizier. These soldiers received different types of textiles on such occasions. It is difficult to understand if there was a distinctive cloth for soldiers, if they all wore the same cloth and how they can distinguish a soldier of their own army from the enemy.

In fact, although the uniform is attested in Europe only from the 17th century AD, it is well known that ancient armies in Egypt, Assyria, Persia, Greece and Rome had distinctive elements or insignia.[56]

The texts list the textiles given for the military campaign. It is not clear if they were garments, but it seems quite possible. The two people are going on a military campaign against the city of Armi.

75.G.1249+ARET III 937+10082 (King Išar-damu, Vizier Ibbi-zikir final years, month *za-'à-tum*) obv. IV 13–19: 4 íb+II-TÚG-sal *I-bí-zi-kir I-ti-lum* lú EN-*zu si-in* níg-kas_4 *Ar-mi*^{ki}.

In the following text, soldiers of Armi are back from a military campaign and they receive various types of textiles as renumeration or as gifts.

75.G.1381 (King Išar-damu, Vizier Ibbi-zikir year 6, parallel to AAM 75.G.10074, month *ḫa-li*) obv. X 6–XI 4: 60 aktum-TÚG 2 *mi-at* 10 lá-3 sal-TÚG 2 *mi-at* ⌜6⌝ [...]⌜60⌝ lá-3 gu-mug-TÚG 2 *mi-at* íb+III-TÚG-gùn túg-túg *Ar-mi*^{ki} al_6-tuš *Gi-za-an*^{ki} *in Bar-ru*$_{12}$^{ki} šu ba_4-ti *áš-du* ì-ti *mi-nu* níg-kas_4 *Na-bù*^{ki}, "60 aktum-textiles, 207 sal-textiles, 206 ... textiles, 57 gu-mug-textiles, 200 íb-textiles for people of Armi who are in the town of Gizan, received in the town of Barru when they went back from the military campaign against Nabu".
75.G.1708 (King Išar-damu, Vizier Ibrium, month *ga-sum*) obv. III 14–IV 8: (⌜textiles and objects of silver⌝ and gold) 1 ugula *A-ḫa-da-mu*^{ki} 27 gu-dùl-TÚG 1 *mi-at* 81 sal-TÚG 40 *gu-zi*-mug-TÚG 20 KIN siki íb+III-gùn-TÚG guruš-guruš-*sù*.

This seems like a supply of textiles and wool to make fabrics for soldiers under the guide of a sheikh (ugula). They have probably participated in a military expedition or are ready for one.

75.G.1790 (King Išar-damu, Vizier Ibrium, month *gi*-NI) rev. XII 10–ff. 2 *mi-at* íb+III-TÚG-gùn to two hundred personal names of those who were probably participating in the military campaign against Armi.
75.G.1744 (King Išar-damu. Vizier Ibbi-zikir, month lost, text type ARET I, 1–8) obv. III' 2–9: 7 *'à-da-um*-TÚG 7 aktum-TÚG 7 íb-sa_6-gùn-TÚG *Šu-ra-gàr-ru wa* šeš-*sù* 1 gu-dùl-TÚG 1 sal-TÚG 1 íb-gùn-TÚG ABxÁŠ-*sù* 20 aktum-TÚG 20 íb+III-TÚG-gùn guruš-*sù*.

Here a translation of "guruš" as "soldiers" is most likely. Šuragarru arrived in Ebla with his brothers, one elder and 20 soldiers.

[56] I thank Prof. Virginio Ilari, professor of Military History, for his advice and suggestions.

4.11. *Textiles given as funerary gifts*

Several texts quote the death of some person of the court and the funerary gifts accompanying with him to his tomb. Several examples of funerary gifts that include rich textiles are already published.[57]

Various groups of fabrics are attested and it should be noted that also coloured textiles are given as funerary gifts. Important women of the court receive almost the same textiles they received during their lifetime, especially at marriage. Even if we can observe that modest people receive small quantities of textiles or only one fabric, sometimes some of them receive fabrics such as zara₆ that are usually considered by us as symbols of high status.

> 75.G.2632 (King Išar-damu, Vizier Ibrium, month *gi*-NI) obv.. IV 6 – V 3 : 1 zara₆- TÚG ú-ḫáb 1 pad-TÚG ⌜1⌝ ⌜*du*⌝-*ru₁₂*-[*rúm*] [.....][.....][.....][2 *bu-di*] [kù]-⌜gi⌝ ⌜šir⌝-za sag-*sù* 1 1/2 gín DILMUN kù-gi 1 kù-sal *Da-ti*-ᵈTu *ma-lik-tum Lum-na-an*ᵏⁱ ÉxPAP *I-šar* ur₄ [šu mu-t]ak₄,
> "1,1,1textiles .... 2 golden brooches whose heads are decorated with one and an half sicles of gold, 1 jewel to Dati-Tu, queen of Lumnan), for her tomb, Išar, the "ur₄"-functionary delivered ".
> 75.G.1416 (King Išar-damu, Vizier Ibbi-zikir, month lost) rev. IX 11'-14': 1 gu-mug-TÚG *Ra-ba si-in* ÉxPAP, "1 gu-mug-textile for Raba for his tomb." Raba is a well known flute-player.

4.12. *Textiles delivered on the occasion of purification rituals*[58]

After a death, members of the family do not receive particular types of textiles. Instead almost all the most common types of textiles are distributed to them. It is possible that the types of fabrics received depend on the status of the deceased and his family.

It was noted by Pasquali[59] that the colour black is the colour of death. In fact, in some cases, for a purification ritual after a death, some black fabrics are delivered. However, more often, the textiles are not black and sometimes even multicoloured textiles are delivered on such occasions. We would expect that, at least one black fabric was given to a family member, but in fact this is not normally the case. We cannot thus conclude that black was the colour of mourning. In other cases, some black textiles are given, even if it is not the occasion of a funeral or a ritual of purification.

> 75.G.1748 (King Išar-damu, Vizier Ibrium, month *za*-[*'à*]-*na*-⌜*at*⌝) obv. IV 9–12: 1 ʾ*à-da-um*-TÚG 1 aktum-TÚG 1 íb+III-TÚG-sa₆-gùn ì-giš-sag *Ib-dur-I-šar* ugula *Iš-da-mu-gú*ᵏⁱ.
> 75. G.1830+ (King Išar-damu, Vizier Ibbi-zikir year 9, month MAxGÁNA*tenû*-úgur) rev. III' 6'-11': 1 zara₆-TÚG 1 gíd-TÚG TAR bar₆:kù 2 *bu-di* 2 sag-*sù* kù-gi ì-giš-sag *ma-lik-tum* NI-*ra-ar*ᵏⁱ *Ḫal-zu-um* šu mu-tak₄.
> 75.G.1775 (King Išar-damu, Vizier Ibbi-zikir year 4 because parallel to 75.G.1918, month *za-'à-tum*) obv. III 8–13: 1 zara₆-TÚG ì-giš-sag ama-gal *Téš-má-zi-kir* dam en.
> 75.G.1775 (King Išar-damu, Vizier Ibbi-zikir year 4 because parallel to 75.G.1918, month *za-'à-tum*) rev. III 10–14: 1 *à-da-um*-TÚG-II 1 aktum-TÚG 1 íb+III-TÚG-sa₆-gùn 1 íb-lá 1 *si-ti-tum* 1 gír-kun 30 (gín) DILMUN) kù-gi [ì-giš]-sag en *Kak-mi-um*ᵏⁱ.

The text 75.G.1962+ was written in the month (*gasum*, as we know from another text) of the death of the great mother of the king, Lady Dusigu, and in the text, there is a long list (regrettably, partially broken) of textiles and objects accorded to her on the occasion of her funerary service,

---

57 See Archi 2002b, 161–199; Biga 2007–2008, 258–262.

58 For this ritual and the gifts on the occasion of a purification ritual, see Biga 2007–2008, 265–266.

59 Pasquali 2005a, 169.

and then the textiles and some jewels given on the occasion of the purification ritual of some members of her family, her son, the king, the queen, two brothers and three of her sisters. None of them received black textiles.[60]

> 75.G.1962+ (King Išar-damu, Vizier Ibbi-zikir year third, month lost) rev. III' 7 ff.: 1 túg-NI.NI 1 íb+III-TÚG-gùn níg-ba-*sù* ì-giš-sag en *wa ma-lik-tum* šè ug_7 ama-gal en 2 [...end of the column broken];
> rev. IV 2' ff. *I-ib-ma-lik wa Ti-la-ì* šeš ama-gal en 1 $zara_6$-TÚG 1 gíd-TÚG 2 *bu-di* TAR kù-gi ì-giš-sag *A-ma-ga* 1 $zara_6$-TÚG šú+ša bar_6:kù 2 *bu-di* ì-giš-sag *Ra-ù-tum* 1 $zara_6$-TÚG 2 *bu-di* šú+ša bar_6:kù ì-giš-sag [...

After the purification, we could expect a delivery of white textiles as customary after a ritual of purification, well attested in Greece and Rome.

> 75.G.1765 (King Išar-damu, Vizier Ibbi-zikir, month *ḫa-li*-NI) obv. VII 9–VIII 13: 1 $zara_6$-TÚG 1 dam *Na-am₆*-ì-giš en in-na-sum 2 gada-TÚG 2 šu *I-bí-zi-kir* 1 aktum-TÚG 1 dam *I-bí-zi-kir* 1 íb-lá gi_6 1 bu-di 3NI kù-gi *Ar-mi*[ki] ì-na-sum *A-zi-mu* ama-gal *I-bí-zi-kir* 1 íb+III-TÚG-*Ar-mi*[ki] *Za-a-šè* dumu-mí *I-bí-zi-kir.*

4.13. *Textiles for deities*

Every type of the most commonly delivered textiles is offered to the deities. It seems that there was not a specific type of textile always offered to the same deity.[61]

> 75.G.1885 (King Išar-damu, Vizier Ibrium, month lost) rev. VIII 10–12: 1 *zi-rí* siki íb+III-TÚG [d]KU-*ra.*
> 75.G.1262 (King Išar-damu, Vizier Ibrium, month *ga-sum*) obv. VIII 13–IX 1: 1 aktum-TÚG 1 gír mar-tu ka-ak níg-ba [d]*A₅-da-bal* ʾ*À-ma-at*[ki] 1 túg-NI.NI 1 *zi-rí* siki níg-ba [d]*Ga-na-na in* ud NAM.NAM-*sù*;
> rev. VIII 16–22: 1 sal-TÚG mu_4. mu 1 $zara_6$-TÚG [d]KU-*ra wa* [d]*Ba-ra-ma* lú ʾ*À-ti*-NI[ki];
> rev. X 5–7: 1 KIN siki *ba-ra-i* 1 siki na_4 *Ma-rí*[ki] gu-dùl-TÚG [d]*Ba-ra-ma* .
> 75.G.1870 (King Išar-damu, Vizier Ibrium, month lost) obv. I 1–3: 1 túg-NI.NI níg-ba [d]KU-*ra.*
> 75.G.1888+11723 (King Išar-damu, Vizier Ibrium, month *ga-sum*) obv. III 8–12: 2 sal-TÚG [d]KU-*ra wa* [d]*Ba-ra-ma* lú ʾ*À-ti-du*[ki].
> 75.G.1761 (King Išar-damu, Vizier Ibrium, month lost) obv. III' 8–13: 1 aktum-TÚG 1 gír mar-tu [KA]-ak níg-ba [d']*A₅-da-"bal"* '*À-ma-du*[ki].
> 75.G.1772 (King Išar-damu, Vizier Ibrium, month lost) obv. III 4–9: 1 aktum-TÚG 1 gír mar-tu ka-ak níg-ba [dʾ]*A₅-da-"bal"* ʾ*À-ma-du*[ki] *in* ud NE-*da-'à-tum.*
> 75.G.1792 (King Išar-damu, Vizier Ibrium, month lost) rev. VI 17–19: 1 *zi-rí* siki íb+III-TÚG [d]KU-*ra.*
> 75.G.1789 (King Išar-damu, Vizier Ibrium, month lost) rev. IX 6–10: 1 túg-NI.NI 1 *zi-rí* siki níg-ba [d]*Ga-na-na in* ud *gi*-NE-*a-sa* [...
> 75.G.1382 (King Išar-damu, Vizier Ibbi-zikir, month *za-ʾà-tum*) obv. VII 8–17: 1 $zara_6$-TÚG 1 gír-mar-tu [dʾ]*A₅-da-"bal"* ⌜1⌝ *du-ru₁₂-ru₁₂* 1 ma-na ša-pi kù-gi 1 *du-ne-za* [d]Be-mí [dʾ]*A₅-da-"bal" Lu-ba-an*[ki] en šu mu-tak_4.
> 75.G.1755 (King Išar-damu, Vizier Ibbi-zikir, month *gi*-NI) obv. V 6–10: 2 *gú-li-lum* kù-gi *wa* 1 túg-NI.NI níg-ba [d]KU-*ra;*
> obv. V 11–VI 3: 1 aktum-TÚG 1 gír mar-tu KA-ak níg-ba [d']*A₅-da-"bal"* ʾ*À-ma-at*[ki].
> 75.G.1763 (King Išar-damu, Vizier Ibbi-zikir beginning, month lost) obv. III' 11'-16': 1 ʾ*à-da-um*-TÚG-II 1 aktum-TÚG 1 íb+III-TÚG-sa_6-gùn 5 KIN siki du_{11}-ga nídba [dʾ]*A₅-da-bal Lu-ba-an*[ki].

5. Textiles in the texts of years 1–3 of Ibbi-zikir as vizier

The study of the texts in order of their relative chronology, year by year, is important not only to reconstruct the sequence of events in the Eblaite kingdom and in several other kingdoms in

[60] For the text and its parallel with the AAM 75.G.10088+, see Biga 1996, 48–50; for the funerary array of the Queen Mother, see Archi 2002b, 178.

[61] Many examples of fabrics offered to the gods are published in Pomponio & Xella 1997; See also Pasquali 2005a, 165–184.

Syria and Upper Mesopotamia, but also to better understand the role of the textiles in the Eblaite economy and commerce. Thus, we can ascertain the number of times some people received textiles *per annum*, the types of textiles they received in the entire year, and so forth.

I am attempting to verify this with the texts written in the period of King Išar-damu's reign from the death of Vizier Ibrium, the beginning of the period of his son and successor Ibbi-zikir and the death of the Queen Mother Dusigu that took place in the third year of Ibbi-zikir as vizier (month *gasum*).

For this period, covering a little more than three years in the life of Ebla, we have different kinds of documents, all mentioning textiles: six "mu-DU" texts, four annual accounts of metals and, so far, more than 30 monthly accounts of textiles.

With this group of texts, it is possible to determine such questions as:

- the number of occasions some people received clothes;
- the frequency with which the king received aktum -TÚG for his head;
- the number and types of textiles the same person received in one year;
- which foreign countries received textiles;
- the frequency and number of textiles that were removed from the storerooms of the palace to be sent to foreign countries or given to the Ebla court;
- how many entered with "mu-DU";
- which emporia and where Eblaite textiles were received, that in that period lay under Eblaite control;
- who the workers were;
- the number of people who were involved in the production of textiles at the time.

However, for these tablets, some broken, I am still looking for joins. I am still in the process of arranging these monthly accounts of textiles in more precise chronological order, month by month, for every one of the three years. This task is not yet complete and therefore I prefer to delay the results of this study to a future article.[62]

Acknowledgements

I would like to thank Marie-Louise Nosch very much for having so kindly involved me in her wonderful group of hard-working textile scholars of different periods. The Centre for Textile Research created by her in Copenhagen is an exemplary place, where researchers can work

[62] I would like to make two observations. The close inventory number of tablets pertaining to two close months is more proof of the fact that in the great archive L.2769, the monthly accounts of textiles were kept in chronological order. The fragments joined to these close tablets have close inventory numbers too, because the tablets fell, were broken and the fragments fell close by. Close inventory numbers correspond to tablets from the same year and successive months. It was possible to conclude that the tablets were ordered both by type, that is to say content or kind of record, and by date in chronological order. Sometimes, in one month, two monthly accounts of textiles were written. This observation was already made by Pomponio and myself some years ago, studying the texts on the birth of the first son of the royal couple. The texts ARET I 15 and ARET IV 7 both registered the deliveries for that important occasion. See Biga & Pomponio 1993, 108–111. In the texts of the first three years of Ibbi-zikir as vizier, 75.G.10137+11703 is a monthly account of textiles written in the same year and in the same month *za-'à-tum* in which text 75.G. 2356 was written. This is proved by prosopographic data and mainly because they both quote the fact that the queen of Ḫarran, the Eblaite princess Zugalum married two years earlier, gave birth to a child. The news was brought to Ebla by a messenger of the kingdom of Ḫarran who received some fabrics as a gift.

together, taking advantage of each others' skills in a cooperative way, and which is also open to discussion and new scholars. I learned much during my stay in Copenhagen in March 2009 and I have more to learn from all the researchers working at the Centre.

And then I must thank both Marie-Louise Nosch and Cécile Michel very much for organizing the meeting in Copenhagen, inviting me and offering me the opportunity to discuss with such a group of scholars and friends, experts on different millennia BC, issues regarding the textiles quoted so often in the Ebla texts.

Abbreviations

AAM	Annual Account of Metals.
ARET	Archivi Reali di Ebla.Testi.
ARES	Archivi Reali di Ebla.Studi.
LL	Lexical Lists.
MEE	Materiali Epigrafici di Ebla.
MAT	Monthly Account of Textiles.
ʾà.	*ʾà-da-um*-textile.
t.	textile/s.

Bibliography

Archi, A. (1980) Allevamento e distribuzione del bestiame ad Ebla, *Annali di Ebla* I, 1–33.

Archi, A. (1993) Trade and Administrative Practice: The Case of Ebla, *Altorientalische Forschungen* 20, 43–58.

Archi, A. (1998) The Regional State of Nagar According to the Texts of Ebla, *Subartu* IV 2, 1–15.

Archi, A. (1999) Clothes in Ebla. In Y. Avishur and R. Deutsch R. (eds.), *Michael. Historical, Epigraphical and Biblical Studies in Honor of Prof. Michael Heltzer,* Tel Aviv-Jaffa, 45–53.

Archi, A. (2000) The "Lords", lugal-lugal, of Ebla. A Prosopographic Study, *Vicino Oriente* 12, 19–58.

Archi, A. (2002a) šeš-II-ib: a Religious Confraternity, *Eblaitica* 4, 23–55.

Archi, A. (2002b) Jewels for the Ladies of Ebla, *ZA* 92, 161–199.

Archi, A. (2003) In margine. In A.A.V.V. 2003, *Semitic and Assyriological Studies Presented to Pelio Fronzaroli by Pupils and Colleagues,* Wiesbaden, 27–43.

Archi, A. & Biga M. G. (2003) A Victory over Mari and the Fall of Ebla, *JCS* 55, 1–44.

Biga, M. G. (1988) Frauen in der Wirtschaft von Ebla. In H. Waetzoldt and H. Hauptmann (eds.), *Wirtschaft und Gesellschaft von Ebla,* Heidelberg, 159–171.

Biga, M. G. (1992) Les vêtements neufs de l'Empereur. *Nouvelles Assyriologiques Brèves et Utilitaires* 19.

Biga, M. G. (1996) Prosopographie et datation relative des textes d'Ébla, *Amurru* I, 29–72.

Biga, M. G. (1998) The Marriage of Eblaite Princess Tagriš-Damu with a Son of Nagar's King, *Subartu* IV 2, 17–22.

Biga, M. G. (2002) Les foires d'après les archives d'Ébla, *Florilegium marianum* VI, *Memoires de NABU* 7, 277–288.

Biga, M. G. (2003a) The Reconstruction of a Relative Chronology for the Ebla texts, *Orientalia* 72, 345–366.

Biga, M. G. (2003b) A Ritual from Archive L.2712 of Ebla. In A.A.V.V. 2003, *Semitic and Assyriological Studies Presented to Pelio Fronzaroli by Pupils and Colleagues,* Wiesbaden, 54–69.

Biga, M. G. (2003c) Feste e fiere ad Ebla. In C. Zaccagnini (ed.), *Mercanti e politica nel mondo antico,* Roma, 55–68.

Biga, M. G. (2006) Operatori cultuali a Ebla. In M. Rocchi, P. Xella and J. Zamora (eds.), *Gli operatori cultuali, Studi Epigrafici e Linguistici* 23, Verona, 17–37.

Biga, M. G. (2007–2008) Buried among the Living at Ebla? Funerary Practices and Rites in a XXIV Cent. B.C. Syrian Kingdom. *Scienze dell'Antichità* 14, 125–151.

Biga, M. G. (2008) Au-delà des frontières: guerre et diplomatie à Ebla, *Orientalia* 77, 289–334.

Biga, M. G. (2009a) Dovecotes in the Kingdom of Ebla, Syria, XXIV Cent. B.C. In P. Negri Scafa and S. Viaggio (eds.), *Dallo Stirone al Tigri, dal Tevere all'Eufrate,* 11–16.

Biga, M. G. (2009b) The "Marriage" of Eblaite King Išar-damu with Queen Tabur-damu as documented by the Contemporary Administrative Texts, *in press.*

Biga, M. G. (2010) War and peace at Ebla in the years 1–3 of Ibbi-zikir as vizier, Fs. Moyen, Rome, 39–57.

Biga, M. G. & Pomponio, F. (1990) Elements for a Chronological Division of the Administrative Documentation of Ebla, *JCS* 42/2, pp. 179–201.

Biga, M. G. and Pomponio, F. (1993) Critères de rédaction comptable et chronologie relative des textes d'Ebla. *M.A.R.I* 7, 107–128.

Breniquet, C. (2008) *Essai sur le tissage en Mésopotamie*, Paris.

Civil, M. (2008) The Early Dynastic Practical Vocabulary A (Archaic HAR-ra A), ARES IV, Rome.

Conti, G. (1997) Carri ed equipaggi nei testi di Ebla. *Miscellanea Eblaitica* 4, *Quaderni di Semitistica* 19, Firenze, 23–71.

Derksen, J. G. (2004) *Old Assyrian Institutions*, Leiden.

Durand, J.-M. (2009) *La nomenclature des habits et des textiles dans les texts de Mari*, Paris.

Gleba, M. (2006) *Auratae vestes*: Gold Textiles in the Ancient Mediterranean. In C. Alfaro and L. Karali (eds.), *Purpureae Vestes*. II *Symposium Internacional sobre Textiles y Tintes del Mediterráneo en el mundo antiguo*, 61–77.

Liverani, M. (1979) La dotazione dei mercanti di Ugarit, *Ugarit-Forschungen* 11, 495–503.

Otto, A. (2006) Archaeological Perspectives on the Localization of Naram-Sin's Armanum. *Journal of Cuneiform Studies* 58, 1–26.

Pasquali, J. (1997) La terminologia semitica dei tessili nei testi di Ebla", *Miscellanea Eblaitica* 4, *Quaderni di Semitistica* 19, Firenze, 217–270.

Pasquali, J. (2005a) Remarques comparatives sur la symbolique du vêtement à Ébla. In L. Kogan, N. Koslova, S. Loesov and S. Tishchenko (eds.), *Memoriae Igor M. Diakonoff, Babel und Bibel 2,* Winona Lake, 165–184.

Pasquali, J. (2005b) Innovazione e continuità nel lessico dell'artigianato nella Siria del III millennio A.C. *Quaderni di Semitistica* 25, 267–299.

Pettinato, G. & D'Agostino, F. (1996) *Thesaurus Inscriptionum Eblaicarum*, Roma, Vol. A, I,2.

Pinnock, F. (1992) Le "Turban royal éblaite". NABU 1992/18.

Pomponio, F. (2008) Note sui tessili della documentazione di Ebla. Il periodo di Arrugum. In M. Perna and F. Pomponio (eds.) *The Management of Agricultural Land and the Production of Textiles in the Mycenaean and Near Eastern Economies*, Napoli, 101–109.

Pomponio, F. & Xella, P. (1997) *Les dieux d'Ebla,* Münster.

Porter, B. & Moss, R. L. B. (1934) *Topographical Bibliography of Ancient Egyptian Hieroglyphic Texts, Reliefs, and Paintings.* Volume IV, Oxford.

Roccati, A. (1970) Una tabella lignea inscritta da Gebelein. *Rivista degli Studi Orientali* 45, 1–10.

Rositani, A. (2001) I tessili della documentazione eblaita come elemento per la datazione interna. *Aula Orientalis* 19, 261–270.

Sallaberger, W. (2009) Von der Wollration zum Ehrenkleid. Textilien als Prestigegüter am Hof von Ebla. In B. Hildebrandt and C. Veit (Hg.), *Der Wert der Dinge-Guten im Prestigediskurs*, München, 241–278.

Veenhof, K. R. (1972) *Aspects of Old Assyrian Trade and Its Terminology.* Leiden.

Waetzoldt, H. (2001) *Wirtschafts - und Verwaltungstexte aus Ebla. MEE* 12, Roma.

Waetzoldt, H. (2007) The Use of Wool for the Production of Strings, Ropes, Branded Mats, and Similar Fabrics. In C. Gillis and M.-L. B. Nosch (eds.), *Ancient Textiles. Production, Craft and Society*, Ancient Textiles Series 1, Oxford, 112–124.

9. Les noms sémitiques des tissus dans les textes d'Ebla

Jacopo Pasquali

Parmi les activités artisanales les plus importantes d'Ebla au milieu du IIIe millénaire av. J.-C., il y a sans aucun doute la production textile, en témoignent les nombreuses tablettes trouvées au cours des fouilles de cette ville et qui comportent les enregistrements mensuels des attributions des tissus. Ces attributions concernent toute la période des archives et seulement une portion a, pour l'instant, été publiée. Dans ces textes on y trouve plusieurs termes concernant les tissus. La plupart sont constitués par des sumérogrammes dont seulement une partie a son correspondant sémitique, connu par quelques passages parallèles et par des études comparées avec d'autres textes administratifs, des listes lexicales et des textes de rituels. À côté des sumérogrammes on trouve donc un certain nombre de termes sémitiques, sujets de cet article.

1. La laine et le lin

À Ebla les étoffes étaient tissées avec de la laine (siki) ou du lin (gada-túg). Les termes sémitiques attestés peuvent indiquer la qualité de ces fibres.

Parmi les termes sémitiques des textes administratifs concernant la laine, les plus fréquents sont: *ba-ra-i* et *ni-za-ù*. *Ba-ra-i* (var. *ba-ra-u*$_9$) rappelle le sém. **bry*, "choisir ; inspecter", indiquant ainsi une variété de laine d'excellente qualité, qui (d'après les nombreuses attributions) était employée pour la fabrication de tissus destinés aux personnages les plus importants de la cour éblaïte.[1] *Ni-za-ù*, dont la lecture est certaine grâce à la rare variante *nu-za-a-tum*[2], correspond à une variété de laine arrachée de bonne qualité, dans la mesure où ce terme tire son origine du sém. **nzʿ*, "ôter; arracher",[3] une racine qui est successivement connue en arabe et en éthiopien. Il s'agirait donc de la laine obtenue après le premier arrachage, tandis que le sumérogramme si-udu-ur$_4$, qui désigne toujours une variété de laine arrachée, lui est parfois opposé, renvoyant à

[1] Pasquali 1997, 220–222. Cela exclut donc toute référence au sém. **brḥ*, "(être) mauvais" bien que ce soit phonétiquement correct.

[2] Pasquali 1997, 258–261.

[3] Voir aussi la racine parallèle **nsʿ*, avec la même signification, qui est présente en akkadien, ougaritique et hébreu. Le mot éblaïte pourrait être expliqué aussi par la racine **nṣʿ*, "(être) blanc, candide", attestée en arabe, mais cette interprétation semble moins probable.

une laine de qualité inférieure probablement obtenue d'un deuxième arrachage, tout comme le terme néo-sumérien plus récent siki-udu-bala-a ur$_{4}$-ra.[4] La lecture sémitique du terme est assurée ; elle est connue par la liste lexicale bilingue : *VE* 1121 si-ur$_{4}$ = *me-a-gu-um*, /malāq-um/, du sém. occ. **mlq*, "arracher; enlever".[5] On trouve également cette racine en hébreu, en araméen, en arabe et en éthiopien. De même, le mot éblaïte *gàr-ti-um* (connu aussi dans les variantes *gàr-ti*, *gàr-ti-ì*, *gàr-ti-ù*) semble faire référence à une qualité de laine obtenue par arrachage, si on l'explique par l'akkadien *qerdu*, "laine arrachée", de *qarādu*, "arracher la laine". Dans les textes édités, ce mot est exclusivement employé pour qualifier des "tissus bariolés" (túg-gùn).[6] La graphie sémitique *ma-za-lum* pourrait, elle aussi, indiquer une autre variété de laine. Dans un texte administratif, elle semble qualifier une certaine quantité de cette fibre textile destinée à la préparation de feutres et de coussins.[7] Cette unique attestation nous oblige à faire preuve de prudence ; il est toutefois possible de faire une comparaison entre ce mot et le sém. occ. **mzr*, connu par les textes de Mari et d'Ugarit, qui concerne une méthode particulière de fabrication des fibres textiles.[8] On peut probablement rapporter le terme hébreu de la Mishna *māzar*, "filer", à la même racine ; celle-ci, en akkadien, est connue seulement par les listes lexicales, où en effet nous trouvons les équivalences ḪI = *mazārum* (dans un contexte lexical concernant la laine) et siki AL.ḪI = *mazrātum*. Nous disposons d'une seule attestation du terme sémitique *gú-bù-rúm* (ARET 15, 38 [40]). Ce mot désigne, lui aussi, une certaine quantité de laine attribuée à une dame de la cour, et renvoie vraisemblablement à la qualité de la laine. Ce mot peut être comparé à l'akkadien *kubburum*, "épais, gras", à propos de laine et d'étoffes.

Toujours dans le lexique de la laine, on trouve la graphie *zi-rí* qui, avec les termes sumériens kin, na$_{4}$ et giš-bal, indique une des unités de mesure qui servait, à Ebla, à quantifier cette fibre textile. On peut rapporter ce mot à la racine sémitique **zwr*, "tourner", attestée dans l'akkadien *zāru*, "tordre, tresser", ainsi que dans l'adjectif dérivé *zēru*, "tressé".[9] L'attestation éblaïte peut être comparée à trois autres termes: $^{(túg)}$*zīrum*, "un mode de tissage ou de finition",[10] gú-ḫa *zīrāti* des tablettes de Tell al-Rimah et siki *zīrtu*(*m*) d'Emar.[11] Cette dernière attestation est très significative, car elle désigne une bande de laine qui constitue une offrande pendant beaucoup de rituels, et elle est posée sur la tête de la déesse Aštarte à l'occasion d'une cérémonie particulière. À Ebla aussi les unités de mesure *zi-rí* concernent presque exclusivement les rituels.

Les termes sémitiques relatifs au lin sont moins nombreux. Il faut rapporter le mot éblaïte *dam-ša-lu*(-túg) aux graphies paléo-akkadiennes gada *dam-šè-lum* et gada-gin *dam-ši-lum* indiquant probablement quelques variétés de lin. Il s'agit peut-être de substantifs ayant *ta-* comme préfixe, du sém. **mṯl*, "(être) pareil, ressembler à". Cela serait en accord avec les règles phonétiques du syllabaire éblaïte où les syllabogrammes de la serie ŠA sont employés pour rendre les interdentales.[12] On peut donc traduire le mot *dam-ša-lu* comme "(le tissu) ressemblant au (lin)". En

---

[4] Voir Waetzoldt 1972, 30.

[5] Conti 1997, 33–34.

[6] Pasquali 1997, 236.

[7] Pasquali 1997, 255–256.

[8] Durand 1990b, 661; 2009, 143.

[9] Pasquali 1997, 267–268.

[10] Durand 1983, 415–416; 2009, 140–141.

[11] Fleming 1992, 26 et 148, n. 57.

[12] Conti 1990, 10–11.

outre, il est presque certain que ce terme indique une variété d'étoffe ou de fabrication étant donné son emploi pour qualifier d'autres tissus, par exemple aktum-túg.[13]

2. Les produits textiles finis

On peut diviser les noms des produits finis en vêtements ou tissus principaux et en accessoires pour s'habiller, comme les ceintures ou les bandes. D'autres mots font ensuite référence aux accessoires du mobilier domestique ou bien à l'équipement des chars et des équidés.

2.1. *Les éléments principaux d'un trousseau*

Il existe deux termes sémitiques relatifs aux tissus principaux : *gu-zi-tum*-túg et *ʾà-da-um*-túg. Le premier, dérivé du sém. **ksy*, "couvrir",[14] est sûrement un mot de tradition akkadienne, ainsi que l'indique la graphie en *gu* pour /k/, alors que les normes phonétiques du syllabaire éblaïte demandent l'emploi de ce signe uniquement pour rendre le /q/ étymologique.[15] Il s'agit donc d'un emprunt oriental ou akkadophone. Le second est un terme occidental, qui a probablement son origine de la racine **ḥtl*, "couvrir", et que l'on retrouve à Ugarit et en hébreu, bien que les substantifs dérivés de cette racine soient assez rares dans ces deux langues plus récentes.[16] Etant donné la signification générique de ces deux racines ainsi que le manque d'autres données, il est très difficile de fixer les différences entre ces deux types d'étoffes. En revanche, il est certain que ces deux termes renvoyaient aux éléments principaux d'un trousseau. On peut leur associer la graphie *du-za-mu* qui est souvent employée pour qualifier les étoffes des trousseaux les plus importantes et de qualité supérieure. On peut interpréter ce substantif comme ayant le schéma *ta12ā3-*, et signifiant "exceptionnellement décoré", du sém. **wsm*, "(être) orné, (être) élégant", une racine déjà connue en paléo-akkadien et attestée par la suite en arabe.[17] On peut aussi trouver la graphie *du-za-mu* utilisée de manière indépendante, mais il faut alors toujours sous-entendre le nom du tissu qu'elle qualifie. Un terme *sí-mi*, provenant du sém. **wsm*, est encore employé dans l'akkadien de Mari pour désigner un type de décoration textile ("dorures" selon les éditeurs)[18] que l'on pouvait appliquer dessus (*šakānum*) ou bien introduire (*šakākum*) dans l'étoffe elle-même.[19]

2.2. *Les accessoires du vêtement féminin*

Deux mots sémitiques désignent des accessoires du vêtement féminin éblaïte. Ils sont importants, surtout à cause de leurs valeurs symboliques. Les termes sont *du-ru$_{12}$-rúm*, "étole-péplum", et *ma-ga-da-ma-tum*, "voile". Ces deux accessoires étaient attribués aux dames de la cour éblaïte à l'occasion des noces et des cérémonies funèbres, confirmant ainsi que le mariage et la mort dans l'Antiquité étaient, pour la femme, deux rites de passage comparables. Ces tissus, à côté d'un vêtement zara$_6$-túg, font partie de la tenue de cérémonie de la reine d'Ebla pour ses noces comme on l'apprend par les rituels d'ARET 11. Le mariage était, de fait, considéré comme une

[13] Pasquali 1997, 222–223.
[14] Pasquali 1997, 240–241, avec bibliographie.
[15] Conti 1990, 39–40.
[16] Pasquali 1997, 218–220.
[17] Fronzaroli 1993, 34–35; Pasquali 1997, 231–233.
[18] Voir par exemple Durand 1983, 137 et n. 42.
[19] Pasquali 1997, 232–233; 2005a, 289. La même étymologie est proposée maintenant par Durand 2009, 140.

transition, surtout pour la femme, qui changeait de statut à l'intérieur de la communauté une fois célébré ce rite.[20] Ce changement était symbolisé par le port de vêtements particuliers. Quant à l'homme, cette transition était plutôt soulignée par l'éducation militaire et l'exercice de la guerre.[21] À Ebla, le nécessaire pour l'habillage des hommes de haut rang, ainsi que des divinités masculines, est toujours constitué d'un ceinturon (íb-lá) muni d'un poignard (gír kun) décoré d'un métal précieux. L'opportunité de s'habiller avec ces objets était donc un signe de distinction pour les hommes. Dans la Rome archaïque, et en général chez les peuples qui habitaient l'Italie dans l'Antiquité, on peut citer à ce propos l'opposition entre *viri cincti* et *non cincti*. Les premiers représentaient les membres du pouvoir politique, détenteurs de la *patria potestas* dans le milieu familial.[22] Malgré cela, il n'y a aucun doute que le mariage conférait aussi un rôle politique à l'homme et à Ebla, comme ailleurs, la célébration des noces était liée à l'intronisation.

Le premier tissu cité ci-dessus, attesté aussi dans les graphies *du-ru*$_{12}$*-ru*$_{12}$ et *du-rúm*, peut être expliqué par l'akkadien *ṭurrum*. Cette "étole-péplum" était destinée seulement aux femmes mariées et elle était souvent "bariolée", gùn. Après le mariage, la femme éblaïte avait la possibilité de s'habiller avec l'"étole-péplum", symbole du passage de célibataire à épouse, le port de ce vêtement étant un signe de distinction.[23]

Le deuxième tissu, que l'on peut interpréter comme un substantif avec schéma *ma12a3-* du sém. **ktm*, "couvrir (avec un voile)", est présent essentiellement dans les textes les plus anciens, tandis que dans les tablettes qui remontent à la phase finale des archives d'Ebla, nous trouvons à sa place le sumérogramme correspondant PAD-túg. Il intervient par exemple dans le plus récent des rituels royaux :

> ARET 11, 2:82–84, *wa* / al$_6$-BAD / dam / [*mu*]-⌜*a*⌝-[*bí-iš-tum*] / *wa* / ⌜*du*⌝*-a-ba-áš* / PAD-túg / *ma-lik-tum* / ⌜7⌝ / *ap* / sag / ⌜šu⌝-⌜šu⌝ / [*ma-lik-tum*] / [PAD-tú]g / [e]n / w⌜a⌝ / *ma-lik-tum* / al$_6$-tuš,
>
> "Et la '[lin]gè[re]' récite la bénédiction. Et elle revêt sept fois la reine avec le voile. Ensuite, elle [couv]re la tête (et) les mains de la [reine], (lorsque) le [r]oi et la reine sont assis",

alors que dans le plus ancien nous trouvons la graphie phonétique *gú-du-mu* qui correspond à l'akkadien *kutummu*, toujours du sém. **ktm*, utilisé à Mari pour indiquer le voile de l'épouse :[24]

> ARET 11, 1:78–80, *wa-a* / *ti-ig-da-ra-ab* / dam / *mu-a-bí-iš-tum* / *wa-a* / PAD / *ma-lik-tum* / *ba-na-sa* / ⌜šu⌝-*sa* / *wa-a* / *du-a-ba-áš* / *gú-du-mu* / *ma-lik-tum* / 7 / dam / NE-*na-áš*^{ki} / en / ù / *ma-lik-tum* / tuš,
>
> "Et la 'lingère' récite la bénédiction. Et elle met le voile à la reine : sur son visage (et) sur ses mains. Et la femme de NE-*na-áš*^{ki} revêt sept fois la reine avec le voile, (lorsque) le roi et la reine sont assis".[25]

Cette variante du lexique a peut-être été introduite dans le texte du rituel le plus ancien à cause de l'influence de la culture de Mari jusqu'au remplacement du mot éblaïte *ma-ga-da-ma-tum*, ayant ainsi les caractéristiques d'un emprunt savant. En outre on doit remarquer que le mot *ma-ga-da-ma-tum* est attesté aussi avec la variante graphique *má-da-ma-tum*, qui indique un

---

[20] Pour la Grèce ancienne, Schmitt 1977.

[21] Voir par exemple Loraux 1981, 40–45.

[22] Camporeale 1991, 61; Torelli 1997, 22; Bartoloni 2003, 159–184.

[23] Pasquali 1997, 224–230; 1998; 2005c, 173–175.

[24] Durand 1988, 103; 2009, 55–56; Michel 1997. Pour ce terme à Ebla, Pasquali 2005c, 174–176; Tonietti 2005, 251–252.

[25] Cela nous rappelle la statuette représentant une femme assise couverte de la tête jusqu'aux mains avec un grand voile découverte dans les fouilles d'Ebla (voir le catalogue *Ebla. Alle origini della civiltà urbana*, Milano 1995, 317).

affaiblissement de la consonne /k/ devant la dentale.[26] Cette graphie nous rappelle ce que Gelb 1961, 27, a remarqué à propos de l'introduction de *má* et d'autres signes analogues dans l'écriture sémitique. Une troisième variante, *má-ga-da-ma-tum*, jusqu'à présent attestée une seule fois, peut être expliquée comme une sorte d'hypercorrection graphique.[27]

Dans les textes concernant le rituel royal, le "voile" est posé sur la reine de façon à la couvrir du visage jusqu'aux mains. Une femme était chargée de cet acte cultuel, la dame de NE-*na-áš*^{ki}, qui préparait aussi les vêtements nécessaires à la célébration des noces et de l'intronisation des souverains d'Ebla. Le rôle de cette femme est précisé par le terme avec lequel elle était désignée dans les textes du rituel royal, c'est-à-dire *mu-a-bí-iš-tum*, participe 0/2 du sém. **lbš*, "s'habiller", interprétable comme "lingère".[28] La dame de NE-*na-áš*^{ki}, selon le cérémoniel, couvre "sept fois" la reine avec le voile, un chiffre qui avait une valeur magique et symbolique importante dans le Proche-Orient ancien.[29]

L'"étole-péplum" est fréquemment offerte aussi aux statues des divinités féminines,[30] mais seule la parèdre (^{d}BAD-mí) du dieu ^{dʾ}A$_5$-*da-bal* (c'est-à-dire Hadda-Baʿal, une hypostase du dieu de l'orage)[31] reçoit l'"étole-péplum bariolée" (*du-ru*$_{12}$*-rúm* gùn) et le "voile" (PAD-túg) ensemble dans :

> ARET 3, 3 :r. 7'-14', 1 túg-gùn 1 ti-túg ú-ḫáb 1 gín DILMUN kù-sig$_{17}$ / níg-ba / ^{dʾ}A$_5$-*da-bal* / *A-ru*$_{12}$*-ga-du*^{ki} / 1 PAD-túg 1 *du-ru*$_{12}$*-rúm* gùn / 1 gín DILMUN kù-sig$_{17}$ / 1 kù-sal / níg-ba / ^{d}BAD-mí / 1 ti-túg ú-ḫáb / am / ^{dʾ}A$_5$-*da-[bal...*,

et

> TM.75.G.10160 :r. VII:9–VIII:5[32], 1 zara$_6$-túg 1 *du-ru*$_{12}$*-rúm* 1 PAD-túg (anep.) 15 kù:babbar 10 kù-sig$_{17}$ 1 *du-rúm* níg-ba ^{d}BAD-mí ^{dʾ}A$_5$-*da-bal ma-lik-tum* ì-na-sum lú du-du *si-in* giš-gál-taka$_4$ *A-ru*$_{12}$*-ga-du*^{ki}.

À notre avis on peut considérer la célébration d'un rite de hiérogamie de ce couple d'importantes divinités éblaïtes.[33] La déesse ^{d}BAD-mí, proprement la "Dame", se révèle, donc, une divinité "matronale", patronne et protectrice des épouses et pour cela vénérée par les femmes de la famille royale qui lui offrent souvent des étoffes précieuses. Tout comme Héra, la νυμφευομένη, dans la Grèce ancienne, l'épouse par excellence, parèdre du dieu suprême Zeus, qui préside au mariage et protège les femmes mariées.[34] Elle reçoit l'offrande d'un πέπλος, par exemple à Olympie, pendant une fête solennelle qui lui est consacrée (Pausanias V, 16, 2: διὰ πέμπτου δὲ ὑφαίνουσιν ἔτους τῇ Ἥρᾳ πέπλον αἱ ἓξ καὶ δέκα γυναῖκες). La péplophorie, à savoir l'offrande du péplum nuptial au simulacre de la divinité, est une cérémonie bien documentée dans les sanctuaires d'Héra en Grèce et en Grande Grèce[35] et la hiérogamie de Zeus et Héra est représentée, entre autres, dans la frise orientale du Parthénon où l'on peut remarquer la déesse assise à côté de son époux pendant

---

[26] Pasquali 2009b. La traduction "coperta" qui a récemment été proposée par Pomponio 2008, 169, n'est pas appropriée aux contextes.

[27] Voir aussi la graphie *ga-du-ma-tum* toujours dérivée de la racine **ktm* dans le texte rituel édité par Biga 2003.

[28] Fronzaroli 1993, 46; Pasquali 2005c, 179–180.

[29] Voir déjà Liverani 1967.

[30] Pasquali 1998.

[31] Voir Fronzaroli 1997, 288–289.

[32] Cité par Archi 2005, 99.

[33] Pasquali 2005b, 64–65; 2005c, 174; 2009b.

[34] Malagardis 1997.

[35] Neuman 1965, 64–66; Cremer 1982; Fridh-Haneson 1988; Greco 1997.

qu'elle est en train d'ôter le voile matrimonial de son visage.[36] La fonction caractéristique d'Héra était vraisemblablement la même qu'avait à Ebla ^{d}BAD-mí, parèdre du dieu $^{d?}$A$_5$-*da-bal*, dont la statue, habillée avec le voile et l'étole-péplum représente l'image-même de l'épouse, alors que c'était la déesse ^{d}*Ba-ra-ma* qui était chargée de la protection de la femme au moment des noces, donc encore *nubenda*. Cette fonction était pareille à celle qu'avait Aphrodite en Grèce.

La cosmogonie archaïque transmise par le mythographe Phérécide de Syros (ou Syrie), ayant vécu au VIe siècle av. J.-C., connue uniquement par les fragments cités par d'autres auteurs, met en évidence le tissage et la décoration du ample et somptueux vêtement nuptial. Ce vêtement est mentionné avec le terme φᾶρος, synonyme de πέπλος, et il est donné par Zas à Chthonie-Ge (sans doute une hypostase d'Héra) le troisième jour du rite afin que leur *hieros gamos* puisse être célébré (F68, col. I):[37]

> κἀπειδὴ τρίτη ἡμέρη γίγνεται τῶι γάμωι, τότε Ζὰς ποιεῖ φᾶρος μέγα τε καὶ καλόν, καὶ ἐν αὐτῶ [ι] ποικ [ίλλει Γῆν] καὶ Ὠγη[νὸν καὶ τὰ Ὠ]γηνοῦ [δώματα...

Il faut considérer le mariage entre Chthonie-Ge et Zas comme la toute première cérémonie nuptiale, son archétype, et grâce à cela elle aura la possibilité de protéger tous les mariages futurs ainsi que les épouses. On peut remarquer l'emploi du verbe ποικίλλω se référant à la décoration du vêtement de Chthonie-Ge. Dans la Grèce antique le péplum relatif au mariage était toujours bariolé, exactement comme à Ebla. Le mot grec ποικίλος et le sumérien gùn, qui correspond à l'akkadien *barmu* (sém. **brm*), ont la même signification.[38]

La polychromie est synonyme de beauté et séduction et les tissus bigarrés et multicolores ont un pouvoir qui leur est attribué directement par Aphrodite. C'est bien cette déesse qui donne à Héra (dans le XIVème livre de l'*Iliade*) sa magnifique ceinture bariolée pour séduire Zeus, le distraire du champ de bataille et célébrer ainsi leur union dans l'épisode de la Διὸς ἀπάτη.[39] En Grèce les vêtements de l'épouse sont ποικιλόμορφοι, un adjectif qui indique aussi bien la polychromie que l'habilité manuelle indispensable pour réaliser un objet artisanal qui était capable d'attirer vers les inextricables filets d'Aphrodite ποικιλόθρονος. Cette épithète lui est attribuée par Sappho et elle peut être interprétée comme "Aphrodite au vêtement décoré de plusieurs couleurs".[40] En effet le terme θρόνα, d'après Hésychios, signifie "fleurs", mais aussi "trames de différentes couleurs". Tout cela, à notre avis, donne une explication à propos de la signification de ^{d}*Ba-ra-ma*, la "Bariolée", comme protectrice de la reine d'Ebla à l'occasion des noces.[41]

L'utilisation de ces tissus très décorés a probablement été introduite en Grèce à travers le Proche-Orient comme l'on peut déduire par quelques allusions faites par Homère dans ses œuvres. Il attribue constamment à l'habilité des femmes sidoniennes la fabrication des péplums bariolés.[42] Euripide aussi, dans le Ion (vv. 1059 et s.tes) décrit comme βαρβάρων ὑφάσματα les tissus richement décorés qui font partie du trésor du temple parmi lesquels on remarquait les péplums pillés par Héraclès aux Amazones dont les références à l'Orient sont très bien connues.

---

36 Mayo 1973, 220.

37 Shapiro 1980, 267; Carter 1988, 93–98; Schibli 1990, 50–51; Scheid & Svenbro 1996, 63–66.

38 Cassin 1968, 116–117.

39 Rissman 1983, 4–5; Borghini 1988; Golden 1989; Faraone 1990; Scheid & Svenbro 1996, 53–61; Bottini 2000, 278–279; Andò 2005, 180–190; Breitenberger 2007, 71–86.

40 Rissman 1983, 5; Svenbro 1984; Scheid & Svenbro 1996, 53–61.

41 Pasquali 2005c, 175; 2008b.

42 Pasquali 2005a, 267–268.

La fabrication de tissus élaborés dans la production artisanale d'Ebla au IIIème millénaire av. J.-C. a beaucoup d'importance parce qu'on dirait confirmer ce qui a été écrit par les auteurs classiques à propos du Proche-Orient.

2.3. *Les vêtements de la royauté*

Pour que l'intronisation des nouveaux souverains puisse s'accomplir il est indispensable que la dame de NE-*na-áš*^{ki} confectionne, pendant la période que les souverains passent au Mausolée (é *ma-tim*, **bayt-i mawt-im*, littéralement, la « maison du mort »),[43] tous les vêtements liés à la royauté et c'est la *ma-lik-tum* qui lui donne la laine nécessaire pour cela (voir par exemple ARET 11, 2:f. VII2'-11', *wa* / túg-nu-tag / *ma-rí-a-tim* / *ma-lik-tum* / ḫi-mu-DU / siki / 2 udu / *maš-da-bù* / *ma-rí-a-tim* / túg-nu-tag ; ARET 11 2:r. XII27–XIII13, [*in*] / al$_6$-tuš / é *ma-tim* / *ša-ti* / *sa-ba-ti-sù-ma* / 4 *ma-rí-a-tum* / túg-nu-tag / ^{d}KU-*ra* / *wa* / ^{d}*Ba-ra-ma* / *wa* / en / *wa* / *ma-lik-tum*). Le rôle de ce personnage est donc encore plus important parce que lié au tissage aussi. Comme on l'a déjà noté ailleurs,[44] le tissage, le mariage et le pouvoir royal sont strictement liés dans leur symbolique. Le tissage, en fait, avec le fuseau et le métier qui permettaient de fabriquer les vêtements, fait partie d'un langage « féminin » qui conduit à la séduction, à l'acte sexuel ainsi qu'au mariage.[45] Les vêtements royaux sont directement liés aux vêtements nécessaires pour le mariage. À ce sujet, dans la Rome archaïque, on peut citer l'exemple paradigmatique donné par la reine d'origine étrusque Tanaquil/Gaia Caecilia qui tisse et introduit l'usage de la *tunica recta*, portée par les épouses le jour de leurs noces, et de la toge royale, portée par Servius Tullius lors de son intronisation, et avec laquelle était également habillée la statue de la déesse Fortuna protectrice de la royauté.[46]

À Ebla, les vêtements royaux sont rarement attestés dans les textes administratifs jusqu'à présent édités. Ils sont cités dans les rituels d'ARET 11 avec la graphie *ma-rí-a-tim* (et variantes) qui peut être interprétée comme /mār-īy-āt-im/, "vêtements à la manière de Mari", comparable à l'akk. *mārītu*, "mariote", que nous trouvons dans les textes paléo-babyloniens de Mari.[47] Bien que dans les textes administratifs nous trouvions quelques tissus appelés *Ma-rí*$^{(ki)}$ et d'autres désignés par des noms de villes parfois difficiles à situer comme *Ar-mi*$^{(ki)}$ et *Ù-ra*$^{(ki)}$,[48] les attestations de ce vêtement cérémoniel (les textes des rituels exceptés) sont limitées à deux passages parallèles relatifs au considérable trousseau[49] donné à *Di-ne-íb-du-lum* à l'occasion de son investiture de prêtresse du dieu ^{d}'A$_5$-*da-bal* de *Lu-ba-an*^{ki} et à un inventaire de tissus apparemment insignifiant et encore inédit (TM.75.G.1504 f. I:5–8: 1 gada-túg / 1 sal-túg / 1 *ma-rí-a-du* / 1 íb-iii-túg).[50] Dans les rituels d'ARET 11 ces vêtements sont liés au terme *maš-da-bu*, qui est interprété comme "bande", d'après le sém. **štp*, attesté en akkadien avec la signification de "couper (des étoffes)". On a donc supposé que dans ce contexte le mot indique les volants qui constituaient les vêtements à la

---

[43] Fronzaroli 1993, 39.

[44] Pasquali 2005c, 177–180; 2008b.

[45] Boëls-Jannsen 1993, 241–252; Scheid & Svenbro 1996, 92; Bartoloni 2003, 117–122; Papadopoulou-Belmehdi 2004; Andò 2005; Menichetti 2007, 130–131.

[46] Champeaux 1982, 274–333; Torelli 1984, 125–131; Martin 1985; Coarelli 1992, 301–328.

[47] Fronzaroli 1993, 27.

[48] Pasquali 2009a.

[49] Pour cela Pasquali 2005b, 15–20.

[50] Je remercie le prof. A. Archi qui a eu l'amabilité de me fournir ce passage inédit.

manière de Mari portés par les souverains et par le couple de divinités qui protégeait la royauté éblaïte, ^{d}KU-*ra* et ^{d}*Ba-ra-ma*.[51] Dans un autre texte le même mot désigne l'"habillage" des souverains à NE-*na-áš*^{ki} pendant un rite en communion avec les ancêtres défunts et leur protectrice ^{d}*Ga-na-na* (ARET 12, 313:r. VII2'-14', 1 *zi-rí* siki / *maš-da-bù* / en / *wa* / *ma-lik-tum* / *si-in* / NE-*na-áš*^{ki} / 1 *zi-rí* siki / lú kin siki / giš-gál-taka$_4$ / ká / ^{d}*Ga-na-na* / *Du-ba-a-du* / [šu]-ba$_4$-⌈ti⌉) et alterne, dans les passages parallèles traitant le même événement, avec les formes du datif-directif:[52] *sa-da-bí-iš* et *ša-dab-tíš* (la seconde n'est pas phonétiquement correcte à cause de l'emploi d'une syllabe de la série ŠA pour rendre /š/ étymologique). Il est bien possible que ce rite avait lieu tous les ans après l'intronisation[53]. Cette signification est certaine grâce à un extrait de la liste bilingue éblaïte (*EV* 0213) où la graphie sémitique *si-dab-tum* glose le terme sumérien ^{mu}mu$_4$, "assignation de tissus, habillage".[54]

L'usage de vêtements à la manière de Mari mis à côté de l'emploi d'un mot typique de cette ville comme *kutummu* pour indiquer le voile de l'épouse, fait imaginer une influence de Mari même sur les symboles de la royauté éblaïte. Tout comme les trônes (giš-uštin) qui sont fréquemment décrits, dans les textes administratifs, "à la façon de Mari".[55] Comme nous n'avons aucune idée sur l'identité de la reine du premier rituel qui épousa le roi *Ìr-kab-da-mu*, on peut supposer comme hypothèse de travail qu'elle venait de Mari et qu'elle introduisit à Ebla quelques symboles royaux d'origine typiquement mariote.

2.4. *D'autres éléments de l'habillement éblaïte*

Un autre élément important de l'habillement éblaïte est indiqué par le mot *ti-ba-ra-nu* (var. *ti-ba-ra-núm*). On peut expliquer ce terme comme un substantif ayant un préfixe *ti-* et un suffixe *-ān* du sém. **'pr* (et var.tes), "couvrir (la tête)", et donc /ti'par-ān-u(m)/, "turban".[56] Généralement produit en lin (gada-túg) avec des appliques en métal précieux ce turban est presque exclusivement destiné au roi et, à une seule occasion, au dieu ^{d}KU-*ra* (MEE 12, 37:f. I 20–27, 5 gín DILMUN kù:babbar / šu-bal-aka / 1 gín DILMUN kù-sig$_{17}$ / *ti-ba-ra-nu* / sag-kešda / ^{d}KU-*ra* / *ma-lik-tum* / in-na-sum). Encore une fois donc on suppose un lien avec la symbolique de la royauté.[57] Il faut considérer que même les personnages féminins pouvaient le porter étant donné qu'il était attribué aux princesses de la cour à l'occasion de leurs noces ou de leurs investitures sacerdotales. L'emploi du lin et du métal précieux pour la fabrication d'importants turbans est également connu à Mari, comme ce passage d'ARMT 23, 202 le prouve: *ṭi-wi-it* gada *a-na* sagšu-ḫá ša ḫuš-a, "fils de lin pour turbans (décorés avec motifs) en or rouge".[58] Les fils de lin étaient en effet très résistants et par conséquent très appropriés pour tenir les nombreux et lourds éléments de ces vêtements élaborés.

Un autre élément fabriqué en lin est un objet indiqué par la graphie *kir-na-nu*, *kir-na-núm*, avec variantes *kir-a-nu* et *kir-na-an*. L'interprétation de ce terme est très difficile. La seule comparaison possible avec l'assyrien *kír-na-a-a*, ayant une attestation uniquement lexicale, n'aide pas à

[51] Fronzaroli 1993, 27–28; Pasquali 1997, 248–253.

[52] Pour le datif-directif à Ebla, Catagnoti 1995.

[53] Pour ces passages Pasquali 1997, 248–253; 2005b, 29–30; 2005c, 168; Pasquali & Mangiarotti 2005.

[54] Pasquali 1996; 1997, 251–253.

[55] Pasquali 2005b, 53.

[56] Pasquali 1997, 262–266.

[57] Les doutes d'Archi 2002, 197, à ce propos ne sont pas justifiés.

[58] Voir Joannès 1984, 185.

comprendre quel objet il pouvait désigner. D'après les contextes où il se trouve, ce tissu était quand même assez précieux. Il arrivait à Ebla grâce surtout aux riches apports (mu-túm) de certains règnes de la Syrie:[59] *Du-lu*[ki], *Du-gú-ra-su*[ki] et *Ra-ʾà-ag*[ki], souvent avec fils (gu) de pierres dures.[60]

La "bande" ou "sangle", représentant un objet d'une certaine valeur, est désignée par le mot *gàr-su* que l'on peut comparer avec l'akkadien *qeršu*(*m*), *qiršu*(*m*), du sém. *qrš*, "couper", qui désigne une sangle en tissu, cuir ou métal.[61] À Ebla, nous avons la certitude, selon un unique texte, que cet objet pouvait être fabriqué en laine et lin (voir TM.75.G.2190:f. III 1, 3 *gàr-su* siki gada).[62] Sa préciosité est compréhensible de par la contre-valeur en métal qu'on lui attribuait pour son achat (níg-sa_{10}) aux "foires" (KI:LAM_7), et le porter était souvent un privilège du roi d'Ebla qui le met une fois à l'occasion des rites de « purification » (ì-giš-sag)[63] lors de la mort d'une des princesses de la cour.[64] L'interprétation du terme est confirmée par son équivalent sumérien aussi : e_6(SU-SU), "sangle", que l'on détermine par quelques passages parallèles.

Un autre objet fabriqué en laine et/ou métal précieux, indiqué par la graphie *ḫar-zu-ba-tum* (var. *ḫar-zú-ba-tum*),[65] est plusieurs fois destiné au roi. Il était utilisé comme ornement pour les heaumes (níg-sagšu) ou bien pour le harnais des équidés (*ba-a-nu*). Ce terme on le compare avec l'ougaritique *ḫrṣbt*, indiquant les "ligaments" des jambes dans un texte concernant les présages liés aux naissances et l'hébreu *ḥrṣbwt*, "chaînes", employé avec sens moral par Isaïe 53:8 et avec valeur méthonimique de "tourments, douleurs" dans Psaumes 73:4. À Ebla, ce mot aura désigné une sorte de petite "chaîne" ou "ficelle"; une signification sémantique correcte si on considère l'attestation où l'objet est lié à un heaume.

D'autres termes sémitiques ayant des rapports avec les vêtements sont : *ga-zi*, "lacet", explicable come un participe au génitif /kāsiy-i/, du sém. **ksy*, "lier", bien connu en akkadien.[66] Cet objet, remis par paires dans les passages où est attesté jusqu'à présent, est toujours mentionné avec le sumérogramme níg-lá-DU, "bande pour les jambes". Il s'agissait donc des "sangles" qui permettaient aux bandes de bien adhérer au corps. Il existe un cas où ce mot est remplacé par son équivalent sumérien, šu-kešda. Et encore *rí-ga-zi* (variantes *rí-gi-za*, *rí-ga-zú*) du sém. **rks*, "lier", toujours cité pour qualifier des tissus gu-mug-túg en précisant probablement une façon de travail.[67]

2.5. *Objets en laine et lin pour le mobilier ou l'équipement de chars et équidés*

Comme cela a déjà été dit, il y a de nombreux termes sémitiques qui se rapportent aux textiles, mais qui ne se réfèrent pas aux vêtements. Ils désignent des objets qui concernent le mobilier ou l'équipement de chars et équidés.

---

59 Pasquali 1997, 243–245.

60 Pour ces fils de pierres dures, voir Pasquali 2005b, 22–24.

61 Pasquali 1997, 233–235. Récemment Archi 2003, 34, a proposé pour ce mot une traduction "bag, hunting bag" sur la base d'une difficile comparaison avec le mot hittite *kursa-*. Cette interprétation ne tien pas compte de l'équivalence du terme éblaïte avec le mot sumérien e_6(SU-SU), "bande; courroie". Pour ce terme à Mari, voir Durand 2009, 108–109.

62 Cité par Archi 2003, 34.

63 Pour ce mot Archi 1996, 17–18.

64 Pasquali 2005c, 169.

65 La variante présumée *ḫar-*ru_{12}*-ba-tum* (voir Pomponio 1982, 210–211) n'existe pas. Le terme doit être lu *ḫar-*<*zu*>*-ba-tum* en. Même la comparaison proposée par Pomponio 2008, 78, avec la graphie *ḫar-ì-ba-tum* d'ARET 15 est phonétiquement difficile.

66 Pasquali 1997, 237–238; 2008a.

67 Pasquali 1997, 261–262.

La graphie *ma-za-bù* indique un objet dont la réalisation nécessite de laine et il est parfois donné avec des "feutres" (túg-du$_8$). C'est un objet lié au giš-uštin, "trône", et il s'agissait peut-être d'un "coussin". D'après les passages parallèles on découvre son équivalent sumérien: (ú-)su. Le mot sémitique est explicable comme un substantif avec schéma *ma12a3-* du sém. **ʾsp*, "recueillir", indiquant le sac ou la taie dans laquelle la laine était mise pour faire le coussin.[68] C'est un terme de tradition occidentale: à Ugarit, dans un texte administratif qui enregistre la sortie de plusieurs genres de tissus, on lit, entre autres, *ḫmš mʾispt qṭ*, "5 coussins en lin". Dans l'inscription punique KAI 122:1 on trouve encore cité *mʾsp*, "coussin", en relation avec *ksʾh*, "trône".[69]

Les termes désignant les accessoires des harnais et des chars réalisés en laine sont beaucoup plus nombreux. *Ḫu-lu* est sans doute le plus attesté. Il est connu dans sa variante *ḫu-li* et dans sa forme double *ḫu-li-ḫu-li* exprimant le pluriel du duel;[70] cet objet était donc utilisé par paires et on peut le traduire "anneau du joug", substantif dérivé du sém. occ. **ġll*, "introduire; enfiler". Selon les règles établies pour le syllabaire d'Ebla les signes de la série ḪA sont régulièrement utilisés pour rendre /ġ/ étymologique.[71] Ce mot éblaïte correspond donc à l'hébreu *ʿl*, "joug" (souvent employé dans le sens moral) et au *ḫullum* des lettres d'Amarna, où il est utilisé comme glose de l'akkadien *nīru*, "joug".[72] Le fait que la racine **ġll* soit employée dans la région sémitique occidentale jusqu'au IIIe millénaire av. J.-C. pour indiquer la partie du joug réalisée en étoffe aide à mieux comprendre l'étymologie et la signification du terme de Mari *ḫullum* (*ḫu-ul-li*, *ḫu-li*, *ḫu-lim*), attribué en général par paires (*ta-pa-al*) et en relation avec d'autres objets concernant l'équipement des chars et des équidés. La même étymologie et les similitudes des contextes indiquent qu'il s'agit du même objet qu'on trouve dans la documentation d'Ebla.[73]

Il y a encore d'autres mots indiquant les parties en étoffe des chars et des harnais: *ma-da-lum* (attesté, en principe, au duel *ma-da-ra*, *ma-da-la* et parfois dans la forme du datif-directif *ma-da-rí-iš*, *ma-da-rí-ša*), qui est toujours obtenu du fractionnement (ḫaš-ḫaš) d'un tissu aktum-túg et destiné aux équidés IGI.NITA du roi (en), de la mère du roi (ama-gal en) et d'autres personnages importants. On le compare avec le rare terme akkadien *maṭrû*,[74] probablement une sorte de lanière ou d'attache,[75] réalisé quelquefois en laine pourpre,[76] connu dans les textes d'Emar et dans les lettres d'Amarna. Pour les équidés IGI.NITA est rarement attestée l'attribution de *bù-a*-LUM et *na-da-lum*, peut-être fabriqués en lin (gada-túg). L'étymologie du premier n'est pas évidente alors que le second est comparable au mot akkadien *nadullu(m)*,[77] attesté à Mari dans ARMT 24, 192:2 et traduit par l'éditeur "couverture de selle".[78] Dans les lettres d'Amarna on le trouve attribué aux chars et aux animaux et on spécifie qu'il est fabriqué en cuir.[79] La lecture sémitique de níg-anše-aka, « brides », c'est-à-dire *lu-bù-gu*, *lu-bù-gú*, connue aussi à la forme duelle *lu-bù-ga*, *lu-*

[68] Pasquali 1995b; 1997, 253–255.
[69] Comme maintenant Watson 1999, 40, n. 12, justement le propose.
[70] Pasquali 2008a.
[71] Conti 1990, 18.
[72] Pasquali 1995a; Conti 1997, 41–42.
[73] Pasquali 1995a; 2008a.
[74] Conti 1997, 30–31.
[75] Durand 1990a, 81 ; 2009, 184.
[76] Adler 1976, 302–303, mais avec traduction "Gehenk".
[77] Conti 1997, 31.
[78] Talon 1985, 107 et 243.
[79] Adler 1976, 310.

ba-ga est beaucoup plus documentée. La variante *ru*$_{12}$*-bù-ga* impose une étymologie de premier /r/ tandis que la graphie ayant *gu* indique que la troisième radicale doit être /q/. L'explication la plus appropriée de ce mot nous l'avons par le sém. *rbq, "atteler (animaux)".[80] Il y a deux autres objets relatifs aux brides qui sont indiqués par des graphies sémitiques. L'un est *ba-a-nu*, toujours réalisé en laine, dont la signification constitue encore un argument de discussion. Il s'agissait peut-être d'un élément décoratif, une bande qu'on posait sur la tête des mulets et en général des équidés. Il faut considérer la comparaison avec l'équivalence *ba-a-nu* = *a-gu-ú* qu'on trouve dans la liste des synonymes An III 234, c'est-à-dire une sorte de bande pour la tête.[81] Dans la liste lexicale bilingue éblaïte *ba-a-nu* glose le sumérien n í g - a n š e - t ú g, dont l'interprétation n'est pas claire. Finalement le terme *ʾa*$_5$*-na-bù*, encore en laine et relatif aux brides, généralement expliqué par l'akkadien *anabu* de An VII, 230 qui indique une bande ou un tissu pour les hanches.[82] À notre avis on peut envisager une comparaison avec le terme de Mari $^{(éš)}$*na-na-ap-tum*, une forme *ma12a3-* sur une racine *ʾ*NP*, désignant un objet fabriqué en laine et utilisé pour l'équipement des chars (par exemple dans ARMT 21, 420; 22, 129; 23, 204), que les éditeurs pensent indiquer une sorte de cordon ou ficelle pour les brides.[83]

Remerciements

Je remercie Mr Gianluca Montanelli pour l'aide apportée à la traduction en français.

Abréviations

ARET Archivi reali di Ebla: Testi. Rome.
ARM(T) Archives Royales de Mari (Traduction). Paris.
KAI H. Donner & W. Röllig, *Kanaanäische und aramäische Inschriften*, I-III. Wiesbaden, 1962–1964.
MEE Materiali epigrafici di Ebla. Neapel 1979.
TM Ebla (Tell Mardīkh), numéros d'inventaire des tablettes.

Bibliographie

Adler, H.-P. (1976) *Das Akkadische des Königs Tušratta von Mitanni*. Alter Orient und Altes Testament 201. Neukirchen - Vluyn.

Andò, V. (2005) *L'ape che tesse. Saperi femminili nella Grecia antica*, Roma.

Archi, A. (1996) Chronologie relative des archives d'Ébla. Dans J.-M. Durand (éd.), *Mari, Ébla et les Hourrites dix ans de travaux, Amurru* 1, 11–28. Paris.

Archi, A. (2002) Jewels for the Ladies of Ebla, *Zeitschrift für Assyriologie* 92, 161–199.

Archi, A. (2003) In margine. Dans P. Marrassini *et alii* (éd.) *Semitic and Assyriological Studies Presented to Pelio Fronzaroli by Pupils and Colleagues*, 27–43. Wiesbaden.

Archi, A. (2005) The Head of Kura - The Head of ^{d}Adabal, *Journal of Near Eastern Studies* 64, 81–100.

Bartoloni, G. (2003) *Le società dell'Italia primitiva. Lo studio delle necropoli e la nascita delle aristocrazie*, Rome.

Biga, M. G. (2003) A Ritual from the Archive L.2712 of Ebla. Dans P. Marrassini *et alii* (éd.), *Semitic and Assyriological Studies Presented to Pelio Fronzaroli by Pupils and Colleagues*, 54–69. Wiesbaden.

Boëls-Jannsen, N. (1993) *La vie religieuse des matrones dans la Rome archaïque*, Rome.

[80] Conti 1997, 45, avec bibliographie.
[81] Conti 1997, 49.
[82] Pettinato 1980, 224; Conti 1997, 53.
[83] Joannès 1984, 139–140; Durand 2009, 185.

Borghini, A. (1988) Gli ornamenti lunari di Afrodite (Inni omerici V 89–90), *Annali della Facoltà di Lettere e Filosofia di Bari* 31, 59–75.

Bottini, A. (2000) Kestos himas poikilos, *Ostraka* 9, 273–279.

Breitenberger, B. (2007) *Aphrodite and Eros. The Development of Erotic Mythology in Early Greek Poetry and Cult*, New York - London.

Camporeale, G. (1991) Eroi e signori nelle prime scene narrative etrusche, *Mélanges de l'École Française de Rome - Antiquité* 104, 57–69.

Carter, J. B. (1988) Masks and Poetry in Early Sparta. Dans R. Hägg, N. Marinatos & C. Nordquist (éd.), *Early Greek Cult Practice. Proceedings of the Fifth Symposium at the Swedish Institute at Athens, 26–29 June, 1986*, 89–98. Stockholm.

Cassin, E. (1968) *La splendeur divine. Introduction à l'étude de la mentalité mésopotamienne*, Paris – La Haye.

Catagnoti, A. (1995) The Suffix -iš in the Ebla Texts, *Quaderni del Dipartimento di Linguistica dell'Università di Firenze* 6, 155–164.

Champeaux, J. (1982) *Fortuna. Le culte de la Fortune à Rome et dans le monde romain des origines à la mort de César. I: Fortuna dans la religion archaïque*, Rome.

Coarelli, F. (1992) *Il Foro Boario*, Rome.

Conti, G. (1990) *Il sillabario della quarta fonte della Lista Lessicale Bilingue eblaita*. Quaderni di Semitistica 17. Florence.

Conti, G. (1997) Carri ed equipaggi nei testi di Ebla. Dans P. Fronzaroli (éd.), *Miscellanea Eblaitica* 4, 23–71. Florence.

Cremer, M. (1982) Hieros gamos im Orient und in Griechenland, *Zeitschrift für Papyrologie und Epigraphik* 48, 283–290.

Durand, J.-M. (1983) *Textes administratifs des Salles 134 et 160 du Palais de Mari*. ARMT 21. Paris.

Durand, J.-M. (1988) *Archives épistolaires de Mari* I/1, ARMT 26/1. Paris.

Durand, J.-M. (1990a) Recension de D. Arnaud, *Recherches au Pays d'Aštata, Emar VI : Textes sumériens et akkadiens*, Paris 1986, *Revue d'Assyriologie* 84, 49–85.

Durand, J.-M. (1990b) Recension de S. Ribichini & P. Xella, *La terminologia dei tessili nei testi di Ugarit, Collezione di Studi Fenici* 20, Roma 1985, Mari Annales de Recherches Interdisciplinaires 6, 659–664. Paris.

Durand, J.-M. (2009) *La nomenclature des habits et des textiles dans les textes de Mari, Matériaux pour le Dictionnaire de Babylonien de Paris*, tome 1, ARMT 30, Paris.

Faraone, C. A. (1990) Aphrodite's κεστός and Apples for Atalanta: Aphrodisiacs in Early Greek Myth and Ritual, *Phoenix* 44, 219–243.

Fleming, D. E. (1992) *The Installation of Baʿal High Priestess at Emar*, Atlanta.

Fridh-Haneson, B. M. (1988) Hera's Wedding on Samos: A Change of Paradigms. Dans R. Hägg, N. Marinatos, C. Nordquist (éd.), *Early Greek Cult Practice. Proceedings of the Fifth Symposium at the Swedish Institute at Athens, 26–29 June, 1986*, Stockholm, 205–213.

Fronzaroli, P. (1993) *Testi rituali della regalità, ARET* 11, Roma.

Fronzaroli, P. (1997) Les combats de Hadda dans les texts d'Ébla, *MARI* 8, 283–290.

Gelb, I. J. (1961) *Old Akkadian Writing and Grammar*, Chicago.

Golden, L. (1989) Διὸς ἀπάτη and the Unity of Iliad 14, *Mnemosyne* 42, 1–11.

Greco, G. (1997) Des étoffes pour Héra. Dans J. de la Genière (éd.), *Héra. Images, espaces, cultes. Actes du Colloque International du Centre de Recherches Archéologiques de l'Université de Lille et de l'Association P.R.A.C. Lille, 29–30 novembre 1993*, Napoli, 185–199.

Joannès, F. (1984) Produits pour le travail du bois, du cuir, et du tissu. Dans J.-M. Durand & D. Charpin (éd.), *Archives Administratives de Mari* 1, ARMT 23, Paris, 133–191.

Liverani, M. (1967) Ma nel settimo anno... Dans *Studi sull'Oriente e la Bibbia offerti al p. G. Rinaldi da allievi, colleghi, amici*, Genova, 49–53.

Loraux, N. (1981) Le lit, la guerre, *L'Homme* 21, 37–67.

Malagardis, N. (1997) Héra, la sans pareille ou l'épouse exclue? A travers l'image. Dans J. de la Genière (éd.), *Héra. Images, espaces, cultes. Actes du Colloque International du Centre de Recherches Archéologiques de l'Université de Lille et de l'Association P.R.A.C. Lille, 29–30 novembre 1993*, Napoli, 93–111.

Martin, P. M. (1985) Tanaquil, la «faiseuse de rois», *Latomus* 44, 5–15.

Mayo, M. E. (1973) The Gesture of Anakalypsis, *AJA* 77, 220.
Menichetti, M. (2007) Metis e regalità nell'iconografia dei principi tirrenici di età arcaica. L'esempio di Dedalo e Medea. Dans P. Scarpi & M. Zago (éd.), *Regalità e forme di potere nel Mediterraneo antico*, Padova, 117–136.
Michel, C. (1997) *Un témoignage paléo-assyrien en faveur du port du voile par la femme mariée*, *NABU* 1997/40.
Neumann, G. (1965) *Gesten und Gebärden in der griechischen Kunst*, Berlin.
Papadopoulou-Belmehdi, I. (2004) *Le chant de Pénélope*, Paris.
Pasquali, J. (1995a) Ḫullum a Ebla e Mari, *NABU* 1995/59.
Pasquali, J. (1995b) Ma'sapu, «cuscino», nei testi di Ebla, *NABU* 1995/99.
Pasquali, J. (1996) La «vestizione» della statua della dea ^{d}TU ad Ebla, *NABU* 1996/129.
Pasquali, J. (1997), La terminologia semitica dei tessili nei testi di Ebla. Dans P. Fronzaroli (éd.), *Miscellanea Eblaitica* 4, Firenze, 217–270.
Pasquali, J. (1998) *Su* ^{d}ga-na-na *e* ^{d}*BAD* ga-na-na-im *ad Ebla*, *NABU* 1998/1.
Pasquali, J. (2005a) Innovazione e continuità nel lessico dell'artigianato nella Siria del III millennio a.C. Dans P. Fronzaroli & P. Marrassini (éd.), *Proceedings of the 10th Meeting of Hamito-Semitic (Afroasiatic) Linguistics*, Firenze, 267–299.
Pasquali, J. (2005b) *Il lessico dell'artigianato nei testi di Ebla*, *QdS* 23, Firenze.
Pasquali, J. (2005c) Remarques comparatives sur la symbolique du vêtement à Ébla. Dans L. Kogan *et alii* (éd.), *Memoriae Igor M. Diakonoff*, *Babel und Bibel* 2, Winona Lake, 165–184.
Pasquali, J. (2008a) La duplicazione dei sostantivi come indicazione del duale nei testi di Ebla, *NABU* 2008/49.
Pasquali, J. (2008b) Une hypothèse à propos du rôle de ^{d}KU-ra dans le rituel royale éblaïte, *NABU* 2008/50.
Pasquali, J. (2009a) Les tissus *ú-ra*$^{(ki)}$ dans les textes d'Ebla, *NABU* 2009/4.
Pasquali, J. (2009b) Les équivalents sémitiques de PAD-túg, « voile », dans les textes d'Ebla, *NABU* 2009/11.
Pasquali, J. & Mangiarotti, P. 2005, Il rito dell'«ipogeo» (*a-ba-i*) ad Ebla, *NABU* 2005/20.
Pettinato, G. (1980) *Testi amministrativi della biblioteca L.2769*, *MEE* 2, Napoli.
Pomponio, F. (1982) Note su alcuni termini dei testi amministrativi di Ebla, *VO* 5, 203–215.
Pomponio, F. (2008) *Testi amministrativi: assegnazioni mensili di tessuti. Periodo di Arrugum*, *ARET* 15/1, Roma.
Rissman, L. (1983) *Love as War. Homeric Allusion in the Poetry of Sappho*, Königstein.
Scheid, J. & Svenbro, J., 1996, *The Craft of Zeus. Myths of Weaving and Fabric*, Cambridge – London.
Schibli, H. S. (1990) *Pherekydes of Syros*, Oxford.
Schmitt, P. (1977) Athéna apatouria et la ceinture, *Annales ESC* 32, 1059–1073.
Shapiro, H. A. (1980) Jason's Cloak, *TAPA* 110, 263–286.
Svenbro, J. (1984) La stratégie de l'amour. Modèle de la guerre et théorie de l'amour dans la poésie de Sappho, *QS* 19, 57–79.
Talon, Ph. (1985) *Textes administratives des Salles Y et Z du Palais de Mari*, ARMT 24, Paris.
Tonietti, M. V. (2005) Symbolisme et mariage à Ébla. Aspects du rituel pour l'intronisation du roi. Dans L. Kogan *et alii* (éd.), *Memoriae Igor M. Diakonoff*, *Babel und Bibel* 2, Winona Lake, 245–261.
Torelli, M. (1984) *Lavinio e Roma. Riti iniziatici e matrimonio tra archeologia e storia*, Roma.
Torelli, M. (1997) *Il rango, il rito e l'immagine. Alle origini della rappresentazione storica romana*, Milano.
Waetzoldt, H. (1972) *Untersuchungen zur neusumerischen Textilindustrie*, Roma.
Watson, W. G. E. (1999) Some Rare Lexical Items in Ugaritic, *SEL* 16, 39–43.

10. New Texts Regarding the Neo-Sumerian Textiles

Francesco Pomponio

The 21th century BC in the history of Near East is the period of the birth, flourishing and fall of the last Sumerian state, the so-called Neo-Sumerian empire, otherwise defined as the Third Dynasty of Ur, following a denomination derived by the Sumerian King List.[1] This political structure included the Southern Mesopotamia, that is, the ancient regions of Sumer and Akkad, the Central Mesopotamia till Assyria, and the modern Khuzistan, the region of the ancient Susa. Moreover, it had under its influence part the western territory of the present Iran (with the name of Elam), northern Mesopotamia occupied for the most part by Hurrian people, and probably a part of the north-eastern Syria.

This state was characterized by a mighty development of the bureaucratic system. As regards the written documentation, that is the cuneiform tablets, produced by its administration, the most imposing in the ancient world, we have about 4,250 texts coming from the capital Ur, 30,000, 24,000 and 3,500 texts respectively from provincial capitals of Umma (including the 1,527 tablets from the rural estate of Garšana of most recent edition[2]), Ĝirsu, and Nippur (but it should be noted that the provinces of the Neo-Sumerian empire were about twenty), 13,800 from the redistributive centre of livestock Puzriš-Dagan, and some hundreds produced by private archives, often of unknown origin.[3] Of course, we are referring to published tablets, but museums and private collections of the Near East, Europe and United States of America are swarming of unedited Neo-Sumerian texts, itching to see the light of the publication. So, in the course of 2010, D.I. Owen will edit about 300 tablets belonging to Ithaca University and coming from another province of the Neo-Sumerian empire, that is Irisagrig.[4]

---

[1] The tablets presented in this article have been transliterated in the ambit of a research project aimed to the publication of the Neo-Sumerian texts from Ĝirsu and Umma, kept in the British Museum (PRIN 2006–2008). The writer wishes to express his gratitude to Dr Farouk Al-Rawi and Dr Lorenzo Verderame who collaborate in this research. The photos of BM 110313 and BM 110615 are published by kind permission of the Trustees of the British Museum. The abbreviations employed in the present article are supplied by the list of the site of M. Molina, *Base de Datos de Textos Sumerios* (http://bdts.filol.csic.es). The date of the Neo-Sumerian tablets is indicated by the name of the king and his relative year of reign, with the following abbreviations: Š = Šulgi; ŠS = Šu-Sîn; IS = Ibbi-Sîn.

[2] Owen & Mayr 2007.

[3] For these data see Molina 2008, 19–53.

[4] The volume is intended to be the 15th issue of the series *Nisaba. Studi Assiriologici Messinesi.*

The economic entries on which the Neo-Sumerian state relied may be subdivided into three basic sectors:

1) The taxes called bala, "service", and gú-un ma-da, "tribute of the country", respectively, delivered by the core of the empire (the about twenty provinces already mentioned) and by its periphery;[5]
2) The tribute and the booty, respectively supplied by the vassal states and by the recurrent raids that regarded a very vast zone, which extended from the valley of the Khabur river to Khuzistan;[6]
3) The income of the industrial complexes of economic centres managed by the administration of Ur and by that of its provinces.

In all probability, the most important of these establishments concerned the production of textiles, and textiles were also one of the dominant assets in the field of the Neo-Sumerian export. The textile industry employed many thousands of female workers (géme-uš-bar is the name of this category), belonging to the lowest stratum of the Neo-Sumerian society and in addition to them female prisoners, provided by the military campaigns of the Neo-Sumerian armies, young "castrates" (amar-ku$_5$), and children were employed in the "houses of weaving" (é-uš-bar). As a matter of fact, the most numerous category of female workers was that of the géme-ús-bar, "female weavers", followed by the géme-kikken, "female millers",[7] géme-UN-íl, "female porters(?)', and géme-ì-sur, 'female oil pressers'.[8] The monthly salaries of these labourers amounted to 30 or 40 litres of barley and their days off were one per five or, more often, six days.

Our data on the production of textiles are supplied by the very high number of tablets regarding both amounts of wool and quantities of textiles which come from the archives of the capital city Ur and of the most important provinces of the empire, Umma and Ĝirsu. In the ambit of this impressive documentation, I have picked out a few unpublished tablets belonging to the collections of the British Museum and coming from the provincial capital of Umma. Their analysis will supply us with glimpses of that enormous production of textiles and meanwhile will illustrate the kind of texts upon which our knowledge of the Neo-Sumerian textile industry bases itself. Their number and the abundance of data probably will arouse the envy of the researchers of other kinds of texts.

The first tablet is a monthly balanced account. This important category of texts is as a rule subdivided in three sections: the first one registers the capital, that is the income; the second the expenditures, that is the redistribution of textiles, of course from the point of view of the provincial administration, that is the office of the ensik, the "governor", who was the highest political authority of the province; the third section establishes if there was a remainder or an overdraft, that is if the deliveries were superior or inferior to the issues. The period covered by these balanced accounts normally varies from a month to a year.

---

[5] See Steinkeller 1987, 19–41.

[6] For the most recent analysis of the objectives of the Neo-Sumerian campaigns in this area, see Steinkeller 2008.

[7] In order to meet the requirements of the provincial administration, workers belonging to a category were provisionally employed in the activity of another category: so, e.g., in Nisaba 18, 146 an amount of flour is supplied by the 'castrates of the House of the weaving' (amar-ku$_5$ é-uš-bar).

[8] See Studevent-Hickman 2008, 145.

BM 110453 (dimensions: 9 × 4.7 × 2.5 cm.)

obv.

60x4 + 6 túg ĝìri / ensik-ka	246 textiles: the governor is the conveyor;
10x2 + 6 túg ĝìri Ur-^{d}Nisaba	26 textiles: Ur-Nisaba is the conveyor;
7 túg ĝìri A-kal-la	7 textiles: Akalla is the conveyor;
1 túg é-kišib-ba é-maš-ta	1 textile from the 'House of the tablets' of the Emaš.
blank	
túg mu-ku$_{x}$-ensik	Textiles: delivery of the governor.
60x2 + 5 túg ki-Nin-me-lám-ta	125 textiles from Nin-melam;
10x2 lá-1 túg-didli ki-ensik-ta	19 various textiles from the governor;
10 + 7 túg ki-Ur-^{ĝiš}gigir ázlag-ta	17 textiles from Ur-gigir, the fuller;
10 + 7 túg ki-Uš-ĝu$_{10}$-ta	17 textiles from Ušĝu;
10x2 + 2 túg ki-Ur-^{d}Iškur-ta	22 textiles from Ur-Iškur;
1 túg ki-Lugal-nì-lagar-e-ta	1 textile from Lugal-nilagare.
blank space	

rev.

šu-níĝin 60x8 + 1 túg-hi-a	Total: 481 different textiles.
šà-bi-ta	From inside it (= the sum of the preceding capital)
60x6 + 10x2 túg é-kišib-ba	380 textiles to the 'House of the tablets';
60 + 10x3 + 3 túg má-a ĝá-ra	93 textiles placed in the boat(s);
1 túg Ur-^{d}EN.ZU	1 textile to Ur-Sîn;
2 túg kišib ensik	2 textiles on the sealed tablet of the governor;
3 túg Lugal-^{ĝiš}kiri$_{6}$	3 textiles to Lugal-kiri;
2 túg Ur-^{d}En-líl-lá	2 textiles to Ur-Enlilla.
blank space	
šu-níĝin 60x8 + 1 túg-hi-a	Total: 481 different textiles
zi-ga-àm	have been expended.
nì-kas$_{7}$-aka túg-bi dé-a	Balanced account of its distributed textiles.
iti ^{d}Li-si$_{4}$	Month of the goddess Lisi (= IX month of the calendar of Umma)
mu-ús-sa bàd / ma-da ba-dù	Year after that in which the Fortress of the Land was built (= 37th year of the king Šulgi).

These accounts are balanced by comparing the total of the income with that of the expenditure. Of course, a remainder (left over after the operations of expenses) or an overdraft incurred may result, or rather as rule results, from these accounts: the former case is indicated by the term lá-ì, the latter by diri, and these differences were strictly recorded in the balanced accounts relating to the immediately following period or, in the case of the remainder, it was returned by the official who had received it. However, in our account there is an exact equivalence between deliveries and expenses. It should also be noted that here there is no information about the typologies of the registered textiles, defined only in an issue as didli, "various", and in the total of expenditure as hi-a, "mixed". The same official may be mentioned both in the former section, that of the suppliers, and in the latter, that of the recipients: it is the case, in our text, of the ensik. It may derive from the fact that an official is the provider of a kind of textiles and the consignee of another kind, but also by the different functions that especially a high official may perform, e.g. he may be at the same time the responsible both for a manufacturing sector that produces textiles and for a team of workers, who receive clothes. At this regard, it should be noted that, as pointed out by H. Waetzoldt, amounts of wool or textiles (the latter only when available) were issued once a year as wages of craftsmen and service personnel, also of very low rank.[9]

In the second text, a yearly document, the number of registered textiles is by far larger, in all 10,251 pieces, and in this case a dozen of different kinds of textiles are specified, and besides this they belong to about seventy sub-categories defined by various characterizing elements of each kind of textiles. Nevertheless, there is an element, which relates the two tablets. The latter text registers, indeed, a long list of textiles that result to be the remainder (33 kinds of textiles) or the overdraft (35 kinds of textiles) of balanced accounts. In other terms, the scribe has here gathered the results of a good deal or the totality of the balanced accounts of textiles drawn up by the Umma administration during the span of about a year. More precisely, the period which regards this circulation of textiles are 11 months (from the first to the eleventh) of the 8th year of the king Amar-Sîn, the son and successor of Šulgi, to whose reign the above-presented text belongs. From the amount of these leftovers or overdrafts (more than 10,000 textiles) we might assume how high must be the number of the textiles involved in the relative operations of deliveries and issues, in all probability all those that have been carried out in the space of time in question in the ambit of the province of Umma.

The official responsible of these thousands of textiles is Ikalla, whose seal is impressed on the tablet.

BM 110313 (dimensions: 13.1 × 7.5 × 3.4 cm.),
see *infra*, Figs. 10.1–10.2.

obv.
col. I

⸢lá-ì⸣ 1 ^{túg}bar-túg-4-kam-ús
60x7 + 8 ^{túg}guz-za-ĝen
600x4 + 60x3 + 10x3 + 3 15 gín / ^{túg}uš-bar
600 + 60x2 + 10x4 + 2 ^{túg}uš-bar-tur
10x3 ^{túg}uš-bar-tur ka-ah
10 ^{túg}šà-ga-dù-ĝen
10 ^{túg}sa-sa
8 ^{túg}uš-bar-tur-gi$_6$
10x4 + 4 ^{túg}ú-gi$_6$-kal
60 + 5 ^{túg}ú-gi$_6$-murub
60x9 + 6 ^{túg}mug-murub
10 + 7 ^{túg}mug-tur
10 + 7 ^{túg}uš-bar-sumun
60 + 10x4 + 9 ^{túg}uš-bar-tur-sumun
10x4 + 5 ½ ^{túg}gú-anše
10 + 2 ^{túg}gú-anše-sumun
3 ^{túg}bar-si-ár-ra-4-kam-/ús
10x3 + 5 ½ ^{túg}ú-kal
2 ^{túg}šà-ga-dù ka-ah
2 ^{túg}*da-ba-tum*-tur
7 ^{túg}ú-sumun

[9] Waetzoldt 1987, 125–126.

col. II

10x2 + 2 ^{túg}[ ]
10 + 3 ^{túg}[ ]
1 ^{túg}uš-[bar]
3 ^{túg}ú-t[ur]
4 ^{túg}šà-ga-dù-ĝ[en]
6 ^{túg}šà-ga-dù ĝen nì-dára$_2$
8 1/2 gada-ĝen
10 + 3 gada šà-ga-dù-4-kam-ús
1 gada ĝen-sumun
3/30 5/6 sìla 5 gín ì-šáh
3,600 + 600x2 + 10x3 + 8 2/3 5 gín / túg-hi-a lá-ì-àm — 4,838 and 25/36 mixed textiles are the remainder.
diri 10x2 + 8 ^{túg}nì-lám-3-kam-ús
3 ^{túg}guz-za-3-kam-ús
10 + 6 ^{túg}nì-lám-4-kam-[ús]
600 + 60x2 + 10x4 + 3 ^{túg}guz[-za-4-kam-ús]
10x3 + 8 ^{túg}nì-lám-[ĝen]
10 ^{túg}guz-za ĝen [ ]
1 ^{túg}guz-za-ĝen [ ]
3,600 + 60x5 + 10 lá-1 [-kam]-/ús

rev.
col. I

above 12 lines lost
⌜60 + 10x3 + 4 + [^{túg}]ú-murub
10x4 + 6 ^{túg}ú-[ ]-tur
10x3 + 1 ^{túg}mug-kal
10x3 + 6 ^{túg}mug-gi$_6$
4 ^{túg}ú nì-dára$_2$
1 ^{túg}nì-lám-ĝen ka-ah
6 ^{túg}nì-sal-lá-saĝ
10 + 2 ^{túg}bar-si-nì-lá
10x2 + 5 ^{túg}bar-si-ĝen
1 ^{túg}bar-si ĝen-sumun
4 gada-3-kam-ús
10 + 2 gada šà-[ga-dù]
10x2 + 8 [ ]
10x5 + 4 + [gú] / 5/6 ma-[na síg]

col. II

3,600x7 + 60x8 + [ ] / 3 ĝur[uš]

blank space with the seal impression:
Ì-kal-la — Ikalla,
dub-sar — the scribe,
dumu Lú-sa$_6$-ga — 'son' of Lusaga.

3,600 + 600x3 + 10 + 3 2/3 5 gín / túg-hi-a diri-ga-àm — 5,413 and 25/36 mixed textiles are the overdraft.
Ur-ĝi$_6$-par$_4$ — Ur-ĝipar

ù Šà-nin-ĝá<<-ĝá>> íb-gi-in	and Ša-ninĝa established.
diri lá-ì Ì-kal-la	Overdraft and remainder of Ikalla.
iti-10 + 1–kam iti še-kin-ku$_5$-ta	11 months are (the period): from the month of the harvesting (= I month of the calendar of Umma)
iti Pa$_4$-ú-e-šè	to the month of the god Pa'ue (= XI month of the calendar of Umma)
kišib Ì-kal-la	Sealed tablet of Ikalla.
[mu en E]ridu^{ki} ba-huĝ	Year after that in which the High priestess of Enki was installed (= 9th year of the king Amar-Sîn).

The textiles in question are the following and they comprehend almost all the textiles mentioned in the Umma texts (the ^{túg}aktum, most frequent in the documentation of Ĝirsu, are mentioned only in a half dozen of texts of Umma):

^{túg}**bar-túg**
4-kam-ús (number: 1)

^{túg}**uš-bar**
(2,613 ¼)
tur (762)
tur-gi$_6$ (8)
tur ka-ah (30)
sumun (17)
tur-sumun (109)
[ ] (1)

^{túg}**sa-sa**
(10)

^{túg}**ú**
kal (35 ½)
gi$_6$-kal (44)
gi$_6$-murub (65)
murub (94 + x)
sumun (7)
tur (3)
[ ] tur (46)
nì-dára$_2$ (4)

^{túg}**mug**
murub (546)
tur (17)
kal (31)
gi$_6$ (36)

^{túg}**gú-anše**
(45 ½)
sumun (12)

^{túg}**bar-si**
-ár-ra-4-kam-ús (3)
ĝen (25)
ĝen sumun (1)

^{túg}**šà-ga-dù**
ka-ah (2)
ĝen (10)
ĝen nì-dára$_2$ (6)
ĝen [ ] (4)

^{túg}***da-ba-tum***
tur (2)

gada
ĝen (8 ½)
šà-ga-dù 4-kam-ús (13)
šà-[ga-dù] (12)
ĝen-sumun (1)
3-kam-ús (4)

^{túg}**nì-lám**
3-kam-ús (28)
4-kam-ús (16)
ĝen (38)
ĝen ka-ah (1)

^{túg}**guz-za**
3-kam-ús (3)
4-kam-ús (763)
ĝen (428)
ĝen [ ] (10)
ĝen [ ] (1)

^{túg}**nì-sal-lá-saĝ**
(6)

All these names are preceded by the determinative túg, which characterizes the items belonging to the category of textiles. But in this category an important difference is to be stated between the textiles which are to be interpreted as clothes and those which are piece-goods; of course, only the latter may be counted in fractions (one fourth, one third or a half). To this second category belong surely ^{túg}ú, ^{túg}uš-bar, ^{túg}gú-anše, gada.

The by far most frequent sequence of characterizing elements of textiles regards the quality. In this regard, we have the following indications, listed from the best to the worst:

> lugal, that is "royal"
> ús, "following (the quality lugal)"
> 3-kam-ús, "of the third class"
> 4-kam-ús, "of the fourth class"
> ĝen, "poor".

This range of values is attributed to the main clothes, which are ^{túg}nì-lám and ^{túg}guz-za, and, in addition to them, to the ^{túg}bar-túg/si, the ^{túg}bar-dul$_5$, the ^{túg}aktum and to the linen gada, even if with these last groups of textiles the categories 3-kam-ús and 4-kam-ús are exceedingly rare.

It should be noted that the first-rate textiles may be indicated in the Ĝirsu texts, and only exceptionally in the Umma texts, also by the term sig$_5$, "of good quality",[10] and more rarely, and exclusively in the Ĝirsu texts, by the title ensik, "of the governor", most likely a variant of the above-cited lugal, "of the king, royal". In some cases, instead of lugal, the personal name of one of the kings of Ur (more precisely, at the best of our knowledge, only of the last two, Šu-Sîn and Ibbi-Sîn)[11] is mentioned, in order to indicate not the recipient of the textile, but its quality. ^{túg}aktum is characterized by the term uru "of the city", the quality of which escapes us, in only three texts from Ĝirsu.[12]

Another category of characterizing elements is indicated by the sequence: KAL – murub – tur. The meaning of the adjective tur is "small", and of murub is "middle"; so, this category must refer to the dimensions. Now, the sign KAL, that has to designate the first class clothes in this respect, may have two values: kal with the meaning of "rare, valuable" and kalag, "mighty, robust", and then "thick". Thus, it is probable that this sequence of adjectives refers to the thickness of the textiles. It is often attributed to two categories of textiles: ^{túg}ú and ^{túg}mug. The meaning of the former name is unknown and that of the latter is "cloth made from waste wool". Each category of textiles is characterized by a sequence of the two above discussed characterizing elements, the quality or, if our hypothesis is correct, the thickness. So, ^{túg}nì-lám and ^{túg}guz-za are never defined as murub or kal, and only the former, and only exceptionally, as tur. A third set of characterizing elements is attributed to the ^{túg}aktum: gal, "great", and tur, "little", but only in the tablets from Ĝirsu.

---

[10] In a few cases the characterizing elements lugal and sig$_5$ are united: see 1 ^{túg}aktum sig$_5$ lugal (ITT 4, 7396:rev. 13; 7522:rev. 11).

[11] Both the title ensik in the more ancient texts (ITT 5, 6713:obv. 3: Š 8; 6810:obv. 1: Š 11; MVN 6, 327:obv. 1': date lost; RTC, 204:rev. I 11. II 10: without year name) and the name of Šu-Sîn or Ibbi-Sîn in the more recent ones (TCTI 2, 3487:obv. 1: ŠS 5; TÉL, 283:obv. 1: ŠS 6; ITT 5, 6797:obv. 1 and TCTI 2, 3468:obv. 1: ŠS 7; UNT, 35:obv. I 1, II 3; UNT, 37: obv. I 1, II 3; ITT 2, 3413:obv. 1; TCTI 2, 2547:obv. 1 and TCTI 2, 3421:obv. 1: ŠS 8; ITT 2, 696:obv. 1 and MVN 13, 804:obv. 1: IS 1) are exclusively related to the textile ^{túg}nì-lám. The cloth in question is always mentioned in the first place of the lists of textiles.

[12] RTC, 208 obv. 3, rev. 1; 276 obv. I 4, rev. I 14; 285 obv. 9.

In contrast, adjectives referring to the colour (only gi$_{6}$, "black", in our text) and state of preservation (sumun, "old") may be assigned both to clothes (^{túg}nì-lám, ^{túg}guz-za, ^{túg}mug, ^{túg}aktum, ^{túg}bar-dul$_{5}$, and ^{túg}bar-si), and to the above-listed piece-goods. It remains to touch on the meaning of the term ka-ah, which is attributed above all to the piece-goods ^{túg}uš-bar, and much less frequently to the piece-goods ^{túg}šà-ga-dù and to the clothes ^{túg}mug and ^{túg}guz-za. Concerning ka-ah, which also refers to hides of sheep and oxen, a meaning of "spoiled", as a consequence of a bad treating or storage, has been attributed. Lastly, the term nì-dára$_{2}$, which in our text is related only to ^{túg}ú and ^{túg}šà-ga-dù, but in other tablets to nearly all the other textiles (^{túg}uš-bar, gada, ^{túg}nì-lám, ^{túg}guz-za), and besides is cited separately as a product, indicates an employ: "rag, sanitary towel". It must refer to textiles in very poor condition to be employed only for the most humble uses.

Unfortunately, we have very few data about the price, that is, the value of the various kinds of textiles in the Neo-Sumerian documents. The piece-goods ^{túg}ú and ^{túg}uš-bar and the poor-quality cloth ^{túg}mug have in two Neo-Sumerian tablets their value indicated in barley, with an equivalent in silver which varies from 1/5 to 2/5 of shekels of silver (that is from about 1.6 to 3.2 grams), while a long ^{túg}nì-lám has a value of 1.5 shekel (about 12.5 grams), but in a text that is over a century older.[13]

All the above-mentioned textiles are made of wool, with the exclusion of the linen gada.[14] Thus, in many texts we have the mention of the textiles' weight in wool or amounts of wool are registered for their manufacture. See e.g. the following tablet:

BM 110320 (dimensions: 12.4 × 7.2 × 2.5 cm.).
obv.
col. I

[síg ^{túg}nì-lám]-/[3-kam-ús]	x of wool for ^{túg}nì-lám of third class;
[síg] ^{túg}nì-lám-/4-kam-ús	x of wool for ^{túg}nì-lám of fourth class;
[m]a-na síg ^{túg}/guz-za-3-kam-/ús	x minas of wool for ^{túg}guz-za of third class;
1 gú lá-1 ma-na síg / ^{túg}guz-za-4-kam-ús	59 minas of wool for ^{túg}gu-za of fourth class;
2 gú 10 ma-na síg ^{túg}guz-za-ĝen	130 minas of wool for ^{túg}guz-za of poor class;
1 gú 8 1/2 ma-na síg ^{túg}/uš-bar	68.5 minas for piece-goods ^{túg}uš-bar;
3 1/3 ma-na síg ^{túg}uš-/bar-gi$_{6}$	3,33 minas of wool for piece-goods ^{túg}uš-bar black.
4 gú 33 ma-na 10 gín	(Total:) 273 minas and 1/6.
Šeš-kal-la na-gada	Šeškalla, the herdsman (who supplied them).
4 ma-na síg ^{túg}nì-lám-/3-kam-ús	4 minas of wool for ^{túg}nì-lám of third class;
4 2/3 ma-na síg ^{túg}nì-lám-/4-kam-ús	4,66 minas of wool for ^{túg}nì-lám of fourth class;
18 ma-na síg ^{túg}guz-za-/3-kam-ús	18 minas of wool for ^{túg}guz-za of third class;
43 ma-na síg ^{túg}/guz-za-/4-kam-ús	43 minas of wool for ^{túg}guz-za of fourth class;
1 gú 10 + 3 1/2 ma-na /síg ^{túg}guz-za-ĝen	73,5 minas of wool for ^{túg}guz-za of poor class;
1 gú 10x3 + 5/6 ma-na síg / ^{túg}uš-bar	91,83 minas for piece-goods ^{túg}uš-bar;

col. II

[ ] ma-na síg ^{túg}uš-bar-gi$_{6}$	x minas for piece-goods ^{túg}uš-bar black.
3 gú 10x5 + 8 ma-na	(Total:) 238 minas.
Lú-^{d}Utu na-gada	Lu-Utu, the herdsman (who has supplied them).
3 1/2 ma-na síg ^{túg}nì-/lám-3-kam-ús	3,5 minas of wool for ^{túg}nì-lám of third class;
7 ma-na síg ^{túg}nì-/lám-4-kam-ús	7 minas of wool for ^{túg}nì-lám of fourth class;

[13] See Waetzoldt 1981, 24, §9.

[14] Exceptionally also the ^{túg}bar-dul$_{5}$ may be made of linen (see ITT 3, 6605:obv. I 4').

10x5 ma-na síg ^{túg}guz-za-/3-kam-ús	50 minas of wool for ^{túg}guz-za of third class;
1 gú 10 + 7 ma-na síg ^{túg}guz-za-/4-kam-ús	77 minas of wool for ^{túg}guz-za of fourth class;
5 gú 10x4 lá-1 ma-na síg / ^{túg}guz-za-ĝen	339 minas of wool for ^{túg}guz-za of poor class;
2 gú 6 1/2 ma-na síg ^{túg}uš-/bar	126.5 minas for piece-goods ^{túg}uš-bar;
6 ma-na síg ^{túg}uš-bar-/gi$_6$	6 minas for piece-goods ^{túg}uš-bar black.
10 gú 9 ma-na	(Total:) 609 minas.
Ur-ru na-gada	Urru, the herdsman (who has supplied them).
10 ma-na síg ^{túg}nì-lám-/3-kam-ús	10 minas of wool for ^{túg}nì-lám of third class;
10x4 + 2 1/2 ma-na síg ^{túg}/nì-lám-4-kam-ús	42.5 minas of wool for ^{túg}nì-lám of fourth class;
rev.	
col. I	
1 gú 3 ma-na síg / ^{túg}guz-za-3-kam-ús	63 minas of wool for ^{túg}guz-za of third class;
3 gú 10x2 + 5 ma-na /síg ^{túg}guz-za-4-kam-ús	205 minas of wool for ^{túg}guz-za of fourth class;
9 gú 7 ma-na /síg ^{túg}guz-za-ĝen	547 minas of wool for ^{túg}guz-za of poor class;
5 gú 1 5/6 ma-na síg / ^{túg}uš-bar	301.83 minas for piece-goods ^{túg}uš-bar;
10 + 9 gú 10x2 + 9 1/3 ma-/na	(Total:) 1169.33 mine,
Lú-Eb-gal ù / Šà-kù-ge	Lu-Ebgal and Šakuge (have supplied them).
mu 2–bi Lú-^{d}Ha-/ìa-ke$_4$ tùm-dam	For two years Lu-Haia has to deliver.
rest: blank space	
col. II	
blank space	
10x3 + 8 gú 9 ½ ma-/na síg-kur-ra	2289.5 minas of mountain (sheep) wool,
síg-kur-ra mu-ku$_x$	mountain (sheep) wool delivered.
mu *Si-ma-núm*^{ki} ba-/hul	in the "year in which Simanum was plundered"
ensik-ke$_4$ šu-ba-ti	(= 3rd year of the king Šu-Sîn) the governor has received.
iti mìn-èš	Month of the sixth sanctuary (VIII month of the calendar of Umma).
[mu]-ús-sa *Si-ma-núm*/^{ki} ba-hul	Year after that in which Simanum was plundered (= 4rd year of the king Šu-Sîn).
rest: blank space	

In this text a certain number of administrative data are supplied: e.g. three herdsmen for the three first items (which register in all 910 minas of wool) and two high functionaries for the fourth item (which, with its 1,169.33 minas, exceeds the total of the preceding amounts) deliver the wool while the governor of Umma is its recipient. It would remain to establish the position of Lu-Haia in this transfer of wool for textiles: it is written that this important official[15] has to bring something for the period of two years (rev. I 6). More interestingly for our topics, the wool in question is defined as "wool of mountain (sheep)", which is the most prized wool, also 9.97 (MVN 15, 117) and 9 shekels of silver (Nebraska, 53 obv. 1–2) per talent, that is, 60 minas (= about 30 kg) of wool, and then 7.5 (AUCT 1, 562; Nebraska, 53 obv. 5–rev. 1; AAICAB 1/1, Ashm. 1911–212, "new"), 7 (AUCT 2, 392; Princeton 1, 300), and 6 (AUCT 3, 251; 313; 373 "new"). This wool is followed for value by the "wool of sheep (uli-)gi": the price of which maybe of 7.5 shekels of

[15] Lu-Haia was for a long period one of the principal subordinates of Ur-e'e, the head of the Central Office of the province of Umma: his seal (with the legend: Lú-^{d}Ha-ìa / dub-sar / dumu Ur-e$_{11}$-e šùš) is impressed on tablets dated from Š 47 (Princeton 2, 191) to IS 2 (AAICAB 1/1, Ashm. 1911–218).

silver (UTI 6, 3778.) and 6.66 (Snell 1982, pl.VI, N.4 obv. I 1–2), but also of 4.98 (D.C. Snell, *op.cit.*, pl.XXI, N.12 obv. 3–4) and 4.75 (AUCT 1, 444 obv. 1–2, "old") per talent of wool. All these data come from texts of Umma. For a comparison, in the Neo-Sumerian texts the value of the normal sheep wool (or at least of the wool of unspecified type) varies from 4 to 6.5 shekels of silver, and that of the goat wool from 1.5 to 2 shekels (see *UNT*, p.73).

In our tablet the number of the various kinds of clothes (^{túg}nì-lám and ^{túg}guz-za) and piece-goods (^{túg}uš-bar) to which the wool is intended is not specified, since the scribe was interested only to the record of the amounts of wool supplied. But the weight of wool per cloth is indicated in numerous lists of textiles, as those presented in the two following tablets:

BM 110568 (dimensions: 5.8 × 4.2 × 2 cm.)

obv.	
6 ^{túg}nì-lám-3-[kam] ús	6 textiles-nì-lám of third quality:
ki-lá-bi 9 5/6 ma-na	their weight is 9.83 minas (= 4.91 kg);
⌜10 lá-1⌝ ^{túg}nì-lám-4-kam ús	9 textiles-nì-lám of fourth quality:
⌜ki-lá-bi⌝ 10 + 8 ma-na	their weight is 18 minas (= 8.99 kg);
10 + 2 ^{túg}guz-za-4-kam ús	12 textiles-guz-za of fourth quality:
ki-lá-bi 1 gú 4 1/2 ma-na	their weight is 64.5 minas (= 32.23 kg);
10 + 7 ^{túg}guz-za-ĝen	17 textiles-guz-za of poor quality:
ki-lá-bi 1 gú 8 1/2 ma-na	their weight is 68.5 minas (= 34.23 kg).
síg kur-ra ^{ĝiš}ga-rík aka	Mountain (sheep) wool combed.
10 + 6 ^{túg}ú-kal	16 thick textiles-ú;
10 + 4 ^{túg}ú-tur	14 light textiles-ú;
rev.	
ki-lá-bi 1 gú 8 1/3 ma-na	their weight is 68.33 minas (= 34.15 kg).
16 ^{túg}mug-kal	16 thick textiles-mug:
ki-lá-bi 56 ½ ma-/na	their weight is 56.5 minas (= 28.23 kg).
2 + [ ] gada-ĝen	2+x linen textiles.
[ki-lá-bi] ⌜tag-ga⌝	Textiles weighed after the weaving.[16]
[ki Ur-^{d}]Nin-tu	From Ur-Nintu
Ni-⌜da⌝-la šu-ba-ti	Nidala has received.
iti Pa$_4$-ú-e	Month of the god Paue (= XI month of the calendar of Umma),
mu bàd-mar-tu ba-dù	Year in which the Wall of Martu was built (= 4th year of Šu-Sîn).
left edge	
^{d}Utu-da-ré pa$_4$-ú-e-ka ba-lá	Utu-dare in (the month) of Pa'ue has weighed.

BM 110615 (dimensions: 4.8 × 4 × 1.8 cm.), see *infra*, Fig. 10.3.

obv.	
2 ^{túg}nì-lám-4-kam ús	2 textiles-nì-lám of fourth quality:
ki-lá-bi 3 ma-na 10 + 8 gín	their weight is 3.3 minas (= 1.64 kg);
4 ^{túg}guz-za-4-kam ús	4 textiles-guz-za of fourth quality:
ki-lá-bi 10 + 9 1/3 ma-na	their weight is 19.3 minas (= 9.65 kg);
10 + 1 ^{túg}guz-za-ĝen	11 textiles-guz-za of poor quality:
ki-lá-bi 10x4 + 5 ma-na 10 gín	their weight is 45.16 minas (= 22.57 kg);

[16] For this operation see Verderame 2008, 112–113.

2 ^{túg}nì-lám-ĝen	2 textiles-nì-lám of poor quality:
ki-lá-bi 3 1/3 ma-na	their weight is 3.3 minas (= 1.66 kg).
síg-kur-ra ^{ĝiš}ga-rík aka	Mountain (sheep) wool carded.
2 ^{túg}ú-ka[l]	2 thick textiles-ú:
rev.	
[ki-lá]-bi 7 2/3 ma-na	their weight is 7.66 minas (= 3.83 kg);
[x] gada-ĝen	x linen textiles.
ki Šeš-sig$_5$-ta	From Šešsig
Ir$_{11}$ šu-ba-ti	Ir has received.
Ur-e$_{11}$-e-ke$_4$ / in-lá	Ur-e'e has weighed.
blank	
iti é-iti-àš	Month of the shrine of the sixth month (= VIII month of the calendar of Umma),
mu ^{d}Amar-^{d}EN.ZU / lugal	Year in which Amar-Sîn became king (= 1st year of Amar-Sîn).

So, limiting ourselves to the textiles mentioned most frequently in the Umma and in the Ĝirsu texts, we have a good deal of indications of weights. It should be considered that some typologies of textiles are rarely mentioned both in the texts of Umma and in those of Ĝirsu (so the ^{túg}aktum for the former documentation and the ^{túg}ú for the latter) and those few mentions don't indicate the weights. In other cases, the comprehensive weight of textiles belonging to different categories (^{túg}nì-lám and ^{túg}guz-za) or sub-categories (textiles-kal, murub, and tur) is supplied, and therefore these data have not been included in the following list:[17]

^{túg}nì-lám lugal	Ĝirsu	1 kg (ITT 2, 909)
^{túg}nì-lám Šu-Sîn	Ĝirsu	1.12 (TCTI 1, 696; TCTI 2, 3421) > 1.14 kg (ITT 2, 3413)
^{túg}nì-lám Ibbi-Sîn	Ĝirsu	0.45 kg (MVN 13, 804)
^{túg}nì-lám ensi	Ĝirsu	0.58 kg (ITT 5, 6713)
^{túg}nì-lám-sig$_5$	Ĝirsu	0.61 kg (MCS 8, 89 BM 100462)
^{túg}nì-lám-ús (lugal)	Ĝirsu	0.82 (HSS 4, 101) > 1.43 kg (TCTI 2, 2564)
^{túg}nì-lám-3-kam-ús	Umma	0.61 (MVN 16, 755) > 0.91 kg (AnOr 1, 292; UTI 5, 3223)
	Ĝirsu	0.8 (HSS 4, 101) > 1.44 kg (TCTI 2, 3398)
^{túg}nì-lám-4-kam-ús	Umma	0.63 (BPOA 1, 461) > 1 kg (UTI 2, 2193; BM 110568)
	Ĝirsu	1.08 (TCTI 2, 2564) > 1.24 kg (ITT 5, 9597)
^{túg}nì-lám-ĝen	mma	0.73 (MCS 8, 89 BM 105406) > 1 kg (DC, 181)
	Ĝirsu	1.08 kg (TCTI 2, 2566)
^{túg}guz-za lugal	Ĝirsu	1.54 (HSS 4, 101) > 3.74 kg (MTBM 199)
^{túg}guz-za-sig$_5$	Umma	0.91 kg (SUL, 403)
	Ĝirsu	1 (RTC, 273 obv. 3) > 2 kg (*RTC*, 273 obv. 1)
^{túg}guz-za-ús lugal	Ĝirsu	1.5 (HSS 4, 101) > 7.24 kg (ITT 4, 7522)
^{túg}guz-za-ús	Ĝirsu	1.49 (ITT 5, 6713 rev. 5–6[18]) > 4.08 kg (TCTI 2, 387)
^{túg}guz-za-3-kam ús	Umma	2.6 (MVN 18, 319; BPOA 1, 461, 616; Nisaba 9, 82 *et passim*) > 3.3 kg (TCL 5, 6054)
	Ĝirsu	1.7 (MTBM, 207) > 4.49 kg (TCTI 2, 3490)

[17] The basic values of the Neo-Sumerian weight system are the gín, "shekel" (= 8.33 g.) and the ma-na = 60 gín (about ½ kg). The highest unit of weight is gú(-un), 'talent', which is equivalent to 60 ma-na. The sub-multiple of the gín, of course rarely employed for the wool, is the še, 'grain', which is equivalent to 1/180 gín.

[18] But the immediately precedent item records 3 ^{túg}[guz]-za-ús-sig$_5$, which must be a variant of ^{túg}guz-za-ús, and their weight (2 ma-na), with the exceedingly low value of 0.33 kg per textile.

^{túg}guz-za-4-kam ús	Umma	2.12 (MVN 16, 679) > 3.2 kg (MVN 16, 959)
	Ĝirsu	1.78 (UNT, 43) > 2.61 kg (UNT, 45 obv. II 10'–11')
^{túg}guz-za-ĝen	Umma	1.98 (BPOA 1, 461) > 2.93 kg (Nik. 2, 394)
	Ĝirsu	2.01 (SET, 274) > 4.12 kg (UNT, 45 obv. III 10'–11')
túg-uš-bar	Umma	1.29 (túg-uš-bar-gi$_6$) (SNAT, 306 obv. 13-rev. 1) > 1.83 kg (SNAT, 306 obv. 11–12)
	Ĝirsu	2.02 (TCTI 2, 2564) > 3 kg (ITT 5, 6773)
túg-uš-bar-sig$_5$	Ĝirsu	2.74 (TCTI 3490) > 2.76 kg (TCTI 2, 3534)
^{túg}bar-dul$_5$	Umma	0.6 (SAT 2, 110) > 1 kg (YOS 18, 100)[19]
^{túg}mug-kal	Umma	1.38 (BPOA 1, 496) > 1.89 kg (BPOA 1, 672)
^{túg}mug-murub	Umma	1.74 (TCL 5, 6054; SANTAG 6, 143) > 2.79 kg (Nik. 2, 394)
^{túg}mug-tur	Umma	1.33 kg (SANTAG 6, 234)
^{túg}mug	Ĝirsu	1.99 (AOS 32, N29 obv. I 3–4, II 8–9) > 2.54 kg (UNT, 45 rev. I 7–9)
^{túg}ú-kal	Umma	1.63 (AfO 40–41, p. 55, N.3) > 1.93 kg (HUCA 29, p. 95, N. 18)
^{túg}ú-murub	Umma	1.52 (SNAT, 317) > 1.9 kg (SANTAG 6, 157)
^{túg}*da-ba-tum*	Ĝirsu	1 (CT 9, pl. 31 BM 20018 rev. 1–2) > 2.3 kg (BPOA 1, 4 obv. 6'–8')[20]

From these data we infer that:

1) In the ambit of one type of textiles the variations of the weight are sizeable, and the same is valid in the ambit of a sub-type.
2) These variations don't result in being proportional to the quality class, and neither to the thickness class. Of course, in the latter case, the difference in weight must be the result of different dimensions. These dimensions are rarely indicated, and they occur exclusively in the Ĝirsu tablets. In these few cases, different terms referring to the length (gíd), to the width (dagal), to the right side (bar-si-sá), or to both sides (bar-šu-nígin) are employed, or some more elements (ki(-tag)) are united to characterize the linear measures (see UNT, pp. 144–148). Also regarding the dimensions, there is an ample span of fluctuations for the same type and sub-type of clothes: e.g. in the same text (BM 14311; see UNT, p.147) the ^{túg}nì-lám lugal is related both to 2.33 kùš (= 1.16 m.) and to 5.66 kùš (= 2.66 m.) and the ^{túg}guz-za-4-kam-ús both to 3.66 kùš (= 1.83 m.) and to 6.5 kùš (= 3.25 m); more precisely, in BM 14311 the bar-si-sá (right side) of the ^{túg}guz-za lugal varies from 1.66 kùš (= 0.83 m.) to 7 kùš (3.5 m.). Thus, we cannot determine with any precision the dimensions of the various kinds of clothes.
3) Notwithstanding the great variations of weights in the ambit of the same typology, the wool intended for the ^{túg}nì-lám and for the ^{túg}bar-dul$_5$ is inferior to that for the ^{túg}mug and, more decidedly, to that for the ^{túg}guz-za. The ^{túg}aktum-guz-za, that must unite the characteristics of two different clothes, presents both the highest weight and the widest span of variations from 4.54 (MVN 6, 498 obv. 7–8) to 15 kg (MVN 6, 493 rev. 7–8).[21] Also the weight of the simple ^{túg}aktum, even if only mentioned in a few texts, has a wide span of variations from 0.55 (MTBM, 207) to 5 kg (MVN 22, 188 obv. 2, 12, rev. 5; 218 obv. 1).
4) As rule, the weight of the textiles results to be higher in the Ĝirsu than in the Umma documentation.

[19] Even if the mentions of the ^{túg}bar-dul$_5$ in the texts from Ĝirsu are definitely more numerous than those in the texts from Umma, in no passage it is possible to draw indications about its unit weigh.

[20] In this case, as well in a few others, the wool employed for the textile is divided in two items: its šid-gú-ku$_5$ (see UNT, p. 114)'and its síg-mug, 'waste wool'.

[21] The weigh of the ^{túg}bar-dul$_5$-guz-za is by long lower: 1 kg (ITT 5, 6711; 6713 obv. 1–2).

Fig. 10.1. BM 110313 obverse (Courtesy of the Trustees of the British Museum).

Fig. 10.2. BM 110313 reverse (Courtesy of the Trustees of the British Museum).

Fig. 10.3. BM 110615 (Courtesy of the Trustees of the British Museum).

In the texts concerning the weighing of textiles, three categories of officials are cited:[22]

1) The official who supplies the product (wool or textiles): he is a foreman of the female weavers. Eight of them are mentioned with particular frequency;
2) The official who weighs it: the governor or, in his place, an official of very high rank (particularly the šùš Ur-e'e or Lukalla, a direct subordinate of his);
3) The official who receives it: Ikalla who recurs in the majority of texts in question. This official, whose seal is impressed on the second tablet we have presented above, is the recipient of the textiles in a series of about 90 tablets from Umma that register the delivery of textiles after their weighing, and in some texts he is also the responsible official for this operation. Successively, Ikalla must deliver these textiles to the fullers for their ultimate elaboration, and the same official supplies the same workers with the raw material for their activity (the herb naga, the oil, and the gypsum im-babbar). In other texts the fullers receive direct delivery of textiles. In both cases, the textiles, properly starched, return under the control of Ikalla, who has the duty to consign them to the Central Office of Umma. And at long last these textiles – invisible for the archaeologists, and also almost invisible for the assyriologists[23] – are ready to begin their journey in the ambit of the province of Umma, outside the province (toward the capital Ur, the Holy City Nippur, the cradle of the royal family Uruk, the redistributive centre of Puzriš-Dagan/Esaĝdana) or also beyond the boundaries of the empire.

Abbreviations

AAICAB	J.-P. Grégoire, *Archives administratives et inscriptions cunéiformes de l'Ashmolean Museum et de la Bodleian Collection d'Oxford*, I-IV. Paris, 1996–2002.
AfO	Archiv für Orientforschung. Wien.
AnOr	Analecta Orientalia. Rome.
AOS	American Oriental Series. New Haven.
AUCT	Andrews University Cuneiform Texts. Berrien Springs, Mich.
BM	British Museum. London. Tablet inventory number.
BPOA	Biblioteca del próximo oriente antiguo. Madrid.
CT	Cuneiform Texts from Babylonian Tablets in the British Museum. London.
DC	J.-M. Durand, *Documents cunéiformes de la IVe section de l'École pratique des Hautes Études*. Paris, 1982.
HSS	Harvard Semitic Series. Cambridge, Mass.
HUCA	Hebrew Union College Annual. Cincinnati.
ITT	F. Thureau-Dangin, H. De Genouillac & L. Delaporte, *Inventaire des tablettes de Tello, conservées au Musée Impérial Ottoman*, tomes I-V. Paris, 1910–1921.
MCS	Manchester Cuneiform Studies. Manchester.
MVN	Materiali per il vocabolario neosumerico. Roma.
MTBM	M. Sigrist, *Messenger Texts from the British Museum*. Potomac, 1990.
Nik. 2	M.V. Nikol'skij, *Dokumenty chozâjstvennoj otčetnosti drevnej Chaldei iz sobraniâ N.P. Lichačeva II: Epocha dinastii Agade i epocha dinastii Ura*. Drevnosti Vostočnyâ V. Moscow, 1915.
RTC	F. Thureau-Dangin, *Recueil de tablettes chaldéennes*. Paris, 1903.
SANTAG	SANTAG. Arbeiten und Untersuchungen zur Keilschriftkunde. Wiesbaden.
SAT	Sumerian Archival Texts. Bethesda.
SET	T.B. Jones & J.W. Snyder, *Sumerian Economic Texts from the Third Ur Dynasty. A Catalogue and Discussion of Documents from Various Collections*. Minneapolis, Minn., 1961.
SNAT	T. Gomi & S. Sato, *Selected Neo-Sumerian Administrative Texts from the British Museum*. Chiba, Japan, 1990.

[22] See Verderame 2008, 114–120.

[23] It should be noted that wool and the textiles are nearly the only commodities, which are never mentioned in any of the more than 850 bala texts of Umma (see Sharlach 2004, 289–311).

SUL	M. Sigrist, *Textes économiques néo-sumeriens de l'Université de Syracuse*. Paris 1983.
TCL	Textes cunéiformes. Musée du Louvre, Département des Antiquités Orientales. Paris.
TCTI	B. Lafont & F. Yıldız, *Tablettes cunéiformes de Tello au Musée d'Istanbul: datant de l'époque de la IIIe Dynastie d'Ur*, 1–2. Istanbul, 1989–1996.
TÉL	Ch. Virolleaud & M. Lambert, *Tablettes économiques de Lagash (époque de la IIIe dynastie d'Ur)*. Paris, 1968.
UNT	H. Waetzoldt, *Untersuchungen zur neusumerischen Textilindustrie*. Studi economici e tecnologici 1. Roma, 1972.
UTI	Die Umma-Texte aus den Archäologischen Museen zu Istanbul. Rome – Bethesda.
YOS	Yale Oriental Series. Babylonian Texts. New Haven – London – Oxford.

Bibliography

Molina, M. (2008) The Corpus of Neo-Sumerian Tablets: an Overview. In S. J. Garfinkle & J. Cale Johnson (eds.), *The Growth of an Early State in Mesopotamia: Studies in Ur III Administration*. BPOA 5, 19–53. Madrid.

Owen, D. I. & Mayr, R. H. (2007) *The Garšana Archives*, CUSAS 3, Bethesda.

Sharlach, T. M. (2004) *Provincial Taxation and the Ur III State*. Leiden-Boston

Snell, D. C. (1982) *Ledgers and Prices. Early Mesopotamian Merchant Accounts*, New Haven – London.

Steinkeller, P. (1987) The Administrative and Economic Organization of the Ur III State. The Core and the Periphery. In Mc. Gibson & R. D. Biggs (eds.) *The Organization of Power. Aspects of Bureaucracy in the Ancient Near East*, 19–41. Chicago.

Steinkeller, P. (2008) The Grand Strategy of the Ur III Empire: Exquisite Design, Perfect Failure. *Paper presented at the 54. Rencontre Assyriologique Internationale, Würzburg 20-25, July 2008.*

Studevent-Hickman, B. (2008) The Workforce at Umma: Some New Questions. In S. J. Garfinkle & J. Cale Johnson (eds.) *The Growth of an Early State in Mesopotamia: Studies in Ur III Administration*. BPOA 5, 141–147. Madrid.

Verderame L. (2008) Il controllo dei manufatti tessili a Umma. In M. Perna & F. Pomponio (eds.) *The Management of Agricultural Land and the Production of Textiles in the Mycenaean and Near Eastern Economy*, 101–133. Napoli.

Waetzoldt, H. (1987) Compensation of Craft Workers and Officials in the Ur III Period. In M. A. Powell (ed.) *Labor in the Ancient Near East*, 117–141. New Haven.

Waetzoldt, H. (1981) Kleidung, *Reallexikon der Assyriologie* VI/1–2, 18–31. Berlin – New York.

11. The Colours and Variety of Fabrics from Mesopotamia during the Ur III Period (2050 BC)

Hartmut Waetzoldt

In Mesopotamia, around 2050 BC, the annual production of textiles was huge, certainly significantly more than 60,000 pieces. Most of them were made from wool. Flax was rarely used. All workers were given a piece of cloth or the corresponding quantity of wool by their masters. This piece of cloth or wool, from which they could produce a piece of cloth by themselves, formed part of their wages. The workers mostly received fabric made from the cheaper sorts of wool, amongst which were black fabrics, too.

The good varieties of wool were classified into five qualities, with the best two or three qualities reserved for the members of the royal family and the upper class.

In this chapter, we will firstly discuss the colours of the textiles and secondly attempt to determine, which of the Sumerian and Akkadian terms denote a fabric, and which denote a piece of clothing. In the final section, we attempt to determine the sorts of weave used in these textiles.

1. The Colours

Most strikingly, the colour of the textiles is almost never mentioned. We therefore suppose that most textiles were made of light-coloured wool. Sometimes we find textiles explicitly called "white" (babbar), but these white textiles were apparently not considered to be better than those without an indication of colour.

In contrast to the situation described above, dark or black textiles were often included at the end of lists, meaning that they were less appreciated. These dark garments were, for example, given to an old man, to water drawers (a-bala) and to slaves (of both sexes).[1] According to some texts from the provincial capital of Umma, up to 10% of those in receipt of cloth rations were given black textiles.[2]

One has to be cautious, however, concerning the evaluation of the colour black. Divinities, too, possess black clothes. According to a literary text, Enkimdu, the god of the farmers possessed

[1] AOS 32 L 5; Ontario II 487:1ff.; UTI 4, 2900: 9+32; cf. UTI 3, 3350:1ff. (for a gendarme of the énsi).

[2] MVN 16, 746, 958:9ff.; cf UTI 4, 2900:9. At Garšana only a few black woven fabrics, caps or ribbons are mentioned (not more than 3: CUSAS 3, 662, 663, 678, 757, 758, 760, 764, 782, 783, 796, 804; number not preserved: 651, 753, 833).

black as well as white textiles, and Dumuzi, the god of the flocks wished to have these textiles.[3] In addition, in economic texts, the mother goddess Ninhursag received a black garment and black wool was also used for the production of a heavy cloth or a carpet (túg-maḫ) for the goddess Inanna.[4]

Red/brown wool (su$_4$-a) was used for the production of shoes, sandals and belts. What exact colour was meant by this term is not yet known.[5] This sort of wool was mostly used to produce braided and not woven cloths.

Yellowish? textiles (sig$_7$) are mentioned extremely rarely. The translation to yellowish is not certain; sometimes the Sumerian word also designates a greenish colour. For the production of a ribbon, black and yellowish wool was mixed.[6] The texts do not mention who wore those garments.

Obviously, the most precious wool was the shining-yellow one (ḫuš-a), since this is also the colour of pure gold. It is attested only in the capital Ur and is only mentioned in the texts in small quantities of about 1–7 kg. The best and finest wool is used exclusively for this colour. From this wool was produced a particular type of fine fabric (^{túg}nì-lám). Fabric made from this wool was probably reserved for the use of the king.[7]

Numerous multicoloured textiles are mentioned in the recently published texts from the royal estate of Garšana. We do not know whether the Sumerian word for multicoloured (gùn-a) meant two- or three-coloured fabrics or fabrics in which colours were combined. The most likely option is that light and dark sorts of wool were combined, also possibly with a third naturally pigmented wool. This argument is supported by the wall painting from Mari, although this is about 250 years younger than the texts. Important too is the fact that for the multicoloured textiles, third and fourth class qualities of wool were also used, which means that upper class people used these textiles.[8] Most of the multicoloured textiles were accompanied by a multicoloured ribbon.[9] Multicoloured textiles were also destined for lower deities. As these textiles were comparatively heavy – they weighed more than 4.5 or 5.5 kg – it remains uncertain whether they served as clothing.[10]

The colours discussed here could have been natural pigmentation or dyed colours. While studying the texts more in depth, we found that normally only naturally pigmented wools were used. This is assured for more than 99% of the production, including the light-coloured, white, dark, reddish brown and multicoloured products. The very precious shining yellow wool (ḫuš-a) was most probably dyed. Concerning the yellowish wool the question must be left open, since the exact shade of colour cannot be identified at present.

These facts clearly demonstrate that, in Mesopotamia around 2050 BC, wool or textiles were only dyed in exceptional cases. The natural colour of wool and clothing was light and white.

[3] Sefati 1998, 326:43–46; 332.

[4] Túg-gi$_6$: CUSAS 3, 787; BIN 5, 19:7f. 40 ma-na (*c.*20 kg) síg-gi$_6$ túg-maḫ-^{d}Inanna-Zabalam$_3^{\text{ki}}$; cf. 251:8–13 five female weavers needed 1½ months for the production. According to MVN 3, 349:4 42 ma-na (*c.*21 kg) were used for it.

[5] CUSAS 3, 782: 2–7, 833:3, 939:1.

[6] Waetzoldt 1972, 51 c (^{túg}bar-síg-gi$_6$-ḪI-sig$_7$-sig$_7$, UET 3, 1702: 5).

[7] Waetzoldt 1972, 51 d.

[8] CUSAS 3, 563, 564, 565, 568, 577: 1–4, 786.

[9] CUSAS 3, 564, 565, 568, 577: 1–4, 786.

[10] CUSAS 3, 584: 4.

Occasionally, the wool of differently coloured animals was used to achieve colour effects. Obviously the shining yellow dyed clothing was reserved for the king, but we do not know the dyestuff used for this process.[11]

For the deities, different pieces of clothing were produced. They were probably used to dress the statues of the gods. Usually there is no information to be found concerning the colours of the garments of the gods. Occasionally, we see references to white or dark garments, but multicoloured pieces of clothing are noted, too. Obviously, the colours appear to have had a certain significance, but what this was is unfortunately unknown. The texts do not reveal any information about this. The *Enmerkar* Epic, for example, describes the sacred marriage. During this ritual, the king in the role of the god Dumuzi puts on a multicoloured cap and a priestess in the role of the goddess Inanna dresses in a white garment.[12] According to the myth of *Inanna and Enki*, the God of Wisdom Enki gave Inanna a black and a multicoloured garment. These garments were said to possess numinous powers, indicating that even the gods were more powerful with these coloured garments.[13]

2. Is it possible to determine the different weaves according to the cuneiform texts?

We have two bases for the discussion:

1. Cases in which two different terms for textiles are used side by side, and
2. Indications of the weight of the wools used for the warp and weft threads respectively.

In the economic texts on textiles, usually only one name for a textile is found per line. For the purpose of this paper, however, we collected cases where two different textile terms appeared in the same line.

The hypothesis is that the first term denominates the piece of clothing. The second term refers to the sort of fabric, or more precisely, to the weave. In the following section, we try to verify this hypothesis.

The term ^{túg}gú-è denominates a kind of "cape or shawl for the shoulder".[14] ^{túg}gú-è occurs with 4 different designations. These designations describe the piece of clothing in more detail:

^{túg}gú-è ba-tab *du$_8$-ḫu-um*[15]
^{túg}gú-è guz-za[16]
^{túg}gú-è nì-lám[17]
^{túg}gú-è *tá-ki-ru-um*[18]

These four terms occurring after gú-è (ba-tab *du$_8$-ḫu-um*, guz-za, nì-lám and *tá-ki-ru-um*) can also denominate wraparound garments. In this context, however, this makes no sense, for gú-è is only a smaller piece of cloth. Its weight is always less than 300 g.[19]

[11] On the natural pigmentation of wool and the dyed colours in the Linear B Inscriptions see Nosch 2004, 32–33, 35.
[12] Waetzoldt 1972, XXI; Cohen 1973, line 578 and 591.
[13] Farber-Flügge 1973, 28: 25.
[14] Waetzoldt 1980–83a, 22 § 7e.3.
[15] CUSAS 3, 586: 2, 587: 2, 609: 5, 618: 6, 748: 7, 834: 7; UET 9, 234: 3, 7.
[16] CUSAS 3, 617: 7, 631: 2, 647: 3f., 649: 7, 9, 731: 1, 742: 6, 743: 7, 758: 5, 773: 4, 786: 8, 803: 4, 804: 3.
[17] CUSAS 3, 579: 16, 618: 5, 639: 2, 649: 2, 748: 6, 793: 3, 6, 821: 1, 829: 5', 1162:18'.
[18] CUSAS 3, 824: 1, TMHNF 1–2, 230: 3; cf. MSL 10, 155 No. 2 V 28.
[19] Waetzoldt 1980–83a, 22 § 7e.3; cf SAT 3, 2070: 1–6.

If we consider the texts recording the delivery of wool to the weavers, we find precisely these terms. We have wool for ba-tab *du*$_8$*-ḫu-um*,[20] wool for guz-za,[21] wool for nì-lám[22] but coincidentally no wool for *tá-ki-ru-um*.[23] These kinds of fabrics would have been woven by the female weavers from the above mentioned wool.

Another example occurs in the following:

^{túg}sagšu ba-tab *du*$_8$*-hu-um*[24]
^{túg}sagšu nì-lám[25]
^{túg}sagšu *tá-ki-ru-um*[26]

Here are only three sorts of fabric mentioned. ^{Túg}sagšu is a cap.[27] These terms can be translated by "cap from ba-tab *du*$_8$*-ḫu-um*-fabric", "cap from nì-lám-fabric", and "cap from *takirum*-fabric".

A further example is:

^{túg}šà-ga-dù ba-tab *du*$_8$ *-ḫu-um*[28] and
^{túg}šà-ga-dù nì-lám.[29]
^{túg}šà-ga-dù is a kind of loincloth, skirt or kilt made from ba-tab *du*$_8$*-ḫu-um* and nì-lám-fabric.

The pieces of clothing mentioned here could therefore be made from two, three or four different sorts of fabric and these could be produced with different weaving techniques. According to the hypothesis suggested here, this is the only possible interpretation, as nì-lám or ba-tab *du*$_8$*-ḫu-um*, for example, never occur with other fabric denominations following them.[30]

In the following section we examine the question of which kind of weave was used to produce the ba-tab *du*$_8$*-ḫu-um*-, guz-za-, nì-lám- and *tá-ki-ru-um*-fabrics. Concerning guz-za and nì-lám we have relatively much information. For the terms ba-tab *du*$_8$*-ḫu-um* and *tá-ki-ru-um*, however, we find only few indications.

In Mesopotamia in the years between 2100 and 2000 BC textiles were usually weighed after weaving. Therefore we possess comparatively many indications of the weight of the different fabrics. Compared to this, the texts inform us only rarely about the size of the fabrics. We are well informed about the guz-za-fabrics. They have a weight of between 1.7 and 3.35 kg and they were 3.5 to 4 m in length and up to 4.5 m in width.[31] A piece of guz-za-fabric, however, could

[20] CUSAS 3, 625:14.
[21] AnOr 7, 201: 1–3 = Molina 1996, 201: 1–3, 518 I' 1'-3', II'2'-3'; CUSAS 3, 596: 5–6, 665: 1.
[22] CUSAS 3, 596: 3–4, 750: 1; AnOr 7, 201: 4–6, 518 II' 1'.
[23] The *tá-ki-ru-um*-fabric is rarely mentioned in the texts, see notes 49–51 and CUSAS 3, 608: 5, 739: 2, 766: 1, 796: 4; UET 3, 1705 III 29, IV 18f. It is normally produced from the best and once from 3rd class wool.
[24] CUSAS 3, 511: 121', 578: 16, 645: 2, 647: 5, 692: 2, 700: 2, 701: 5, 708: 3, 6, 709: 5, 742: 3, 744: 2, 745: 2, 746: 7, 747: 7, 773: 9, 793: 5, 799: 2, 801: 2, 803: 1, 806: 4, 835: 4', 1254: 41'.
[25] CUSAS 3, 646: 4, 669: 5, 674: 3, 1254: 12'; *JCS* 54, 2002, 14 Rev. V 16.
[26] CUSAS 3, 563: 2, 569: 16, 579: 17, 591: 8, 617: 2, 636: 2, 669: 2, 753: 2, 764: 2, 789: 2, 793: 1, 796: 1, 814: 3, 1162: 14'.
[27] Waetzoldt 1980–83b, 200 § 8a.
[28] CUSAS 3, 511: 120', 569: 8, 579: 14, 618: 4, 738: 4, 742: 2, 743: 1, 3, 746: 6 and see CUSAS 5, 176 in CUSAS 4, 176.
[29] CUSAS 3, 569: 9, 579: 13, 637: 1, 662: 3, 673: 4, 674: 2, 738: 1f. and see CUSAS 4, 176.
[30] In CUSAS 3, 579: 18 the authors read 4 [^{túg}níg-lá]m ba-tab *tuḫ-hu-um*-3–kam-ús, but according to the parallel text 578: 17, the correct reading is 4 [^{túg}sagš]u ba-tab *tuḫ-hu-um*-3–kam-ús. In SAT 3, 2070: 3 we find 20–lá-3 ^{túg}nì-lám-guz-za. This is all in all a very unusual text; is it possible to translate "17 nì-lám- and guz-za-fabrics"?
[31] Waetzoldt 1980–83a, 21 § 7e.1; Seri 2007, 28 I 5–6, 14–15, II 1–2, 12–15, 24–25, III 4–7, IV 3–6, 14–17, 21–24, V 3–6, 16–20, Rev. I 22–23, 28–29, II 3f. (between 4.4 and 5.5 ma-na = *c*.2.2 and 2.75 kg).

have a length of 6.8 m.[32] Generally, the almost square guz-za-fabric was doubled and used as a wraparound garment (Wickelgewand).

The production of a guz-za-fabric from fourth-class wool is described in a text as follows:[33]

> 1 guz-za-fabric from fourth-class wool,
> the mixed wool for it (weighs) 4 kg,
> 1 woman cleans and combs 125 g in a day (and)
> 1 woman 'mingles' (HI.HI) 1 kg in a day (possibly production of roving/slubbing);
> the warp threads for it (weigh) 333 g (and)
> 1 woman spins 8.3 g strongly twisted threads (for the warp);
> the weft threads for it (weigh) 1.66 kg (and)
> 1 woman produces 61 g (of them) in a day (for the weft);
> (the) length (of the guz-za-fabric is) 3.5 m (and)
> (the) width (is) 3.5 m;
> 3 women warp in 3 days (and)
> 2 women weave 50 cm in a day.

Such detailed descriptions of the production of textiles are very rare. From this example we learn that warp threads have a weight of 333 g and the weft of 1.66 kg. Hence the total weight was 2 kg and the ratio of warp and weft threads is 1:5. In one day, a female weaver could produce only 8.33 g of warp-thread, but 61 g of weft thread. The production of the warp thread needed 40 days, while weft thread was produced over only some 16 and a half days.

Three women needed three days to warp the ground loom. Two women could weave 50 cm of cloth a day. Therefore to produce the complete guz-za-fabric they needed only seven days. The rate of 50 cm a day was comparatively high since for nì-lám-fabrics only 25 cm could be produced a day.

One text describes the warp thread in even more detail: "241 g of warp thread for the third-class guz-za-fabric, the UB for it (are) 8 (and) the *qin-nu-um* for it are 2".[34] In this context, the meanings of UB and *qin-nu-um* are totally vague. Both terms could relate to the guz-za-fabric, too. But then, for me they don't make any sense at all.[35]

We have only few hints indicating what sort of cloth guz-za was. These refer to the weaving rate as well as to the relation of warp and weft threads. In the example above, the ratio of warp to weft was 1:5, in other texts it is 1:4 and 1:4.2.[36] Hence there is about four times more weft yarn than there is warp thread. I wonder whether this is because we are dealing with "shaggy cloth". This was how guz-za was translated in the Pennsylvania Sumerian Dictionary.[37] Regrettably, we are unaware of the time it takes to weave such a cloth.

Concerning nì-lám-fabrics, the ratio of the weights of warp and weft yarn is between 1:0.94 and 1:2.25.[38] The weight of the warp for a single piece of cloth was about 333 g for a medium-weight fabric. The weight of the weft yarn can be estimated by comparing the weight of the nì-

[32] Waetzoldt 1972, 144–145, 148.
[33] Waetzoldt 1972, T.32 III 4–Rev. I 4.
[34] Waetzoldt 1972, 125.
[35] UB normally means "corner, angle and nook" and *qinnum* can be translated as "nest, lair and family" (CAD Q 257; p. 260 qinnu B meaning unknown).
[36] Waetzoldt 1972, 124; cf. ibid. T.23: 1–3 = MVN 19, 89: 1–3 1:2.1; UET 9, 337:7–9.
[37] Pennsylvania Sumerian Dictionary, Vol. 2 B, 120.
[38] Waetzoldt 1972, 124.

lám-fabrics, which were between 0.666 and 1.3 kg per piece.[39] This makes the weft between about the same weight as the warp and approximately double the weight of the warp. This leads me to the assumption that nì-lám-fabrics were produced in plain weave. The nì-lám-fabric must have been comparably light in spite of its size of 3.5 × 3.5 m, 4 × 3.5 m or 4.5 × 6 m (but it could have up to 7.5 m in length).[40] In addition, during excavations cloths were found which were woven in plain weave as demonstrated by Elisabeth Völling.[41] Finally, another fact supporting a smooth fabric is that ni-lam-fabrics were used for the king's bed.[42]

In only few cases we found the following combination: (síg-)^{túg}nì-lám šà-$^{(túg)}$bar-dul$_5$,[43] which means nì-lám-fabric with weft from the bar-dul$_5$ (-type). This means that weft threads originally produced for the bar-dul$_5$-fabric were woven into a nì-lám-fabric. According to this, the weft threads for nì-lám- and bar-dul$_5$-fabrics must therefore have differed from each other. The difference must have been visible to specialists after weaving.

The ba-tab *du$_8$-ḫu-um*-cloth was highly regarded and mostly produced from very good wool.[44] Another argument for its very good quality is the fact that the female weavers needed 240 days to produce a second-quality ba-tab *du$_8$-ḫu-um*-cloth. This working time comprises all tasks starting from the raw wool up to the complete piece of cloth. To produce fabrics from the finest wool, the female weavers needed 480 to 960 days.[45] Unfortunately, the weights of these fabrics are not attested because of the accidental nature of excavations and text publishing. The ratio between the weights of warp and weft threads is 1:4.27. This value, however, is not exact, because it comprises the wool for other textiles, too.[46]

Based on part of the word ba-tab, the Sumerian verb tab may mean: "to double, to lay double" or "to lay something parallel". This interpretation leads to the surmise that several threads of the weft were put into one shed. This would mean that ba-tab *du$_8$-ḫu-um* is a denomination for twill or rep weave (German: Rips).[47] The second part of the word *du$_8$-ḫu-um* cannot be interpreted at present. Therefore, the proposal for the translation "rep" is rather hypothetical. The second part of the word could theoretically be connected with the Akkadian term *tuḫḫum* "bran and draff". This interpretation, however, makes little sense here.[48]

Regarding the fourth cloth denominated *tá-ki-ru-um* we do not know much. A large piece of *tá-ki-ru-um*-fabric of the best quality weighs either 2.6 or 3.5 kg.[49] Once we see 3 kg, but the text is not completely preserved.[50] A "small piece of *tá-ki-ru-um*-fabric" had a length of 2.59 m and a width of 1.25 m. Its weight is not indicated in this text.[51]

---

39 Waetzoldt 1972, 125 (weight of warp) and UET 9, 337: 7–8 (weight of weft not fully preserved); BPOA 2, 1815: 1–2, 1835: 1–6; CUSAS 3, 582: 1–2; MVN 16, 679: 1–2, 7–8, 755: 1–4, 959: 1–4; MVN 22, 121: 1–2; Nisaba 11, 5: 7-8; Seri 2007, 28 I 1–4, 10–11, 20–23, II 22–23, 29–III 3, IV 1–2, Rev. I 20–21.

40 Waetzoldt 1972, 129, 148.

41 Völling 2008, 203–204, 206, 208–209, 212, 215–218, 222, 227–228 etc.

42 Nisaba 11, 25 III 24 – Rev. I 1.

43 Waetzoldt 1972, 127; MVN 3, 188: 3; Nisaba 6, 33 II 11.

44 CUSAS 4, 25; Pennsylvania Sumerian Dictionary, Vol. 2 B, 21.

45 Waetzoldt 1972, 139.

46 Waetzoldt 1972, 124.

47 Völling 2008, 204, 209, 211, 213, 234, 240.

48 CAD T 452; Pennsylvania Sumerian Dictionary, Vol. 2 B, 21.

49 CAD T 78 a textile; UET 3, 1673:1–2, 5–6.

50 UET 3, 1697:1–2.

51 Widell 2003, 104 = Sallaberger 1993–94, 60, 6.

Takkirum-fabrics were only woven in small quantities. Obviously it was a precious fabric, for it was mainly produced in the capital of Ur and in the palace of the princess Simat-Ištaran at Garšana.[52] The deity Ninsianna, too, received a piece of it on several occasions. S/he was the deity of the evening star meaning Venus.[53]

^{túg}bar-dul$_5$: bar-dul$_5$ is on one hand a sort of fabric and on the other hand "a coat or a long cloak".[54] Its production is described in the same text as the guz-za-fabric:[55]

> "1 bar-dul$_5$-U$_2$–fabric from fifth-class (wool),
> the mixed wool for it (weighs) 2 kg;
> 1 woman cleans and combs 125 g in a day (and)
> 1 woman 'mingles' (HI.HI) 1.5 kg in a day (possibly production of roving/slubbing);
> the warp yarns for it (weigh) 666 g (and)
> 1 woman spins 16.6 g strongly twisted threads (for the warp);
> the weft yarns for it (weigh) 833.3 g (and)
> 1 woman produces (of them) 41.6$^{??}$ g in a day (for the weft)";

The rest of the text is lost.

From this we see that the total weight of the warp thread was 666 g and that of the weft thread 833.3 g. Hence, we have a ratio of warp to weft thread of 1:1.25. One piece of bar-dul$_5$-U$_2$–cloth weighed 1.5 kg. In other cases we have found weights for 1 bar-dul$_5$-cloth of up to 1.8 kg or only 604 g.[56] To spin the warp thread a woman needed 40 days and only half of this was needed for the weft thread, meaning 20 days. The length of a bar-dul$_5$-cloth could be up to 4 m.[57] The width is not mentioned in any of the texts known to me.

According to one text, the total working time for the production of a bar-dul$_5$-cloth from the best and finest wool comprised 1080 days.[58] In the text discussed above, we admittedly only find a total of 77 working days, but this text is only half preserved and the quality of the wool is worse, being 5th class.

* * *

In this paper we have discussed five words for different kinds of cloth: guz-za, nì-lám, ba-tab *du$_8$-ḫu-um*, *tá-ki-ru-um* and bar-dul$_5$. Through this investigation it became clear how difficult it is to find arguments for how a certain cloth was woven. Quite rarely, the relationship between the weights of warp and weft yarn is known. But it still does not become obvious how a cloth was woven. For guz-za was proposed the interpretation of "shaggy cloth", for nì-lám "plain weave" and for ba-tab *du$_8$-ḫu-um* "rep". But next to these interpretations we have to add question marks.

The textiles found during excavations can give us an idea of which kind of weave was popular

52 See notes 22, 49–51; CUSA 4, 546.

53 CUSAS 3, 608: 5, 684: 1, 739: 2, 766: 1, 796: 4: for Nisiannna see *Reallexikon der Assyriologie* Vol. 9, 1998–2001, 487–288.

54 Waetzoldt 1980–83a, 21 § 7e.1; *Pennsylvania Sumerian Dictionary*, Vol. 2 B, 119–121 (also made of linen).

55 Waetzoldt 1972, T.32 Rs. I 6–14.

56 CUSAS 3, 724: 1–2; SAT 1, 110.

57 Waetzoldt 1972, 147

58 Waetzoldt 1972, 139.

in a certain period and what was possible to produce on the ground loom. These textiles can stimulate philological research. And hopefully, help to solve the problems.

It was a surprise, that relatively many textiles could be identified by Elisabeth Völling as rep weaves (Rips). Therefore it was necessary to look for a Sumerian term for this weave which we may have identified in ba-tab du_8-*ḫu-um*. But we have to wait for new texts to prove this hypothesis right or wrong.

In the Ur III-Period between 2100 and 2000 BC, it is remarkable that, very often three or four times as much wool was used for the weft yarn as for the warp yarn. This is even the case for the very cheap fabric called túg-mug. For this fabric, the weft yarn was often from five to seven times heavier than the warp yarn: Although the warp was produced from better wool the weft yarn was made from noil.[59] Fabrics with more weft yarn that warp yarn are termed "weft-faced" fabrics.

All these different kinds of fabrics were woven on the horizontal loom. In a newly published text wooden parts of the loom are listed, indicating that the looms in Mesopotamia could be more than 5 m wide.[60]

Acknowledgements

My colleague and friend David Owen, Ithaca (USA) and Rachel Fenton, Cambridge (Great Britain) were so kind to correct and improve my English. I am very grateful for their great help.

Abbreviations

AAS	J.-P. Grégoire, *Archives administratives sumériennes*, Paris 1970.
AnOr 7	N. Schneider, *Die Drehem- und Djohatexte im Kloster Montserrat (Barcelona).* Analecta Orientalia 7, Roma, 1937.
AOS 32	A. L. Oppenheim, *Catalogue of the Cuneiform Tablets of the Wilberforce Eames Babylonian Collection in the New York Public Library*. American Oriental Series, Vol. 32, New Haven 1948.
BIN 5	G. G. Hackman, *Temple Documents of the third Dynasty of Ur from Umma.* Babylonian Inscriptions in the Collection of J. B. Nies, Vol. V, New Haven
BPOA 2	T. Ozaki/M. Sigrist, *Ur III Administrative Tablets from the British Museum, Part two.* Biblioteca del Próximo Oriente Antiguo 2, Madrid 2006.
CAD	*The Chicago Assyrian Dictionary*, Chicago/Glückstadt 1956ff.
CUSAS 3	D. I. Owen/ R. H. Mayr, *The Garšana Archives*, Cornell University Studies in Assyriology and Sumerology, Vol. 3, 2007.
CUSAS 4	A. Kleinerman/D. I. Owen, *Analytical Concordance to the Garšana Archives*, Cornell University Studies in Assyriology and Sumerology, Vol. 4, 2009.

[59] Waetzoldt 1972, 124f. The warp yarn is produced from 5th class wool, whereas the weft yarn is made from noil (ibid. T. 44: 8–10; CUSAS 3, 744: 11–13, 747: 13–15), or the warp yarn is produced from wool of the uli-gi-sheep (*AAS* 135 IX 5–9), cf. Waetzoldt 1972, T. 45 II 1–3, III 16–18, IV 7–9.

[60] CUSAS 3, 635:1–5 (^{giš}sag-du-túg-ga gíd 12/10/8/7/6 kùš-ta): the beams have a length of 6, 5, 4, 3.5, and 3 m. During the Ur III-period (*c.*2100–2000 BC) horizontal looms were used in Mesopotamia. There is good evidence for this in the fact that the main parts of the loom described were two beams (up to 6 m long; weaver's beam and cloth beam), uprights (^{giš}dar-kak) fixed to the beams with ropes. These ropes are called éš-túg-gíd in Sumerian "rope to make the cloth long", which means "to stretch the cloth (and the warp)". Instead of ropes woven ribbons (^{túg}bar-si-túg-gíd) could also be used for this purpose. The beams are called ^{giš}sag-du and ^{giš}gi-na (Waetzoldt 1972, 133f.). ^{giš}dar-kak was discussed most recently by M. Civil (1989, 57) and for ^{túg}bar-si-túg-gíd see CUSAS 3, 793: 11 (and Waetzoldt in preparation).

JCS	*Journal of Cuneiform Studies*
MSL 10	B. Landsberger, E. Reiner, M. Civil, *The Series HAR-ra = hubullu Tablets XVI, XVII, XIX*. Materials for the Sumerian Lexicon, Vol. 10, Roma 1970.
MVN 3	D. Owen, *The John Frederick Lewis Collection*. Materiali per il Vocabolario Neosumerico, Vol III, Roma 1975.
MVN 16	H. Waetzoldt, F. Yildiz, *Die Umma-Texte aus den Archäologischen Museen zu Istanbul*. Materiali per il Vocabolario Neosumerico, Vol XVI, Roma 1994.
MVN 19	P. Mander, *Testi Economici Neo-Sumerici del British Museum (BM 12600–12750)*. Materiali per il Vocabolario Neosumerico, Vol XIX, Roma 1995.
MVN 22	M. Molina, *Testi amministrativi neo-Sumerici del British Museum BM 13601–14300*. Materiali per il Vocabolario Neosumerico, Vol XXII, Roma 2003.
Nisaba 6	F. al-Rawi, F. D'Agostino, *Neo-Sumerian Administrative Texts from Umma kept in the British Museum, Part One (NATU I)*. Nisaba 6, Messina 2005.
Nisaba 11	F. N. H. al-Rawi, L. Verderame, *Documenti amministrativi neo-sumerici da Umma conservati al British Museum (NATU II)*. Nisaba 11, Messina 2006.
Ontario II	M. Sigrist, *Neo-Sumerian Texts from the Royal Ontario Museum II. Administrative Texts Mainly from Umma*. Bethesda 2004.
SAT 3	M. Sigrist, *Texts from the Yale Babylonian Collections, Part II*. Sumerian Archival Texts, Vol. III, Bethesda 2000
TMHNF 1–2	A. Pohl, *Rechts- und Verwaltungsurkunden der III. Dynastie von Ur.* Texte und Materialien der Frau Professor Hilprecht Collection, Neue Folge I/II, Leipzig 1937.
UET 3	L. Legrain, *Business Documents of the third Dynasty of Ur*. Ur Excavations Texts III, London 1937.
UET 9	D. Loding, *Economic Texts from the third Dynasty*. Ur Excavations Texts IX, Philadelphia 1976.
UTI 3	F. Yildiz & T. Gomi, *Die Umma-Texte aus den Archäologischen Museen zu Istanbul*. Bd. III, Bethesda 1993.
UTI 4	T. Gomi & F. Yildiz, *Die Umma-Texte aus den Archäologischen Museen zu Istanbul*. Bd. IV, Bethesda 1997.

Bibliography

Civil, M. (1989) The Statue of Šulgi: The Inscription. DUMU-E$_2$-DUB-BA-A. *In Studies in Honor of Ake W. Sjöberg*, Occasional Publications of the Samuel Noah Kramer Fund 11, Philadelphia, 49–64.

Cohen, S. (1973) *Enmerkar and the Lord of Aratta*, Ph.D. University of Pennsylvania.

Farber-Flügge, G. (1973) *Der Mythos "Inanna und Enki" unter besonderer Berücksichtigung der Liste der me*, Studia Pohl 10.

Molina, M. (1996) *Tablillas Aministrativas Neosumerias de la Abadía de Montserrat (Barcelona)*, Aula Orientalis – Suppl. 11.

Nosch, M.-L. (2004) Red Coloured Textiles in the Linear B Inscriptions. In L. Cleland & K. Stears (eds.) *Colour in the Ancient Mediterranean World*, BAR International Series 1267, 32–39.

Sallaberger, W. (1993–94) Keilschrifttexte einer Privatsammlung. *Archiv für Orientforschung* 40/41, 53–63.

Sefati, Y. (1998) *Love Songs in Sumerian Literature*, Bar-Ilan Studies in Near Eastern Languages and Culture.

Seri, A. (2007) *The Mesopotamian Collection in the Kalamazoo Museum*. Cuneiform Digital Library Journal 2007:1. http://cdli.ucla.edu/pubs/cdlj/2007/001.html.

Völling, E. (2008) *Textiltechnik im Alten Orient. Rohstoffe und Herstellung*. Würzburg.

Waetzoldt, H. (1972) *Untersuchungen zur Neusumerischen Textilindustrie*, Studi Economici e Tecnologici I, Rome.

Waetzoldt, H. (1980–83a) Kleidung. A. Philologisch. *Reallexikon der Assyriologie* Vol. 6, 18–31.

Waetzoldt, H. (1980–83b) Kopfbedeckung. A. Philologisch. *Reallexikon der Assyriologie* Vol. 6, 197–203.

Widell, M. (2003) *The Administrative and Economic Ur III Texts from the City of Ur.*

12. The Textiles Traded by the Assyrians in Anatolia (19th–18th centuries BC)

Cécile Michel and Klaas R. Veenhof

The records of the Old Assyrian traders found in the archives of their houses in the commercial quarter (*kārum*) in the lower town of the ancient Anatolian city of Kaneš, modern Kültepe (*c.*20 km northeast of modern Kayseri), which date from the 19th and 18th centuries BC,[1] contain an enormous amount of references to a large variety of textiles. The traders imported great quantities of primarily woollen textiles from their hometown Aššur into Anatolia, conducted some business in textile products in northern Mesopotamia on their way there, and engaged in a brisk trade in locally produced woollen textiles within Anatolia. Since most of the names and designations of these textile products do not appear in contemporary sources from elsewhere, and since the excavations of Kaneš have yielded no textile remains, their identification is not easy. Most appear in purely commercial contexts, which mention their purchase, packing[2] and transport, sale, and the taxes levied on them in Anatolia, but hardly ever describe their nature, which was of course known to those involved in the trade. Only rarely, in a few private letters written by or to women, do we obtain some information on the production of certain textiles, when traders state their preferences and the women who made them react to such wishes or criticism of their products.[3] We must of course study their names, some of which allow an etymological analysis or are attested in other sources, while others link a textile with a particular town, people or land by being a nisbe, e.g. "Abarnian (textile)" or by means of the relative pronoun *ša*, "(that) of",

Together the two authors collected the data from the sources, discussed the organization and presentation of the material, wrote the introduction, read each other's contributions several times; Michel wrote § 2 and 4, Veenhof § 1, 3 and 5. The manuscript was completed in spring 2009, before the new book by J.-M. Durand, *La nomenclature des habits et textiles dans les textes de Mari* (Durand 2009), had become available. Some references to it have been inserted at the last moment between brackets.

[1] These texts are quoted by their excavation numbers that start with Kt (=Kültepe), followed by the mention of the excavation year (a = 1948 until z = 1971, continued by 72ff.), a slash (/) and k (= *kārum*, the commercial quarter in the lower town) and the number of the individual tablet. For a recent list of published or quoted tablets with these numbers, see Michel 2003, 60–140, continued in Michel 2006, 438–445 and Michel (in press a), where the interested reader can find the data on the tablets we only quote here by excavation number.

[2] Usually in bales or "bags" (called *naruqqum*), note e.g. the small text LB 1269 (quoted Veenhof 1972, 38) that lists the contents of four bags with in all 30 textiles of 11 different types and qualities.

[3] For such letters, see Veenhof 1972, 103–115; Michel 2001, nos. 302–304, 307, 317–319 and Michel 2006b.

e.g. "(textile) of Šubarum". We can also use data on their prices and numbers (the former reflect their production costs, expensive textiles usually occur in small numbers), their quality (ranging from "royal quality" to "poor quality") and on certain characteristics, occasionally revealed by qualifying adjectives (e.g. "thin" or "heavy"). Moreover, the sequence in which they are listed, combinations of two textiles, and especially occasional 'categorizations' are helpful, such as statements like "x textiles of type a, among which/including y of type b", or "textile a (made/consisting) of (*ša*) textile b", where the latter presumably denotes a particular type of fabric.

We are of course not the first to study the Old Assyrian textiles that are so prominent in the trade. Ignoring scattered earlier observations, usually in the comments on particular texts, we mention here the studies in Garelli 1963, 284–293 ('Les étoffes') and Veenhof 1972, part II, 79–216 ('Textiles and Wool')[4] and take the systematic collection of data and their analysis by the latter as our point of departure, while also referring to the treatment of textile names in the *Chicago Assyrian Dictionary,* where especially the later volumes contain a wealth of references.[5] These earlier studies demonstrated that part of the textiles shipped by the Assyrians to Anatolia were imported into Aššur from "Akkad" or Babylonia, while others were produced by women in Aššur, and, that they were woollen products. Uncertainties remained about the nature of the various textile products, their origin, the trade in Anatolian textiles, and the important question of whether they were (predominantly) untailored fabrics or (also) ready-to-wear garments; it seems that this last category has been overestimated in the past. Since the publication of the investigations mentioned above, many new textual sources have become available, especially now that the first archives excavated at *kārum* Kaneš by Turkish archaeologists since 1948 are becoming available.[6] They offer the possibility of critically assessing and supplementing the current data and insights in the hope of solving at least some of our problems. We will do so by treating the various issues mentioned above, starting with the basic question of the material from which the textiles were made.

1. Materials

1.1. *Wool (síg, šaptum)*

Today, it is clear that most textiles traded by the Assyrians were made of wool. Confusion had been caused when the most frequent textile product, *kutānum*, only attested in Old Assyrian,[7] was at

[4] See also Veenhof 1988, 254–257, on the purchase prices of the textiles in Aššur, and Larsen 1967, 97–140: "Caravan Accounts" (Ch. III, c), on the purchase and sale of textiles.

[5] Julius Lewy, the early expert in the study of the Old Assyrian texts, contributed many insights in the comments in his text editions and in various articles, especially in the copious footnotes, but offered no systematic analysis.

[6] For the present situation see Veenhof 2008, 68–75: "Work on texts excavated since 1948". AKT 4 was published in 2006 (see Veenhof 2009) and volumes with editions by K. R. Veenhof (AKT 5 = *Kuliya*) and M. T. Larsen (AKT 6) of archives excavated in 1992 and 1994 are in press in Ankara. Many small groups of and even single texts have been published in a great variety of articles in congress volumes, Festschriften and journals (among which *Archivum Anatolicum*, inaugurated in Ankara in 1995, and *Anadolu Medeniyetleri Müzesi Yıllığı*, the annual of the Museum in Ankara, where the excavated tablets are preserved). A few thousand unpublished texts circulate among Old Assyrian specialists and we are grateful to be able to quote or refer to words and lines in them that are important here, thanks to those who deciphered them and will publish them in due course.

[7] It was considered to occur in Mari as well, but the only references are in the closely related letters, A. 2881:13 and ARMT 13, 101:28, sent there from Aššur, in which an Assyrian trader promises to send such a textile to his colleague in Mari; see Durand 2001, 119–120. Another occurrence in an Old Babylonian text is in al-Rawi & Dalley 2002, 97:32–33,

first rendered as "Chitons-Stoffe" by Landsberger (1925, 20) and as "*kutānu*-Stoffe" in *EL* (passim), but interpreted as "linen" by Garelli, in which he was followed by Larsen. This identification was supported by the fact that an Ugaritic bilingual text renders the logogram for linen, túg-gada by *ktn.*[8] Later, von Soden in AHw 930a, suggested a new etymology, interpreting the word as a *purās*-form, a nominal formation used for "deverbale Vergegenständlichungen".[9] It made him parse the word as *qutānum*, meaning "das Dünne", derived from *qatānum*, "to be thin", whose first consonant, emphatic /*q*/ instead of /*k*/, would rule out a connection with comparable terms in other languages, and AHw adds "wohl nicht *kutānum* zu hebr. *kuttonet*; > χιτών?" However, **qutānum* is superfluous alongside túg*raqqutum*, "thin textile", well attested in Old Assyrian, and there is probably a text where the adjective *qatnum*, "thin", is applied to *kutānum* itself, which would be a tautology for **qutānum.*[10] The letter TC 3, 17:21–23 (below § 3.4.1) requires that one side of a "thin textile" (*ṣubātum qatnum*), "if it is still hairy should be shorn like a *kutānum*", and this would seem strange if the latter textile itself was a **qutānum*, "thin textile."[11] That *kutānum* was a woollen product is clear from the statement by the writer of TC 2, 7:24–26, that he is unable produce the *kutānū* requested "because there is no Šurbu-wool (*šaptum šurbuītum*) available".[12] Furthermore, in TC 2, 14:6–9 (Michel 2001, no. 108), textiles designated as *šurbuīum* and apparently made of this type of wool are included in the category of *kutānum*. *Kutānum*, as argued by Oppenheim (1967, 158, note 82), was probably a "Kulturwort" of unknown origin, related to Hebrew *kuttonèt* and Greek χιτών, but curiously absent in other periods of ancient Mesopotamia. It most probably was a "woollen cloth" with, according to TC 3, 17:21–22, a flat and smooth surface, achieved by shearing (*qatāpum*, see § 3.4.1). This meaning fits the fact that, in several cases, the words *ša kutānim*, "(made) of *kutānum* (fabric)" or "of *kutānum* type", qualify other textiles or garments, such as *nahlaptum* (OIP 27, 11:11–13), *namaššuhum* (Benenian 5:2–3), *nibrārum* (Kt 94/k 1686:9–10, courtesy of Larsen), and *šitrum* (*RA* 59 [1965] no. 14:16).[13] It also explains why *kutānum* can be used as a designation of a specific kind of fabric under which other textiles can be subsumed, e.g. *kusītum*, in AKT 4, 23:1–2 and Prag I 616:18–19 (see below § 4.1). The meaning of the combination 1 *kutānum e-pì-ší* in Kt 87/k 452:3–4 (courtesy of Hecker) is not clear, but it again suggests a woollen product (see below § 3.3 *s.v. ēpišum*). That the bulk of the traded textiles was made of wool also

túg *ku-ta-a-ni*, *ša* 1 *ma-na* kù-babbar; it occurs in a letter dealing with trade, alongside "20 minas of tin".

[8] Garelli 1963, 288: "On s'accorde généralement à voir dans le mot *kutānum* une forme élargie de *kitūm*, le "lin", dérivé du sumérien GAD, qui aurait donné naissance à l'hebreu *kuttōnet*, au grec χιτών et au latin *tunica*". Larsen 1967, 152: "linen-cloth". [This incorrect meaning is repeated in Durand 2009, 599 (Index), *s.v.*] See for Ugaritic *ktn* (plural *ktnt*), Van Soldt 1990, 332, Conclusions, 3, where he describes it as "a cloth made of linen. It is not a finished garment but a piece of cloth which can be used to manufacture garments".

[9] *GAG* § 55k, 15, e.g. the textile name *ṣubātum*, from the root *ṣabātum*, "to seize", therefore a woven fabric that "holds" or "is attached to" the body.

[10] Kt 91/k 356:28–29, 2 *kutāni ša-hu-šu-x-ri qatnūtim*, "2 thin *kutānu* of *h.*" (or "*kutānū* of thin *h.*"; meaning of *h.* unknown).

[11] Note also TC 3, 72:24–26, [x] *kutānī* [x x x x] ù! 1 túg *qat*[*nam* x x x] *ikla.*

[12] This nisbe is to be derived from the place-name Šurbu (see § 2.1.1, *s.v.*) and refers primarily to wool from that area, cf. 'Tablet Rendell' (unpubl.), lines 6 and 16, túg *ša šu-ur-bu-i-a-tim*, where the fem. plural form must refer to *šaptum*, "wool". But it is also used for textiles made of it. TC 2, 14:6–9, quoted above, shows that this type of wool was expensive, since textiles made of it cost *c.*25% more than normal *kutānū*. Note the statement by the writer of TC 2, 7:25–28, that, for lack of wool from Šurbu he will buy a "heavy textile" (*ṣubātam kabtam*).

[13] The reading "25 pounds of refined copper, the price of wool of a *kutānum*", in lines 4–6 of this text, accepted by CAD K, 608, cannot be correct because of the strange writing síg-*tí-e-em* for *šaptim*, "wool". See below § 3.3, *s.v. lud/ṭûm.*

explains the fear that they might be "affected by moth" if they remained too long in storage or packed in bales, a danger forestalled by regularly airing them.[14]

While textile production in Aššur must have consumed large quantities of wool, information on it is rare (cf. Michel 2006b, 290–293). We may assume that certain inhabitants or perhaps institutions (e.g. temples) of the city owned herds of sheep and/or that wool was acquired from pastoral nomads grazing their flocks to the east or west of the city, but the only indications date from about 1770 BC. Two texts from Mari mention that Suhu-nomads, who grazed their herds along the Middle Euphrates and in the area of the Wadi Tarthar, would normally go to Aššur to pluck their sheep and presumably sell their wool there.[15] In the heyday of the trade, about a century earlier, the situation must have been similar, but there would have been no need to mention it, unless there were problems that interfered with the production of the textiles, to be reported to Kaneš. The purchase of wool is only mentioned in Kt 93/k 325:10–13, where an Assyrian woman writes "Send me silver so that we can buy wool and make a garment (*ṣubātum*) for you to wear (*ana litabšika*)". Letters occasionally mention problems in the supply of wool, e.g. TC 2, 7:24–26, quoted above, which reports that wool from Šurbu was not available in Aššur. The letters by Lamassī, the wife of the prominent trader Pūšukēn, also mention wool. In BIN 4, 9:3–6 she complains about not having received the wool (twice 5 pounds) sent to her, and in lines 18–20 and in BIN 6, 7:16–18 she asks: "When you send me the purse, add wool to it", which must mean the same as suggested by Garelli (1965), 158, no. 25:13–16, "When you prepare 1 mina of silver (for transport to Aššur), put it inside wool", a request argued in both texts by mentioning that "wool is expensive in the City".[16] Prag I 554:9–10 mentions a shipment to Aššur of primarily silver that includes "3 pounds of wool for Waqurtum", a lady active in the production of textiles in Aššur. For reasons unknown to us, wool was apparently occasionally in short supply and hence expensive, so that even small amounts of wool sent from Anatolia were welcome, but they cannot have helped much, considering the number of textiles produced by some women there.[17] There is no evidence of large-scale shipments of wool from Anatolia, which anyhow would have been too expensive considering the cost of the transport. However, for the women who wove textiles and through their sale in Anatolia tried to earn silver for themselves, small amounts of wool too were at times welcome.

The most explicit evidence for the use of wool is found in TC 3, 17 (see below § 3.4.1), a letter addressed to the above-mentioned Waqurtum, in which she is asked to process 1 pound of wool

---

[14] For this feature, see Michel 1998; we can now add the following references Kt h/k 18:15–16, Kt n/k 717:11–15, Kt 91/k 290:24–27, Kt 92/k 174:11–12, and Kt 94/k 1257:13, all of which use the expression *ṣubātū sāsam laptū* and the first text states "I keep airing his textiles every day" (*ūmešamma ṣubātīšu uttanappaš*). Note also Kt 94/k 823:5–6, "we aired your textiles and your textiles are in good shape" (*šalmū*) and Kt 94/k 1131:36–41, which states that textiles kept in a storeroom (*huršum*) have to be aired because they are "weary" (*anhum*), presumably by having been kept there too long.

[15] Charpin & Durand 1997, 377 and 387–391. The first letter (A 2459 rev.: 3'-6') describes a situation of war in which the sheep of the Suhu have to be plucked where they live, so that the Assyrians are forced to go there to obtain their wool, while the second (A 4535-bis, rev. 2'–5') mentions the complaints by the Assyrians that their traders and the sheep and wool (of the Suhu?) are held back. [For wool, see the texts from Mari, now Durand 2009, 142–155, and for texts recording the purchase of wool from the Suhu see M. 11269:1–7, and 11281:1–9].

[16] For these letters, see Veenhof 1972, 112–113; Michel 2001: nos. 299–311; Michel 2006b.

[17] *ATHE* 44 mentions 17 pieces produced by Waqqurtum; the various letters written by Lamassī (including CCT 6, 11a) together mention more than 60 textiles sent by her to Anatolia.

more per piece of textile, but to make sure that the weave stays "thin" (*qatnum*), which implies the use of thin threads, woven densely, because "the warp has to be much/numerous" (*šutûšu lū mādât*). The evidence from the Ur III period, analyzed in Waetzoldt 1972, demonstrates that the thickness of the threads used for various types of textiles and the difference between the threads used for the warp and the weft was extremely important and conditioned the quality and the labor costs of the woven fabric. It must have been similar in the Old Assyrian period, when (as will be shown below, in § 3.2) the same classification of qualities existed, but we have almost no evidence of spinning and weaving. Apart from the wish in TC 3, 17, to weave with a dense warp, we can only mention a reference in a contract found in Kaneš (Kt 91/k 388:5–7) recording the claim of the wife of an Assyrian trader on another Assyrian for "30 pounds of soft wool for making the weft" (*šapātim narbātim .... šakākiš*), to be delivered within two months.[18] She had apparently provided him with money to supply her with this type of wool and this may indicate that she, or women or slave-girls in her household, did engage in textile production; we have few further evidence for such activities by Assyrian women in Anatolia.[19]

The evidence of Anatolian wool and its trade by Assyrians is abundant, amply documented in several archives, including those excavated in 1993 and 1994, which will be edited by C. Michel and M. T. Larsen. The topic deserves a separate investigation and here we only mention a few basic facts.[20] The goal of this trade, in which some Assyrians apparently were much more active than others, was also to earn silver, which means that wool was bought, shipped elsewhere and sold, either directly for silver, or first for copper, which was then converted into silver according to the local opportunities and the 'market'.[21] Local palaces could also be involved, as sellers of wool (we have a reference to a large amount of "wool of Kaneš, of the palace", see below, note 32) and they could also derive income from it by levying the 5% *nishatu*-tax on it, as was customary for textiles.[22] Important transactions could be joint enterprises, in which various traders had shares and in which also the Assyrian trading organization (the *kārum*) played a role.[23] This trade could handle large quantities of wool. CCT 4, 47a:30–33 requests to convert 80 talents of white and 20 talents of red wool (*c.*3 tons) into copper and BIN 6, 76:13 mentions 60 talents for the same purpose, on which the trader "reached an agreement with our own people and with the retailers" (*pāširū*, local traders).[24] A group of records in the archive of Šalim-Aššur (excavated in

[18] *Šakākiš* is an infinitive with terminative ending of a verb known to mean "to string (beads, a rope), to harrow", and this is the first occurrence with the meaning "weaving the weft".

[19] In the excavation reports there are some mentions of loom weights found in the houses of the Assyrian traders, see for example N. Özgüc & Tunca 2001, 247.

[20] See already Lewy 1958, 97–99; Veenhof 1972, 130–139 (also on prices, organization, woollen fleeces); Dercksen 2004, 183–190, 'Wool trade in Anatolia'.

[21] A nice example is the letter Prag I 768:4–12, where Imdīlum is told: "I hear that much wool has now entered Wahšušana (a city northwest of Kaneš) and when I arrive in town I will sell the wool at any price and send you the silver".

[22] See ICK 1, 97:3–6, "I. brought here 680 pounds of wool, of it the palace levied 34 pounds as tax" (*issuh*); CCT 6, 19b:4–9, "Over there U. must declare to you both the amount of the *nishatu*-tax [and ...] and whatever wool is cleared you must sell cash".

[23] See Veenhof 1972, 134–139, and Dercksen 1996, 125–127, 145, 160, and 173. In the letter Kt 93/k 721, the individual shares amount to 4 2/3 talents of wool; Kt n/k 539 reports about a settlement of accounts, whereby a trader had deposited more than 23 pounds of silver in the *kārum*-office and acquired (the right to collect) *c.*120½ talents (*c.*4 tons) of wool; also Kt n/k 1475:20–23, see note 26.

[24] See Veenhof 1972, 137–138 with footnote 237, and also CCT 6, 19b. BIN 4, 181 mentions nearly 68 talents (more than 2 tons) of wool, and AKT 4, 58:4–10 copper acquired in the *kārum* office alongside wool sold for copper.

1994) documents the purchase of in all *c.*50 tons of wool, acquired from a high Anatolian official in exchange for many tons of copper and sold for silver.[25] In several transactions, wool also figures alongside hides, in particular "fleecy hides" (*maškū šapātim*, "hides of /with wool"), also designated as "thick hides" *(maškū šapiūtum)*, and the local Anatolian woollen textile product called *pirikannum*.[26] The latter is probably also meant when in CCT 2, 18:4–5 a trader somewhere in Anatolia reports enthusiastically: "Textiles and wool are available (here)!"

The quality of the wool was important, as is demonstrated in the request in TC 3, 65:18–22 to buy "soft, long, extremely good wool" *(naribtam araktam damiqtam ūtartam)* of Mamma, and the promise in AKT 4, 52: 6–8, "I will give you (for 1 shekel of silver) 6 minas of fine, soft wool", because fine wool obviously yields textiles of better quality. Apart from the rather frequent adjectives "good" and "soft"[27] and rare references to red and white wool,[28] we find wool qualified by the adjectives *wašium/ušium* and *lahum* or *lahhum*, but the meaning of these terms is unknown.[29]

The price of 6 minas of (soft) wool for 1 shekel of silver recorded in AKT 4, 52, in *EL* 243:5 (100 shekels of silver for 10 talents of wool) and in Kt n/k 860:15–17,[30] seems to be fairly normal. It appears similar to the price attested for Babylonia in the Old Babylonian period (6 pounds for 1 shekel of silver according to § 1 of the *Laws of Ešnunna*), but wool was cheaper during the Ur III period (usually 10 pounds for 1 shekel of silver).[31] However, a comparison is difficult, because in Anatolia, silver, used as a standard of value, had much less buying power than in Mesopotamia. And in Anatolia we also find higher prices, e.g. 5 pounds for 1 shekel of silver in TPAK 1, 36:5–6 (12 shekels per talent) and *c.*3¾ pounds in TPAK 1, 35:4–15, and it is likely that these differences were determined both by the quality of the wool as well as by the geographical situation. In CCT 6, 19b:14–16 wool is sold in exchange for copper at an exchange rate of 2:1 (*ana itaṭlim šanā'um*), that is 2 pounds of wool for 1 pound of copper, which equals *c.*4 to 5 pounds of wool for 1 shekel of silver.

---

25 See Larsen 2008, 86.

26 See already Lewy 1958, 97–98, and Veenhof 1972, 124–125, and 132–134 for fleeces, where it is argued why *maškum* alone may frequently mean "fleece", i.e. the hide with the wool still attached to it. Additional examples of the co-occurrence of wool and hides/fleeces or *pirikannu*-textiles are CCT 4, 27a:11, *POAT* 8:28, *Kuliya* 57:12–14, CCT 6, 7a:5–9, Kt f/k 128:8–10, Kt n/k 1475:20–22 (wool and fleeces of Luhusaddiya) and Kt 93/k 781:7–9//505:9–11. In Kt 93/k 721, wool occurs together with a large number of *ukāpū*, a blanket-like woollen saddlecloth, a combination also attested in AKT 1, 7:32–34 and ICK 1, 37B:16–19.

27 It is more frequent than the few references in CAD N/1, *s.v.* suggest; see also kt a/k 572:3–8, kt f/k 123:16, Kt n/k 860:17, Prag I 740:4, and AKT 4, 53:6–7 (in line 27 "good quality wool"). See for *narbum*, "soft", used for textiles (also once for *pirikannu*-textiles), see below § 3.4.3.

28 CCT 4, 47a:30–33, quoted above; CCT 4, 27a, l.e. 1; *šaptum makrītum*, a type of red, occurs in OIP 27, no. 7:3–4, and probably in 48B:3'; "dyed wool" (*šaptum šinītum*), in BIN 4, 54:15.

29 Kt 87/k 545:15–16 (courtesy of Hecker), "wool, half of it soft and half of it *ú-ší-tum*", and Kt 93/k 84:3–4, "in Luhusaddiya they gave/sold me 7 talents 10 minas of *šaptam la-hu-tám*, and also there 1 talent 50 minas of *šaptam ušītam*". Equally unclear is the qualification *nu-ha-tum* in Kt 93/k 239:10, which in Kt 93/k 253:45 is used for copper, but it might perhaps be connected with the adjective *nuhhutu*, used in Neo-Babylonian texts to qualify silver and perhaps linen (see CAD N/2, 318 *s.v.*).

30 *AMMY* 1992, p. 54, no. 2; it stipulates that as interest on a silver loan the debtors will "give for 10 shekels of silver 6 minas of soft wool per (shekel)". Note AKT 4, 53:9–14, "I gave you 4½ shekels of *tiri*-silver for which you gave me 10 pounds of wool, (but) at that time it(s rate of exchange) stood at 6 pounds per (shekel of silver)", which implies that the writer paid far more than normal! See for the price of wool in Anatolia, Michel 2006b, 291; its price in Aššur is unknown.

31 See Veenhof 1972, 131, and for the Ur III period now also Snell 1982, 178–181, 16. The price of 15 pounds of wool for 1 shekel of silver, mentioned in a building inscription of Šamšī-Adad I (*c.*1800 BC; see RIMA 1, 49–50, lines 66–67), as obtained during his reign in the market of Aššur, is clearly too favorable and propagandistic.

Flocks of sheep must have grazed throughout Anatolia and there are indications that the palaces too had them, e.g. the palace of Kaneš.[32] Assyrian traders sold their wool in many areas, but the wool in our texts seems to have originated especially from the more southern areas, notably from the cities of Luhusaddiya[33] and Hurama, but also from Hahhum, Kaneš, Mamma and Timilkiya,[34] and there are some references to wool acquired in Balihum.[35] The massive trade in Anatolian wool implies a well-developed local textile industry that must have produced the woollen textiles called *pirikannum, sapdinnum* and *tisābum* (see for these textiles § 3.3, *s.v.*), in the trade of which the Assyrians were heavily involved. Several of the towns from which these textiles occasionally are said to originate play an important role in the wool trade, which implies the existence of a local textile industry, and it cannot be accidental that we have attestation of both red wool (CCT 4, 47a, mentioned above) and red *pirikannu*-textiles.[36] Regrettably our texts provide no information on this local textile production, although there are a few occurrences of fullers (*ašlākum*) with Anatolian names.[37]

1.2. *Linen (kitā'um)*

While it is now clear that *kutānum* is a woollen and not a linen product, there are about a fifteen occurrences of *kitā'um* (plural *kitā'ātum*), the word for flax and linen, which may refer to the threads and the fabrics made of them (CAD K, s.v. *kitû*), but in Old Assyrian it is only attested as referring to fabrics. This word must be distinguished from *kitītum*, rarely attested in Old Babylonian under the form ^{túg}*kitītum*, the designation of a garment, apparently not of linen but of wool, because of the occurrence of síg, "wool", in the corresponding Sumerian logograms in lexical texts, where it frequently appears alongside *raqqatum* and *itqum*.[38] CAD K, 466 maintains the etymological link with *kitûm*, "linen", by proposing a meaning "fine (lit. linen-like) wool". ^{túg}*kitītum* is not a combination of noun and adjective, since túg/*ṣubātum* is masculine, nor does it mean "textile of

[32] See for occurrences of a "chief of the shepherds" and "a shepherd of the queen", Veenhof 2008a, 223 *s.v. rē'ē/rē'im*, and in general for husbandry in ancient Anatolia, Michel 1997, 108–111 and Dercksen 2008, 152–154. He mentions a text, Kt 94/k 1024:15–16 (courtesy of Larsen), which records that an Assyrian will pay "21 talents of wool of Kaneš, of the palace" in the city of Kuburnat, in the north.

[33] See for data Veenhof 1972, 131, 2, and for Luhusaddiya also Kt 93/k 84:3–9 (cited in note 29) and Kt n/k 1475 (quoted in note 26).

[34] Hahhum, OIP 27, 7:6, kt b/k 27:5–6; Hurama, *EL* 243:14–15 (sic!); Kaneš, Kt 94/k 1024:15–16; Mamma, TC 3, 65:18–22; Timilkiya, Kt m/k 114:1–2 (courtesy of Hecker). In some cases place names mentioned in connection with wool (or textiles) document trade there, but not necessarily the origin of the wool.

[35] BIN 6, 176 (// ICK 2, 277, see Veenhof 1972, 134–135), f/k 185 (courtesy of Umur), Kt c/k 922 and 944 (published in Albayrak 2008). While the first (see Veenhof 1972, 134–135) and third texts deal with the acquisition of wool only, the other texts mention both wool and fleeces. It is doubtful whether Balihum (which in these texts seems to denote a town or region) is to be connected with the well-known river of that name in the western part of the Jazira, within the bend of the Euphrates. Luhusaddiya must be located at least 200 km north of the Euphrates, in the general area of the plain of Elbistan, and this makes an enterprise to acquire wool and hides in both towns rather unlikely.

[36] For the trade in Anatolian textiles, see below § 2.3.2, and for red *pirikannū*, TC 1, 43:24–27, "Buy red *pirikannū* and send them to me, the *pirikannū* that you acquire must be red!"

[37] See Dercksen 2001, 62 with note 130; Kt 94/k 833:31–32, mentions "the fuller of the ruler" *(ašlakum ša rubā'im)*.

[38] In Old Babylonian "Proto-Diri", síg-bu = *sulumhû, itqum, kitītum* (*MSL* 15, 46: 422–424) and *ibidem* 172, Diri V:131–136 (cf. Hh XIX:153–158); s u l u m h i = túg-síg-sud is equated with *sulumhû, itqu, kitītu, raqqatum, lubuštum, lamahuššû. Sulumhû* according to CAD S, *s.v.*, is 1. "a long-fleeced breed of sheep", 2. "a garment", but it occurs only in lexical lists. The sheep, which occurs in a few Neo-Babylonian texts, is listed in Hh 13:16: dud-síg$^{\text{su-lu-hu}}$sud = ŠU-*u*. *Itqu* (CAD I/J, *s.v*) is "fleece", "tuft of wool" and "a garment made of fleecy wool".

kitītum wool", since *kitītum* in the Old Babylonian occurrences is in the nominative form. It must be a substantivated feminine adjective, like *raqqatum*, which also figures as the name of a textile.[39]

Kitā'um occurs in very small numbers (between 3 and 1), and twice in the plural without numbers; in TC 3, 271:9 they belong to the contents of a trader's house (alongside silver, gold, silver cups and tablets). The letter Kt 89/k 252 reports that an Anatolian palace (the location of which is not mentioned) "needs linens" and that its ruler puts pressure on the agent of the Assyrian owner to fix their price so that he can acquire them.[40] That Anatolian palaces owned linen textiles and apparently attached value to them is demonstrated by the single occurrence of the Anatolian title "head of linens" (*rabi ki-ta-a-tim*) in BIN 4, 160:7–8, where he figures as the debtor of an Assyrian trader.[41] Although no prices are mentioned, linens apparently belong to the more expensive textiles. In kt 89/k 266:10, linen figures (among textiles brought to an Anatolian official) alongside 1 fine *raqqutum*, 1 *kutānum*, 1 *kusītum* and 1 *ṣubatum damqum*, and the list Kt n/k 152:7–9 (courtesy of Bayram) mentions 11 Abarnian textiles, 1 fine *kutānum* and 3 *ki-ta-a-tum*.[42]

A few texts provide more information and there are several cases where linens are sent from Anatolia to Aššur. "The *kitā'um* and the belt/scarf (*išrum*) for the god Amurrum", brought to Aššur according to CCT 3, 25:27–28, may well be a set of clothing, and Kt 93/k 196:5–8, a letter probably sent to Aššur,[43] mentions a shipment of silver, some gold, one "linen of Tuttul" (*ki-ta-am ša Tuttul*) and 3 pounds of carnelian". Shipments to Aššur are also mentioned in Kt 93/k 241:21–23, where Lamassatum (in Aššur) writes to Iddin-Sîn: "Send me tin, *nabrītum* and 2 linens", and in AKT 3, 79:26–28, where Nuhšatum in Kaneš is asked: "Send me *nabriātim ša i-lá-tim* and a large linen" (*ki-ta-a-am rabītam*). According to *RA* 81 (1987) 59 no. 71:36–37 "one *supannum*, one linen and two *samālu*-cups" were sent from Anatolia to Lamassī in Aššur.[44]

CCT 4, 44b:17–22 gives the order to buy (apparently in Anatolia) and send "one *kitā'um* of fine quality of 15 or 20 cubits?", probably referring to the length of this (strip?) of linen.[45] In the enumeration of CCT 5, 12a:9–10, among the Anatolian textile products entrusted to a traveling agent, we find two *ki-ta-a-tum ba-li-lu*, but the meaning of the latter qualification, although attested a few times more as name of a garment or textile, is unknown. We have to conclude that the

[39] See for Mari, where Durand translates "pièce de lin", in addition to the references given in CAD K, *s.v.*, also ARMT 21, 219:22, "1 ^{túg}*kitītum*, its value in silver 5 shekels", 318:2, 349:11, 2 ^{túg}*ki-ti-tum*, 383 VII:14", 2 bar-si *ki-ti-[tum]*, "2 châles en lin", 369:8, 1 gú-è-a *ki-ti-tum*, 383 II: 4) [and now Durand 2009, 159–160]. In the last two references *kitītum* apparently designates a type of fabric from which the textiles it qualifies have been made [cf. Durand 2009, 160 note c)]. In Babylonia, also in O 342 (unpubl., Old Babylonian Kiš), I:9–10, 2 ^{túg}guz-za, 1 ^{túg}*ki-ti-tum*, together stored in one box.

[40] Lines 3–8, "Here I asked him about the linens that are with Zumana, saying: 'The palace needs linens' " (*kitā'ātim ekallum hašah*). Kt 89/227:17–19, a letter to the same addressee, mentions "3 linens that you sent to Zumana", one of which has been sold, while two are still available in the latter's house. Kt 89/k 266:10–11, a memo from the same archive, lists "1 linen his servant brought him".

[41] This does not prove that he had become indebted by buying the linens, although this is a possibility, since four lines above another official, "the head of the guard", is said to owe a similar amount of copper as the price of an Abarnian textile. Dercksen 2008, 144 takes this title as evidence that flax was grown in the area of Kaneš and that linen garments were produced, but we do not know what exactly he supervised, cultivation, production, storage or distribution.

[42] Kt n/k 216:7–9, in a letter to Uṣur-ši-Ištar, "If in addition to the *ki-ta-a-tim* [you have] a *kutānum*, give it to me to dress myself in".

[43] Letter by Uṣur-ša-Aššur sent to Ali-ahum s. Iddin-Sîn, Lamaša-Aššur and Aššur-ṭāb.

[44] See also Prag I 488:8–9, 1 *karpatam šarašrānam, a-lá-nu ki-ta-um tamalakkū*, and Kt 93/k 196:6–8, 1 *ki-ta-am ša Tù-tù-/ul*, 3 *mana* $^{na}{}_{4}$gug, A. *naš'akkunūti*.

[45] The length of the *īdum* is not certain, but presumably something like a cubit, see Veenhof 2007. Line 4 also mentions a linen: "1/2 mina of silver [x x x] / *ki-tám* 4 *ša* x x [x], with/due from Š."

Assyrians did not import linens into Anatolia and that they played only a minor role in the trade in locally produced linens, although local palaces had an interest in them, as was also the case later during the Hittite empire period.[46] Their origin is unknown and the single reference to a "linen of Tuttul", a city on the Middle Euphrates, does not prove that the few others mentioned also originated in that area.

2. Geographical aspects

The Old Assyrian tablets mainly document the long distance trade organised by Assyrian merchants between their home city Aššur and Anatolia. Among the textiles they exported to Anatolia, many had previously been imported to Aššur, others were locally produced in Aššur and some in Northern Mesopotamia, the area crossed by the caravans. In addition, the Assyrians also traded textiles which were produced in Anatolia itself. A study of textile terminology needs to make a distinction between the different production areas. This can be done first by analyzing the textiles named after (the so-called nisbes) or connected with (by means of *ša*, "of, from") toponyms. Secondly, some documents, mainly letters, give indications about the origin of various textiles and such data allow us to draw up a map of the production areas of the main textile types mentioned in the texts.

2.1. *Textiles named after geographical names*

The provenance of a textile may be indicated by the name of the textile itself if it is a nisbe, by the construction *ša* + geographical name, or by a simple genitive relation.[47]

2.1.1. *Nisbe qualifying textiles*

Some textiles are referred to by a nisbe.[48] In Old Assyrian, nisbes derived from a place-name or noun ending in a consonant have the ending *-īum*, those formed from place-names on *-a*, we render as *-aīum*, notwithstanding a variety of spelling.[49] Some of the nisbes dealing with textiles correspond to well-known toponyms, while others might be interpreted as a nisbe, but the town has not yet been identified. For example, the textiles quoted as *takkušta'um* and *šilipka'um* (also attested elsewhere in Mesopotamia) might be nisbes, but their origin and the corresponding place names are unknown and there is a great variety in

---

[46] See Klengel 2008, 76–77.

[47] Veenhof 1972, 189–191.

[48] This is not specific to Old Assyrian. For example, many textiles mentioned in the Mari tablets are referred to by a nisbe: Yamhadû is frequent in Mari (Durand 1983, 401), but we also find Akkadû (ARMT 18, 28:10–11), Elamûm, from Elam [Durand 2009, 67, 100], Gublāyum, from Byblos [Durand 2009, 100], Haššûm, from Haššum [Durand 2009, 69], Kakmûm, from Kakmum [Durand 2009, 141], Nurrugayum, from Nurrugum [ARMT 22, 110], Parahšu, from Marhaši [Durand 2009, 71], Suhûm, from Suhu [Durand 2009, 507, n. 100], Šubarûm, from Subartu (ARMT 21, 318, 5; 23, 617, 1) and Tuttubayum, from Tuttub [Durand 2009, 111, 130]. Of doubtful identification are Buššurum, Kišihhu, Laharû and Mar(a)tû, [see Durand 2009, 56, 86 and 106–107].

[49] We do not write the latter as *-ājum*, the hypothetical character of which is indicated when *GKT* § 57b writes "scheint die Nisbenendung die Form *-āj* anzunehmen". We prefer rendering *-a-i-um* as *-aīum*, without indicating the presence of a glide or aleph, which seems to be the "classical" Old Assyrian form. Rare writings as *Ca-um*, without *-i* may, as N. J. C. Kouwenberg suggests to us (personnal communication 2009), render a spoken *-ājum* or *-ajjum* and be the precursors of the Middle Assyrian form of the nisbe. And rare spellings with additional vowels – e.g. *A-bar-ni-ú-um* (Kt 93/k 253:47 and Kt f/k 39:19), *Ba-ad-na-e-em* and *Tí-mì-il$_5$-ki-a-e-em* (Kt 00/k 10 III:26'-27', from the later level Ib), are probably attempts to render the intervening glide or aleph unambiguously, but they are exceptions. See the next footnote for the contracted endings of Šilipkûm and Takkuštûm.

spelling. Therefore we refer to them by these citation forms, unless a transliteration is necessary.[50] A list of nisbes is given below in alphabetical order.

Abarnīum. This nisbe, treated as a noun (plural *Abarnīū*), is among the most commonly used for textiles exported from Aššur to Kaniš, and thus it must refer to a place located somewhere in Upper Mesopotamia, but the corresponding city has not yet been identified.[51] *Abarnīum* is attested in an Ur III text from Drehem.[52] It mentions men and officials from l. 5: Marhaši, l. 6: Ebla, l. 9: Mari, l. 12: Abarnīum (*A-ba-ar-ni-um*^{ki}). The town also occurs in a inscription of king Šu-Sîn of Ur, in "a passage listing the peripheral regions of the Ur III empire,"[53] in the sequence (5'–7'): Ebla, Mari, Tuttul, Ma......, Urkiš, Mukiš!?, x-x-x, x x-la, *A-bar-nu-um*^{ki}... This again might suggest a town in Northern Mesopotamia.

a-li-ú-tum (?). One text discovered in 1993 mentions some textiles which are qualified as *a-li-ú-tum*. This adjective could be a nisbe of *ālum*, "the city", which refers to Aššur in the Old Assyrian tablets, and thus could mean "from Aššur/made in Aššur/according to the Aššur fashion";[54] it would then be a synonym of the qualification *ša ālim* describing some textiles (see under § 2.1.2). See for another more likely interpretation, below p. 246, d), with footnote 199.

Alkuaīum. It occurs twice as the qualification of a *kusītu*-garment,[55] in both cases mentioned as a possible alternative to other types of garments, in TC 3, 169:10–12, "1 *kusītu*-garment, either (*lu*) one of Alkuwa, or else (*u lu*) a *šilipka'u*-garment", while VS 26, 74:37–43 asks to buy and send from Aššur to Anatolia "either white *kusītu*-garments or one from Alkuwa, or thin garments of good quality, or white *lubūšu*-garments." The place name Alkuwa is unknown and the nisbe might be a variant (or mistake?) of Malku(w)aīum, see below.

Gasurīum(?). The unique adjective GA-ZU-*ri-im* has been interpreted as a nisbe of Gasur, a city east of the Tigris, called Nuzi during the middle of the 2nd millennium BC, since a person is twice designated in this way.[56] If not a nisbe of Gasur, it could alternatively be taken as *kaṣṣurum*, a verbal adjective of the D stem of *kaṣārum*, perhaps referring to a tightly knotted textile. According to the CAD K, 261–262, there is no D

---

[50] *Takkušta'um* never has the nisbe ending *-īum*, so that the underlying word may end in *-ta*. If so, one would have expected the common Old Assyrian nisbe ending *-a-i-um*, but it is never spelled with inserted *-i-*. That *šilipka'um* is also frequently written *šulupka'um* and also has forms ending in *-kīum* (the normal nisbe ending after final consonant) and *-ka'um* (which suggests a final *-a*), shows that the writers themselves hesitated about its correct spelling (and perhaps derivation). In addition it exhibits plene writings with additional vowel, *ší-li-<ip>-kà-ú-um* (CCT 4, 5a:6), *šu-lu-up-ki-ú-um* (BIN 4, 148:11), *ší-li-ip-ki-e-ú* in Kt 94/k 829:10 (but *-ki-ú* in lines 15 and 20!), and *ší-li-ip-kà-e-a/kà* (*RA* 60, 96:5, 8), alongside contracted forms, *ší-li-ip-ku-um* (KTS 55a:11), *šu-lu-up-kà-am* (91/356:25). The latter is also (but more rarely) the case with *takkušta'um*: *ta-ku-uš-tù-um* (CCT 5, 34c:11), *ta-ku-uš-tù-ú* (AKT 3, 59:19), and *ta-ku-uš-tum* (Yale 13092:11; cf. *ta-ku-uš-té-kà* in CCT 5, 46b:17). Contraction is normal in the later Mari texts, *-tu(-ú)* [Durand 2009, 121 *s.v.*] and probably in *ši-li-ip-ki-im*, the only occurrence elsewhere, in Old Babylonian Kisurra 177:20, unless we consider it the genitive form of *-ki-um* and not of *-ku-um*.

[51] See Veenhof 1972, 156–158 and below § 3.3 s.v. *Abarnīum*. They are expensive textiles, qualified as "good" and "extra good". Correct AKT 2, 24:4–5 to: 1 túg *a-bar*$^{!}$*-a-ni-a-am*, sig$_5$ *wa-at-ra-am*. Note the construction with *ša*: *šitrē ša abarnīē* (TC 1, 19:12) and *šitram ša abarnīim* (Kt 93/k 75:7).

[52] CST 468, see Owen 1992, p. 144, no. 17; this text mentions "the messenger of the ensi of Abarnium" which may indicate this town was further away. I. J. Gelb, cf. RGTC 4, 2, tentatively identifying it with classical Abarne, "half-way between Malatya and Amida", proposed a location in Eastern Turkey, at modern Çermük, which does not correspond to our sources.

[53] Civil 1967, 37.

[54] Kt 93/k 765:13–14, šà-ba 20 túg sig$_5$ *a-li-ú-tum*, *tardīūtum*. The Assyrians were creative in this respect, as shown by the recently published first occurrence of *ekalliyum*, "of palatial quality", AKT 4, 28:7, 2 túg^{há} sig$_5$-*tim* diri é-gal-*li-ú-tim*, sent from Aššur.

[55] TC 3, 169 and VS 26, 74, which is a copy of a letter sent to Aššur.

[56] See RGTC 4, 40; Veenhof 1972, 189–190 and CCT 4, 2a:31.

stem of this verb with a meaning related to textile production; but the *kāṣirum* (CAD K, 264) is described as a "craftsman producing textiles by a special technique."

Hahhīum. Textiles from Hahhum are usually designated as *ša Hahhim*,[57] but the nisbe is used a few times, both for wool[58] and for textiles (in lists), in the plural and dual feminine form.[59] This well-known city, located on the Euphrates in the area of Samsat, is the site of the main crossing used by the Assyrian caravans on their way to Anatolia.[60]

Malku(w)aīum. We have two certain occurrences of this nisbe, which may derive from an otherwise unknown place-name *Malku(w)a. In both cases it qualifies a *kusītu*-garment, spelled *ma-al-ku-a-i-tám* (Kt 91/ k 360:22) and *ma-al-ku-a-tim* (Kt 94/1686:17, plural, with the variant spelling *ma-lu-ki-a-tum* in Kt 94/k 1687:32, courtesy of Larsen). These occurrences suggest the correcting of CCT 2, 3:15–16 to *kusītum ma-al-<ku>-a-i-tum*, also because the place name Mal'a does not seem to exist,[61] and in *RA* 81, 14 no. 3:7, we might perhaps also read 6 *kusiātum* [*m*]*a*!-*al-ku-i-a-tum*. These expensive textiles (more than one pound of silver according to CCT 2, 4:15!), were exported from Aššur to Anatolia, which suggests the location of Malku(w)a somewhere in Upper(?) Mesopotamia. There is reason to assume that this nisbe was confused with or was an alternative writing of Alku(w)aīum (see above), which sounds similar and both occurrences of which also apply to *kusītu*-garments. Moreover, the letter *POAT* 7:8–9, which deals with the same issue as CCT 2,3, omits the nisbe and calls the *kusītum* "white", which recalls VS 26, 74:38–40 where such garments "of Alkuwa" are also an alternative for "white *kusītu*-garments".

Susēium. A document found in 1962 mentions a garment (*lubūšum*) with the qualification *sú-sé-e-a-am*.[62] If this corresponds to a nisbe Suse/ē, it could perhaps concern the city of Susā, well-known from the Mari royal archives and situated in the Ida-Maraṣ, not far from Šubat-Enlil.[63]

Šarzuaīum. There is only one reference to textiles named by means of this possible nisbe, to be derived from a place name Šarzu(w)a, which is not attested elsewhere thus far.[64] It has sometimes been interpreted as corresponding to Arzua, a geographical name mentioned in Hittite documentation, but this seems unlikely.[65]

Šilipka'um. The *šilipka'um* or *šulupka'um*[66] textile, quite frequent (*c*.40 times) in the texts, appears together with many different kinds of textiles that are known as fabrics made in Mesopotamia; in at least two occurrences, this type of textile is bought in Aššur and exported to Anatolia.[67] This nisbe occurs once in a Kisurra letter.[68]

[57] Veenhof 1972, 129 and see below § 2.1.2.

[58] BIN 6, 136:14.

[59] *RA* 58, 60:5, 2! *ha-hi-ta-an* ; VS 26, 123:8, *pūh ha-hi-té-en*$_6$. Since it is always in the dual form, it could correspond to shoes, or to pieces of garments which cover legs or arms.

[60] Veenhof 2008b, 7–10.

[61] The text numbered Mat. II, 4a, x+15, quoted by Bilgiç 1951, 34, Veenhof 1972, 159 and RGTC 4, 81, cannot be found; it probably is an error, since all the Prague texts have now been published.

[62] AKT 2, 24:10, *lu lubūšam sú-sé-e-a-am*.

[63] Charpin & Ziegler 2003, 266.

[64] ICK 1, 81:18, 9 túg^{hi} *ša-ar-zu-a-i-ú*, alongside 5 *kutānū*, together (line 27) 14 sold textiles.

[65] RGTC 4, 13–14, already rejected by Veenhof 1972, 190.

[66] See above, note 50 and below § 3.3 *s.v. šilipka'um*.

[67] See TC 1, 47:4, 2 túg *ší-li-ip-ki-ú* and *RA* 60 [1966], 111, n°43:6, 19 (Larsen 2002, no. 82) where this textile is listed among *kutānum* and *kusītum*. An unpublished tablet preserved in New Haven, Yale 13092:8–12 (courtesy of Larsen) lists this type of textile together with Akkadian pieces: 7 *lubūšū* 6 *kusiātum* 1 *šulupka'um* 1 *takuštûm* šu-nigin 15 *ša A-ki-di-*NI*-im*, at 9,1 shekels of silver apiece.

[68] FAOS B. 2, 177:20.

Šurbuīum. This nisbe applies primarily to wool,[69] identified as originating from a town Šurbu, which occurs in some 3rd millennium sources and in a geographical list from the early 2nd millennium BC found at Tell Harmal (MSL 11, 58:164).[70] Secondly, it is in one instance used for textiles made from this type of wool, in TC 2 14:6, where 27 túg sig$_5$ *šu-ur-bu-i-<ú>-tum* are bought in Aššur.[71] Šurbu is located in the Hamrin mountains, Southeast of Aššur, an area well-known for sheep breeding. The *šurbuīum* wool is used to produce the *kutānu*-textiles exported to Anatolia.[72]

Takkušta'um. This word presents the same ending as *Šilipka'um*.[73] This type of textile (*c*.15 occurrences) occurs mostly in relatively small numbers (11 in VS 26, 11:11–12, 10 in CCT 5, 46b:8) alongside textiles exported to Anatolia such as *kutānu*-textiles.[74] It is also attested about ten times in the Mari royal archives, where it is usually written *ták/ta-ak-ku-uš-tu-ú*. These textiles come from Babylon, Kurdā and Karanā.[75] Thus, if it is a geographical name, Takkušta should be located south of Aššur.

Talhatīum. This nisbe corresponds to the city of Talhat, well-known from the Mari archives and located west of the Habur triangle.[76] It primarily qualifies *ēpattum* garments,[77] which are not bought in, but sent to Aššur. The *ēpattum* seems to be a specific, local product made in this Northern Mesopotamian town. Less often, this nisbe is used of *išrum* "belt".[78] Once, a *sapdinnum* textile is said to come from Talhat, and once, perhaps, a *kusītum* garment.[79]

This list suggests that most of the textiles named by means of a nisbe are exported to Anatolia; they mainly originate from places located east of the Euphrates, in "Northern Mesopotamia."

2.1.2. *Geographical designations added by means of ša*

These designations fall into two categories, those where *ša* is followed by a nisbe and those where it is followed by the name of a country or town.

ša Akkidīē (Akkadīum). To the first category belongs the best-known qualification *ša Akkidīē*, lit. "of the Akkadians", which contains a nisbe of "Akkad" in the plural (with vowel harmony), used as an adjunct after the names of textiles.[80] These textiles clearly come from Southern Mesopotamia according to a letter that describes problems of supply: "As for the Akkadian textiles you wrote about, since you left, Akkadians have not entered the city, their country is in revolt. If they arrive before the winter and there is a possibility to buy for you with profit we will buy them for you".[81] An important text is Kt n/k 1228 (courtesy of

[69] TC 2, 7:25–26 (Michel 2001, no. 108), *šaptum, šu-ur-bu-i-tum lašu*.
[70] The earlier interpretation, a textile made of wool consisting of four twisted threads, the word being derived from *arbe*, "four" found in CAD Š/3, 342b, *s.v. šurbuītu*, most probably has to be abandoned in favour of a nisbe, see Dercksen 2004, 16, note 32.
[71] This emendation is necessary since the adjective is added to the masc. plural of túg = *ṣubātū*.
[72] For this type of wool, see § 1.1.
[73] See above, note 50 and below § 3.3 *s.v. takkušta'um*.
[74] To the texts quoted by Veenhof 1972, 166–167, add for example : Kt 93/k 344:20, *lu ta-ku-uš-ta-ú* ; AKT 3, 52:19, 12 túg *ta-ku-uš-tù-ú*, same form in Yale 13092:11 (1 *ta-ku-uš-tum*).
[75] [Durand 2009, 121–122].
[76] See for Talhat, C. Michel's review of RGTC 4 (*WO* 24, 1993, 176) and Veenhof 2008a, 18–21.
[77] See below § 3.3 *s.v. epattum*. To the references given by Veenhof 1972, 190–191, add Prag I 686, 17.
[78] Prag I 488:6. See below § 3.3 *s.v. išrum*.
[79] Kt 94/k 1672 (courtesy of Larsen), see below § 3.3 *s.v. sapdinnum*; kt n/k 1452:10–14 (courtesy of Çeçen) *rādiam lu kusītam ša Mamma lu <ša>? ta-al-ha/(text:A)-at labbiššu*, "clothe the escort in a *kusītu*-garment from Mamma or one <from> Talhat".
[80] In Prag I 616:3–4, it is written exceptionnally *ša A-ki-dí-im*: túg *ša ṣuhrim ša Akkidīim*; note also 1 túg *ša Akkidīim* in Kt 94/k 368:18 (courtesy of Larsen). The singular *Akkidīim* is perhaps caused by the singular of *ṣubātum*.
[81] VS 26, 17:4–11: *aššumi šīm* túg *ša a-ki-dí-e, ša tašpuranni, ištu tuṣ'u a-ki-dí-ú, ana ālim*^{ki} *ula ērubūnim, māssunu sahi'atma,*

Çeçen):13–17, *ina ṣubātī ša Akkidīē, (...) ibašši lubūšum, kusītum ú šulupka'<um>*, which suggests that the last three types of garments fall under *ša Akkidīē*. There are various garments which are said to be Akkadian: *burā'um, kutānum, kusītum* (several times), *lubūšum, nibrārum, šitrum, šilipka'um, šūrum, takkušta'um.*[82]

ša Šubirim. This qualification (with Assyrian vowel harmony), which means "of Šubarum", contains the name for the Hurrian speaking area located north of Aššur, along the Tigris river (called S/Šubartum in Old Babylonian sources). It qualifies textiles sent from Aššur to Anatolia[83] and is added to the generic term *ṣubātum* (túg), to *nibrārum* and to *kusītum.*[84] Three times it occurs alongside *ša ālim*, "of the City". [85]

More often, textiles are qualified as originating from a specific town by the adjunct: "*ša* + geographical name". Some of the towns named belong to Upper Mesopotamia, while others are located in Anatolia. Of the many place names located between Aššur and the Euphrates, the following ones are used to describe textile products.

Ālum, the City = Aššur. "Of the City" is added to the generic term *ṣubātum*[86] and to specific garments such as *nibrārum*[87] or *šitrum.*[88] This adjunct would be a synonym of the once attested adjective *a-li-ú-tumi* that is a nisbe from *ālum.*[89] Textiles "of the City" occur a few times alongside textiles from Šubartu.[90]

Apum. There are a few references to one or two pieces of textile originating from Apum (Tell Leilan), in the Habur triangle,[91] including the as yet unpublished text (LB 1268:13–14) that mentions 2 *nibrārum* garments from Apum.[92] "Of Apum" probably signifies that these textiles were bought *en route*, on the way to Anatolia.

Hahhum. Alongside the nisbe *hahhīum* (see § 2.1.1), Hahhum itself also appears in the formula "*ša* + geographical name" applied to wool or textiles,[93] including *tisābum ēpišum*[94] and *pirikannum* said to be "from the land of Hahhum" (*ša māt Hahhim*),[95] both of which are in fact Anatolian products.

Haqqa. Textiles as well as *tisābum* and *pirikannum* of good quality are said to be "of Haqqa." The town might be located on the road to Anatolia, between Eluhut, North of the Habur triangle, and Zalpa, which should be located on the northern bend of the Euphrates.[96] However, according to M. Forlanini, it could also be

---

šumma a-kuṣṣi imtaqtūnimma, šīmum ša balāṭika, ibašši niša'amakkum. See also TC 1, 11, BIN 6, 75 and TC 2, 7.

[82] See Veenhof 1972, 99, 158–159 and add to the references: *burā'um ša Akkidīē* (Kt 94/k 966, courtesy of Larsen; AKT 2, 44), *kusītum ša Akkidīē* (KTS 2, 22:5, Prag I 686:21), *nibrārum* (kt n/k 524:10) and for unspecified túg *ša Akkidīē*, AKT 3, 52:61, 91. Note the wrong writing in Yale 13092, 12 (courtesy of Larsen), where *lubušū, kusiātum, Šilipka'um* and *Takkušta'um* are said to be *ša a-ki-dí-NI-im*.

[83] RGTC 4, 108–109 and Veenhof 2008c, 17–19, see also Michel in press b.

[84] See Veenhof 1972, 173 and Prag I 686:19–20: *nibrāram, lu ša šu-bi-ri-im lu ša a-lim*[ki]. FT 4:34–35 (Larsen & Möller 1991, 231, 239): 1 túg *kusītum, ša Šu-bi$_4$-ri-im*.

[85] Túg[(há)] *ša Šu-bi/i$_4$-ri-im*, see AKT 3, 16:2–4; FT 4:5 (Larsen & Möller 1991, 231, 239).

[86] AKT 3, 16:3; Kt 93/k 887:27; FT 4:6 (Larsen & Möller 1991, 231, 239).

[87] Prag I 686, 19.

[88] Kt n/k 437 (courtesy of Günbattı):3–4, 2 *šitrē, ša ālim*[ki].

[89] See § 2.1.1 *s.v. Ālīum* and note 54.

[90] See above, note 85 for occurrences alongside *ša ālim*.

[91] Kt 93/k 344:21, 32.

[92] LB 1268:14–15: 2 *nibrārū, ša Apim*.

[93] Túg *ša Hahhim* : Kayseri 25 (Landsberger), *KUG* 13:20 = *EL* 332:20, KT c/k 695:11–12 (courtesy of Dercksen). Wool *ša Hahhim* occurs in Kt b/k 27:5–6 and OIP 27:7, 6 + 46b, dated to the period of *kārum* Kaneš level Ib, see Dercksen 2001, 47, note 44.

[94] Kt 94/k 1672:19 (*tisābum*) and Kt c/k 729:3, 43 pieces (*ēpišum*).

[95] Kt n/k 518:89–90 (courtesy of Günbattı).

[96] BIN 4, 43:29, túg[há] *ša Ha-qá-ma* ; Kt c/k 753:7 (courtesy of Dercksen), *tisābam ša Ha-qá*; Kt 93/k 60:1–2, 34 *pirikannī ša Ha-qá*.

situated north of the Euphrates,[97] a proposal that accords with the fact that the *pirikannum* textile is a typical Anatolian product, whose production therefore should start beyond the Euphrates.

Nihriya. An unpublished tablet quotes a *tisābum* textile from Nahriya, which is commonly written Nihriya in the Old Assyrian texts.[98] This city is located on the Upper Balih, north of Harrān.

Qaṭṭara, probably Tell Rimah, is given as the origin of a *nibrārum* textile in a document recovered in 1993.[99]

Talhat (see § 2.1.1 s.v *Talhatium*), *ṣubātū ša Talhat*, Kt c/k 709:2 (courtesy of Dercksen) and Kt 94/k 1395:16–17 (courtesy of Larsen); *sapdinnū* sig$_5$ *ša Talhat*, Kt 94/k 1387:19–20 (see footnote 79).

Tuttul. Linen textiles (*kīta'um*) were produced in Tuttul, on the Middle Euphrates according to an unpublished document.[100]

Zalpa. There are several towns called Zalpa at the beginning of the 2nd millennium BC, but the one connected with textiles is located on, or beyond, the Euphrates, North of Hahhum. Apart from the generic textile name, túg,[101] four different types of textiles are said to come from (the land of, kt n/k 457:34–35) Zalpa: *šitrum*, *nibrārum*, *tisābum* and *pirikannum*.[102]

Beyond the bend of the Euphrates, in Anatolia, many towns are also connected with textiles by means of the expression "*ša* + geographical name". Thus, unspecified textiles (*ṣubātum*) could come from the towns of Burušhattum,[103] Hurrama,[104] Šalatuwar,[105] Timilkiya[106] or Tuhpiya;[107] it is, however, not always clear whether they were produced there or simply traded. Typical Anatolian *pirikannum* textiles originated from Kaniš[108] and Mamma, a town also known for its fine wool production.[109]

2.2. *The origin of textiles*

2.2.1. *Origin of the textiles exported to Anatolia*

Apart from the textiles produced in Anatolia, which are also traded by the Assyrian merchants, there are many place names from Northern Syria and Upper Mesopotamia connected to textiles in the Kaniš archives.

The textiles from southern Mesopotamia were bought by Assyrians, to all appearances in Aššur, in order to be exported to Anatolia. The textiles originating from small places, designated by

---

[97] See Forlanini 2004, 426 and the discussion by Veenhof 2008b, 11–13.

[98] Kt 94/734:2 (courtesy of Larsen): 3 túg *tí-sà-b[i] ša Na-ah-ri-a*.

[99] Kt 93/k 75:13–14: 1 túg *ni-ib-ra-ra-/am, ša Qá-ṭá-ra*; for the bibliography about the location of Qaṭṭarā, see Michel 2006c.

[100] Kt 93/k 196:6: 1 *kitā'am ša Tù-tù-/ul*.

[101] *ATHE* 63:17; KTS 2, 4:6; Kt 93/k 517:20.

[102] *Šitrum ša Zalpa*: BIN 6, 184:19; CCT 1, 50:6 and Kt 92/k 239. *Nibrārum ša Zalpa*: Kt 94/k 734:3. *Tisābum ša Zalpa*: Kt 93/k 891. *Pirikannum ša Zalpa*: Kt 93/k 891:5–6 and Kt 93/k 59:1–2.

[103] Cole 9:36.

[104] *ATHE* 63:17; KTS 2, 4:6.

[105] BIN 4, 148:8.

[106] TC 1, 3:30.

[107] Kt 93/k 517:21–22; Kt 93/k 522:4.

[108] CCT 5, 12a:7; Kt 93/k 505:10; Kt 93/k 708:23; Kt 93/k 781:8; Kt 94/k 734:6–7.

[109] TC 1, 43:4. For wool from Mamma, see TC 3, 65:18–22 and § 1.1.

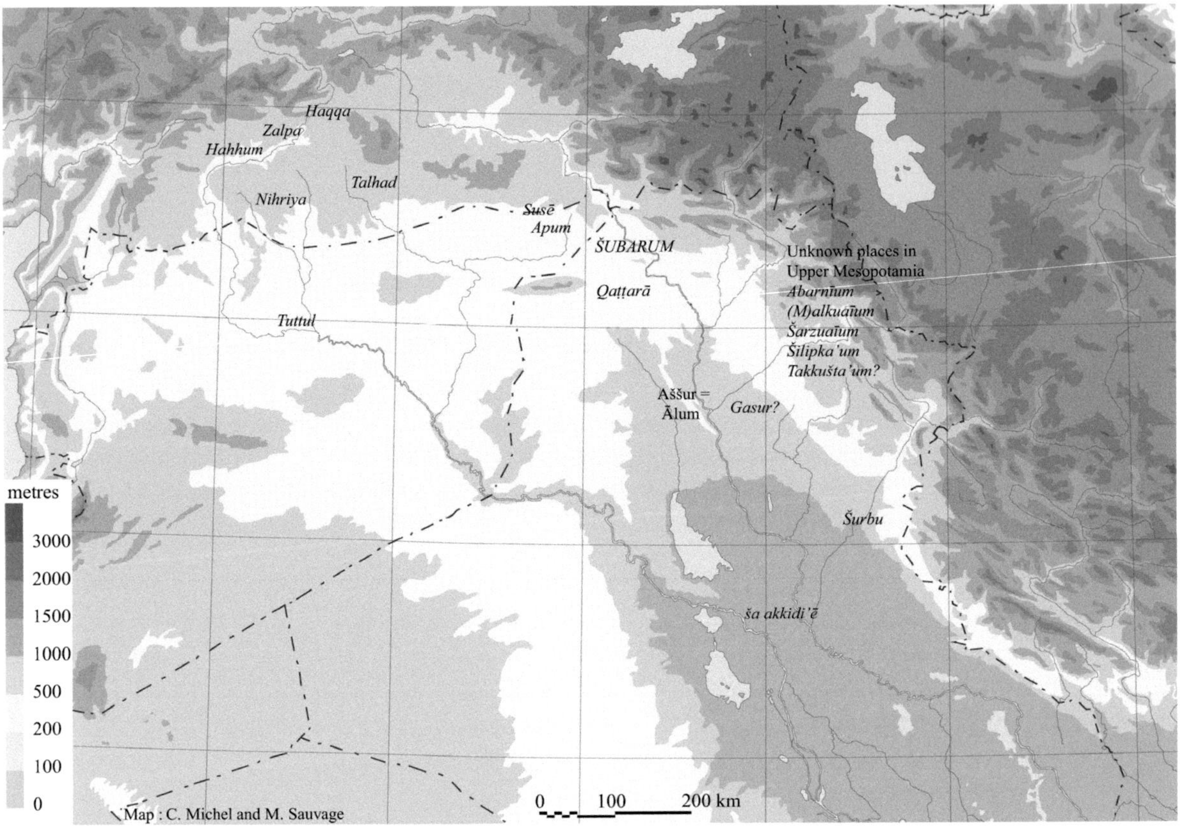

Fig. 12.1. Map of Upper Mesopotamia with geographical names connected to textiles.

their nisbes such as Abarnīum, (M)alkuaīum, Šarzuaīum, Šilipka'um and Takkušta'um, were also transported to Kaniš in order to be sold there for as much profit as possible. The various place names situated north of Aššur and between Aššur and the Euphrates, quoted in connection with textiles, correspond to stations on the road followed by the Assyrian caravans going to Kaniš: Qaṭṭarā, Apum, Nihriya, Hahhum, Zalpa and Haqqa. Some textiles might thus have been bought en route in these towns.

2.2.2. *Distinction between origin and fashion of manufacture*

This conclusion, based on a simple link between a textile and its geographical qualification, interpreted as "made in + geographical name", must be qualified and discussed. It is not always clear whether "*ša* + geographical name" signifies that the textiles in question originate from a particular town, because they could have been acquired there by trade, or (which seems to be true in many cases), were local products, manufactured in that town and perhaps exhibiting specific local or regional features. The textiles called *ṣubātum ša Akkidīē*, « Akkadian textiles », have clearly been made in Babylonia. In fact, the term "Akkadians" does not refer to inhabitants of the city of Akkad, but to Babylonians who are always referred to in that manner in Old Assyrian documentation.

The association of a geographical name with a textile could also refer to specific techniques that are reproducible somewhere else, or to particular material as in the case of *šurbuīum wool*, used by Assyrian women in Aššur. The *abarnīum* textile is an expensive item exported to Anatolia; in some occurrences, it must originate from Aššur, for it is woven with the greatest expertise by Assyrian women who are able to reproduce its typical features: "About the Abarnian textile which you sent me, you should not send me a similar one again. If you want to make one, make one like the one I wore there."[110] In this case, a translation "textile from Abarna" is unacceptable, and a meaning "textile according to the fashion/technique of Abarna" is better. The same applies to *šilipka'um* and *takkušta'um* textiles, which are said to be "Akkadian" products.[111]

2.3. *Geographical areas of textile production according to letters*

Frequently textile names are not associated with nisbes or place names, and we therefore need other criteria to identify their production areas. In order to classify the many different textile types traded by the Assyrians, we can primarily distinguish two geographical zones separated by the Euphrates: Upper Mesopotamia including northern Syria, and Anatolia. While private notices or accounts only provide the names of the textiles and, sometimes, their prices, the letters are much more informative. From the names of the writers and recipients, we can often deduce the origin of the textiles mentioned in them, especially if they ask to buy them or to ship them.

2.3.1. *Textiles exported from Aššur to Anatolia*

Many letters deal with the shipment of merchandise from Aššur to Anatolia, among which various textile types are mentioned. These documents enable us to draw a list of the textiles made or bought in the area of Aššur by the Assyrians. For example, in a message he addresses to Aššur-nādā and Aššur-taklāku, Ilī-ālum announces the shipment of "16 *kutānu*-textiles, 18 *ṣubātum*, 5 *šurūtum*, 2 *raqqatum* textiles, 1 *lubūšum* garment, 1 *šilipka'um* textile and 2 *kusītum* textiles".[112] In another letter sent to Imdīlum, a well-known Assyrian merchant living in Kaniš, the inventory of the merchandise shipped lists: "4 textiles for wrapping and 221 *kutānu*-textiles including those for wrapping, 6 *kusītum malkuīātum*, 6 heavy *burā'um* textiles, among which are 3 soft *burā'um* textiles, [x] white *lubušum* garments, [x] *šilipka'um* textiles, 1 fine *kusītum* textile of extra good quality".[113]

All these textiles are thus produced in the vicinity of Aššur. The *kutānu*-textile is the most common type woven by the women in Aššur, the *burā'um*, *kusītum* and *šūrum* textiles are also made there or imported from southern Mesopotamia (see § 3.2).[114] The *(m)alkuaīum* and *šilipka'um* might have been woven in these places or, like the *abarnīum*-type, produced by Assyrian women according to the fashion or technique of these small towns. The *lubūšum* is a generic term for

[110] See TC 3 17:23–28 § 3.4.1.
[111] Yale 13092:12 (courtesy of Larsen), quoted above note 67.
[112] *RA* 60, 111, no. 43:4–7: 16 túg *kutānū* 18 túg, 5 *šurūtum* 2 túg *raqqatān*, 1 túg *lubūšum* 1 túg *ší-li-ip-ki-um*, 2 túg *kusītān*.
[113] *RA* 81 [1987], 13–15, no. 3:5–12: 4 túg *liwītim u* 2! *meat* 21 túg!, *kutānū qadum*! *ša liwītim*, 6 *kusiātum* [*m*]*alkuīātum*, 6 túg *burā'ū kabtūtum*!, [šà]-ba! 3 túg *burā'ū narbū*, [x tú]g *lubūšū paṣīūtum*, [x túg š]*u-lu-up-<ki>-ú*! 1 túg *kusītum*, [1 *raqqutu*]*m* sig$_5$ diri.
[114] Note the occurrence of *kusītum ša akkidīē* in KTS 2, 22:5.

garment (see § 5) and *raqqatum* is a substantivated adjective which means "fine". These two textile names are not linked to a geographical area.

Other textile types too, connected with geographical names, belong to the exported products. For example, the *nibrārum* textile comes from Aššur, Šubarum, Apum, Qaṭṭarā and even Zalpa.[115]

2.3.2. *Anatolian textiles*

The Assyrians also traded in Anatolian products, which were always cheaper than those imported, but allowed the Assyrian traders to make some profit. Letters allow us to make a list of textiles traded only in Anatolia. Best-known are those called *pirikannum*, *sapdinnum* and *tisābum*,[116] but we also have references that connect Anatolian textiles with a particular town, such as "textiles of Zalpa", "*pirikannū* of Kaneš/Mamma"... Thus, the *menuniānum* textile, also woven in Anatolia, is cited alongside textiles from Tuhpiya.[117] Apart from their names, which do not seem to be Akkadian, the context in which they occur helps to establish in nearly all cases those that are Anatolian and accordingly they never occur in the caravan reports of textiles bought in Aššur and shipped to Anatolia. In general, the names of these native Anatolian textile products do not reveal to us what they were and data gathered from the context are meagre, although it is clear that *pirikannū* were made of wool, rather cheap and weighed less that the imported textiles.[118]

3. Names and qualifications of the textiles

3.1. *Etymology*

One method of identifying textiles is by linguistic analysis of their names. However, many names of textiles exported by the Assyrians are etymologically unclear, while those of Anatolian textiles are not transparent because of our limited knowledge of the early languages of Anatolia. If a convincing etymology can be suggested, the resulting meaning however is often too general or vague to be of much help. The most frequent term for a textile or garment, *ṣubātum*, must be a *purās*-form from the verb *ṣabātum*, "to seize, to grasp", used for "deverbale Vergegenständlichungen" (*GAG* § 55k, 15), and therefore denotes a fabric that "holds" or perhaps "is attached to" the body.[119] But this is true of most garments, and in fact this derivation does not even help us to choose between a textile or untailored garment and a ready-to-wear one, because *ṣubātum* became a generic term and figures as determinative with all kinds of textile names. If *kusītum* is a *purīs*-form from the root *kasûm*, "to bind", it should, according to *GAG* § 55k, 16, be used for "substantive deminutiver

[115] See below § 3.3, *s.v. nibrārum*.

[116] See below § 3.3, *s.v.* Notwithstanding the fact that a verdict of the City of Aššur, at some time, prohibited trade in *sapdinnum* and *pirikannum* textiles; see VS 26, 9, edited in Veenhof 1972, 126–127 and Michel 2001, no. 199. Note that in the later period the trade in *pirikannū* was accepted and even figured in the treaty between the Assyrians and the ruler of Kaneš (see Veenhof 2008a, 193, § h).

[117] Kt 93/k 517:19–23: 63 túg *ibaššiū*, šà-ba 8 túg *ša Zalpa*, 55 túg *lu menuniānū lu ša, Tuhpiya*. Kt 93/k 522:25–27: 71 túg *piri[kannū]*, šà-ba 30 túg *menuniānū*, 41 túg *ša Tuhpiya*. Or together with *pirikannum* textiles: Prag I 740:13–15: 1 *me-nu-ni-<a>-nu-um*, 1 *piri[kann]um*, 2 *na[hlapāt]im*; Kt 94/k 463: 21 *pirikannū* 14 *menuniānū u* 21 *maškū šapātim*.

[118] See for the *pa/i/arakannū*, below § 3.3 *s.v.* In ICK 1 53:4–5, one donkey carried 38 pieces. We may also derive some information from occasional classifications, such as "10 *sapdinnum* textiles, 2 of which are *tisābū*" (Kt f/k 117: 5–6; same CCT 5, 12a:8–9). Note also "1 *sapdinnum* textile of extremely good quality for me to wear" (*ana litabšia*, Kt 94/k 1373: 18–19).

[119] A better example would be *lubāšum*, from *labāšum*, "to wear, to dress oneself in", "something which one wears", which occurs once in Old Assyrian (CCT 1, 27a:4 = 5, 48d:4), but not as garment but to store items in! (*ina lubāšim šaṣṣer*).

oder pejorativer Bedeutung", but this does not yield a suitable meaning and one would expect it to denote a textile or garment bound around the body. The dangers of etymology are clear from attempts to fix the meaning of *kutānum*, which also looks like a *purās* formation, mentioned above (§ 1.2).

Etymology is more helpful for *nahlaptum*, from *halāpum*, "to slip in or through, to cover, to cloth" (CAD H, 35),[120] but its actual meaning is more specific, according to CAD N/1, 138 *s.v.* "a wrap, outer garment (worn by soldiers and as festive apparel)". According to the Sumerian logogram túg-gú-è(-a) it would be "a piece of clothing from which the neck sticks out".[121] CAD N in most cases translates "cloak" and it may serve as outer or upper garment, as in ARM 10, 17:10, where the wife of king Zimrī-Lîm asks him "to put on his shoulders the *ṣubātum* and the *nahlaptum* I made", and in a text from Ugarit the person who breaks a contract "will hang his *nahlaptum* on the doorbolt and go out in the street".[122] Durand (1983, 397), referring to texts from Alalakh and Mari, which list sets of clothing comprising several items of the series túg/*ṣubātum* = "garment", túg-bar-si/*paršigum* = "sash, headdress", gada-šà-dù/*misarrum* = "girdle, belt", and túg-gú-dè-a/*nahlaptum*, suggests the meaning "chemise".[123] A set must also be meant in Old Assyrian, KBo 9 rev. 8', where a *sakkum*-garment, a *nahlaptum* and a pair of shoes are delivered.[124] This might explain the small weight (*c.* half a pound) of a *nahlaptum* according Ur III texts (Waetzoldt 1972, 52, note 118), but in some Old Babylonian texts (CAD N/1, 139, c) they weighed between 2 and 2 2/5 pounds. This suggests a difference in quality (see CAD N/1, 139, e, and the occurrence of "2 extremely fine, soft *n.*" in the Old Assyrian text Kt 87/k 378:16–18, courtesy of Hecker) and perhaps in size.[125] This also explains the differences in price, which in Old Assyrian range from 10 shekels to *c.*5¾ shekels of silver, but there may also have been different shapes or applications, such as the (túg) gú-dè-a *riksi*, attested at Tell Rimah (*OBTR* 59:13 and 80:4, "with ties?"), which, moreover, are distinguished as "long and not long" (sud-a *ù la* sud-a). Note that in Nuzi (HSS 14, 607:14) a *nahlaptum* is also used as a bedcover (*ša majāli*), but the same is the case with *lubuštum* (HSS 15, 139:18).

A complication is created by the logogram túg-gú, which occurs at Mari (ARMT 21, 383 ii:3–4; 384:2–6), and according to CAD N/1, 138 and Durand 1983, 397 note 12 and 405 is also the equivalent of *nahlaptum*. Note also the spelling gú-du-a in ARMT 23, 39:3–6, where, as its author suggests, DU is an abbreviation of UD.DU = è. Eidem (1992, 24) shares this view for the Šušarrā

[120] The *mapras(t)* formation is used inter alia as *nomen instrumenti* (*GAG* § 56 b/c), cf. *nalbašum*, a kind of cloak, from the verb *labāšum*. A derivative of *halāpum* is also *hulāpum*, for which CAD H, *s.v.* registers only one occurrence and proposes a meaning "a bandage", adding "possibly a free variant of *ulāpu*, "bandage". This can now be corrected, since the meaning clearly is "rags", "tatters", in which a slave is wrapped (CCT 4, 45b:31). Additional occurrences confirm this meaning: KTS 34b:14–15, "the girl is clad in rags (*hulāpam labšat*) and is starving", and Kt 92/k 152:4–5, "the tablet is wrapped in a rag" (*ṭuppum ihhulāpim lawi*).

[121] A curse known by Neo-Assyrians was that the moon god will "clad people with leprosy as with a *nahlaptum*".

[122] See Van Soldt 1990, 328, note 50, who defines a *nahlaptum* as "a cloak, i.e. a loose outer garment". A similar clause attested in Boğazköy and Emar uses simply túg, "garment".

[123] [Cf. Durand 2009, 67, where he gives the following translations: "habit de dessus, chemise, casaque, côte de maille".] See also CT 45, 36 II, 2–4, in a summary of textiles delivered by the weavers: 242 túg^{há}, 488 túg-gú-è^{há}, 79 túg-bar-si^{há}, 31 túg-bar-si-gal, where the number of *nahlapātum* is the double of that of the *ṣubātū*, and on a more domestic level, in the dowry listed in BE 6/1,84:7–8, 10 túg^{há} 20 túg-bar-si^{há}, 1 túg guz-za 2 túg-gú-è.

[124] See for this text Dercksen 2001, 52 with note 69.

[125] See ARM 18, 11 for an order of hundreds of *nahlapātum* in five different colors.

texts and both logograms also occur at Tell Rimah.[126] This appears convincing and also explains the writing túg-gú *Hurri*, "Hurrian cloaks/shirts" in the peripheral text EA 22 (from Mittani), while contemporary Hittite texts write túg-gú-é-a *Hurri*. However, texts from Babylonia proper only write túg-gú-è(-a), and since in a letter by Hammurabi (AbB 2, 44:5) túg-gú and túg-gú-è-a occur side by side – together with headdresses (*paršigum*), sandals, leather containers (or hides) and oil, as equipment for troops – they must be different items.

Another etymologically clear term is ^{túg}*raqqatum* (in Old Assyrian with vowel harmony *raqqutum*), in Sumerian túg-sal-la, "a thin textile', well attested in the Old Babylonian period,[127] whereby *raqqum* qualifies the fabric as such as "thin" (its opposite is *šapium* "thick"), which is to be distinguished from *qatnum* = sig, "thin", primarily applicable to the yarn (and to hair). But, like other qualifications of wool, it is also used for textiles made from such thin threads (see especially TC 3, 17:6–7, below, § 3.4.1, on the qualities required for a *ṣubātum qatnum*).[128] This textile – whose name is a substantivated feminine adjective, **ṣubātum raqqum* is not attested – occurs in many periods, just as the adjective *raqqum* is applied to various textile products, notably to *kusītum* (see § 3.2, on túg-bar-dul$_5$).

Finally, *lubūšum* must be mentioned, the Akkadian word for "1. clothing, wardrobe, 2. (a specific piece of apparel), 3. clothing allowance" (CAD L, 236), derived from the verb *labāšum*, "to put on clothing".[129] Again, the etymology does not answer the question of the type of clothing or garment represented. For Old Assyrian, as we will see below in § 5.3, the question is whether túg *lubūšum*, in lists of textiles exported to Anatolia, was a ready-to-wear garment or not. Moreover, we have to distinguish it from its feminine counterpart, *lubūštum*, originally *a nomen unitatis*, for which CAD L, *s.v.* gives the same meanings as for *lubūšum*, but the term is very rare in Old Assyrian, and occurs only as "clothing (allowance)" and does not figure among textiles exported and traded.

3.2. *Occurrences in other corpora and periods*

Occurrences of a textile name in other periods and text corpora can be helpful by their context, contemporary lexical data and occasional logographic spellings. In Old Assyrian, however, logographic spelling, apart from the ubiquitous túg = *ṣubātum*, is extremely rare and there are only two exceptions.

The first is túg-bar-dul$_5$, the logogram for *kusītum*, already used in Presargonic times and recorded in the lexical tradition, which occurs only once in Old Assyrian, in CTMMA 85A:12: 2

---

[126] Cf. *OBTR* nos. 57:4–5 (wool for 50 gú^{há} *zi-ra-ti*), 59:13, 80:4 (gú-è-a *riksi*), 60:4 (túg^{há} *ù* gú^{há}).

[127] Cf. Durand 1983, 408 [and now Durand 2009, 87–90, "un des items les plus courants à Mari"]; its logogram sal-la, added both to gú and to túg, is occasionally spelled with LÁ (ARMT 21, 386bis:6), a spelling also attested at Acemhöyük, see Karaduman 2008, 287 (Ac.i.920), alongside túg *raqqatum* (Ac.i.890, 915, 923, 1092), túg-sal-la (Ac.i.907), and simple sal-la (Ac.i 1199), not surprising because these bullae were attached to packets of different origin. In Old Babylonian this textile figures in dowries, e.g. Bruxelles O 342 I:1–3, ([x+] 2 túg-sa[l-la] 2 túg-sal-la [...] 1 túg-sal-la [...], and BM 12645 II:4, 1 túg-sal-la *sūn ramanišu* ("with a tassel/fringes/flounce of its own material") 2 túg-sal-la *ša la sūnim* (Dalley 1980, 69). Cf. also *OBTR* no. 133:26.

[128] See also Veenhof 1972, 214–216, Excursus IV, on the differences and confusion between *qatnum* and *raqqum*, and 154, where earlier interpretations ("loincloth", on the assumption that sal-la stands for gal$_4$-la = *biṣṣuru*, "vagina"; "netlike fabric", based on *raqqum* as a name for the turtle) are refuted.

[129] There is also a derivative *lubāšum*, attested twice, once in Old Assyrian, in CCT 5, 48d:4, = CCT 1, 27a:4, where a man is instructed to preserve some oil and small items in a *lubāšum*". Hardly a garment, but perhaps a pouch or sack made of a piece of textile.

túg-bar-dul$_5$ *raqqetēn*, "2 thin *kusītu*'s". While the occurrence of this logogram does not add to our knowledge, we note that the qualification "thin" is also attested in Presarg., Sarg. and Ur III texts and in lexical lists (Hh 19, 106: túg-bar-dul$_5$-sal-la = [*kusītum*] *raqqatum*), which indicates that this quality was apparently typical for a *kusītu*-garment.[130] According to CAD K, *s.v.*, a *kusītum* was "an elaborate garment". It is thus far absent from Mari texts and fairly rare in OB, where it is nearly always written logographically, see CAD K, 586, c.[131] During the 1st millennium BC, according to Babylonian sources, it is a precious and coloured outer garment that belongs to the vestments of goddesses.[132]

Túg-nì-lám, the logogram for *lam(a)huššûm*, a well-known name for an expensive, fine garment from the Ur III period,[133] rare in Old Babylonian,[134] has recently turned up in Old Assyrian, in AKT 4,24:1–3, 22⅓ túg *damqūtum watrūtum* (3) šà-ba 1 túg-nì-lám, "22⅓ textiles of extra fine quality, among which one *lamahuššûm*", shipped from Aššur to Anatolia (see Veenhof 2009, 194). This logogram enables us to identify the Old Assyrian textile name *namaššuhum* as a variant of *lamahuššûm*, a conclusion supported by other spellings with the initial *n*, such as túg-na-ma-huš-a at Mari (ARMT 21, 257:22–23; 386bis:19'), *nab/waššuhum* in two Ur III texts and *namanšu'um* in TCL 10, 100:34 (Old Babylonian), spellings showing that early scribes had some problems with the Akkadian rendering of the name of this textile product.[135]

Of the Old Assyrian textile names *kitā'um* ("linen", see above § 1.2), *kusītum, lubūšum, nahlaptum, namaššuhum* (=*lamahuššûm*), *palīlum, paršigum* (rare and only for personal use), *raqqatum, šilipka'um* and *takkušta'um*,[136] apparently all made of wool, are also attested in other periods, but the last two are extremely rare outside Old Assyrian sources. Information on them (including the lexical data, especially in Hh 19 and its forerunners) and the context in which they appear there (production, use, prices, etc.) at times help us to understand what they are, as shown above in connection with *kusītum* and *nahlaptum*. What is salient is mentioned below in § 3.3, under their names.

It is rather surprising that, in the Old Assyrian texts, many of the well-known textiles appearing

---

[130] In Hh 19, it is the first textile product treated after "wool", but in the Old Babylonian Forerunner from Nippur it comes only in line 99 (*c.*60 lines after the section on "wool" has ended) and the qualification sal-la, "thin", does not occur. In the younger Forerunner from Ras Shamra it appears in line 68, as the first specific textile name after 15 lines consisting only of túg with an adjective (but note already in the wool section, in lines 15–16, síg-bar-dul and síg-bar-dul-sal-la), which foreshadows the sequence in the canonical Hh 19.

[131] It only mentions AbB 9, 16:35, *kusīt ši-ka-ti-im* (for veiling a girl). CAD Š/2, *s.v. šikkatu* B, takes it as "a tassel or edging on textiles", and note also AbB 1, 134:13–14, "PN brought me a cloak, a headdress and 5 *kiššātum*" [see also Durand 2009, 152, *s.v.*], and a túg-bar-dul$_5$ weighing 10 pounds, assigned to various priests in an account of ritual expenses from Old Babylonian Larsa. Additional occurrences are in Edzard *Tell ed-Dēr* no. 107:18, MDP 18, 100:6, Bruxelles O 342, I:13 (dowry), and YOS 14, 310:16 (followed by specifications, cf. CAD N/1, *s.v. napāšu* B, 1), and it must be meant in the text edited in Lackenbacher 1982, passim, where it is written túg-bar-dib, qualified as "thin" (sig), MA-IM-TE-NA (=?), *laharītum*, and "of second quality" (ús), and where its various finishing treatments are listed.

[132] See Zawadzki 2006, 117–118.

[133] According to Šulgi Hymn A:43, the king ran from Nippur to Ur with his hips covered by a níg-lám-bàn-da, a combination also found in Hh 19:115–116. Hh forerunners equate *lamhuššû* also with túg-ZIxZI-lagab, see CAD *s.v.*

[134] CAD L, *s.v.* lists no Old Babylonian occurrences, apart from those in lexical texts, but we find it in AbB 11, 170:14 (in a school letter), whose exact parallel, AbB 11, 179:17, has *lubūšum*! The logogram also occurs in an Old Babylonian letter found at Haṣor (perhaps originating from Qaṭnā and listing goods to be sent to Mari), see Horowitz & Oshima 2006, Haṣor 12:8', 20 túg-nì-lam *bu-re-e* (cf. túg-na-ma-huš-a *bu-re-em* in ARMT 21, 257:22).

[135] CAD L, *s.v. lamahuššû*, lists no syllabic spellings apart from those in lexical lists.

[136] Long unknown elsewhere, but now attested at Mari, see below. One might add *mardatum*, but it is attested only once in Old Assyrian, see § 3.3. *s.v.*

in Old Babylonian sources (including those of Mari and the bullae of Acemhöyük), do not appear, such as *ha/ururum, kitītum, laharītum, taddītum, taktimum* (túg-an-dul), *uṭba, uṭ/tublum, yamhadûm, zakûm*, túg-bar-kar-ra, túg-guz-za, túg-nì-bar and túg-si-sá. Not to mention the many other, presumably more specific textile products and pieces of apparel, figuring in the records from Mari as goods given out (usually as gifts or remunerations), coming in, or produced,[137] as well as some textiles mentioned in the administrative texts from Šušarrā (northeast of Aššur, dating to shortly after 1800 BC; see Eidem 1992, 24). Particularly remarkable is the absence of túg-guz-za, prominent in Ur III texts and attested in Old Babylonian,[138] which still figures in the Forerunner from Ras Shamra, lines 179–187, but has disappeared from the canonical Hh 19.

The explanation for this state of affairs, apart from temporal (Assyrian texts are about a century older than the texts from Mari), regional and dialectical differences in terminology, which are a universal feature, must be that the Assyrians imported fairly standardized woollen textile products into Anatolia, mainly untailored fabrics, presumably of cloth, rather than a variety of ready-to-wear garments. This resulted in a limited vocabulary for the main textile items imported, of which, apart from the generic term *ṣubātum*, only *raqqatum, kusītum* and *lubūšum* are well-known from other sources, as well as the specific, but in Old Assyrian rare *kitā'um*, "linen". *Nahlaptum* does occur, but nearly always only one or two pieces,[139] not among the items exported from Aššur and rather for private use than as an article of trade. The Old Assyrian textile repertoire also included a few specific products, usually in small numbers, such as *namaššuhum = lamahuššûm, šilipka'um* (*šulupka'um*) and *takkušta'um*. *Šilipka'um* occurs once in an Old Babylonian letter from Kisurra, quoted in CAD Š/2, 444 s.v. b), and *takkušta'um* – whatever the origin of its name – has now turned up in texts from Mari (see § 2.1.1 and § 3.3, *s.v.*). It is interesting to see that in ARMT 24, 188:1 this textile was a gift a man from Mari had received on a visit to Babylon,[140] and this suggests that the *takkušta'um* mentioned in Assyrian caravan records also originated from Babylonia, and this may therefore also be the case with *šilipka'um*, since the letter from Kisurra mentions that it was made in Babylonia.[141] The one called *makūhum*, unknown from Mesopotamia and not among

[137] Such as *aguhhum, dabadum, guššum, nalbašum, šušippum*. However, note that Mari too knows textile products qualified as "Šubarian" (*šubarûm*), see ARMT 21, 318:5 (see § 2.1.2). For an overview and analysis, see Durand 1983, 393–427, with texts nos. 318–386. We can now add ARMT 22, nos. 108–182, ARMT 23, nos. 8–50, 225–230, 444–451, 571–576; ARMT 24, 181–220 [and see now Durand 2009, *passim*].

[138] Rare in Mari (ARMT 22, 139:7, qualified as *bērum*, "select"; 164 rev. 1'–7', qualified as gíd-a, "long"), once at Haṣor (Horowitz & Oshima 2006, Haṣor 12:11'), better attested in Babylonia, e.g. AbB 9, 16:36, and in dowries, usually one or two pieces, see BAP 7:12, BE 6/1, 84:6, Bruxelles O 342, I:9, CT 8, 2a:4, CT 45, 46:8, TLB 1, 229:13, YOS 13, 91:3', BM 16978:5' (5 pieces; see Dalley 1980, 73). Also in CT 45, 36:I:15 and III:9, and in Lackenbacher 1982, col. I:5', 23' and III:13, where it qualified as "royal" (lugal), "thin" (sig) and *šikimtum*; in VI:9 such a garment is meant for the goddess Nanāya. Note in BM 16465 II:9–10 (dowry, Dalley 1980, 69), 4 túg-guz-za šà! 2 *ana kimāhim*, "two of which are for the tomb", to be used as shrouds? (Cf. Ziegler 1999, 196 no. 25:6', an *uṭublu* ús [*ana k*]*imāhim ša* PN, a musician in the harem of Mari). This textile still occurs much later, e.g. in the dowries listed in El Amarna no. 22, col. IV:12, 15, and in PRU III (MRS VI) 184 (RS 16.146+161):12–13 (written túg-sig$_4$-za), both as a garment for the wardrobe and as a cover for a chair or throne. One wonders what the Akkadian equivalent of túg-guz-za was (see CAD I/J, *s.v. i'lu* = túg-sig$_4$-za, which does not list Old Babylonian and earlier concurrences). [Durand 2009, 35, notes a–b, now suggests that its Akkadian equivalent was *gizzum*].

[139] The only exception is the mention of 10 *nahlapātum* in the younger text OIP 27 no. 11:8; *paršigum* occurs only three times, see CAD P, *s.v.* a, 2'.

[140] 1 túg *ta-ak-ku-uš-tu-um ša qīšti* Z. *inūma ana Babilim illiku.*

[141] Cf. also Yale 13092:9–12 (courtesy of Larsen), cited above note 67, and the combination 1 *šulupka'am, u kusītam ša Akkidīē, u* 2 *kutānī* in Kt 91/k 356:25–27.

the textiles exported from Aššur, might be considered an Anatolian product, but it occurs also in a letter from a Phoenician king found at Ugarit (see CAD M/1, 141 *s.v.*, b), which suggests the possibility that some names of textiles are of western origin. The frequent "textiles of the Akkadians" (*ṣubātū ša Akkidīē*) are designated by what is not a real name and this designation covers various textile products that share particular traits or are made from a particular fabric (see above § 2 and 3.3 *s. v.*).

The names of Anatolian textile products such *pirikannum, sapdinnum, tisābum* and *memuniānum* remain elusive, since they occur only in Old Assyrian and do not appear in later Hittite sources. In texts from the younger period of *kārum* Kaneš level Ib (first half of 18th century BC), where *kutānum* and *kusītum* still occur, ^{túg}*kuššatum* (only attested in TC 3, 61:3 during the earlier period), which also appears at Mari,[142] becomes more frequent. And we now also meet ^{túg}*sakkum*, which is well attested at Old Babylonian Mari (see Durand 1983, 411–12) and also occurs on the bullae from Acemhöyük, see § 3.3, *s.v.*[143] The appearance of *kuššatum* and *sakkum* in later Old Assyrian texts indicates changes in the assortment of textiles, probably due to increased contacts with the area to which Mari belonged.

3.3. *The names of textiles in alphabetical order*

The large number of attestations of the main textiles traded implies that references have to be selective, restricted to what is more informative; for more data the reader is referred to Veenhof 1972, 144–180, and to the entries in CAD. For rare and less well-known textiles all or most occurrences are given. For names that are nisbes or are connected with the name of a town or region by means of *ša*, see also § 2.

abarnīum (Veenhof 1972, 156–158), a nisbe derived from the town of Abarn(i)um, usually treated as a noun (2 túg *abarnīū*), see above § 2.1, *s.v.*, and 2.2.2. Expensive textiles (in BIN 4, 4:4–5 sold for 25 shekels of silver apiece!), also worn by the traders themselves;[144] their price in Aššur, where they were also produced, is once 10 shekels of silver, and in CCT 6, 25d:4'-6', 3 such textiles of good quality cost 23 shekels apiece. They usually occur in modest quantities (1 to 6 pieces), but in Kt 94/k 1687:13, as part of a very large caravan carrying more than 600 textiles,[145] we have 46 *abarnīū lu kamsūtum lu nibrārū ša kutāni damqūtim watrūtim*, "46 Abarnian or *kamsu*- or *nibrāru*-textiles (made) of fine *kutānu*-cloth".[146] They are qualified as *damqum*, "of good quality" (CCT 1, 25:26) and as *damqum watrum*, "of extra good quality" (AKT 2, 24:4–5 – read *a-bar*$^{!}$*-ni-a-am*; BIN 4, 185:2–4; CCT 5, 44a:1–2; Kt 93/k 288:5–6) and are regularly lumped together with other expensive and fine textile products, e.g. in CCT 4, 29b:4 (together with *kutānū* and Akkadian textiles) and in Kt 94/k 1697 quoted above. Twice we meet a *šitrum* (see below *s.v.*) qualified as "Abarnian" (TCL 1, 19:12, 2 pieces *ša abarnīē*; Kt 93/k 75:7, one piece *ša abarnīim*), where the use of *ša* + genitive instead of simple *abarnīum* could mean "belonging with an Abarnian garment."

---

[142] See ARMT 22, 164:1–7 and 23, 375:11–19 [and now Durand 2009, 54].

[143] Kt b/k 21:7–8, from this same period, mentions túg^{há} *šapti*, "woollen textiles", but it is not clear whether this is a descriptive designation or a new name.

[144] Kt 86/k 193:15–17, "select a heavy, soft Abarninan textile for me to wear" (*allitabšia*); also BIN 4, 94:12 (*abarni'am ana litabšia lūšēliam*). See also § 5, notes 267–268, for evidence of women in Aššur who sent single Abarnian textiles to traders in Anatolia, to sell or perhaps rather to wear them.

[145] The 92 pieces in Kt 94/k 1446:6 (alongside *kutānū*), according to Larsen, may belong to the same caravan.

[146] Somewhat different in the parallel text Kt 94/k 1676:13–15, 46 *ṣubātū lu abarni'ū lu kutānū damqūtum watrūtum lu kamsūtum lu nibrārū*, "46 textiles, as well Abarnian ones, as *kutānū* of extra fine quality, *kamsu*- and *nibrārū*-textiles".

b/palītum, TPAK 1, 59:17–18, *šīm* BA-*li-tim lu-bi*$_4$*-ri-im* .... *luptāni(m)*, "write down the price of the ... textile"; *ATHE* 47:24, "While I said: 'Buy and send me textiles to be used as clothing for the servants', you have kept sending me BA-*li-a-tim* of 1 and 1½ shekel apiece". Veenhof 1972, 182–183, rejecting AHw 816a s.v., "Palā-Gewand", and quoting CCT 4, 45b:29 "*jattum* BA-*li-*<....> has fallen from me" (followed in line 43 by "send me (from) there whatever BA-*li-tám* there is, so that I can depart)", considered the possibility of restoring *ba-li-*<*at*>, "mine is worn out", but CAD P, s.v. *palītu*, restores *jattum pá-li-*<*tum*>, "my own *palītu*". The combination with *lubēru* (only occurrence in Old Assyrian, but attested a few times in Middle Assyrian texts and considered a by-form of *lubāru* by CAD L, *s.v.*) is not helpful.

burā'um (Veenhof 1972,173–174), only twice without the determinative túg, occurs in small numbers, frequently only one piece, e.g. in caravans carrying textiles, tin and one *burā'um*, cf. ICK 1, 189:14', *EL* 110:2, CCT 1, 20b:2 (alongside 26 *kutānū*), CCT 1, 36a:10 (with a *nibrārum*), TuM 1, 2c:5 (alongside 46 *kutānū*). CCT 1, 38a:2 mentions that one *b.* was sold en route to supplement the income. The biggest number in *RA* 81 no. 3:8–9, "6 heavy (*kabtūtum*[!]) *burā'ū* 3 of which[?] are soft." It could be worn: BIN 4, 160:12–13, "I clothed him in a *burā'um* of extra fine quality", similarly in Kt m/k 43:2 (courtesy of Hecker), and the writer of Kt 94/k 966:12–14 (courtesy of Larsen) demands: "Give me garments I can wear (*ša litabšia*), either a *šūrum* or a *burā'um* in Akkadian style (*ša Akkidīē*)". This latter type also occurs in AKT 2, 44, 3–5, "4 Akkadian textiles, among which one *burā'um*", and in Prag I 709:24, 1 túg *bu-ra-ú-um ša Akkidīē,* alongside a few *kusītu*-garments, which is also the case in Kt 94/k 1446:12 (courtesy of Larsen) and Kt n/k 524:42. In Benenian 5:7, it appears alongside 1 fine *kutānum*, in Kt c/k 710:8 (courtesy of Dercksen), 1 túg *burā'um* sig$_5$ diri, LB 1268:13–16 one bag contains 3 *burā'ū 2 nibrārū ša Apim,* 1 *tisābu ú,* 1 *pirikannum.* It was probably not one of the types of textiles traded and may thus have been the personal property of the traders, who did wear it, when necessary also during caravan trips.

DU-DU-*ru*?, only BIN 6, 186:7, 5 túg DU-DU-*ru ša* PN, meaning unknown; CAD M/1, 141, s.v. *makūhu* reads *tutturū*, but this word is not registered in CAD T, *s.v.*

epattum (Veenhof 1972, 128–129 and above § 2.1.1 s.v. *Talhatīum*), plural *epadātum*, occurs a dozen times, frequently qualified as *Talhatītum*, "of Talhat", a city in Northern Mesopotamia,[147] which also produced a specific type of *išrum*, "belt". This is confirmed by Kt n/k 391 (courtesy of Günbatti), where a man (probably the ruler) of Talhat swears that he will deliver 22 túg *epadātim*. The use of the determinative túg (only once), the (rather doubtful) identification with the Hebrew *'ēpōd* and Syrian *peḏtā* (see CAD E, 183 *s.v.)* and its occurrence alongside *raqqutum* in CCT 1, 32c:14–15, indicate a textile product, perhaps a kind of cloak. They were apparently appreciated in and shipped to Aššur in small numbers, and in the letter CCT 4, 6e:6–8, the son of a well-known trader, who was undergoing scribal training in Aššur, asks his father to send him an *epattum* as a gift for his teacher. However, the *epattum* was also traded in Anatolia, since OIP 27, 62:43 mentions 20 *epādātum Talhatiātum* deposited in Kuburnat. The specific features of this product remain unknown.[148]

ēpišum (Veenhof 1972, 171–172, and earlier Lewy 1958, 98 note 65), first vowel mostly written as *e-*, but once as *i-* (BIN 4, 78:6). The reading of the first consonant as *p* assumes a connection with the verb *epēšum* I, "to make", or perhaps rather (according to Landsberger) *epēšum* II, "a type of weaving". However, CAD B, *s.v. ebišu*, which lists as alternative readings *ebiššu, ibi(š)šu* and *ip/bi(š)ša*, considered it a native Anatolian appellative, "a low-priced textile, a subcategory of *pirikannum.*"[149] It was a woollen product that may also

[147] For this important city, see Veenhof 2008, 18–21, also connected with other textile products, and see below § 2.1.1.

[148] The mention of an *epattum* worn as a garment, in Veenhof 1972, 96, quoting VAT 9237 (from a photo), should be cancelled; read with VS 26, 40:9, *ṣú-ba-tám.*

[149] For details, see Veenhof 1972, 172. The letter VS 26, 9 (see Veenhof 1972, 126–127) mentions a verdict of the City

have served as a kind of blanket, as suggested by the combination *pirikannī ēpišī šapātim*, "*pirikannū* (in the shape of/worked into?) *ēpišū* of wool", in VS 26, 9: 6–7, comparable to the combination *maškū šapātim*, "hides with wool" = "woollen fleeces."[150] Perhaps also usable as a saddlecloth, since AnOr 6, 15:3–5 mentions a transport consisting of one donkey, 27 *kutānū* and one [*ē*]*pišum*, but it does not belong to the standard donkey harness, as reconstructed in Dercksen 2004, 270–277. Mentioned alongside Anatolian textile products in BIN 4, 78: 6–10, "buy *pirikannū* as/for clothes for the servants, or *menuniānu*-textiles or *i-pí-šu* or *pirikannū* that are strong enough to serve as clothes....", and also in Kt 89/k 421:4, "209 *pirikannū*, 41 *ēpišū*". They occur alongside *makūhum* in TC 3, 132:11, where they are sold at 9½ shekels (of silver) apiece, which demonstrates that they were not a cheap product, cf. the price of *c.*7 shekels apiece in TC 3, 91:33–34 and Kt 91/k 481:12. Kt n/k 127:3–4 mentions 15 túg *e-pí-ší* alongside 20 *kutānū*, and Kt 87/k 452:3–4 (courtesy of Hecker) has the combination 1 *kutānum e-pì-ší*, (not *ēpišim*!), perhaps made op *epišu*-fabric? AKT 3, 16:10, mentions 9 túg *e-pì-ší* deposited in Wahšušana.[151]

g/kasistum, Prag I 428:33, 2 túg GA-*sí-is-té-en*$_6$, unclear and perhaps a mistake, since the partial duplicate published in *JCS* 41 (1989) 40:28, writes 2 *kusītēn*. Attested as adjective in Kt c/k 710:14-15 (courtesy of Dercksen), 1 *išram*, GA-*sí-is-tám damiqtam*.

hirurum? Kayseri 4695:11–13, 3 túg *abarnni'ū* 1 túg *lubūšum*, 1 túg *hi*?-*ru-ru-um ù raqqutum*.

illūkum, BIN 4, 168:16, 121 túg *i-lu-ki*; TC 3, 192:8, 4 ANŠE *ša pirikannī i-lu-ki*, perhaps to be connected with lexical túg-níg-sal-íl-ŠÌR, túg-du$_8$-du$_8$, and túg-gu-za = *ullūku* = *lubār sāmu*, see CAD I/J, 86 *s.v.*, where no Old Assyrian references are mentioned.

išrum, "belt" or "scarf", plural *išrātum*, occurs *c.*20 times. The meaning is indicated by VS 26, 40:13–14 and Kt 88/k 625:11–12, both mentioning an "*išrum* for my waist" (*ana qablia*), and CAD I/J, 261 *s.v.* interprets ICK 1, 88:16–18, 1 túg *išram kīsam* ... PN *naš'akkum* as "PN is bringing you one *išru*-belt (with an attached) money bag". See also the sequence in BIN 4, 88:4–6, "Give 1 *išrum* to yourself, 1 *išrum* to [x x], 4 *šakkukātum* to ...", where the last word is another term for a belt or girdle. ICK 1, 88:16 and CCT 6, 3a:1 (5½ túg *išrātum*) are the only cases where the determinative túg is used.[152] In *POAT* 42:10–13 an *išrum* is sent to somebody together with a pair of sandals and some oil, TC 1, 19:19–20 mentions it alongside a butcher's knife. It figures as a gift to a local ruler in OIP 27, 58:26 (3 pieces) and in CCT 3, 25:27, together with a piece of linen, it figures as a (votive) gift for the god Amurrum. Single items are sent to people, apparently for personal use, e.g. *KTH* 7:34 and TC 3, 210:8; CCT 5, 41a:29–31 mentions 5 pieces, "4 for my representatives, 1 for PN". Like *epattum* it is frequently qualified as *talhatīum*, "of Talhat", see the examples quoted in CAD I/J, 261 and Veenhof 1972, 176–8, especially BIN 4, 160:6–8, "I paid 20 pounds and 32 shekels of copper for 16 *išrātum* to the people of Talhat", which makes it a very cheap item. *Išrātum* of Talhat also occur in Kt k/k 46:9 (3 pieces), Prag I 488:6 (14 pieces). Prag I 740:2 mentions 10 *išrātum* after *kutānū* and *šūrūtum* textiles and before 1 *šitrum ša lubūšim*.

kitā'um, "linen", see under "Materials", § 1.2.

---

of Aššur that prohibited Assyrians to trade in the native Anatolian textile products *saptinnū u pirikannū*, and their qualification as *e-pí-ší šapātim* could also be rendered as "woollen products", revealing that the reason for this measure was protection of the Assyrian import of Mesopotamian woollen products.

150 Note also the combination in Prag I 429:17–18, "of the red *pirikannū* 4 are *ša i-pì-ša*", and perhaps TC 3, 65:3–5, "I gave 15 shekels of silver *ša i-pí-ša* to PN".

151 This textile product may also be meant in Kt 94/k 297:4, *lu-tí-a-am ša e-pí-ší-im ša Apim*.

152 It should not be confused with túg *išrātum*, which means "textiles as tithe", e.g. in Kt m/k 45:6 (courtesy of Hecker) and presumably in CCT 6, 36a:1–3, 5½ textiles *išrātum*, at 13½ shekels per 1½ textile they balanced to you".

kusītum (Veenhof 1972, 159–161), whose etymology has been mentioned in § 3.1, must have been a garment made of woollen cloth (it could range under the category *kutānum*),[153] occasionally qualified, as in other periods (see § 3.2), as *raqqum*, "thin", as "of good quality" (VS 26, 51:18) and a few times as "white".[154] We also find *kusītu*-garments *ša Akkidīē*, "of Akkadian make/style",[155] or qualified by the nisbe *(m)alkuaīum*, and once a *kusītum* qualified as such is described as white in a parallel text.[156] What a *kusītum mardātum* (only occurrence in CCT 1, 29:6) was, depends on the meaning of the latter term, discussed in Durand 1983, 409–11, where the single Old Assyrian occurrence is not mentioned.[157] A *kusītum* was not cheap: in Aššur one paid 7 shekels for it (CCT 1, 35:15) or more.[158]

kuššatum (in Assyrian with vowel harmony), only once attested in a text from *kārum* Kaneš level II, TC 3, 61:3 (in Anatolia, 20 shekels of silver *šīm kuššitim*), more frequent in texts from the younger level Ib, in OIP 27,11:4–7 (after *kutānū* and *kusiātum: 2 kuššātum damqātum 2x-na 5 kuššātum 1 kuššutum ša lubūšti*), 36:7–8, and 37:5', and now also a few times attested in Mari, see above § 3.2.

kutānum, see above § 1.1, on its etymology and meaning, where evidence is presented for its nature as a woollen fabric, of more or less standardized size (probably *c.*4.5 by 4 m.), whose finishing treatment had turned it into a woollen cloth. That it was a large, untailored fabric, explains that texts mention fractions of a *kutānum*, usually 1/2 or 1/3.[159] *Kutānū* were traded but could also serve as material from which specific textile products and pieces of apparel could be made, designated as *ša kutānim*, "made of *kutānu*-cloth/of *kutānu*-type". This qualification occurs with *nahlaptum* (OIP 27, 11:11–13), *namaššuhum* (Benenian 5:2–3), *nibrārum* (Kt 94/k 1686:9–10), and *šitrum* (*RA* 59 [1965] no. 14:16).[160] Note also the request in Kt n/k 216:7–9, "If in addition to the linens there is a *kutānum* (available), give me that *kutānum*".[161] In summaries, other textiles or garments made from this material could be subsumed under it, e.g. *kusītum*, in AKT 4, 23:1–2 and Prag I 616:18–19, "x *kutānum*, thereof y *kusiātum*", see below § 4.1.1 Since *kutānum* was the main type of cloth and obviously served as material for making various types of garments, the word was occasionally also used with the meaning "garment made of *kutānum*", e.g. in *kutānum ša ṣuhrim*, "a *kutānum* for a child" (Kt 93/k 75:6–7). The meaning of the combination *kutānum e-pì-ší* in Kt 87/k 452:3–4 (courtesy of Hecker) is not clear, but also suggests a woollen product (see above *s.v. ēpišum*). It was the most frequent textile product, thousands of which were shipped to Anatolia. The whole scale of qualifications listed below in § 3.4.1 can be applied to *kutānum*, and we also find the qualifications *kabtum*, "heavy" (kt 86/k 193:19–20), and perhaps *sāmum*, "red" (KTS 2, 35:30, 6 *kutānī sà*[!]*-mu-tim*). What "thin and yellow *kutānū ša hu-šu-x-ń*"

[153] On the basis of statements such as "x *kutānū* of which (šà-ba) y are *kusiātum*", AKT 4, 23:1–2, Prag I 616:1–3, see § 4.

[154] CTMMA 1, 85b:12 (quoted in § 3.2), Kt 89/k 257:15 (1 *kusītum ra-qú-*[*tum*]), *KTB* 7:5, and perhaps VS 26, 74:41, cf. CAD R, s.v. *raqqu*, b.

[155] See above § 2.1.2. In Yale 13092:8–12 (courtesy of Larsen), 7 *lubūšū* 6 *kusiātum* 1 *šilipka'um* and 1 *takkušta'um* are added up in line 12 as "together 15 *ša A-ki-dí-*NI-*im*", which presumably means "of Akkadian style/make"; they were purchased for an average of *c.*9,2 shekels of silver. Kt 91/356:32–33 also mentions a *kusītum* together with a *šilipka'um* and it occurs alongside a *burā'um* in Kt 94/k 1446:12–13 and in Prag I 709:23–25 (2 *kusitān u* 1 *burā'um ša Akkidīē*).

[156] See for *ša Akkidīē*, Kt c/k 323:9–10 (courtesy of Dercksen), Kt m/k 22:5 (courtesy of Hecker), and 91/k 356:26, and for *(m)alkuaīum*, above § 2.1.1.

[157] Durand 1983, 409–411: according to the Mari texts "clairement comme originaire de l'Ouest" ... "pouvait être un habit très orné" ... "une ample pièce de tissu". [See now Durand 2009, 61–65: tapisserie]. CAD M/1, 277 s.v., "fabric woven with several colors in a special technic", well attested at Nuzi.

[158] In Kt n/k 199:18–20 (courtesy of Bayram), 6 textiles for wrapping and 6 *kusiātum* together cost 69 shekels of silver.

[159] See below, § 5.1.

[160] The reading "25 pounds of refined copper the price of wool of a *kutānum*", in lines 4–6 of this text, accepted by CAD K, 608, cannot be correct, because of the strange writing síg-*tí-e-em* for *šaptim*, "wool". See below, *s.v. lud/ṭûm*.

[161] *šumma aṣṣēr kitā'ātim*, [túg] *kutānam allitabšia*, [túg *k*]*utānam dinam*.

of Kt 91/k 356:27–28, and a *kutānum ša ša-da-dim* of Prag I 741:8' were, is not clear, but it should be noted that nowhere does a **kutānum ša Akkidīē* occur.

kutinum, only two occurrences, but, as shown by RC 1749:8–9, 1 túg *ku-ta-num ša* Ṭ., 1 *ku-tí-num ša* I., not a rare variant of *kutānum*. Imported from Aššur, according to *KTB* 2:3, 5 túg *ku-tí-nu*, "which in the City cost 13 minas (of copper) apiece", they are sold for 25 shekels of silver apiece in Anatolia.

lubērum, only twice, in *RA* 60, 140–41 no. 8:3–4, "You sent me 10 túg^{há} *šurūtim lu-bi-ri*", which cost 8 shekels of silver apiece, and in CTMMA I, 79:19–23, "Send me also garments to wear (*ṣubātī ana litabš*[*ia*]). I am staying (here) without garments, I am clothed (*labbušāku*) in his *lubērum* ... " According to CAD L, 232 *s.v.*, which takes the word as a variant of *lubārum*, "clothing, garment", the same word is attested a few times in Middle Assyrian. In the first text, it is in apposition to and qualifies *šūru*-textiles, in the second, it seems to be worn for lack of a proper garment and here a derivation from the root *labārum*, "to be old" might fit, but its meaning remains obscure.

lubūšum, "garment", appears independently, in enumerations, "but (túg) *lubūšū*", can also be qualified by a following genitive, e.g. *lubūš ṣuhārim* or *lubūšum ša ṣuhārim*, and we meet *ša lubūšim* qualifying other textile products, e.g. *šitrum ša lubūšim* (see below § 5.3).

***lud/ṭium**, perhaps a textile product, attested in Kt 94/297:4–5 (courtesy of Larsen), *lu*-DÍ-*am ša ēpišim ša Apim*, "a *l.* of a rug/blanket(?) from Apum", followed by: "3 *kurušnanū*, a yoke, for the yoke *ša lu*-DÍ-*i*, if (there is) 1 hide (1 <*ma*>*aškum*?) of an ox". The reason to list it is the occurrence in Garelli 1965, 35 no. 14:5–6 of "25 pounds of copper payment for *lu*-DÍ-*e-im*, *ša ku-ta-ni*" (where Garelli's reading síg$^{!}$-*tí-e-im*, "of wool", is excluded). However, what *lud/ṭûm* means is not clear (cf. the remarks of K. R. Veenhof in AbB 14, 210, *s.v.*).

makūhum, not exported from Aššur, but traded in Anatolia, possibly an Anatolian product, but it also occurs in a broken text from Ugarit, a letter from the king of Sidon, alongside other textiles called túg *sa-ga-li*^{meš} (see CAD M/1, 141 *s.v.*, and Veenhof 1972, 169–170). Etymology and meaning are not clear. Trade in Anatolia is documented in Kt n/k 1689:23–29, "I hear that *makūhū* are expensive/in demand in Burušhattum, buy for 10 minas of silver *makūhū* and send them to me, so that you may earn 1 or 2 pounds of silver!" The writer of ICK 1, 190:8'-9' states "I am entitled to/ have a share of 18 túg *makūhū* in the palace in Burušhattum", and in KTS 18:5 they cost 20 minas of copper apiece. Kt 91/k 436:1–8 lists tin, "24 *makūhū*, 1 *pirikannum* and 2 donkeys belonging to me, 21 *makūhū*, 1 donkey belonging to I.". They occur together with *ēpišū* in TC 3, 132:1 ([x+]40 *makūhū lu ēpišū*, sold at 9½ shekels of silver apiece) and in Kt 87/k 423:1–2 (courtesy of Hecker, 24 túg *lu makūhū lu* túg *ēpišū*). In Kt 93/k 277:1–3, 55 pieces occur alongside a large number of *šulhu*-textiles, and Kt 91/k 344:23–24 also writes "either *šulhū* or *makūhū*". *Makūhū* are never said to be made of a particular fabric (*ša* ...), or to belong to a specific category ("thereof/including x *makūhū*"), but they can be summarized under the general category of "textiles" (*ṣubātū*), e.g. in CCT 1, 15a (= *EL* 132):1–6, where 90 *kutānū* 50 *ma-ku-hu* 3 *lubūšū* 3 *namaššuhū* and 2 *šulupka'ū* are added up as 148 *ṣubātū*. In AKT 3, 52:17–20, 162 *makūhū* appear alongside 15 *kutānū* 12 *takkuštû* 1 *šulupkûm*, and 2 túg *ša Šubirim*.[162]

mardatum, occurs only in CCT 1, 29:6–7: 1 túg *kusītam*, *ma-ar-da-a-tám u sahertam* given to the chief (*barullum*) of a town. Also attested at Mari, where, in ARM 6, 67:13 they probably figure as a gift to two commanders and are described by Durand 1983, 409–411, as "un habit très orné", apparently originating from the west; later also denoting a kind of carpet. (For this word, see also A. Wisti Lassen in this volume). [See now Durand

[162] See further: BIN 4, 113 (= *EL* 261): 4–5, copper of (*ša*, earned by selling?) "his *makūhū* or his *kutānū*, which are his votive gifts" (*ikribū*); BIN 6, 186:7, 5 túg *makūhū u* DU-DU-*ru*. According to TC 3, 132:4 "they took 1 *makūhum* during the journey", and in AKT 3, 61:23–25, 98 túg *makūhū* figure as price of an *emūqum* (= ?), and 28 pieces in line 35.

2009, 61–64, who mentions its embroidered decoration and notes that in inventories it is listed after the textiles and appears alongside fabrics used on furniture, which suggests a meaning "cover" (also used for a sun-shade), "carpet" and when worn by men perhaps a "cape"].

***maškunum**, an item probably made of a textile fabric, perhaps a kind of cover or pouch, as suggested by BIN 6, 84:16, where, in a damaged context, "a double *maškunum* for/of a textile/garment" (*[aššini]šu maškunam ša [ṣu]bātim*) occurs after wool, a *[tisā]bum* and a *šitrum*. In BIN 6, 184:4, ⅓ *ṣubātum, 1 kusītum maš-ku-num*, as part of a trader's share, the absence of a numeral before *maškūnum* suggests that it is in apposition and qualifies the *kusītu*-garment, which served as *maškunum*. Its function is indicated by AKT 3, 82:25, which requires one to carefully pack a valuable tablet and "to put it in a *maškūnum* of good quality" for shipment overland.[163] The three *ma-áš-ku-nu*, costing 7 shekels of silver, mentioned in BIN 6, 140:1–3, together with some tin given for expenses for the transport of a load of textiles, may also have been used for protecting or packing goods. Not informative are ICK 1, 98:7, 8 *maš-ki-ni*, mentioned after a large amount of wool, and Kt n/k 1385:25, which mentions the availability of copper, the price paid for "my *maškunū*" (*šīm maš-ki-ni-a*). A possible identification with *maškanum*, which in later texts can also mean "tent", suggested by CAD M, *s.v. maškūnu*, is unlikely.

menuniānum (Veenhof 1972, 171), an Anatolian textile product, never exported from Aššur. In CCT 4, 27a:11–13 *menuniānū* appear alongside wool, woollen fleeces and *pirikannū*, in Kt 94/k 463:1–4 (courtesy of Larsen), 21 *pirikannū*, 14 *menuniānū* and 21 woollen fleeces are transported on 2 donkeys. In Prag I 740:13–14, 1 *menunênum* figures together with 1 *pirikannum* and 2 *nahlapātum*, and in *KTH* 1:17–19 alongside *kusītu*-garments of Mamma and woollen fleeces. In BIN 4, 78:6–9, "*pirikannū* for clothing of the servants (*ša lubūš ṣuhārī*), either *menuniānū* or *ēpišū*", they are considered to be suitable as (or for making) clothes, which may indicate that they were made of *pirikannu*-fabric. Here they cost only 1 shekel of silver apiece, as in *KTH* 6:6–8 (3 pieces, for clothing servants), in Kt n/k 190:1–2, 1¼ shekel. Only Kt n/k 214:29 adds a qualification, "thin" (*qatnum*). Kt n/k 162:6 writes *ma-nu-ni-a-ni-im* and in CCT 1, 16b (= *EL* 131):26, and Prag I 740:13 there is contraction, *me-nu-né-nu-um*; the plural in Kt 93/k 522:26 is spelled *me-nu-ni-a-e*. In *EL* 131, together with other items, including textile products of Talhat, it seems to have been shipped to Aššur. What a *menuniānum* was remains unclear (see also § 2.3.2 with note117).

nahlaptum, whose etymology and meaning have been discussed above in § 3.2, occurs a few dozen times in very small numbers (but 10 pieces in OIP 27, 11:8), frequently as personal property (e.g. TC 3, 193:6 and Kt 88/k 71:46). It features occasionally in a list as an article of trade (Prag I 616:9, 2 pieces, after 2 *raqqātum*; Prag I 740:15, 2 pieces, deposited in a house together with a *menuniānum* and a *pirikannum* to be sold in Kt n/k 437:6). It served as a gift to a "lord of the town" (*bēl ālim*) in Kt 91/k 548:3–4, and in AKT 4, 30:11 it comes from (had been made by?) "our bride-in-spe" (*kallutum*). It could be made of *kutānu*-cloth (OIP 27, 7:11.13, *ša kutānim*), and together with a piece of linen (*kitā'um*) and a pair of sandals apparently made a full set of clothing in KBo 9, 9 rev. 8'–9'. According to the marriage contract Kt 94/141:9–10, if the wife misbehaves, her husband will strip her of "her *ṣubātum* and her *nahlaptum*", together apparently a normal set of clothing of a woman.[164] That it was worn on the body is shown by Kt k/k 2:24, "a *nahlaptum* for your breast" (*ana irtika*[165]); according to ARM 10, 17:10 it is placed around a person's shoulder. There is, as with most textiles, variety in quality, the best are "extremely fine, soft" (*nahlaptēn damiqtēn watartēn naribtēn*; Kt 87/k 378:16–18, courtesy of Hecker). Differences in quality (and style?) are reflected in differences in weight, as registered in CAD N/1, 139, c (no weight attested in Old Assyrian), ranging between 5 (at Nuzi), 2½ (Old Babylonian) and 2 pounds (Nuzi) apiece. Prices also vary, due to quality, size and to where they

[163] *ina maš-kà-nim* sig$_5$ *šuknā*, where, however, the absence of vowel harmony may indicate a different noun.

[164] *ṣubāssa u nahlaptaša ihammaṣṣima.*

[165] Cf. the occurrence in Mari, in T 108:9 (Ziegler 1999, 56 note 359), of a [túg-bar]-si *irtim*, a gift to a woman.

were paid, and some *nahlapātum* (especially those mentioned in the somewhat later texts from Alişar and Boğazköy) were probably Anatolian products. In ICK 2, 83:13' one *nahlaptum* can be purchased for 10 shekels of silver, in Kt u/k 5:5 two(?) pieces cost 7½ shekels, in KBo 9, 26:3–4, one *nahlaptum c.*5¾ shekels, in Kt 88/ k 71:46, some pieces 4½ shekels. That a *nahlaptum* was smaller and lighter than a normal garment is also clear from the use of strips of complete garments for making them, e.g. ARMT 21, 354, where 4 of them could be made from one *uṭublu*-garment cut into strips (*ana šerṭi ša* 4 *nahlapātim*; Durand translates "pour le rapiéçage"; see CAD Š/3, 113 *s.v. širṭu*).

namaššuhum, the Old Assyrian equivalent of *lam(a)huššû* (see above §3.2), was an expensive item, produced in and exported from Aššur. Most references are to one single piece, but CCT 5, 44a lists "10 *namaššuhum* of very good quality", alongside 20 Abarnian and 10 Akkadian textiles, also of extra fine quality; 6 pieces occur in Kt c/k 449:7//458:10 (courtesy of Dercksen). According to Benenian 5:2–3, 4 túg *namaššuhum ša kutāni*, they could be made of *kutānu*-cloth and note also the listing in CCT 1, 39a:13–17 of five times 5 *kutānū u namaššuhum*, as if they belong together. According to the letter Kt 89/k 221:10–22 (courtesy of Kawasaki) an Anatolian ruler had taken a *namaššuhum* and used it as a garment: "As for the *namaššuhum* about which you wrote me, up to five times we went up to the ruler saying: 'We will give you one mina of tin and then you must give us (back) the *namaššuhum*'. He answered: 'At that time (when I took it) I asked you for tin, but you did not give it to me. Since you did not give me tin I have dressed myself in that textile and I have now worn it out' (túg *altabašma u ultabbiršu*)".

nasistum, perhaps a textile product, but never written with the determinative túg. OIP 27, 55:8–9 // BIN 6, 162:12 mention "10 *na-sí-sà-tum*, their price 3 shekels of silver", KTS 2, 53:5–6 mention 1 shekel of silver as the price of 2 *na-sí-sà-tim* (after a long list of *pirikannū*), and Kt 87/k 45:8 (courtesy of Hecker) lists 11 *na-sí-sà-tum* after wool, tin, *palīlu*-textiles, a *kutānum, ṣubātū waDiūtium* and *šapiūtum* and before items belonging to the harness of the donkeys. The context of the last two texts suggests that it may have been a textile product.

nibrārum, a name without a convincing etymology for a textile product that appears more than 20 times, both in Aššur, Northern Mesopotamia and in Anatolia, see Veenhof 1972, 172–173. It could be the same product as that mentioned in a Presargonic text from Mari, according to which 130 *na-ab-ra-ru*.túg are in a storeroom (MARI 5, 78, no. 18). Prag I 686:19–21 distinguishes between *nibrārū* of *Šubarum* and of the City." In BIN 4, 10:35 Lamassī promises that she will send one from Aššur, according to *KTB* 7:12, 2 túg *nibrārū* are supplied to a transporter together with "loose tin". In CCT 1, 36a:11, one *nibrārum* figures alongside tin for expenses, oil and a *burā'um* textile, and in *POAT* 19:16–18 one *nibrārum* is shipped by Dān-Aššur together with 6 *kusiātum*. In Kt c/k 443:12 6 pieces feature among textiles exported from Aššur, but the parallel texts Kt c/k 449:7 and 458:10 show that this is an error for 6 túg *namaššuhū*. Kt 93/k 75:14–15 mentions one of Qaṭṭarā, Kt 94/k 734:3 (courtesy of Larsen), 5 *nibrārū* of Zalpa (cf. AKT 4, 30:9–11, "PN of Zalpa brought 1 *nibrārum*"), and they could be identical to "the *nibrārū* of Šubarum" of Prag I 686:19–21. We find them among Anatolian textiles, e.g. in *Kuliya* 57:16, with a *sapdinnum* and two *šitrū* of Zalpa, and LB 1268:14–15 mentions 2 *nibrārū* of Apum; in Prag I 588:2–3 and Kt 91/k 372:5, they occur alongside *tisābū*. They are qualified as "Akkadian" (Kt n/k 524:10), "of good quality" (*damqum*, in Kt 91/k 372:6, where they occur alongside *tisābū* and *šitrū*), and Kt 94/k 1686:8–10 and 38–40 list "46 Abarnian textiles or/either *kamsu*-textiles or *nibrārū* of extra good *kutānu*-cloth" (*ša kutānī damqūtim watrūtim*), but the classification remains difficult because the parallel text Kt 94/1687:13–16 writes "46 textiles, either Abarnian ones, or extra fine *kutānū* or *kamsu*-textiles or *nibrārū*". Prag I 686:19–21 asks to buy "a *nibrārum*, either of Šubarum or of the City, or an Akkadian *kusītum*, paying 10 or 12 shekels (silver), to be worn by me", which demonstrate that they were valuable products, which were or could be made into garments worn by Assyrian traders.

pālilum, also attested in Mari (ARMT 22, 136:25 and 321:2–5, where the writing with the sign PA identifies the first consonant) and Nuzi, see CAD P, *s.v.*, equated with níg-šu-gur-ra. In TC 3, 164:12, 2 shekels of silver are paid for 2 túg *pālilū*, in Kt 94/k 823:25–26 (courtesy of Larsen), 5 *pālilū* cost 10 2/3 shekels of silver, and in Kt 94/k 1302:15–16, 2 pieces for 6 shekels of silver. TPAK 1, 37:5 mentions 3 *pālilum ša qātim*, "of normal quality", and Kt 94/k 823:27–28 reports that "your servants are bringing you both *pālilum* and wool". Further attestations: Prag I 768:13, *pālilum mādiš šarruṭū*, "the *pālilum* are completely torn into shreds"; Kt n/k 97:3, lists 16 *pālilū* between silver, tin and donkeys; in Kt c/k 355:10–12, 2 *pālilū*, mentioned alongside *pirikannū*, cost 2½ shekels 15 grains (of silver); Kt 87/k 45:4 (courtesy of Hecker), "tin, 23 *pālilū* 1 *kutānum usmum*, 10½ túg *waDiūtum*; Kt 94/k 843: 3–4 (courtesy of Larsen), "22 fine textiles, a donkey with its harness, 2 *pālilū*; Kt 94/k 1302:15–16 (courtesy of Larsen), 2 *pālilēn*, alongside Anatolian textiles. The combination *kita'ātum pālilū*, in CCT 5, 12a:10, "linens being/serving as *pālilum*", suggests that it denotes a specific funtion or shape, and this is comparable to ARMT 22, 321:3–4, 57 túg-šà-ha *pa-li-lu*, *ša* síg šà-ba *za-am-ra*, "57 *šahhu*-textiles being *pālilū*, into which wool has been ..." [Durand 2009, 179, *s.v.* now suggests that it denotes a type of container on the basis of ARMT 22, 321:2 (p. 240), 35 *palilū* šà 10–àm túg šà-ba, "35 housses-*pâlilum* contenant chacune 10 étoffes."].

paršigum, rare, CAD P, *s.v.* 'a sash, often used as headdress"; wearing one is expressed by the verb *apārum* in the stative. In Old Assyrian: CCT 3, 31:34, one, brought to a woman (followed by a *nahlaptum*), BIN 6, 122:11–12, "either a *šitrum* or a *paršigum*"; Kt n/k 1459:27, 2 *pár-ší-ge ša šārtim* (of goat hair?; after 2 ropes of palm-fiber), Kt 94k 938:7, 2 *pá-ar-ší-ge*. Not for trade but for personal use.

pirikannum (also *parakannum)* is by far the most frequent and numerous Anatolian woollen textile product, attested in large numbers, see Veenhof 1972, 124–6 and CAD P, *s.v.* By means of *ša*, a *pirikannum* can be connected with place names (attested are Hahhum, Haqqa, Kaneš, Mamma and Zalpa), which indicates its origin or specific local style of weaving, see above § 2.3.2. The largest number, 317 pieces, occurs in the broken letter CCT 6, 7a:5–6,[166] and trade in these textiles is the subject of the emotional letter CCT 6, 14:47–5, whose writer wonders whether the trade in these cheap items is worth all the trouble and will yield enough profit.[167] However, AKT 3, 19:7–11 mentions the sale of 300 *kutānū* and 300 *pirikannū* to a local Anatolian palace. Kt n/k 1385:15'–18' implies that "10 soft *pirikannu*-textiles belonging to Ištar-bāštī" (produced by her or was she involved in their trade?) were converted into "refined copper of Taritar". Texts record prices ranging between ¾ and 4 shekels apiece, which implies differences in quality (and size?) and we meet *pirikannū* that are said to be "extremely good" (Kt 94/k 364:15, courtesy of Larsen). Several times "soft" *pirikannū* are preferred (see § 3.4.3 s.v. *narbum*). In TC 2, 60:1–8, 25 shekels are paid for 10 pieces to an Assyrian, and 15 shekels for 4 pieces to a native Anatolian. According to OIP 27, 55:1–4 63 *pirikannū* for garments for servants cost 110 ¼ shekels of silver, and 40 other *pirikannū* 86 2/3 shekels, i.e. prices of *c.*1¾ and 2 shekels apiece. These textiles become more prominent during the later period of *kārum* Kaneš level Ib and a text from this period, Kt n/k 30:4–8, mentions 90 *parakannū* that have been deposited in the house of a Assyrian trader because of 2¼ minas of silver, i.e. at *c.*1 1/2 shekel apiece. [168] *Pirikannū* were used as (or for making) clothing for personnel, *ana lubūš ṣuhārī*, cf. OIP 27, 55 (quoted above), BIN 4, 78:6–8 ("*pirikannum* that are strong enough to serve as clothing"), and TC 2, 49:19 (*ana lubūšti bētim*). Prag I 429:17–18 mentions red *pirikannū*, four of which are *ša i-bi-ša*, TC 1, 43:16 demands *pirikannū* that are of

[166] CCT 6, 7a:3–11: 22 túg^{hi}-[*kà*], *ša Šubirim* 1 x [.....] [5]3 *me'at* 17 túg^{hi}-*k*[*à*] , *pirikan*[*nū*] 22 túg^{hi}-*k*[*à*], *maškunū* 11 *maškū šap*[*ātim*], 22? *pirikannū*, [x]-x-*rūtum*, [.......] [10] [šu-nígin?] 3 *me'at* 91 túg [...], [*ša Š*]*ubirim* [...] (broken).

[167] Lines 49–54, "What is the profit on *pirikanū* that I would trade them? May (the gods) Aššur and Šamaš trample it to dung! Are 30 donkey loads worth 30 pounds of silver? How many donkey drivers, how much harness and what journeys do they have to make?"

[168] 90 túg *parakannī ša ina bēt E. ... A. aššumi 2 mana* 15 gín kù.babbar *iššiknuniāti*. AKT 3, 91:6–10 mentions that two traders have established a claim on 60 *parakannū* and 12 Akkadian textiles.

good quality and large (*lu damqū lu rabû*), and Kt 94/k 364:14–16 qualifies *pirikannū* as being of extremely fine quality and mentions that "fine purchases" (*šīmū ṭābū*) are possible. That they are made of wool is likely because they are frequently listed together with woollen fleeces (*maškū*, at times qualified as *šapātim* or *šapiūtum*), cf. VS 26, 30:4, CCT 4, 27a:11, CCT 6, 7a:5–7, BIN 6, 10:10, OIP 27, 55:19, and *POAT* 8: 28 and 34. Of interest is the occurrence, in KTS 10:4–6, of túg *kutānī pirikannim*, which suggests that they could be worked into cloth of the *kutānu*-type, which explains the verdict of the City of Aššur quoted in VS 26, 9:4–11, that forbade Assyrians to engage in their trade, to all appearances in order to protect the Assyrian import of woollen textiles into Anatolia: "Here a court case arose concerning *sapdinnu*- and *pirikannu*-textiles and many people have been fined. You too have been ordered to pay 10 pounds of silver". It led to the advice (lines 20–23), "Please do not get involved in *sapdinnu*- and *pirikannu*-textiles and do not buy them!".[169] Yet later, during the period of level Ib, this was no longer a problem and in the treaty with the ruler of Kaneš (line 69–70) it was even stipulated that he would receive 10% of the *parakannū* imported into his town as tax.[170] TC 3, 192:7 mentions 4 donkey loads of *pirikannī i-lu-ki*, where the last word, also a name for a textile, could be a qualification of *pirikannum*.

pūkum, a rare designation. We have both túg *pūkum* (*RA* 60 [1966] 119:24 and FT 4:6, in Larsen & Möller 1991, 231), between a *ṣubātum* of Šubarum and one of the City, and three references to a *šitrum ša pūkim*, in Kt 91/k 466:1, Kt 93/k 542:9 and Kt 91/k 501, in the last text qualified as "of extremely fine quality", which cost 4 shekels of silver apiece. It could be a specific type of weave, from which *šitrū* were made, or a type of garment with which it had to fit.

raqqatum, "thin textile/garment", a substantivated adjective (singular in Old Assyrian also written with vowel harmony, *raqqutum*) with and without the determinative túg. It has a clear etymology (the adjective *raqqum* is occasionally also added to other textile names, see below § 3.4). It is well attested at Mari,[171] and is frequent among the textiles exported from Aššur[172] and traded in Anatolia, where it figures as one of the more expensive products, which occurs in small numbers, bought in Aššur for prices ranging from 5 to 10 shekels silver (of course depending on quality)[173] and sold in Anatolia for up to three times that price. In TC 1, 39:7–8, it occurs together with a *lubūšum*, offered as a gift to a ruler.[174] Several times qualified as "good" (*damqum*; CCT 2, 32a:17, TC 3, 269:5, worth 30 shekels of silver apiece) and in CCT 4, 48b:18–19 as "good and thin" *(damqum qatattum)*, where *qatnum* is the result of using "thin yarn". It occurs in enumerations alongside *lubūšum, šilipka'um, kusītum, kutānum* and *šūrum*,[175] in Kt c/k 458:10–13 together with *kamsu*-textiles, *namaššuhū, kusiātum*, and *nibrārum*, and it appears regularly alongside *lubūšum*.[176] Other combinations are: *lubūšū, raqqutum, kutānū* (Kt 91/k 299:10–12), *raqqatum, šilipka'um, kutānum* (Prag I 74916–19) and note VS 26, 11: 26–30, "110 textiles, thereof 40 *kutānū*, 11 *takkušta'ū* and 11 *raqqātum*, including 2 *šilipka'ū* (*qādum šilipkên*), which could imply that the latter were a specific type of "thin garments" (see below Kt 94/k 1751, courtesy of Larsen). It also features with Abarnian textiles (*KUG* 6:3–4) and *kutānum* alone (KTS 2, 29:4–5; CCT 1, 28b:2–3; in Kt n/k 469 *kutānum damqum*, Kt 89/k 266:5). Twice *raqqātum* are qualified as *ikribū*, "votive gifts" (and as such property) of a god, of Aššur in RA 60, 111, no. 43:22, and of Ilabrat in VS 26, 11:20. It is

[169] For this letter, see Veenhof 1972, 126–127 and for its background Veenhof 2003, 90–94. AKT 3, 52:6–4 mentions among the many items belonging to a certain Azu, alongside copper, antimony, oxen, donkeys, *ēpišū*, *makūhū* and Akkadian textiles, also three bales of *pirikannū*.

[170] See Veenhof 2008a, 193, h).

[171] [See now Durand 2009, 87–90].

[172] However, note the statements in TC 2, 7:29 and CCT 5, 5b:26, that there are no *raqqātum* available in Aššur, which could imply imports from Babylonia.

[173] Kt 91/k 491:14, "10 shekels of silver for a *raqqutum* of (=made by?) Lamassī"; 6 shekels in AKT 4, 17:7.

[174] See Veenhof 1972, 152–4 and CAD R, *s.v. raqqatu* A.

[175] *RA* 60 [1966] 116, no. 43:4–7 and 18–20; CCT 1, 41a:1–8; CCT 5, 28c:6–8; VS 26, 74:37–42.

[176] TC 1, 39; 6–7; CCT 5, 18b:9–10; ICK 1, 92:2–4; *KTH* 11 rev. 8–10; Kt 94/k 1701:29'-30'.

possible that *raqqatum* was a ready-to-wear product, because in Kt 94/k 1751:5–7, 2 *šulupka'u*-textiles and 2 *raqqātum* of good quality are qualified as *ša lubūšī ša abini*, "for/as garments for our father". This would also make them appropriate as a gift, e.g. in KTS 57a:6, where two pieces are offered to a local ruler.

sakkum occured first only in texts from the later period of *kārum* Kaneš level Ib: in *UF* 7 (1975) 318, no. 3:4 (alongside *kutānū*), OIP 27, 11:9 (read: 3 *sá-ku*), and KBo 9, 8:13 and 9:8 (*sá-kà-am*). But it has now turned up already in a level II text from *kārum* Kaneš, in Kt c/k 866:7 (courtesy of Dercksen), where 17 túg *sà-ku* sig_5 occur in a list of exported textiles. Also known from Mari [see Durand 2009, 54–55] and from inscribed bullae from Acemhöyük (Karaduman 2008), it is written túg *sa-kum*, but *sa-ak-kum* in Ac.i.1085. It is distinguished for quality into sag, "top quality", and ús, "second quality", and the former, according to ARM 10, 19:5, was worn by Mari's king. In Acemhöyük and Mari, we also have the combination *sa-ak bu-re-(e)-em* (Ac.i 1097, ARM 7, 253:5, 270:6, and ARMT 21, 257:18 [cf. Durand 2009, 92–93, 'tissage (serré) pour un matelas"]), where the second word is perhaps rather the textile *burûm/burā'um* (see above *s.v.*) than *burûm*, "reed mat". *Sakkum* also qualifies other textile products (see CAD S, *s.v. saqqu*), e.g. bar-si and gú-è-a, which suggests that it was a particular type of fabric or cloth.

sapdinnum, after *pirikannum*, alongside which it occurs frequently, the most important native Anatolian textile product (see Veenhof 1972, 170, no. 15). Both are also mentioned together in a verdict of Aššur that prohibits Assyrians to trade them (see above *s.v. pirikannum*). The meaning of the word is unknown. Lewy proposed, tentatively, "fleecy cloth", connecting it with Syrian *spudnā* (with an Anatolian ending *-innum*?), which is better than Garelli's interpretation "woollen textiles", who read the word as *saptinnum* and connected the word with túg^{há} *šapti(m)*, "textiles of wool", which occur in a few texts from *kārum* Kaneš level Ib (see above § 3.2, end, with note 143). According to KTS 36c:9–10, *sapdinnū* of good quality were bought in Hahhum, Kt 94/k 1672:20 and Kt 94/k 1387:19–20 (courtesy of Larsen) mention *sapdinnū* of Talhat, and *Kuliya* 57:16 lists *sapdinnam nibrāram 2 šitrē ša Zalpa*. They were not expensive, 5⅓ pounds of copper was paid for one in BIN 6, 227:8–9, but TC 1, 81:5–6 registers a price of 5⅓ shekels of silver as a debt. *Sapdinnum* was a category of textiles that comprised products called *tisābum* according to CCT 5, 12a:8–9 and Kt f/k 117:5–7 (respectively 14 and 10 *sapdinnū* of which 4 and 2 were *tisābū*), but Kt n/k 141:4–5 lists them alongside each other (1 túg *sapdinnum* 1 túg *tisābum* 4 *maškē*). The biggest number is 23, in Prag I 434:3, where, together with another lot they will be sold piecemeal.[177]

Many textiles are named or qualified by adding to *ša* + noun in the genitive to túg/*ṣubātum* in order to indicate its origin or specific nature.

ša Akkidīē, "of the Akkadians", also without túg, because the combination is very frequent; see § 2.1.2 and Veenhof 1972, 98–103 and 158–159, where the texts VS 26, 17:4–14 and TC 1, 11:9–18 are quoted, which mention that they were brought to Aššur by "Akkadians", that is inhabitants of Babylonia. Other items qualified as "of the Akkadians" in Old Assyrian texts are *musārum*, a type of belt or girdle (CAD M/2, 110–111, *s.v. miserru*), and a kind sheep (udu^{há} *ṣuppū raqqūtum ša Akkidīē*, "thin Akkadian *ṣuppu*-sheep"). See for the rare writings *ša A-ki-dí-im* above, § 2.1.2, *s.v.* These textiles occur frequently, usually in restricted numbers, but 80 pieces in KTK 39:7'; 34 in BIN 4, 51:5; 15 in BIN 6, 54:4, CCT 5, 36a:8, and TC 3, 128A:5, etc. They belong to the more expensive products (in Kt v/k 151:4–5 one pays 5 minas of tin for one piece) and some are qualified as "extra fine, of royal quality" (CCT 5, 44a:4), "extra fine" (TC 1, 72:5) and "fine" (TC 3, 36:22). The term refers to a woollen fabric of a particular type or style, which apparently could (also) be used to make specific textiles or garments (see above § 2.1.2, note 82), such as *kusītum* and *burā'um* (Prag

[177] 23–25, *sapdinnī išti tamkārī alqema išti, sapdinnīšu uš-ta-ta-tim sapdinnī ašar, ataddinu* ...; KKS 2, 29:4 records an agreement about a trader's liability for, among others, two *sapdinnī*. Additional occurrences: Kt 94/k 1302:9 (28 pieces at 1⅚ shekels of silver apiece); Kt 94/k 1605:81-82.

I 709:23–25), *šitrum* (Kt n/k 437:3–4), *nibrārum* (Kt n/k 524:10) and those mentioned in Yale 13092:8–12 (quoted in notes 67 and 282).

ša ālim, "of the City", scil. of Aššur, see § 2.1.2. It is used of *ṣubātum* (Kt 93/k 887:27), in FT 4:25, and AKT 3, 16:3 it occurs alongside "a *kusītum* of Šubarum", and Prag I 686:10–20, mentions a *nibrārum*, either of Šubarum or of the City, or an Akkadian *kusītum*". In Kt n/k 437:3–5, 2 *šitrū ša ālim* occur alongside a *šitrum ša Akkidīē.*

ša b/pard/titi, read *ša maštīti* in Veenhof 1972, 181, because the first syllable was written in OIP 27, 55:63 with the sign MAŠ or BAR, to be corrected on the basis of a duplicate of this text, Prag I 429:33, which writes "1¼ shekel *ana* túg *ša* BA-*ar-dí-tí*", followed by a payment for *pirikannū*. However, the meaning remains unclear.

ša liwītim, "for wrapping" (see Veenhof 1972, 28–30, for the various combinations), a functional designation, usually added to *ṣubātum* and occurring in the combination "x textiles, including those for wrapping" (*qādum ša liwītim*), but it is also regularly added to the textile called *šūrum*. Note also *kutānū ša liwītim* in CCT 4, 23a:14 and 2 *ṣubatēn kabtēn*, "two heavy textiles", used for this purpose in *ATHE* 51:2'. Used primarily for wrapping the plaques of tin, but occasionally also for packing other textiles and it is clear that one normally used less expensive, perhaps somewhat coarser fabrics for this purpose. This is ordered in Prag I 718:21, "do not use a fine textile for wrapping" (*ṣubātam damqam illiwītim lā talawwiā*),[178] but occasionally one also used a few textiles of good quality for this purpose (HG 74:10), which is understandable because textiles that had served this purpose were also sold. However, in *EL* no. 123:4 and 6 *ṣubātū ša liwītim* are said to be part of a large group of textiles of good quality (*damqum*), and Kt 93/k 304:5 has the puzzling statement "81 *kutānū* of medium quality, including those for wrapping (*qādum ša liwītim*), of which 20 are *kutānū* of good quality".[179]

ša ma'ēšu, "of its water", meaning unclear, hardly "waterproof"; two occurrences in Veenhof 1972, 181 and Kt 94/k 723:9, 3 túg *ša ma-e* sig$_5$-*tum*.

ša rab/pād/tim, see for two references Veenhof 1972, 182, and now also Kt a/k 532:5–7, [x]+3 *abarni'ū* [x] *kutānū* 1 túg *kamsum*, each qualified as *ša ra*-BA-TIM, given to a man for transport. The presumably long middle vowel (no vowel harmony) suggests the infinitive of the verb *rapādum* "to run", although the adjective *kabtum*, "heavy", added in VS 26, 58:59 does not favor that, but CAD R, 148, 1, a, 2', accepts it and translates "for traveling", but this is not exactly what *rapādum* means; uncertain.

ša qātim, "of the hand" = "of current/normal quality", also abbreviated to *qātum*, used as a noun in apposition. Extremely frequent, especially alongside textiles qualified as "good" (*damqum*), see below § 3.4.1.a.

ša ṣuhrim, "of/for children (*ṣuhrum* is a collective). Qualifies various textile products, such as *ṣubātum*, Kt 75/k 78:2, AKT 2, 52:10, etc.; *kutānum*, Kt 94/k 75: 8; *lubūšum*, BIN 6, 84:30, but it is occasionally also qualified itself by an added *ša* + genitive: 1 *ṣubātum ša ṣuhrim ša lubūšim*, KTS 2, 31:3, and 3 *ṣubātū ša ṣuhrim ša Akkidīē*, *Kuliya* no. 22:9–10 (in Prag I 616:4 written *ša Akkidīim*). In the last case, the addition must indicate from which type of fabric it was made or in which style it was fashioned, but the adjunct *ša lubūšim* does not mean "for (wearing as) a garment", but rather made from the fabric used for a garment. In most cases

---

[178] See for the corresponding verb CAD L, 73 *s.v. lamû*, 3, a, and also Kt n/k 1466:7–9, "4 talents of tin, 8 textiles of you, they wrapped" (i.e. the tin in the textiles?), and VS 26, 149:4–7, "9 bags with tin, of them 6 are wrapped and 3 not" (6 *lawiā* 3 *ulā lawiā*), that is 6 bags contained tin wrapped in textiles and 3 unwrapped tin. *POAT* 19:33 asks that the tablet with the last will of a trader be wrapped in reed (*ina qanu'ē lawwiā*) before being entrusted for transport, and Kt n/k 405:10–11 asks to wrap a debt-note "carefully in a hide" (*ina maškim damqiš lawwiā*) for sending it overland.

[179] In *EL* no. 143:18 one encounters "bags for wrapping/packing" (*naruqqum ša liwītim*). See also footnotes 218–220.

it is simply listed, but occasionally, the context shows that it was actually meant to be worn by a child, e.g. Kt 75/k 78:2–4, 1 *ṣubātam ša ṣuhrim mer'assu ulabbiš*, "I gave his daughter a garment for a child to wear".

ša šadādim, twice, Prag I 741:8', [x] *kutānī ša ša-da-dí-im*, and Kayseri 4698:46–47, "after the textiles had come down from the palace, A. took a textile (*ṣubātam*) *qá-du ša ša-da-dim*." Identifying *šadādim* as a gentive of the corresponding verb, "to draw, to haul", does not yield a suitable meaning. Cf. perhaps Kt 87/k 434:1–3 (courtesy of Hecker), "23 *kutānū*, of which 5 are túg *ša-da*[!]*-im* and 18 *šūru*-textiles", which confirms that they can be made of woollen cloth, but is equally unclear.

šaddum(?). CCT 1, 37b:9 mentions 1 túg *kutānum damqum ša*-DU-*um* and TC 2, 37:26–28 writes "Take a decision on the *ṣubātī ša-du-tim* that are here". CAD Š/1, 42 *s.v. šaddu*, 3 "(uncertain meaning)", in the second text transliterates *ša-tù-tim* and translates "delayed(?)" (which requires a reading *šaddūtim*!), starting from a special meaning of the verb *šadādum* attested in Old Assyrian (CAD Š/1, 30, 6). The first reference is too laconic to decide whether its fits. The younger text KBo 9, 21:1–10 lists a series of túg *ša*-DU (unless one emends into *ku*[!]*-ša-tù*) at different prices (ranging from 4 2/3 shekels to 9 shekels apiece), to end with "1 *kutānum* for 10 shekels". This *ša*-DU probably is a different word, since it lacks the adjective ending and "delayed" does not fit in this list.

ša šarruttim, abbreviated from *ša lubūš šarruttim*, "of royal wear", top quality, see below § 3.4.1.a. Qualifies *ṣubātum* (BIN 6, 23:16), *abarnīum* (Kt n/k 533:18), *ša Akkidīē* (CCT 5, 44a:4–5) and *kutānum* (Kt m/k 8:22–23, courtesy of Hecker).

ša Šubirim, "of Šubarum (Šubartum)", or simply "Šubarian", see § 2.1.2 and Veenhof 2008, 18–19.[180] It qualifies "textiles" (*ṣubātū*) in general (22 pieces in CCT 6, 7a:2–3), but also specific ones, e.g. a *kusītu*-garment, in Larsen & Möller 1991, 231 no. 4:34–35 (alongside "2 textiles of the City"). In Prag I 686:19–22 a garment to be sent to the writer to dress himself in (*ana litabšim*) could be a *nibrārum*, a Šubarian one, one of the City, or an Akkadian *kusītum*. This shows them to be fine products, qualified as "good" (*damqum*) in *RA* 58, 117–118:4, and Kt 91/k 344:10–12 mentions a shipment of "6⅓ heavy and good Šubarian textiles" (*ṣubātī kabtūtim damqūtim ša Šubirim*). We do not know what kind of a product it was, it could be in the style or weaving technique used in Šubarum, but it could perhaps also refer to the wool it was made of, since a letter from Šušarrā (Eidem & Laessøe 2001, 50:7) mentions the existence of "Šubarian sheep" (udu[há] *Šu-ba-ri-i*). Here the real nisbe is used, also attested in tug *Šubarûm* in Mari, see ARMT 21, 318:5, 2 túg *šu-ba-ru-ú*, perhaps also in 23, 617:1 (1 *šu-ba-rum*, without túg).[181]

šiknum, only in Prag I 429:63–64, 2 DU-KU-DU 2 túg *kusiātum* 1 *ši-ik-nu-um* 1 *raqqutum*, listed in CAD Š/2, *s.v. šiknu* A, 439, but on the basis of an older edition, where the numeral before *šiknum* is missing so that it might be taken as a qualification (in the singular?) of the preceding *kusītu*-garments. As a separate name it could mean a spread or cover, as suggested by Durand 1983, 407, for *RA* 64 (1970) 33 no. 25:1–2, 1 túg *ha-li šiknum ša* [giš]ná, on the basis of the final words "of/for a bed".

šilipka'um (šulupka'um), see Veenhof 1972, 165–166 and CAD Š/2, 444 *s.v.*, presumably a nisbe, see § 2.1.1. Belongs to the more expensive textiles exported from Aššur to Anatolia and usually occurs alongside *raqqutum, kusītum, takkušta'um* and *kutānū* of good quality, nearly always only one or two pieces, but four in *KTB* 16:4, and possibly made in Babylonia (see above § 3.2 on its occurrence in an Old Babylonian text from Kisurra). We know nothing of its characteristics, but according to VS 26, 11:27–30, "42 *kutānū* 11

[180] We may add to what is mentioned there in footnote 24, that the nisbe in Old Assyrian not only appears as *šubrīum* (in *amtam šu-ub-ri-tám*, CCT 3, 25:35), but also as *šubirīum*, in Kt n/k 213:28–29 (courtesy of Bayram), *wardam šu-bi*$_4$*-ri-a-am*, and Kt 79/k 101:21, *ana ... šu-bi-ri-im*, "to a Šubarian". See also Michel in press b.

[181] [Durand 2009, 111 points out that, as the mention of "Šubarian sheep" at Šušarrā confirms, Šubartum was rich in wool].

takkušta'ū 11 *raqqātum*, including 2 *šilipkī'ū*", they seem to belong to the category of "thin textiles".[182] In the letter Kt 93/k 505:12// 781:9–10 a single *šilipka'um* turns up as part of a shipment inside Anatolia, for the rest consisting of wool, *pirikannu*-textiles and woollen fleeces, which the addressee is expected to sell for copper. According to Yale 13092:8–12 (see notes 67 and 282), a group of such textiles, including 1 *šilipka'um*, on purchase in Aššur cost on average 9.2 shekels of silver, but a similar group of 3 *raqqātum*, 2 *šilipkī'ū*, 4 *takkušta'ū* and 12 *lubūšū* in TC 1, 47:3–6 together cost only *c.*1 mina of silver or on average *c.*5 shekels. In Anatolia, one paid 35 shekels of silver for one *šilipka'um* in BIN 4, 218:7, in CCT 3, 45b:4–9 one was given in commission to an agent for 22½ shekels, in *RA* 60 no. 35:8 one cost 50 pounds of good quality (*damqum*) copper, but according to Kt n/k 118:12–14 it was sold there among Assyrians for 30 minas copper, and for 15 minas of refined (*masium*) copper in Kt 91/k 330:2.

šitrum, according to CAD Š/3, *s.v.*, "possibly an undergarment, if the lex. ref. - tu-un = tùn = *šuppulu, šitru* - pertains to this word". See also *HUCA* 27 (1956) 33, note 116, where a meaning "cover" or "veil" is proposed, and the word is related to the Hebrew verb *sātar*, "to cover"; cf. Veenhof 1972, 174–6. It is worn in particular by women, cf. Kt 94/k 208:21–25, "I invited Š. and his wife and I gave him a fine *kutānum* and his wife a fine *šitrum* of *kutānu*-cloth". Kt 88/k 71:47–9 mentions the payment of 4 shekels of silver for some *šitrū* for a woman (alongside one for "her cloaks", *nahlapātum*), followed by one of 3 shekels for *šitrū* "which I 'tied' to the slave-girls" (*amātim arkus*), and Kt m/k 121:7 (courtesy of Hecker) registers "one *šitrum* of my lady". When in Kt 2000/k 325b: 36–38, brothers during three years have to give their sister a *ṣubātum* and a *šitrum*, this must be a set of clothes. Note also the association of a *šitrum* with a "belt" (*šakkukum*) in TPAK 1, 28:10–12 and 18, the statement "neither a *šitrum* nor a *paršigum* (a headdress)" in BIN 6, 122:13–14, and the sequence 3 *šitrē* 5 *išrātim* 2 *raqqāti šitta nahlapātim* 5 *musarrētim*, in Prag I 616:3–8. It was usually made of wool, cf. Kt n/k 214: 24–29 (in a letter to a woman), "Buy for me soft wool for *šitrū* and send it at the next opportunity, since I have no *šitrum* for my trip".[183] It was a fairly cheap item, in *KUG* 19:8–10 the price of some *šitrū* was 1 shekel of silver, in Kt 88/71:47–48, 4 and 3 shekels were paid for *šitrū*, in Kt 94/k 432:12–15 (courtesy of Larsen) amounts of 3/4 and 1 1/8 shekel of silver, but CCT 1, 50 (= *EL* 296):6–8 mentions 6 *šitrū* of Zalpa sold for 20 shekels of silver. A *šitrum* made of *pūku*-fabric cost 4 shekels in Kt 91/k 466:1–2 and must have been of fine quality, cf. the request in Kt 91/k 501:6–8 to buy "*šitrū ša pūkim* of extremely fine quality", as may have been the one that was a votive gift for the gods Sîn and Šamaš (together!) in Kt 91/372:5–8.[184] Some occurrences relate it to a town: *šitrū* of Zalpa in AKT 4, 4:5, BIN 6, 184 rev. 10', *Kuliya* 57:13, CCT 1, 50:6; "of the land of Nawar" in Kt 94/k 432:13–15. More frequently it is linked with various types of textiles by means of *ša*, presumably referring to the fabric from which it was made or the type it should match. TCL 1, 19:10–13 asks to send 2 *šitrē ša lubūšē*, 2 *šitrē ša sapdinnī* and 2 *šitrē abarnīē*, and such specifications occur more often: *ša abarnīē* (also in Kt 93/k 75), *ša Akkidīē*[185] (also in KTS 4, 28:13; TPAK 1, 28:10, Kt a/k 253:12–13 and Kt n/k437:4), *ša lubūšī/lubūšim* (OIP 27, 58:25, TC 1, 19:10–11, *RA* 60 [1966] no. 43:33, and Prag I 740:3), *ša kūtānim* (*RA* 59 [1965] 35:16 and Kt 94/k 208:23, *damqum*), *ša pūkim* (Kt 91/k 501:6–7, Kt 91/k 466:1, and Kt 93/k 542:9), and *ša ālim*, "of the City (of Aššur)" (Kt n/k 437:3).

šulhum a textile product, usually written with the determinative túg, that appears *c.*15 times, but neither etymology nor context illustrate what kind of product it was. See for references CAD Š/3, 239–240 *s.v.* and Veenhof 1972, 168–169, 13, where LB 1293:15–18 is quoted, which mentions that "56 *šulhū* and Akkadian textiles cost 7⅓ shekels of silver apiece", and indicates purchase in Aššur. This price agrees with CCT

[182] Note that in the description of the individual lots, the 2 *šilipkī'ū* in line 9 are mentioned alongside 3 *raqqātum*, which suggest that they were a specific type of "thin textiles".

[183] *šaptam*! *naribtam ša šitrī šāmamma*sic *išti panîmma šēbilānim ana alākia šitram, lā išû.*

[184] [x] *šitrū* [š à]-b a 1 *šitrum* [*š*]*a ikribī* [*š*]*a Sú-in ú* ᵈu t u.

[185] As other examples indicate, this is in fact an abbreviation of **ša ša Akkidīē*, "*šitrū* made of/belonging to (*ša*) a textile of (*ša*) the Akkadians". See also footnote 255.

5, 38b:5 (64 shekels of silver for 7 *šulhū*) // KTS 48c:5 (64 shekels for 8 pieces). Kt 94/k 1517 mentions a price of 41 shekels for 5 *šulhū* and 2 textiles of normal quality, owed by somebody in Anatolia, and sale in Anatolia is also at stake in ICK 2, 296:43–4 (cf. 18), where 2 *šulhū* cost 70 minas of copper. Kt 93/k 277:1–4 mentions 152 *šul[hū]*, alongside 55 *makūhū* and 10 *maškinī*, in TC 1, 109:1–2 11 pieces figure alongside 31 *kutānū*, and an unpublished text in Ankara mentions 38 pieces. KTS 2, 4:13–14 mentions the shipment of textiles from Zalpa and Hurama, presumably to Kaneš, "together with the *šulhū* you will buy", which may refer to purchase in Anatolia, as does Kt 91/344:23–5, "look out for *šulhu*- or *makūhu*-textiles that please you there".[186] Cf. Kt 94/k 1775:6–7 (courtesy of Larsen), which mentions "30 túg *šulhū* which one will acquire in Ullama".[187]

šūrum, an adjective (notwithstanding a few plural forms *šu-ru-ú*) not attested outside the Old Assyrian sources, whose meaning is unclear (see Veenhof 1972, 154–6 and CAD Š/3, *s.v.*), usually preceded by túg, but occasionally without it, nearly always treated as adjective, but a few times túg *šu-ru(-ú)* (BIN 4, 189:19; 6, 60:17). A clue for its meaning is perhaps offered by ICK 1, 172:13, where *šūrum* qualifies udu^{há} *ṣu-pú-tim*. According to CAD Ṣ, 249, *s.v. ṣuppu* A (where its lexical equation with udu-babbar, "white sheep", is recorded), such sheep were probably characterized by a white and curly fleece, in which case the added *šūrum* might indicate a darker color (brownish?), which then might also apply to these textiles.[188] Túg *šūrūtum* were exported in great numbers from Aššur and in caravan reports they frequently occur alongside the 30 to 50% more expensive *kutānū*. They were a cheaper and presumably somewhat coarser textile (made from coarser wool or threads?), but they could nevertheless be subsumed under the *kutānū*, as in Kt/87 k 434:1–3, "23 *kutānu*-textiles, thereof 5 túg *ša-da*$^{!}$*-im* (and) 18 *šūru*-textiles". However the summary of various bales of textiles in VS 26, 11:27–39 lists them separately from the *kutānū, takkušta'ū* and *raqqātum*, but BIN 6, 60:13–15, "We counted 85 textiles, thereof 24 Abarnian ones, including (*qādum*) one *šūru*-textile" surprisingly ranges one among the Abarnian textiles. We frequently meet túg *šūrūtum ša liwītim*, "for wrapping" (see above under *ša liwītim*), regularly used for packing the slabs of tin[189] and occasionally also for other merchandise. "Heavy (*kabtum*) *šūru*-textiles for wrapping" occur in CCT 3, 4:7 and *ATHE* 51:2–3 (read: *kabtēn*, [*ša li-wi*]*-tim*. These textiles were regularly sold in Anatolia.[190] In the records Kt 92/k 98, 110, 113, and 121 many bales of these textiles are said to be of/belong to (*ša*) a number of persons, presumably their owners or the agents who had shipped them, and they specifiy how many of them had been paid as tax, had been pre-empted (no doubt by the local palaces), had been used for wrapping, and how many remained available for sale.[191] The large memorandum CCT 5, 36a, which lists substantial numbers of textiles, summarizes in lines 17–20: "In all 335 *kutānū*, 128 *šūru* or Akkadian textiles, 24 thereof for wrapping" (túg *liwītum*). Unclear are *RA* 60 (1966), 141 no. 8:3–4, 10 túg^{há} *šūrūtim lu-bi-ri*, according to CAD L, *s.v. lubēru*, "as clothing", and Kt c/k 173:1–3, 20 *kutānū* 17 túg *šūrūtum ša*-DU-*um šu-ri-im*. That they were appreciated and could be valuable is clear from *EL* 145:5; where they figure as a gift to an Anatolian queen, and BIN 6, 186 rev.:5', where they

186 Lines 24–25: *ša a-ma-kam ṭá-áb-ú-ku-*<*um*>.

187 CCT 6, 7c:8–10, [....] *ša Zalpa* [x x] *lu šu-ul-hi i-*[x x] *ta-áš-a-am*, is unclear.

188 Michel 1997, 109, with note 156, has doubts, observing that *ṣú-pu-tim* is an adjective, while the plural should be *ṣuppū*, as in CCT 5, 32a:13–15, 14 udu^{há} *ṣú-pu-ú raqqūtum ša Akkidīē*, "14 thin Akkadian *ṣuppu*-sheep" (another parallel between designations of sheep and textiles). Yet, since no other explanation has been suggested, it seems possible to consider *ṣuppūtim* in ICK 1, 272:13 a mistake, understandable since the word follows the noun udu, and perhaps caused by the following *šūrūtim*. See for *ṣuppum* also Kt 94/k 462:1–4 (courtesy of Larsen), which record a payment of 2½ shekels of silver for a sheep (udu) and of (only) 1 shekel 15 <grains> for a *ṣú-pí-im*.

189 For the system, see Veenhof 1972, 30–32: one donkey load of tin weighed 130 pounds, 65 in each "half pack", designated as a "(standard) weight" (*šuqlum*), which comprised two packets of *c.*32.5 pounds, each wrapped in one textile.

190 Note in Kt c/k 110:4'–5', the sequence *kutānū, šūrūtum, kabtūtum*.

191 E.g. Kt 92/k 121:1–11: 30 lá 1 túg *šu-ru-tum, ša Šu-Ištar*, 15 túg *ša En-na-Sú-in*, šu-nígin 44 túg *šu-ru-tum, ša Šu-Ištar ù En-na-Sú-in*, šà-ba 4 túg *i-na, li-wi-tim*, 4 túg *ni-is-ha-tum*, 4 túg *i-ší-mì-im*, 32 túg (remain available).

are a votive gift to the goddess Ištar. This is confirmed by TC 1, 19:16–18, which mentions two such textiles of fine quality intended as garments for a trader (2 *šūrēn damqēn ana litabšia*).

ta-dí-im, meaning unknown, in KTS 1, 54b:1–4, *ina* 22 *kutānī*, 1 túg *ta-dí-im*, 1 túg *kamsum*, 2 túg *sapdinnū*, which suggests that the final *-im* is not a genitive ending. See also the mention of its price in Kt c/k 811b:14–15 (courtesy of Dercksen), 1 mina of silver *ša šīm, ta-dí-im*.

takkušta'um, an expensive textile, presmuably a nisbe, see § 2.1.1, bought in Aššur (TC 1, 47:43–8), exported from there into Anatolia, but also attested at Mari, where a text shows that it had been given to a man from Mari in Babylon (see above § 3.2). It occurs *c.*15 times, in modest quantities, rarely more than ten pieces (11 in VS 26, 11:28) and according to CCT 5, 46b:6–8 (see Veenhof 1972, 166) the "counter value/ equivalent" (*mehrātum*) of ten pieces was 2 minas of silver or *c.*12 shekels apiece, but this may not reflect a normal sale.[192]

tisābum (the spelling is conventional, always spelled with DÍ-ZA-BA/U/I, which leaves the nature of the consonants unclear, while first vowel could be *e* or *i*), a native Anatolian textile product whose etymology is unknown and which has not turned up outside Old Assyrian sources, where it is thus far attested *c.*15 times (see Veenhof 1972, 170–171). CAD T, 371 *s.v. tēṣābu* lists it together with a few occurrences of *te/i-ṣa-bu* in Neo-Assyrian sources, which is tentatively translated as "leftovers", but this is not convincing, because such a meaning does not fit in Old Assyrian, where it is an Anatolian textile product. Kt 94/k 1373:18–19 (courtesy of Larsen) mentions one of very good quality to be used as a garment (1 túg *tisābam damiqtam ūturtam ana litabšia*) and shows that the noun was feminine. This may also be the case in *Kuliya* 54:4–6, where an Anatolian owes "1½ shekel of silver (remainder) from the price of a textile, that (for which?) he bought a *tisābum* for his wife".[193] The relation of *tisābum* to the other main Anatolian textile products, *pirikannum* and *sapdinnum*, is not very clear. Kt 94/k 1672:19–20 (courtesy of Larsen) asks the purchase of "either *tisābū* of Hahhum or *sapdinnū* of Talhat", while in CCT 5, 12a:9 (which keeps them separate from *pirikannū* of Kaneš) and Kt f/k 117:5–7 *tisābū* are said to belong to the category of *sapdinnū*,[194] and Kt 93/k 891:5–9 writes, "I gave you 9 *pirikannū* of Zalpa under my seals, 3 thereof are *tisābū*". In Kt c/k 141:4–5 a *sapdinnum* and a *tisābum* occur together, while in Kt c/k 102:3–5, "4 *tisābū* of Timilkiya" figure alongside *pirikannū* and woollen fleeces. In Kt a/k 626:1–6 a shipment consists of tin, 61 *kutānū*, 1 *tisābum* and three woollen fleeces, BIN 4, 51:39–40 mentions 7 pieces alongside 3 woollen fleeces. In BIN 6, 84:5–6 and 14–15 twice 1 *tisābum* occurs alongside wool and other items; Prag I 588:2–3 mentions one together with a *nibrārum*, as is the case in Kt 91/k 372:4–5, while LB 1268:13–17 lists as the contents of one bag: 3 *burā'ū*, 2 *nibrārū* of Apum, 1 *tí-sà-bu* and 2 *pirikannū*. The biggest number, 18, occurs in *KTB* 7:5–6.[195]

tudīqum (*d/tudiqqum*?), only in TC 3, 49:30, DU-*dí-qam* ... *ša* 3 *inammitim alqe*, "I acquired a *tudīqum* measuring 3 cubits", perhaps a textile product if to be connected with the verb *edēqum*, "to don a garment".

z/ṣ/sirum (quality of the first consonant uncertain), attested in KTS 2, 17:20', 1 túg ZI-*ru-um*, in damaged context, after other textiles, and in Kt 94/k 1023:8 (courtesy of Larsen), 2 túg ZI-*ri* 1 túg *abarnīum*. This textile, plural ZI-*ra-tum*, is known from Old Babylonian texts from Susa and from Mari, where, in ARM 18, 47:1, 48:1, and 21, 338:1, the words "x *ziratum lawû*" head identical lists with pieces of apparel presumably meant for one person, which suggest a use for wrapping or packing goods; see Durand 1983, 450 note 13.

[192] The mention of "counter value /equivalent" suggests that the situation was not one of normal sale in Anatolia, but that they were taken over by another trader (line 8, "I gave it to you"), who paid a standard price on receipt and it may even have been a preliminary payment, since line 16 mentions (in broken context) "the outstanding claim (for payment of) your *takkušta'u*-textiles".

[193] *Ina šīm ṣubātim ša ana aššitišu tí-sà-ba-am iš-ú-mu-ú.*

[194] CCT 5, 12:8–9, 14 *sapdinnū* šà-ba 4 *tisābū*, Kt f/k 117:5–6, 10 túg^{há} *sapdinnum* šà-ba 2 túg *tí-sà-bi*$_4$*-im*.

[195] Additional occurrences in TPAK 1, 200:13, a debt consisting of 13 shekels of silver and 2 *tisābū*, and in Kt 93/k 253:33.

[Durand 2009, 140–141, documents its use as saddlecloth (on a donkey, a wagon or a litter), for wrapping and packing, and for making a coat, *z/s. ša nalbaši*]. See CAD Z, 137, *s.v. zīru* B, which suggests a possible link with the adjective *zēru*, "braided, plaited".

3.4. *Qualifications*

3.4.1. *Indications of quality*

The quality of textile products is indicated by the following adjectives, to which are added those attested in Ur III texts, both for wool and for woollen textiles, especially for the more expensive textiles called guz-za, nì-lám and bar-dul$_5$:

Old Assyrian	Ur III[196]
a) *ša (lubuš) šarruttim*, "royal class", "of royal wear"	šàr (lugal)
b) *damqum watrum* (sig$_5$-diri), "of very fine quality"	(sag)
c) *damqum* (sig$_5$), "of fine quality"	sig$_5$ / ús-šàr
d) *tardium*, "of next good quality"	ús-sig$_5$ / 3-kam-ús
e) *ša qātim / qātum*, " of normal/current quality"	gin
f) *maṭium*, "of inferior quality"	(murgu$_x$ / egir$_x$)

Ur III only knows a), the quality sag, "top quality", is not used, but occurs e.g. with túg-nì-lám-ma in Hh 19, RS Forerunner (MSL 10, 74ff.), alongside ús, but the standard recension in lines 114–118 distinguishes túg-nì-lám bàn-da lugal, sag (equated with *rēštû*) and ús (=*terdennu*). In Mari, textiles called *uṭublum* and si-sá are distinguished as sag and ús, and on the Acemhöyük bullae (Karaduman 2008, 283–285) *raqqatum*, *sakkum* and *baratû* are qualified as sag, a *sakkum* once as ús (Ac, i, 903). At Šušarrā we find the sequence sig$_5$-ga – *terdennu* (Eidem 1992, no. 138:1–6), and at Tell Rimah sag – ús (*OBTR* no. 70:6–7). Instead of sag one occasionally uses *bērum*, "select" (= igi-zàg/zag-ga, frequent with wool), e.g. ARMT 22, 139:7, túg-guz-za *bé-rum*; 23, 375:1, *uṭuplu be-ru*). In general the quality depends both on the type of weave and the quality of the wool and therefore wool qualified as ús yields a textile qualified as ús (ARMT 23, 376:1–2).

a) ***ša (lubuš) šarruttim***: nearly always in combination with b); – CCT 5, 44a:4–5 ("Akkadian textiles"); Kt m/k 8:22–23 (*kutānū*, courtesy of Hecker); Kt n/k 533:17–19 (*ṣubātū lu lubūšum paṣium lu abarnīum*) – but alone in BIN 6, 23:7 (*ṣubātī* 5 *ša šarruttim*).[197] A variant *of ša šarruttim* is the nisbe *ekallium*, "of 'palatial' quality" (AKT 4, 28:7 and Kt c/k 676:21', said of sandals, courtesy of Dercksen). A textile (túg *šušinnu*) qualified as lugal, appears at Šušarrā (Eidem 1992, no. 134:1).

b) ***damqum watrum***: several times of *kutānū*, but occasionally also of Abarnian (BIN 4, 185:3–4, CCT 5, 44a:1–2) and Akkadian textiles (CCT 5, 44a:4–5), of *burā'um* in BIN 4, 160:12, of *nahlaptum* in Kt 87/k 378:16 (courtesy of Hecker, + *naribtum*, "soft"), of *namaššuhum* (^{túg}nì-lám) in AKT 4, 24:1–3 and CCT 5, 44a:2–3, of *pirikannum* in Kt 94/k 364:15–16 (courtesy of Larsen), of *tisābum* in Kt 94/k 1373:1–19, and of unspecified *ṣubātū*. Note also Kt m/k 35:11–12 (courtesy of Hecker): 11 *ṣubātī damqūtim watrūtim qaqqad ṣubatī*, where "extra fine textiles" are called "top textiles". "Extra fine" of course means expensive, cf. the order in Kt 94/k 729:12–15 to send from Aššur *kutānū* of extra fine quality that cost at most 12 shekels of silver apiece (*ša* 12 gín-ta kù-babbar *ù*, *šapliš*).

[196] See Waetzoldt 1972, 47–48, who distinguishes for wool an older system (with c, d, ús, and e) and a younger one (with a, c, 3-kam-ús, 4-kam-ús, and e). Occasionally 5-kam-ús is added before e) and in Lagaš (Gudea) šàr may alternate with or appear alongside sig$_5$.

[197] Note its use of iron, in Kt 94/k 1455:26–27 (courtesy of Larsen): *ašiam zakku'am ša šarruttim*, "pure, top class iron".

c) ***damqum***: frequent, used of unspecified *ṣubātū* and of *abarnīum, kutānum, raqqutum* (Bursa 3773:5, Kt 94/k 131:5–6), *kusītum* (TPAK 1, 173:6–7, KTS 2, 25:16), *šulupka'um* (Kt 94/k 131:5–6), *kitā'um* in CCT 4, 44b:17–18, *nibrārum* in Kt 91/k 372:6, *sapdinnum* in KTS 36c:9 and Kt k/k 29:4, *kuš(š)atum* in OIP 27, 11:4. Note the reference Chantre 10:6–8: x túg sig_5 thereof ...... 4 $túg^{há}$ *ša liwītim* and passim in the sequence "x túg sig_5 y túg".

d) ***tardium***: "of next, following (quality)",[198] presumably after *damqum,* cf. the sequences sig_5 – ús and sag – ús in various texts (Mari, Acemhöyük, Tell Rimah, see above). In Šušarrā (see above) and at Nuzi (see CAD T, 227, b, 2'), we have the syllabic spelling *ter/tdennu*. In Ur III in the combinations ús-šàr and ús-sig_5 the relative notion ús, "following after", is specified and this yields a continuous series: šàr, ús-šàr = sig_5, ús-sig_5=3-kam-ús, 4-kam-ús, 5-kam-ús. What *tardium* means can become clear when it occurs in a sequence, especially after *damqum*, as in 1 *kutānum damqum* 1 *kutānum tardium* 1 *raqqutum* 1 *ša Akkidīē* (Kt 94/k 981:1–3, courtesy of Larsen), in "164 textiles, thereof 20 textiles of good quality, the others (*allīūtum*?)[199] *tardiūtum* (Kt 93/k 765:12–14), and in "400 textiles, thereof 25 of good quality, including the Abarninan ones, 18 textiles *tardiūtu* ... 17 *maṭiūtu* (Kt c/k 443:5–8). However, from "603 *kutānū*, thereof 100 *kutānū tardiūtum*" (Kt 94/k 1687:1–4) we cannot simply conclude that the other 503 pieces were of good quality. And "82 textiles of good quality shipped by A., thereof 24 *tardiūtum*, packed in 18 bags ... 94 textiles of good quality shipped by L., thereof 25 *ṣubātū tardiūtum* (and) 10 textiles for wrapping, packed in 18 bags" (Kt 91/k 106:1–6, 9–15), too is unclear. These last occurrences may indicate that many of the good quality textiles could include a number of textiles which were of less good quality, but still too good to be qualified as "current/standard quality" (*ša qātim*). This may explain the use of *tardium* as an independent mark of quality, as happened in Nuzi, where *tertennu* figures as a noun without gender congruence, e.g. in *iltennūtu hullānu tertennu damiqtum* (HSS 19, 79:16), "one *nahlaptu tertennu* (HSS 15, 201:13, cf. CAD T, 227b, 2').

e) ***ša qātim***, "of the hand", "current, normal quality", also abbreviated to simple *qātum*, used as an apposition with case congruence, e.g. in BIN 4, 221:6–8, 110 *kutānī qá-tám* 8 *kutānī damqūtim*, and CCT 6, 3a:22–23, *ṣubātī damqūtim qabliūtim ù qá-tám*.[200] When used in a sequence it always occurs alongside *damqum*, cf. VS 26, 53:10–11, *POAT* 28:20–22, etc. Note BIN 4, 65:16–17, *ina damqūtika* 2 *maṭiūma allibbi ša qātim nad'*[*ū*], "of your good quality textiles 2 pieces are of less quality and they have been added to those of normal quality". Old Assyrian *ša qātim* is comparable to Ur III/OB gin(= *alākum*), which has the notion of "being current", cf. *mahīrat illaku*, "the current rate of exchange". Cf. also síg-gin, "normal/ current wool", alongside síg-igi-sag-ga, "selected wool".

f) ***maṭium***, "lacking in quality", a relative notion, rare in enumerations, but 16 *kutānu damqūtum* 10 *kutānū maṭiūtum* (Kt b/k 198:14–15) and 400 *ṣubātū* šà 25 *ṣubātū damqūtu* ... 18 *ṣubātū tardiūtu* ... 17 *maṭiūtu* (Kt c/k 443:5–8). More frequently used to single out some textiles that are below standard and are included in a

[198] In Old Assyrian, the adjective is also used of persons, "second son of" (Kt 94/k 1233:16, DUMU A. *ta-ar-dí-im*) and of objects, to refer to their size (*Kuliya* 59: 2, 4, used of wooden beams and boards, after *rabium*, "big").

[199] We doubt whether in this sequence *a-li-ú-tum* is a nisbe, "of the city (of Aššur)". See for the meaning "others", *EL* 274B:14–15, where a committee of three "outsiders" (*ahiūtum*) has to accompany persons to inspect the archive of a dead trader. If some of those asked to do so refuse, *alliūtum errubū*, "others will enter". We do not accept the interpretation in *EL* I, p. 310, note a, followed by CAD A/1, 210, 1, a, and 390, s.v. **alû*, a), but identify the adjective with *allû*, used as "the other" in Nuzi texts, cf. CAD A/1, 358 *s.v.* b.

[200] If followed by "of PN", this *qātum* can be wrongly interpreted as "the share of PN" (equated with *qāti* PN), e.g. in BIN 4, 185:7–8, 1 *ṣubātum qá-tum ša* PN, but its meaning is clear from lines 1–3, 162 *ṣubātū qá-tum* 20 *ṣubātū damqūtum watrūtum*, "162 textiles of standard quality, 20 textiles of extra good quality". In Mari the qualification šu/*qātim* of the textile *šušippum* (ARMT 21, 318:12 alongside ARMT 22, 324 III:51) has again a different meaning, referring to the part of the body this strip of textile has to cover, as shown by *šušip birkim* (ARMT 22, 324 III:50), "for the knee" [see now Durand 2009, 117–120].

lower category, cf. BIN 6, 65:16–17, quoted under e). They are used for various purposes, such as paying taxes, making a deposit in the *kārum*, or for wrapping (Kt 94/k 848:3–4, courtesy of Larsen, "20 *kutānū*, including 12 *kutānū maṭiūtim* for wrapping"), but they still could be sold: "The rest (of them), 3 pieces of less quality among them, were sold at 16 shekels apiece" (Kt u/k 3–9–11, *šitti* 3 *ṣubātū maṭ'ūtišunu*). Note also "I took 2 *kutānū*, the less good ones have been sold for 36 shekels of silver" (CCT 4, 14a:8), and TC 3, 73:337, "he took one textile of good quality and the one of less quality he rejected" (1 túg sig_5 *ilqema maṭi'am iddi*).

In this list we have not included:

g) ***qablium***, a nisbe derived from *qablum*, "middle", whose meaning is not clear. It might refer to quality, "of middle/mediocre quality", coming after "good", or refer to the size of a piece of textile. Textiles with this qualification are among the textiles bought in Aššur for shipment to Anatolia: TC 3, 69:16: "for half of the silver arriving in Aššur *kutānū qabliūtum* must be bought"; CCT 4, 46a:8–9, "x tin, 50 *kutānū qabliūtum* and 2 donkeys". The statements in Kt c/k 158:15, "your textiles are *qabliū* as to ..., they do not appeal to me" (*ṣubātūka šu-wu-ur qabliū ēnī lā mahrū*), where the meaning of *šuwur* is unknown, and Kt 93/93:2–3, "I have looked for *ṣubātī qabliūtim* whose inside is perfectly finished" (*ṣubātī qabliūtim ātamarma ša qerbam šalmūni*) are interesting, but not explicit enough. Since *qablium* expresses a relative notion, we have to look at enumerations,[201] where we see the following sequences: *damqum, qablium, ša qātim/qātam; damqum (watrum), tardium, qablium; damqum watrum, qablium, ša Akkidīē; damqum, Abarnīum, qablium;* and *Abarnīum, qablium, ša qātim*. It is difficult to draw a conclusion from them, unless one takes Kt 93/k 304:5–9 literally: "81 *kutānū qabliūtum* including those for wrapping, of which 20 are *kutānū* of good quality", where *kutānū* of good quality seem to range under *kutānū* of *qablium* quality, which would only make sense only if *qablium* referred to size or style and not to quality. However, the sequences *damqum - tardium - qablium* (2x) and *qablium - ša qātim/qātum* (2x) point in the other direction. Note also Kt m/k 9:4 (courtesy of Hecker) where 14[0 *k*]*utānī qá-*[*áb-li-ú-tim*], if this restoration is correct, are identical to 140 *kutānū wasmūtim* in the parallel text Kt m/k 8:5 (courtesy of Hecker).

h) ***wasmum*** (also *usmum*), "fair, proper", "of decent quality", a positive qualification attested a dozen times. Also used of a packet of tin, *šuqlam wasumtam*, probably referring to its full weight (ideally 65 minas) rather than to its quality. Used alone: Kt m/k 8:5, 140 *kutānu wa-as-mu-tim*; TC 3, 161:4, 1 túg *ús*!*-mu-um* (*akkārim labbušim*); CCT 6, 25d:1–2, [x+]1 *kutānū wasmūtum* cost 100 shekels of silver; Kt 93/k 288:23, *ahamma* 20 túg *ús-mu-tim*; Kt 94/k 1675:21–22, *pirikannū narbūtim wasmūtim*. In a sequence, after *ša qātim*: Sadberk 11:6–7, 97 *ṣubātī ša qātim u* 47 *ṣubātī wasmūtum*, but it also qualifies *ša qātim*: Kt 94/k 415:6–7, *kutāni ša qātim wassumūtim*; perhaps also Kt 94/k 503:21 (courtesy of Larsen), *kutānī ša qātim ú-sí-mu-tim*. Alongside *damqum (watrum)*: Kt 93/k 497:8–9, 10 *ṣubātū wa-as-mu-tim u* 10 *ṣubātī damqūtim*; Kt 93/765:21–22, 20 túg *wa-sú-mu-tim u* 3 *kutānī* sig_5 diri; Kt m/k 22:9–10 (courtesy of Hecker), 20[+x *kutānī*] *ús-mu-tim* 5 *ku*[*tānī*] sig_5 (cf. lines 1–3: 22 *kutānī* sig_5 25 *kutānī ša qātim*). This suggests the meaning "slightly better than *ša qātim*, but not *damqum*", and Kt m/k 9:4, quoted at the end of g) may indicate that *wasmum* can be the same as *qablium*, "of medium quality". The adjective is also used of the Anatolian *pirikannū*, in Kt 94/k 1675:21–22, *pirikannī narbūtim wasmūtim*, "soft *pirikannū* of fair quality".

---

[201] 30 *ṣubātū* [*damqūtum*?], 20 *qabliūtum*, 11 *ša qātim*, Chantre 14 rev.:2'-4"; [1]7 *ṣubātū damqūtum* 23 *ṣubātū tardiūtum* 30 *ṣubātū qabliūtum*, Kt 93/k 308:1–3; 10 *ṣubātū qabliūtum* 10 *ṣubātū ša qātim*, Kt 89/k 257:10–11; 1 *ṣubātum abarnīum* 1 *ṣubātum qablium*, 1 *ṣubātum ša qātim*, TPAK 1, 145:1–3; 21 *damqūtum watrūtum* 7 *ṣubātū tardiūtum* 4 *ṣubātū qabliūtum*, Kt f/k 40:17–19; *ṣubātī damqūtim qabliūtim u qātam*, CCT 6, 3a:22; [x *ṣubāt*]*ū damqūtum watrūtum*, [y *ṣubāt*]*ū qabliūtum* [z *ṣubāt*]*ū ša Akkidīē*, BIN 6, 90:4–6; *ina šalištim naruqqim* 4 *damqūtum* 2 *qabliūtum darkū*, LB 1268:9–11; 77 *ṣubātū damqūtum*, 6 *ṣubātū abarnīū* 32 *ṣubātū qabliūtum qādum ša liwītim*, Chantre 10:1-4. Note Kt 94/k 829:3-4, 10 *kutānū qabliūtum* for wrapping the tin.

The uncertainty about what quality *tardium, qablium* and *wasmum* denote has two reasons. These terms express relative notions and the Old Assyrian system is less rigorous than the Ur III one, where "next good quality", ús, is part of a fixed sequence, as mentioned above under d), *tardium*. In Mari, too, there are traces of such a sequential ranking, e.g. *RA* 64 (1970) 32, no. 20:4–5, 1 túg sal-la ús 2 túg-sal-la 3-kam, "one *raqqatum* of second rank quality, 2 of third rank quality", which implies the existence of "first rank quality" (sag or sig_5). Old Assyrian texts in general offer no concrete information on what a particular quality means, only the prices and the numbers provide a clue and they show - not surprisingly - that the "extremely good", "Akkadian" and "Abarnian" textiles occur in small numbers and are more expensive. In Aššur they can cost up to twice as much as normal textiles,[202] and the price of course relates to the production costs, that is, the amount of work required, the type and quantity of wool and the thickness of the yarns used for the warp and the weft, their thread count and the finishing procedures. The Ur III texts analyzed in Waetzoldt 1972 and the Old Babylonian tablet AO 7026, edited in Lackenbacher 1982, demonstrate that there were remarkable differences in the amount of labour invested in spinning, preparing the loom, weaving and in the quality and amount of yarn used.

Some Old Assyrian letters show concern for the finishing of the textiles. TC 3, 17:6–22 (see insert) gives instructions about the treatments of both surfaces or sides ("faces", *pānum*) of a woollen textile, a concern also expressed in three letters insisting that the "inside" (*qerbum*) of textiles has to be well finished (*lū šalim*).[203] TC 3, 17 probably also speaks of "striking/beating" (*mašādum*) and "shearing/cropping" (*qatāpum*)[204] the weave, the latter treatment being necessary to create a smooth, flat surface that is not "hairy" (*šārtam išûm*), which characterizes a *kutānum* assumed to be woollen cloth.[205] In the unpublished letter "Rendell" lines 5–13, Lamassī in Aššur writes to her husband in Anatolia: "As for the textiles made of wool from Šurbu, about which you wrote me, saying: 'Send me a garment to dress myself in', the garment has indeed been made, but it is now with the fuller, so that I have not yet sent it up to you".[206] This is the first reference to the activity of a fuller in Aššur – the other references to this profession refer to people in Anatolia – and it shows that the finishing treatment by a fuller was a normal procedure for such woollen fabrics. Waetzoldt writes "Das Walken der Stoffe ist bisher nur für einige Sorten belegt, doch dürften fast alle Gewebe so behandelt worden sein", because it is necessary to make woven fabrics suitable for garments.

[202] The writer of TC 3, 17:23–24 is not happy with the Abarnian textile a woman sent him and asks her to make one "like the one I wore over there" (i.e. in Aššur), but we do not know what that meant.

[203] BIN 4, 63:19–20, 5 *ṣubātī damqūtim ša qerbam šalmūni*; Kt 93/k 497:11, dealing with *ṣubātū wasmūtum* and *damqūtum*, and Kt 93/93:2–3, dealing with *ṣubātū qabliūtūm*

[204] We prefer "to shear" over "to pick off bits of wool from the surface of a textile" of CAD Q, *s.v.* 1, d.

[205] ICK 2, 299:10'f. distinguishes between 1 túg *kutānam* and 1 túg *lá qá-at-pá-am.*

[206] Lines 5–13: *aššumi, ṣubātim ša šu-ur-bu-i-a-tim, ša tašpuranni, umma attāma*, 1 túg *ana litabšia šēbilim, ṣubātum wadde, epiš, išti ašlākimma, adini ulā ušēlišu.*

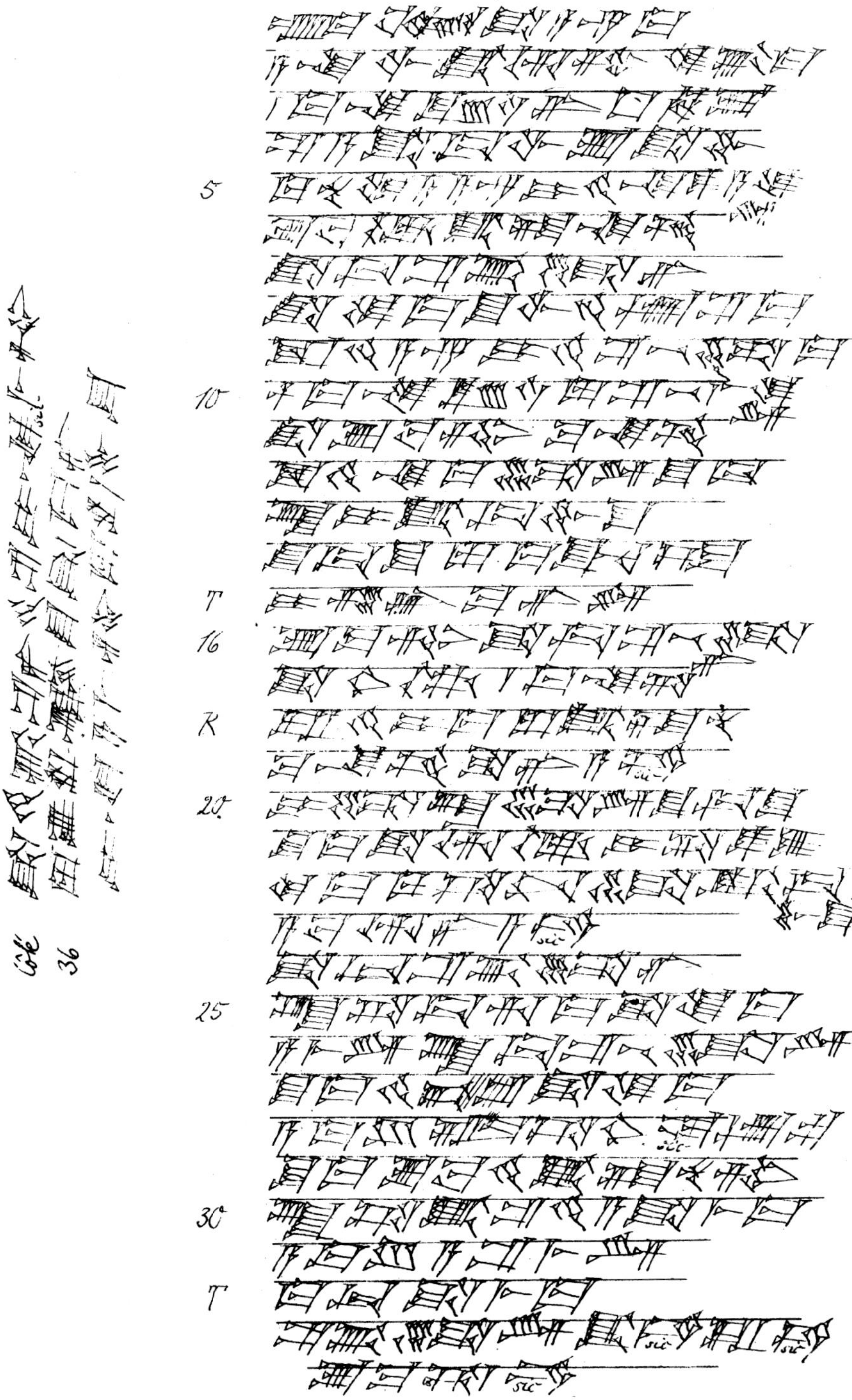

Fig. 12.2. Cuneiform copy of TC 3, 17. Source: Lewy 1935, plate XIV.

TC 3, 17

Umma Puzur-Aššur-ma		Thus Puzur-Aššur,
ana Waqurtim qibima		say to Waqurtum:
1 mana kaspam nishassu		"1 mina of silver – its excise added,
watar šaddu'assu šabbū		with the transport fee he is satisfied –
kunukkia Aššur-idī naš'akkim		Aššur-idī brings you under my seal.
ṣubātam qatnam		The thin textile
ša tušēbilinni		you sent me, such ones
ša kīma šuwati epšīma		you must make and send me
išti Aššur-idī šēbilimma		with Aššur-idī and I will send
½ mana kaspam lūšēbilakkim		you ½ mina of silver (apiece).
ša ṣubātim pānam		One must *strike* the
ištēnamma limšudū		one side of the textile,
lā iqattupūšu		and not shear it,
šutûšu lu mādat		its warp should be close.
iṣṣēr panîm		Add per piece one pound of wool
ṣubātim ša tušēbilinni		more than you used for
šaptam 1 mana-ta		the previous textile you sent me,
raddīma lu qatnū	rev.	but they must remain thin!
pānam šaniam		Its second side one must
i-li-la limšudū		*strike only lightly.*
šumma šārtam itaš'û		If it proves still to be hairy
kīma kutānim liqtupūšu		let one shear it like a *kutānum.*
abarni'am		As for the Abarnian textile
ša tušēbilinni		you sent me,
lā taturrīma ša kīma		such a one you must not
ammîm lā tušēbilim		send me again.
šumma teppišī ša kīma		If you make (one), make (it)
ammakam altabšu epšī		like the one I dressed myself in there.
šumma ṣubātī qatnūtim		If you do not manage (to make) thin
lā takaššidī ašammema		textiles, I hear that there are
ammakam aššīmim		plenty for sale over there,
mādū šāmīma		buy (them) and send them
šēbilim gamram		to me. A finished textile
ṣubātam ša tepišīni tiše inammitim		that you make must be nine
lu urukšu šamānē		cubits long and
ina ammitim lu rupuššu		eight cubits wide".

This letter (see Veenhof 1972, 104–109 and Michel 2001: no. 318) contains the most detailed information on various kinds of textiles women in Aššur made for export to Anatolia. The main problem are the instructions on the finishing of both "sides" (*pānum*), obviously the inside (elsewhere called *qerbum*, see §3.4.1) and the outside, of a "thin textile".

Qatāpum (lines 13 and 22), "to pick, to crop, to shear" ("glattstutzen"), to be applied after *mašādum*, removes raised hairs, the nap (see Lackenbacher 1982, 144 and CAD Q, *s.v.*, 1. d). It is forbidden for one side (which?) and for the other has to be done "lightly", if the fabric is still hairy (see § 3.4.1 end). *Mašādum* (lines 12 and 20), "to comb", is applied to wool and hair, not to a fabric (also not in AO 7026 = Lackenbacher 1982), although Landsberger claims this, stating that it has the same effect as *mašārum*, carried out with thorns and thistles (*OLZ* 60, 1965, col. 158, on no. 299). It seems attractive to follow AHw 623a, who starts from the basic meaning "to beat" ("schlagen, walken"), a treatment applied by the fuller in order to create the smooth surface of a cloth, typical of a *kutānum* (l. 22). Then "lightly" and "cropping" after "striking" fit, but what the technical difference with *kamādum*, "foulage à la main" (Lackenbacher 1982, 141–142) consists of is not clear. *Šumma šārtam i-ta-áš-ú*, in line 21, is translated by CAD Q, *s.v. qatāpum*, 1, d, as "if it still has loose hairs", which probably is an implicit correction of CAD K, 608, s.v. *kutānu*, d), and CAD N/2, s.v. *našû* A, 1, f), which translate "if it has (lit. has raised) a nap". Since final *-u* with a singular subject is impossible with *naša'um*, the form must be a perfect tense of *išû*, "if it (still) has", in agreement with CAD Q, *loc. cit.* Note that this text uses the word *šārtum*, "hair", primarily used for "hair" of animals and humans, in particular "goat hair", but here clearly referring to the hairs of a woollen fabric. In the only other occurrence of the word in Old Assyrian, Kt n/k 1459:27, "2 headdresses of *šārtum*" (2 *paršigē ša šārtim*), it could mean goat hair.

3.4.2. *Colours*

Old Assyrian texts offer very little information on colours. Only once (in TC 3, 69:22) a 'multicolored and dyed textile" (*ṣubātam barrumam u šinītam*) is mentioned, but the request in this letter not to buy such a piece implies that they were made and for sale, but the "caravan reports" never mention such specification for exported textiles.

Yellow, warqum/erqum. We have 3 references for *warqum/erqum*, "yellow/green": 7 *lubūšī erqūtim*, (after 6 white ones, ICK 1, 92:5), 13 túg *war*(BAR)-*qú-tim* (*RA* 58, 64, 7:5), 2 *kutānī ša hu-BA-na-ri qatnūtim u er-qú-tim*, "2 thin and yellow *kutānū* ......" (Kt 91/k 356:27–29). Note also Kt 94/k 1686:19–21, 600 *samru'ātim* šà-ba 86 *wa-ar-qá-tim*.

White, paṣium. There are about a dozen references for "white" (*paṣium*) textile products, most numerous are "white *lubūšū*" (Veenhof 1972, 164, and also KTS 2, 26:15', Kt/n k 533:170), but we also have a "white *kusītum*" (*POAT* 7:8). A reference to large amounts (80 and 20 talents) of "white and red wool", in CCT 4, 47a:30–31 applies to Anatolia, where the Assyrians were involved in the wool trade.

Red, samum. A few occurrences refer to *pirikannu*-textiles, Anatolian products, BIN 4, 162:4, 35 (40 pieces at 2/3 shekel apiece), Prag I 429:17, and TC 1, 43:24–26 ("buy red *pirikannū* ... the *pirikannū* you acquire must be red!"), but KTS 2, 35:30 perhaps mentions 6 *kutānī sà*?(copy A)-*mu-tim*. "Red wool" in CCT 4, 47a:31 is an Anatolian product, according to this text available in the towns of Durhumit and Tišmurna.
There is one reference to *makrûm*, "reddish", used of wool in OIP 27, 7:4.

Black, ṣalmum. One occurrence, AKT 2, 24:23–26, "Let them give you 100 *ṣubātū* and send them here with the lady, do not .... black ones!" [207]

Dyed, šinītum. The fem. adjective *šinītum*, "dyed", cf. CAD Š/3, *s.v.*, is used once of wool, *šīm* síg^{há} *šinītim*, BIN 4, 54:15, and it occurs a few times as a noun, a name of a textile, in TC 3, 69:22 alongside *barrumum*, "multicolored". Note also 2 túg^{há} *šiniātum* in CCT 3, 49b:24, and "He offered us 30 textiles but they are affected by the moth (16) *ù ší-ni-a-tum*; since the textiles have lost value (*batqū*) *ù ší-ni-a-tù-ni* we refused to

[207] *ṣalmūtim lá ta-sá-hi-a* in line 26 is difficult, since the verb *saha'um*, "to become troublesome, rebellious", is not construed with an accusative object and requires a preposition (*ana, aššumi*).

handle them" (Kt n/k 717:13–19, courtesy of Albayrak). The combination with damage by moth suggests a negative meaning, but it is uncertain whether the writers want to say that the effect of the moth is more damaging on dyed textiles or that such textiles were anyhow less attractive."[208]

Other adjectives that refer to the outward appearance of textiles are:

waršum, "dirty": three references in Veenhof 1972, 188, j, and one in *Or* 52, 197, no. 2:5': 66 [fine(?) *kutānū*] 7 *a*[*barnīū*], 1 túg *šilipka'um* 1 túg *waršum*.[209] Its opposite,

zakium, "pure, clean", once in BIN 4, 23:4, "I gave to K. 2 *kusītān*, 1 túg *za-ki-am* 2 *šulupka'ē*.

3.4.3. Other qualifications

ad/tmum, Cole 6:4–5, 90 túg, sig_5 *ú* 93 túg *ad/t-mu-tù-um*, meaning unclear, but perhaps the verbal adjective of the verb *adāmum*, "to invest, have a share in", rather than to be connected with níg-bàra-ga = *atmu*, a kind of "spread for the bed" (Civil 1964, 80).

anhum, "weary, old", Kt 94/k 1106:4-5 (courtesy of Larsen), "there are old textiles" (*ṣubātū anhūtum, ibaššiū*), that have to be aired (see also footnote 14).

dannum, "strong", *KUG* 29:12, 6 túg[há] *dannūtim ana nišī bētim*.

eDium, see *waDium*.

kabrum, "thick", used of wool and garments: Prag I 487:1–2, "túg GA-*ar*-ZA-*am kabram* PN is bringing you".

kabtum, "heavy" [see for Mari, Durand 2009, 104, s.v. *ṣubātum kabtum*], more than 30 references. Used as adjective with *abarnīum* (Prag I 435:11; Kt 86/k 193:15, with added *narbam*, "soft"), túg *šūrum* (CCT 3, 4:7, 4 túg *šūrūtum kabtūtum ša liwītim*, used for packing merchandise, cf. *ATHE* 51:2', 2 túg *kabtēn* [*ša liwī*]*tim*). In most cases, túg *kabtum* is used independently, as a heavy textile, according to CCT 3, 20:19–20 "for (wearing on) a wagon" *(ana narkabtim)*, and the writer of TC 2, 7:25–28 states that for lack of wool from Šurbu he will buy a "heavy textile" (*ṣubātam kabtam*) in the market in Aššur. Note AKT 3, 73:13–18, 40 túg[há] šà-ba 20 túg *kabtūtum*.....32 túg šà-ba 10 túg *kabtūtum damqūtum;* Kt c/k 110+:4–5, 35 *kutānū* 10 *šūrūtum* 5 túg *kabtūtum*; AKT 3, 65: 4–6, 94 túg šà-ba 5 túg *kabtūtum* 15 túg *tardiūtum* ..., 17–18, 11 túg *kabtūtum* 10 túg *ša qātim*. "Heavy textiles" are qualified as *damqum*, "of good quality", e.g. AKT 3, 73:18, as "of Šubarum" (6⅓ túg[há] *kabtūtim* sig_5 *ša Šubirim*, Kt 91/k 344:10–11), as *ša liwītim*, "for wrapping" (*ATHE* 51:2'), and as *narbum*, "soft" (Kt 93/k 350:10–11). Heavy garments may have been appreciated during the Anatolian winter.

kamsum, a (verbal) adjective of uncertain meaning. AHw's "etwa mit Appretur versehen" is a guess, possibly by deriving the adjective from *kamāsum* A, "to finish, complete".[210] Qualifying a textile as finished (and hence expensive), is a meaning that would fit BIN 6, 165 (see below). There are *c.*20 references, usually small numbers, and they are also produced by women in Aššur, cf. Veenhof 1972,184, d, and especially *ATHE* 31:7–8, where Pūšukēn's wife entrusted 3 *kamsu*-textiles of good quality and 7 *kutānū*, which she must have produced herself, for transport to Anatolia, and according to BIN 4, 9:24 (cf. line 6) she sent 3 *kamsu*-textiles and 5 *kutānū* there. According to BIN 6, 165:1–6 (note the sequence!), 15 *kutānū* of extra good quality, 5 *kamsūtum*, 20 *kutānu* of good quality, 20 túg *kutānū tardiūtum* and [x] Akkadian textiles" were sent to Anatolia. This suggests that *kamsum* (without determinative túg!) describes a type of *kutānum* of fine quality, less than "extremely good", but better than simply "good".

[208] There is one Old Assyrian occurrence of *ṣarpum*, "dyed", used of woollen fleeces in Kt 93/ k 915: 7–10, 1 *maškum ṣarpum* 2 *maškē šūrūtim*, " I sold him one dyed fleece and 2 dark coloured(?) fleeces".

[209] Wilcke, who edited the text, pointed out that the same lot of textiles is mentioned in CCT 4, 5a:5–6, where, however, túg *waršum* is omitted.

[210] Cf. *ATHE*, p. 46, 7; in Old Babylonian it is used once of finishing an object (AbB 3, 34:19, a reed door).

In lists, where the sequence has informative value, we find *kamsūtum* – *kutānū* (*POAT* 15:13–14), *ṣubātū ša qātim* - *kamsūtum* - Akkadian (Kt 94/ k 204:1–3), Abarnian – *kamsūtum* (Kt 94/k 218:19), Akkadian – *kamsum* – Abarnian (at 45 shekels apiece, followed by *kutānū* at 30 shekels; BIN 4, 4:3–6), *kutānū qātum* - *damqūtum* - *kamsūtum* (BIN 4, 221:6–8, shipped off from Aššur). See also Kt c/k 443:7–8 (courtesy of Dercksen): 400 textiles of which 25 *damqūtum* (including the Abarnian ones), 18 *tardiūtum*, 17 *maṭiūtu* ..., and in lines 12–13, 5 *kamsūtu* 6 *nibrārū* 2 *kusītān* 1 *raqqutum* 1 *nibrārum*. Qualifications are rare, we have *damqum*, "fine", in Kt 94/k 1446:15–17, "for a child" in BIN 4, 68:10 (*ša ṣuhrim*), and in Kt 94/k 204:1–3 they are qualified as "of Akkadian make/style" (2 túg^{há} *ša qātim kamsūtum ša Akkidīē*). Note for its relation to other textiles especially Kt 94/k 1686:36–40 (courtesy of Larsen): "Of my 100 ... *kutānu* 35 are of extra good quality 46 *abarnīū*$^{!}$ [*ka*]*msūtum ú nibrārū ša kutānī damqūtim*, in all 181 *kutānū*", where "(made) of good *kutānu*-cloth" qualifies the *nibrārū*, but perhaps also the Abranian ones of *kamsu*-quallity. However, the parallel Kt 94/k 1697:13–16 (courtesy of Larsen) gives the second series as "46 *ṣubātū*, either Abarnian ones or *kutānū* of extremely good quality or (*lu*) *kamsūtum* or *nibrārū*", where this group seems to comprise four different types of textiles and the qualification *ša kutānī damqūtim* has become a separate category!

karsum (?) meaning unknown, see Prag I 487:1–2, quoted under *kabrum*.

lahhub/pum, "?", ATHE 62:8, 3 túg *lá-hu-b/pu-tim*; Kt c/k 675:15, 3 túg *lá-hu-b/pu-tim*, followed by 1 túg *raqqutum*. From a verb *lahhub/pum*, of unknown meaning, see CAD L 239a, with the statement in ICK 1, 15:18-19, "The rest of the textiles I will *lahhub/pum* so that they can wear them (*ulahhab/pma iltabaššūniššunu*). Compare Kt 94/ k 966:17-18, whose writer asks to provide him, from Hurrama, with a set of garments (1 túg$^{\text{hi-a}}$) for his personal dress, either a *šūrum* or an *Akkadian burā'um*; "let one *lahhub/pum* and bring it into the town" (*lu-lá-hi-b/pu lušēribuniššu*).

narbum, "soft", usually said of wool (see above, § 1.1) but also applicable to textiles made of soft wool, Kt 86/k 193:15ff., túg *abarniam kabtam na*[*rba*]*m a-litabšia*, "a heavy (but) soft Abarnian textile for me to wear"; Kt 92/k 112:14–15, a *ṣubātam narbam* sent by the writer's sister *ana litabšia*, "for me to wear". "Soft" implies better quality, cf. Kt 87/k 378:16–17 (courtesy of Hecker), 2 túg *nahlaptēn damqātēn watartēn naribtēn*, "2 extremely good, soft cloaks", and does not exclude "heavy", *RA* 81, 14:8–9, "6 heavy *burā'u*-textiles, including 3 soft ones". Since *narbum* is also used of Anatolian wool, we are not surprised to meet "soft *pirikannū*": Kt n/k 1385:16 (10 pieces) and Kt 94/k 1675:21–22, "soft *pirikannū* of fair quality" (*narbūtim wasmūtim*).

qatnum, "thin", refers to the textile as such, also in Old Babylonian, including Mari (CAD Q, 174a, b), because it is made of thin threads ("thin wool", CAD, *loc.cit.*, c, attested for Old Babylonian but not for Old Assyrian). According to the locus classicus TC 3, 17:6–22 (above, § 3.4.1 end) a *ṣubātum qatnum* to be made by an Assyrian woman, who has to process 1 pound of wool more apiece than she did, should have a dense warp (*šutûšu lū madât*), but "the textiles must be thin" (*lū qatnū*, line 18). See also Veenhof 1972, 214, Excursus on the difference between *qatnum* and *raqqum*. *Qatnum* is used as an adjective in CCT 4, 48b:18, túg *raqqatam damqam qatattam*, and Kt n/k 391: 4, 1 túg *qatnum* has to yield ½ mina of silver. Note Kt 91/ k 356:28–29, 2 *kutānū ša hu-BA-na-ri qatnūtim u erqūtim*, "(send me) 2 thin and yellow *kutānu* of ......", and CCT 5, 39b:18–20, "I took 12 Akkadian textiles and from these I took 1 thin textile". *Qatnum* is used independently, without túg, in LB 1201:14–15, 1 *Abarnīum* 2 *qatnūtum* 4 *ša qātim*. Note that in § 182 of the Hittite Laws, a túg-sig is the most expensive quality.

raqqum, "thin", used as substantivated adjective and spelled ^{túg}*raqqutum* (not *ṣubātum raqqum*!), denotes a light and thin garment. Cf. above § 3.3, *s.v. raqqatum*. Occasionally *raqqum* is used as a real adjective: *CTMMA* 1, 85a:12, 2 túg bar-dul$_5$ *raqqitēn*, "2 thin *kusīātu*-garments", and *KTB* 7:4–5, 4 túg *kusīātum raqqātum* (see under *kusītum* in § 3.3.3).

šapium, "padded, thick", regularly used of wool and woollen fleeces *(maškū šapiūtum)*, and saddlecloths (*ukāpum*), but occasionally also of textiles, *JCS* 14, no. 2:18–20: "bring together the rest of my textiles, 7 *kutānū*[!] *šapiūtim ù eDiūtim*; Mixon 10:4–5, 13½ túg *šapiūtum* 9½ túg *waDiūtum.*

waDium, meaning unknown, also written *eDium*, usually alongside *šapium*, "thick", said of fleeces, saddlecloths and garments, e.g. Kt 94/k 734:51, 4½ túg *wa-dí-ú-tum*; see under *šapium* and the comments on *Kuliya*, no. 57:4.

3.4.4. *Format and size*

The adjectives "small" (*ṣahrum*) and "big" (*rabium*) in general are not used to qualify textiles. There is only one possible reference to *ṣahrum* in Kt n/k 469:2 (courtesy of Günbatti), 10 túg *ša li-tab-ší-im*, 18 túg *ṣa*[!]*-hu-ru-tum* PN *ilqe*, "PN took 10 textiles to be worn and 18 small textiles".[211] In a letter to his wife (BIN 4, 10:14–15, see below) Pūšu-kēn states that the textiles she sent him "are (too) small, are not good". "Big" *(rabium)* is used only once in TC 1, 43:16, an order that *pirikannū* to be acquired should be big (*lu ra-bu-*[*ú*]). "Long" (*arkum*), occasionally attested elsewhere (ARMT 22, 164 rev. 1', [túg-gu]z-za gíd-a; *OBTR* 80:4–5, gú-è-a *riksu* sud-a *ù la* sud-a), does not occur in Old Assyrian. This suggests that the textiles in Old Assyrian trade had standard sizes, known to the parties involved, which need not be mentioned in caravan accounts and lists. Accordingly we only have very few indications of size, only mentioned in letters for particular reasons: *a*) the request in TC 3, 17:33–38 that a finished woollen *ṣubātum qatnum* should measure 8 by 9 cubits *(tiše inammitim lu urukšu šamāni ina ammitim lu rupuššu*; see § 3.4.1, end), or *c.*4.5 by 4 m; *b*) the statement in Kt 94/kI350: 31–32 (courtesy of Larsen), *raqqatam arbē ina ammitim rupussa u ešar urukša*, "a thin textile, 4 cubits wide and 10 long", or *c.*2 by 5 m.; and *c*), in CCT 4, 44b:17–18, a reference to the size of a piece of "linen" (*kitā'um*; see above § 1.2). While the size in a) is in a request to the woman who produced the textile, apparently because such textiles sold well, the purpose of the linen that has to be bought and sent in c) is unknown and we do not know what size "linens" in general had. In b) the measures of the *raqqutum* shipped to the addressee may have been mentioned because they deviated from the standard. In this connection also the letter BIN 4, 10:14–19 is interesting (edited Veenhof 1972, 111–112), where Lamassī complains that her husband had written: "They (the textiles) are (too) small, they are not good", to which she reacts with: "Did I not reduce their size at your own order? And now you write: 'Add half a pound (of wool) to each of your textiles!' – I have done so!" This is comparable to the request made in TC 3, 17:15–18 (above, § 3.4.1) to process in each *ṣubātum qatnum* 1 pound of wool more than before. These pieces of information show that there could be important differences, but they do not reveal how much wool was processed on average, e.g. in a *ṣubātum qatnum* or a *kutānum*. If we take BIN 4, 10 at face value, processing more wool yields not only a heavier, but also a larger textile and this may have been implied in TC 3, 17 too, and be the reason why the required size is stated at the end of the letter. The adjectives used, on the one hand "thin" (*raqqum* and *qatnum*), and on the other hand "heavy" and "thick" (*kabtum, kabrum, šapium*), suggest different weights, which could be based on the nature and/or the amount of wool processed and on the thickness

---

[211] The reading of ZA is not fully certain, Günbatti writes "maybe A", which would yield *ahhurūtum*, "still due, still to be delivered", which is also unique. *Ṣahhurūtum* is the well-known plural of *ṣahrum*, a pseudo-D-stem, with an added vowel.

of the threads and density of the weave.[212] In Veenhof 1972, 89–90, considering both the data provided by texts from Ur III and Nuzi and the carrying capacity of donkeys (one usually carried 25 textiles or a few more), a weight of *c.*5 minas apiece for the most current types, *kutānū* and *ṣubātū* was suggested, but a "thin" textile (*raqqutum*) and a *kusītum*, also regularly qualified as "thin" (also in lexical texts), must have been lighter and perhaps also (see reference c), above) smaller. This is no problem since the lists of textiles shipped by caravan to Anatolia comprise only small numbers of these textiles. However, we admit that the argument derived from the carrying capacity of donkeys is not very strong, because the textiles' volume rather than their weight may have determined the amount an animal could carry.

Puzzling and disturbing, finally, is the small text Kt n/k 200 (courtesy of Bayram), which reads: 226½ *ṣubātū šuqultašunu* 7 GÚ 20 *mana*, "226⅓ textiles, their weight 7 talents and 20 minas", which yields a weight of less than 2 minas apiece. The type of textiles is not specified, but considering the large number it may well refer to those current in the trade.

4. Categorisation

Apart from the generic word corresponding to "textile", usually written with the logogram túg and less often with the Akkadian term *ṣubātum*, many of the textiles cited in the Old Assyrian tablets belong to specific categories. We have texts with enumerations of various types of textiles that list certain types together or in a particular, probably not coincidental sequence, e.g. by quality/price, in an ascending or descending order. Other enumerations may state that particular types of textiles belonged to or ranged under another, larger category, or that certain types of textiles were considered as alternatives. There are a number of references where a particular type of textile is qualified as *ša* another textile, in particular *ša kutānim*, which most probably means that such a textile was made of a particular type of fabric.[213] The study of all these combinations provides an understanding of some connections between the many textile types.

4.1. *Textile type included in another type*

The most informative combination between two categories is by means of an inclusive link, where we have "X textiles of type A, among which Y are of type B". This can be expressed either by the logogram šà-ba (*iqqerbim*), "among which", or by the Akkadian word *qādum*, "including". Also, some accounts give the total number (šu-nígin) of textiles of a certain type which comprises several other textile categories.

4.1.1. šà-ba *(ina qerbim), "among which"*

Many occurrences of šà-ba do not offer much information when the first group is simply "textiles" (túg), without any further details. For example, a document presents the following inventory: "94 textiles among which 5 *kabtūtum* textiles, 15 *tardiūtum* textiles, 30 *kusiātum* textiles, 1 *šulupkum* textile, 1 *lubušum*, 2 *nibrarān*".[214] We merely learn that all these items are textiles.

---

[212] Note that when Pūšu-kēn, in BIN 4, 10, qualifies a textile he considers "(too) small" as "not good", "good" (*damqum*) here, is not the same as when in a list textiles are qualified as "good"; "not good" meaning that they are not appreciated, do not sell well.

[213] See for example 3 *šitrim ša kutānim* (RA 59 1965, n° 14, 15) and below § 4.3.

[214] AKT 3, 65 (Michel 2001, no. 167):4–7, 94 túg šà-ba 5 túg, [*k*]*abtūtum* 15 túg *tardiūtum*, 30 túg *kusiātum* 1 túg *šilipka'um*,

The enumerated textiles may be of a specific quality. Thus a *namaššuhum* garment is counted amoung extra fine quality textiles.[215] An *abarnīum* textile can be included among extra good quality textiles.[216] An unpublished tablet lists in all 58 bags in which various textiles are packed. In the first 18 bags we find 82 textiles of good quality (túg sig$_5$) among which are 24 *tardiūtum*.[217] This contrasts with the traditional idea that *tardium* refers to items of secondary quality; it seems that *tardium* could be used as an independent mark of quality and means "of next good quality", after *damqum*, "of very fine quality" (see above § 3.4.1 d). The next 18 bags include 94 good quality textiles transported by Laqēp(um), among which are 25 *tardiūtum* textiles and 10 textiles for wrapping (*ša liwītim*).[218] Again, in addition to those called *tardium*, textiles *ša liwītim*, used for wrapping other textiles and usually cheaper, are included here among good quality pieces.[219]

A link between two textile categories, besides being an indication of quality, may suggest the geographical provenance of a particular type. Among Babylonian textiles qualified as *ša Akkidīē*, we find textiles for wrapping and a *burā'um* piece.[220] This shows that a *burā'um* could be made in southern Mesopotamia, but it does not rule out the possibility that such textiles could also be woven in another geographical area. The same observation can be made for excellent quality *kutānu*-textiles, among which some are said to be *abarnīum*.[221]

More informative are the connections between various textile categories where quality or provenance do not play a role. The *kutānum* category, which is the main type exported from Aššur to Anatolia, includes several other categories: *kusītum*,[222] *šūrum* and once *ša šadā'im*.[223] The *pirikannum* category, which is the main type of textiles produced in Anatolia, also comprises other types: *tisābum*[224] and *menuniānum*,[225] and the *sapdinnum* category also includes *tisābu*-textiles.[226] More difficult to interpret is the combination illustrated in the text CCT 5, 12a, where a bale of 14

---

1 *lubūšum* 2 *nibrārān*. See also AKT 3, 61:12–13, 1 *me'at* 3 túg^{há} : *ina* é-gal^{lim} *izkūnim*, šà-ba 66 *kutānī u* 31 túg *ša a-ki-dí-i*; AKT 2, 34:8–9, 12 túg *ša tamkārim*, šà-ba 5 *abarnīū*; Kt 93/k 765:12–13, 164 *ṣubātū* šà-ba 20 túg sig$_5$ *alīūtum tardiūtum*.

[215] AKT 4, 24:2–3, 22⅓ túg *damqūtum watrūtum*, šà-ba 1 túg nì-lám, "22⅓ textiles of extra fine quality, among which one *namaššuhum*", shipped from Aššur to Anatolia (see above §3.2. *s.v.* túg-nì-lám).

[216] Kt 93/k 288, 5–6: 8 túg^{há} sig$_5$ diri šà-ba, 1 túg *abarnīum*.

[217] Kt 91/k 106, 1–5: *ina* 82 túg^{há} sig$_5$, *ša šēp Ali-abim*, šà 24 *tardiūtim*, *ina* 20 lá 2 *naruqqātim*, *darkū*. The same remark applies to Kt 93/k 304 where *qablium* garments, usually translated as "medium quality textiles", include good quality *kutānums* l. 5–8: 81 *kutānū*, *qablīūtum*, *qadum* : *liwītim*, šà-ba 20 *kutānū* sig$_5$^{tum}.

[218] Kt 91/k 106, 9–15: 94 túg sig$_5$ *ša šēp*, *Laqēpim*, šà 25 túg *tardiūtum*, 10 túg *ša liwītim*, *ina* 20 lá 2, *naruqqātim*, *darkū*. The same observation can be made from Chantre 10, 5–7: šu-nígin [1 *me'at*] 15 túg *ša* dam-gàr, 92 túg^{há} sig$_5$ š[à-ba] 4 túg^{há}, *ša liwī*[*tim*].

[219] See also *EL* no. 123:4 and 6 where *ṣubātū ša liwītim* are part of a large group of textiles of good quality (*damqum*) and Kt 93/k 304:5 81 *kutānū* of medium quality, *qādum ša liwītim*, of which 20 are *kutānū* of good quality.

[220] Textile *ša liwītim*: CCT 5, 36a:18–20, 128 túg *lu šūrūtum*, *lu ša Akkidīē*, šà-ba 24 túg *liwītim*. *Burā'um* textile: AKT 2, 44:3–5, 4 túg^{tí}, *ša Akkidīē*, šà-ba 1 *burā'u*[*m*].

[221] Kt c/k 174 (courtesy of Dercksen): 6 túg *kutānū* sig$_5$^{tum} *watrūtum* šà-ba 3 *abarnīū*.

[222] AKT 4, 23:1–2, 21 túg *kutānī*, šà-ba 5 *kusiātim*; Prag I 616:1–2, 18 túg^{há} *kutānū*, šà-ba <x> túg *kusiātum*, followed by few pieces of various types of textiles and we do not know if they are counted among the 18 *kutānū*, l. 3–8, 4 túg *ša ṣuhrim*, *ša Akkidim*, 3 *šitrē*, 5 *išrātim*, 2 *raqqātim*, 2^{šitta} *nahlapātim*; if we suppose that the missing number of l. 2 is <2>, then we obtain a total number of 18 textiles. Kt c/k 174 (courtesy of Dercksen): 175 *kutānū* šà-ba 12 *kutānī qabliūtum* 5 *kusiātum ša Akkidīē* 3 *šūrūtim*.

[223] Kt 87/k 434:1–3 (courtesy of Hecker), 23 túg *kutānī*, šà-ba 5 túg *ša ša-da-im*, 18 túg *šūrūtim*.

[224] Kt 93/k 59:1–3, 9 *pirikannī* /sig$_5$, *ša Zalpa*, šà-ba 3 *tisābū*; Kt 93/k 891:5–7, 9 túg *pirikannī*, *ša Zalpa kunukkia*, *addinakkum* šà-ba 3 *tisābū*; Kt 93/k 60, 1–3: 34 *pirikannī*, *ša Haqqa* šà-ba 12 *tisābū*.

[225] Kt 93/k 522:25–27, 71 túg *piri*[*kannī*], šà-ba 30 túg *menuniāē*, 41 túg *ša Tuhpiya*.

[226] CCT 5, 12:8–9, 14 *sapdinnū* šà-ba 4 *tisābū*; Kt f/k 117:5–6, 10 túg^{há} *sapdinnum* šà-ba 2 túg *tí-sà-bi$_4$–im*.

sapdinnū includes 4 *tisābū* (and?) 2 *kita'ātum pālilū*. We know that *tisābu*-textiles may belong to both the *pirikannum* and *sapdinnum* categories, which consist of woollen textiles, but it is impossible to decide whether the *kitā'um pālilum*, which should be translated as "linen being/serving as *pālilum*", could also belong to the *sapdinnum* type.[227] Perhaps it is better to suggest that only the 4 *tisābū* textiles belong to the *sapdinnum* category, and that the 2 *kita'ātum pālilū* are counted apart, but were added to the same donkey load.

4.1.2. *Qādum "including"*

The same remarks apply to the less common expression "type A *qādum* (including) type B". Beside the most common expression *qādum ša liwītim* "including those for wrapping"[228] and the examples referring to túg in general, we find combinations of specific categories with qualities[229] or with geographical provenance: Akkadian textiles including dirty (*waršūtum*) textiles,[230] *šūrum* ranging among Abarnian textiles[231] or *raqqātum* textiles including *šilipka'um*.[232] More interesting is the reference to *kutānu*-textiles of *qablium* quality including textiles for wrapping (<*ša*> *liwītim*).[233]

4.1.3. *šu-nígin (ištēniš) "total amount"*

Contrary to the situations outlined above where first the total number of textiles is given, followed by a specification about some categories included in this total, many documents list the number of textiles belonging to each category separately and sum up all the textiles (šu-nígin), thus again combining several types.[234] In such a combination, textiles made of wool from Šurbu (see § 1.1) belong to the *kutānum* type.[235] *Lubūšu*-garments, *kusītu*-textiles and two types named by a nisbe, *šilipka'u*- and *takkušta'u*-textiles, are totalled as Akkadian textiles.[236] This combination of two different geographical qualifications is difficult to understand: one name could refer to the real provenance while the second to the technique used to weave the textile or to its shape if dealing with a garment. Again the *kutānu*-textiles appear as a large category containing several others: *šūrum* and *takkušta'um*[237] or *abarnīum kamsum* and *nibrārum*.[238]

To sum up, there are two main textile categories which include several others: the *kutānum* type, made in Upper Mesopotamia and the Anatolian *pirikannum* type. Two terms which were

---

[227] CCT 5 12a:6–10, 50 lá 1 túg *pirikannī*, *ša Kaneš*^{ki}, 14 *sapdinnū* šà-ba, 4 *tisābū* 2^{šitta}, *kita'ātum pālilū*. For the categorisation of Anatolian textiles, see above § 3.3 *s.v. tisābum*.

[228] See for example LB 1268:9–11, 77 *ṣubātū damqūtum*, 6 *ṣubātū abarni'ū* 32 *ṣubātū qabliūtum qādum ša liwītim*.

[229] kt c/k 443:6 // Kt c/k 449:4 (courtesy of Dercksen): 25 túg sig$_5$ *qādum abarniē*.

[230] CCT 5, 36a:26–27, 3 túg *ša Akkidīē*, [*qā*]*dum waršūtim*.

[231] BIN 6, 60:13–15, "We counted 85 textiles, thereof 24 Abarnian ones, including (*qādum*) one *šūru*-textile".

[232] VS 26, 11:26–30, "110 textiles, thereof 40 *kutānū*, 11 *takkušta'ū* and 11 *raqqātu*, including 2 *šilipka'ū* (*qādum šilipkên*); this could imply that the latter were a specific type of "thin garments".

[233] Kt 93/k 304:5–7, 81 *kutānū*, *qabliūtum*, *qādum* : *liwītim*.

[234] CCT 5, 36a:1–9, 1 *me'at* 57 *kutānū*, 43 túg *šūrū*<*t*>*um*, 4 túg *kutānū ša ikribīšu*, 2 túg *namaššuhū*, šu-nígin 2 *me'at* 6 túg *ša abini*, 43 túg *kutānū* 8 túg *abarnīū*, 32 túg *šūrūtum* 25 túg *ša abuni iddinanni*, 34 túg *ša Akkidīē*, šu-nígin 1 *me'at* 42 túg *ja'ūtum*.

[235] TC 2 14:6–9, 27 túg sig$_5$ *Šu-ur-bu-i-*<*ú*>*-tum*, kù-bi 3⅓ *mana* 1 2/3 gín, 63 *kutānū* kù-bi, 6 *mana* 1⅓ gín šu-nígin 90 *kutānū*.

[236] Yale 13092 (unpubl. courtesy of Larsen), mentions textiles bought in Aššur, 8–12: 7 *lubūšū*, 6 *kusiātum*, 1 *šulupka'um*, 1 *takkuštûm*, šu-nígin 15 *ša A-ki-dí-*NI*-im*.

[237] CCT 5 36a:10–17, 38 túg *ša Ahā*, 45 túg *lu kutānū lu šūrum*, 7 túg *takkušta'ū*, *ša Ali-ahim*, dumu *Enah-ilī*, 5 túg *ša kārim*, 14 túg *ša Ali-ahim*, 6 túg *ša Laqēp*, šu-nígin 3 *me'at* 35 túg *kutānū*.

[238] kt 94/k 1686:36–39 (courtesy of Larsen), *ina* túg^{há} 100 *kutānī* [*tardi*]*ūtim*, 35 *kutānī* sig$_5$-diri 46 *abarnīē* [*ka*]*msūtum u nibrārū ša kutānī* sig$_5$ *watrūtim* šu-nígin 181 *kutānū*.

considered as referring to medium or second quality textiles, *qablium* and *tardium*, may rather belong to good or top quality textiles. We can imagine, for example, that *qablium*, "medium", does not refer to quality but to size and would indicate a medium sized piece (see above § 3.4.1. g). This explanation is not completely satisfactory because *kutānum qablium* can be used for wrapping, and one would imagine that this operation requires larger size textiles. Finally, categories defined by a nisbe or geographical name do not necessarily refer to the provenance of the textile, but it could indicate a shape, or a special weaving technique.

4.2. Alternative textile types

Certain types of textiles were considered alternatives, either by the Akkadian expression *lu ... u lu*, or they were interchangeable in parallel documents.

4.2.1. *Lu ... u lu*

The expression "(x textiles a,) either textiles b or textiles c", used sometimes, implies that some categories could be exchanged and thus are considered equivalent. The choice may occur between textiles of two different geographical origins. For example, a *kusītum* garment may be either *(m)alkuaīum* or *šilipka'um*;[239] a *nibrārum* garment could come either from the Šubarum, from Aššur or be exchanged with an Akkadian *kusītum*.[240] Most often, we observe that a "geographical" type and another category can be alternatives. Among textiles exported from Aššur to Anatolia:

- *kutānu*-textiles may be either *šurūtum* or Akkadian,[241]
- good quality textiles may be either *kutānum*, *abarnīum* or Akkadian,[242]
- *raqqutu*-textiles may be either *lubūšum* or *takkušta'um*,[243]
- a *lubūšum* from Susē and a *kutānum* of fine quality are alternatives for an extra fine Abarnian textile to be worn by a man.[244]

Among the Anatolian textiles:

- textiles may be either *menuniānum* or from Tuhpiya,[245]
- *tisābum* from Hahhum may alternate with *sapdinnum* from Talhat.[246]
  Since we do not know the criteria along which the Assyrian merchants distinguish or compare these categories, it is a difficult task to choose between these alternatives. The trade being the main purpose of this documentation, it is most probable that the first criterion is the commercial value of the textiles (Kt 93/k 344:9–21).

4.2.2. *Parallel texts*

Some caravan accounts were written in several copies to be kept by the sender, the transporter and the recipient of the merchandise. In few cases, we observe small variations between

239 TC 3 169:10–12, 1 túg *kusītum, lu alkuaītum, ù lu*[!] *šilipka'um.*
240 Prag I 686:19–21, *nibrāram, lu ša Šubirim lu ša ālim*[ki], *lu kusītam ša Akkidīē.*
241 CCT 5, 36a:18–19, 1 *me'at* 28 túg *lu šūrūtum, lu ša Akkidīē.*
242 CCT 4 29b:3–4, túg[há] : sig_5 *lu* túg *kutānū, lu abarnīē lu ša Akkidīē.*
243 Kt 93/k 344:19–20, 18 túg *raqqātum lu lubūšū*[!], *lu takkušta'ū.*
244 AKT 2 24:4–5, 10–11, 1 túg *abarnīam*, sig_5 *watram ... lu lubūšam suseīam, lu* túg *kutānam* sig_5.
245 Kt 93/k 517:21–22, 55 túg *lu menunianū lu ša, Tuhpiya.*
246 Kt 94/k 1672:19–20, *lu* túg *tisābī ša Hahhim, lu sapdinnī-ma ša Talhat.*

duplicates, either in the number of items counted or in the name of the item. For example, the three parallel documents Kt c/k 443, 449 and 458 (courtesy of Dercksen) show the following variations (underlined).

Kt c/k 443:12–14	Kt c/k 449:7–9	Kt c/k 458:9–13
5 túg *kamsūtum* 6 túg ***nibrārū***	**6** túg *kamsūtum* 6 túg *namaššuhū*	5 túg *kamsūtum* 6 túg *namaššuhū*
2 túg *kusītān* 1 túg *raqqatum*	2 túg *kusītān* 1 túg *raqqutum*	2 túg *kusītān* 1 túg ***a-ra-qá-tim***
1 túg *nibrārum*	1 túg *nibrārum*	1 túg *nibrārum*

The 6 *nibrārū* textiles listed in the first text are changed into 6 *namaššuhū* textiles in the two other documents. Either it is a confusion made by the author of the document, or those two types are in fact very alike. Note also the variation between 1 túg *raqqutum* into 1 túg *a-raqqatim*, "1 piece of textile for a *raqqatum*", which expresses a purpose rather than a textile type.

Another example is given by two texts excavated in 1994 (courtesy of Larsen).

Kt 94/k 1686:3–6, 8–11, 36–40	Kt 94/k 1687:1–7, 10–11, 13–15
603 *kutānū*, šà-ba 100 *kutānū tar<di>ūtum*,	603 túg *kutānū*, ... ***ina qerbim***, 100 túg *kutānū*, *tardiūtum*,
ahamma 35 ***kutānū*** sig$_5$ **diri**,	35 **túg** sig$_5$ ***watrūtim***,
šà-ba 3 *kutānū* ...	***ina qerbim***, 3 *kutānū* ... **šu-nígin 600,** ***u*** **38 túg** ...
ahamma 46 túg *abarnīū*, *lu kamsūtum*	*ahamma* 46 túg ***lu*** *abarnīū*, ***lu*** *kutānū* sig$_5$ *watrūtum*,
lu nibrārū, ***ša*** *kutānī* sig$_5$ *watrūtum*,	*lu kamsūtum lu nibrārū*
šu-nígin 684 túg^{há} ...	-----------------

(again l. 36–39):

ina túg$^{\text{hi-tí}}$-a 100 *kutānī*, [*tardi*]*ūtim*
35 ***kutānī***, sig$_5$ **diri** 46 ***abarnīē***
[*ka*]*msūtum*, ***ú*** *nibrārū* ***ša*** *kutānī* sig$_5$, *watrūtum*
šu-nígin 181 *kutānū*

The interpretation of these two tablets is not clear. In lines 8–11//13–15, should we read:

> "*abarnīū* textiles either *kamsūtum* or *nibrārū* of good quality *kutānū*",
> "*abarnīū*, *kamsūtum* and *nibrārū* of good quality *kutānū*",

or should we understand:

> "*abarnīū* textiles, either good quality *kutānū*, or *kamsūtum* or *nibrārū*"?

This example demonstrates the complexity of such statements. In the first two translations, the *nibrārum* type may be made of *kutānum* fabric, but not in the third one. In addition, according to the first and the third interpretations, *abarnīum* is the broad category that comprises *kamsūtum*, *nibrārum* and perhaps *kutānum* textiles, but not in the second version.

4.3. *Textile "ša" another type of textile*

Another very informative categorial link is provided by the expression "textile A *ša* textile B", which we may understand as "textile A made from/in the style of textile B". One of the two items again may be a geographical name or nisbe:

- *kusītum ša Akkidīē*[247]
- *burā'um ša Akkidīē*[248]
- *nibrārum lu ša Šubirim lu ša Ālim*,[249] *ša Akkidīē*[250]
- *šitrum ša abarnīē*,[251] *ša Akkidīē*,[252] *ša Ālim*,[253] ša Zalpa.[254]

Several textile types can be produced in different places or made in different shapes or techniques, but it does not help to identify each category.

When the word combinations do not include a toponym, one of the main issues is the question of whether the items mentioned were textiles or garments. For example, the *šitrum* is said to be of (*ša*): *kutānum*, *sapdinnum*, *pūkum* and *lubūšum*.[255] We can imagine that the *šitrum* was a piece of clothing that could be made of several types of textile that are different as regards to their material or weaving technique. This definition nicely fits the first three types, *kutānum*, *sapdinnum*, *pūkum*, but not the last one, *lubūšum*, which has usually been interpreted as a piece of clothing. In this case, a *šitrum ša lubūšim* may be translated "a *šitrum* belonging with/of the same type as a *lubūšum*".[256]

Both *nahlaptum* and *namašuhhum* are made of *kutānu*-fabric and must correspond to garments.[257] The *nahlaptum* is usually translated as an outer garment, a coat or a shirt (see above § 3.3 *s.v.*). The case of the *nibrārum* type is more complicated. It belongs either to the categories of well-known Anatolian textiles, *pirikannum* and *sapdinnum*,[258] or it can be made from a woollen *kutānum* fabric, usually produced in Aššur.[259]

5. Textiles or garments?

5.1. *Context, names and fractions of textiles*

Did the Assyrians trade in woollen fabrics in the shape of large sheets or textiles of standard sizes or in ready-to-wear garments? It is a more general problem when dealing with the names of ancient textiles, also encountered in connection with the texts from Mari, where it has been clearly formulated by Durand 1980, 394–395. The names of the "textiles" frequently are not informative enough to answer this question and we also have no pictorial evidence to help us, while information on textile production is almost completely absent. The preference for "soft" textiles (§ 3.4.3, s.v. *narbum*) and for those whose "inner side has been well finished"[260] may apply to both. Occasional information on garments worn, e.g. in the letter TC 3, 17:23–28, "As for the

---

[247] *ATHE* 46:7–8, 4 *kusiātim*, [*š*]*a Akkidīē*; Kt m/k 22:6–7 (courtesy of Hecker), 1 *kusītam ša*, *Akkidīē*.
[248] Kt 94/k 966:14–15: *lu burā'um*, *ša Akkidīē*; AKT 2, 44:3–5, 4 túg[tí], *ša Akkidīē*, šà-ba 1 *burā'u*[*m*].
[249] Prag I 686:19–20, *nibrāram*, *lu ša Šubirim lu ša ālim*[ki].
[250] Kt c/k 524:10, 1 túg *nibrārum ša Akkidīē*.
[251] TC 1, 19:12, 2 *šitrē ša abarnīē*; Kt 93/k 75:7 *šitram ša abarnīim*.
[252] BIN 6, 64:7, 1 *šitrum ša Akkadî*; Kt n/k 437:4–5, *šitram*, *ša Akkidīē*.
[253] Kt n/k 437:3–4, 2 *šitrē*, *ša ālim*[ki].
[254] BIN 6, 184:r9–10, 1 *šitram*, *ša Zalpa*; CCT 1, 50:6, 6 *šitrī ša Zalpa*.
[255] 3 *šitrim ša kutānim* (RA 59 1965, n° 14:15); *šitrum ša sapdinnim* (TC 1, 19:11); *šitrum ša pūkim* (Kt 91/k 501:6; Kt 93k 542:8); *šitrum ša lubūšim* (Prag I 740:3; *RA* 60, 1966, 113, n° 43:33; OIP 27, 58:25).
[256] The correct interpretation of the *lubūšum* is given below § 5.3.
[257] *Nahlaptum ša kutānim*: OIP 27, 7:11, 13; *namaššuhū ša kutānī*: Benenian 5:2–3 (unpubl.).
[258] Kt 93/k 891: 9, túg *pirikannum ša Zalpa* including 3 *tisābum*; CCT 5, 12a:9, 14 *sapdinnū* including 4 *tisābū*.
[259] Kt 94/1686, 9–10: *nibrārū*, *ša kutānū* sig$_5$ *watrūtum*. Note the writing: Kt 87/k 452:9 (courtesy of Hecker), 1 túg *kutānu ti-sà-bu*.
[260] In Assyrian *ša qerbam šalmūni*, see footnote 203.

Abarnian textile you sent, a similar one you must not send again. If you make one, make it like the one I wore there (= in Aššur)", may apply to the woollen fabric from which it was made or to the finished garment.[261] Even the distinction been a woollen sheet and a garment is not always clear, since some of the latter were hardly tailored and more of the type of a (large) "wrap-around" garment. Of course, tailored garments of various types and shapes, at times provided with fringes and tassels, did exist and there was also a distinction between undergarments and shirts, and outer or upper garments such as cloaks, the more ceremonial "toga-garments",[262] and coats.

Above (§ 1.1) we concluded that the most frequent textile product exported to Anatolia, *kutānum*, was not a garment but a woollen fabric, presumably a kind of cloth, which could be used for making garments and specific pieces of apparel qualified as *ša kutānim*, "(made) of *kutānum*".[263] Therefore *kutānum* may also occur as a type or category of woollen fabric under which other textile products or garments, made from it, could be subsumed, e.g. *kusītum*, see above § 4. The absence of a statement of the type "x *ṣubātū*, of which y are *kutānū*" suggests that, in many cases, textile products designated by the generic term *ṣubātū* may have been *kutānū*-textiles and not garments. Besides, what is true for *kutānum* is most probably also valid for *šūrum* (see § 3.3, *s.v.)*, essentially a somewhat cheaper and possibly coarser alternative, much used for wrapping merchandise, for which a tailored garment would not have been used.

A further argument for considering most common "textiles" woollen fabrics of standard sizes is the occurrence of parts or fractions of them. While some of these fractions may reflect shared ownership (like the occasional occurrence of "half a donkey") or be the outcome of a balancing of accounts, most are real and imply that certain woollen textiles could be cut into pieces still retaining a commercial value. This is understandable considering the (few) data we have on their large size: a "thin textile" (*ṣubātum qatnum*) could measure *c.*4 by 4.5 m and a *raqqutum c.*2 by 5 m (see above § 3.4.4). Fractions usually occur when an Anatolian palace levies a tax (*nishatum*) of 5% on imported textiles and uses its right to pre-empt 10% (or a the tithe, *išrātum*) of the remaining ones. This regularly yields odd figures and in such cases the Assyrians hand over parts of textiles, regularly a half, one third and even occasionally one fourth of a textile, which for the above mentioned items means pieces of between 9 and 2.5 square meters, which could still be used for making garments.[264] There are even a few cases where pieces of textile are exported from Aššur, "6⅓ heavy Šubarian textiles of good quality" in Kt 91/k 344:10–12, and "31 textiles and 1/3 Akkadian textile" in Prag I 704:8–9.[265] For still smaller pieces or fractions due, a "balancing payment" (*nipiltum*) in silver is made.[266] That such fractions only occur in *ṣubātum, kutānum* and

[261] For the Assyrian text, see § 3.4.1.

[262] See the contribution by B.R. Foster in this volume.

[263] As shown here, this applies to *nahlaptum, namaššuhum, nibrārum,* and *šitrum*.

[264] Note also TC 3, 164:21–22, "I paid two shekels of silver for one textile and I used them to provide clothing to two servants" (2 *ṣuhārē ulabbiš*).

[265] The curious writing 31 *ṣubātū* ù ⅓ túg *ša Akkidīē* suggests that the last item was an addition to the load, still saleable in Anatolia, which is feasible, since an "Akkadian textile" is not a tailored garment, but a woollen fabric or type of textile from which other, more specific pieces of apparel could be made (see above § 2 and 3.3. *s.v.*).

[266] The tax paid may consist of a fraction of a textile plus some silver, and if, in order to avoid fragmentation of textiles, the palace receives a little more, it compensates by paying some silver back, but occasionally tiny fractions are ignored. For examples, see Larsen 1967, 122–134 and 156–159 (the edition and analysis of his texts "type 3" nos. 11–130) and also Veenhof 1972, 85–86 and 94–95. To give some examples: in BIN 4, 61:5–10, the tax on 85 *kutānū* amounts to 4 ¼ pieces and after deduction of the tithe of 8 *kutānū*, there remain 72¾ textiles; in CCT 5, 39a: 9 of the presumably 9[1] textiles,

šūrum, never in textiles called *abarnīum, kusītum, namaššuhum, šilipka'um*, and *takkušta'um*, suggests that the latter were either tailored garments or textiles of specific shapes and styles, that could not be cut into pieces.

A further argument for considering many textiles as being woollen fabrics of standard types is that there is almost no evidence of specific textiles meant as garments for women.[267] There are several references that link a *šitrum* with a woman,[268] but men also wear it (see § 3.3, *s.v.*), and there must have been differences in the headdress (e.g. *paršigum*), but a *nahlaptum* was worn by persons of both sexes. The two cases (see below § 5.2) of women claiming to have given up a garment (*ṣubātum*) of their own to add it to the merchandise sent to Anatolia, suggest that their garment was of a standard type that could be sold in Anatolia, not necessarily only to and for women. According to KTS 50c (= *EL* 150):1–7, the queen of Wahšušana, on the occasion of her arrival (in town), received as a gift one *kutānum* and one *šūrum* which apparently were suitable for her wardrobe. And the statement in *RHA* 18, 37:15–16, "I clothed his wife and him in two garments of good quality" (see below § 5.2) does not suggest that they received two different textile products. The same is true of the Anatolian textile called *tisābum*, since *Kuliya* 54:4–6 mentions that an Anatolian bought one for his wife (see § 3.3, *s.v. tisābum*). This picture is confirmed by data from other periods. In Mari, the textiles allotted to women in the royal harem are not different from those given to male personnel of a more or less similar rank, and both usually receive a túg-si-sá (perhaps to be read *išār(t)um*, "ordinary, normal textile"), whose name indicates the type of weave and fabric rather than its function or characteristics as a garment.[269] Among the "textiles" listed as part of a dowry during the Old Babylonian period (we lack dowry lists for the Old Assyrian period), we also do not find specific garments tailored for women. Those listed are the usual items that occur in administrative and economic texts such as túg, túg-bar-si, túg-guz-za and occasionally túg-gú-è(-a), túg-sal-la, *uṭublum, laharītum,* and *kitītum.*[270] In the text published by Lackenbacher 1982 (III:1 and VI:23), a túg-guz-za can belong both to the god Enki and to the goddess Nanaya, and also the garments in the wardrobe of the goddess Ištar of Lagaba, as far as identifiable, do not seem to be typically feminine.[271]

---

after deduction of the tax of [4½ pieces] and the tithe of [$8\frac{2}{3}$ pieces] there remain $77\frac{5}{6}$ textiles; in AnOr 6 no. 15, the tax on 27 *kutānū* amounts to 1⅓ piece, the tithe to 2½ pieces, so that (line 14) $23\frac{1}{6}$ pieces remain.

[267] We must disregard cases where the woman mentioned is not the recipient of the textile, but the one who had made it or sent it to a trader in Anatolia, e.g. in CCT 1, 25 (*EL* no. 166):26–28, "one textile of good quality and a Abarnian one of the lady I entrusted to A."; TC 1, 105:5–6, "one Abarnian textile of his wife N., I. and E. brought to Suejja"; TC 3, 158:27–30, "I gave you one Abarnian textile of the daughter[!] of A."; AKT 4, 30:11–13, "One *nahlaptum* of our bride-to-be (*kallutum*) P. brought me".

[268] AKT 4, 29:21–23, in a letter to an Assyrian woman in Kaneš, mentions a gift to her from her daughter (in Aššur) : "One *šitrum* of Akkadian make with the seals of Bēlatum, your daughter, Š. is bringing to you".

[269] See Ziegler 1999, 193–196. Higher ranked women may receive an *uṭublum* (text no. 25:3'), which is also given to men; Durand 1980, 405–406, for túg-si-sá and the fact than an *uṭublum* is characterized by a specific kind of weave (result of "tissage de la serge", shared by the textile called *raqqatum*).

[270] See the data presented above in § 3.1, with note 7 and 11, and for several texts Dalley 1980 nos. 3–6, 10 col. II (where the meaning of túg *ú*-ZU-*um* in line 8 remains unclear) and 11, and Bruxelles O 342, col. I. An exception could be the túg-sal-la *sūnim* in BM 16465 II:5 and 12, and a túg-bar-si *irtim*, attested once at Mari (see § 3.3 note 43, but the *nahlaptum ana irtika* in Kt k/k 24:24–25 is meant for a man).

[271] See Leemans 1952,1–2, where apart from various kinds of *paršigū*, we also find the rare *aguhhum, gadamahhûm*, túg *tuqnātum* (meaning unknown), túg *taktīmum* (perhaps a fine blanket, cf. CAD T, *s.v.*) and túg-sal-la = *raqqatum.*

Taking into account comparative data as well, we may conclude that a *kusītum* was a type of garment (which could be "thin", made of *kutānu*-fabric, or "of Akkadian make/style", see § 3.1 and 3.3 *s.v.*), although possibly not much tailored and rather of the "wrap-around" type. The same may be true of *raqqutum*, "thin textile", because in Kt 94/k 1751:5–7, 2 *šulupka'u*-textiles and 2 *raqqātum* of good quality are qualified as "for/as garments for our father" (*ša lubūšī ša abini*). In the Old Babylonian period too, this textile was worn as a garment and belonged to a person's or a god's wardrobe (see CAD R, *s.v. raqqatu* A, 169, b–c), which made it appropriate as a gift, e.g. in KTS 57a:6, where two pieces are offered to a local ruler. Products with specific names, such as *namaššuhum* (=*lamahuššûm*), *šilipka'um* and *takkušta'um*, which figure in small quantities among the items exported, probably also were garments. Babylonian sources show that *lamahuššûm* was a rather sumptuous garment and its Old Assyrian counterpart was worn as a garment by an Anatolian ruler according to Kt 89/k 221 (quoted in § 3.3, *s.v. namaššuhum*). That a *šilipka'um* was a garment is suggested by Kt 94/k 1751, quoted above, and for *takkušta'um*, it may be inferred from the fact that in Mari it figures as a gift received by a man at the court in Babylon (see § 3.3, *s.v.*), since kings and courts (as is well attested at Ebla, Mari and Babylon) used to hand out (sets of) garments as gifts to important visitors. It is, however, impossible to say what the nature of these garments was: perhaps a specific type of weave, with finishings and perhaps colours, rather than extensive tailoring.

5.2. *The use of the verb labāšum*

Further evidence for the issue of "textiles or garments" can be found by studying the occurrences of the verb *labāšum*, "to put on clothing", in Old Assyrian attested in the reflexive Gt-stem (*litabšum*), "to clothe oneself in ...", and in the D-stem (*labbušum*), "to clothe, to provide somebody with clothing". This raises the question of the meaning of the derivative noun (túg) *lubūšum*, which in Old Assyrian occurs in different contexts and with different meanings.

The D-stem occurs several times (also in the stative) and when the impersonal object is *ṣubātum*, the latter apparently means "clothing, garment", see CAD L, 19 b, 1', and apart from the two examples quoted below, also *RHA* 18, 37:15–16, "I clothed his wife and him in two garments of good quality" (2 túg^{há} sig$_5$ ... *ulabbiš*), and Kt 94/ k 486:10–12, "you failed to give me the price of the garment in which I dressed you". Note also Kt 75/k 78:2–3 and 29–30, "I provided his son / the daughter of Š. with one garment for a youngster" (1 túg *ša ṣuhrim ulabbiš*), where the garment given matches the age, gender and size of the recipient. However, TC 3, 164:21–22, "I paid 2 shekels of silver for a *ṣubātum* and clothed (with it) two servants" (2 *ṣuhārī ulabbiš*), suggests that *ṣubātum* here was a large sheet of textile that could be cut into two to make garments.

The Gt-stem, *litabšum*, occurs about two dozen times, inter alia in letters where traders in Anatolia ask to send them, usually from Aššur,[272] textiles "to be worn, to be put on by me" (*ana litabšia*). In several letters they ask to send or buy an unspecified *ṣubātum* for that purpose, provided either by a fellow trader[273] or by a female relative. In Kt 91/k 508:13–15, Ummī-Išhara writes to her brother, "With the next caravan I will send two garments (*ṣubātū*) of good quality for you

[272] However, in VS 26, 40:8–8–10, "a textile for me to put on" (*ṣubātum ana litabšia*) has to be bought for 6 shekels of silver in Anatolia, since it is followed by a request to buy a *pirikannum* and a belt.

[273] CCT 4, 45b:27–29 (meant for women, *allitabšišina*); CCT 6, 3a:24–26 (will be sent when the road is open again); CTMMA 1, 79:19–21("send me *ṣubatū* to put on, I am staying here without *ṣubātū*"); Prag I 477:19–25 (may cost 20 shekels of silver); see also ICK 2, 210:2 and Prag I 440:45–47 ("the oil you left behind for me to anoint myself and the *ṣubātum* to put on...").

to wear", Kt 92/k 102:18–20 asks "Why did you hold back the garment (*ṣubātum*) my sister sent for me to wear?", and TC 3, 210:3–6 mentions the shipment of 12½ shekels of silver (to Aššur) to the address of two women "for buying *ṣubātū* for me to wear".[274] That a *ṣubātum* was worn as a garment is also clear from Kt 91/k 543:25–27, where a woman writes from Aššur, "You know very well that I stripped the garment from my shoulders to give it to you".[275] In Prag I 440:3–6 (also a letter from a woman in Aššur to a trader in Anatolia) such a *ṣubātum*, worn by a woman, is put on a pair with the *ṣubātū* that to all appearances had been bought in Aššur for export to Anatolia: "Together with your own *ṣubātū* (plural) one is bringing you a *ṣubātum* that is my own garment of which I stripped myself!" Other texts also mention that some of the textiles traded could be worn or used as garments: ICK 1, 15:18–19, "The rest of the textiles I will ......[276] and (so that?) they can wear them", and Kt 91/k 449:1–8, "When we counted the textiles (*ṣubātū*) of my transport, the palace took 2 *sapdinnū*, ... 5 textiles they (had) put on" (*iltabšū*). In CCT 3, 20:17–20, "Since the girl has now grown up, I have now made a few heavy textiles for (on) the wagon" (túg *ištên u šina kabtūtim ana narkabtim ētapaš*), it is not clear whether she had to wear them as garments or they were used as covers or blankets on the wagon.

Other texts mention particular types of garments in which people dressed:

- *Abarnīum*, apart from TC 3, 17:23–28, quoted above, also Kt 86k 193:15–25, "If (you have?) a heavy and soft Abarnian textile for me to dress in, wrap it and also select two heavy *kutānū*, either from those of mine or from those belonging to the trader ...". Perhaps also in AKT 2, 24:4–12, "Š. owes me one Abarnian[277] textile of very fine quality. Ask it from him and when you ... Š., make him wear it" (*labbišāšu*), and the letter adds that if Š. refuses, the addressees should give Š. a *lubūšum* of Susē (? *sú-sí-e-a-am*, see § 2.1.1, *s.v.*) or a fine *kutānum* for that purpose.
- *burā'um*, BIN 4, 160:11–13, "Š. owes 11 minas of refined copper (because) I dressed him (*ulabbiššu*) in a *burā'u*-textile of good quality", and Kt m/k 43:1–2 (courtesy of Hecker), "When the textile(s) came up from the city, I dressed A. in a *burā'u*-textile (1 túg *burā'am A. ulabbiš*). Also Kt 94/k 966:12–20 (courtesy of Larsen), "Get yourself out of there! Give one of the textiles that I can wear, either a *šūru*-textile or a *burā'um* in Akkadian style ... to an independent trader and let one ... (it) and bring it into the town. There are (here) no textiles I can dress myself in".[278]
- *kutānum*, CCT 5, 33:9–15, "I gave him a *kutānum* to put on (*allitabšišu*) and will send up for him from the City an Abarnian textile"; with the D-stem of the verb, Kt 94/k 1226:18, "I provided the Hattians[279]

[274] In Kt 93/k 325:10–13, a woman writes from Aššur: "Send me silver so that we can buy wool and then we will make one garment for you to wear" (1 túg *ana litabšika lu nēpušakkum*). When this was impossible, garments could also be bought, as in TC 3, 17:29–33, "If you don't manage to make thin textiles (*ṣubātū qatnūtum*), as I hear there are plenty for sale there, buy and send them!", and this could be done on the local market, cf. TC 2, 7:25–28, "Because there is no *šurbu*-wool (for making a textile) available, we will buy a heavy textile on the market (*ina mahīrim*)"

[275] *Attāma tidê ṣú-ba-tum*[sic] *bu-dí-a* [*ah-m*]*u-úṣ-ma addinakkum*. Cf. ARM 10, 17:10–14, where the queen of Mari writes to her husband, "Let my lord put the garment and the cloak I made on his shoulders (1 *ṣubātam u nahlaptam ... ana budišu liškun*).

[276] *Šitti ṣubātī ú-lá-ha-áp-ma iltabbušunišsunu*. CAD L, 18, c, 1', suggests for *lahhupum* a meaning "to set aside" or "to select" (p. 239, *s.v. luhhupu*, "to treat textiles in a particular way"); AHw *s.v. la''upum* II, referring to Arabaic *lḥf*, proposes "in ein Tuch hüllen", but this action is always expressed by the verb *lawā'um* in Old Assyrian. For this verb, see also line 18 in footnote 278.

[277] Sic? The edition has túg *A-ni-a-ni-a-am*.

[278] Lines 12–19, *ramakkunu šulhā*, 1 túg[há] *ina litabšia, lū šūrum*[sic] *lu burā'am, ša Akkidīē ištu, Hurama ana mer ummiānim, dimma lu-lá-hi-pu, lūšerrebūniššu*, túg[há] *allitabšia lā išû*.

[279] Courtesy of Larsen; *Hattīum*, presumably a nisbe derived from Hattum, probably also in KTS 1, 8a:4–5, *i-na a-lim*[ki] (written *a ki lim*) *ša* [*H*]*a-tí-e*.

with two *kutānū* as clothing" (2 *kutānī Ha-tí-e ulabbiš*), and see above, under *Abarnīum*, for such a use of a (fine) *kutānum* in Kt 86/k 193:19–20 and AKT 2, 24:11.

- *lubūšum*, see AKT 2, 24:10, quoted above, under *Abarnīum*.
- *palīlum*, TC 3, 164:11–13, "I paid 6 shekels of silver for two túg *pá-li-li* and the slave-girls have put them on" (*iltabšāšunu*).
- *raqqutum*, Kt 93/k 93:6–9, "The thin textile that A. promised you – I saw it and it is not fit for you to wear, it would be a shame" (*lā ša litabšika magriat*).
- *šūrum*, TC 1, 19:17–21, "Send two *šūru*-textiles of good quality to me to wear (and) 2 *sapdinnū* of good quality"; also Kt 94/k 966, quoted under *burā'um*.
- *tisābum*, Kt 94/k 1173:18-21, "A. promised me one extra fine *tisābum* to dress myself in" (1 túg, *tisābam* sig_5 *uttu<r>tam, ša ana litabšia*).

These data show that different varieties of textiles could be and were used as garments, not only those called *abarnīum, burā'um, palīlum* and *raqqutum*, but also *kutānum* and *šūrum*, textiles whose names, as indicated above, probably refer primarily to a particular type of woollen fabric. They may have been appropriate to use as "wrap-around garment" or perhaps their fabric, after some tailoring or cutting may have been turned into a garment. This would put them on a par with the unspecified *ṣubātū*, which various texts show to have been worn as garments, although there is also evidence for cutting them into pieces to yield more than one garment.

5.3. *lubūšum*

This bring us, finally, to the noun *lubūšum*, which etymologically means "what one wears, is put on", according to CAD L, *s.v.*, "1. clothing, 2. piece of apparel, 3. clothing allowance". In Old Assyrian it occurs in various constructions and contexts.[280]

The third meaning is frequent in connection with the hiring of caravan personnel who, in addition to a wage, receive a "clothing allowance", normally in the form of some silver (*c.*1½ to 2 shekels per person). This could be the price paid by a trader for buying (cheap) garments for his employees or the silver he had actually handed over to them.[281]

The first meaning, not always well distinguishable from the third one, occurs in general statements, such as "we will spend it (the copper) for clothing (*ana lubūš*) for your wife and son and for their food" (BIN 6, 187 rev. 13'-16'), and in combinations where a textile is qualified as "clothing of/for" a particular person either by a construct state, as in textiles *ša lubūš šarruttim*, "as clothing of kings" (see § 3.4.1,a) and *ša lubūš ṣuhārī*, "for clothing of servants" (Prag I 429:1, *pirikannu*-textiles), or by means of *lubūšum ša*, e.g. *lubūšam ša ṣuhrim*, "a garment/clothing for a youngster" (*AAA* 1, 2:5–6). We also find textiles qualified as *ša lubūšim ša*, e.g. 1 túg *ša lubūšim ša ṣuhrim*, "one textile as garment for a youngster" (BIN 6, 84:30), or "2 *šulupka'ū* ... (and) 2 thin textiles (*raqqātān*) as garments (plural!) of our father" (*ša lubūši ša abini*, Kt 94/k 1751:1–6). However, *ša lubūšim* is also used more independently, as in 1 túg *ša ṣuhrim ša lubūšim* (KTS 2, 31:3) and 1 *šitrum ša lubūšim* (OIP 27, 58:25, Prag I 740:3, alongside *kutānū* and *išrātum*). Since a translation "as garment" would amount to tautology and the contexts make "as (part of a) clothing allowance"

[280] Old Assyrian also knows *lubūštum* (see § 3.1), "clothing, clothing allowance", but it is very rare, see CAD L, 233, d, 1'.

[281] See Larsen 1967, 150–151, table, under "clothes", and Veenhof 1972, 97 with note 160. We regularly meet in descriptions of caravan expenses the phrase "x silver the wages of the harnessors, together with their clothing allowance" (x *kaṣṣārū qadum lubūšišunu*), e.g. in *KTB* 17:8–9. Cf. also *POAT* 41, 17–18, "send 5 minas of wool, her clothing ration" (*lubūšiša*, plural or mistake for *lubūšša*?)

unlikely, we take it as referring to a particular (kind of) garment to which the item mentioned belonged or which it had to match, to make a set.

In such cases, *lubūšum* may be used because it was the standard type of garment and this may also apply in cases where a number of *lubūšū* are listed together with small numbers of specific types of textiles or garments, exported from Aššur.[282] Here a meaning "garment" is not distinctive enough and would make all other exported textiles more or less automatically not garments, which is difficult to accept, as indicated above. CAD L, *s.v.*, 2 therefore takes it as "a specific type of apparel" and the occasional use of the determinative túg before *lubūšum* supports this view and distinguishes it from textiles with more specific names that appear alongside it, but what its characteristics were is difficult to say. In the listing in Yale 13092 (see footnote 282), together with other textiles, it is qualified as "Akkadian" and this is also the case in Kt n/k 1228:13–17 (courtesy of Çeçen), "among the Akkadian textiles, both yours and mine, there are a *lubūšum*, a *kusītum* and a *šulupka'*[*um*]". There are a number of references to white and one to yellow *lubūšū* (see § 3.4.2) and they could be of fine quality,[283] expensive products according to Kt n/k 533:16–22 (courtesy of Günbattı), "if the textiles, either a white *lubūšum* or an Abarnian one, both of extremely fine, royal quality, seem right to you, give one to her".[284] In TC 3, 161:1–3, 3 *lubūšū*, among which a white one, are a gift for a queen,[285] and in AKT 2, 24:10–12 a *lubūšum* of Susē (*sú-sí-e-a-am*) and a *kutānum* of fine quality are alternatives for an extra fine Abarnian textile to be worn by a man.

These data indicate that túg *lubūšum* was a fine garment, presumably of a standard type, a valuable product exported in small quantities, which could be provided with a matching *šitrum* and of which also a smaller version, fit for children, existed.

* * *

The Old Assyrian archives contain an important number of textile names, but the data concerning them are restricted to particular features, such as: the numbers, qualities and prices of the textiles traded, their geographical origin and, sometimes, also the material they are made of. The production techniques and the textile usages are rarely mentioned. The Assyrians and the Anatolians used only two materials: wool and linen, goat hair being mentioned only rarely. However, the terminology dealing with textiles is rich: there are about seventy different words referring to the various kind of fabrics, material, shape, use of textiles, and a few data on their weights. We found also twenty different words to specify the colour, quality or appearance of the textiles. The great majority of this vocabulary is peculiar to the Old Assyrian corpus; in fact,

---

282 We have the following enumerations (cf. CAD L, *s.v.* 2, a): 5 túg *lubūšū* 1 túg *kusītum* 2 *raqqātān* (Prag I 623:1–3); 5 *šū*[*rūtum*] [x] *lubūšū*, [y] *raqqātum* (Prag I 672:4–6); 7 *lubūšū* 6 *kusiātum*1 *Šilipka'um* 1 *takkuštûm*, "in all 15 Akkadian textiles" (Yale 13092:8–10); 5 *kusiātum* 1 túg *Abarnīum* sig$_5$ 2 túg *lubūšū* 1 túg *nibrārum ša Akkidīē* (Kt n/k 524:8–10); 10 *kusiātum* 2 túg *burā'ū* 5 *lubūšū* (Kt n/k 524:32–33, exported from Aššur); 1 *kutānum* 5 túg *lubūšū* 1 túg *raqqutum* (Kt 91/k 299:11–12), etc.

283 Note *RA* 60, 139:18–21, "As for the garment (*lubūšum*) for (the woman) A., I looked for a fine *lubūšum*, but I could not find one to send her".

284 *Šumma ṣubātū, lu lubūšum paṣium, lu Abarnīum, damqūtim watrūtim, ša šarruttim, išširūnikkum, diššim.*

285 Lines 1–6, 1 túg *lubūšum paṣium, ša A.* 2 túg *lubūšī, ša P. a-rubātim iššī,* 1 túg *paṣium inūmi ana* [5] *bēt mūṣîm iqrûšuni, akkārim labbušim*, "one white *lubūšum* of A., two *lubūšū* of P. he brought as a gift to the queen; one white garment when they invited him to the house of the *mūṣium* to dress the *kārum*".

there are relatively few words dealing with textiles, which are also found in cuneiform sources from other areas and periods. Thus, textile crafts have a strong regional tradition compared to other crafts. The important geographical aspect of the production is confirmed by the twenty geographical terms referring to textile types.

The study of textile terminology, in the absence of archaeological remains, rests on etymology, categorization and the combination of various terms, apart from a few more detailed descriptions. One can see that, most often, it is difficult to give an accurate and appropriate translation for each term, but, at least, we could distinguish textiles from garments. The translation of the technical terms will be enhanced by a good knowledge of weaving techniques, and the workshop, whose results are published in this volume happened to be very successful in this respect[286].

Abbreviations

AAA — Annals of Archaeology and Anthropology. Liverpool.
AbB — Altbabylonische Briefe in Umschrift und Übersetzung. Leiden.
AHw — W. von Soden, *Akkadisches Handwörterbuch*. Wiesbaden, 1958–1981.
AKT — Ankara Kültepe Tabletleri = Ankaraner Kültepe Texte. Ankara.
AMMY — *Anadolu Medeniyetleri Müzesi Yıllığı*. Ankara.
AnOr — Analecta Orientalia, Roma.
ARM(T) — Archives Royales de Mari (Traduction). Paris
ATHE — B. Kienast, *Die altassyrischen Texte des orientalischen Seminars der Universität Heidelberg und der Sammlung Erlenmeyer*. Berlin, 1960.
BE — The Babylonian Expedition of the University of Pennsylvania. Philadelphia
BIN — Babylonian Inscriptions in the Collection of J. B. Nies. New Haven.
CAD — *The Assyrian Dictionary of the Oriental Institute of the University of Chicago*. Chicago (1956–).
CCT — Cuneiform Texts from Cappadocian Tablets in the British Museum. London.
Chantre — E. Chantre, *Mission en Cappadoce 1893-1894*. Paris, 1898.
CST — T. Fish, *Catalogue of the Cuneiform Tablets in the John Rylands Library*, Manchester 1982.
CT — Cuneiform Texts from Babylonian Tablets in the British Museum. London.
CTMMA — *Cuneiform Texts in the Metropolitan Museum of Art*, vol. 1. New York, 1988 (nos. 71–98, 'Old Assyrian Texts', by M. Trolle Larsen).
EL — G. Eisser & J. Lewy, *Altassyrische Rechtsurkunden vom Kültepe*, I–II. Mitteilungen der Vorderasiatisch-Aegyptischen Gesellschaft 30, 35/3. Leipzig, 1930, 1935.
FAOS — Freiburger Altorientalische Studien. Stuttgart.
GAG (Erg.) — W. von Soden, *Grundriss der Akkadischen Grammatik*. Analecta Orientalia 33. Roma 1952 or later); *Ergänzungsheft*. Analecta Orientalia 47. Roma 1969.
GKT — K. Hecker, *Grammatik der Kültepe-Texte*. Analecta Orientalia 44. Roma 1968.
HSS — Harvard Semitic Series. Cambridge, Mass.
HUCA — *Hebrew Union College Annual*. Cincinnati.
ICK — Inscriptions cunéiformes de Kültepe. Prag.
JCS — *Journal of Cuneiform Studies*. New Haven – Boston – Ann Arbor.
KBo — Keilschrifttexte aus Boghazköi. Berlin.
KKS — L. Matouš & M. Matoušová, *Kappadokische Keilschrifttafeln mit Siegeln aus der Sammlung der Karlsuniversität in Prag*. Prag, 1984.
KTB — J. Lewy, *Die Kültepetexte der Sammlung Rudolf Blackertz*. Berlin, 1929.
KTH — J. Lewy, *Die Kültepe-Texte aus der Sammlung Frida Hahn*. Berlin, 1930.
KTS — Keilschrifttexte in den Antiken-Museen zu Stambul. Istanbul, 1926.

286 See also the contribution of A. Wisti-Lassen in this volume.

KUG	K. Hecker, *Die Keilschrifttexte der Universitätsbibliothek Giessen*. Berichte und Arbeiten aus der Universitätsbibliotheker 9. Giessen, 1966.
Kuliya	K. R. Veenhof, *The Archive of Kuliya, son of Ali-abum (Kt 92/k 188–263), Kültepe Tabletleri*, Ankara, 2010, quoted by text number = AKT 5.
LB	Texts in the collection de Liagre Böhl. Leiden.
MARI	Mari. Annales de recherches interdisciplinaires. Paris.
MDP	Mémoires de la Délégation de Perse. Paris.
MSL	Materialien zum sumerischen Lexicon. Roma.
NABU	*Nouvelles Assyriologiques Brèves et Utilitaires.* Quoted by year + number of contribution.
OBTR	S. Dalley, C. B. F. Walker & J. D. Hawkins, *The Old Babylonian Texts from Tell al Rimah*, London, 1976.
OIP	Oriental Institute Publications. Chicago.
POAT	W. C. Gwaltney Jr., *The Pennsylvania Old Assyrian Texts* (*HUCA* Supplement 3). Cincinnati, 1983.
Prag I + no.	Siglum of texts in K. Hecker, G. Kryszat & L. Matouš, *Kappadokische Keilschrifttafeln aus den Sammlungen der Karlsuniversität Prag*. Prag, 1998.
RA	*Revue d'assyriologie et d'archéologie orientale.* Paris.
RGTC	Répertoire géographique des textes cunéiformes. Wiesbaden
RHA	*Revue hittite et asianique*. Paris.
RIMA 1	A.K. Grayson, *Assyrian Rulers of the Third and Second Millennia BC (to 1155 BC)*. Royal Inscriptions of Mesopotamia, Assyrian Periods, vol. 1. Toronto, 1987.
Sadberk	V. Donbaz, *Sadberk Hanım Müzesi'nde Bulunan Çiviyazılı Belgeler.* Istanbul, 1999.
TC	Tablettes cappadociennes du Louvre. Paris.
TCL	Textes cunéiformes du Louvre. Paris.
TLB	Tabulae Cuneiformes a F.M.Th. de Liagre Böhl Collectae Leidae Conservatae. Leiden.
TPAK	C. Michel & P. Garelli, *Tablettes paléo-assyriennes de Kültepe*, 1 (*Kt 90/k*). Paris 1997.
TuM	Texte und Materialen der Frau Professor Hilprecht Sammlung. Leipzig.
UF	*Ugarit-Forschungen*. Internationales Jahrbuch für die Altertumskunde Syrien-Palästinas. Neukirchen-Vluyn.
VS	Vorderasiatische Schriftdenkmäler der (Königl.) Museen zu Berlin. Berlin.
WO	*Die Welt des Orients*. Göttingen – Tübingen.
YOS	Yale Oriental Series. Babylonian Texts. New Haven.

Bibliography

Al-Rawi, F. N. H. & Dalley, S. (2000) *Old Babylonian Texts from Private Houses at Abu Habbah, Ancient Sippir*. É.DUB. BA.A 7. London.

Bilgiç, E. (1951) Die Ortsnamen der « kappadokischen » Urkunden im Rahmen der alten Sprachen Anatoliens, *Archiv für Orientforschung* 15, 1–37.

Charpin, D. & Durand, J.-M. (1997) Aššur avant l'Assyrie, Mari Annales de Recherches Interdisciplinaires 8, 367–392. Paris.

Charpin, D. & Ziegler, N. (2003) *Mari et le Proche-Orient à l'époque amorrite. Essai d'histoire politique*, Florilegium marianum 5, Paris.

Civil, M. (1964) A Hymn to the Beer Goddess and a Drinking Song. In R. M. Adams *et al.* (eds.), *From the Workshop of the Chicago Assyrian Dictionary. Studies Presented to A. Leo Oppenheim*, 67–89. Chicago.

Civil, M. (1967) Šū-Sîn's Historical Inscriptions: Collection B, *Journal of Cuneiform Studies* 21, 24–38.

Dalley, S. (1980) Old Babylonian Dowries, *Iraq* 42, 53–74.

Dercksen, J. G. (1996) *The Old Assyrian Copper Trade in Anatolia.* Publications de l'Institut historique-archéologique néerlandais de Stamboul 75. Leiden.

Dercksen, J. G. (2001) "When we met in Hattuš". Trade according to the Old Assyrian texts from Alishar and Boğazköy. In W. H. van Soldt (ed.), *K. R. Veenhof Anniversary Volume*, 39–66. Publications de l'Institut historique-archéologique néerlandais de Stamboul 89. Leiden.

Dercksen, J. G. (2004) *Old Assyrian Institutions*. Publications de l'Institut historique-archéologique néerlandais de Stamboul 98. Leiden.
Dercksen, J. G. (2008), Observations on Land Use and Agriculture in Kaneš. In C. Michel (ed.), *Old Assyrian Studies in Memory of Paul Garelli*, 139–158, Old Assyrian Archives Studies 4, Leiden.
Durand, J.-M. (1983) *Textes administratifs des salles 134 et 160 du palais de Mari*. Archives Royales de Mari 21. Paris.
Durand, J.-M. (2001) Une alliance matrimoniale entre un marchand assyrien de Kanesh et un marchand mariote. In W. H. van Soldt *et al.* (eds.), K. R. *Veenhof Anniversary Volume*, 119–132. Publications de l'Institut historique-archéologique néerlandais de Stamboul 89. Leiden.
Durand, J.-M. (2009) *La nomenclature des habits et des textiles dans les textes de Mari*. Archives Royales de Mari 30. Paris.
Eidem, J. & Laessoe, J. (2001) *The Shemshara Archives*. Vol. 1. The Letters. Copenhagen.
Forlanini, M. (2004) Dall'Alto Habur alle montagne dell'Anatolia nel II millennio A.C., Note sulla Geografia storica di una regione poco conosciuta. In Ch. Nicolle (ed.), *Nomades et sédentaires dans le Proche-Orient ancien. Compte rendu de la XLVIe Rencontre Assyriologique Internationale (Paris, 10–13 juillet 2000)*, 405–426. Amurru 3. Paris.
Garelli, P. (1963) *Les Assyriens en Cappadoce*. Paris.
Garelli, P. (1965) Tablettes cappadociennes de collections diverses (suite), *Revue d'Assyriologie* 59, 19–48, 149–176.
Horowitz, W. & Oshima, T. (2006) *Cuneiform in Canaan. Cuneiform Sources from the Land of Israel in Ancient Times*. Jerusalem.
Karaduman, A. (2008) Acemhöyük Sarıkaya Sarayı'nda bulunmuş olan etiketlerden bir gruo. In T. Tarhan, A. Tibet & E. Konyar (eds.), *Muhibbe Darga Armağanı*, 283–290. Istanbul.
Klengel, H. (2008) Studien zur hethitischen Wirtschaft 4, *Altorientalische Forschungen* 35, 68–85.
Lackebacher, S. (1982) Un texte vieux-babylonien sur la finition des textiles, *Syria* 59, 129–149 (the cuneiform copy of the text was published in *Syria* 58, 97–98).
Landsberger, B. (1925) *Assyrische Handelskolonien in Kleinasien aus dem dritten Jahrtausend. Der Alte Orient* 24/IV.
Landsberger, B. (1967) Über Farben im Sumerisch-Akkadischen, *Journal of Cuneiform Studies* 21, 139–173.
Larsen, M. T. (1967) *Old Assyrian Caravan Procedures*. Publications de l'Institut historique-archéologique néerlandais de Stamboul 22. Leiden.
Larsen, M. T. (2002) *The Aššur-nādā Archive*, Old Assyrian Archives 1, Leiden.
Larsen, M. T. (2008) Archives and Filing Systems at Kültepe. In C. Michel (ed.), *Old Assyrian Studies in Memory of Paul Garelli*, 77–90, Old Assyrian Archives Studies 4, Leiden.
Larsen, M. T. (in press) *The Archives of the Šalim-Aššur Family*, vol. 1, Ankara Kültepe Tabletleri, Ankara = AKT 6.
Larsen, M. T. & Möller, E. (1991) Five Old Assyrian Texts. In D. Charpin & F. Joannès (eds.), *Marchands, Diplomates et Empereurs, Etudes sur la civilisation mésopotamienne offertes à Paul Garelli*, 227–252. Paris.
Leemans, W. F. (1952) *Ishtar of Lagaba and her Dress*. Studia ad tabulas cuneiformes a de Liagre Böhl collectas pertinentia I/1. Leiden.
Lewy, J. (1935) *Tablettes cappadociennes du Louvre*. Tablettes cuneiformes du Louvre 19. Paris.
Lewy, J. (1958) Some Aspects of Commercial Life in Assyria and Asia Minor in the Nineteenth Pre-Christian Century, *Journal of the American Oriental Society* 78, 89–101.
Michel, C. (1991) *Innāya dans les tablettes paléo-assyriennes*, vol. 1–2, Paris.
Michel, C. (1997) À table avec les marchands paléo-assyriens. In H. Waetzoldt & H. Hauptmann (eds.), *Assyrien im Wandel der Zeiten, Compte rendu de la XXXIXe Rencontre Assyriologique Internationale*, 95–113. Heidelberger Studien zum Alten Orient 6. Heidelberg.
Michel, C. (1998b) Les mites d'Assyrie: Moths in the Assyrian Texts of the Second Millennium B.C., *Journal of the American Oriental Society* 118, 325–331.
Michel, C. (2001) *Correspondance des marchands de Kaniš au début du IIe millénaire avant J.-C.*, Littérature anciennes du Proche-Orient 19. Paris.
Michel, C. (2003) *Old Assyrian Bibliography*, Old Assyrian Archives Studies 1. Leiden.
Michel, C. (2006a) Old Assyrian Bibliography 1 (Febr. 2003 – July 2006), *Archiv für Orientforschung* 51, 435–449.
Michel, C. (2006b) Femmes et production textile à Aššur au début du IIe millénaire avant J.-C., *Techniques et culture* 46, 281–297.
Michel, C. (2006c) Qaṭarā. In M. Streck (ed.), *Reallexikon der Assyriologie*, vol. XI/1–2, 154–156.

Michel, C. (in press a) Old Assyrian Bibliography 2 (July 2006 – April 2009), *Archiv für Orientforschung* 52.
Michel, C. (in press b) Šubartum. In M. Streck (ed.), *Reallexikon der Assyriologie,* vol. XII/1–2.
Oppenheim, A. L. (1967) Essay on Overland Trade in the First Millennium B.C., *Journal of Cuneiform Studies* 21, 236–254.
Özgüç, T. (2003) *Kültepe Kaniš/Neša. The earliest international trade center and the oldest capital city of the Hittites*, The Middle Eastern Culture Center in Japan. Istanbul.
Özgüç, N. & Tunca, Ö. (2001) *Kültepe-Kaniš. Sealed and Inscribed Clay Bullae,* Türk Tarih Kurumu Yayınları V. Dizi – Sayı 48, Türk Tarih Kurumu. Ankara.
Owen, D. I. (1992) Syrians in Sumerian Sources from the Ur III Period. In M. Chavalas *et al.* (eds.), *New Horizons in the Study of Ancient Syria*, Bibliotheca Mesopotamica 25, Malibu 1992.
Snell, D. C. (1982) *Ledgers and Prices. Early Mesopotamian Merchant Accounts.* Yale Near Eastern Researches 8. New Haven-London.
Van Soldt, W. H. (1990) Fabrics and Dyes at Ugarit, *Ugarit-Forschungen* 22, 321–357.
Veenhof, K. R. (1972) *Aspects of Old Assyrian Trade and its Terminology.* Studia et Documenta 10. Leiden.
Veenhof, K. R. (1988) Prices and Trade. The Old Assyrian Evidence, *Altorientalische Forschungen* 15, 243–263.
Veenhof, K. R. (2003) Trade and Politics in Ancient Aššur. Balancing of Public, Colonial and Entrepreneurial Interests. In C. Zaccagnini (ed.), *Trade and Politics in the Ancient World*, 69–118. Saggi di Storia Antica 21. Roma.
Veenhof, K. R. (2008a) The Old Assyrian Period. In M. Wäffler (ed.), *Annäherungen 5. Mesopotamia. The Old Assyrian Period.* Orbis Biblicus et Orientalis 160/5, part I, 13–264. Freibourg-Göttingen.
Veenhof, K. R. (2008b) Across the Euphrates. In J. G Dercksen (ed.*), Anatolia and the Jazira during the Old Assyrian Period* 2–29. Old Assyrian Archives Studies 3. Leiden.
Veenhof, K. R. (2008c) Some Displaced Tablets from *kārum* Kanesh (Kültepe), *Altorientalische Forschungen* 35, 10–34.
Veenhof, K. R. (2009) A New Volume of Old Assyrian Texts from *kārum* Kanesh, *Jaarbericht van het Vooraziatisch -Egyptisch Genootschap Ex Oriente Lux* 41, 179–202.
Waetzoldt, H. (1972) *Untersuchungen zur neusumerischen Textilindustrie*. Roma.
Zawadski, S. (2006) *Garments of the Gods. Studies on the Textile Industry and the Pantheon of Sippar according to the Texts from the Ebabbar Archive.* Orbis Biblicus et Orientalis 218. Freibourg-Göttingen.
Ziegler, N. (1999) *Le Harem de Zimrî-Lîm*. Florilegium marianum 4. Paris.

13. Tools, Procedures and Professions: A review of the Akkadian textile terminology

Agnete Wisti Lassen

Craft production was of prime importance in the ancient Near East, and both rich and poor needed clothes, furniture, tools, pots, and weapons. Unfortunately, not much evidence is available for the study of ancient crafts in Western Asia, and only few workshops have been firmly identified archaeologically.[1] A few investigations on the basis of the Sumerian written material have been undertaken,[2] but the Akkadian terminology is often left as an unexploited resource of information about technology and the evolution of crafts in ancient society.

The perishable nature of archaeological evidence means that certain aspects of some crafts are completely lost, and it is often not possible to reconstruct processes and social or religious aspects of the ancient crafts on the basis of physical remains alone. Studies in terminology can therefore corroborate both the archaeological evidence we possess, and shed light on issues not illuminated by archaeology at all. An example of the latter is the possible symbolic connotations of certain raw materials. For instance, we learn from the written evidence, that a particular fibre known as ú-ki-kal, perhaps nettle, was used only for religious garments in the city of Ur in the late 3rd millennium BC.[3]

It is possible to identify a textile term in Akkadian on the basis of the word's context, of the use of the determinatives túg and gada, and on the basis of etymological evidence, such as the occurrence of the word in later Semitic dialects, e.g. Arabic and Hebrew. Often, however, these criteria will not reveal any precise technical information about the term, and this presents a great obstacle for a better understanding of the Mesopotamian textile production, namely the changing degree of accuracy in the extant translations of the specific technical terms. Often, our dictionaries[4] place great emphasis on a correct understanding of etymology or context, whereas the specific translations seem to be based on a limited technical knowledge, and translations

---

[1] A good example is the textile workshop at Gordion, Burke 2005. For a general introduction to ancient Near Eastern crafts and technology, see Sasson 1995, chap. 7: Technology and artistic production.

[2] See e.g. Waetzoldt 1972 and van de Mieroop 1987.

[3] Lassen 2008, 42.

[4] W. von Soden: *Akkadisches Handwörtebuch* I–III, Wiesbaden 1965–1981 (*AHw*), and to a lesser degree, the *Chicago Assyrian Dictionary*, Oriental Institute, Chicago 1956– (*CAD*). Note also *A Concise Dictionary of Akkadian* Black J., A. George, and N. Postgate (eds.), Wiesbaden 2000 (*CDA*) with updated and improved translations of textile terminology.

of textile terms are left broad and open. Thus, when reviewing the Akkadian textile vocabulary in the dictionaries, a striking number of words are simply translated: "a textile", "a garment" or "a tool". Many of these terms occur only in lexical lists, or in texts written in the common uninformative tone of self-informedness. For this reason, it is well-nigh impossible to reach any clear understanding of what these objects actually were.

It is striking just how often textiles are mentioned in the cuneiform sources. Plainly, regardless of period, geographical area and sphere of society, textiles had enormous economical and social importance. This just makes it all the more problematic that the translation of the terminology is so vague. In the present article I hope to demonstrate that it is often possible to offer more precise translations of words in this technical vocabulary with a knowledge of the actual physical procedures involved in ancient craft production.

There are a number of problems to be faced during this exercise. First is an issue of representativeness. Although an overwhelming number of texts written on clay tablets in the Akkadian language has come down to us, there are numerous aspects of society that are less than well illuminated by the written evidence. There is, so to speak, a periphery to the literate world. Textile production often belongs to this periphery, as it is frequently associated with the private sphere and the female gender. Also, as in many other ancient societies, Mesopotamia was home to a large textile production administered by palaces and temples and recorded by bureaucrats. Yet, the terminology of administrative records kept in such large organisations tends to be generalised and focus on raw materials and products rather than on the actual work procedures and tool repertoire. Plainly, there existed a rich vocabulary of technical expressions and terms beyond those noted in our texts, and we cannot simply assume that the extant textile terminology preserves a complete record of ancient processes and production. In addition, the types of texts that have come down to us from each individual period and region are not uniform, which can present a problem of comparison.

The methodology is very simple: the presence of a linguistic term of a given procedure or tool implies its existence in the society where the language was spoken. It follows that a study of a particular technical vocabulary in a diachronic perspective can inform us about the nature and evolution of craft and production. Whenever a new technique or technology was invented it needed a name. Often, it seems, this new name was provided by just adding another meaning to a word already present in the vocabulary. In addition to a general sense, the word would adopt a technical meaning. In such cases only the new usage of the word reveals its new specific and technical meaning.

Loanwords from foreign languages were also adopted, most importantly Sumerian, which has provided Akkadian with a great number of words basic and fundamental to the textile craft, e.g. *qû*, "yarn, thread", and *kītum*, "linen".[5] A less apparent feature of the Akkadian textile vocabulary is the large number of terms that occur in texts from the site of Nuzi and/or are loanwords from Hurrian. Oppenheim suggested already in 1966 that a revolution in dyeing technology took place at Nuzi.[6] Indeed, the great variation in the textile vocabulary, especially with regard to work procedures, indicates that several techniques may have been developed here in addition to dyeing.

[5] Note, however, that it cannot be excluded that these words were borrowed into Sumerian from Akkadian and not the other way around.

[6] Oppenheim 1966.

1. Textile terminology in private letters

From the Old Assyrian private letters found at the site of Kültepe in Central Turkey we catch a glimpse of the diverse and complex Akkadian textile terminology as employed by private craftspeople. The women of this period, as presumably in most other periods, spun and wove fabrics in their own home. Their husbands or male relatives sold some of these textiles on their behalf, and the women were thus able to earn some extra cash for themselves.[7] The travelling merchants selling goods in Anatolia seem to have had a thorough knowledge of the techniques of weaving, and they were able to send home very detailed specifications of what was in demand on the markets in Anatolia. Among the most discussed and interesting of these texts is TC 3, 17, which is a letter from the merchant Puzur-Aššur to the lady Waqqurtum[8]:

ša ṣú-ba-tim pá-na-am	Let them full
iš-té-na-ma li-im-šu-dù	one side of the textile.
la i-qá-tù-pu-šu	They should not teasel it.
šu-tù-šu lu ma-da-at	Its warp should be close, so compared
i-ṣé-er pá-ni-im	to the textile you sent previously
ṣú-ba-tim ša tù-šé-bi$_4$*-li-ni*	process one more mina
ša-áp-tám 1 *ma-na-ta*	of wool into each textile
ra-dí-i-ma lu qá-at-nu	– but keep them fine/thin.
pá-na-am ša-ni-a-am!	The other side
i-li-la li-im-šu-du-šu	they should full *lightly*.
šu-ma ša-ar-tám i-ta-áš-ú	If they want to raise the nap,
ki-ma ku-ta-nim li-iq-tù-pu-šu-nu	they should teasel them like a *kutānum*

The text is an instruction in how Waqqurtum should make her textiles in order to sell them on the market. Lines 11–13 seem to be concerned with the finishing treatments of one side of such a textile, lines 14–18 with the warping, lines 19–22 with the finishing treatments of the other side of the textile, and lines 34–37 with the size of the textile.

More wool can be added to a textile by either using thicker threads or increasing the number of threads. If the threads are thicker the textile will become coarser, whereas if the thread count is higher, the textile will become denser.[9] In lines 14–18 Puzur-Aššur requests that the fabric should be denser, but that it should retain its fineness by increasing the thread count.

After weaving, each side of the textile is exposed to two basic treatments: *mašādum* (l. 12 and l. 20) and *qatāpum* (l. 13 and l. 22). A third effect (l. 21 *ša-ar-tám i-ta-áš-ú*) is attainable if the textile is exposed to *qatāpum*. The translation of these processes is disputed, and the standard dictionaries offer incompatible suggestions.

It is agreed that the verb *mašādum* has the basic meaning "to beat, to hit". In technical contexts CAD *s.v.*, 352 adds the technical meaning "to comb wool", connecting *mašādum* with the noun *muštum* (gišga-zum), "a comb". The dictionary even refers to TC 3, 17 specifically, although it provides no translation of the passage. However, the connection between *muštum* and *mašādum* is uncertain, and a *muštum* is only used to comb wool – not finished textiles. Also, it is never mentioned alongside woven textiles. Finally, it is difficult to see how exactly the abovementioned

[7] Veenhof 1972, 103–123.

[8] For lines 1–10 and 23–37 see Michel & Veenhof pp. 249–251, in the present volume.

[9] Mårtensson, Andersson, Nosch, Batzer 2007.

textile could be "combed", and the suggested translation seems unlikely in the context.[10] Instead of "to comb", the AHw II, 623 translates *mašādum* as: "walken?". Fulling is indeed a standard procedure in finishing a woollen textile, and this translation seems preferable.

The second treatment the textile was exposed to is *qatāpum*, a verb which CAD *s. v.* translates as "to pluck, pick", and in relation to textiles, d) "to pick off bits of wool from the surface of a textile." In CAD Ṣ, 222–223, however, we find the translation of *qatāpum*: "to teasel." AHw II, 907 has "heraus, -abpflücken", and in regard to textiles: 1) "Haare aus Stoffen." In their treatment of the text, Michel & Veenhof disagree with both dictionaries, and translate *qatāpum* "to shear".[11]

The suggestion "to pick off bits of wool from the surface of a textile" seems unlikely. Cleaning a finished textile for any bits of wool, threads or dirt is standard procedure and does not change the textile in any particular way. It would thus not be a feature qualifying a textile. Instead, *qatāpum* is connected to the third effect mentioned in line 21: *šartam našā'um*. The meaning of *našā'um* (to lift, carry) seems to acquire a technical meaning in connection with *šartum* (hair). The verbatim translation "to lift the hair" points to a technical process known from many other contexts as "to raise the nap".[12] If this is indeed the correct understanding, it means that *qatāpum* must be a process similar to "teaselling", as suggested by the CAD vol. Ṣ.[13]

In line 22, TC 3, 17 states that the *kutānum* textile was also teaselled. The *kutānum* is the most common type of textile mentioned in the Old Assyrian sources – more than 30,000 individual pieces are attested in some 800 texts. Scattered references to *kutānus* used for clothing, similar to TC 3, 17 make it possible to get a vague idea of what a *kutānum* could have been, but no certain conclusions can be drawn. The etymology of the word is debated.[14] The most obvious connection is to Akkadian *kitū* and Aramaic *ktn*, "linen", and perhaps also Phoenician *ktn*, "coat". Much emphasis has been placed on the fact that the *kutānum* could be made of wool,[15] leading to the rejection of an etymological relation between *kutānum* and a word for linen *kt(n)*.[16] In this context it is interesting to note the apparent similarities between Akkadian *kutānum*, Hebrew *kuthnoh* (cotton), Arab *quṭun* and Hebrew *kuttoneth* (tunic, coat, shirt), and its relation to Arab *kattan* (flax, linen), W. Semitc *ktn* (flax, linen), and Akkadian *kitūm* (flax, linen). This root was adopted into Greek *khitōn* (Linear B *ki-to*), Latin *tunica* and perhaps also Hittite *kattanipu*.[17]

A modern equivalent to this shift in meaning over time is the usage of the word *linen* in English, which today only rarely refers to a textile that is specifically made of flax. More commonly, it denotes cotton textiles, as in "bed linen" or "table linen".[18] What characterises "linen" in this

[10] Note, however, that textiles with the adjective "combed" also appear in Linear B texts from Mycenae. See Del Freo, Nosch & Rougemont in this volume.

[11] Michel & Veenhof in this volume, p. 251.

[12] The writing *i-ta-áš-ú* is ambiguous. I follow Veenhof's (1972, 104) understanding of the inflexion as a Gt durative, but on the basis of the suggested meaning of *šartam našā'um* as a *terminus technicus* describing the raising of the nap, I prefer a different reading than Veenhof's: "If it is still hairy".

[13] CAD Ṣ, 222–223. Teaselling is a treatment of a woven textile in which the surface is rubbed with a thistle/teasel pulling out the fibre ends to give it a soft and slightly fluffy surface.

[14] See Michel & Veenhof in this volume pp. 216–217.

[15] E.g. TC 2, 7:24–26: *a-na ku-ta-ni ša ta-áš-ta-na-pá-ra-ni ša-ap-tum šu-ur-bu-i-tum lá-šu*: "Concerning the *kutānus* that you keep writing about – there is no *šurbuītum*-wool available."

[16] Cf. CAD K, 608; Landsberger 1967, 158.

[17] Mallory & Adams 2006, 237.

[18] *Oxford English Dictionary*: *s.v.* "linen", 3: "Garments or other articles made of linen; often by extension applied to garments normally or originally made of linen, even when other materials are actually used."

context is not that it is made from flax, even though this is indicated by the etymology, but rather its dense tabby weave (also called "linen weave"), its white or light (bleached) colour, and its stiff quality. Fabrics of other fibre types, but with the same qualities, fall into the same category. It is possible that the same principle was valid in the case of *ktn/kutānum*; perhaps it was originally used to refer to a type of textile made of flax, but that its properties – weave, colour and texture – eventually became a more important characteristic than the fibre it was made of.

Meaning shifts of this type presents an obstacle in the translation of technical terminology. Etymology becomes a very unsecure source of information and placing too much emphasis on it can at times even be misleading. It is necessary to consider the usage and context of these terms in the different periods and combine it with the technical knowledge we possess.

2. Procurement of wool and spinning

The textile tool repertoire of the Akkadian-speaking cultures as represented in the written evidence is limited. In the category of tools for obtaining raw materials we have only 3 words attested. The first is *magzazu*, which occurs only in Standard Babylonian lexical lists. The word derives from the verb *gazāzum*, which I will return to later; a tentative translation would be "shearing blade". In addition, there is the *serpu* or *sirpu*, "shears", which occur in lists of tools, among items given as a dowry etc. In the texts is often stated that shears were made of iron and were for shearing (*gizzu*). So far, the term occurs exclusively in Neo-Babylonian texts. Finally, we have *muštu šipāti*,[19] "wool comb", occurring from Old Assyrian and Old Babylonian times onwards. Other words for comb are attested, but none seem to appear in contexts that connect them to wool or textile production.

In spite of its limited size, this vocabulary of tools coincides well with what we know of shearing in ancient times: in the Bronze Age wool was plucked, combed, and shaved off the back of the sheep, and only in the Iron Age, with the arrival of iron shears, could wool also be sheared.[20]

The verbal act that goes along with the shears is *gazāzum* – a word that occurs already in Old Akkadian texts, but which seems to have undergone a shift in meaning at some point. In the Old Babylonian texts *gazāzum* is used only of goats, whereas *baqāmum*, "to pluck", is the verb preferred for the action of obtaining wool from sheep. In Neo-Babylonian texts, however, *gazāzum* appears with the shears, and is used instead of *baqāmum*. It seems likely that *gazāzum* initially meant "to cut hair" e.g. wool off goats, but that later, with the invention of shears, it came to be used with the meaning "to shear".[21]

Only the name of a single tool used for spinning fibres is attested in the vocabulary, namely *pilaqqum* "the spindle". In the texts spindles occur as prestigious items made of gold, silver, lapis and rare types of wood given as royal gifts, and such spindles have indeed been found in excavations.[22] Furthermore, lexical lists record words for different parts of the spindle, for instance, the spindle whorl (*qaqqad pilaqqi*), words for different types of spindles, such as the hooked spindle (*pilaq qarni*), and a word for spindle container (*bīt pilaqqi*).

[19] The comb, the spindle and other "women's items" often occur together also in magical texts.

[20] Barber 1991, 29.

[21] Similarly, Barber 1991, 261 suggests that the mistake of not differentiating between plucking and shearing was made by Liddell and Scott in their Greek-English dictionary in the translation of πεκειν (*pekein*) as "to comb, card; shear". Note also that there seems to have been some confusion of *gazāzum* and Sumerian giš-ga-zum "comb".

[22] See Barber 1991, 56–65.

Likewise, a group of words refers to the process of spinning. Most common is *ṭamûm* or *ṭawûm*, which denotes spinning in general, while *patālum* seems to refer only to the spinning of plant fibres and intestines, and could thus perhaps be more accurately translated as "to twine". Also the word *eṣēpum* is usually translated "to twine", but from the contexts it appears in, I would suggest that the precise technical meaning should be "to ply". The basic meaning of the word *eṣēpum* is "to double", and beside contexts related to textile production, it occurs also in mathematical texts. An example of its use comes from the text known as *Šurpu*: "she plied with white wool and black wool a double-stranded thread with a spindle, an excellent thread, a great thread, a double-coloured thread".[23] In Sumerian the equivalent to *eṣēpum* is tab, a sign that consists of just two parallel wedges. Tab appears in a sequence of textile procedures: nu-nu: to spin, tab: to ply, zé: to prepare the warp, and tag-tag: to weave.[24]

In some texts, three instead of two strands are plied into a thread. This presumably made the CAD suggest an additional and rather different meaning of the verb, namely "to plait".[25] There is, however, no compelling evidence to suggest that the word should have had this connotation, and of course, plying can be done with more than two strands.

3. Warping and weaving

Before weaving can commence, the yarn must be set up on the loom in a process called warping. Two words document this process, *dêpu* and *šatû*. Both are known from bilingual lexical lists that do not offer any details of the process. The warping process is naturally unattested in the archaeological record, but there are pictorial representations on cylinder seals that might depict it.

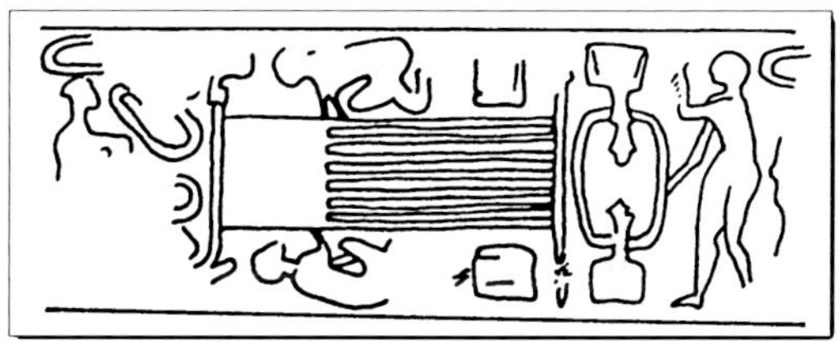

Fig. 13.1. Cylinder seal from Susa (from Barber 1991, p. 84)

The seal shown above seems to illustrate the process of warping on what has been suggested to represent a ground loom.[26] If this interpretation is correct, the ground loom was in use in Elam, and probably also Mesopotamia, from very early on. Another type of loom in use in ancient Mesopotamia was the vertical two-beam loom, which Elisabeth Barber has suggested was invented at Ebla in the 3rd millennium on the basis of this town's supposed outstanding position as a textile production centre.[27] The Akkadian loom terminology is relatively sparse and not very well understood, and we cannot say for sure when and where the various types of looms were used. Words for looms and loom parts come mostly from lexical lists, which makes accurate translations of the terms difficult. Nevertheless, there is occasionally some room for progress in our understanding and identification of an Akkadian loom terminology. In the CAD we find

[23] Reiner 1958, *Šurpu* V–VI 150–153: *ši-pa-a-te pi-ṣa-a-te ši-pa-a-te ṣal-*[*ma-a-te*] *qa-a iṣ-pa ina pi-lak-ki iṭ-me qa-a ṣi-ra qa-a* gal-*a qa-a bit-ru-*[*ma*...].

[24] van Dijk 1953, 65.

[25] CAD Š/3, 383.

[26] See e.g. Breniquet 2008, 295.

[27] Barber 2000.

the term *asû*, which is translated simply as: "a wooden part of a loom."[28] However, a closer look reveals further details. Firstly, *asû* is a word also used for a part of a door. Secondly, an upper, *elû*, and a lower, *šaplû*, *asû* is attested. Finally, the *asû* tend to come in pairs in the texts. On the basis of these three circumstances, it seems plausible to suggest a more precise translation and interpretation of the term: that the *asû* is a part of the frame of a loom – as well as the door – probably the transversal beams of the loom and the threshold and the lintel of the door. Such beams make up the frame of both the horizontal and the vertical loom, but he existence of an "upper" and a "lower" *asû* indicates that the loom which the *asû* are part of is in fact be the vertical two-beam loom.[29]

The word is first attested first already in Old Akkadian texts, and it occurs also in Old Babylonian and Middle Babylonian sources, at Nuzi, and in Neo-Assyrian lexical list. As already mentioned, this does not suggest that the loom type was used only in these periods, but it does demonstrate its existence, and probably over a large geographical area. This is more than could be deduced from the archaeological evidence. Plainly, it also means that we must reconsider Barber's proposal that the vertical two-beam loom is primarily a Syro-Palestinean tool invented at Ebla.

An important aspect connected to the two-beam loom is its use in the weaving of tapestries and knotted pile. Evidence for these textile types comes from a Middle Assyrian inventory text from Kār-Tukultī-Ninurta, which records precious objects brought into the palace of that city.[30] All the objects are carefully described, and figuring on the list, immediately below the royal throne and some illegible object, are two valuable textiles.

The first word of the text, which gives the name of the textile, is unfortunately broken. It reads:

[1 túg x] *ša iš-pár ka-ṣi-ri ù* [x	1 [*x*-textile] (the work) of the knotter ...
[x x x] *lu-ri-*DU*-e i-na ša-*[*ap-li-ti*]	[... depicting] a pomegranate tree ...
ù ú-ma-am-tu ta-ra-ha ù [x	a female ibex un[der a tree?]
te-qi-a-tu-šu zi-qu ia-ú-r[*u*	is its decoration.[31] A decorated edge of
gu-ha-ṣu-šu ša ṣir-pa-ni eš-r[*u* ...]	rosettes ... The fringe of colored wool is ...
----------------------------	----------------------------
1 ^{túg}*mar-du-tu ša* 5 *pi-*x [x	1 *mardātu*-textile with 5 [x] (the work)
ša ši-pár uš-bar *bir-mu-šu* [x	of a 'weaver of colourful textiles' ...
*ni-še ù ú-ma-ma-ni a-*x-[x	of people and wild animals...
ša uru-didli *du-un-nu ù* [x	of different towns and
ṣa-lam lugal *i-na* ^{giš}*ma-as-*[x	fortifications, and...
^{giš}x*-a-hu* [*n*]*a-ha-*x-[x	the image of the king on a pedestal(?)...
ṣa-lam lugal *i-na* [x	the image of the king on a x...

Köcher, who published the text, restored [1 ^{túg}*mar-du-tu*] in line 27, presumably on the basis of the *mardātu*-textile being mentioned in the second section of the text, line 32. However, the craftsmen who produced the two textiles clearly were of different professions – one was a knotter, the other a type of weaver – and the detailed description of the two textiles shows that they were of a different nature. Köcher's restoration thus seems unconvincing. The first textile is said to

[28] CAD A/2, 347.

[29] For an image of the vertical two-beam loom and the warp-weighted loom, see Andersson Strand, this volume.

[30] Köcher 1958, 300–313.

[31] For the translation of *tēqītu*, see Barrelet 1977, 57(b).

Fig. 13.2. Neo-Assyrian gypsum door-sill designed to appear as a carpet (BM 124962).

have been produced by a *kāṣiru*, "knotter". Our earliest attestation of the word appears in an Old Assyrian text,[32] but it becomes frequent from the 16th century BC onwards.[33] Only a few texts state clearly which types of fabric the knotter produced, but in some instances *lamhuššu*-textiles are mentioned.[34] The only connection between the *mardatu* and the *kāṣiru* is the unconvincing restoration made by Köcher. Regrettably, no knotted carpets have survived in Mesopotamia, but they are known from pictorial representations, for instance from the rugs carved in stone in the doorways of Neo-Assyrian royal palaces.[35]

The second textile mentioned, the *mardatu,* is woven by an "*išpar birmi*", a type of craftsman who wove textiles in multiple colours. I would like to suggest that the *mardātu* is in fact a type of pattern weaving, perhaps tapestry, and that the *išpar birmi* was a tapestry weaver. As mentioned above, words, even technical terms, can change their meaning over time, and *išpar birmi* seems to be an example of this phenomenon. In the present context *išpar birmi* denotes a weaver of a specific type of textile. In the Neo-Babylonian period, however, the term simply means "textile worker".[36] The term thus developed from having a very specific meaning to a more general meaning.

The craftsman producing the *mardatu* is a specialised weaver holding his own occupational term. Textual evidence from Mari mentioning *mardatu* shows that it was an expensive type of

---

[32] CCT 5, 32b.

[33] CAD K, 264: "a craftsman producing textiles by a special technique." AHw I A-L, 458: "Knüpfer, Gewandschneider(?)" CDA, 151: "knotter".

[34] E.g. Camp. 245:4, 7

[35] Dalley 1991.

[36] I thank Stefan Zawadzki for pointing this out to me.

Fig. 13.3. Clay model of an Assyrian king from Nineveh (British Museum: WA 93011) and detail of relief of protective spirit from Nimrud (British Museum: WA 124561).

textiles exchanged as royal gifts. For instance, one *mardatu* carrying the image of a *lamaštum* was given by a certain Mukannišum to the king of Babylon.[37] There are also records of *mardatu* sashes, *mardatu* loin-cloths, *mardatu* cushions, *mardatu* curtains, "large *mardatus*", and *mardatu* garments.[38] The *Akkadisches Handwörterbuch* translates *mardatu*, as: "*ein Teppich?*".[39] However, it is important to note that a textile woven in tapestry technique is not necessarily a carpet, or another heavy type of textile for that matter. This weaving technique is used also for producing garments and other light fabrics. Unfortunately, the suggestion offered in the AHw has later led to some curious translations, such as the: "1 *ku-sí-tám ma-ar-da-a-tám*" in an Old Assyrian letter: "ein *kusîtum*-Gewand in Teppich Art?".[40] The *kusītum* is a garment wrapped around the body, presumably a type of robe or mantle, which in the first millennium BC. could be elaborately decorated, and was worn by gods, goddesses and kings on special occasions.[41] *Kusītus* could be purple, red and 'multicoloured' (*birmu* and gùn), and a list from Nuzi records wool of many different colours used in the making of *kusītus*. If we accept the translation of *mardatu* as a 'tapestry woven textile', the meaning of the word becomes much clearer. Then the *kusītu mardatu* is what we see on Neo-Assyrian reliefs as worn by the king and his elite.

In the present paper I have used some of the extant Akkadian textile vocabulary to review the

[37] Durand 1983, 410.

[38] Durand 2009, 63 and 107.

[39] AHw II M-S, 611.

[40] CCT 1, 29: 1 túg *ku-sí-tám ma-ar-da-a-tám ù sà-he-er-tám a-na ba-ru-lim ša Pá-ha-tí-ma*. (note also the occurrence of an equivalent at Nuzi: *ku-sí-tù ša mar-dá-ti-im-ma* in HSS 13, 431, 46). Durand 1983, 409 n. 58.

[41] Postgate 2001, 378–381.

different stages of the textile production in ancient Mesopotamia. These terms, however, make up only a tiny sample of a rich vocabulary concerned with textiles and textile production. With a basic understanding of the technology involved in the making of textiles, I have attempted to show that the translations offered by our dictionaries at times leave room for improvement and specification. And such specifications can in turn reveal information about the history of textiles. Behind a word, such as *asû* – the wooden part of a loom, – lies information about the existence of a specific technology. Behind *mardātu* a concretisation of the images of dress we see in wall paintings and on stone reliefs.

Etymology can provide hints to the original meaning, but meaning shifts occur, and it is important to look at the terms in their different usages from period to period to get a fuller understanding. At times, placing too much emphasis on etymology can even be misleading.

It is interesting to note that so many different translations of the text TC 3, 17 occur. Several suggestions for each of the techincal terms mentioned in the text can be found in the dictionaries and in the literature, as is the case in this volume. It is in some way symptomatic of the difficulties we face when dealing with textile terminology.

When most of the primary evidence is missing – the tools as well as the textiles – we must be creative in order to reconstruct the past. We must tease out bits and pieces of information from different sources and put them together with the hope of getting the broader picture. Hopefully, this article shows how a combination of linguistic evidence and practical knowledge of the actual technology can be used in the reconstruction of the history of textiles.

Abbreviations

AHw W. von Soden, *Akkadisches Handwörterbuch*. Wiesbaden, 1958–1981.
AKT Ankara Kültepe Tabletleri = Ankaraner Kültepe Texte. Ankara.
BIN Babylonian Inscriptions in the Collection of J. B. Nies. New Haven.
CAD *The Assyrian Dictionary of the Oriental Institute of the University of Chicago*. Chicago (1956–).
CCT Cuneiform Texts from Cappadocian Tablets in the British Museum. London.
CDA J. Black, A. George & N. Postgate, *A Concise Dictionary of Akkadian*, Wiesbaden, 2000.
HSS Harvard Semitic Series. Cambridge, Mass.
KBo Keilschrifttexte aus Boghazköi. Berlin.
RA *Revue d'assyriologie et d'archéologie orientale*. Paris.
TC Tablettes cappadociennes du Louvre. Paris.

Bibliography

Barber, E. J. W. (1991) *Prehistoric Textiles: The Development of Cloth in the Neolithic and Bronze Ages with special reference to the Aegean*. Princeton.

Barber, E. J. W. (2000) *The beginnings of tapestry weaving*. Second Annual GFR Lecture.

Barjamovic, G. (2005) *A Historical Geography of Ancient Anatolia in the Assyrian Colony Period*. University of Copenhagen, Unpublished PhD thesis.

Barrelet, M.-T. (1977) Un inventaire de Kār-Tukultī-Ninurta: Textiles décorés Assyriens et antres. *Revue d'Assyriologie* 71, 51–92.

Bilgi Ö. (2001) *Metallurgists of the Central Black Sea Region. A New Perspective on the Question of the Indo-European Original Homeland*. Istanbul.

Breniquet, C. (2008) *Essai sur le tissage en Mesopotamie des premieres communautes sedentaires au milieu du IIIe millenaire avant J.-C.* Paris

Burke, B. (2005) Textile production at Gordion and the Phrygian Economy. In L. Kealhofer (ed.) *The archaeology of Midas and the Phrygians: recent work at Gordion*, 62–81. Philadelphia.
Curtis, J. E. and Reade, J. E. (eds.) (1995) *Art and Empire. Treasures from Assyria in the British Museum*. London.
Dalley, S. (1991) Ancient Assyrian Textiles and the Origins of Carpet Design, *Iran* 29, 117–135.
van Dijk, J. J. A. (1953) *La Sagesse suméro-accadienne*. Leiden.
Durand, J.-M. (1983) *Textes Administratifs des Salles 134 et 160 du Palais de Mari*. Archives royales de Mari 21. Paris.
Durand, J.-M. (2009) *La nomenclature des habits et des textiles dans les textes de Mari*. Archives royales de Mari 30. Paris.
Forbes, R. J. (1956) *Studies in Ancient Technology* 4. Leiden.
Köcher, F. (1958) Ein Inventartext aus Kār-Tukultī-Ninurta. *Archiv für Orientforschungen* 18, 300–313.
Landsberger, B. (1967) Über Farben im Sumerisch-Akkadischen. *Journal of Cuneiform Studies* 21, 139–173.
Lassen, A. K. W. (2008) *Textiles and Traders: A study of the technology, organisation and socioeconomics of textile production in Anatolia in the Middle Bronze Age, c. 2000–1700 B.C.* Unpublished MA-Thesis. University of Copenhagen.
Mallory, J. P. & D. Q. Adams (2006) *The Oxford Introduction to Proto-Indo-European and the Proto-Indo-European World*. Oxford.
van de Mieroop, M. (1987) *Crafts in the Early Isin Period: A Study of the Isin Craft Archive from the Reigns of Išbi-Erra and Šu-Illišu*. Orientalia Lovaniensia Analecta 24. Leuven.
Mårtensson, L., E. Andersson, M.-L. Nosch, A. Batzer (2007) *Technical Report. Experimental Archaeology: Part 3 Loom weights* Centre for Textile Research, Copenhagen. Available at:
http://ctr.hum.ku.dk/upload/application/pdf/f51d6748/Technical%20report%203,%20experimental%20archaeology.PDF
Oppenheim, A. L. (1966) Mesopotamia in the Early History of Alchemy. *Revue d'Assyriologie* 60, 29–45.
Postgate, J. N. (2001) Assyrian Uniforms, In van Soldt, W. H. (ed.) *Veenhof Anniversary Volume. Studies presented to Klaas R. Veenhof on the occasion of his sixty-fifth birthday* Publications de l'Institut historique-archéologique néerlandais de Stamboul LXXXIX, 373–388. Leiden.
Reiner, E. (1958) *Šurpu. A Collection of Sumerian and Akkadian Incantations*, Graz.
Sasson, J. (ed.) (1995) *Civilizations of the Ancient Near East*. London.
Veenhof, K. R. (1972) *Aspects of old Assyrian trade and its terminology*. Studia et documenta ad jura Orientis antiqui pertinentia 10. Leiden.
Waetzoldt, H. (1972) *Untersuchungen zur neusumerischen Textilindustrie*. Studi Economici e Tecnologici I, Rome.

14. Les textiles du Moyen-Euphrate à l'époque paléo-babylonienne d'après un ouvrage récent

Anne-Claude Beaugeard

L'histoire du textile à l'époque paléo-babylonienne dans la région du Moyen-Euphrate a été renouvelée par J.-M. Durand grâce aux informations fournies dans son ouvrage *La nomenclature des habits et des textiles dans les textes de Mari, Matériaux pour le Dictionnaire de Babylonien de Paris,* tome 1, Archives Royales de Mari 30, 2009, Paris. Grâce à sa longue expérience du corpus mariote, J.-M. Durand est à même de proposer un lexique complet des termes désignant les textiles, et nombre des traductions qu'il propose s'écartent sensiblement des traductions fournies par les grands dictionnaires d'akkadien actuellement disponibles.

On trouve dans ce livre un inventaire des termes désignant toutes les catégories des textiles à usages divers présents dans le corpus des textes de Mari, avec un certain nombre d'éléments provenant d'autres sites syriens de l'époque paléo-babylonienne. L'auteur souligne dans son avant propos qu'il ne s'agit pas d'une synthèse sur l'industrie textile ou sur les habitudes vestimentaires à Mari. Néanmoins, ce livre est plus qu'un simple lexique, car l'auteur présente ses réflexions au fil des pages et, grâce à cette méticuleuse recherche, les historiens disposent maintenant d'un matériau particulièrement fiable. Le corpus de base ne représente pas l'ensemble des textes exhumés à Mari mais les textes dont disposait l'auteur, en fonction de la constitution des lots de textes pour leur publication. Néanmoins, l'échantillon de textes est suffisamment important pour pouvoir proposer un lexique que l'on présume complet.

Malheureusement pour l'historien des textiles, les textes dont nous disposons ne sont pas les archives de gestion de la production textile, mais les archives de la gestion des stocks textiles : le vocabulaire spécifique de la production manque presque totalement. L'allocation générale de vêtements distribuée par le palais se dit túg-ba/*lubuštum*[1]. Il peut aussi y avoir une allocation de laine (síg-ba). En dehors des vêtements, le palais gère beaucoup d'étoffes qui servent uniquement à l'"ameublement" (en général). Dans ce cas, les fonctions des différents textiles repérées par J.-M. Durand sont les suivantes :

- la literie :
  burê'um : la toile à matelas (destinée à être rembourré de roseaux), qui peut être faite de différents tissus, y compris les plus luxueux.

[1] Le terme donné en premier correspond à l'idéogramme sumérien utilisé dans les textes de Mari, et le terme écrit en italique correspond au mot akkadien attesté dans ces textes.

lîṭum : le drap? de lit.
nalbašum : la couverture en lin.
nâsisum : un type de couverture
túg-(níg)-barà/*uṣûm* : un type de couverture.

- l'équipement des équidés, des chars, des chariots, des bateaux et tentes diverses :
  duzurum : le tapis de selle.
  gaba-gál/*gabagallum* : le pectoral, le décor frontal d'un char.
  kammakum : la bâche du chariot, en laine ou lin.
  hayyû : le tapis de siège.
  hurpatum : le tissu en laine servant de velum dans une cour ou sur un bateau. Il faut imaginer une pièce de tissu d'une dimension considérable (son poids est de plus de 50 kg).
  kuštârum : la tente de bédouin.
  nakbasum/nakbisum : la carpette.
  napdû : la toile de protection (?)
  sîrum/zîrum : un tissu épais servant pour garnir un chariot, un palanquin ou une tente.

Le vocabulaire collecté dans cette étude est celui qui est utilisé dans le palais, et non celui des textiles fabriqués à domicile par les particuliers, à une exception près semble-t-il.

Les tissus de laine peuvent être travaillés de façon diverse, mais les termes qui servent à désigner les textiles sont souvent peu éclairants sur l'aspect fini qu'ils pouvaient avoir. On connaît les termes techniques utilisés pour désigner la chaîne (síg-ná-àm ou *bittum*) et la trame (*nîdum*). Les termes relatifs aux techniques de tissages et de finition des tissus relevés par J.-M. Durand sont les suivants :

aguhhum : le tissu de style "kaunakès" (tissage avec de longues mèches de laines qui ressortent). Il s'agit d'un tissage démodé au début du 2ème millénaire, qui n'est plus couramment porté et qui est réservé à la garde-robe des statues de dieux.
túg-bar-kar-ra/*barkarrû/barkarrîtum* : un tissu épais et résistant de qualité commune ou "tissu grossier".
sûnum : un tissu servant à faire un galon ou une ganse pour "ourler" les tissus. (Sur cette question, voir plus bas).
halûm : un tissu avec frange amovible, parfois tressée (?) de style *ṣipparim*.
hamdû : un tissu de petite taille et d'un tissage très fin.
huršânim/huršanîtum : un tissage "à la mode montagnarde".
si-sá/*išartum* : un tissu naturel ou "écru". Le palais gère de très grandes quantités de textiles non teints. Il semble que la majorité des textiles utilisés à Mari aient été de ce type.
ilbum : un tissu à poils longs
itqum : une frange, réalisée probablement par torsade de fils pour créer un "effet ressort". La frange est réalisée séparément de la réalisation du tissu et peut être ajoutée ou enlevée au gré des envies.
kap lurmim : un tissu décoré de plumes d'autruches : la question de la technique utilisée pour fixer les plumes d'autruches sur le tissu reste ouverte.
kabtum : un tissu "lourd", de grand luxe, à la mode uniquement à l'époque du début des archives de Mari.
mazzû : un tissu élamite.
narbâtum : un tissu fait de laine fine.
hurrurum ou *harrurum* : un textile rasé, très courant (équivalent du terme *gizzum*).
túg-guz-za/*gizzum* : un textile fabriqué de laine de piètre qualité (laine de mouton mort?) et rasé ensuite. Ce textile est utilisé pour l'ameublement.
lamahuššû : un tissu de laine assez précieux, qui peut servir comme toile à matelas.
sakkum : le tissage typique de Mari et de Haṣor, peut-être un tissage très serré. Ce tissu est utilisé aussi bien pour le vêtement que pour l'ameublement.
uṭba : le tissu royal et princier typique de Mari, servant à faire des capes pour hommes et femmes.

ša panûm : le tissu double face, réversible (pour le tissu de couverture).
qaršum : J-M. Durand n'a pas de proposition définitive pour ce terme. Peut-être s'agit-il d'un tissu où l'on a aménagé une fente permettant de le nouer facilement au niveau de la ceinture (fermeture style "portefeuille"). La mode n'est connue qu'au début des archives de Mari.
túg-sal-la/*raqqatum* : un tissu sergé assez léger. C'est l'un des textiles les plus courant à Mari.
šahhû : une toile résistante qui peut servir pour faire une tente ou une voile de bateau.
ša šûṣupum : un tissage avec un fil double (pour le tissu de couverture).
terdennûm : un tissu de seconde qualité.
ša tûšurum : un tissage lâche ou ajouré (pour le tissu de couverture).
uṭublum : un tissu sergé, assez rêche, c'est la qualité d'étoffe la plus courante à Mari, produite en grande quantité. Ce textile est toujours originaire de haute Mésopotamie et est probablement fabriqué dans les maisons. Les coupons semblent avoir une taille plus ou moins standard.
zakûm : un tissu lustré ou qui brille (?)
zîmum : une broderie figurative(?) effectuée sur un tissu destiné à servir de voile de luxe pour les femmes.

Le corpus couvre trois générations et semble indiquer des modes vestimentaires et textiles, car certains termes tombent en désuétude. Ces évolutions correspondent aussi à l'importance de telle ou telle ethnie sur le palais de Mari. Le vocabulaire listé est spécifique à Mari et ne semble guère avoir de correspondance dans les textes plus récents de la même zone géographique comme le corpus d'Emar. Beaucoup de termes de "technique textile", font en fait référence aux pratiques textiles des régions avec lesquelles le palais de Mari est en contact économique et politique. Dans la liste précédente, on aura par exemple déjà relevé le *mazzum,* tissu à la mode élamite. Dans de nombreux cas les qualificatifs textiles nous informent simplement sur le fait que ces textiles étaient considérés comme "exotiques" à Mari, sans que l'on sache si c'était en raison de leur couleur, du type de tissage, de leur dimension ou de la forme des vêtements auxquels ils servaient. Parfois simplement, on sait qu'il s'agissait de vêtements/tissus de prestige, qui devaient être représentatifs de l'artisanat de luxe de chaque région.

À défaut d'autre information, on apprend donc que les traditions textiles sont très spécifiques à chaque zone géographique ou à chaque ethnie. Le tissage comporte une forte dimension identitaire. Les toponymes et les qualificatifs ethniques attestés correspondent aux régions qui avaient des contacts économiques et politiques avec Mari. On a relevé les mentions suivantes :

ša bišri : l'étoffe à la manière du Bishri.
elamûm : l'étoffe de style élamite.
gublâyum : le tissu à la mode de Byblos.
huršânû : l'étoffe de Huršânum.
huššîtum : l'étoffe de Haššum.
iamhadu : l'étoffe du Yamhad.
kiššihhu : l'étoffe de Kiš.
laharû : l'étoffe de Lahâru.
lullûm : l'étoffe des Lullû.
maratû : l'étoffe de la côte Méditerranéenne.
nurrugayu : l'étoffe du Nurrugayu (Nord-est de la Mésopotamie).
paramšitum : l'étoffe à la mode de Marhaši.
sûhum : l'étoffe souhéenne.
šubarûm/šubarîtum : l'étoffe à la mode du Šubartum.
tuttubum : le tissu de Tuttub.

La compréhension des différents termes relatifs aux textiles est extrêmement difficile car on ne possède ni données archéologiques ni corpus figuratif suffisant qui permettrait d'éclairer le lexique présent dans les textes. De plus, le vocabulaire textile actuel signifie des réalités et des classifications tellement différentes de celles qui sont signifiées par le vocabulaire mariote que chaque ébauche de classification ou de traduction pose plus de problèmes qu'elle n'en résout.

Le premier problème est celui de l'emploi du mot actuel "habit". Dans quelle mesure est on en droit de considérer qu'un objet textile est spécifiquement un vêtement? Doit-on se contenter du vocable général "tissu"? J.-M. Durand propose comme définition minimale du terme "habit", "celle d'une pièce d'étoffe dont la destination principale est d'être portée par le corps humain, et qui est par là même dotée d'une certaine forme ("coupe"). Si la première partie de la définition peut faire consensus, la seconde partie de la définition est beaucoup plus problématique. La notion de "coupe" est extrêmement ambiguë. La notion de "coupe", suppose l'utilisation d'une géométrie complexe visant à créer des volumes tout aussi complexes qui ne peuvent être obtenus par simples pliures, drapés, serrages et fixations au moment de l'habillement lui-même, et il est peut-être difficile d'utiliser cette terminologie pour une première approche du lexique textile d'une période aussi reculée. La forme générique d'une pièce de tissu est le rectangle, plus ou moins parfait, de grande ou de petite taille. Il peut exister des techniques de tissage créant d'autres formes, mais cela reste à établir. On devrait s'en tenir à une notion de dimension et de qualité de l'étoffe, même si, précisément, les textes mariotes ne nous donnent pas d'indications de dimensions, ni très souvent de poids des tissus, ce qui ne facilite pas l'enquête.

Le terme générique à Mari pour dire l'étoffe est túg/*ṣubâtum.* Deux sumérogrammes servent à classer les textiles : le classificateur gú-è-a s'oppose au classificateur túg. La distinction essentielle est donc entre le tissu "par lequel passe la nuque" et les autres tissus qui servent ou non à l'habillement. Le premier est traduit par "chemise" et le second par "étoffe", mais les deux sont portés.

Une question importante dans le cas d'étoffes-vêtements, est celle de l'existence ou non d'une *façon permanente* sur ces étoffes. La *façon permanente* peut provenir certes d'une coupe de la pièce initiale de tissu, mais concerner également la création d'un volume permanent. La "forme" du vêtement est alors liée à une pratique d'assemblage du morceau d'étoffe sur lui même, ou de plusieurs morceaux d'étoffes entre eux de façon permanente. C'est à proprement parler la technique de la couture. Des pièces assemblées peuvent faire l'objet de désassemblage pour récupération. La question du décor surajouté au tissu est également importante car le textile est avant tout un élément de *parure*, au moins pour les parties supérieures de la société. Il n'y a que quelques indices de formes particulières pour les vêtements de Mari (voir ci-dessous la liste des vêtements particuliers recensés). Pour le reste, J.-M. Durand utilise le vocable général de "finitions", et l'on pense surtout à des broderies ou à des techniques qui concernent l'obtention de boucles ou de fils qui sortent du tissu et qui sont coupés ou rasés en complément du simple tissage. Il semble même que l'on pouvait enfiler des perles sur le tissu au fur et à mesure du tissage.

Un élément important de cette enquête lexicale est l'existence, selon J. M. Durand, d'une profession particulière, celle de "couturier" (lú-túg, "tailleur" selon sa terminologie), à côté du simple tisserand (lú-túg-du$_8$). Le premier terme ne serait pas la simple abréviation du second. Cette information est importante car cela laisse entrevoir que les tissus utilisés en guise de vêtement ne sont pas de simples étoffes rectangulaires de taille ou de qualité différentes, mais peuvent recevoir un début de "façon" (assemblage par couture, coupe ?...). On note que des éléments tels que les coiffes peuvent être confectionnées avec plusieurs pièces de tissus qui ne sont pas forcément cousues entre elles mais simplement arrangées directement sur la personne.

La technique du ravaudage des vêtements par ajout de petits morceaux de tissus (*pirsum*) sur l'étoffe trouée est l'affaire de spécialistes (lú-túg-kal-kal a*mukabbû*). Les vêtements peuvent donc être ravaudés (*kubbûm*). Le tissu peut être donné à repriser (*ana šûšurim*). Les vêtements sont parfois donnés à laver (*ana mesîm).*

En fin de vie, il y a de multiples attestations de la coupe des tissus afin de fabriquer différents bandages, langes etc... Le tissu est déchiré (*šarâṭum*) pour former des bandes (*ṣertum*).

Une des questions importantes quand on aborde les techniques de couture est celle de la finition des bords du tissu. Selon les exemples réunis par J.-M. Durand, il semble que les bords des vêtements étaient traités avec une ganse surajoutée (le galon-*sûnum*). Il semble que ces galons n'étaient pas considérés vraiment comme des tissus (une seule attestation de galon prêt à l'emploi). Souvent ce galon était classé parmi les laines. L'aspect de ce que nous traduisons par "ganse" ou "galon" n'est donc pas certain. Par ailleurs, la technique de l'ourlet simple semble être attestée à l'époque paléo-babylonienne (*sûn ramânišu* : "bord" pris dans l'étoffe elle-même). Dans la comptabilité du palais, il est très souvent fait mention de tissus qui sont *lâ sûnim* (sans ourlet). La pratique du gansage en bordure de vêtement est, selon J.-M. Durand, typique de la région du Yamhad : il ne s'agit donc pas, là non plus, d'une technique universelle, mais d'une tradition ethnique particulière.

Selon J.-M. Durand, il semble établi que certaines *formes* de vêtement existaient bel et bien, qui n'étaient pas simplement des rectangles de différentes dimensions. Néanmoins, on a en fait très peu d'indications sur les formes, et ce terme reste problématique en l'état de nos connaissances. On sait qu'un ensemble d'habillement complet se compose d'un petit nombre de pièces essentielles : la "chemise" (gú-è-a), le manteau, la ceinture, éventuellement le voile pour les femmes ou la coiffe pour les hommes, et bien sûr les sandales.

Le terme générique que l'on traduit par "chemise", (gú-(è-a)/*nahlaptum*) ne concerne probablement pas un vêtement ayant reçu un quelconque façonnage particulier, mais un tissu qui est principalement utilisé pour jouer le rôle de "chemise". La dimension du tissu doit entrer en ligne de compte, et probablement la possibilité ou non de draper l'étoffe sur la personne. Par contre, on ne sait pas si ces "chemises" faisaient l'objet d'un quelconque procédé de couture. En effet le tissu *nahlaptum* peut être parfois utilisé comme textile d'ameublement, en guide de "housse", ce qui serait difficile s'il avait reçu une véritable façon. J.-M. Durand semble penser que les "chemises" ont au moins des coutures latérales (d'où peut-être sa traduction par "housse" lorsqu'il s'agit de tissu d'ameublement), mais pour l'instant cela n'est pas prouvé. Il existe des gú-(è-a)/*nahlaptum* en lin (gad/*kitîm*). Le fait que ce terme gú-è-a/*nahlaptum* serve de classificateur à toutes sortes de textiles indique simplement l'usage prioritaire de ces textiles.

J.-M. Durand remarque que le terme utilisé pour signifier "se vêtir" (*labâšum/lubbušum*) correspond au terme akkadien qui rend le sumérien túg, terme générique désignant l'étoffe, et non au classificateur général gú-(è-a)/*nahlaptum*) que l'on traduit par "chemise". De ce fait on peut déduire que la notion de "chemise" (gú-è-a) est secondaire par rapport au simple fait de nouer un tissu autour de son corps, túg désignant le "manteau" mais aussi simplement le pagne.

Les types de vêtements attestés se répartissent en plusieurs groupes :

- celui des pagnes qui peuvent servir de "sous-vêtements" divers :
  maksûm : les langes, faits de tissus récupérés.
  nahramum : le pagne masculin.
  nâsisum šu iški : le cache sexe.
  túg-bar-si/*parsikkum* : c'est le vêtement le plus courant à Mari. Cela doit être une sorte de foulard que l'on peut nouer à différents endroits du corps, y compris autour des reins en guise de pagne. C'est en tout cas une pièce de tissu assez légère.
  túg *šusippum* : bande qui peut servir soit à faire un pagne, soit à enrouler autour des mains pour les protéger.

- celui des voiles féminins :
  kutummû : les voiles de la mariée (mot toujours utilisé au pluriel).
  mardatum : le voile brodé typique de la Syrie.

- celui des "chemises" :
  gú-(è-a)/*nahlaptum* : ce terme est un classificateur, et désigne en général toute "chemise", comme nous venons de le voir.
  nahZaBum : une chemise ornée de pierres précieuses.

- celui des ceintures :
  patinnu : la ceinture rigide.
  naṣbadu : la ceinture, le châle plié en bandeau, parfois en lin.

- celui des accessoires vestimentaires divers :
  ahatum : les manches (?). Le terme n'est attesté qu'une fois, dans le cadre de pièces de tissus récupérées. Cette réalité textile peut donc exister mais n'est visiblement pas courante à Mari.
  kaballu : en laine, typiques des montagnes, ce sont des sortes de bandes molletières ou des "chaussettes". Il en existe une variante en cuir.
  karikkum : les chaussettes (?). J.- M. Durand suggère que ce terme fasse référence à un tissage fermé sur lui même pour créer un tissu tubulaire. Le mot n'est attesté qu'une fois à Mari : il s'agit d'une réalité étrangère.
  rittum : des gants (?).
- les différents "manteaux" sont les suivants :
  dabadu : est un mot générique pour désigner un vêtement à la fin du 3ème millénaire et qui est encore en usage dans les archives de Mari les plus anciennes.
  kimura : une étoffe servant de manteau précieux, peut-être d'un blanc brillant, attesté seulement au début des archives de Mari.
  kussatum : une sorte de manteau, attesté seulement au début de l'époque des archives de Mari.
  kutânum : vêtement cappadocien en laine. La question de savoir si ce mot a un rapport avec le grec *chiton* reste problématique.
  lullumtum : un vêtement fait de laine babylonienne, connu à la fin du 3ème millénaire, mais passé de mode à l'époque de Mari et employé seulement en contexte liturgique.
  taddiâtum : un vêtement du couronnement, recouvert de pierreries, d'or et d'argent.

Enfin, beaucoup des mots du lexique désignent des morceaux de tissus, des bandages divers pour former d'autres formes textiles (coiffes). Il semble que l'on puisse fabriquer également des perruques en laine (*hupurtum*). Les cordages, sangles et les liens peuvent, outre le cuir, être en lin ou en poil de chèvre.

Les tissus sont conservés dans des sacs en cuir, des corbeilles en roseau, et mis dans des housses de lin destinées à les protéger.

Certains vêtements du type "chemise" (gú-è-a) sont en peau (kuš), et même formés de plusieurs peaux (kuš^{há}), ce qui signifie que ces morceaux de cuirs étaient cousus ensembles pour former des vestes. De même, les sandales (*mešênum)* étaient en cuir, garnies ou non de divers accessoires.

En général, les tissus attestés sont en laine, mais quelques vêtements en lin (gad/*kitûm*) sont connus également. On trouve :

- le tissu de lin *hîrum* qui est une bande de lin de dimension modeste, destinée à toutes sorte d'usages, aussi bien pour le vêtement (autour des reins, sur la tête...), que pour servir de lien ou pour empaqueter des produits.
- le šà-ga-dù/*nêbahum,* la ceinture. Le terme désigne un tissu de lin, probablement plus long que large qui sert en général de "ceinture", mais qui peut aussi être utilisé enroulé autour de la tête pour faire une sorte de turban (*agûm*).
- différentes sortes de liens (*riksum).*
- des bandelettes (*šusippum*) qui servent pour les reins (*birki*) ou pour les mains (*qatim*).
- quant au terme *kitîtum*, il semble bien désigner un tissage particulièrement fin plutôt qu'un simple tissu de lin.

Au total, l'usage du lin est bien moins courant à Mari que celui de la laine, et semble ne se présenter que sous la forme de tissus de petites dimensions.

15. Linen in Hittite Inventory Texts

Matteo Vigo

1. Linen terminology and trades: a general overview from the Ancient Near East

Linen[1] makes up the largest part of the Egyptian economy and is also its main export produce.[2] Egypt has, since the Old Kingdom, had a thriving craft industry based both on the cultivation of raw flax as well as the spinning, weaving and trading of linen cloths.[3] Egyptian linen is referred to in writings from the middle Babylonian period as a cloth used to dress statues of deities.[4]

Linen, like patterned textiles, is undoubtedly a high quality product and can rightly claim a place alongside luxury goods,[5] since it is always included in the exchange of gifts between sovereigns along with precious metals such as gold and silver.[6]

[1] Sumerogr. GADA (for all the transcriptions of the Sumerogram the updated HZL reading has been followed); Akkad. *kitû(m)*; Hitt. *kattanipū-*(?). Cf. HW, 105, 272; Tischler 2001, 76; HEG "A-K", 544; cf. Siegelová 1986, 376, note 14; 378, but with the example of Siegelová (1986, 621), it appears obvious that the Sumerogram has been interpreted wrongly (^{GAD}*tanipu-*). Cf. Alp 1940, 33, with note 2; Güterbock 1930, Vorwort 1, No. 1; Košak 1982, 125, rev. 28, for the forms ^{GAD}*tanipuliš* and ^{GAD}*tanipu* BABBAR. The still valid study by Forbes (1964, 27–43) has been referred to for a history of linen in antiquity. In general, for the most recent study on the trade and manufacture of textiles in the ancient Near East, Völling 2008 has been referred to with its extensive bibliographic references.

A list of attestations (GADA ideographic form with and without phonetic complements), updated to Siegelová 1986, with corrections: KBo VII 23, obv.$^{?}$ 2; KBo VII 26, obv.$^{??}$ right col. 1; KBo IX 89, rev. col. VI 5, 11; KBo IX 90, col. V 7; KBo IX 91, left edge 3$^{???}$; rev. B 5 (^{LÚ}GAD.TAR(?); Cf. Košak 1982, 29; HZL, 174, No. 173); KBo XVIII 154, rev.$^{?}$ left col. 9; KBo XVIII 170, obv. 7–9; KBo XVIII 175, obv. col. I 3–6; col. II 4$^{?}$, 6–7; col. V 7$^{?}$, 11$^{?}$, 13, 15; KBo XVIII 178, obv. 4; KBo XVIII 179, obv. col. II 6; KBo XVIII 180, rev. 5; KBo XVIII 181, obv. 2, 4–6, 13, 19–20, 24–26, 31; rev. 3–4; 13, 16–18$^{?}$, 25–28; KBo XVIII 183, obv.$^{?}$ left col. 2; KBo XVIII 184, rev. 5, 8, 9; KBo XVIII 185, obv.$^{?}$ 3; KBo XVIII 186, left edge 1, 3; KBo XVIII 187, obv.$^{?}$ 6–7; KBo XVIII 198, rev.$^{?}$ 3, 8; KBo XXXI 52 + KBo VII 25, obv.$^{?}$ right col. 8; KUB XII 1, rev. col. III 26, 31; col. IV 9; KUB XL 96 + Bo 1016, right edge 2; KUB XLII 11, obv. 7; KUB XLII 13, obv. col. II$^{?}$ 2, 4$^{?}$; rev. col. V 4; col. VI$^{?}$ 7; KUB XLII 14, rev.$^{?}$ col. IV$^{?}$ 3, 5; KUB XLII 15, rev. left col. 6; KUB XLII 16, rev. col. IV$^{?}$ 6, 8; col. V 11; KUB XLII 17 obv. left col. 4$^{?}$-5$^{?}$; right col. 3$^{??}$; KUB XLII 31, obv.$^{?}$ 2; KUB XLII 34, obv.$^{?}$ 10, left edge 1; KUB XLII 42, obv. col. II 9, rev. col. III 7$^{??}$ (GADA SAR; Cf. Košak 1982, 60); KUB XLII 47, obv.$^{?}$ 10; KUB XLII 49, obv. 2, 4, 8, 11; KUB XLII 51, obv. 5; rev. 4; KUB XLII 52, obv. 2$^{?}$, 5$^{??}$; KUB XLII 53, obv.$^{?}$ 8, rev.$^{?}$ 1; KUB XLII 54, obv.$^{?}$ 3, 5; KUB XLII 55, obv. col. I 8, 11; KUB XLII 56, obv. 4–6, 9, 11, 13, 14; KUB XLII 58, obv. 4, 9; KUB XLII 59, obv. 7, 8, 12, 14, 16, 17; rev. 8, 11, 12$^{?}$ 14–22, 26, 28; KUB XLII 61, obv. col. I 4, 9; KUB XLII 63, rev. 4; KUB XLII 69, rev. III$^{?}$ 1, 4; KUB XLII 75, obv. 1–2; KUB XLII 106, obv. 1, 10–11, 12, 17, 20; rev. 9, 11; IBoT I 31, obv. col. I 3–4, 8, 10, 20, 26; NBC 3842, rev. 7.

[2] Bibliography is too vast to provide a complete list. For a recent in depth study, see Jones 2008; Kemp & Vogelsang-Eastwood 2001, 25–34, 53–55, 438, but also Vogelsang-Eastwood 1995, and before that, Vogelsang-Eastwood 1992.

[3] For an overview of the first archeological finds relating to Egyptian linen (Neolithic – Predynastic Period – Old Kingdom) see *supra*. Now also Jones & Oldfield 2006, 33–35; cf. van de Mieroop 2007, 157–161.

[4] CT 2, 2: 8: GADA *ša* ^{URU}*Miṣir*.

[5] For the concept of luxury goods, see below note 117.

[6] Cf. Waetzoldt 1980, 30–31; van de Mieroop 2007, 162: "Their appearance alongside gold and silver objects, and the fact that Hittite king explicitly demanded those gifts in a treaty he concluded with the king of Ugarit, demonstrate

Egyptian terminology for linen is as varied as its uses.[7] There is no doubt however that the so-called 'king's linen' (Sumerian GADA LUGAL) is the finest of all. The Egyptian term for linen (*šś nsw*) is compared to the Greek word βύσσος (byssus),[8] found on the Rosetta stone. This was the term used by Greek merchants to refer to prestigious dyed linen and it is clear that the expression was already used in Akkadian as *kitū šarri*, meaning precisely 'king's linen'.[9] The same term is also often found in the correspondence between the Egyptian and Hittite courts.[10]

The treatment and dyeing of linen seem to be a characteristic of traditional Egyptian manufacturing.[11]

However, trading in dyed linen increased during the New Kingdom and probably already involved foreign trade industries.[12]

In fact, interesting data emerges from close examination of the materials listed in the so-called Hittite 'inventory texts' (CTH 240–250),[13] also with regards to the origin of certain luxury goods such as linen.

It is difficult, for instance, to establish if a cloth mentioned in certain inventory texts and linked to the name of a specific place or city, actually originates from that place or whether it was simply crafted elsewhere using the traditional process, method and custom of the place that it has been named after. With regards to this, the observation by Košak concerning the so-called 'Egyptian chests',[14] "It is often impossible to distinguish whether an object is simply 'from Egypt, in Egyptian style' or 'from Egypt (directly A.N.)'",[15] can be applied to all other objects linked directly to the name of a specific place.

There are exceptions however that may question of this assumption. Such is the case of IBoT I 31,[16] which refers to linen cloths linked to the land of Amurru and to the city of Alašiya.[17]

The expression *ŠA* KUR *A-MUR-RI-kán* GADA$^{\text{ḪI.A}}$ *an-da*, and those alike, but not identical, such as SÍG *ḪA-ṢAR-TUM* A.AB.BA-*kán an-da* (line 17), SÍG *ḪUR-RI-kán an-da* (line 25), and above all BI-IB-RI$^{\text{ḪI.A}}$-*kán an-da* (line 18), clearly demonstrate that it is difficult to establish whether items

the great value of these textiles."

7 See Gardiner 1931, 161–183; Hannig & Vomberg 1999, 471.

8 Already Forbes 1964, 43.

9 Cf. CAD "K", 474, 3. See also the Akkadian form *būṣu*. Cf. CAD "B", 350.

10 Cf., for example, KBo XXVIII 47, rev. 3–6; KBo I 29 + KBo IX 43, rev. 1–4; KUB XXXIV 2, rev. 3–5.

11 See, among others, a recent study by Germer 1992. Alternatively Cochavi-Rainey & Lilyquist 1999, 215 who uphold that, "Patterning and color were not part of Egypt's traditional linen industry".

12 Cf. Forbes *loc. cit.* See also KBo XVIII 14, rev. 3–14; KUB IV 95, obv. 7; KBo XVIII 36, rev. 11.

13 For the principle editions: Košak 1982; Siegelová 1986.

14 E.g. KUB XLII 11, obv. col. I 13.

15 Košak 1982, 37.

16 The text has been widely discussed. Collected bibliography in Košak 2002, selecting the reference text: CTH 241.1. See in particular Košak 1982, 4–10; Siegelová 1986, 74–86.

17 IBoT I 31, obv. 2–4: $^{2'}$ ^{GIŠ}PISAN SA$_5$ GAL GÌR UR.MAḪ IGI.DU$_8$.A *ŠA* KUR *A-MUR-RI-kán* $^{3'}$ GADA$^{\text{ḪI.A}}$ *an-da IŠ-TU* GIŠ.ḪUR *gul-aš-ša-an* $^{4'}$ EGIR-*an-da-ya-kán* 37 GADA ^{URU}*A-la-ši-ya*, "one large, red basket, (on) lion feet, (fit for) a gift. Contains: Amorite linen. As jotted down on a wooden board $^{(?)}$. Furthermore 37 items of linen (from$^{??}$) Cyprus". The impossibility of a link between the term IGI.DU$_8$.A and the following sentence (*ŠA* KUR *A-MUR-RI-kán*) for reasons of syntax had already been put forward by Goetze (1956, 33, and note 9). This interpretation was accepted by later editors of the text. The term IGI.DU$_8$.A, also, translated by Goetze as "show piece $^{(??)}$", has been variously interpreted. Cf. Goetze *loc. cit.*; Košak 1982, 8; Wilhelm 1992, 503. The translation "gift" seems to be the currently acceptable one. For this see Tischler 2001, 233; Mora 2006, 135 with the bibliography collected in notes 17–18; *Eadem* 2007, 537.

were simply channeled through the area or whether they are typical of that place because they were, for instance, crafted there.[18]

Using comparison with other documents it is, however, possible to draw up a fairly detailed assumption of the linen trade, at least as far as Cyprus is concerned.[19] In a quote from 'Ezekiel's Prophecy on Tyros',[20] which can be dated between 580–570 BC, the materials forming the allegorical ship that represents Tyros include embroidered linen from Egypt and dyed blue and purple linen from the island (or coast) of Cyprus (*'iyyê(y) 'ĕlišâh*). The correspondence between Pharaoh Akhenaton and the King of Alašiya provides interesting information on the island's linen trade.[21] It is precisely from the testimonies in the el-Amarna Letters that one learns of the difficulty that Cyprus had in getting hold of this material. In EA 34 the King of Cyprus, after reminding the Egyptian Pharaoh that he had sent 100 talents of copper,[22] asks in exchange for the delivery of 42 bolts of linen, 50 linen shawls and two linen robes as well as four bolts and four shawls in 'king's linen type' (byssus).[23]

In EA 40 the "governor"[24] of Cyprus promises gifts to the Egyptian Pharaoh consisting of five (ingots) of copper (for a total of three) talents of refined copper, one elephant's tusk, one (beam)

[18] It seems evident that in this context the particle *-kan* cannot have a local value. Košak (1982, 5) translates "contains..."; Siegelová (1986, 81) produces "darin...". On the value of the particle *-kan*, see, among others, Friedrich 1940, 85–86; Neu 1993, in particular 145. It has been suggested for quite some time (Boley 2001), that the particle *-kan* had, at least during the Ancient and Middle Hittite periods, an absolutely non-local value (*Contra* Hoffner & Melchert 2008, 374). The value of *-kan* varies depending on whether it is inserted into a nominal sentence or followed by a motion verb or by a non-motion verb. Thus, for example, the particle *-kan* on line 4 (with an anaphoric value) does not so much refer to the name (*Alašiya*) but rather to the fact that "within" (*anda* + *kán*) the basket, there are also items of 'Cypriot linen'. Cf. Goetze 1956, 33, note 5; Neu 1993, 147, § 12.7. See *infra*. It must be stressed that the use of this particle is particularly typical of 'New Hittite', usually substituting other local particles in New Hittite texts. Cf. Hoffner & Melchert 2008, 374, 28.82. The text can be dated to the 13th century BC. According to Carruba (1968, 20), based on the *ductus* and on the presence of grammatical elements characteristic of so-called 'classic Hittite', it should be possible to date the text further back.

[19] References are made in other contexts of textiles also coming from Amurru. See for example: RS XVI 146 + 161 (a list of the worldly goods belonging to the queen mother Aḫat-Milku). Transcribed and translated by Nougayrol 1955, 182–183. See also the transcriptions and translations provided by Cochavi-Rainey & Lilyquist 1999, 180–181. Obv. 10–11: 20 TÚG.ME.MEŠ SAL.LA *ša* KUR *Ḫur-ri* 20 TÚG.ME.MEŠ SAL.LA *ša* KUR.MAR.TU.KI 20 ^{TÚG}*ša-bat-tu*$_4$ *ša* ^{KUR}*Ḫur-ri* 20 ^{TÚG}*ša-bat-tu*$_4$ *ša* KUR.MAR.TU.KI. However no mention of linen is made in this passage, even though the expression ^{TÚG}*nalbašu raqqatu* (logographic form: TÚG.ME.MEŠ SAL.LA), translatable as "the finest capes$^{?}$", hints at a high quality textile. For the meaning of the word *raqqatu(m)* one refers to CAD "R", 168; Nougayrol 1950, 19; Oppenheim 1947, 128. Cf. also AHw "M-S" Band II, 958: (feine Wolle?). For the textile *šabattu(m)*, see CAD "Š" Part I, 8. Even if, from the example given, it can be deduced that a textile industry existed in the land of Amurru, we cannot establish for certain whether linen too was either produced or processed there. Cf. Klengel & Klengel 2009, 205–206.

[20] The Book of Ezekiel 27. 7. This passage has been studied in relation to Phoenician commerce by Moscati 1966, 108–110; cf., among others, also Smith 1953, 97–110. For a general approach to the problems that can arise from the reading of the document, see van Dijk 1968, 65–66 with the relative bibliography. References are also found in Goetze 1956, 36.

[21] Apart from the classic editions of the letters of el-Amarna by Knudtzon 1915; Moran 1987 and Liverani 1998; 1999; for the *corpus* of the Alašiya texts (EA 33–40) see Cochavi-Rainey 2003, 5–42.

[22] EA 34, obv. 18. A talent of copper corresponds to approximately 30 kg of metal made into ingots in the typical 'oxhide' shape. For a preliminary study of 'oxhide' ingots and the trading of them in the Mediterranean during the 2nd millennium BC, see Parise 1968; Muhly 1977. More recently Yalçın *et al.* 2005, 560.

[23] EA 34, obv. 22, 23, 25.

[24] For the Hittite term (^{LÚ}*pidduri-*) reference is generally made to CHD "P", 368 with the relative bibliography; Tischler 2001, 134; HEG "P", 638–639, with the relative bibliography. See also, among others, Carruba 1968, 29, with note 65; Steiner 1962, 135–136; Kühne 1973, 85–86; Moran 1987, 209, note 1.

of boxwood,[25] and a ship's mast.[26] There is no mention of linen cloths.

These same goods are also found in the Annals of Thutmosis III[27] (summary of the 34th year of reign), which mentions a tribute made by the 'chief' (*wr*)[28] of *'Asiya* (Cyprus) which that year included 108.5 ingots of smelted (?) copper ore (lit. "heavy") 2.400 *deben*,[29] five (+X) ingots of lead/tin,[30] 110 *deben* of lapis lazuli, one ivory tusk and two *śigu*-wood logs. The summary of the 38th year of reign again lists the tribute made by the 'chief' of *'Asiya*, which includes a certain quantity of copper ore and two horses.

A further reference to goods channeled through Cyprus found in the Annals of Thutmosis III is in the summary of the 39th year of the Pharaoh's reign in which the tribute made by the 'chief' of *'Asiya* includes two ivory tusks, 40 copper ore ingots and one ingot of lead/tin.

Thus even in EA 36–37[31] the King of Cyprus is able to guarantee enormous quantities of copper and some horses, but there is never a mention of 'Alašiya linen'.

According to the information provided by the texts it would be reasonable to assume that Cyprus had no thriving industry for the manufacturing of linen but that it brought it from Egypt and crafted it into undoubtedly prestigious goods, which were then channeled into the Syrian inland and to Anatolia.[32] The expression GADA ^{URU}*Alašiya* documented in the Hittite inventory texts can therefore be translated as "Cypriot linen", in the sense of: 'crafted in the "Alašiyan" way'.[33]

2. Some notes on Hittite terminology for linen textiles

Cypriot linen in the form of finished goods is also mentioned in another Hittite inventory text together with a range of other materials.[34] In the third line of the obverse there is even a mention

[25] *Buxus sempervirens*, commonly known as boxwood, is an evergreen shrub which grows up to five meters high and is still widely found today in an area stretching from the Atlantic coast of the Iberian Peninsula to the Dodecanese islands. It is a slow growing shrub preferring a sandy, lime rich soil but which also adapts well in woodland as undergrowth. It is a highly valued and heavy hard wood which is still today widely used in the skilled crafts. During the period in question it was probably essentially used for making refined furniture and not for constructing boats. In fact, in the letter quoted above, boxwood is differentiated from the wood needed for a ship, since the specific weight of dry boxwood is larger than that of water. However it is quite interesting to underline that the only wooden diptych found in the Eastern Mediterranean until now (along the seashore of South Anatolia – Uluburun Peninsula), and dated to the end of the second millennium BC is made of boxwood. Cf. Warnock & Pendleton 1991. The boxwood is referred to in the Egyptian texts written in Akkadian as *taškarinnu(m)* and compared with the Egyptian *śi-gu*. See also CAD "T", 280–282; AHw "Ṣ-Z" Band III, 1336–1337.

[26] EA 40, obv. 10–15.

[27] Copy of this text in Sethe 1906, 707–708. Translation in Endesfelder & Priese 1984, 211, 216, 219. See also, in general, Helk & Drenkhahn 1995, 98–99.

[28] The Egyptian term *wr* means literally "the Great", but it is generally used by the Egyptian chancellery of the 14th and 13th centuries BC to designate all foreign sovereigns, not necessarily Egyptian vassals. Cf. Erman & Grapow 1955a, 329; Moran 1987, 35–36, with note 71.

[29] The Egyptian *deben* corresponds to approximately 91 grams of metal.

[30] It was decided, for various reasons, to keep two translations for the Egyptian term *ḏḥty*. Initially it was thought that this term referred only to lead. Apropos this see Erman & Grapow 1955b, 606; Helk & Drenkhahn 1995, 102.

[31] EA 36, obv. 5–14; EA 37, obv. 9.

[32] The correspondence between ancient Near Eastern courts and Ḫatti proves that Cyprus was not, at the time in question, the only exporter of fine, skillfully processed textiles. See, for example Singer 2008, 29. Some archeological remains could perhaps reinforce the validity of the hypothesis given here and of the data collected. See, particularly, Reese 1987, 205.

[33] See Carruba 1968, 20.

[34] KBo XVIII 175, obv. col. I 5. See Košak 1982, 10; Siegelová 1986, 410.

of linen from Ḫurri,[35] whereas in the fourth line it refers to 39 linen shirts of which two are in red linen and one in *šuḫru* linen.[36]

Lines 6–7 of the obverse even describe the route traveled by the produce: by caravan (*ŠA* KASKAL*[NI]*) to the Palace ([GIŠ]PISAN *ŠA* É.GAL*[LIM]* lit. "the chest of the Palace").[37]

The linen traded in the Near East during the 2nd millennium BC. was subjected to a range of processes. In fact in KBo XVIII 175 (CTH 241.2) one also reads: "a red linen robe in the 'Tapašpa style'".[38] In this case too it is a reference to the method of crafting the cloth.[39] It should be stressed

[35] [1 [GI]PI]SAN SA$_5$ LÚGUD.DA 40 [TÚG]GÚ *ḪUR-RI* GADA, translated by Košak 1982, 13, "one small red basket: 40 linen Hurrian shirts". In this case too the translation "Hurrian shirts (made of linen)" is preferred, rather than "linen Hurrian shirts". 'Hurrian shirts', that is to say in Hurrian fashion, are in fact well documented in the Hittite texts (cf. lastly Klengel & Klengel 2009, especially in the inventories (Cf. Siegelová 1986, 650), as well as in the correspondence between Tušratta of Mittani and Amenhotep IV. (Cf. most recently Cochavi-Rainey & Lilyquist 1999, 51–164, in particular p. 144), therefore the statement "Hurrian linen" remains a ἅπαξ. See also CTH 250.1=KBo VII 23, obv. 2: [TÚG]G]Ú *ḪUR-RI* GADA *ŠU* [*-UḪ-RU* (?). Thus Siegelová 1986, 516; Košak 1982, 160, however, renounces this integration. See in particular Goetze 1955, 53, with note 53. Also it should not be forgotten that within the Hittite court there were real "corporations" of skilled craftsmen who worked for the palace, producing items that were typical of their native lands. See, among others, Klengel 2008, 72; already Hoffner 1992, 93–94. On the mobility of artisans in the ancient Near East during the second half of the 2nd millennium BC see, in general, Moorey 2001, with previous bibliography. For a study of Hurrian textiles, with particular reference to the textile industry in Nuzi, as is evident in the texts, see Zaccagnini 1981, with the preceding bibliography.

[36] 39 [TÚG]GÚ GADA 2 GADA SA$_5$ 1 GADA *ŠU-UḪ-U-RU*. For the meaning of *šūru(m)*, here given as *šuḫru*, see Košak 1982, 14, with the list of attestations and *ibidem*, 293. Cf. also Siegelová 1986, 339, note 6. Generally reference is made to CAD "Š" Part III, 367–368. Veenhof (1972, 154–156) accepts the interpretation "black cloth" by Landsberger (1925, 14), who, in TC I, 27: 2 translated, "zehn schwarze Stoffe". Also, again, AHw "Ṣ-Z" Band III, 1287 translates *šūru(m)* as "schwarz, grau", accepting the equation with *šaḥōr* "schwarz werden" from Hebrew and Aramaic. As Veenhof (1972, 30, note 59) rightly observes, the cloth accompanied by this adjective is indicated in the texts as a material for packing goods. Therefore Košak (1982, 14) suggests that it could refer to an untreated (rough[?]) cloth, in apposition to the term [TÚG]SIG, which would hence refer to a fine cloth. The possibility should not be excluded here that the term *šuḫru*, used as an adjective, could simply indicate an untreated and undyed cloth. Cf. Košak 1982, 202. Neither does this interpretation exclude the possibility of translating the term as "grey, black". It would therefore appear that the adjective serves to indicate the quality of the textile rather than its colour. In fact, in the inventory texts, it often occurs next to [TÚG]SIG ("fine fabric"), almost as if highlighting the difference. See the statements in Košak 1982, 14. Cf. MVAeG 33 No. 155: 4, *a-na* 1 [TÚG]*šu-ri-im ú ra-qí-tim*, translated, "für 1 schwarzen Stoff oder dünnen (Stoff)", but, "for an untreated textile or (a) fine one" would be preferable.

Košak (1982, 202, 260) has also suggested that the Sumerogram ḪI.ḪI linked to textiles (in the forms [TÚG]ḪI.ḪI, [GADA]ḪI.ḪI, [TÚG]ḪI.ḪI-*natar*) might indicate a type of treatment, even though "not coloured" (undyed[?]) is often given in translation. Cf. however *ibidem*, 16, according to the interpretation suggested in KBo XVIII 175, obv. col. I 21. Siegelová (1986, 654) on the other hand translates it as "meliert[(?)]". The term literally means "storm, thunderbolt" and it is often a divine attribute. Cf. Tischler 2001, 232, "auch eine Farbbezeichnung?". The principally accepted meaning of the term probably alludes to something with dark, opaque tones; in indicating a cloth it possibly refers to poor quality or to the fact that it has undergone no treatment or dyeing process. Even if a similarity with the Akkadian term *šūru(m)* seems farfetched, one cannot however preclude it from belonging at least to the same semantic area. See, conversely, the equivalence of the Akkadian *raqqatu(m)* and the Sumerian [TÚG]SAL.LA or [TÚG]SIG. Cf. Veenhof 1972, 153. See also the considerations by Siegelová 1986, 204–206, note 2; HZL, No. 335, "Farbbezeichnung(?)".

[37] One would assume that in this context É.GAL[LIM] refers to a particular structure located in the Tempel I in Ḫattuša, or to the worship - administrative centre as a whole. Cf. Güterbock 1974, 305-306; Siegelová 1986, 155, note 8, with previous bibliography.

[38] KBo XVIII 175, rev. col. V 11. Cf. the interpretation by Siegelová 1986, 418 in KUB XLII 14 rev. IV[?] 11: [TÚG]*ta-piš-pa*. See Košak 1982, 17, 22; HZL, No. 244.

[39] This could be a collective term (*-a*), derived from the name of the city ([URU]*Ta-ap-pa-aš-pa-aš*). See also Košak 1982, 22 and Siegelová 1986, 338, note 1. Cf. HEG "T,D/1", 124–125. In order to locate the city of Tapašpa refer to del Monte & Tischler

that this document always refers to goods that have only been provisionally inventoried since they come from convoys that are waiting to be channeled to a destination.[40]

In IBoT I 31, obv. 10 the expression 1 GADA *eḫ-li-pa-ki* appears. It seems plausible that this expression may also refer to a type of linen. Goetze claims that the term *eḫlipaki* is undoubtedly to be linked to *eḫlipak(k)u*,[41] which is often quoted in the Qatna inventories[42] and translated by Bottéro as "possiblement améthyste".[43] As Bottéro himself recalls,[44] the term is often cited in the el-Amarna inventories as the *ḫi-li-p/ba* stone.[45] It is highly likely to be a precious stone, as Laroche rightly observed;[46] the expression 1 GADA *eḫ-li-pa-ki* could therefore be translated as "*eḫlipaki* stone (amethyst?) coloured linen",[47] meaning dyed in a shiny bright purple colour.

In a passage from CTH 243.2.A[48] another reference is made to processed linen. In KBo XVIII 170+, obv. 7 'bolts/rolls?'[49] of linen are referred to, as are 'sheer linens'.[50]

A number of minas of linen and (for)[51] four tunics[52] with refined embroidery[53] are also

---

1978, 400, even though one cannot exclude its geographical vicinity to Konya (^{URU}*Ik-ku-wa-ni-ya*), from which six garments originate (KBo XVIII 175, rev. col. V 8) and are placed in the same container (KBo XVIII 175, rev. col. V 5).

[40] Even though the problem is extremely complex and undoubtedly requires in-depth studies, the stocking of goods should consist of three stages: provisional storing (as in this case), the listing of the goods on wooden tablets (Cf. for example IBoT I 31, obv. 3), storing of the goods at their final destination (e.g. CTH 241.12 A-B) and the drawing up of an inventory. See, among others, Archi 1973, 215–216; Košak 1982, 52; Mora 2007, 538.

[41] Goetze 1956, 36.

[42] Bottéro 1949, 138–168, lines 96–131–215–218–235–252–299–329. For a revised and updated version of the Qatna inventories, see Fales 2004, 83–128.

[43] Bottéro 1949, 18, with note 2.

[44] Bottéro *loc. cit.*

[45] Cf. Goetze 1956, 36, note 52.

[46] Laroche 1976, 76; compared to Košak 1982, 8.

[47] Already Košak 1982, 6; Siegelová 1986, 80, with note 6; 81, 63, 65, with note 4. For other interpretations see, in particular, Polvani 1988, 13–14, with the bibliography suggested in the notes. Cf. *ḫilibû* in CAD "Ḫ", 186.

[48] KBo XVIII 170+ and its copy KUB XLII43 (CTH 243.2B). See Košak 1982, 109–110; Siegelová 1986, 482–488. In this context the listed products suggest a private destination. In fact in CTH 243.2.A, obv. 1–6 e rev. 1–4 objects have been listed which are already present in the so called 'Inventory of Manninni'. See Košak 1978, 99–123; Siegelová 1986, 441–451. See also KBo XVIII 170+, rev. 5: "Šuppiluliuma, head of the *ilana*-house". Cf. Košak 1982, 110; Siegelová 1986, 487–488, note 15. Cf. Pecchioli Daddi 1982, 512; HW, No. 271. For an archeological comparison see Naumann 1971, 415, 454.

[49] In the Hittite documents the logogram PAD is mainly attested in connection with metals, designating a metal ingot; here ^{TÚG}PAD-*me-eš*, could be a strip, a bolt of untreated fabric (linen?). Compare Košak 1982, 112; Siegelová 1986, 338, note 1, 667; HZL, No. 295; Tischler 2001, 251. See, more recently Singer 2006, 252–258, in particular 253, note 45.

[50] As far as the term ^{GADA}*ḫu-u-wa-an-d/ta-r(a)* is concerned, it is probably such finely worked linen that it is sheer, almost transparent. It appears to be a neuter, abstract noun with the function of adjective. A purely homophonic association could link it back to the Hittite term *ḫuwant-* "wind". Compare Tischler 2001, 58; HEG "A-K", 328–331. The same suggestion is made by Puhvel 1991, 430. The term also occurs in KUB XLII 43, obv. 11 (Košak 1982, 111–112; Siegelová 1986, 482–488), in KUB XLII 34, left edge, line 1 (Košak 1982, 54–55; Siegelová 1986, 56–59), also in KUB XLII 60, obv. 7 (Košak 1982, 186; Siegelová 1986, 524–525). See, for example, the adjective derived from it, attested in the neuter, plural form, *ḫuwandaruwanda*. Cf. Beckman 1985, 141.

[51] Like Siegelová 1986, 225, "[x +] 1 Minen Leinen (für) 4 gemusterte, erstklassige Gürtel".

[52] KBo IX 90, obv. 7. It could have been a loose fitting tunic tied at the waist (E.ÍB), to create a skirt which was slightly longer at the back. The tunic in fact often ended with a kind of tail at the back (KUN). Cf. KUB XLII 48, obv. 3(?), 11, 15; KUB XLII 44 (+) KBo IX 89, obv. 1–2, 7, 9. The garment E.ÍB could be associated with the Hittite ^{TÚG}*gapari-*. Cf. Goetze 1955, 56; Tischler 2001, 72; HEG "A-K", 490, "tunic with kilt", maybe not to be confused with TUG ÍB.LAL, Hitt. ^{TÚG}*putalliya-*, "leichtes Marschkleid der Soldaten", thus wrongly Ünal 1978, 126. Cf. the valid observations by Goetze 1955, 56, "tunic with kilt which the 'soldier gods' from Yazılıkaya are wearing". See also what is claimed by Siegelová 1986, 213–216.

[53] The Akkadian word *mašlu* often used in the inventory texts as Akkadogram, could have the meaning "trimmed" or

mentioned in a list of gifts (IGI.DU$_8$.A)[54] from different parts of Anatolia.[55]

Other types of linen also existed such as *karnaša* linen.[56] This was, in all likelihood, a special type of cloth[57] used for covering precious pieces of furniture.[58] In the same text[59] one also comes across the expression GADA *tiyalan* which possibly refers to a type of cloak for special occasions worn as an over-garment.[60] Speaking of which, it is interesting to note the term used without

"embellished", also in this context. The word in fact occurs nearly always next to insertions of gold and silver in the clothing. Cf. Goetze 1955, 53, note 55; CAD "M" Part I, 318, b; HZL, No. 20; Košak 1982, 202, with the relative bibliography. Lastly Klengel & Klengel 2009, 207.

[54] KBo IX 90. For the category of these texts see, in general, Siegelová 1986, 213–217. Transcription and translation in Košak 1982, 116–117; Siegelová 1986, 224–225.

[55] The textiles in this group of texts (CTH 243.3 = KBo XVIII 197 (+) 197a (+) 197b + KBo IX 89 + 90 (+) KUB XLII 44), come from the city of Paniša (KUB XLII 44, rev. col. V 5, 8), Parnaša (KUB XLII 44 + KBo IX 89, rev. col. II 1–2, 5–6, 8, 10) and Zarwiša (KBo IX 90, rev. col. V 3, 8 + KBo IX 89, rev. col. VI 1, 6). As far as the first place name is concerned it should be stressed that it was already a place of worship at the time of Arnuwanda I, initially located near Yerköy, along the old Roman road which now runs past the modern Yozgat (maybe the Hittite Ḫarana(ša); Lat. *Corniaspa?*) from Ḫattuša towards the south (Kaneš?). Cf. Cornelius 1967, 71; del Monte & Tischler 1978, 300–301; Siegelová 1986, 217. According to Forlanini (cited as personal communication), the city would have been located, together with Tiura, in the valley between the Melendiz/Hasan Dağ massif and the middle-Kızılırmak. The city of Parnaša is yet to be accurately located, even though it was an important place of worship. It is likely however to have been located on the north eastern banks of the Tuz Gölü and corresponds to ancient Parnassos and today's Parlasan (lastly Forlanini 2009, 54, with previous bibliography). Already del Monte & Tischler 1978, 306. Lastly, as far as Zarwiša is concerned, it may have been near Tarḫuntašša and Karadağ, in the area between the Taurus mountains, the Ḫulaya river and the modern Ereğli. As in Forlanini 1988, 137–138. Cf. del Monte & Tischler 1978, 496; del Monte 1992, 191–192. In the light of these suggested locations, a region of textile production and trade could be hypothesized in the heart of the Hittite Lower-Land between the cities of Pašura (?), Ikkuwaniya (Konya), Wašḫaniya (according to Forlanini [2009, 49, 68] to the west of Kültepe/Kaneš and near the Kızılırmak river), as far as Zarwiša, near Ḫurniya/*Korne* (personal communication by Forlanini). Cf. Košak 1982, 203.

[56] The term occurs, in association with the items on which it is placed, in KBo XVIII 181, rev. 5, 7 (integrated: *kar-n]a-aš-ša* in KBo XVIII 181, rev. 6). In KBo XVIII 186, left edge 6, Košak (1982, 170) suggests for GADA $^{(GIŠ)}$*karnašaš*: "the cloth for the karnasa-table", while Siegelová (1986, 379) transcribes directly ^{GIŠ}*karnašaš*, even though it is evident both in the drawing of KBo XVIII, and in the photo of the fragment, that the determinative GADA appears before the noun (not GIŠ$^!$), and with no possibility of being mislead by the various spelling of the two symbols (cf. HZL, Nos. 174, 177), since, just before (line 3: GADA ^{GIŠ}*šar-pa-aš*) the two words appear side by side and can be clearly distinguished. The term is accompanied by the determinative indicating items made of wood (GIŠ) in KUB LII 96, obv. 3.

KBo XVIII 181 and KBo XVIII 186 have been catalogued by Siegelová (1986, 363–387) in the category 'Textilen aus persönlichem Besitz'; KUB LII 96 in 'Persönliche Zuweisungen'. Cf. Siegelová 1986, 328–362.

[57] Cf. Siegelová 1986, 359, note 1; *ibidem*, 375, note 12.

[58] It is not clear if the linen takes the name of the object on which it is placed (^{GIŠ}*karnaša-*). From the contexts in which the term occurs it cannot be excluded that it refers to a kind of throne. Compare Tischler 2001, 74. It is very probably not the same kind of linen described in KBo XVIII 181, rev. 28: GADA ^{GIŠ}ŠÚ.A (linen [for a] chair). The term *karnaša-/karnaši-* could be formed by the Luwian term *karna-* – of unknown meaning – with the Luwoid appurtenance suffixes *-aš(š)a-/-aš(š)i-* (as in Puhvel 1997, 91). For *karna-* see CLL, 101, '?'; Tischler 2001, 73 provides, with some reserve, the motto '*kar-na-an ma-ar-na-an e-eš-ša-ú*' of KUB I 1, col. IV 80: "schalten und walten". Otten 1981, 29, "...soll ein jeder für die Gottheit *nach besten Kräften* schaffen!". According to HEG "A-K", 513; Puhvel *loc. cit.*; Košak 1982, 124, it is possibly a piece of furniture on which objects can be placed. In fact, in KBo XIV 33, obv. col. I 9–10, for example, it is said that the head of the palace employees brings a precious club and places it on the king's *karnaši*. Compare Klinger 1996, 482–483, 775.

[59] KBo XVIII 181, rev. 14 and may be also in obv. 2, 29 and rev. 11.

[60] The term is in fact almost definitely retraceable to the verb *tiyalai-* ("to cover", "to wear", "to put over"), from which the noun *tiyala-*. In the Hittite inventories the term is clearly used as an adjective, meaning it serves to qualify the workmanship of linen and wool (that is to say the manufacture of an item of clothing). Compare Tischler 2001, 176; HEG "T,D/3", 364. See the interesting suggestion by Siegelová 1986, 365, note 3; *ibidem*, 370–371, note 4.

its determinative GADA, in KUB XLII 106, obv. 4:]x 1 ^{TÚG}GÚ.È.A[61] *ḪUR-RI ŠU-RI-PU ḪAŠ-MAN-NI ti-ya-la-*⸢*a*[*n*$^{?}$.

It should be noted that in this passage the *tiyalan* robe is tailored to be put on over the head through a specific hole in the cloth (^{TÚG}GÚ.È.A). This garment is almost always linked to the land of Ḫurri. It is therefore assumed that the GÚ.È.A cloak originates from Hurrian lands.[62]

The element that appears to confirm the link $^{TÚG/GADA}$*tiyalan* = cloak, is in the interpretation of the term *šurīpu.*[63] In the Hittite texts it seems to have been used as an Akkadogram referring to the "freezing winter" or simply to "winter",[64] despite the fact that the equivalent Hittite word is known.[65] Therefore Košak's interpretation of the expression ^{TÚG}GÚ.È.A *ḪURRI ŠURIPU tiyalan* appears very convincing.[66]

The Luwian term[67] *lakkušanzani-*, often preceded by the Sumerogram for linen,[68] is of unknown etymology. This lexeme, widely attested in the Hittite inventories, always occurs in association with the term ^{GIŠ}NÁ ("bed").[69] The textile referred to is evidently used in connection with beds.[70] The most viable hypothesis is that the expression $^{TÚG/GADA}$*lakkušanzani-* refers to a canopy.[71]

In rev. 27 one comes across the expression GADA IGI, "linen (for) the eyes".[72] It is quite plausible that this refers to fabrics worn for specific occasions, particularly since other garments are mentioned in the lines that follow.[73]

---

[61] For this type of garment, which is characterised by the way it is put on, see Goetze 1955, 52–54.

[62] Cf. lastly Klengel & Klengel 2009, 206.

[63] For the meaning of the word see CAD "Š" Part III, 347–348; Landsberger *apud* Bauer *et al.* 1934, 157–159.

[64] Güterbock *apud* Freydank 1968, 317, was the first to give a specific meaning to *šurīpu*, attested as Akkadogram in the fragments of letters 453/w (left edge, line 3) and 2236/c (rev. 5). The 'winter context' in which this expression is inserted, '*kinuna ŠURIPU mekki*' (KBo XVIII 35, left edge 3; cf. KBo XVIII 79, rev. 30: *ŠURIPU-wa kuitman wa*[*kkari*$^{?}$]) translated, "und jetzt ist hier viel Eis" (Hagenbuchner 1989, 166–167, 179–180), suggests to us the possibility of translating the Akkadogram *ŠURIPU* as "winter". Compare AHw "Ṣ-Z" Band III, 1284. 6a.

[65] *gim(a)-*. HW, 109 and bibliography; HEG "A-K", 571–573; Tischler 2001, 79.

[66] Košak 1982, 139, "Hurrian winter coat?". Cf. Klengel & Klengel 2009, 207, "... wohl auf ein Kleidungsstuck hinweist, das während der kalten Jahreszeit getragen wurde".

[67] The presence of the inflected Luwian form (nominative plural) in KBo XVIII 175, rev. col. V 15: *la-ak-ku-ša-an-za-ni-en-zi* leads us to a Luwian origin of the term. Compare Košak 1982, 17; CLL, 121; Neumann *apud* Puhvel 2001, 40; Kloekhorst 2008, 515. In KBo XVIII 154, rev.$^{?}$ left col. 9, the reading (*la-ak-ku-ša-an-za-ni-uš*) by Güterbock in Inhaltübersicht VI of KBo XVIII, as an accusative plural, would demonstrate that the forms $1^{NU<-TIM>}$ in fact anticipate *Pluralia Tantum*. Cf. also Goetze 1931; *Idem* 1955, 53, note 54. In any case, according to Starke (1982, 361) it should be read as GADA *la-ak-ku-ša-an-za-ni-iš*$^{!}$, even though it is evident that the final syllable is *-uš*, both in the drawing in KBo XVIII and in the photo of the fragment of tablet. Compare HEG "L-M", 19.

[68] For the list of statements in the Hittite inventory texts, see Košak 1982, 224; Siegelová 1986, 604.

[69] The term rarely occurs in conjunction with other objects. Cf. CHD "L-N", 21, b.

[70] As in, among others, Otten 1982, 285; CHD "L-N", 20–21; Puhvel 2001, 39; Siegelová *loc. cit.*

[71] Compare valid considerations by Košak 1982, 17.

[72] The same expression is to be found in KUB XLII 59, rev. 20, in KUB XLII 56, obv. 4, 14, also with the phonetic ending (acc. plur. *-aš*) in KBo XVIII 187, obv.$^{?}$ 6.

[73] Košak 1982, 125. Cf. Klengel 2008, 76.

3. A propos colours of fabrics in the Hittite world

In a Hittite medical text[74] there is reference to a remedy for an ophthalmic complaint which probably causes profuse lachrymation.[75] The treatment procedure is described in detail. A compress made with an ingredient (possibly obtained from Alašiya)[76] which is mixed with wine in a bronze vessel and then applied to the patient. This procedure is repeated day and night for some time, during which the patient must wear a special bandage.[77] The main ingredient in this preparation could be chrysocolla (hydrated copper silicate),[78] a substance probably used for the same purpose also in the Akkadian ritual AMT 9, 1: 39[79] and mentioned again by Pliny in a passage from his *Naturalis Historia* as an excellent remedy for "*initia glaumatum*".[80] It is however difficult to establish how *mar(ru)wašḫa-* should be accurately translated. In 1919 Emil Forrer, in reference to the 'kanesic' rendering of words originating from Luwian and *vice versa*, suggested linking the hydronym *Maraššand/ta – Maraššantiya* to the ideographic form (^{ÍD}SI-A) attested in the

[74] KUB VIII 38 + XLIV 63, rev. col. III 10–21. *Editio princeps* by Burde 1974, 30–34.

[75] Unfortunately the text is damaged in this part. As suggested by Haas 2003, 117, "Rezept gegen vermehrte Sekretion von Tränenflüssigkeit und Brennen liegt in KUB 8.38 + KUB 44.63 vor.", it is more likely that it is a matter of an over secretion from the eye, rather than a lack of secretion. As in CHD "L-N", 202, "or (if) he (the patient) [lacks?] tears". The pathology to which the text refers to is in all probability trachoma (compare Haussperger 2000, 443–444; Starke 1986, 163–164), a *chlamydia trachomatis* infection which still occurs in certain parts of the world today. A form of contagious conjunctivitis with a pronounced tendency to scar the upper eyelid, it causes in-growing eyelashes, the inability to close the eye, infections, perforation of the eyeball and blindness. The infection also affects children. It may refer to a simple conjunctivitis, much less contagious and frequent among peasants because of wounds, even only superficial ones, caused by the vegetation (the so-called harvesters' ulcer). It is improbable that it refers to blockages of the lachrymal duct which usually only occurs for persons over 60 and therefore at an age which was at that time rarely reached. "The salts of heavy metals and the tannin from wine are still used, particularly the former. Until recently Zincometil was used (a zinc eye wash). Colloidal silver is still a treatment for gonorrheal conjunctivitis in newborns. Yellow mercuric oxide cream is used for parasites such as lice and *phtirius pubis*, which, occasionally, still today can be seen on the eyelashes. All heavy metals are disinfectants, but I cannot remember ever having come across the use of copper". This was a personal communication (01/13/2009) by Dr. Roberto Bellucci, head physician of the ophthalmology department in the 'Borgo Trento' hospital in Verona. Starke (1986, 163) claims that, "Nun wird aber seit alters das Trachom (Conjunctivitis granulosa), das vor allem in subtropischen Klimazonen auftritt, gerade durch Ätzung der erkrankten Lidbindehaut mit sogenannten Blaustein oder Kupferstift behandelt (Heute verwendet man stattdessen Sulfonamide)". So does Puhvel 2004a, 195, "Before sulfa drugs, granulation and corneal scarring caused by trachoma and conjunctivitis were treated by cauterizing diseased tissue with copper sulfate". For a more detailed technical account of ophthalmic pathologies in the ancient Near East, see Fincke 2000 and more in general the bibliography collected by Haas 2003, 117, note 520.

[76] KUB VIII 38 + XLIV 63, rev. col. III 11: ^{U}]^{RU} *A-la-ši-ya-az*. Compare CHD "L-N", 202; Polvani 1988, 66.

[77] $^{\text{(SÍG)}}$*ḫandala-*. Compare Tischler 2001, 38; HEG "A-K", 154; HW$_2$ Band III/1, 168. The Egyptian language too has specific words to indicate bandages (particularly in linen), used in traditional medicine. Compare Forbes 1964, 43, with note 399.

[78] The Hittite term *mar(ru)wašḫa-* seems to be borrowed from Luwian. Compare CLL, 142; HEG "L-M", 152. It is important to stress that the mineral is indicated by the omission of the determinative for stone (NA$_4$). It cannot be excluded, in fact it is very possible and certainly logical, that in this case, as in others, it was necessary to indicate that the mineral did not come in the form of a stone. It also was unlikely to refer to a vitreous paste made with pigments derived from it, since it was used for medical purposes. For the complex problem related to the interpretation of mineral terminology with or without the determinative NA$_4$, see, in general, Polvani 1988, 56–57, and the relative bibliography. For more in depth etymological research, see *infra*.

[79] Cf. Polvani 1988, 66, notes 6–7.

[80] Pliny, *Naturalis Historia*, 33. 92; already cited in Polvani 1988, 66–67, notes 8–10 and by Puhvel 2004a, 195–196. It should be remembered that Strabo (Γεωγραφικά XIV, 6. 5) also cites chalcanthite (copper sulphate) and "copper rust" – from Cypriot mines – used for their medicinal purposes.

Hittite texts (the modern ^{ÍD}SA$_5$ transcription), "red river" (the old Halys, today called Kızılırmak, which in fact means "red river" in Turkish).[81] Later Güterbock too, following Goetze,[82] suggested placing the ideographic form ^{ÍD}SA$_5$ with the same hydronym ^{ÍD}*Maraššanda*, a noun possibly formed by the Luwian word **mar(ru)wa*, "red?",[83] together with the participle stem (*-ant*) from the Hittite verb derived from it.[84] Laroche was the first to suggest seeing in the term *mar(ru)wašḫa* a noun derived from the Luwian **mar(ru)wa(i)-* (rougir).[85]

The suffix *-(°a)sha-* has been interpreted in many ways. Gusmani has suggested for some time seeing 'Verbalabstrakta' in nouns ending in *-(a)sha*.[86] From a morphological point of view it is basically a matter of abstract names, often *nomina actionis*.[87]

Howard Berman,[88] who was not convinced by Gusmani,[89] tried to explain why the suffix *-asha-*[90] added to a verb root would form words denoting names of actions or states.[91]

This question has since been tackled again[92] and other scholars have reached similar conclusions, leaving aside the morphological issues[93] and the debate on the nature of the suffix and its function when linked to verbs or nouns.[94] The term *mar(ru)wašḫa-* is a nominal deverbative

---

81 Forrer 1919, 1039. Compare del Monte & Tischler 1978, 538–539; del Monte 1992, 207. Today the identification is not so certain. Compare now HEG "L-M", 148–149, "Angesichts der umfangreichen Lücken nach bzw. vor diesen beiden Textstellen ... ist die Identität von ÍD.SA$_5$ ^{ÍD}*Marassantiya* keineswegs gesichert; später publizierte Texte zeigen im Gegenteil, daß es sich um verschidene Flüsse handelt ... Und zwar scheint der ÍD.SA$_5$ ein Zufluß des ^{ÍD}*Marassantiya* zu sein, wie das Ritualfragment KBo XII 94 zeigt, ... Bezüglich der Identifizierung des *Marassantiya* läßt sich sagen, daß er an Nerik vorbeifließt ... Der *Marassanta*-Fluß floß früher nach Westen(?) oder nach Nordwesten". See also Košak 1982, 202. Already Gurney 1977, 207.

82 Goetze 1930, 25, note 50.

83 As in Laroche 1959, 69.

84 According to Güterbock himself (1956b, 116, note b), the Hittite 'passive participle' *-ant/d* is equivalent to the Luwian 'passive' participle *-ma* attested in the gloss *marušam(m)a-* (KUB XXII 70, rev. 11), designating a colour other than white (*ḫarki-*), probably found precisely together with this term (in the genitive form *ma-ru-ša-aš-aš*) also in IBoT I 31, obv. 16. Here Košak 1982, 6 translates, "white and red (textiles)"; Siegelová 1986, 83, "weisse *marušaša*". Regarding this, see reserves by Melchert in CLL, 141. Also in IBoT III 110, obv. 6, one could suggest the integration *ma-]ru-ša-ma*[. As also in col. IV, 5–6. Compare Siegelová 1986, 394, 416–417. The relationship between the hydronym Maraššanta and the term indicating a colour could be explained phonetically with the vowel alternation (a/u) confirmed, for example, also in the cases of *idalu-/adduwali-, innarawant-/annarummi-*. Cf. HEG "L-M", 149. Güterbock thus corrected the first reading of *marušam(m)a-*, as a neuter plural adjective indicating a colour. Cf. Güterbock 1956a, 122–123, "Either 'black, dark' in contrast to 'white'" Cf. Puhvel 2004a, 194. On questions relating to colour, see *infra*.

85 As in Laroche 1958, 113. Cf. now CLL, 141, **mar(ru)wa-* "blacken, darken". See, more recently HEG "L-M", 151. For more in depth study, see *infra*.

86 Gusmani 1972, 254–266; in particular 265, with note 39. Now however, for a purely phonetic and phonological insight, see Melchert 1994, 122 and the examples of deverbative forms provided on pp. 69–71. On the complexity of PIE laryngeal derivatives in PA and then in strong and weak fricatives see the more recent suggestions by Kloekhorst 2008, 75–82.

87 Clear examples of these are *armuwalašḫa-* ("moonlight") from the verb *armuwalāi-* ("rising like the moon"), *maliyašḫa-* ("approval") from the verb *malāi-* ("allow, approve"); *ḫarnamniyašḫa-* ("restlessness") from the verb *ḫarnamniya-* ("incite, stir up"); *etc.*

88 Berman 1977, 231–239; in particular 234–237.

89 Berman 1977, 234, note 8.

90 On the basis of an example, which is not even very significant, Berman (1977, 237) claims that the suffix is *-asha-* and not *-sha-* or even *-sh-*.

91 Berman 1977, 235, "...*-asha-* is added to a verb stem and forms names of actions or states corresponding in meaning to the English gerund.".

92 Starke 1979.

93 Compare, for instance, Starke 1979, 261, note 64.

94 With regards to this the three scholars draw different conclusions. According to Gusmani 1972, 262, the suffix is

form from the Luwian **mar(ru)wa-*.[95]

If the association of ^{ÍD}SA$_5$ (lit. "red river") with ^{ÍD}*Maraššanda* is arguable at a purely 'topographical' level,[96] it is even more so from a lexical point of view.

There are texts in which the verb **mar(ru)wa-* and therefore the nouns linked to it, including *mar(ru)wašḫa-*, can hardly be seen to qualify the colour red.

KBo VI 29, obv. col. II 10–13 states the following:

> *numu* ^{D}*Ištar* ^{URU}*Šamuḫa* GAŠAN-*YA warriššišta nu šarazzi katterraya anda :maruwāit nu nepiš tekanna katkattenut.*

In this passage is described the epiphany of Ištar, who, on coming to the aid of Ḫattušili III, "darkened" above and below[97] (all the lands of Ḫatti); shaking the sky and the earth.

In the same way the deities in KUB LIV 78, rev. 6[98] and in KUB VII 38, obv. 6,[99] are clearly "dark divinities".[100]

The term *mar(ru)wašḫa-*, cited in the inventory text KUB XLII 18, obv. col. II 5,[101] probably doesn't refer to a precious stone, but to a mineral which is difficult to identify; perhaps to its actual pigment.

The underlying problem lies in the fact that previously the term *mar(ru)wašḫa-* was interpreted based on the derivation from the verb **mar(ru)wa-*, where **mar(ru)wa-* = "to be red" > *mar(ru)wašḫa-* = "redness".

Nevertheless, from the examination of the Hittite texts, but above all from reading the fragments just cited, it can be deduced that the colour of the mineral is difficult to define. It also seems possible that the perceived colour of this mineral (or of colours in general) and therefore, the colour classification criteria the Hittites used, were different from ours.[102]

It is more logical to think that the cited minerals in the text have a blue tone rather than red

present in other Indo-Germanic languages and is hence a 'pre- or proto-Anatolic' phenomenon. According to Berman 1977, 234, it is a phenomenon which is also characteristic to the more recent Indo-European languages such as Greek and Latin. According to Starke 1979, 256, the suffix is shared by the Anatolian languages of the 2nd millennium BC, such as Hittite and Luwian. See also Weitenberg 1984, 232, § 600; 460, note 626.

95 Among others Weitenberg 1984, 295–296; § 800; CHD "L-N", 202–203; HEG "L-M", 152; CLL, 142; Puhvel 2004a, 195; *Idem* 2004b, 91.

96 Cf. Košak 1982, 202.

97 *katterraya anda :maruwāit.*

98 $^{\text{DINGIR.MEŠ}}$*mar-ku-wa-ya[-aš.*

99 ^{D}*Mar-ku-wa-ya-aš.*

100 The *markuwayaš* deities (dat. nom. plur. neuter of *mar(k)uwaya-?*) quoted in the Hittite text should correspond to ^{D}*Marwayanza* of KUB XXIV 11, col. II 8 and KUB XXIV 9, col. II 27 (cf. $^{\text{DINGIR.MEŠ}}$*Marwāinzi*, nom.plur. neuter of KUB XLIV 65, col. II 11. Cf. Starke 1986, 162–163, with previous bibliography. See the list of attestations in van Gessel 1988, 299–300. For more in depth study, see CLL, 142, and the relative bibliography; HEG "L-M", 139; Puhvel 2004b, 77–78 and lastly Kloekhorst 2008, 563.

101 [^{NA}]$_4$$^{?}$*mar-ru-wa-aš-ḫa*(-)x[. Cf. Košak 1982, 30; Siegelová 1986, 50. Looking at the drawing in KUB XLII the integration of the word is not so certain. As also Rosenkranz 1965, 247. *Contra* Burde 1974, 34. Within this context (a list of precious stones kept in some baskets, like a tribute), it seems probable that pigments of this substance were used to perhaps make objects from a vitreous paste. It is important to underline that in the Hittite texts the term never appears as associated with the determinative for stone. The restoration made by the editors in the reference text determine, as such, a ἅπαξ difficult to explain, but easily questionable.

102 As already noted by Landsberger (1967, 139) and recently confirmed by Polvani (1988, 174), the definition of certain colours by the Mesopotamian people, as with later civilisations, were somewhat different to those of today. Such is the case, for example, for the colour blue. Cf. Landsberger *loc. cit.*, note 7. An explanation of this is given further below.

because all the other stones and objects identified are of the same colour, which does not seem to be a coincidence.[103] If we try to apply the same logic that scholars used in trying to identify the mineral on the basis of its colour,[104] it can be deduced that the mineral in question cannot be chrysocolla, based on the fact that it came from the island of Cyprus.[105] Chrysocolla [$(Cu,Al)_2H_2Si_2O_5(OH)_4{\cdot}n(H_2O)$] is characterised by a very irregular, amorphous structure; it is formed by oxidation in copper ore deposits; it is found in crusts or very softly, green, earthy masses, and can be of varying colours from green to bright blue. Since it does not have a uniform crystalline structure, (it is amorphous and therefore a tendency to be described as a 'mineraloid', rather than a mineral), it is not often used in jewellery. This fact strongly hinders the possibility of integrating the determinative for stone (NA_4) into the text KUB XLII 18, obv. col. II 5, or to identify $^{(NA_4)}$*mar(ru)wašḫa-* as chrysocolla. One option automatically excludes the other.[106]

The minerals that are often associated with chrysocolla are quartz, limonite, cuprite, malachite and azurite.

Malachite [$Cu_2(CO_3)(OH)_2$] is a mineral with a distinctive green colour, in the form of acicular clumps more or less densely united; it is rare that they form single crystals. This mineral is usually a compact mass or encrustation with a velvety texture. It is a basic copper carbonate, but is principally used as an ornamental stone. Malachite is actually one of the best-known semi-precious stones in the world. Its name derives from the Greek word for mallow.[107] Many malachite stones contain particular combinations of other minerals. It includes azurite, mottramite, chrysocolla and limonite. Moreover, malachite is usually associated with almost all secondary copper minerals, whether they are carbonate minerals or not. Malachite is often closely associated with azurite, not only for the obvious colour of both minerals (usually green) but also because the two minerals have very similar chemical formulae.[108] Malachite can sometimes substitute azurite, giving origin to what is usually defined in mineralogy as a 'pseudomorph'; in this case an exact structural copy of the azurite crystals.[109]

---

[103] Although this observation is irrelevant to the definition of the meaning of the verb **mar(ru)wa-* and the noun maybe semantically linked to it (*marwāy(a)-*), perhaps related to Hittite in the form *mar(k)uwaya-*, cited in the texts as mentioned above (KUB LIV 78, rev. 6; KUB VII 38, obv. 6). It is also difficult to think of a 'red deity', just as it is easier to imagine a climatic disturbance (a storm for example), characterised by dark chromatic tones, created by the goddess cited in KBo VI 29, obv. col. II 10–13.

[104] It has been essentially based on the equation: *mar(ru)wa-* (a verb that indicates a colour): *mar(ru)wašḫa-* (abstract noun of this colour) = *mar(ru)wašḫa-* (abstract noun of this colour): ^{NA_4}*mar(ru)wašḫa-* (stone X of this colour).

[105] So proposes, with reservation, Polvani 1988, 65, "crisocolla"; "verderame".

[106] The possibility of creating a stone which imitates and substitutes precious stones, which are often difficult to find or very costly, involves the "agatisation" of chrysocolla with chalcedony quartz, which guarantees the hardness of the stone. The quartz crystals in the stone, if well polished, can result in the natural colour of chrysocolla. Such an operation is typical in specialised industries, which have sophisticated instruments to work hard stones. It cannot be excluded therefore that a stone with these characteristics can be identified in ^{NA_4}*YAŠPU*, a term rarely used in Hittite documents, but always provided with the determinative. Cf. Polvani 1988, 123–125, with preceding bibliography.

[107] μαλάχης with the variants μολόχης/μολάχης. Also in Pliny, *Naturalis Historia*, 37. 8: μολοχῖτις λίθος. Cf. Hebrew: *malūăḥ*.

[108] Chemical formula for azurite: $Cu_3(CO_3)_2(OH)_2$; Chemical formula for malachite: $Cu_2(CO_3)(OH)_2$.

[109] The charges on the copper ions are the same for both minerals at two positive; each hydroxide has a charge of one negative and each carbonate has a charge of two negative. The change in colour between malachite and azurite, considering that the charges in the copper are the same, is determined by the higher oxidation of malachite compared to azurite. This means that malachite reflects a later stage of oxidation.

Malachite has low solubility and the ethyl alcohol (wine) cited in the text KUB VIII 38 + XLIV 63, rev. col. III 14, is certainly not a reagent which would dissolve it.[110]

Azurite [$Cu_3(CO_3)_2(OH)_2$] is a basic copper carbonate, produced by weathering of copper ore deposits with a unique dark blue colour. The chromatic tone of this mineral is due to the high presence of copper.[111] It can often be found in the form of clumps or joined together in aggregations with radial texture. Azurite is highly soluble.[112]

From this re-examination so far one can deduce that the mineral (not stone) often cited in KUB VIII 38 + XLIV 63, rev. col. III, 11–16, is azurite.[113] It is easily ground to a powder, is highly soluble using reagents like ethyl alcohol, and is one of the principal secondary copper minerals. The term *mar(ru)wašḫa-* cited in KUB XLII 18, obv. col. II 5 probably refers to the colour, similar to that of azurite, of the missing object.

However, the identification of minerals leaves the colour of the fabrics described in the inventory text IBoT I 31, obv. 16, open to debate. When azurite, reduced to powder and mixed with water and other reagents, it can take on a range of pigmentations. On the other hand, even the grinding of the mineral can strongly influence the tone of the final pigment that can vary from dark blue (coarse powder) to light blue (fine powder). Moreover if not ground sufficiently, the powder can become too sandy and granular, making it unsuitable to be used as a pigment. Apart from this consideration, the translation "to become dark, black(?)" from the Luwian verb **mar(ru)wa-* – from which the term *mar(ru)wašḫa-* could have been derived – seems completely acceptable.[114]

Research into the dyeing of the fabrics cited, even in the Hittite inventory texts, would require too deep a study to be covered here.[115] It should also be stressed, as already mentioned, that there is the risk of translating terms that refer to colour in ancient Near Eastern languages, based on words used in the treatise writings on painting belonging to the Renaissance period.[116]

---

[110] It is interesting to note that malachite is used in Egyptian medicine, along with *asa foetida* and naphtha, to heal cataract problems. Cf., for example, Daglio 2008, 79.

[111] It needs to be highlighted that, unlike lapis lazuli with which it is often confused and substituted, once azurite is reduced to powder and heated or mixed with acid reagents, it undergoes chromatic changes and becomes very dark. This process is defined as decarbonisation.

[112] For an overview of the definition and use of minerals in the ancient Near East refer in general to Thompson 1936; Forbes 1963, 67–93 and more recently Moorey 1994, 79–103.

[113] Starke (1986, 163) had already reached the same conclusions, today reconfirmed (personal communication). Cf. Beckman 1996, 35; Puhvel 2004a, 195.

[114] Cf. previously Güterbock 1956a, 122–123 and more recently Starke 1986, 163; CLL, 141–142; Puhvel 2004a; *Idem* 2004b, 89–91; Kloekhorst 2008, 562–563.

[115] The bibliography on this subject is vast. The following are some summary bibliographical references. About colours in the ancient Near East see in general *RlA* 3 (1957–1971), 19–26; Landsberger 1967 and recently Verderame 2004, in particular p. 327. On colours in the Hittite world Riemschneider 1957 and the bibliography collected by Singer 2008, 29, note 89. For the colour of fabrics Goetze 1956, 34–35; Waetzoldt 1980, 20–21 and recently Klengel 2008, 7–14; Völling 2008, 151–157, with preceding bibliography. For major studies on the chromatism of fabrics in the ancient Near East, with particular reference to purple, see Singer 2008, above all 22–27, with bibliography in notes; van de Mieroop 2007, 162–166.

[116] For this reason, even today, there are many expressions connected to the literature of the ancients that seem to be too suggestive, like, for example, the "wine-dark sea", the "hyacinth hair" of Ulysses or "black blood", often cited in the Odyssey. With reference to the colours of fabrics cited in Hittite inventory texts see Košak 1982, 201, "The perception of colours in various cultures differs greatly and it is therefore difficult to find exact equivalents". The starting point for this discussion can be seen in Geiger 1871, followed by Magnus 1877 and Marty 1886. See, more

4. Linen as a luxury good. A case of study: ^{GADA}*lupan(n)i*

The definition of linen as a luxury good[117] is confirmed by the presence of the fabric in series of inventory texts which cite the royal couple, the name of a king or a queen and objects denoting royalty or events connected to the enthronement of a sovereign.[118]

In the left edge of KBo XVIII 186 there is a list of cloth items. Among these there are 14 linens$^{(?)}$,[119] perhaps related to a chair; two (cloths?) of linen presumably for a "royal seat".[120] The most difficult fabric to identify is definitely ^{GADA}*lupanneš*,[121] quoted in line 2.

The term $^{\text{TÚG/GADA}}$*lupan(n)i-*[122] seems to describe a type of textile craft. This, however, recurs in Hittite inventory texts with or without determinatives.[123]

The evidence of the term in this passage is significant because of the presence of the expression "royal *lupan(n)i-* (of linen)".[124] A similar expression appears in another inventory text (KUB XLII 22, obv. col. II, 11).[125] The tablet, related to KBo XVIII 179 (duplicate of KUB XLII 27), is of great interest, because it seems to contain an inventory of goods listed on the occasion of the enthronement of King Ḫattušili III.[126]

The preciousness of the object, or the material from which it is made, seems to be confirmed by the 'substitute-king ritual' where the substitute was dressed in royal clothes, evidently including a *lupan(n)i-*.[127]

Moreover, in a statue – literally "image" = ALAM – dressing ceremony, representing the king, it states as follows: "A *lupan(n)i-* is placed on his head".[128]

The object in question seems to be a distinctive attribute of the Sungod, as evident in a Hittite ritual.[129] Even women could wear a *lupan(n)i-*, as seems to be evident in a passage of the ritual

---

recently Berlin & Kay 1969.

[117] The concept of luxury goods related to the complex ceremony of gift exchange between the courts of the ancient Near East defines precious objects which usually recur as the possessions of powerful sovereigns in the late Bronze Age like Egypt, Ḫatti, Assyria, Kassite Babylonia and Mittani. For a general overview of gift exchange in the ancient Near East, of the concept of luxury goods, and of international diplomatic relations, see the fundamental works of Mauss 1923–1924; Zaccagnini 1973; *Idem* 1989–1990; Liverani 1990, 205–282; Thomason 2005. Most recently, Giorgieri & Mora (in print).

[118] Following, as an example, are a few passages of the many texts refer specifically to linen in association with kingship.

[119] Left edge 1$^{?}$: GADA *ta-ni-pu-ú.*

[120] Left edge 3: 2 GADA ^{GIŠ}*šar-pa-aš*. For the controversial meaning of the term $^{\text{(GIŠ/KUŠ)}}$*šarpa-* refer to CHD "Š", 287; Tischler 2001, 146; HEG "Š/1", 928.

[121] ^{GADA}*lu-pa-an-ni-eš*. Nominative singular or plural. Cf. CHD "L-N", 85; Puhvel 2001, 120.

[122] For the meaning of the term see CHD "L-N", 85–86; HEG "L-M", 77–78; Puhvel 2001, 119–121, all with previous bibliography.

[123] A list of references in Hittite inventory texts is available in Košak 1982, 225–226; Siegelová 1986, 605–606.

[124] ^{GADA}*lu-pa-an-ni-eš* LUGAL$^{\textit{UT-TI}}$.

[125] ^{TÚG}*lu-pa-an-ni* LUGAL-*a*[*n*$^{?}$-*na-aš*]. As perhaps also in KUB XLII 98, obv. col. I, 10–11: ^{TÚG}NÍG.LÁM^{MEŠ} LUGAL$^{\textit{UT-TI}}$ *wa-aš-ši-ya-an-zi wa-aš-š*[*i-ya-an-zi-ma*]1$^{\textit{NU-TIM}}$ ^{TÚG}GÚ.È.A *ḪUR-RI* 1 ^{TÚG}E.ÍB 1 ^{TÚG}*lu-*[*pa-an-ni-*$^{\text{(?)ME}}$]^{Š}, "(the substitute) is dressed in royal clothes: one Hurrian overcoat (coat$^{?}$), one light tunic and one *lupan(n)i-*".

[126] Cf. Siegelová 1986, 35; Mora 2007, 539.

[127] KUB XXIV 5 + IX 13, obv. 21–22: $^{\text{[TÚG]}}$*lu-pa-an-ni-na-wa-kán ke-e-da-ni ši-ya-n*[*u-u*]*n*. Cf. Kümmel 1967, 10.

[128] KBo XV 15, rev.$^{?}$ 5: [*nu-uš-š*]*a-an* ^{TÚG}*lu-u-pa-*[*an-ni-in A-NA* SAG.DU-*ŠU ši-ya-an-zi*]. Cf. Kümmel 1967, 136.

[129] KUB XVII 15, obv.$^{?}$ col. II 13 and duplicate KUB XXXV 145, obv. col. II 17: [(*na-aš-ta an-da*)] *ŠA* ^{D}UTU *lu-u-pa-an*[*-ni-i*(*n*) KI.MIN *na-aš-ta an-d*(*a*) *ŠA* ^{D}U]. Cf. Starke 1985, 231.

called 'the expansion of the cult of the Deity of the Night'.[130]

In the inventory text IBoT I 31 (obv. 9) we find a blue *lupan(n)i-* from Ašpunawiya.[131]

In CTH 250.36[132] it is cited again *lupan(n)i-* in a fragmentary passage.[133] Even if the tablet is extremely damaged, the presence of a number of personal names,[134] such as the heir to the throne and the queen,[135] provides interesting information regarding the date of the text as well as an interpretative hypothesis on the characteristics of the *lupan(n)i-*. When comparing this document and other trial reports for embezzlement (e.g. the so-called Gerichtsprotokolle (KUB XIII 35+ and KUB XXVI 49),[136] useful information emerges about the identity of the receivers (?) of the goods cited.[137] As already pointed out by Siegelová,[138] the concomitant presence of some key members

---

[130] The new title is subject to a historical revision of the divine cult in question. For a comprehensive overview refer to Miller 2004, 259. KUB XXIX 4, obv. col. I 44–46: $^{44'}$ 1 TÚG *ša-ra-a ḫu-it-ti-ia-an-za* 1 ^{TÚG}E.ÍB *MAŠ-LU* 1 ^{TÚG}*ka-ri-ul-li* $^{45'}$ 1 ^{TÚG}*lu-pa-an-ni-iš* 1 ^{TÚG}*ka-lu-up-pa-aš* 1–*NU-TIM* ^{TÚG}E.ÍB$^{?!}$ *TA-ḪAP-ŠI* $^{46'}$ 1–*NU-TIM TU-DÌ-IT-TUM* KÙ.BABBAR *ki-i ŠA* MUNUS-*TIM* {1 TÚG} 1–*NU-TIM* ^{TÚG}GÚ.È.A. See lastly Miller 2004, 277; *Idem* 2005b, 225.

[131] 1 ^{TÚG}*ka-lu-pa-aš* ZA.GIN 1 *l[u]-pa-ni-iš* ZA.GIN SUM ^{f}*Aš-pu-na-wi-ya*, "one blue vest, one blue *lupanni*: (present from/to$^{??}$) Ašpunawiya". The name is a ἅπαξ λεγόμενον in the Hittite texts (Laroche 1966, 45, No. 177), but seems to refer to the Luwian onomastics. Cf. Goetze 1956, 38. It is important to highlight that in the Hittite texts the term *lupan(n)i-* is often associated with *kaluppa*: the correct meaning, "women's underclothing". Refer in general to Goetze 1955, 61; *Idem* 1956, 36; HEG "A-K", 471; Puhvel 1997, 32–33; Tischler 2001, 70.

[132] KUB XLII 51.

[133] KUB XLII 51, obv. 4: x +] 6$^{(?)}$ ^{TÚG}*lu-pa-an-ni-i[š*. Probably also in line 8: ^{TÚG}*lu-pa-a]n-ni*$^{(?)}$ ZA.G[ÌN$^{(?)}$.

[134] KUB XLII 51, rev. x+1:]x ^{f}*Ḫé-en-t[i*; rev. 5$^{?}$:]x ^{m}EN-LUGAL-*ma* ^{m}*Ne-ri-ik[*.

[135] KUB XLII 51, obv. 2: ^{L}]^{Ú} *tu-ḫu-kán-t[i*; obv. 6:]x MUNUS.LUGAL [.

[136] To study these documents refer to the bibliography of Košak 2002, selecting the texts referred to (CTH 293 e CTH 297.6).

[137] The Hittitie inventory texts in which officers of the Hittite court are mentioned cartainly deserve a more in-depth research. From the study of these texts the possibility emerges that the listed luxury goods are re-assigned and distributed to prominent members of the Hittite court. The reasons for these procedures presently remain difficult to justify, but future study could open new scenarios on the redistribution mechanisms and hoarding of goods within the central Hittite power and could help us to clarify the relationships within the complex palatine system of the Hittite Empire at the end of the 13th century BC. In some of the texts within this group, such as KUB XXVI 66, which unsurprisingly makes constant reference to the heir to the throne ($^{\text{(LÚ)}}$*tuḫkanti*) and the queen (MUNUS.LUGAL), the context of the redistribution of treasures is not clear. Cf. Košak 1982, 71. The treasurer Eḫli-Kušuḫ entrusts quantities of silver to various officers (Walwaziti, Pupuli, Lullu), in the presence of the queen (KUB XXVI 66, col. III 10; cf. Košak 1982, 68), often within religious contexts (KUB XXVI 66, col. III 3: DINGIR^{LIM} ^{URU}*A-ru-uš-(uš)-na*; col. III 11: MUNUS *ḫar-na-wa-aš*), but even typically administrative contexts (KUB XXVI 66, col. III 15), if one follows the interpretation of the verb *ḫarkanzi* by Košak (1982, 69), "held them outside (i.e. separate from the general account)", and by Siegelová (1986, 107, note 19). Cf. also Tischler 2001, 42; HW$_2$ "H", 195a. For the dating of the texts see Siegelová 1986, 119–121. Some texts, like KBo XVIII 153+, appear as lists of receipts, therefore the officers mentioned could have the function of guaranteeing the redistribution of the items, on behalf of the palatine authority, for various purposes. The text KUB XXVI 66, however, seems almost to be a list of distributed items, wisely disguised as offerings for religious purposes or as tributes to cult centres. (Cf. Košak 1982, 71, "The background of entire complicated transaction is not at all clear. 2 minas of silver are first held by the two officials, Walwaziti and Pupuli, then taken over by the queen and finally, the amount is split: 1 mina is now in the possession of priest Lullu who hands it over to the MUNUS *ḫarnawaš* while the remaining mina is sent for the embellishment of divine statues in Urikina."). As a last analysis, they could be compensation or bribes for the *entourage* of the Palace, not coincidentally composed of important figures. Authorisation of this hypothesis would be: a) the presence of a treasurer who controls the distribution (a person who does not appear in KBo XVIII 153); b) the presence of the queen as witness to the redistribution; c) the sentences, "Formerly (it was) with..." (KUB XXVI 66, col. III 2–5); "Furthermore, 3 minas of silver were taken separately and the chief scribe (i.e. Walwaziti) and Pupuli held them outside (i.e. separate from the general account)" (KUB XXVI 66, col. III 14–15). For in-depth study on this topic refer to Mora 2006, 145–146.

[138] Siegelová 1986, 344–345.

of the Hittite court, historically known and cited in other Hittite documents, suggests that the examined fragment dates to the reign of King Ḫattušili III. Without taking into consideration the possibility of a coincidence of names, the person cited in rev. 5 (EN-LUGAL-*ma*)[139] could be the same leading officer in the Hittite court who appears as witness in the 'Bronze Tablet' treaty and also in other documents in the time of Tudḫaliya IV.[140]

Ḫenti, the female character cited in the reverse, is unfortunately not quoted elsewhere. Yet we know from various sources a woman called Ḫenti linked to the Hittite court. In the inner register of an extremely fragmented *bulla* there remain traces of two syllables[141] of which Güterbock, comparing various texts dated at the time of the reign of Šuppiluliuma I,[142] suggests the name Ḫenti.[143] The interpretation was subsequently accepted by Otten, who suggested to connect the name Ḫenti, to this king – documented in a 'list of offerings' – as being the actual wife of Šuppiluliuma I.[144] On the basis of this supposed royal couple (Šuppiluliuma I – Ḫenti),[145] it is then suggested that the integration of a fragmentary passage of a decree pertaining to the 'kizzuwatnean priesthood' of Telipinu.[146] Assuming the historical credibility of the so called 'royal lists' are accepted with some reservation,[147] the chronological sequence of the queens appearing in the text reconstructed by Carruba (Daduḫepa – Ḫenti),[148] seems to be confirmed from another fragment of a 'royal list'[149] and completed by the contemporary(?)[150] presence of other Middle-Hittite queens[151] (Walanni – Nikkalmadi – Ašmunikkal – Duduḫepa[152] – Tawannanna) in the 'ritual of the Queen' at Taḫurpa (the fifth day of the *nuntarriyašḫa-* festival).[153] The name Ḫenti is also found in a fragmentary letter(?) dated to the Middle-Hittite period,[154] and in a very damaged text,

---

139 Phonetic reading: Ewri-Šarruma.

140 For the prosopography of this figure refer to van den Hout 1995a, 136–138. More recently Marizza 2006, 163–164. The same officer appears in the court proceeding KUB XIII 35+, col. III 8, mentioned above, in the form: ^{m}*Ib-ri*-LUGAL-*ma*. Cf. Werner 1967, 10–11. The text relating to the proceedings 'for embezzlement' (widerrechtliche Verwendung) was recently re-studied by Hoffner 2002, with previous bibliography.

141 [....]-*in-ti-x*[....]. As in the reading of the signs given by Güterbock 1940, 4.

142 *Ibidem*, 4, note 14.

143 *Ibidem*, 5.

144 Otten 1951, 56, "Daduḫepa und Ḫintī sind die Namen von Šuppiluliumas beiden Gemahlinnen"; *ibidem*, 57–58, "Das entspricht der Folge in Liste C: Daduḫepa (Rs. 8), Ḫintī (Rs. 9) und Tavann[anna] (Rs. 10), wobei die beiden ersten als Gemahlinnen Šuppiluliumas zu bestimmen waren".

145 Cf. in particular Güterbock 1940, 5, "... wenn wir dementsprechend Daduḫepa als erste, Ḫinti als zweite Gemahlin des Šuppiluliuma ansetzen ...".

146 KUB XIX 25, col. I 1–2: $^{1'}$ [*UM-MA* ^{D}]UTU^{ŠI} ^{m}*Šu-up-pí-lu-li-u*[*-ma* LUGAL GAL LUGAL KUR ^{URU}*Ḫa-at-ti*] $^{2'}$ [^{MUNUS}*Ḫi-in-*]*ti* MUNUS.LUGAL GAL ^{m}*Ar-nu-an*[*-da* DUMU LUGAL *Ù* ^{m}*Zi-da-a* GAL *ME-ŠE-DI*]. Goetze 1940, 12. Already Cavaignac 1933, 157.

147 Concerning this cf. Carruba 2007, 131–132.

148 KUB XI 7 + KUB XXXVI 121 + KUB XXXVI 122 [(+) KBo XIII 42 (+) KBo XIII 43]. Carruba 2007, 137–139, in particular 139.

149 KUB XXXVI 124, col. I 8–12. Cf. Otten 1951, 49 (text F [Bo 2893]). For the reference bibliography see Košak 2002, selecting the text CTH 661.6.

150 Cf. Goetze 1940, 77, note 312.

151 Cf. already Güterbock 1940, 31–32 (No. 60).

152 The transcription D/Tuduḫepa, doubtful in KUB XXV 14, obv. col. I 46 ([^{f}*D*]*u*$^{(?)}$*-ú-du-ḫé-pa*), seems to be confirmed by KUB XXV 14, obv. col. I 28: ^{f}*Du-ú-d*[*u-ḫé-pa* and by col. III 10: ^{f}*Du-ú-du-ḫé-pa*. For the alternation Daduḫepa/Duduḫepa, see already Otten 1951, 57, notes 4–5 and recently Giorgieri 1999, 70, note 24.

153 KUB XXV 14 obv. col. I 25–30, 42–49; col. III 3–12. Cf., lastly, Nakamura 2002, 192–195, 205. In general refer to Haas 1994, 833.

154 KBo XXXII 201, upper edge 1. Cf. Beal 1992, 171.

where a woman by the name of Ḫenti, is forbidden to eat.[155]

In conclusion, the presence of the name of Queen Ḫenti along with other queens of the Middle-Hittite period, suggests placing her chronologically at the time of King Šuppiluliuma I, as his second wife, in a sequence that lists the following royal couples: Walanni – Kantuzzili (?), Nikkalmadi – Tutḫaliya II (?), Ašmunikkal – Arnuwanda I, Daduḫepa[156] – Šuppiluliuma I (?), Tawannanna II/III(?)[157]– Šuppiluliuma I.[158]

Unfortunately, the late dating of the most of the texts (13th century BC),[159] excluding the fragmentary letter cited above, is not useful for identifying Ḫenti and does not even confirm the validity of the previously suggested royal couple.[160] From the range of references to this name in the Hittite texts it is clear that a precise identity cannot be attributed to the person called Ḫenti cited in the inventory text KUB XLII 51.[161]

The last name that can be matched with some certainty[162] is that of prince Nerikkaili, cited in line 5 of the reverse.[163] If we accept the identity of the person in this text as the son of King Ḫattušili III,[164] we can assume that the queen (MUNUS.LUGAL), cited in line 6 of the obverse, should be identified with Puduḫepa,[165] perhaps Nerikkaili's mother.[166] Staying within the realm of speculative hypotheses it could logically follow that the *tuḫukanti* cited in line 2 of the obverse can be identified as Tudḫaliya IV.[167]

Whatever the true identity of the figures cited in the text, an important fact emerges from the analysis undertaken: the *lupan(n)i-* appears again as a precious fabric to give(?), not only to

155 KUB LX 112, rev. 5–12. Cf. Cohen 2002, 170; Groddek 2006, 109–110. Maybe we are dealing here with a fragmentary inventory text (see in partucular the reference to a basket probably coming from a caravan in obv. 2–4: ŠÀ ⌜GI⌝PISAN x[...]*ŠA* KAŠKAL*NI*[). See Košak 1994, 289.

156 First wife?

157 Third wife?

158 For all suggestions see Carruba 2008, 125–141, 191.

159 Of significant importance is the dating of the text KUB XXV 14, where all queens are listed. Cf. Nakamura 2002, 188.

160 The name of a certain Manninni within the royal lists, next to the name of Arnuwanda and of the queens frequently cited (cf. Carruba 2007, 139), demonstrates the difficulty of using lists of offerings as historically reliable documents for the purpose of reconstructing the sequence of Hittite kings. For evidence of the name refer to Laroche 1966, 113, No. 747. Cf. also Košak 1978, 119. For evidence of the name within the Hittite inventory texts, see Košak 1982, 301; Siegelová 1986, 715.

161 Cf. Siegelová 1986, 344, "Es ist fraglich, ob in der Tafel etwa Opfer des Verstorbenenkultes aufgezeichnet waren, oder ob man vielleicht doch eher eine andere Persönlichkeit darunter suchen soll.".

162 Cf. Košak 1982, 183; Siegelová 1986, 345.

163 KUB XLII 51, rev. 5: ᵐ*Ne-ri-ik*[. For the prosopography of this figure refer to van den Hout 1995a, 96–105.

164 Following Imparati suggestions this identification cannot be proved. Cfr. Imparati 1995, 153.

165 Cf. Imparati *loc. cit.*

166 For this hypothesis see Houwink ten Cate 1996, 43. The maternity of Puduḫepa is a problematic issue. A direct bloodline has been suggested between this queen and Tudḫaliya IV. Otherwise Nerikkaili would have been born from a previous marriage of Ḫattušili III. For an in-depth study refer to van den Hout 1999, 232 and to Frantz-Szabó & Ünal 2006, 107–108, both with previous bibliography. See also Bryce 2005, 272, with bibliography included in note 37. In the inventory text KUB XXVI 66, as previously mentioned, the queen is cited (col. III 10) together with the heir to the throne (col. III 17). Van den Hout (1995a, 98, note 125) identifies them as Puduḫepa and Nerikkaili.

167 The interpretive hypothesis, in no way verifiable, is based on the presence, within the text, of the name of the older brother of Tudḫaliya IV (Nerikkaili), without a royal title. For events relating to the appointment to the kingship (LUGAL-*iz-na-ni tapariya-*) of Tudḫaliya IV, in place of the designated heir to the throne (Nerikkaili) cf. Imparati, 1995, 151–153; see lastly Bryce 2005, 272, 295–297, with previous bibliography in note 1 of p. 471.

a woman (the queen),[168] but above all to people of high status (the crown prince, or other son$^{??}$ of the King).[169]

The presence of the *lupan(n)i-* within the Hittite inventory texts shows that the wool or linen fabric was often decorated after undergoing specific treatments, such as dyeing. Colours of which vary from blue to green-blue.[170]

The presence of the term *lupan(n)i-*, without determinative (TÚG or GADA) in the inventory list of luxury goods, such as weapons or bronze objects and other precious metals, considerably complicates the identification of the fabric.

A clear example is the so-called 'Inventory of Manninni' (CTH 504),[171] inside of which are listed *lupan(n)i-* in association with weapons.[172]

The suggestion to identify the *lupan(n)i-* in these contexts as a textile covering of a hilt or handle of a sword is based on the interpretation of the expression: EME (+ the name of the metal of which it is composed) *lu-pa-an-ni-eš* GAB, often attested in the Inventory of Manninni, like "covering of the handle of the blade (sword, knife?)".[173]

However, the Inventory of Manninni, like the majority of inventory texts, shows an extremely varied list of objects, from weapons to textiles, from votive statues to furnishings.[174]

This fact suggests the possibility of considering the listed objects after the swords as simple metallic utensils, but not necessarily weapons.

The term EME (literally "tongue"), often cited in Hittite inventories and translated as "blade", seems to be an object closely linked to the *lupan(n)i-*. Following a reading of other passages it emerges that *lupan(n)i-* is a fabric worn on the head and it is also an attribute of kingship. Therefore the "metal-tongue" (sometimes made of gold, silver, "black iron"[175] or bronze) could

[168] This element could help us to confirm that the *lupan(n)i-* is actually an object worn by both men and women. As well as the examples indicated above, the *lupan(n)i-* recurs in connection with women in the document KUB XLII 49, obv. 8–9: $^{8'}$ 1[^{TÚG}*l*]*u-p*[*a-*]*an-ni* ZA.GÌN 1 GADA.DAM ZA.GÌN 1x[$^{9'}$ [*ŠA*$^{(?)}$] ^{f}*Ḫé-pát*-ÌR. For the prosopography of woman cited in the fragment see Siegelová 1986, 347, note 1.

[169] It is worth mentioning again the text KUB XLII 22, obv. col. II 11, where, as already pointed out, *lupan(n)i-* of royalty$^{(?)}$ are cited (6 ^{TÚG}*lu-pa-an-ni* LUGAL-x[), together with other precious fabrics, objects and cult weapons made of bronze. The afferent tablets at CTH 241.12A–B (KUB XLII 27 (+) 23 (+) 22; KBo XVIII 179) probably describe objects provisionally stored in the 'royal treasury', as tributes or donations, on the occasion of the enthronement of the royal couple. Refer in general to Siegelová 1986, 32–35 for the relevant discussion regarding the problems that can result from the study of these important documents, and for the hypotheses in identifying the royal couple. Cf. note 126.

[170] Keep in mind the observations made above regarding the rendering of terms, which indicate colours in the ancient languages. For evidence of coloured *lupan(n)i-* see, for example, CHD "L-N", 86, d.

[171] Published by Košak 1978 and re-edited by Siegelová 1986, 441–451.

[172] KUB XII 1, col. III: 6–13: $^{6'}$ 21 GÍR ŠÀ.BA 1 GÍR ^{URU}*Ḫa*[*-at-ta*$^{(??)}$ $^{7'}$ ^{GIŠ}DÙG.GAN GUŠKIN GAR.RA *ŠU-RU-UḪ-TU*[*M* $^{8'}$ 2 EME AN.BAR GE$_6$ GAB *lu-pa-an-n*[*i-eš* $^{9'}$ *A-NA* 1 EME ZABAR *lu-pa-an*[*-ni-eš* $^{10'}$ 11 EME ZABAR *lu-pa-an-ni-eš* GAB $^{11'}$ SAG.DU-*ZU* NU.GÁL *ŠU-RU-UḪ-DU* $^{12'}$ GAB *lu-pa-an-ni-eš* AN.BAR $^{13'}$ 1^{EN} *ši-ik-kiš* AN.BAR GE$_6$ GAB *lu-pa-an-ni-eš* $^{\text{NA}_4}$Z[A.GÌN$^{(?)}$; KUB XLII 42, rev. col. IV 3–6: $^{3'}$]x EME ZABAR *lu-pa*[*-an-ni-eš* GAB $^{4'}$]x*-te-eš* *ŠUM-ŠU* ^{m}*Tu*[*-ut-ḫa-l*]*i-ya*$^{(?)}$ DÙ*-an* DÙG.GAN KUŠ N[U. GÁL] $^{5'}$ 1]^{EN} EME AN.BAR GE$_6$ *lu-pa-an-ni-eš*[ ]*za* $^{6'}$ x+] 1 EME ZABAR *lu-pa-an-ni-eš* AN.B[AR$^{(?)}$ (KUB XLII 42 is linked, with reservation, to CTH 504; cf. Košak 1982, 59; Siegelová 1986, 469 with note 1). The term *lupan(n)i-*, determinative lacking, recurs even in KBo XXXI 54, obv.$^{(?)}$ 6 – the text closely linked to the Inventory of Manninni – like in VBoT 87, obv.$^{(?)}$ 2 and in KBo XVIII 170 (+) 170a, rev. 2. For the form *lu-wa-an-ni-eš*, documented here, see below.

[173] Like, for example CHD "L-N", 86, 2, "EME 'blade' *lupan(n)i-* 'cap' and GAB 'front(?)' are here understood as parts of a GÍR 'dagger'". Cf. Puhvel 2001, 119, "cap(ping), pommel (on a sword's or dagger's hilt)". Cf. Starke 1990, 411, note 1481.

[174] Cf. Košak 1982, 200.

[175] For the interpretation of the term AN.BAR GE$_6$ see Maxwell-Hyslop 1980, 87–88.

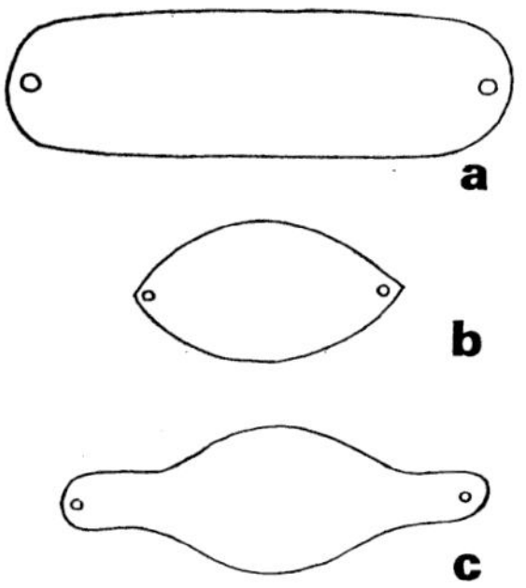

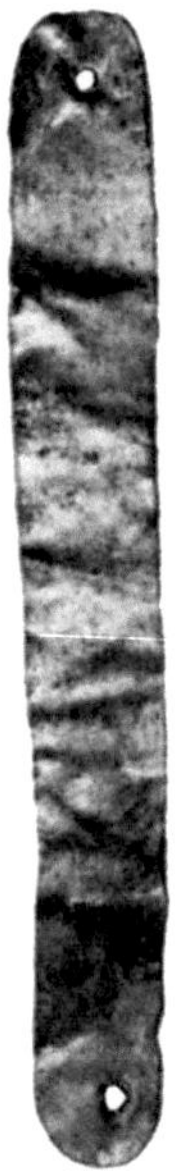

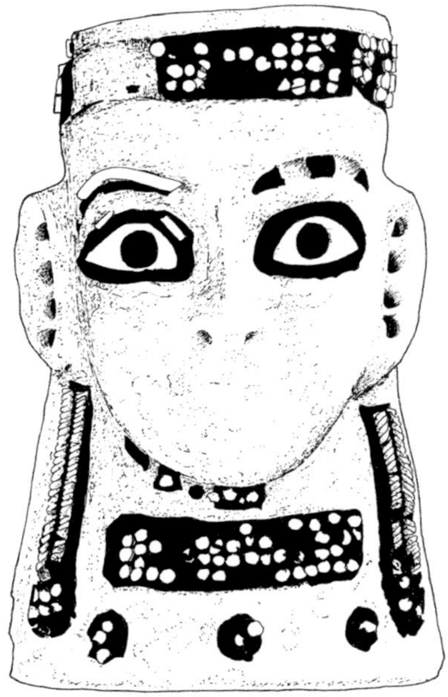

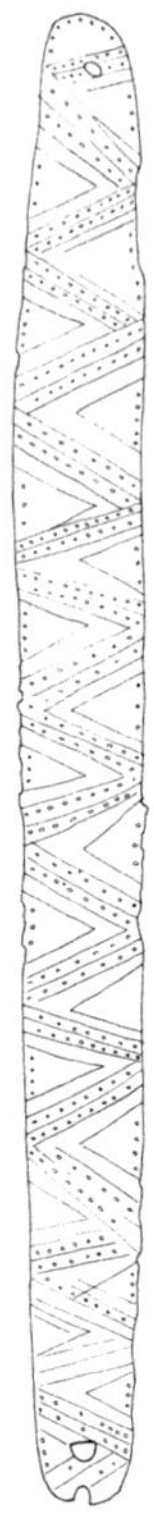

From left to right: Fig. 15.1 Gold frontlets from Ur (Maxwell-Hyslop 1971, 23, Fig. 13); Fig. 15.2. Gold diadem from Tomb 201 in Tell Fara (Maxwell-Hyslop 1971, Pl. 203); Fig. 15.3. Glazed frit mask from Rimah (Maxwell-Hyslop 1980, 87, Fig. 4); Fig. 15.4. Gold frontlet from Ajjul (Maxwell-Hyslop 1980, 88, Fig. 5)

probably indicate a diadem (or a part of it); that is, the metal plate – whether oval, rectangular, spherical, convex or in the shape of a zoomorphic protome – which is used to fasten a piece of cloth onto the forehead.[176] The piece of cloth in question, therefore, could be the *lupan(n)i-*, so the expression: X EME GAB *lu-pa-an-ni-eš*, literally translatable: X (metallic) tongue (that is added) in front of a *lupan(n)i-*, that is, the diadem or part of it.[177] It is feasible that an inventory listing luxury goods destined for a high dignitary of the court, the king or the royal couple, would also include weapons and diadems.

The presence of 'tongues' of precious metal together with the term *lupan(n)i-* seems necessary, even though this fabric that is tied around the head to support the diadem often does not leave archaeological traces. The validity of the suggestion, however, is supported by actual archaeological evidence.[178]

[176] *Contra* Siegelová 1984, 153–154.

[177] For similar conclusions see Kümmel 1967, 31; Goetze 1955, 62; Košak 1978, 107; *Idem* 1982, 201. For a possible archaeological comparison see, in general, Maxwell-Hyslop 1971, 22; *Eadem* 1980, 87–88 and Figs. 15.1–15.4.

[178] See, above all, Frangipane *et al.* 2009, in particular p. 18, with the important indications of remains of fabric under the deceased's diadem. Regarding the Hittite findings see Boehmer 1972, 38–41 and Fig. 15.5; for the gold diadems in the Syrian environment, see above all Parrot 1959, 95 and Fig. 15.6; Hauptmann & Pernicka 2004, 102 and Fig. 15.7. For other bronze diadems found in funeral contexts and positioned exactly under the head of the deceased, see Jean-Marie 1999, 166 (III Z17 SO91–SO109) and Fig. 15.8. For the presence of bronze plates (diadems) in elite funeral contexts (*Šakkanakku*); *ibidem*, 186 (III D1 SO4 bis) and Fig. 15.9. This last example confirms once more the contextual presence of weapons and diadems, as symbols of kingship, acquired in life and preserved after death.

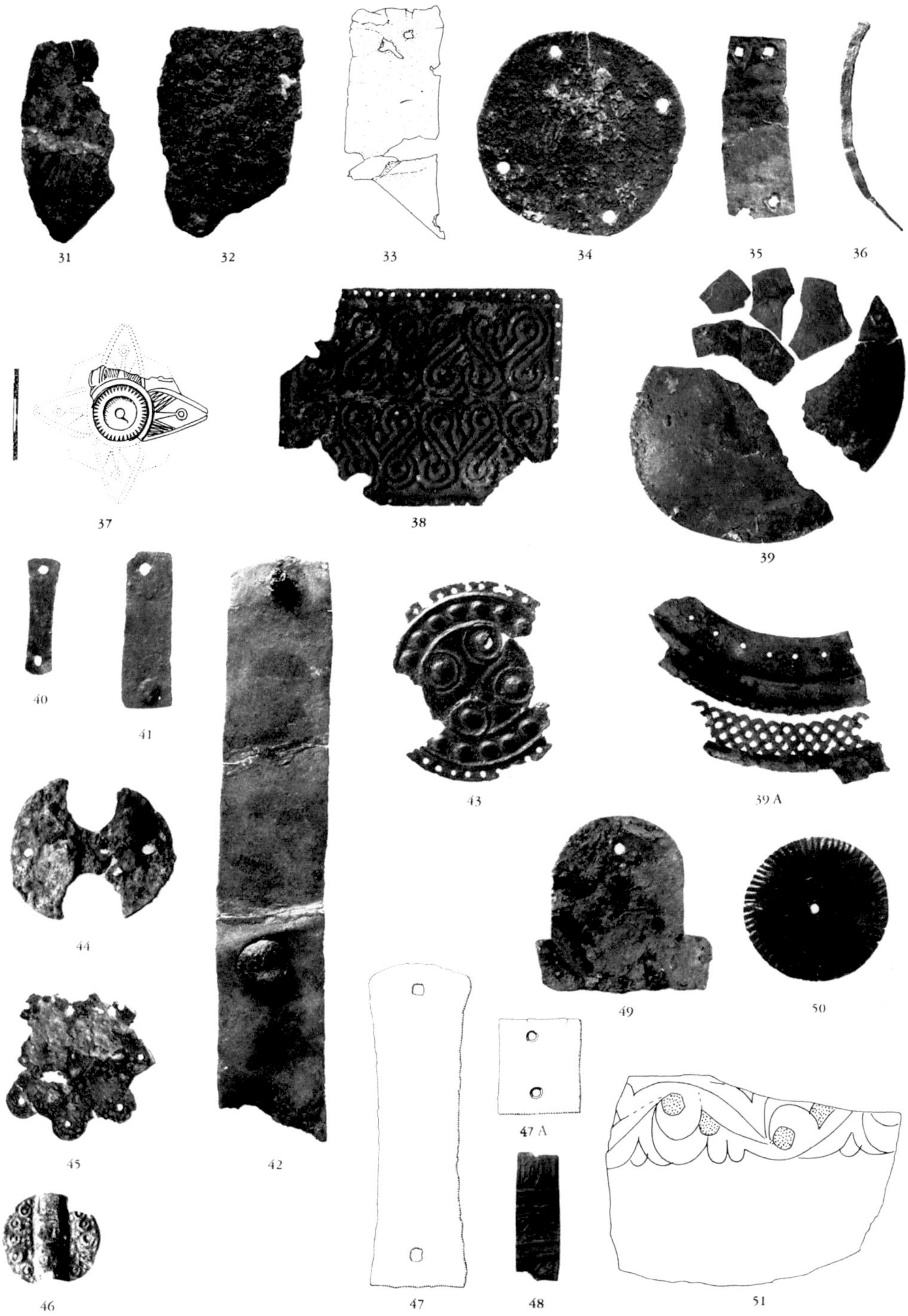

Fig. 15.5. Thin layers from Boğazköy-Ḫattuša. Among them some strips maybe were used as part of 'diadems' (Boehmer 1972, Pl. III)

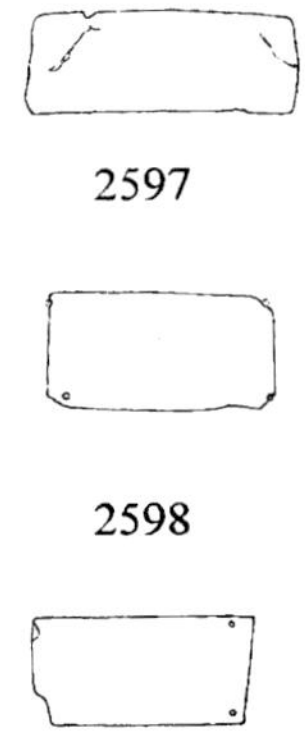

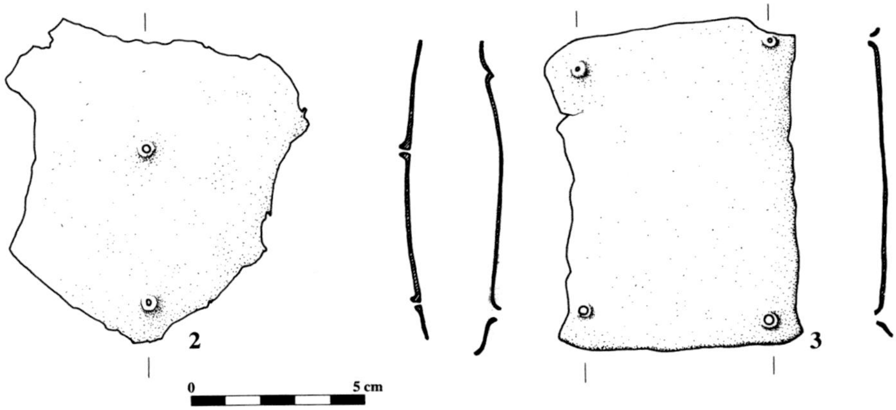

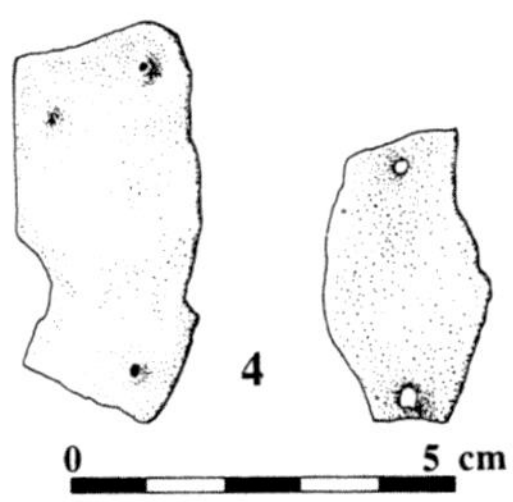

Fig. 15.6 (Top). Gold frontlets from the throne room of the Palace of Zimri-Lim in Mari (Parrot 1959, 95, Fig. 70);

Fig. 15.7 (Top right). Drawing of gold frontlets from the throne room of the Palace of Zimri-Lim in Mari (Hauptmann & Pernicka 2004, Pl. 158, Nos. 2598–2599)

Fig. 15.8 (Middle). Bronze frontlets from Tomb 727 in Mari (Jean-Marie 1999, Pl. 127, Nos. 2–3)

Fig. 15.9 (Left). Bronze frontlet from Tomb 1018 in Mari (Jean-Marie 1999, Pl. 205, No. 4)

Goetze's suggestion,[179] which was rejected for a long time,[180] of identifying the term $^{\text{TÚG/GADA}}$*lupan(n)i-* with the Sumerogram ^{TÚG}BAR.SI (a band for the head), though based on the analysis of a single fragment of text, seems fairly satisfactory, particularly in absence of valid alternatives to justify the impossibility of such a suggestion.[181]

The relationship between *lupan(n)i-* and headband, often intended as part of a diadem, is supported by the possibility of substituting the term with the Sumerogram BAR.SI, as can be deduced from a number of Hittite texts, above all KUB IX 15, col. III 1 ff. The headband, sometimes

[179] Goetze 1955, 62.

[180] CHD "L-N", 86.

[181] We agree with the observation of Košak 1982, 200, "It is true that the equation of lupanni- with BAR.SI is based on insufficient evidence but such objection applies even more to the new proposal advocated by CHD".

adorned with jewels or thin metal layers fixed onto the forehead, is a canonical element of clothing for priests or priestesses, perhaps even worn by rulers to legitimise their own high ranking position as "High Priests".[182] There is less of a tendency to consider the object as an article of clothing used exclusively by men.[183]

Little can be deduced from an etymological perspective. The term *lupan(n)i-* does not appear to be a word of Hittite origin. The form *lu-wa-an-ni-eš*, attested in KBo XVIII 170 (+) 170a, rev. 2,[184] could suggests a word of Hattic origin,[185] but this is invalid due to the presence of the 'Glossenkeil' in the derivative adjective *lūpannawant-* documented in KUB XXXVIII 1, col. II 7–8.[186] Even a Hurrian derivation of the term[187] seems to be excluded due to the fact that there is very little evidence of words that begin with the consonant *l-* in the Hurrian vocabulary.

The comparison of the term with other later Indo-European languages,[188] although suggestive, raises many doubts. Thus, Indo-European origin is very difficult to support.[189] It is therefore tricky to think of Luwian word, as Rosenkranz has already postulated[190] and subsequently been accepted also by Melchert.[191] The term *lupan(n)i-* could be a loan word or a 'Hittite' adaptation of a foreign word of non-Anatolian origin.[192]

In the recent past Kammenhuber has highlighted the probable link between the term ^{TÚG}*lu-pa-ri*, ἅπαξ λεγόμενον attested in the fragmentary oracular text KUB XVIII 29, col. IV 6,[193] and the Akkadian ^(TÚG/GADA)*lubāru*.[194] The term generally indicates a fabric or cloth, often of high quality,[195] and appears frequently in the Old-Babylonian and Neo-Babylonian tablets. It is also found in documents written in Akkadian from the archives of Alalaḫ and Nuzi, as well as the Akkadian texts of Mari and of Egypt (el-Amarna). However, in some texts the fabric was perhaps used as a shawl, like in EA 14, the letter between Amenḫotep IV and Burna-Buriyaš of Babylon,[196] but above all as a headdress or headgear, such as in the ritual BRM 4.[197] Even if the examples given are interesting, it is difficult to ascertain that the term *lupan(n)i-* could be an 'Anatolian adaptation'

[182] Cf. Boehmer 1980–1983, 198. Cf. the preliminary study of Goetze 1947.
[183] See Košak's doubts 1982, 200.
[184] Cf. Siegelová 1986, 488.
[185] Cf., for example, Puhvel 2001, 121.
[186] Cf. del Monte 1985, 154.
[187] Cf. Košak 1982, 110.
[188] Suggested examples can be seen in HEG "L-M", 78 and Puhvel 2001, 121.
[189] The only reconstructed Indo-European root for the term is **reup-/rup-* "break, smash", with the alternation of the liquid (*l/*r). Cf. Latin *ru-m-pô* "I break", Vedic *rop-* "to break" (*rûpa-* "form, structure" and for thoroughness "beauty"). Cf. *rûpin-* (literally: "to have form"), but even in Old English *rêofan* "break". The proposal is nevertheless clearly unfeasible from a semantic point of view.
[190] Rosenkranz 1957.
[191] CLL, 129.
[192] Cf. Kloekhorst 2008, 55.
[193] Cf. Košak 2002 (CTH 577). The fragment is to be surely placed in relation to KBo LIII 110. Cf. Miller 2005a, Inhaltübersicht VII.
[194] Kammenhuber 1985, 541.
[195] Cf. AHw I "A-L", 560; CAD "L", 228–231, 1; CDA_2, 184. We agree with the observation of CAD "L", 231, "There is no reason to connect *lubāru* with verb *labāru*", for the translation of the term *lubāru* "altes Gewand" of von Soden 1955, 387.
[196] EA 14, col. III, 19: 1 ME GADA *l[u-bá]-ru* GAL *ták-ti-mu*. Cf., finally, Cochavi-Rainey & Lilyquist 1999, 18, 26.
[197] 6:21, 44: *niši māti ṣubāt qaqqadišunu šaḫṭu ina lubarašunu qaqqasunu katmu*, "the people of the village remove their headdress and cover their heads with their *lubāru*"; *ina lubarušunu nukkusūtu qaqqasunu katmu*, "they cover their heads with cut *lubāru*".

Fig. 15.10 (left). Yazılıkaya: relief of King Tudḫaliya IV carved on the right wall of Kammer A (Ehringhaus 2005, 25, No. 38)

Fig. 15.11 (right). Relief of King Tudḫaliya IV: detail of his cover head (Bittel et al. 1967, Pl. 24, No. 3)

of the Akkadian *lubāru*, above all from a linguistic point of view.[198]

Thus, from the analysis undertaken so far, it can be confirmed that *lupan(n)i-* indicates a headband. Its use seems to be often relegated to religious contexts and is therefore a distinctive object, a luxury good. Furthermore the *lupan(n)i-* seems to be used by the king (or the queen). It also appears in inventory texts as a precious tribute offered on particular ceremonial occasions in the presence of the royal couple, and is even used as an accessory for the Sungod.[199]

All this could possibly suggests an archaeological comparison between the *lupan(n)i-* and the image of the headdress worn by Tudḫaliya IV in the relief No. 64 of Room A of the Yazılıkaya rock sanctuary,[200] just like the image of the royal couple in procession, carved on the Alaca Höyük orthostat,[201] or the image of Muwatalli II engraved in relief 1 of Sirkeli.[202] The same type

[198] The comparison offered by Puhvel 2001, 121, "The phonetic relationship of *luwanni-* to *lupari-* would match that of Hitt. *kuwanna-* to Lat. *cuprum* 'copper'" seems to be insignificant. The alternation r/n, that makes the association of TÚG*lupari* of KUB XVIII 29, col. IV 6 to *lupan(n)i-* possible, could be a result of the stabilization of a Luwian dialect form in the kizzuwatnean area. The consolidated form of the 13th century should therefore be the product of the doubling of the nasal, where truly conjecturable, due to the long vowel, perceived as atonic, from the Akkadian *lubāru* and then the 'Hittite' *lupāri*, with consequent -i motion.

[199] Cf. van den Hout 1993, 11.

[200] Figs. 15.10–15.11.

[201] Fig. 15.12.

[202] Fig. 15.13.

Fig. 15.12 (Top left). Alaca Höyük: stone-block relief of the Hittite royal couple during an offering scene (Ehringhaus 2005, 8, No. 3)

Fig. 15.13 (Middle). Drawing of the relief of King Muwattalli II in Sirkeli (Ehringhaus 2005, 98, No. 176)

Fig. 15.14 (Top right). Boğazköy-Ḫattuša: Relief of the Sungod in Kammer A of Südburg (Seeher 2006, 102, No. 108)

Fig. 15.15 (Bottom). Yazılıkaya: relief of the Sungod of Heaven carved on the right wall of Kammer A (Ehringhaus 2005, 20, No. 27)

Fig. 15.16 (above): Taşcı A: detail of human figures carved on the rock (Ehringhaus 2005, 67, No. 123)

Fig. 15.17 (right): Taşcı B: detail of the image of a prayer(?) carved on the rock (Ehringhaus 2005, 69, No. 127)

of headband, which is fixed to the head with one or more (metallic?) strings wrapped around the forehead, is evidently worn by the Sungod in the well-known representations of Room 2 of Südburg, in Ḫattuša,[203] or relief No. 34 of Room A,[204] also at Yazılıkaya, just to cite some examples from monumental art.[205] Obviously it is only a speculative hypothesis that requires further investigation.[206]

* * *

From this brief study on Hittite inventory texts it is obviously impossible to obtain significant elements to understand the techniques used to produce fabrics in the Hittite world, or even to establish whether the Hittites produced linen or, on the contrary, created trade routes across which the fabric arrived to the heart of Anatolia (other than sporadic examples such as Cyprus or northern Syria[207]).

---

203 Fig. 15.14.

204 Fig. 15.15.

205 Seen in relation to previous considerations of CHD "L-N", 86; Kümmel 1967, 31, with note 94. We could add the images of persons in the Taşcı rock-reliefs (A–B) Figs. 15.16–15.17. They are wearing a small headdress on which a horn is attached. Cf. Kohlmeyer 1983, 75, 79. Unfortunately the reliefs are very damaged or almost completely buried underground. Cf. Ehringhaus 2005, 65–70, in particular p. 66. On the identity of persons 'depicted' on Taşci reliefs see different suggestions collected by Stokkel 2005, 173.

206 For a valid overview of headdress represented in Hittite art, see Boehmer 1980–1983, 207–208. Also see the study of iconography of Tudḫaliya IV, with an in-depth study of royal headdress in van den Hout 1995b, in particular p. 565–568.

207 As rightly pointed out by van de Mieroop (2007, 165): "Northern Syria may have been a source of much desire textiles in this period, as many terms designating special garments have Hurrian names, seemingly indicating their region of origin." Cf. Klengel 2008, 76: "Geographisch wird GAD vor allem mit dem mittelsyrischen Land Amurru sowie mit Zypern verbunden, die hier wohl als wichtige produktionsstandarte gennant werden".

The aim of this text was to provide a general picture of a particular category of luxury goods, 'precious fabrics' and specifically linen, that appear in the Hittite inventory texts.

Given the current state of research and resources available, analysis of the documents has demonstrated that a re-examination of terms referring to objects listed in the Hittite inventories would be highly desirable. We hope that this work can serve as an incentive for future, in-depth studies.[208]

Abbreviations

AHw	W. Von Soden, *Akkadisches Handwörterbuch* – Wiesbaden 1958–1981.
AMT	R. C. Thompson, *Assyrian Medical Texts* – Oxford 1923.
A.N.	Author's note.
BRM	Babylonian Records in the Library of J. Pierpont Morgan.
CAD	The Assyrian Dictionary of the Oriental Institute of the University of Chicago – Chicago 1956–.
CDA_2	J. Black & A. George & N. Postgate (eds.), *A Concise Dictionary of Akkadian – 2nd (corrected) printing* – Wiesbaden, 2000.
CHD	H. G. Güterbock & H. A. Hoffner Jr. (eds.), *The Hittite Dictionary of Oriental Institute of Chicago* – Chicago 1980–.
CLL	H. C. Melchert, *Cuneiform Luvian Lexicon*, (Lexica Anatolica Volume 2) – Chapel Hill 1993.
col.	Column.
CT	Cuneiform Texts from Babylonian Tablets in the British Museum – London 1986–.
CTH	E. Laroche, *Catalogue des Textes Hittites* – Paris 1971.
EA	Texts from el-Amarna, according to J. A. Knudtzon, Die El-Amarna-Tafeln, Vorderasiatische Bibliothek 2 – Leipzig 1915.
HEG	J. Tischler, *Hethitisches Etymologische Glossar*, Innsbrucker Beiträge zur Sprachwissenschaft -Band 20) – Innsbruck 1983–.
HW	J. Friedrich, *Hethitisches Wörterbuch* – Heidelberg 1952.
HW_2	J. Friedrich & A. Kammenhuber, *Hethitisches Wörterbuch. Zweite, völlig neu bearbeitete Auflage auf der Grundlage der edierten hethitischen Texte* – Heidelberg 1975–.
HZL	C. Rüster & E. Neu, *Hethitisches Zeichenlexicon – Inventar und Interpretation der Keilschriftzeichen aus den Boğazköy-Texten* (Studien zu den Boğazköy-Texten, Beiheft 2) – Wiesbaden 1989.
IBoT	Istanbul Arkeoloji Müzelerinde bulunan Bogazköy tabletleri – Istanbul.
KBo	Keilschrifttexte aus Boghazköy – Berlin 1916–.
KUB	Keilschrifturkunden aus Boghazköy – Berlin 1921–.
MVAeG	Mitteilungen der Vorderasiatisch-ägyptischen Gesellschaft – Leipzig.
NBC	The Nies Babylonian Collection – Yale.
obv.	Obverse.
PA	Proto-Anatolic.
PIE	Proto-Indo-European.
rev.	Reverse.
RlA	Reallexikon der Assyriologie. Berlin, New York.
RS	Field numbers of tablets excavated at Ras Shamra – Ugarit.
TC	Tablettes Cappadociennes – Paris.
VBoT	A. Goetze (hrsg.), Verstreute Boghazköi-Texte – Marburg 1930.

[208] The contribution presented here is part of a larger nationally-funded research project aimed at an up-to-date study of the tribute practices and the treasuring and hoarding of goods in the Hittite Empire. The project is directed by Prof. Marcella Frangipane (Università La Sapienza di Roma); Prof. Clelia Mora (Università di Pavia) organizes the research undertaken at the local level, as well as that by the author, by Dr. Mauro Giorgieri (Università di Pavia), by Dr Lorenzo d'Alfonso (Università di Pavia), and by Dr. Maria Elena Balza (Freie Universität – Berlin). By the time this paper was finished a couple of new interesting articles by Horst Klengel were published (here Klengel 2008 and Klengel & Klengel 2009). Where possible they have been referred to. I am glad to read that the author proposed similar hypotheses on specific topics discussed in my paper.

Bibliography

Alp, S. (1940) *Untersuchungen zu den Beamtennamen im hethitischen Festzeremoniell.* Sammlung orientalischer Arbeiten Heft 5. Leipzig.

Archi, A. (1973) L'organizzazione amministrativa ittita e il regime delle offerte cultuali. *Oriens Antiquus* 12, 209–226.

Bauer, Th. & Friedrich, J. & Landsberger B. (1934) Lexikalisches Archiv. *Zeitschrift für Assyriologie* 42, 152–169.

Beal, R. H. (1992)*The Organisation of the Hittite Military.* Texte der Hethiter 20. Heidelberg.

Beckman, G. (1985) Boekbesprekingen-Hethitolgie: Archi, Alfonso. Keilschrifturkunden aus Boghazköi, Heft III. Hethitische Orakeltexte und Texte verschiedenen Inhalts. *Bibliotheca Orientalis* XLII, 138–144.

Beckman, G. (1996) Hittite Documents from Hattusa. In A. B. Knapp (ed.) *Sources for the History of Cyprus: Near Eastern and Aegean Texts from the Third to the First Millennia BC, Volume II*, 31–35. Albany.

Berlin, B. & Kay, P. (1969) *Basic color terms. Their Universality and Evolution.* Berkeley – Los Angeles.

Berman, H. (1977) Greek –σαι: Hittite –asha-. *Kuhns Zeitschrift* 91, 231–239.

Bittel, K. & Naumann, R. & Otto, H. (1967) *Yazılıkaya. Architektur, Felsbilder, Inschriften und Kleinfunde*, Osnabrück.

Boehmer, R. N. (1972) *Die Kleinfunde von Boğazköy aus den Grabungskampagnen 1931-1939 und 1952-1969.* Wissenschaftliche Veröffentlichung der Deutschen Orient-Gesellschaft 87. Berlin.

Boley, J. (2001) Suggestions for the Original Function of -kan. In O. Carruba & W. Meid (eds.), *Anatolisch und Indogermanisch - Anatolico e Indoeuropeo: Akten des Kolloquiums der Indogermanischen Gesellschaft, Pavia 22.-25. September 1998*, 19–28. Innsbruck.

Bottéro, J. (1949) Les Inventaires des Qatna. *Revue d'Assyriologie et d'Archéologie Orientale* 43, 1–40.

Bryce, T. R. (2005)*The Kingdom of the Hittites - New Edition.* Oxford.

Burde, C. (1974) *Hethitische Medizinische Texte.* Studien zu den Boğazköy-Texten 19. Wiesbaden.

Carruba, O. (1968) Contributo alla storia di Cipro nel II millennio, *Studi classici e orientali* 17, 5–29.

Carruba, O. (2007) Per una ricostruzione delle liste reali etee. In D. Groddek & M. Zorman (eds.), *Tabularia Hethaeorum. Hethitologische Beiträge. Silvin Košak zum 65. Geburtstag*, 131–142. Dresdner Beiträge zur Hethitologie 25. Wiesbaden.

Carruba, O. (2008) *Annali etei di Medio Regno.* Studia Mediterranea 18 – Series Hethaea 5. Pavia.

Cavaignac, E. (1933) Remarques sur les «Ahhijavā-Urkunden» de M. Sommer. I. La femme de Subbiluliuma. *Revue hittite et asianique* 2. 12, 155–159.

Cochavi-Rainey, Z. (2003) *The Alashiya texts from the 14th and 13th centuries BCE: a Textual and Linguistic Study.* Alter Orient und Altes Testament 289. Münster.

Cochavi-Rainey, Z. & Lilyquist, C. (1999) *Royal Gifts in the Late Bronze Age. Fourteenth to Thirteenth Centuries B.C.E.* Beer –Sheva, Studies by the Department of Bible and Ancient Near East – Volume XIII. Beer-Sheva.

Cohen, Y. (2002) *Taboos and Prohibitions in Hittite Society. A Study of the Expression natta āra ('not permitted').* Texte der Hethiter 24. Heidelberg.

Cornelius, F. (1976) Neue Arbeiten zur hethitischen Geographie. *Anatolica* 1, 62–77.

Daglio, C. (2008) Le malattie. In S. Malgora (ed.), *Ur Sunu. Grandi dottori dell'Antico Egitto. Malattie e cure nella terra dei faraoni, Casale Monferrato, Palazzo Sanazzaro, 27 Settembre-21 Dicembre 2008*, 65–81. Vercelli.

van Dijk, H. J. (1968) *Ezekiel's Prophecy on Tyre (Ez. 26,1-28,19). A New Approach.* Roma.

Ehringhaus, H. (2005) *Götter, Herrscher, Inschriften. Die Felsreliefs der hethitischen Grossreichszeit in der Türkei.* Sonderbände der Antiken Welt. Mainz am Rhein.

Endesfelder, E. & Priese K. H. (1984) Historisch-bibliografische Urkunden aus der Zeit Thutmosis III = Nr. 201–Nr. 215. In E. Blumenthal & I. Müller & W. F. Reinke (eds.), *Urkunden der 18. Dynastie - Übersetzung zu den Heften 5-16*, 180–241. Berlin.

Erman, A. & Grapow, H. (1955a) *Wörterbuch der Aegyptischen Sprache - Erster Band.* Berlin – Leipzig.

Erman, A. & Grapow, H. (1955b) *Wörterbuch der Aegyptischen Sprache - Fünfter Band.* Berlin – Leipzig.

Fales, F. M. (2004) Rileggendo gli inventari di Qatna. *Kaškal* 1, 83–128.

Fincke, J. C. (2000) *Augenleiden nach Keilschriften Quellen. Untersuchungen zur altorientalischen Medizin.* Würzburger medizinhistorische Forschungen Band 70. Würzburg.

Forbes, E. J. (1963) *Studies in Ancient Technology - Volume VII.* Leiden.

Forbes, E. J. (1964) *Studies in Ancient Technology - Volume IV - Second Revisited Edition*. Leiden.

Forlanini, M. (1988) La regione del Tauro nei testi hittiti. *Vicino Oriente* 7, 129–169.

Forlanini, M. (2009) On the Middle Kızılırmak. In F. Pecchioli Daddi & G. Torri & C. Corti (eds.), *Central-North Anatolia in the Hittite Period. New Perspectives in Light of Recent Research. Acts of the International Conference Held at the University of Florence (7-9 February 2007)*. Studia Asiana 5, 39–69, Roma.

Forrer, E. O. (1919) Die acht Sprachen der Bogazköi-Inschriften. *Sitzungsberichte der Preußischen Akademie der Wissenschaften* 1919, 1029–1041.

Frangipane, M. & Andersson Strand, E. & Laurito R. & Möller-Wiering S. & Nosch, M. L. & Rast-Eicher, A. & Wisti Lassen, A. (2009) Arslantepe, Malatya (Turkey): Textiles, Tools and Imprints of Fabrics from the Fourth to the Second Millenium BCE. *Paléorient* 35. 1, 1–28.

Frantz-Szabó, G. & Ünal, A. (2006) Puduḫepa. A. *Reallexikon der Assyriologie* 11, 106–108.

Freydank, H. (1968) *bīt šurīpim* in Boğazköy. *Die Welt des Orients* 4. 2, 316–317.

Friedrich, J. (1940) *Hethitisches Elementarbuch*. Heidelberg.

Gardiner, A. H. (1931) Two hieroglyphic signs and the Egyptian words for 'Alabaster' and 'Linen', etc. *Bulletin de l'Institut Français d'Archéologie Orientale au Caire* 30, 161–183.

Geiger, L. (1871)*Vorträge zur Entwickelungsgeschichte der Menschheit*, Stuttgart.

Germer, R. (1992) *Die Textilfärberei und die Verwendung gefärbter Textilien im Alten Ägypten*. Ägyptologische Abhandlungen Band 53. Wiesbaden.

van Gessel, A. H. L. (1998) *Onomasticon of the Hittite Pantheon I*. Handbuch der Orientalistik 33/1. Leiden.

Giorgieri, M. (1999) Zu den hurritischen Personennamen in den Amarna-Briefen. *Studi Micenei ed Egeo-Anatolici* XLI, 63–83.

Giorgieri, M. & Mora, C. (in print) Luxusguter als Symbole der Macht: in den "Schatzkammern" der altorientalischen Königen. In G. Wilhelm (ed.), *Akten des LIV Rencontre Assyriologique Internationale, 20.-25. Juli 2008*. Würzburg.

Goetze, A. (1930) Bemerkungen zu dem hethitischen Text AO 9608 des Louvre. *Revue hittite et asianique* 1. 1, 18–30.

Goetze, A. (1931) *ištēnūtu* und *tapalu*. *Zeitschrift für Assyriologie* 40, 79–80.

Goetze, A. (1940) *Kizzuwatna and the Problem of Hittite Geography*. Yale Oriental Series-Researches XXI. New Haven.

Goetze, A. (1947) The Priestly Dress of the Hittite King. *Journal of Cuneiform Studies* 1, 176–185.

Goetze, A. (1955) Hittite Dress. In H. Krahe (ed.), *Corolla Linguistica - Festschrift Ferdinand Sommer zum 80. Geburtstag am 4 Mai 1955*, 48–62. Wiesbaden.

Goetze, A. (1956) The Inventory IBoT I 31. *Journal of Cuneiform Studies* 10, 32–38.

Groddek, D. (2006) *Hethitische Texte in Transcription KUB 60*. Dresdner Beiträge zur Hethitologie 20. Wiesbaden.

Gurney, O. R. (1977) Book Review: Keilschrifttexte aus Boghazköi, 23 Heft. by Heinrich Otten and Christel Rüster, Berlin 1976. *Journal of the Royal Asiatic Society* 1977, n. 2, 207.

Gusmani, R. (1972) Keilhethitische Nominalableitungen auf -(a)sha-. *Kuhns Zeitschrift* 86, 255–266.

Güterbock, H. G. (1930) *Keilschrifturkunden aus Boghazköi - Heft XXV (Festrituale)*. Berlin.

Güterbock, H. G. (1940) *Siegel aus Boğazköy. Erster Teil. Die Königssiegel der Grabungen bis 1938. Archiv für Orientforschung* Beiheft 5.

Güterbock, H. G. (1956a) Notes on Luwian-Studies (A propos B. Rosenkranz' Book Beitrage zur Erforschung des Luvischen). *Orientalia. Commentarii periodici Pontificii Instituti Biblici. Nova Series* 25, 113–140.

Güterbock, H. G. (1956b) The Deeds of Suppiluliuma as told by his son Mursili II. *Journal of Cuneiform Studies* 10, 107–130.

Güterbock, H. G. (1974) The Hittite Palace. In P. Garelli (ed.), *Le Palais et la Royauté (Archéologie et Civilisation). XIX Rencontre Assyriologique Internationale organisée par le groupe François Thureau-Dangin, Paris, 29 juin - 2 juillet 1971*, 1974, 305–314. Paris.

Haas, V. (1994) *Geschichte der hethitischen Religion*. Handbuch der Orientalistik I/15. Leiden.

Haas, V. (2003) *Materia Magica et Medica Hethitica - Ein Beitrag zur Heilkunde im Alten Orient Vol. II*. Berlin - New York.

Hagenbuchner, A. (1989) *Die Korrespondenz der Hethiter. 2. Teil: Die Briefe mit Transkription, Übersetzung und Kommentar*. Texte der Hethiter 16. Heidelberg.

Hannig, R. & Vomberg, P. (1999) *Kulturhandbuch Ägyptens. Wortschatz der Pharaonen in Sachgruppen*. Kulturgeschichte der Antiken Welt, 72 – Hannig-Lexica, 2. Mainz.

Hauptmann, H. & Pernicka, E. (2004) *Die Metallindustrie Mesopotamiens von den Anfängen bis zum 2. Jahrtausend v. Chr.* Rahden.

Hausperger, M. (2000) Einige medizinische Anmerkungen zum Text BAM 3. In S. Graziani (ed.), *Studi sul Vicino Oriente Antico dedicati alla memoria di Luigi Cagni*. Istituto Universitario Orientale di Studi Asiatici – Serie Minor LXI, Volume I, 439–450. Napoli.

Helk, W. & Drenkhahn, R. (1995) *Die Beziehungen Ägyptens und Vorderasiens zur Ägäis bis ins 7 Jahrhundert v. Chr.* Erträge der Forschung – Band 120. Darmstadt.

Hoffner, H. A. Jr. (1992) Syrian Cultural Influence in Ḫatti. In M. W. Chavalas & J. L. Hayes (eds.), *New Horizons in the Study of Ancient Syria*, 89–106. Bibliotheca Mesopotamica 25. Malibu.

Hoffner, H. A. Jr. (2002) Records of Testimony Given in the Trials of Suspected Thieves and Embezzlers of Royal Property. In W. W. Hallo & K. L. Younger Jr. (eds.), *The Context of Scripture – Archival Documents from the Biblical World – Volume III*, 57–60. Leiden – Boston – Köln.

Hoffner, H. A. Jr. – Melchert, H. C. (2008) *A Grammar of the Hittite Language. Part I: Reference Grammar*. Languages of the Ancient Near East 1. Winona Lake.

van den Hout, Th. P. J. (1993) *Tudḫalija Kosmokrator. Gedachten over ikonografie en ideologie van een Hettitische koning. Inaugurationsrede an der Universität von Amsterdam, 16. Juni 1993*. Amsterdam.

van den Hout, Th. P. J. (1995a) *Der Ulmitešub-Vertrag. Eine prosopographische Untersuchung*, Studien zu den Boğazköy-Texten 38, Wiesbaden.

van den Hout, Th. P. J. (1995b) Tutḫaliya IV. Und die Ikonographie Hethitischer Großkönige des 13. Jhs. *Bibliotheca Orientalis* LII, 546–573.

van den Hout, Th. P. J. (1999) Nerikkaili. *Reallexikon der Assyriologie* 9, 231–232.

Houwink ten Cate, P. H. J. (1996) The Hittite Dynastic Marriages of the Period between ca. 1258 and 1244 BC. *Altorientalische Forschungen* 25/1, 40–75.

Imparati, F. (1995) Apology of Ḫattušili III or Designation of His Successor? In Th. P. J. van den Hout & J. de Roos (eds.). *Studio Historia Ardens. Ancient Near Eastern Studies Presented to Philo H. J. Houwink ten Cate on the Occasion of his 65th Birthday*, 143–157. Publications de l'Institut historique-archéologique néerlandais de Stamboul 74. Leiden.

Jean-Marie, M. (1999) *Mission Archéologique de Mari. Tomo V. Tombes et Nécropoles de Mari*. Institut Français d'Archéologie de Beyrouth. Bibliothèque archéologique et historique. Tome CLIII. Beyrouth.

Jones, J. & Oldfield, R. (2006) Egypt's earliest linen. *Egyptian Archaeology* 29, 33–35.

Jones, J. (2008) Pre- and Early Dynastic textiles: technology, specialisation and administration during the process of state formation. In B. Midant-Reynes & Y. Tristant (eds.). *Egypt at Its Origins 2. Proceedings of the International Conference "Origin of the State. Predynastic and Early Dynastic Egypt, Toulouse (France), 5th–8th September 2005*, 99–132. Orientalia Lovaniensia Analecta 172. Leuven.

Kammenhuber, A. (1985) Bemerkungen anlässlich eines neuen hethitischen Wörterbuchs. *Orientalistische Literaturzeitung* 80, Heft 6, 534–542.

Kemp, B. J. & Vogelsang-Eastwood, G. (2001)*The Ancient Textile Industry at Amarna*. Egyptian Exploration Society. London.

Klengel, H. (2008) Studien zur hethitischen Wirtschaft, 4: Das Handwerk. Werkstoffe: Wolle und Leder, Holz und Rohr. *Altorientalische Forschungen* 35/1, 68–85.

Klengel, H. & Klengel, E. (2009) „Hurritische Hemden" in der keilschriftlichen Tradition. *Altorientalische Forschungen* 36/2, 205–208.

Klinger, J. (1996) *Untersuchungen zur Rekonstruktion der hattischen Kultschicht*. Studien zu den Boğazköy-Texten 37. Wiesbaden.

Kloekhorst, A. (2008) *Etymological Dictionary of the Hittite Inherited Lexicon*. Leiden Indo-European Etymological Dictionary Series – Volume 5. Leiden.

Knudtzon, J. A. (1915) *Die El-Amarna-Tafeln mit Einleitung und Erlauterungen*. Vorderasiatisches Bibliothek 2 – Band I – Die Texte. Leipzig.

Kohlmeyer, K. (1983) Felsbilder der hethitischen Grossreichszeit. *Acta praehistorica et archaeologica* 15, 7–154.

Košak, S. (1978) The Inventory of Manninni CTH 504. *Linguistica* 18, 99–123.
Košak, S. (1982) *Hittite Inventory Texts: (CTH 241-250)*. Texte der Hethiter 10. Heidelberg.
Košak S. (1994) Review: Klengel H. Texte verschiedenen Inhalts (= Keilschrifturkunden aus Boghazköi, Heft XL). Akademie der Wissenschaften der DDR, Zentralinstitut für Alte Geschichte und Archäologie. Berlin 1990. *Zeitschrift für Assyriologie* 84, 288–290.
Košak, S. (2002) *http://www.hethport.uni-wuerzburg.de/hetkonk/*. Mainz.
Kühne, C. (1973) *Die Chronologie der internationalen Korrespondenz von El-Amarna*. Alter Orient und Altes Testament 17. Neukirchen – Vluyn.
Kümmel, H. M. (1967) *Ersatzrituale für den hethitischen König*. Studien zu den Boğazköy-Texten 3. Wiesbaden.
Landsberger, B. (1925) Assyrische Handelskolonien in Kleinasien aus dem dritten Jahrtausend. *Der Alter Orient* 24, no. 4, 3–34.
Landsberger, B. (1967) Über Farben im sumerisch-akkadischen.*Journal of Cuneiform Studies* 21, 139–182.
Laroche, E. (1958) Etudes de Vocabulaire VII. *Revue hittite et asianique* 16. 63, 85–114.
Laroche, E. (1959) *Dictionnaire de la langue louvite*. Bibliothèque archéologique et historique de l'Institut Français d'Archéologie d'Istanbul Vol. VI. Paris.
Laroche, E. (1966) *Les Noms des Hittites*. Études Linguistiques IV. Paris.
Laroche, E. (1976) Glossaire de la Langue hourrite I: A-L. *Revue hittite et asianique* 34, 13–161.
Liverani, M. (1990) *Prestige and interest. International Relations in the Near East ca. 1600–1100 B.C.* History of the Ancient Near East. Studies. Vol. I. Padova.
Liverani, M. (1998) *Le lettere di el-Amarna – Vol. I: Le lettere dei «Piccoli Re»*. Testi del Vicino Oriente Sez. 2, Letterature Mesopotamiche 3. Brescia.
Liverani, M. (1999) *Le lettere di el-Amarna – Vol II: Le lettere dei «Grandi Re»*. Testi del Vicino Oriente Sez. 2, Letterature Mesopotamiche 3. Brescia.
Magnus, H. (1877) *Die geschichtliche Entwickelung des Farbensinnes*, Leipzig.
Marizza, M. (2006) La carica di GAL DUMU^{MEŠ} É.GAL nel Regno ittita. *Studi Micenei ed Egeo-Anatolici* XLVIII, 151–175.
Marty, A. (1886) Anzeige von R. Hochegger, Die geschichtliche Entwickelung des Farbensinnes. *Göttingische gelehrte Anzeige* 1886, Heft 1. (then in J. Eisenmeier, A. Kastil & O. K. Niemeyer (eds.) *Gesammelte Schriften Vol. I, Sez I*, 1918. Halle).
Mauss, M. (1923–1924) *Essai sur le don. Forme et raison de l'échange dans les sociétés archaïques*. Année Sociologique, seconde série. Paris.
Maxwell-Hyslop, K. R. (1971) *Western Asiatic Jewellery c. 3000–612 B.C.* London.
Maxwell-Hyslop, K. R. (1980) Jewellery listed in the Inventory of Manninni. *Anatolian Studies* 30, 85–88.
Melchert, H. C. (1994) *Anatolian Historical Phonology*. Amsterdam – Atlanta.
van de Mieroop, M. (2007) *The Eastern Mediterranean in the Age of Ramesses II*. Singapore.
Miller, J. L. (2004) *Studies in the Origins, Development and Interpretation of the Kizzuwatna Rituals*. Studien zu den Boğazköy-Texten 46. Wiesbaden.
Miller, J. L. (2005a) *Keilschrifttexte aus Boghazköi – Dreiundfünfzigstes Heft – Texte aus dem Bezirk des Grossen Tempels V*. Berlin.
Miller, J. L. (2005b) Die Erweiterung des Kultes der Gottheit der Nacht (CTH 481). In B. Janowski & G. Wilhelm (eds.), *Omina, Orakel, Rituale und Beschwörungen*, 223–229. Texte aus der Umwelt des Alten Testaments-Neue Folge 4. Gütersloh.
del Monte, G. F. (1985) Review: H. G. Güterbock – H. A. Hoffner, The Hittite Dictionary of the Oriental Institute of the University of Chicago. Vol. 3, fasc. 1 Chicago 1980, 96. *Oriens Antiquus* 24, 148–154.
del Monte, G. F. (1992) *Die Orts- und Gewässernamen der hethitischen Texte: Supplement*. Beihefte zum Tübinger Atlas des Vorderen Orients – Reihe B Geisteswissenschaften No. 7/6; Répertoire Géographique des Textes Cunéiformes Band 6/2. Wiesbaden.
del Monte, G. F. & Tischler, J. (1978) *Die Orts- und Gewässernamen der hethitischen Texte*. Beihefte zum Tübinger Atlas des Vorderen Orients – Reihe B Geisteswissenschaften Nr. 7/6; Répertoire Géographique des Textes Cunéiformes Band 6. Wiesbaden.
Moorey, P. R. S. (1994) *Ancient Mesopotamian Materials and Industries. The Archaeological Evidence*. Oxford.

Moorey, P. R. S. (2001) The Mobility of Artisans and Opportunities for Technology Transfer between Western Asia and Egypt in the Late Bronze Age. In A. J. Shortland (ed.), *The Social Context of Technological Change. Egypt and the Near East, 1650-1550 BC. Proceedings of a conference held at St Edmund Hall, Oxford. 12-14 September 2000*, 1–14. Oxford.

Mora, C. (2006) Riscossione dei tributi e accumulo dei beni nell'Impero ittita. In M. Marazzi & M. Perna (eds.), *Atti del Convegno «Fiscality in Mycenaean and Near Eastern Archives», Naples, 21-23 October 2004*, 133–146. Napoli.

Mora, C. (2007) I testi d'inventario e gli 'archivi' di cretule. In D. Groddek & M. Zorman (eds.), *Tabularia Hethaeorum. Hethitologische Beiträge. Silvin Košak zum 65. Geburtstag*, 271–292. Dresdner Beiträge zur Hethitologie 25. Wiesbaden.

Moran, W. L. (1987) *Les Lettres d'El-Amarna: correspondance diplomatique du Pharaon*. Paris.

Moscati, S. (1966) *Il mondo dei Fenici*. Milano.

Muhly, J. D. (1977) The Copper Ox-Hide Ingots and the Bronze Age Metals Trade. *Iraq* 39, 73–82.

Nakamura, M. (2002) *Das hethitische nuntarriyašha-Fest*. Publications de l'Institut historique-archéologique néerlandais de Stamboul 94. Leiden.

Naumann, R. (1971) *Architektur Kleinasiens: von ihren Anfangen bis zum Ende der hethitischen Zeit (2. erw. Auflage)*. Tübingen.

Neu, E. (1993) Zu den hethitischen Ortspartikeln, *Linguistica* 33, 137–152.

Nougayrol, J. (1950) Textes hépatoscopiques d'époque ancienne conservés au Musée du Louvre (III), *Revue d'Assyriologie et d'Archéologie Orientale* 44, 1–44.

Nougayrol, J. (1955) Textes économiques. In Le Palais Royal d'Ugarit III, 177–211.

Oppenheim, A. L. (1947) Book Review. Mathematical Cuneiform Texts. Edit by O. Neugebauer and A. Sachs (with a chapter by Goetze), *Journal of Near Eastern Studies* 6, 126–128.

Otten, H. (1951) Die hethitischen "Königlisten" und die altorientalische Chronologie. *Mitteilungen der Deutschen Orient-Gesellschaft zu Berlin* 83, 47–71.

Otten, H. (1981) *Die Apologie Hattusilis III. Das Bild der Überlieferung*. Studien zu den Boğazköy-Texten 24. Wiesbaden.

Otten, H. (1982) Bemerkungen zum Hethitischen Wörterbuch IV. *Zeitschrift für Assyriologie* 72, 280–288.

Parise, N. F. (1968) I pani di rame del II millennio a.C. Considerazioni preliminari. In *Atti e Memorie del I Congresso Internazionale di Miceneologia*, 117–133. Roma.

Parrot, A. (1959) *Mission Archéologique de Mari. Volume II: Le Palais. Documents et Monuments*. Institut Français d'Archéologie de Beyrouth. Bibliothèque archéologique et historique. Tome LXX. Paris.

Pecchioli Daddi, F. (1982) *Mestieri, professioni e dignità nell'Anatolia ittita*. Incunabula Graeca 79, Istituto per gli Studi Micenei ed Egeo-Anatolici, Consiglio Nazionale delle Ricerche. Roma.

Polvani, A. M. (1988) *La terminologia dei minerali nei testi ittiti - Parte Prima*. Eothen 3 – Collana di Studi sulle Civiltà dell'Oriente Antico. Firenze.

Puhvel, J. (1984) *Hittite Etymological Dictionary Vol. 2 - Words beginning with E and I*. Trends in Linguistics Documentation 1. Berlin – New York.

Puhvel, J. (1991) *Hittite Etymological Dictionary Vol. 3 - Words beginning with H*. Trends in Linguistics Documentation 5. Berlin – New York.

Puhvel, J. (1997) *Hittite Etymological Dictionary - Vol. 4- Words beginning with K*. Trends in Linguistic Documentation 14. Berlin – New York.

Puhvel, J. (2001) *Hittite Etymological Dictionary - Vol. 5 - Words beginning with L. Indices to Volumes 1-5*. Trends in Linguistics Documentation 18. Berlin – New York.

Puhvel, J. (2004a) Darkness in Hittite. *Zeitschrift für vergleichende Sprachforschung auf dem Gebiete der indogermanischen Sprachen* 117.2, 194–196.

Puhvel, J. (2004b) *Hittite Etymological Dictionary - Vol. 6 - Words beginning with M*. Trends in Linguistics – Documentation 22. Berlin – New York.

Reese, D. S. (1987) Palaikastro shells and Bronze Age purple-dye production in the Mediterranean basin. *Annual of the British School of Archaeology at Athens* 82, 201–206.

Riemschneider, K. K. (1957) Hethitisch "gelb/grün". *Mitteilungen des Instituts für Orientforschung* V, Heft 2, 141–147.

Rosenkranz, B. (1957) Zum Luvischen. *Die Welt des Orients* 2. 4, 369–377.
Rosenkranz, B. (1965) Ein hethitischer Wirtschaftstext. *Zeitschrift für Assyriologie* 57, 237–248.
Seeher, J. (2006) *Hattusha Guide. A Day in the Hittite Capital. 3. Revised Edition*. Istanbul.
Sethe, K. (1906) *Urkunden der 18. Dynastie*. Urkunden des Ägyptischen Altertums IV: 3. Leipzig.
Siegelová, J. (1984) Gewinnung und Verarbeitung von Eisen im hethitischen Reich im 2. Jahrtausend v.u.Z.». *Annals of the Náprstek Museum* 12. Prague.
Siegelová, J. (1986) *Hethitische Verwaltungspraxis im Lichte der Wirtschafts- und Inventardokumente*. Praha.
Singer, I. (2006) Ships Bound for Lukka: A New Interpretation of the Companion Letters RS 94.2530 and RS 94.2523. *Altorientalische Forschungen* 33/2, 242–262.
Singer, I. (2008) Purple-Dyers in Lazpa. In B. J. Collins & M. R. Bachvarova & I. C. Rutherford (eds.), *Anatolian Interfaces. Hittite, Greeks and their neighbours. Proceedings of an International Conference on Cross-Cultural Interaction, September 17–19, 2004 Emory University, Atlanta, GA*, 21–43. Oakville.
Smith, S. (1953) The Ship Tyre. *Palestine Explorations Quarterly* 85, 97–110.
von Soden, W. (1955) Zum akkadischen Wörterbuch. 67–80. *Orientalia. Commentarii periodici Pontificii Instituti Biblici. Nova Series* 24, 377–394.
Starke, F. (1979) Zu den hethitischen und luwischen Verbalabstrakta auf *-šḫa-*. *Kuhns Zeitschrift* 93, 246–261.
Starke, F. *(1982)* Boekbesprekingen-Hethitologie: The Hittite Dictionary of the Oriental Institute of the University of Chicago edited by Hans G. Güterbock and Harry A. Hoffner. Vol. 3, Fasc. 1: -la to -ma. Chicago 1980. *Bibliotheca Orientalis* XXXIX, 356–363.
Starke, F. (1985) *Die keilschrift-luwischen Texte in Umschrift*. StBoT 30. Wiesbaden.
Starke, F. (1986) Boekbesprekingen-Hethitologie: The Hittite Dictionary of the Oriental Institute of the University of Chicago edited by Hans G. Güterbock and Harry A. Hoffner. Vol. 3, Fasc. 2: -ma to miyaḫuwant-. Chicago 1983. *BiOr* XLIII, 157–165.
Starke, F. (1990) *Untersuchungen zur Stammbildung des keilschrift-luwischen Nomens*. Studien zu den Boğazköy-Texten 31. Wiesbaden.
Steiner, G. (1962) Neue Alašija-Texte. *Kadmos* I, 130–138.
Stokkel, P. J. A. (2005) A New Perspective on the Hittite Rock Reliefs. *Anatolica* 31, 171–188.
Tischler, J. (2001) *Hethitisches Handwörterbuch mit dem Wortschatz der Nachbarsprachen*. Innsbrucker Beiträge zur Sprachwissenschaft – Band 120. Innsbruck.
Thomason, A. K. (2005) *Luxury and Legitimation. Royal Collecting in Ancient Mesopotamia*, Aldershot.
Thompson, R. C. (1936)*The Assyrian Dictionary of Assyrian Chemistry and Geology*. Oxford.
Ünal, A. (1978) *Ein Orakeltext über die Intrigen am hethitischen Hof (KUB XXII 70 = Bo 2011) Text und Übersetzung*. Texte der Hethiter 6. Heidelberg.
Veenhof, K. R. (1972) *Aspects of Old Assyrian Trade and its Terminology*. Studia et Documenta ad Iura Orientis Antiqui pertinentia, Volumen X. Leiden.
Verderame, L. (2004) I colori nell'astrologia mesopotamica. In H. Waetzoldt (ed.), *Von Sumer nach Ebla und zurük. Festschrift Giovanni Pettinato zum 27. September 1999 gewidmet von Freunden, Kollegen und Schülern*, 327–332. Heidelberger Studien zum Alten Orient 9. Heidelberg.
Vogelsang-Eastwood, G. (1992)*The Production of Linen in Pharaonic Egypt*. Leiden.
Vogelsang-Eastwood, G. (1995) *Die Kleider des Pharaos; Die Verwendung von Stoffen im Alten Ägypten*. Hannover – Amsterdam.
Völling, E. (2008)*Textiltechnik im Alten Orient: Rohstoffe und Herstellung*. Würzburg.
Waetzoldt, H. (1980) Kleidung. A. Philologisch. *Reallexikon der Assyriologie* 6, 18–31.
Warnock, P. & Pendleton, M. (1991) The Wood of the Ulu Burun Diptych. *American Journal of Archaeology* 41, 107–110.
Weitenberg, J. J. S. (1984) *Die Hethitischen U-Stämme*. Amsterdamer Publikationen zur Sprache und Literatur Band 52. Amsterdam.
Werner, R. (1967) *Hethitische Gerichtsprotokolle*. Studien zu den Boğazköy-Texten 4. Wiesbaden.
Wilhelm, G. (1992) Hurritisch e/irana/i "Geschenk". In H. Otten & H. Ertem & E. Akurgal & A. Süel (eds.), *Sedat Alp'a Armağan – Festschrift für Sedat Alp. Hittite and Other Anatolian and Near Eastern Studies in Honour of Sedat Alp*, 501–506. Ankara.

Yalçın, Ü. & Pulak, C. & Slotta, R. (2005) *Das Schiff von Uluburun. Welthandel vor 3000 Jahren, Katalog der Ausstellung des Deutschen Bergbau-Museums Bochum von 15. Juli 2005 bis 16. Juli 2006*. Bochum.

Zaccagnini, C. (1973) *Lo scambio dei doni nel Vicino-Oriente durante i secoli XV–XIII*. Roma.

Zaccagnini, C. (1981) A note on Nuzi textiles. In M. A. Morrison & D. I. Owen (eds.), *Studies in Honor of Ernest R. Lacheman on his Seventy-Fifth Birthday. April 29, 1981*, 349–361. Studies on the Civilization and Culture of Nuzi and the Hurrians 1. Winona Lake.

Zaccagnini, C. (1989–1990) Dono e tributo come modelli istituzionali di scambio: echi e persistenze nella documentazione amministrativa Vicino-Orientale del tardo bronzo. *Scienze dell'Antichità 3–4*, 105–110.

16. Textile Terminology in the Ugaritic Texts

Juan-Pablo Vita

1. Ras Shamra and the Kingdom of Ugarit

In 1929, Claude Schaeffer carried out the first archaeological campaign in modern Ras Shamra (Syria).[1] The tell is located about 10 km to the North of the modern city of Latakia (Syria) and less than 1 km from the Mediterranean coast.[2] The mound site contained the remains of the ancient city of Ugarit, the capital of the kingdom of the same name. From 1929 to the present, except for the duration of World War II, Ras Shamra-Ugarit has regularly been excavated, almost annually, by a French – later Syrio-French – mission. The long series of archaeological campaigns has brought to light the royal palace and a complex of buildings closely linked to it, several temples, remains of fortresses, important residential areas, several private and palace archives and a great amount of all sorts of archaeological objects. Just about 1/6 of the tell of Ras Shamra has been excavated.

Therefore, almost eighty years of research and more than fifty archaeological campaigns have allowed us to gradually know the history and culture of the ancient kingdom of Ugarit, located in the modern Arab Republic of Syria. Because of its geographical extension, political power and military capacity, it was a middle-sized kingdom within the parameters of Syria in the 2nd millennium BC, but a leading one from an economic point of view. Bordered by the Mediterranean to the west and a mountain range parallel to it to the east, its boundaries suffered several modifications throughout history. Ugarit's foreign policy also had to adapt to the changing political circumstances of each period, determined by the converging interests in Syria-Palestine of the great powers at the time, in particular Egypt, Mitanni and Ḫatti. Ugarit disappeared from history at the beginning of the 12th century BC, swept away by the wave of the so-called "Sea Peoples", a time, which witnessed the collapse of the main Syrian centres of the Final Bronze Age and the disappearance of the kingdom of Ḫatti.

---

[1] This chapter is also one of the results of the Research Project "Bancos de Datos Semíticos Noroccidentales: Desarrollo y aplicación de nuevas tecnologías para el estudio y conservación de la documentación semítico-noroccidental del II y I milenio a. C." (HUM2007–65317), funded by the Spanish "Ministerio de Ciencia y Tecnología" within the National Plan for Scientific Research, Development and Technological Innovation (I+D+I) and by the European Union (Feder Funds).

[2] On Ugarit see *e.g.* Yon 1997; 2006; Watson & Wyatt 1999; Cornelius & Niehr 2004; Al-Maqdissi & Matoïan 2008.

2. Languages and scripts in Ugarit

Most of the texts found in Ras Shamra-Ugarit were written in two main scripts and languages.[3] On the one hand, Akkadian, a Semitic language written in the cuneiform syllabic script of Mesopotamian tradition, which reached the status of the international language of the Ancient Near East throughout the 2nd millennium BC. The Akkadian texts found in Ugarit comprise a great variety of genres, such as school, literary, religious, legal and economic texts as well as letters. On the other hand, Ugaritic, a northwest Semitic language of local use, had an alphabet of 30 signs; Ugarit is the only place in Late Bronze Age Syria, which has yielded abundant literature written in the local language. The Ugaritic corpus is made up of the same genres as the Akkadian, with the addition of some exclusive genres such as the hypiatric texts (veterinary texts dealing with the health of horses) or a sizeable collection of rituals.

Nowadays, Ugarit provides philologists and historians with hundreds of texts dealing with the most varied aspects of life within the kingdom and international relations at the time. As a whole, the archives of Ugarit are crucial to the understanding of the history of the ancient Near East from the second half of the 14th century to the beginning of the 12th century BC.

3. The textile industry in Ugarit

The texts from Ugarit provide information regarding different crafts and trades, such as work with hard stones, precious stones and metals, construction of houses, ships and chariots, production of different weapons, manufacture of pottery and food and the textile industry.[4]

The research carried out in the last twenty-five years has always been interested in the question of textiles in Ugarit from different yet complementary perspectives. Archaeology has allowed us to find in Ras Shamra physical elements regarding textile activity, mainly in the domestic sphere; for example, spindle whorls and loom weights, which indicate that there were looms in many households.[5] In some houses, what can be interpreted as sinks for washing raw materials to make fabrics or for dyeing them, have been found.[6] An area which could be exploited is the iconographic study of the garments depicted in some images of Ugaritic art, in particular, human images which illustrate a great number of cylinder-seals made in Ugarit.

Some other research work focused on the economy of the kingdom of Ugarit has also taken into account the written documents regarding textiles. This work has given us more accurate knowledge of the economic importance of the textile industry of Ugarit and its global dynamics.[7] Some authors have paid particular attention to the question of raw materials – mainly wool

[3] On this subject see *e.g.* Malbran-Labat 1999; Bordreuil & Pardee 2004.

[4] Vita 1999, 484–492; Heltzer 1999, 452–453.

[5] Callot 1994, 190: "des pesons et des fusaïoles qui indiquent que de très nombreuses maisons possédaient des métiers à tisser". See also Mallet & Matoïan 2001, 121.

[6] Callot 1994, 190.

[7] Ribichini & Xella 1985, 15–23; 1988; Heltzer 1999, 446–447, 452–453; Vita 1999, 486–487; Vidal 2005, 101–103; McGeough 2007, 210–211.

(Ugaritic *šʿrt*, RS Akkadian síg = *šīpātu*)[8] and linen (Ugaritic *pṯt*, RS Akkadian gada(.meš) = *kitû*)[9] – their prices and the prices of the manufactured products,[10] the different qualities of the fabrics and cloths[11], the dyeing elements and the types of dyes and colours,[12] as well as the accessories (such as fibulae or belts). Finally, other works have studied the terms used to describe occupations and professions connected with textiles,[13] such as the "fuller and/or washers" (*kbs/ś*, = syll. *kābisu*, RS Akkadian lútúg.me[14]), the "spinners" (*ǵzlm*), the "weavers" (*mḫṣ* = syll. *māḫiṣu*, RS Akkadian uš.bar), or the "purple-dyers" (*ṣāripūtu*).[15]

4. Brief history of research into textile terminology in Ugarit

The most important combined study specifically dealing with terms regarding textiles in the Ugaritic language continues to be the work of Sergio Ribichini and Paolo Xella, *La terminologia dei tessili nei testi di Ugarit*, from 1985. It produces a total of 78 lexical entries, with an annexe which presents, both in transcription and translation, a selection of 22 Ugaritic texts dealing with textiles. This work was completed and expanded by some recensions, in particular, those made by Hartmut Waetzoldt (1987) and Jean-Marie Durand (1990), by specific research work on some particular terms[16] and by the work of Wilfred H. van Soldt in 1990 on fabrics and dyes.

The Ugaritic-English dictionary by Gregorio del Olmo and Joaquín Sanmartín of 2003, can be placed at the other chronological end of the most recent lexicography. It offers the most comprehensive and recent state of the matter regarding Ugaritic textile terminology. Later works by Wilfred Watson, *Lexical Studies in Ugaritic* (2007) and Josef Tropper, *Kleines Wörterbuch des Ugaritischen* (2008) are equally useful in this respect. Likewise, we should highlight the recent work by Durand (2009); although dealing with textiles and garments in Mari, it provides significant comparative material for the Ugaritic lexicography as well as a good number of proposals on the possible meaning of specific Ugaritic terms.

A comparison between the Ugaritic terms which are considered as part of the textile semantic field by the aforementioned work of Ribichini and Xella, on the one hand, and by del Olmo and

---

[8] *šʿrt* (DUL 799, 2) "wollen textile"): < Semitic *ś(š/s)aʿar(t)*- (AHw 1191; Huehnergard 1987, 183), cf. Ugaritic *šʿr* (DUL 798: 1) "hair, hairs", 2) "pelisse, fleece"; Tropper 2008, 117: "Haar, Vlies, Schafsfell"), Hebrew *śaʿarāh* (Gesenius 18, 1295: "Haare, Fell"), Akkadian *šārtu* (AHw 1191: "Haare, behaarte Haut, Fell") (cf. DUL 799; Tropper 2008, 117). Tropper 2008, 117: "Haar, Schurwolle". The term is also attested to in the syllabic text PRU 6, 128:5, 1 túg *šá-ḫar-tu* /šaʿartu/ (Huehnergard 1987, 183; van Soldt 1991, 330 n. 158; del Olmo 1993, 194–195; Sanmartín 1993, 207; CAD Š/1 81: "a garment"). In the Akkadian of Ugarit the term is written síg (Huehnergard 1989, 407; van Soldt 1990, 333; DUL 799).

[9] Huehnergard 1989, 363; van Soldt 1990, 325–329; DUL 688.

[10] Heltzer 1978, 38–50; Stieglitz 1979, 17 and 19; Ribichini & Xella 1985.

[11] For example, *mlk* "royal", that is, "of superior quality", or "fine", in RS 17.111:11 (4.270), *mrbdt mlk*, cf. DUL 554 sub *mlk* 2)b, *contra* Ribichini & Xella 1985, 50 n. 68; 12 túg *qallu* "12 small garment(s)" (PRU 3, 206, RS 15.135:1; Huehnergard 1987, 174; 1989, 386).

[12] The costume *all*, for instance, is presented in RS 15.115:4–6 (4.182) in the colours white, carnelian red and violet purple; another one, *ḫpn*, presents the colours violet and reddish purple, cf. RS 15.082:1 (4.168).

[13] Especially Heltzer 1979; Ribichini & Xella 1985, 18–22; van Soldt 1990, 352–354; Sanmartín 1995, 177, 179 and 181; cf. also Vita 1999, 486–487.

[14] See also Lackenbacher 2002, 1238: "Foulon et/ou blanchisseur, lú túg.lá; dans la tablette inédite RS 94.2519, où les mêmes termes sont écrits en alphabétique et en idéogrammes, dans le même ordre et avec les mêmes chiffres, (lú) túg.lá a comme équivalent *kbsm*".

[15] Lackenbacher 2002, 95 n. 276; Vidal 2005, 101–102; Singer 2008, 22.

[16] E.g. Watson 1990; Sanmartín 1992.

Sanmartín on the other, shows great differences between both of them. Suffice it to say that 37 out of the 78 terms studied by Ribichini and Xella, are not collected as textile or textile-related terms by del Olmo and Sanmartín, (that is, practically half of them). In contrast, at least ten terms considered by del Olmo and Sanmartín as belonging to the textile world, are not studied by Ribichini and Xella. This situation shows, indeed, the reality of our knowledge of Ugaritic textile terminology, with many uncertainties as regards the meaning of a large amount of terms used to describe fabrics, garments, dyes or professions dealing with textile production. On the textile terminology existing in the syllabic texts from Ugarit, see below sub §6.1.

Ugaritic terms regarding textiles appear in all types of genres: literary or economic texts, letters, rituals, etc. Most of them, however, appear in economic or administrative texts. The so-called "East Archive" of the royal palace offers the highest number of documents with references to textiles or dealing entirely with them.[17] Thus, we find texts which register the delivery of raw materials for making fabrics and costumes, the production of garments, the price of them, raw materials and garments for officials, costumes for worship as well as lists of various textiles. In general, however, a comprehensive and detailed study on the exact function of the economic texts dealing with textiles is still pending, a study which will allow us to understand the *raison d'être* of each document within the administration machine of the palace and kingdom, although the first steps have already been taken in this sense (Vita 2007; Rougemont & Vita 2010).

5. Problems in the study of textile terminology in Ugaritic

As we have already pointed out, the dictionary of del Olmo and Sanmartín may be considered the most modern and comprehensive lexicographic study of the Ugaritic language. However, despite the high scientific quality of the work, in a few cases it is not possible to accept the meanings given to certain terms regarding textiles; this is due to the well-known problems posed by Ugaritic lexicography: suffice it to recall that the Ugaritic alphabetic script does not indicate (except for a few special cases) the vocalization of the words or the reduplications of consonants.

An example of this is the term *mṣrr*(*t*), which del Olmo and Sanmartín translate as "A piece of fabric or a garment (?)" (DUL 589). The term occurs only once in an administrative text, in the phrase: *t̠lt̠m t̠lt̠ kbd mṣrrt pt̠t* "thirty three *m.* of linen". In all certainty, it is a Semitic nominal pattern mVq(t)tVl from the root ṢRR "be narrow; lock in, besiege". In Hebrew, the verb *ṣrr* "to tie, wrap" has an intensive participle form (Pual) *mṣrrm* (*mᵉṣorārīm*) with the meaning "bundled". On these grounds, some authors have suggested the meaning "bundle" for the Ugaritic term. On the other hand, in Akkadian, the term *muṣarrirtum*, intensive form of the verb *ṣarāru* "to flow, drip", is attested too, and it denotes a type of vessel (AHw: "ein Tropgefäβ", CAD M/2 241: "a flat dish"). Consequently, *mṣrr*(*t*) could be a) either a type of vessel or box for flax (based on the Akkadian *muṣarrirtum*) or b) some kind of "packs" or "bundles" of flax (based on the Hebrew *mᵉṣorārīm*). In neither of the two cases do the possible etymologies allow us to consider that *mṣrr*(*t*) alludes to a "piece of fabric" or "garment", and therefore, it must be discarded from the question we are dealing with.

[17] Widbin 1985; Vita 2007 and 2008.

Another example which illustrates the ambiguity rendered sometimes by the Ugaritic material is the syllabic term *mašlaḫama*.[18] Nougayrol[19] suggested the meaning "un certain façonnage d'étoffe", CAD M/1 379 includes the term as "adj. (?); (qualifying textiles?)", which would be a northwest Semitic term attested to only in Ugarit. Huehnergard puts forward the general meaning of "a garment" (alternative: "throw") and links it to the alphabetic term *mšlḥ*.[20] The meaning of *mšlḥ*, however, has not been properly determined yet; Xella[21] translated it as "Lieferung" (Semitic root ŠLḤ), Sanmartín[22] suggested (based on the same root) the meaning "battering ram" (of a ship), a meaning which is also shown in DUL 593: "battering ram (?)". DUL (*ibid.*), on the other hand, does not include the possible equivalence syll. *mašlaḫama* = alphabetic *mšlḥ*, which, nevertheless, does seem plausible.[23] However, both instances would be *hapax* (one within the syllabic corpus and the other within the alphabetic corpus) and the contexts of the texts PRU 6, 123 and RS 20.008 (4.689) which mention them do not allow us to state with certainty that these terms denote a type of garment or qualify it in any way.

Another type of problem posed by Ugaritic texts is the actual reading of some of the terms, due to the poor state of preservation of some of the tablets; in this respect, see below (§6.1), for example, the comment of the term *kndpnṯ*.

6. Textile terminology in Ugarit

The following lines offer a state of the matter regarding the vocabulary on cloths and garments produced by the texts from Ugarit.[24] The lexemes are shown in two sections, respectively dealing with the present terms in the alphabetic and syllabic texts. Only the texts which can be deemed as written in Ugarit are taken into consideration.[25]

6.1. *Alphabetic Texts*

Each of the following lexemes is accompanied by its (possible) etymology and a brief comment.[26] In the selection of the terms, which can be identified as designating textiles in the alphabetic texts, we generally follow DUL, although with a critical approach, providing also the relevant information given by Ribichini & Xella (1985). Some other terms, both debated and debatable, could have also been part of the list, and they will be revised in the future. The bibliography for each entry is preferably amongst the most recent available on each term (and refers to earlier previous bibliographies).[27] Precise textual references for each term can be found both in DUL and in CUW.

[18] PRU 6, 123:3; a discussion on the morphology of the term in Sivan 1984, 111; Huehnergard 1987, 181.
[19] Nougayrol 1970, 159.
[20] Huehnergard 1987, 181.
[21] Xella 1982, 33.
[22] Sanmartín 1988, 272–273.
[23] Cf. also Vita 2000, 284–285.
[24] The rest of the aspects concerning textiles in Ugarit, such as, for instance, types of dyes, colours, prices, qualities, garments with leather, ropes, etc., will be the subject of a future global study of the Ugaritic textile industry.
[25] In other words, all the texts in alphabetic script and part of the texts written using the Mesopotamian syllabic system; this latter corpus was, to a great extent, outlined by Huehnergard 1989, 285–311.
[26] The common Ugaritic term for "clothing" is *lbš* (DUL 492; Ribichini & Xella 1985, 44–45; Sanmartín 1992; Tropper 2008, 64; Watson 2007a, 91); add *lbš* in RSO 14, 53:36".
[27] Good use will also be made of the bibliography by Dietrich & Loretz 1996.

all: < Hurrian *alāli* (de Martino & Giorgieri 2008, 55: "ein Gewand"; Wegner 2007, 246: "Gewand"). DUL 56: "Outher garment (type of 'cape, cloak')"; Ribichini & Xella 1985, 28–29: "probabilmente una sorta di mantello ... un tipo di indumento d'uso religioso e cerimoniale"; Watson 2007a, 39: "There is now strong evidence that *all* is a Hurrian loanword"; Tropper 2008, 4: "eine Art Mantel".

ipd: < Semitic ʾPD (DRS 28), cf. Hb. *ʾepôd* (Gesenius 18, 88, "Prachtkleid, Sakralgewand"), Akkadian *epattu* (AHw 222: "ein Gewand aus Talḫad"; CAD E 183: "a costly garment"). DUL 89: "Type of garment (usually: 'tunic')"; Ribichini & Xella 1985, 31: "sorta di corto gonnellino?"; Tropper 2008, 4: "ein prachtvolles Gewand"; Tropper 2000, 183, 286; Pardee 2000b, 329. For Hittite and Egyptian cognates see Watson 2007a, 119, 136–137.

ušpǵt: < Hurrian *us/špaḫḫu* (AHw 1438: "ein Gewand", Nuzi-Akkadian). DUL 118: "A garment"; Ribichini & Xella 1985, 33: "una veste, di possibile impiego cultuale"; Tropper 2008, 14: "ein Gewand"; Tropper 2000, 125; Pardee 2000b, 233–234; Watson 2007a, 126. In Mari, *ušpaḫḫi* denotes a type of rope (Durand 2009, 186).

uṭb: < Semitic(?) ʾṬB(?) (cf. Durand 2009, 131). The syllabic equivalent is, with great probability, túg-gada gal (van Soldt 1990, 354 n. 232). DUL 123: "A garment with passemanterie"; Ribichini & Xella 1985, 33: "si può pensare ad un capo di vestiario indossato a diretto contatto col corpo"; Tropper 2000, 175; Pardee 2000a, 50: "il s'agit ... d'un 'habit de luxe'"; Tropper 2008, 14: "ein Prachtgewand". Perhaps it is the alphabetic equivalent of the garment *uṭba* (a term of yet unprecise meaning) attested to in Mari (Durand 2009, 129–131); but see Watson 2007a, 79 for other possible options.

az: term of uncertain etymology and meaning which qualifies *mrdt* (cf. also Watson 2007, 127). DUL 136: "A textile fabric (?)"; Ribichini & Xella 1985, 27: "un tipo di tessile"; Tropper 2000, 106.

a/izml: Syll. sg. *azamil(l)u* (van Soldt 1991, 283 n. 76: RS 20.134:III:28 (unpublished): *a*(?)-*za*-ME-*lu*), pl. *azamil(l)āte* (PRU 6, 155:8: *a-za*-ME-*la-te*). About the morphology of the term see also Huehnergard 1987, 282 n. 67 and Tropper 2000, 106 and 170. Attested to in Akkadian *azamil(l)u* (AHw 92: "Tragsack, -netz"; CAD A/2 525: "sack, with netlike reinforcement"), term of unknown linguistic origin (cf. AHw 92), *a/izml* is a probable loan from Akkadian to Ugaritic. DUL 137: "A kind of sack"; Tropper 2008, 12: "ein sackartiger Stoff"; Ribichini & Xella 1985, 28: "tessile"; Watson 2007a, 80.

ʿ***rm***: Uncertain etymology (cf. DUL 183; Durand 1990, 661; Watson 2007, 81). DUL 183: "a cloak or blanket"; Ribichini & Xella 1985, 56: "una sorta di copertura, in ipotesi un mantello"; Watson 2007, 81: "a cloak or blanket".

ʿ***rp***: two possible etymologies have been suggested: a) < Semitic ʾ/ʿ/ǴPR "to cover", with metathesis, hence *lbšm ʿprm* "ciertas prendas de vestir dotadas de 'capucha' o de cierto tipo de tocado, esclavina, guardapolvo, etc. ... vestidos dotados de *ʿrp* (capucha(?))" (Sanmartín 1992, 101); b) < Semitic ʿRP "to be dark", cf. Watson 2007a, 41: "This indicates that the Ugaritic term *ʿrp* in connection with clothing does ... refer ... to the shelter (from the sun) it provides. Accordingly, the meaning 'cape' or 'cloak' would be quite suitable". DUL 184: "A type of garment (with a hood ?)".

blḫdr: < Hurrian **pilaḫ(a)=t=are*, cf. Sanmartín 1980, 335: "bestehend aus einem Element **blḫ*, offensichtlich eine Nebenform zu *plk* 'Spindel' ... und aus den Affixen *-t-* (> *d*) (Zustand, Eigenschaft) und *-are*. Die *blḫdrm* dürften somit vorläufig als 'gesponnene Tücher oder Kleidungsstücke' lexikographisch zu bestimmen sein"; DUL 222: "Piece of cloth or garment"; Watson 2007a, 127: "it denotes spun cloth or garments"; Ribichini & Xella 1985, 34: "un panno tessuto o di un capo di vestiario".

ǵprt: Uncertain etymology (cf. DUL 323: "n. f. of a garment"). According to Sanmartín 1992, 101: "es un producto morfoléxico de la ... [semitic] base ʾ/ʿ/ǴPR 'cubrir' [see above], y tiene su isolexia en el ac. *epartu*"; but Akkadian *epartu* does not seem to be an (Semitic) Akkadian term, cf. AHw 222: "u.H, ein Gewand" and

CAD E 183a: "(a coat)...foreign word"). Ribichini & Xella 1985, 57 compare the term to Hittite *ḫuppara*, a type of fabric (cf. also Watson 2007, 120).

ḥtl: < Semitic ḤTL "to wrap", cf. Hebrew *ḥittûl* (Tropper 2008, 48; Ribichini & Xella 1985, 37). DUL 376: "Nappy"; Ribichini & Xella 1985, 37: "una 'benda' o 'fascia' ... Non è tuttavia da escludere che si tratti di un oggetto imprecisato, e non di un tessile"; Tropper 2008, 48: "Windeln (zum Wickeln eines Säuglings"). Watson 2007a, 120 mentions possible alternative etymologies.

ḫlpn: a term etymologically linked to Akkadian *ḫulāpu* (cf. Durand 1990, 66; AHw 353: "eine Binde(?) < *ḫalāpu*, AHw 310: "hineinschlüpfen"; CAD E 228: "bandage"; cf. also Tropper 2000, 148). Add *ḫlpn* in RSO 14, 53:36. DLU 393: "Type of cloak or cape"; Ribichini & Xella 1985, 38: "sorta di mantello o di tunica"; Tropper 2008, 50: "Textilgewebe, Decke" od. ein Gewandtyp ('Umhang, Mantel')"; Tropper 2000, 300; Watson 2007a, 86. It is very likely that ^{túg}gú-è = Akkadian *naḫlaptu* (AHw 715: "Gewand, Mantel"; CAD N/1 138: "wrap, outer garment"; for Mari, cf. Durand 2009, 67–69) is not the equivalent of *ḫlpn* in the syllabic texts from Ugarit (cf. van Soldt 1990, 328 n. 50; diff. DUL 393); see also *ḫpn*.

ḫpn: Uncertain etymology (DUL 400; Tropper 2008, 51: 'Etym. unsicher (vom Lexem *ḫlpn* zu trennen)', cf. also Tropper 2000, 148). The equivalent in the syllabic texts from Ugarit may be ^{túg}gú-è (cf. van Soldt 1990, 328 nn. 50 and 55). DUL 400: "Garment or accessory"; Ribichini & Xella 1985, 39: "verosimile pensare che *ḫpn* designasse una sorta di coperta che, come tale, poteva essere adibita a vari usi, ovvero un indumento per l'abbigliamento esterno"; Tropper 2008, 51: "Mantel, Umhang, Satteldecke (für Pferde)"; Tropper 2000, 148: "*ḫpn* ... das eine phonetische Variante des Lexems *ḫlpn* darstellen könnte ... Wahrscheinlicher besitz *ḫpn* aber eine andere Etym."; Watson 2007a, 87.

kndpnṯ: < Hurrian *kindabašše* (cf. Watson 2007, 88: "A Hurrian loanword [in Ugaritic] possibly borrowed through Akkadian"; Durand 1990, 662: "La base est sans doute formée par un terme hourrite au génitif 'ce qui recouvre le *knd*' "). Variants *knd* (syll. *kanādu*, PRU 6, 163:5', 1 $^{\text{túg(?)}}$*ka-na-dú*, cf. Huehnergard 1987, 137–138) and *kdwṯ* (with the assimilation of the /n/). Cf. in general Tropper 1997; diff. DUL, which present them in three different entries (*kdwṯ* p. 432, *kndpnṯ* p. 449, and *knd* p. 449). However, the proposal of Tropper 1997 involves reading problems and, consequently, in order to be accepted, a careful collation of the relevant tablets will be necessary. Ugaritic version of the term attested to in Middle-Assyrian *kindabašše* (CAD K, 384: "a garment ... foreign word"). DUL 432: *kdwṯ* "garment", 449: *kndpnṯ* "A garment (woman's underclothes?)", 449: *knd* "A type of garment or cloth"; Ribichini & Xella 1985, 40: *kdwṯ* "capo di vestiario", 41: *knd* "veste", 41: *kndpnṯ* "capo di vestiario"; Pardee 2000a, 53: "une sorte de sous-vêtement ... Le fait qu'il était 'pour' *grgyn* peut aussi indiquer que celui-ci était pour ussage masculin – on décrit le *kindabašše* médio-assyrien comme un vêtement intime féminin"; Watson 2007a, 88, 129, 130.

kpṯ: cf. syll. *kupšu*.

kst: < Semitic KSY "to cover oneself with" (DUL 466; Tropper 2008, 466 and 153: "sich bedecken mit; (Kleider) anziehen"; see also Durand 2009, 55). DUL 466: "A type of robe or cloak"; Tropper 2008, 60: "Bedeckung, (Ober-)Gewand"; Tropper 2000, 191; Watson 2007a, 90 n. 209. DUL 466 points out the possibility that the term *kuššatu* (cf. §6.2) may be the syllabic version of Ugaritic *kst*. However, *kst* is in all probability the Ugaritic version of Akkadian *kusītu* (AHw 514: "Gewand"; CAD K 585: "an elaborate garment"; Lackenbacher 1982, 138 n. 10: "un vêtement, et à époque tardive c'est une sorte de manteau d'apparat"), and both terms *kst* / *kusītu* may not be identified with Akkadian *kuššatu* due both to phonetic reasons (it does not seem that the Ugaritic phoneme /s/ may be equivalent to Akkadian /š/, cf. Tropper 2000, 103 sub 32.143.23) and to the geographical distribution of both words (cf. Durand 2009, 54).

ktn: Uncertain etymology (DUL 468; cf. van Soldt 1990, 332: "whatever its etymology and vocalization"). Syllabic ^{túg}gada (van Soldt 1990, 329; DUL 468. In PRU 6, 172:5', a specific type is mentioned, $^{⌜\text{túg}⌝}$gada kud^{meš}, Nougayrol 1970, 158: "chiffon" ?, cf. also Huehnergard 1989, 66; RSO 7, 25:48, 1 ^{túg}gada *še-er-ṭu*$_4$, Malbran-Labat 1991, 60: "couverture"). "Kultur-" or "Wanderwort" attested to in various languages, both Semitic and Indo-European (references in DUL 468). van Soldt 1990, 332: "refers to a cloth made of linen. It is not a finished garment but a piece of cloth which can be used to manufacture garments ... *ktnt* are not weighed but counted, which suggests that they had a standard size"; several texts, both Ugaritic and Akkadian, clearly support van Soldt's conclusions (*contra* DLU 468: "A type of 'tunic'"). Ribichini & Xella 1985, 43: "veste ... usualmente resa con 'tunica', una traduzione che può essere verosimile, anche se risente forse del parallelo con miceneo *ki-to*"; Tropper 2008, 61: "Leinen; Leinenstoff, -gewand"; Pardee 2000b, 233; Watson 2007a, 148. Cf. also *kutānum* in Mari (Durand 2009, 55).

mizrt: < Semitic ʾZR "to wrap, to enfould" (Huehnergard 1987, 105; Tropper 2008, 68; cf. also Durand 2009, 143). Syll. sg. */*maʾazaru*/(?), pl. *maʾzarūma*(?) (PRU 6, 126:4, 2 $^{\text{túg.meš}}$*ma-za-ru-ma*^{meš}; PRU 6, 168:4, 2 *ma-za-r*[*u-ma*(?)], cf. Huehnergard 1987, 105). DUL 519: "Garment, ritual tunic"; Ribichini & Xella 1985, 47: "plausibile pensare ad un capo di vestiario intimo o ad una fascia che doveva cingere i lombi"; Tropper 2008, 68: "eine Art Lendenschurz bzw. eine kurze (zweiteilige) Beinbekleidung"; Watson 2007a, 92: "woollen undergarment".

md: < Hebrew *mad*, Aramaic *md*, Punic *md* (Tropper 2008, 69. Along with the previous terms, Tropper also mentions Akkadian *mudûm* –AHw 666: "ein Gewand", CAD M/2 168: "a garment"–, but see Durand 1990, 661 against this option). Add *md* in RSO 14, 51:5.12.16 (*mdy* "my *md*") and 13 (*mdk* "your *md*"). DUL 524: "Cape, covering"; Tropper 2008, 69: "(Ober-)Gewand"; Ribichini & Xella 1985, 48.

mḫtrt: Uncertain etymology (DUL 538). Ribichini & Xella 1985, 48: "potrebbe essere plausibilmente un tipo di veste traforata o trapunta"; a proposal which Durand (1990, 661) considers "peu crédible" and he suggests: "Puisque une formation en *ma-* peut désigner un nom d'objet, pourquoi ne pas y voir un 'ciseau' assez fort pour être capable de couper du lin?". DUL 538: "Article of clothing". Watson 2007a, 93–94 reviews the possible explanations of the term and concludes: "No clear conclusion can be drawn".

mks: < Semitic KSY "to cover oneself with" (DLU 544; Tropper 2008, 72 and 153: "sich bedecken mit; (Kleider) anziehen"). Possible syllabic version: ^{túg}*ma-ki*(?)-*sú* in RS 15.135:4 (PRU 3, 206; Nougayrol's reading; cf. also DLU 544); but this form poses serious reading and interpretation problems, see Huehnergard 1987, 192 and Durand 1990, 661. DLU 544: "Blanket"; Ribichini & Xella 1985, 49: "una sorta di copertura"; Tropper 2008, 72: "Bedeckung". Cf. also *maksûm* in Mari (Durand 2009, 60–61).

mlbš: < Semitic LBŠ "to get dressed" (cf. DUL 547; Tropper 2008, 73; Akkadian *nalbašu*, AHw 724: "Mantel", CAD N/1 200: "a fine cloak"; Mari, Durand 2009, 74: "étoffe qui revêt"). DUL 547: "Cloak, luxurious cape"; Ribichini & Xella 1985, 49: "probabilmente ... una specie di mantello"; Tropper 2008, 73: "Kleidung; Gewand, Mantel".

mrbd: < Semitic RBD "to spread, to cover" (DUL 573; Tropper 2008, 159: *rbd* Gp "ausgebreitet/bereit werden"). Add *mrbd* in RSO 14, 53:34". DUL 573: "Bedspread, counterpane"; Ribichini & Xella 1985, 50: "tessuto adibito a coperta, usato specificamente per l'allestimento del letto"; Tropper 2008, 77: "Decke(?)".

mrdt: < Semitic RDY(?) "?" (Huehnergard 1987, 177: "cf. perhaps Arabic *r-d-y* Gt and tD 'to put on a garment', *ridā* 'cloak, robe") or MRD / WRD "to go dawn, to descend" (cf. Durand 2009, 64: "le (tissu) qui descend"); cf. also Akkadian *marda/ātu*, AHw 611: "ein Teppich(?)", CAD M/1 277: "fabric woven with several colors in a special technique". Syll. PRU 3, 206:5, ^{túg}*m*]*ar-de*$_4$*–tu* /*mardētu*/ (cf. Huehnergard 1987, 177; van Soldt 1990, 329 n. 59). DUL 573: "A piece of material or a garment (?)"; Ribichini & Xella 1985, 51: "in ipotesi, una coperta, un arazzo, una cortina"; Tropper 2000, 297; Watson 2007a, 95; Durand 1990, 662: "c'est un habit bien connu, souvent caractérisé par des broderies dès la période OB, puis, à époque moyenne, un tapis ou au moins un rideau". In Mari, Durand 2009, 64: "Il n'y a donc pas, en fait, de contexte net où la *mardatum* soit un habit porté par l'individu ... il est possible ... qu'il designe la tapisserie de mur, ou de recouvrement d'une structure".

mšlt: Uncertain etymology (cf. Pardee 2000b, 794 n. 65: "notre vocalisation [*mašallātu*, cf. p. 788] reflète la racine arabe ŠLL"; cf. also *ibid.* p. 1166). DUL 592: "Garment or harness"; Ribichini & Xella 1985, 52: "tipo di veste imprecisabile"; Pardee 2000b, 787: "(vêtements-)*mšlt*"; Watson 2007a, 96. Ribichini & Xella 1985, 52 proposed identifying *mšlt* with the garment-*massilâtum* attested to in Mari (Durand 1990, 662; *id.* 2009, 66: "D'après sa forme extérieure, le nom semble de formation sémitique"; AHw 1573: "u. H."), a possibility which was initially accepted by Durand 2009, 66: "Il faudrait donc considérer que le terme documente une alternance *š* (Ugaritic) / *s* (Akkadien-Mari)".

mṯyn: < Hittite *maššiya* (Watson 2007, 122: "a garment"; cf. also Ribichini & Xella 1985, 52, + Hurrian article *-nni* (Durand 1990, 663; Tropper 2008, 83). Syll. PRU 6, 7 A:7[28], 1 túg : *ma-aš-ši-ia-an-na*; RS 25.131:11, ^{túg}*ma-ši-ia-an-nu* (Lackenbacher 1989, 318): Nougayrol 1970, 158: *maššiyanna* "châle"; van Soldt 1990, 336; AHw 629: "eine Schärpe ?", CAD M/1, 389: "a garment"; Lackenbacher 1989, 318: "châle(?)"). DUL 606: "A garment": 'shawl, sash(?)'", also pointing out the Hebrew *mšy* "feines Gewebe" (Gesenius 18, 752; see in this respect also Watson 2007, 122, as well as p. 142 on the possible link between Ugaritic *mṯyn* and Eg. *msy* "a garment"; Tropper 2000, 112. Ribichini & Xella 1985, 52: "Si tratta quindi, con tutta probabilità, di uno scialle in tessuto leggero e costoso"; Tropper 2008, 83: "Schal".

npyn: < Semitic NPY "to weave" (Tropper 2008, 89). DUL 639: "Tunic"; Ribichini & Xella 1985, 53: "le possibilità che *npyn* designi una veste ci paiono non trascurabili"; Tropper 2008, 89: "Kleid(ung)".

pǵdr, pǵndr: < Hurrian *paḫandari* (GLH 192; Huehnergard 1987, 169: *paǵandarri*; DUL 666). Syll. PRU 3, 206:9, 3 ^{túg}*pa-ḫa-dar*$_6$*-ru* gada (Huehnergard 1987, 169 and 317: *pa-ḫa-dar*$_6$(TAR)*-ru* /*paǵan/ddarru*/; cf. also Huehnergard 1989, 282; AHw 810 and CAD P 20: *pa-ḫa-tar-ru*). Hurrian loan to Ugaritic, perhaps through the Hurro-Akkadian *paḫantaru* (AHw 810 *paḫantaru*: "eine Decke"; CAD P 20 *paḫantarru*: "a blanket or cover"; cf. Watson 2007, 132). DUL 666: "Type of blanket or cloak"; Ribichini & Xella 1985, 61: "tipo di coperta o stoffa".

pld: Uncertain etymology (cf. DUL 670; Tropper 2008, 95; Huehnergard 1987, 168: "Probably related to Hebrew *p*^{e}*lādôt* 'blankets(?)' ... although the ultimate etymology of the term remains obscure"). Add *pldm* in RSO 14, 40:9. Syll. PRU 6, 127:4, 3 ^{túg}*pa-[l]i?-du-ma*; 128:3, 5 ^{túg}*pa-li-du-m[a]*; 129:5', ^{túg}*pa-l[i-du]* (cf. also Huehnergard 1987, 167–168). DLU 670: "Cloth or garment"; Ribichini & Xella 1985, 61: "non è possibile stabilire con precisione se *pld* designi una veste ovvero si riferisca ad un tipo di tessuto"; Huehnergard 1987, 167: "type of garment or other article of fabric"); CAD P 66: "a textile".

psm: vid. Akkadian *pusummu* (AHw 883: "Schleier"; CAD P 537: "veil") (cf. DUL 685; Ribichini & Xella 1985, 60). DUL 685: "veil, gauze(?)"; Ribichini & Xella 1985, 60: "una sorta di velo o di tessuto finissimo"; Watson 2007a, 100–101): "1. 'veil', ... 2. 'sleeved(?)' ... Either solution is possible".

rṭ: < cf. Arabic *rīṭ, rayṭa* (DUL 749; Renfroe 1992, 142). DUL 749: "Cloth or garment"; Ribichini & Xella 1985, 61: "tipo di stoffa o di veste, molto fine e leggera"; although according to Watson 2007a, 54–55, it could be a type of wool (with referente to Akkadian *urṭu*, AHw 1434: "Ginsterfarbstoff, gelb-bläuliche Wolle").

---

[28] Huehnergard 1989, 298 (cf. also *id.* 1999, 387: "proveance unknown") does not consider this letter part of the epistolary corpus from Ugarit; see, nevertheless, Nougayrol's considerations in this respect (1970, 9 n. 1).

sk: < Hebrew SKK "to wrap, to cover (with a dress)" (Huehnergard 1987, 156; DUL 756; Tropper 2008, 110). Syll. PRU 6, 127:6, 2 ^{túg}⌜ZU/SU?-*ku*?-*ma*^{meš}⌝ (reading according to Huehnergard 1987, 156: */*sukku*/?, pl. /*sukkūma*/). DUL 756: "1) Coverlet, cloak 2) Covering, lid(?)"); Ribichini & Xella 1985, 55: "sorta di stola o mantello"; Tropper 2008, 110: "ein Kleidungstyp"; Watson 2007a: 104: "cloak"; Pardee 2000b, 793–794. Durand 2009, 92: "Le terme [*sakkum*, Mari] pourrait se retrouver dans le *sk* ougaritique".

ṣt: Uncertain etymology (DUL 793; Tropper 2008, 115). DUL 793: "A garment"; Ribichini & Xella 1985, 55: "forse ... una sorta di tunica"; Tropper 2008, 115: "ein Gewandtyp"; Watson 2007a, 107.

ṯprt: Uncertain etymology (DUL 926). DUL 926: "Garment"; Ribichini & Xella 1985, 69: "un tipo di veste o di tessuto"; Watson 2007a, 113: "cf. Akkadian *šapru* 'thigh, female pudenda' (based on Arab. *ṯafr*-, *ṯufr*-, 'vagina'). Ug *ṯprt* is the garment covering this anatomical region", *id. ibid.* p. 230: "woman's undergarment".

ṯq: < Semitic ŚAQ (DUL 927; AHw 1027, *s*/*šaqqu*(*m*): "sem. ausser Arabic *śaq* ... Sack, Trauergewand"; Tropper 2008, 135: "wsem. *s^2*aq*"). DUL 927: "A cloth or textile"; Ribichini & Xella 1985, 69: "Non è inverosimile ravvisare in *ṯq* una sorta di stoffa grezza utilizzata nella bardatura di animali"; Tropper 2008: 135: "ein Textilprodukt". Diff. Watson 2007b: 42: "I propose the meaning 'halter, lead-rope' or a similar term, perhaps even 'bridle', and derive the noun from Ugaritic *yṯq*, 'to bind, tie' ... This would provide an inner-Ugaritic derivation for *ṯq*".

6.2. *Syllabic Texts*

The textile terminology in the Ugarit texts written in Akkadian has been less studied. In this respect, we could mention the following statement by Malbran-Labat:[29] "les termes [of garments] notés syllabiquement (...) sont rarement connus en akkadien (...) et leur sens reste souvent mystérieux". Indeed, the syllabic texts from Ugarit produce a number of terms of difficult reading and/or interpretation, as can be noted, for instance in the list of cloths and garments provided by Nougayrol.[30] We try to offer below a state of the matter regarding our knowledge of this terminology, each entry is accompanied by the readings of the term and the corresponding textual references.

azamil(l)u (RS 20.134:III:28, *a*(?)-*za*-ME-*lu*; PRU 6, 155:8, *a-za*-ME-*la-te*): "A kind of sack", cf. alphabetic *a*/*izml*.

iʾlu: (PRU 3, 39a:4, 3 ^{túg}sig$_4$.za; PRU 3, 206c:8, 4 ^{túg}sig$_4$.<za>(?) gal, cf. Huehnergard 1989, 410). AHw 373: "eine Binde ?"; CAD I/J 90: "a garment"; Durand 2009, 35: "Il vaudrait mieux comprendre l'*iʾlum* comme 'à trame serrée', l'habit étant obtenu par une certaine technique de tissage".

kanādu (PRU 6, 163:5', 1 $^{\text{túg(?)}}$*ka-na-dú*): cf. alphabetic *kndpnṯ*.

kimdu (PRU 3, 206c:13, 1 ^{túg}*ki-im-da*): AHw 478: "ein Gewand"; CAD K 372: "cloth woven and prepared in a special way"; Huehnergard 1987 137. See in this respect also the data from Mari, Durand 2009, 128: "*ša kimdim*, une certain sorte de tissage. À Mari, seuls sont dits '*ša kimdim*' l'étoffe *uṣûm* et l'étoffe *hayyû* qui sont tous deux des tissus d'ameublement. Le procedé *ša kimdim* devait donc produire une trame serrée

[29] Malbran-Labat 1999, 92.

[30] Nougayrol 1970, 159. See also, as examples: PRU 3, 206–207:16, 3 *šup*(?)-*pa-ti* (AHw 1280 *sub šuppatu*(*m*) "Binse": "RU-*pa-ti* dazu?"; CAD Š/3, 326 *sub šuppatu* "rush, reed thicket": 3 *šup-pa-ti* "uncert. ... among garments"); ^{túg}LA^{meš} (PRU 6, 126:5, cf. Huehnergard 1989, 66 and 356; *id*., 1987, 191; also Durand 2009, 143 and 184); the possible term *paršīgu* (AHw 836: "Kopfbinde, Mütze"; CAD P, 203: "a sash, often used as a headdress") in Ug 5, 50:5', although, as Nougayrol points out (1968, 138 n. 2): "D'autres restaurations sont possibles".

et un tissu relativement lourd, comme celle d'une tapisserie ou d'un tapis"; *ibid.* p. 51: "*kimdum* ... C'est le tissage qui renvoie au procédé *kamâdum*", whose precise meaning, as Durand states, has been pointed out by Lackenbacher 1982, 141–142: "je pense qu'il ne peut s'agir de tissage ... il doit s'agir du foulage à la main, bien attesté dans les techniques traditionnelles".

kitû **(gada)**: cf. alphabetic *ktn*.

kupšu (PRU 6, 128:4, 2 ^{túg}sagšu?^{meš}; RS 15.076:20 (= CAT 4.165:20 = PRU 6 p. 99c:5), 10 $^{túg.meš}$*ku-up-šu*): Durand 2009, 53 sub *g*): "[*kub/pšum*] n'a pas d'étymologie évidente en akkadien". AHw 497: *kubšu* "Kopfbinde, Turban"; CAD K *kubšu* 485: "headdress, cap"; Nougayrol 1970, 158: "bonnet". It is very likely that the alphabetic version of the term is *kpṯ* (see, for example, Ribichini & Xella 1985, 43: "turbante"; Huehnergard 1987, 140), hence the plausible script *kupšu* of the syllabic version (instead of the more common *kubšu*, see Nougayrol 1970, 158: *kupšu*(?); cf. regarding this Durand 2009, 52: "La transcription des textes ougaritiques pourrait faire abandonner la lecture *kubšum* des dictionnaires, s'il ne s'agit pas d'un traitement phonétique -bš- -> -pš- propre à cette langue, comme *špš* provient de *šmš*"). However, the meaning "turban, etc." of *kpṯ* continues to be debated, see for example DUL 453: "1) 'flor' (> 'firmament, sky'), 2) 'gangplank'", Tropper 2000, 111 and 139, and Watson 2007a, 89 with n. 198. In Mari (Durand 2009, 52–53) it was a typical ornament of the king, decorated with fine gems and precious metals: "On peut s'imaginer le *kubšum* d'après le bonnet que porte Hammu-rabi sur la stèle ou est rédigé son code ... devait donc comporter une structure solide, même si les documents l'attestent en laine et non en lin ... Il devait donc s'agir ... d'un objet tout d'une pièce et assez volumineux". A text from Mari, originating from Haṣor, shows, however, that the *kubšum* was a much more common object in the West "et certainement pas uniquement une marque royale. On songe, dès lors, à une utilisation comme un casque ou une protection de la tête dans les combats". This could have also been the case in Ugarit.

kuššatu (PRU 6, 6:26[31]; Nougayrol: túg : *ku-u*[*š-ši-ti*]; CAD K, 600, based on the notations of the term in other corpora: *ku-u*[*š-ša*(?)*-ti*]). Durand 2009, 55: "Le terme doit être sémitique et typique de l'Ouest et du Nord. On peut ainsi le rapprocher de l'arabe *kasā* (KSW) qui signifie au propre 'vêtir qqu'un' ... KŠŠ ne serait qu'une variante dialectale de KSW". AHw 517: "ein Gewand"; CAD K, 600: "(a garment)". Durand 2009, 54: "Il s'agit donc manifestement d'une réalité occidentale ... vraisemblable que *kuššatum* désigne une forme de vêtement".

maʾazaru (PRU 6, 126:4, 2 $^{túg.meš}$*ma-za-ru-ma*^{meš}; PRU 6, 168:4, 2 *ma-za-r*[*u-ma*(?)]), < Semitic ʾZR, cf. alphabetic *mizrt*.

mardētu (PRU 3, 206c:5, ^{túg}*m*]*ar-de*$_4$*-tu*), < Semitic RDY(?) "?" or MRD / WRD, cf. alphabetic *mrdt*.

martû (PRU 6, 126:8, 3 túg.meš *ma-ar-tu-*[*ú*], cf. Durand 2009, 108; diff. Nougayrol 1970, 100; Huehnergard 1987, 192, 3 túg.meš *ma-ar-*⌜TU?⌝; *id.* 1989: 88, 3 túg.meš *ma-ar-tu*, CAD M/1, 300, 3 túg.m[e]š *ma-ar-tu* [(x)]; PRU 6, 129:11', 2 túg.meš *ma-a*[*r*(?)*-tu*(?), cf. Nougayrol 1970, 102; Huehnergard 1987, 192, with a discussion of the possible etymology). Qualifying adjective of túg = *ṣubātu*. AHw 610: "Stoff nach Marad(ON)-Art"; CAD M/1, 276 (sub *maratû*): "adj; (describing garments)"; CAD M/1, 300 (sub *martu* B): "(a garment)", whose only testimony would be PRU 6, 126:8, and adds: "Possibly to be connected with *mardatu*", a possibility which was rejected by Nougayrol 1970, 158 n. 6. AHw does not compile the examples from Ugarit. As Nougayrol already did (1970, 158 n. 6), CAD considers that the testimonies from Ugarit do not belong to the lexeme *maratû*. But, following Durand 2009, 108, we must accept the identification of *martû* from Ugarit with *mar(a)tû* from Mari: in Mari the most frequent script *ma-ra-tu-*(*ú*) alternates with the variant

[31] The text PRU 6, 6 (RS 17.144) is not part of the corpus of texts written in Ugarit, cf. Huehnergard 1989, 298. However, the passage quotes the contents of the letter from the prefect of Ugarit to which PRU 6, 6 is the answer; in the passage the prefect requested the delivery of a garment *kuššatu*.

ma-ar-tu-(ú) (cf. Durand 2009, 108); dictionaries only knew the script *ma-ra-tu-ú*. It would be a *nisbe*, see Durand 2009, 108: "Le plus simple serait donc de considérer que l'on a affaire ici à 'l'étoffe/l'habit de la Côte méditerranéenne'. À l'Ouest, on devait ainsi distinguer le *Marratum* '(Region de) la Mer', de l'*Amurrum* '(Région des) Eaux saumâtres', soit tout la frange des déserts au sud de l'ancien royaume d'Ebla" (*ibid.* 600: "à la façon de la Côte méditerranéenne").

maššiyannu (PRU 6, 7 A:7, 1 túg : *ma-aš-ši-ia-an-na*; RS 25.131:11, ^{túg}*ma-ši-ia-an-nu*): < Hittite, cf. alphabetic *mṯyn*.

mazru (PRU 6, 123:2, *ma-az-ru*, cf. Nougayrol 1970, 99 and 159, CAD M/1 439; Durand 2009, 143, [^{éš}*mu-r*]*u-ú ma-az-ru-*[*tu*]): < Semitic MZR "to thread (a thread)" (cf. AHw 637: *mazru* "s. mhe. *mzr* spinnen ?, etwa 'verfilzt'"; Huehnergard 1987, 105; Durand 2009, 143). Nougayrol 1970, 159: "feutré"; CAD M/1, 439: "adj.; (mng. uncert.)" (attested to only in Ugarit). Huehnergard 1987, 105: "may be an Akkadian v. adj.". According to Durand 2009, 143 *mazru* from Ugarit could be, however, a type of rope (see *ibid.* 181 on the notations túg and éš): in Mari, *mazrum* could be a "fil tordu par simple cardage" (Durand 2009, 600).

muruʾu (PRU 6, 126:3, 2 $^{\text{túg.meš}}$*mu-ru-ú-ma*^{meš}; sg. */mōruʾu/*, pl. */mōruʾūma/*?, cf. Huehnergard 1987, 134): < Semitic YRʾ (< *WRʾ, see a discussion in Huehnergard 1987, 134). CAD M/2 230: *murūma* s. pl. "(a textile), foreign word(?)" (attested to only in Ugarit). According to Durand 2009, 143, it could be a type of rope (reading ^{éš}*mu-ru-ú-ma*).

naktu (PRU 3, 206c:6, *n*]*a-ak-tu*$_4$ gada; 10, 1 ^{túg}*na-ak-*[*tu*$_4$/*tu*; 14, 10 *na-ak-tu* gada; 15, 3/$^{\text{túg(?)}}$*na-ak-tu*$_4$ šu gada; cf. CAD N/1, 197; Huehnergard 1987, 151; van Soldt 1990, 329). AHw 724: *na-ak??-tum* "aus Leinen ganz. uns."; CAD N/1, 197: "a garment or fabric"; Huehnergard 1987: 151: "?NGD /nagdu/? 'a linen garment' ... It is quite uncertain whether the same word is written in all four lines ... Whether this word (or these words) is Semitic, and if so the composition of the root, remain uncertain".

nēbeḫu (Ug 5, 48:16, 1 íb.lá): Nougayrol 1968, 136: "châle ceinture"); Lackenbacher 2002, 297: "écharpe ceinture"; AHw 773: "Gürtel, Binde ... 1) als Kleidungsstück"; CAD N/2, 143b: "waist sash". Cf. also Huehnergard 1987, 190 *sub* IB-*lu*.

paḫaddarru (PRU 3, 206:9, 3 ^{túg}*pa-ḫa-dar*$_6$*-ru*): < Hurrian, cf. alphabetic *pǵdr*, *pǵndr*.

palidu (PRU 6, 127:4, 3 ^{túg}*pa-*[*l*]*i?-du-ma*; 128:3, 5 ^{túg}*pa-li-du-m*[*a*]; 129:5', ^{túg}*pa-l*[*i-du*]): cf. alphabetic *pld*.

qablītu (PRU 3, 192a: 8.10: ^{túg}murub$_4$^{meš}): Huehnergard 1989, 66: "perhaps for *qablītu* '(a garment)' (see CAD Q, 6a, *qablītu* 5 ...) or *qablu* 'belt' (CAD Q, 11–12)". Term attested to in Mari, Durand 2009, 86: "il pourrait s'agir d'un corsage".

qaqqaru (PRU 3, 206c:12, 3 ^{túg}*qa-qa-ru*): CAD K, 124: "a type of wool or garment".

sukku (PRU 6, 127:6: 2 $^{\text{túg⌜}}$ZU/SU?-*ku?-ma*$^{\text{meš⌝}}$): cf. alphabetic *sk*.

šabattu (PRU 3, 39a:4, $^{\text{t]úg}}$[*š*]*a-bat-tum*): AHw 119: "ein Gewand"; CAD Š/1, 8: "a garment or textile"; Durand 2009, 114: "c'est une fabrique qui existe aussi bien pour le pays hourrite que pour les régions occidentales (Ugarit) ... Le terme est certainement originaire de la culture matérielle du nord de la Mésopotamie ... Le terme, d'après la documentation mariote, pourrait donc être posé *šabâtum* au singulier, et répresenter une forme primitive **šaybatum*. Il s'agirait d'une toile 'blanche' ".

* * *

Ugarit offers thus a variety of terms to refer to fabrics and garments. However, it can be seen that, although in the last few years, certain advances have been made in the understanding of the meaning of some terms, there are still plenty of generic translations such as "a garment", "a cloth or textile", etc. The Ugaritic texts often give too little information regarding a particular term and the only possibility of attempting to understand it is through etymology, which, as is well known, involves plenty of problems and risks when it becomes the only available tool. Due to this, the study of a term regarding textiles or garments in a text from Ugarit frequently requires checking the information provided about that same term in other external archives in order to better understand it.

From a linguistic point of view, it can be noted that, as could only be expected, most of the textile terminology of Ugarit is Semitic. At times Biblical Hebrew and Akkadian are the main Semitic languages, which contribute to fix the meaning of an Ugaritic term. After the Semitic component, Hurrian is the language, which can provide most assistance in the understanding of some Ugaritic textile terms.[32] We must bear in mind that Hurrian was, after Semitic, the second most important cultural component of Ugaritic society, and that Hurrian was a living language in Ugarit until the end of the kingdom.[33] In other cases we can question whether a given term is directly Hurrian, used as such in the Hurrian from Ugarit, or it is a Hurrian term, which arrived in Ugarit as a Hurro-Akkadian loan (for example *pǵdr*, *pǵndr*). Certain terms seem to be exclusive to the Ugaritic vocabulary (for example *pld*).

Finally, we consider that a new and detailed analysis of the Ugaritic textile terminology, that takes into account, for instance, the lexicographic advances made in the last few years (both in Ugaritic and in neighbouring languages) as well as the information offered by other archives on each term, will certainly allow us to make some of the meanings more precise and to propose new ones. In this sense, collaboration with specialists in other archives and other languages could be of enormous interest; suffice it to consider in this respect that several authors have also proposed Eblaite, Hittite or Mycenaean etymologies for several Ugaritic terms, proposals, which should be carefully revised and reassessed.

Abbreviations

AHw	W. Von Soden, *Akkadisches Handwörterbuch*. Wiesbaden 1965–1981.
CAD	*The Assyrian Dictionary of the University of Chicago*. Chicago-Glückstadt.
CAT	M. Dietrich, O. Loretz & J. Sanmartín, *Cuneiform Alphabetic Texts from Ugarit, Ras Ibn Hani and Other Places*. Münster 1995.
CUW	J.-L. Cunchillos, J.-P. Vita, J.-Á. Zamora, *A Concordance of Ugaritic Words*, 5 vols. Piscataway 2003.

[32] On some occasions, Hurrian can indirectly help us to understand a Ugaritic term. This is the case with the term *all*, for which Ribichini and Xella 1985 were not in a position yet to provide a satisfactory etymology. In 1996, Erich Neu published the entire bilingual Hurrian-Hittite text known as "Epos der Freilassung". The following could be read in a passage in the Hurrian version: "(Ist) Teššub nackt, wollen wir (ihn) – jeder von uns – mit einem *alāli*-Gewand bekleiden". The Hittite version of the same passage says: "Wenn der Wettergott aber nackt ist, bekleiden wir – jeder (von uns) – ihn mit einem *kušiši*-Gewand". The parallel with the Hittite word ^{túg}*kušiši*- thus allowed us to know that the Hurrian term *alāli*, unknown until then, denoted a type of garment. From this equivalence, and independently, several authors (Neu 1996, 314) suggested identifying the Ugaritic *all* with the Hurrian *alāli*, which perfectly fits within the context of the Ugaritic passages that mention this term.

[33] Vita 2009.

DRS	D. Cohen, F. Bron & A. Lonnet, *Dictionnaire des racines sémitiques ou attestées dans les langues sémitiques*. Leuven 1994.
DUL	G. del Olmo & J. Sanmartín, *A Dictionary of the Ugaritic Language in the Alphabetic Tradition*, 2 vols. Leiden 2003.
Gesenius 18	W. Gesenius, *Hebräisches und Aramäisches Handwörterbuch über das Alte Testament*, 18 ed. Berlin 1995–.
GLH	E. Laroche, *Glossaire de la langue hourrite*. Paris 1980.
PRU 3	*Le Palais Royal d'Ugarit* III. Paris.
PRU 6	J. Nougayrol, *Le Palais Royal d'Ugarit* VI. Paris 1970.
RS	Ras Shamra
RSO 7	P. Bordreuil (ed.), *Une bibliothèque au sud de la ville*, 27–64. Ras Shamra-Ougarit VII. Paris 1991.
RSO 14	M. Yon & D. Arnaud (eds.), *Études ougaritiques I. Travaux 1985-1995*. Paris 2010.
Ug 5	J. Nougayrol, E. Laroche, Ch. Virolleaud & Cl. F. A. Schaeffer, *Ugaritica* V. Paris 1968.

Bibliography

Bordreuil, P. & Pardee, D. (2004), *Manuel d'Ougaritique*. Paris.

Callot, O. (1994) *La tranchée « Ville Sud ». Études d'architecture domestique*, Paris.

Dietrich, M. & Loretz, O. (1996) *Analytic Ugaritic Bibliography 1972-1988*. Neukirchen-Vluyn.

Cornelius, I. & Niehr, H. (2004) *Götter und Kulte in Ugarit. Kultur und Religion einer nordsyrischen Königsstadt in der Spätbronzezeit*. Mainz am Rhein.

Durand, J.-M. (1990) review of Ribichini & Xella (1985), *Mari* 6, 659–664.

Durand, J.-M. (2009) *La nomenclature des habits et des textiles dans les textes de Mari*. Paris.

Heltzer, M. (1978) *Goods, Prices and the Organizarion of Trade in Ugarit*. Wiesbaden.

Heltzer, M. (1999) The Economy of Ugarit. In W. G. E. Watson & N. Wyatt (eds.), *Handbook of Ugaritic Studies*, 423–454. Leiden.

Huehnergard, J. (2008 rev. ed.) *Ugaritic Vocabulary in Syllabic Transcription*. Atlanta.

Huehnergard, J. (1989) *The Akkadian of Ugarit*. Atlanta.

Huehnergard, J. (1999) The Akkadian Letters. In W. G. E. Watson & N. Wyatt (eds.), *Handbook of Ugaritic Studies*, 375–389. Leiden.

Lackenbacher, S. (1982) Un texte vieux-babylonien sur la finition des textiles, *Syria* 59, 129–149.

Lackenbacher, S. (1989) Trois lettres d'Ugarit. In H. Behrens, D. Loding & M. T. Roth (eds.), *Studies in Honor of Åke W. Sjöberg*, 317–320. Philadelphia.

Lackenbacher, S. (2002) *Textes akkadiens d'Ugarit*. Paris.

Malbran-Labat, F. (1991) Lettres. In P. Bordreuil (ed.), *Une bibliothèque au sud de la ville*, 27–64. Paris.

Malbran-Labat, Fl. (1999) Langues et écritures à Ougarit, *Semitica* 49, 65101.

Mallet, J. & Matoïan, V. (2001) Une maison au sud du 'Temple aux rhytons'. In M. Yon & D. Arnaud (eds.), *Études ougaritiques I. Travaux 1985-1995*, 83–190. Paris.

McGeough, K. M. (2007) *Exchange Relationships at Ugarit*. Leuven.

Al-Maqdissi, M. & Matoïan, V. (2008) *"L'Orient des palais". Le Palais royal d'Ougarit au Bronze récent*. Damas.

De Martino, S. & Giorgieri, M. (2008) *Literatur zum Hurritischen Lexikon (LHL)*, vol. 1. Firenze.

Neu, E. (1996) *Das hurritische Epos der Freilassung I. Untersuchungen zu einem hurritisch-hethitischen Textensemble aus Ḫattuša*. Wiesbaden.

Nougayrol, J. (1968) Textes suméro-accadiens des archives et bibliothèques privées d'Ugarit. In J. Nougayrol, E. Laroche, Ch. Virolleaud & Cl. F. A. Schaeffer, *Ugaritica* V, 1–446. Paris.

Nougayrol, J. (1970) *Le Palais Royal d'Ugarit*. Paris.

Del Olmo, G. (1993) Sheep and Goats at Ugarit: Alphabetical Texts, *Bulletin on Sumerian Agriculture VII. Domestic Animals of Mesopotamia - Part I*, 183–197. Cambridge.

Pardee, D. (2000a) Trois comptes ougaritiques. RS 15.062, RS 18.024, RIH 78/02, *Syria* 77, 23–67.

Pardee, D. (2000b) *Les textes rituels*. Paris.

Renfroe, F. (1992) *Arabic - Ugaritic Lexical Studies*. Münster.

Ribichini, S. & Xella, P. (1985) *La terminologia dei tessili nei testi di Ugarit*. Roma.

Fr. Rougemont & J.-P. Vita (2010) Les enregistrements de chars à Ougarit et dans le monde mycénien: approche comparative sur l'administration au Bronze récent. In W. H. Van Soldt (ed.), *Society and Administration in Ancient Ugarit*. Leiden, 123–150.

Sanmartín, J. (1980) Glossen zum Ugaritischen Lexikon (IV), *Ugarit-Forschungen* 12, 335–339.

Sanmartín, J. (1988) Glossen zum Ugaritischen Lexikon (IV), *Ugarit-Forschungen* 20, 265–275.

Sanmartín, J. (1992) Tejidos y ropas en ugarítico: apuntes lexicográficos, *Aula Orientalis* 10, 95–103.

Sanmartín, J. (1993) Sheep and Goats in the Akkadian Economic Texts from Ugarit, *Bulletin on Sumerian Agriculture VII. Domestic Animals of Mesopotamia - Part I*, 199–207. Cambridge.

Sanmartín, J. (1995) Wirtschaft und Handel in Ugarit: Kulturgrammatische Aspekte. In M. Dietrich & O. Loretz (eds.), *Ugarit. Ein ostmediterranes Kulturzentrum im Alten Orient*, 131–158. Münster.

Singer, I. (2008) Purple-dyers in Lazpa. In B. J. Collins, M. R. Bachvarova & I. C. Rutheford (eds.), *Anatolian Interfaces. Hittites, Greeks and Their Neighbours*, 21–43. Oxford.

Sivan, D. (1984) *Grammatical Analysis and Glossary of the Northwest Semitic Vocables in Akkadian Texts of the 15th–13th C.B.C. from Canaan and Syria*. Neukirchen-Vluyn.

Stieglitz, R. R. (1979) Comodity Prices at Ugarit, *Journal of the American Oriental Society* 99, 15–23.

Tropper, J. (1997) Beiträge zur ugaritischen Lexikographie, *Ugarit-Forschungen* 29, 661–668.

Tropper, J. (2000) *Ugaritische Grammatik*. Münster.

Tropper, J. (2008) *Kleines Wörterbuch des Ugaritischen*. Wiesbaden.

van Soldt, W. H. (1990) Fabric and Dyes at Ugarit, *Ugarit-Forschungen* 22 (1990), 321–357.

van Soldt, W. H. (1991) *Studies in the Akkadian of Ugarit. Dating and GrammArabic* Neukirchen-Vluyn.

Vidal, J. (2005) *Las aldeas de Ugarit según los archivos del Bronce Reciente (siglos XIV-XII a.n.e.)*. Sabadell-Barcelona.

Vita, J.-P. (1999) The Society of Ugarit. In W. G. E. Watson & N. Wyatt (eds.), *Handbook of Ugaritic Studies*, 455–498. Leiden.

Vita, J.-P. (2000) Los antecedentes de la marina fenicia: barcos en Ugarit. In M. E. Aubet & M. Barthélemy (eds.), *Actas del IV Congreso Internacional de Estudios Fenicios y Púnicos*, vol. 1, 281–288. Cádiz.

Vita, J.-P. (2007) Les documents des archives est du Palais royal sur les textiles: une contribution à la connaissance de la procédure administrative à Ougarit. In J.-M. Michaud (ed.), *Le royaume d'Ougarit de la Crète à l'Éuphrate. Nouveaux axes de recherche*, 243–265. Sherbrooke.

Vita, J.-P (2008) Les textes administratifs en langue ougaritique des 'Archives Est' du Palais Royal. In Y. Calvet & M. Yon (eds.), *Ougarit au Bronze moyen et au Bronze récent*, 227–234. Lyon.

Vita, J.-P. (2009) Hurrian as a living language in Ugaritic society, *Aula Orientalis*. In D. A. Barreyra Fracaroli and G. del Olmo Lete (eds.), *Reconstructing a Distant Past. Ancient Near Eastern Essays in Tribute to Jorge R. Silva Castillo*, Aula Orientals-Supplementa 25, Sabadell – Barcelona, 219–231.

Waetzoldt, H. (1987) review of Ribichini & Xella (1985), *ZA* 77, 297–299.

Watson, W. G. E. (1990) Two Ugaritic Terms Concerning Textiles, *Aula Orientalis* 8, 138–140.

Watson, W. G. E. (2007a) *Lexical Studies in Ugaritic*. Sabadell – Barcelona.

Watson, W. G. E. (2007b) More loose ends, *Nouvelles Assyriologiques Brèves et Utilitaires*, 42–43.

Watson, W. G. E. & Wyatt, N. (eds.), *Handbook of Ugaritic Studies*. Leiden

Wegner, I. (2007 second ed.) *Hurritisch. Eine Einführung*. Wiesbaden.

Widbin, R. B. (1985) *The East Archives in Le Palais Royal d'Ugarit: A Structural and Socio-Economic Analysis*. Ph.D. dissertation, Brandeis University.

Xella, P. (1982) Die Ausrüstung eines kanaanäischen Schiffes (KTU 4.689), *Die Welt des Orients* 13, 31–35.

Yon, M. (1997) *La cité d'Ougarit sur le tell de Ras Shamra*. Paris.

Yon, M. (2006) *The City of Ugarit at Tell Ras Shamra*. Winona Lake.

17. The Terminology of Textiles in the Linear B Tablets, including Some Considerations on Linear A Logograms and Abbreviations

Maurizio Del Freo, Marie-Louise Nosch and Françoise Rougemont

Terminology can be defined as (1) the vocabulary specifically linked to a field of expertise, as (2) the academic discipline devoted to the study of this vocabulary, or as (3) the linguistic theory reflecting on these terminologies: their formation, evolution, functioning and use.[1]

As such, it is clear that terminology can only apply to deciphered scripts and languages that are identified and understood, so that its application to the non-deciphered Linear A documents is a bit far stretched. Nonetheless, the nature of this script, as well as the state of research on the phonetic value of some of its signs may allow a few observations on the logograms and adjuncts used to designate textiles. However, it is clearly not intended by the authors as terminology in the full sense of the concept.

Linear B tablets have been found on the island of Crete as well as on sites on the Greek mainland (See Fig. 17.1, map, below). They are dated between *c.*1450 and 1200 BC, according to the site on which they have been unearthed.[2]

How can one study the corpus of Linear B tablets recording textile matters from the point of view of the tools of terminology? In a lexicographical perspective, the words belonging to the Mycenaean textile vocabulary have already been gathered, recorded and classified especially by J. T. Killen, J. L. Melena, Y. Duhoux, A. Morpurgo Davies, E. Barber, M.-L. Nosch and E. Luján.[3] As it emerges from the contributions of these authors, in order to study the textile vocabulary, one can rely on different approaches. Basically one can try to understand the meaning of a Mycenaean word by studying its context or its etymology. It is clear that safe results can only be achieved when it is possible to combine these two methods. The elements of the textile vocabulary can then be grouped together and compared from both a historical and a structural perspective.

[1] See Dury & Lervad, this volume.

[2] At Knossos on Crete, the texts found in the so-called *Room of the Chariot Tablets* (RCT), which is the oldest deposit of Linear B texts known until now, are dated to the Late Minoan (=LM) II period (*c.*1450–1400 BC), whereas the bulk of the archives found in the same palace is dated to LM III A2 or III B (*c.*1375–1325 and 1325–1190 respectively); the Pylos and the Thebes tablets, as well as the Mycenae tablets found in the citadel, are dated mainly to the end of the Late Helladic (=LH) III B2 period (LH III B2 = *c.*1225–1200 BC); the Mycenae tablets found in the so-called House of the Oil Merchant are dated to the end of the LH III B1 period (LH III B1 = *c.*1300–1225 BC).

[3] Duhoux 1976; Morpurgo Davies 1979, 87–108, esp. 99–101 and Appendix; Luján 1996–1997; Luján 1999; J. T. Killen 1962; 1963; 1964; 1966; 1968; 1972; 1974; 1979; 1982; 1984; 1988; 2001; Nosch 2000; Barber 1991, 260–282; Rougemont (forthcoming).

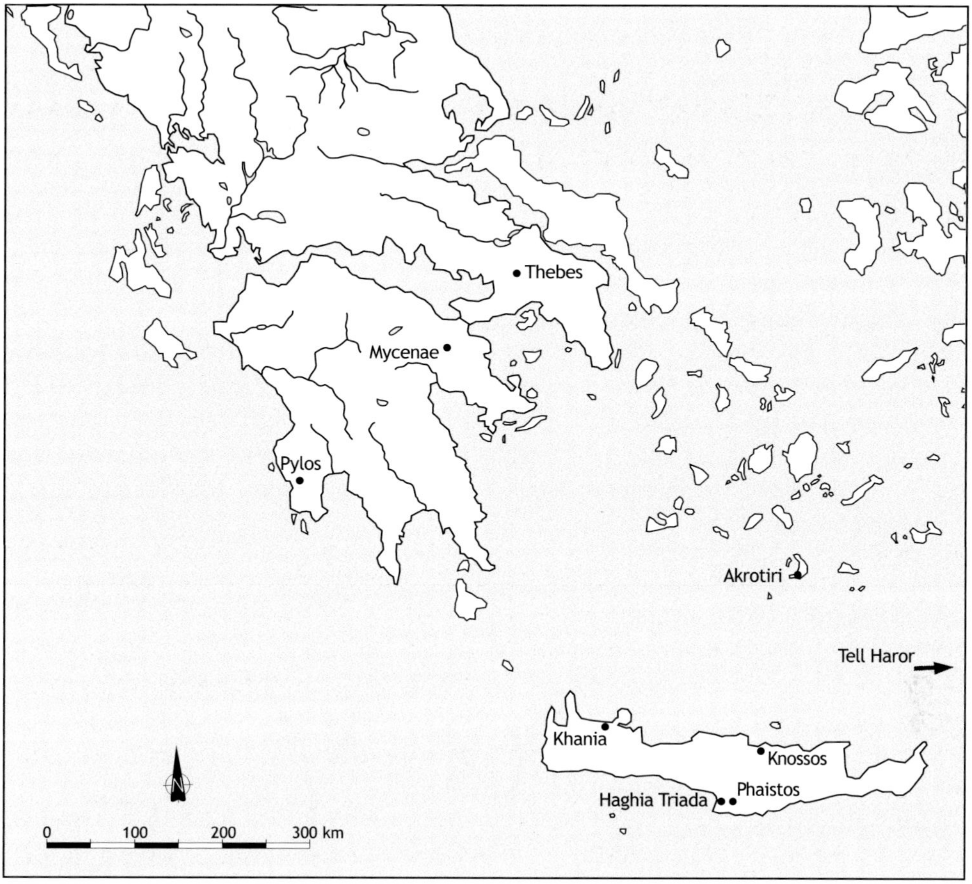

Fig. 17.1. Map with the sites where the Linear A and Linear B inscriptions mentioned in the text have come to light. Courtesy of P. Darcque.

The aim of this chapter is to combine the approaches mentioned above in a new perspective. Firstly, the reader will find below a classification of the extant textile terminology according to the *chaîne opératoire*;[4] second, we present a reflection on the relationship between logographic notations and textile terminology, as can be grasped from the study of Linear B texts, but also going back to the elements provided by the Linear A documents, as far as they can be understood; then the focus is on the terminology of tools and weaving, and its relationship to technical characteristics; and finally a reflection on the use of words in textile terminology (antonyms, synonyms, and what can be inferred from it about the information that was particularly important to the Mycenaean scribes) is presented.

[4] This first section owes a great deal to the reflections already developed in the framework of a workshop on the processing of thread held in Nanterre, see Rougemont (forthcoming). See also Del Freo & Rougemont (forthcoming).

1. From raw fibres to finished textile products

1.1. *Fibres: wool and linen/flax*

Two main textile fibres have been identified in Linear B records: they are, unsurprisingly, wool and linen/flax.

1.1.1. *Wool*

Wool is designated in Linear B by the logogram **145*, conventionally transcribed by the Latin word LANA.[5] The corresponding word is not directly attested, only known through other designations, such as the adjective *we-we-e-a*, meaning "woollen", related to Gr. εἶρος.

The logogram is also used as metrogram, i.e. to indicate a quantity, the value of which is estimated to be 3 kg. It denotes, to the best of our knowledge, exclusively sheepswool.[6] However, it has been suggested by M. Perna that the logogram **142* could represent goatshair, which would fit nicely into the wider range of textile fibres recorded in contemporary Near Eastern documents, for example those found at Nuzi.[7] Bronze Age goatshair textiles are also attested in the archaeological record.[8]

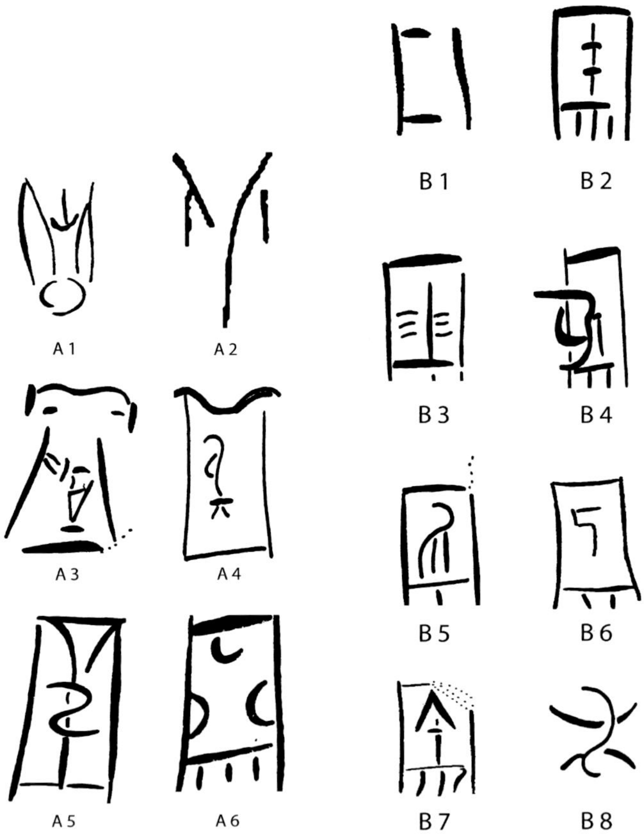

Fig. 17.2. Chart of Linear B wool/flax and textile logograms. Drawings by L. Godart.

Table A

1. *LANA* logogram. From KN Lc(1) 525 by scribe 103
2. *SA*, syllabogram and also logogram for flax. From KN Gs (1) 685 by scribe 135
3. *TUN+KI* logogram. From KN L 870 by scribe 114?
4. *TUN+RI* logogram. From KN L 178 by scribe "124"
5. **146* logogram. From KN M 467 by unidentified scribe
6. **164* logogram. From KN L 520 by unidentified scribe

Table B

1. *TELA* logogram. From KN Lc(1) 534 by scribe 103
2. *TELA+PA* logogram. From KN Ld(2) 787 by scribe 114
3. *TELA+TE* logogram. From KN Lc(1) 526 by scribe 103
4. *TELA+KU* logogram. From KN L(4) 516 by scribe 208
5. *TELA+PU* logogram. From KN L(4) 515 by scribe 208
6. *TELA+PO* logogram. From TH Lf 139 by unidentified scribe.
7. *TELA+ZO* logogram. From KN L 433 by unidentified scribe.
8. **166+WE* logogram. From KN Oa 745 by unidentified scribe.

[5] On this logogram, see Nosch 2007.

[6] Perna 2004, 278–280.

[7] Cf. Schneider-Ludorff 1998, 163–168.

[8] Spantidaki & Moulherat 2009; Frangipane *et al.* 2009.

J. T. Killen has demonstrated that the palatial sheep flocks which were the source of the wool used by palatial workshops were constituted of castrated wethers;[9] they contain in general 100 animals or a multiple of this number; the quantity of wool expected by the palace is 1 unit of wool (3 kg) from 4 adult sheep, meaning around 750 g per animal. Even when a flock is not a wether-only flock, but also contains ewes, old animals and yearlings, the production target for wool is identical. However, in flocks destined for reproduction, the composition of which include ewes and lambs, the wool target is only 10 animals : 1 unit of wool, i.e. *c.*300 g per animal.

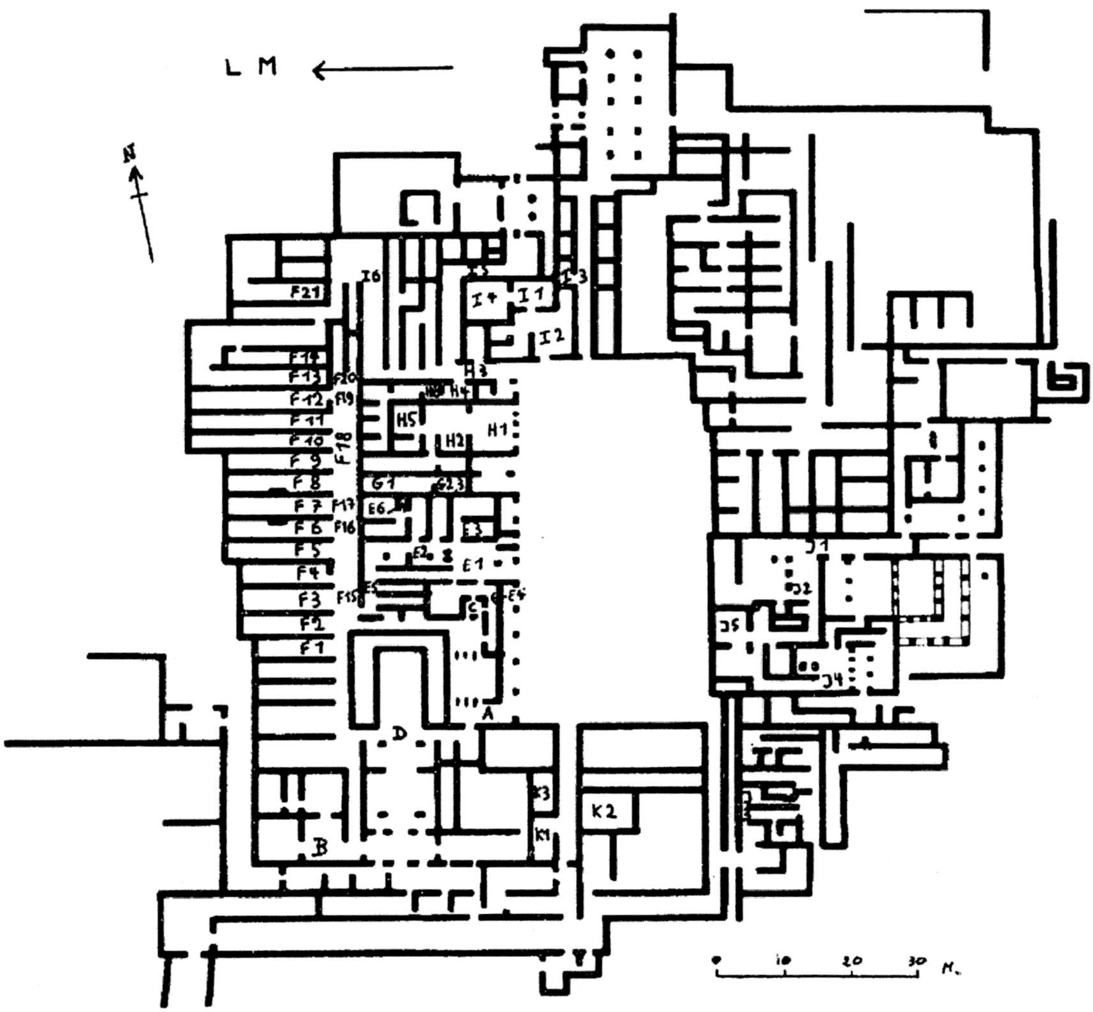

Fig. 17.3. Map of the Knossos palace

[9] Cf. Killen 1964.

The Knossos records include mentions of fleeces designated by the word *po-ka /pokai/*:

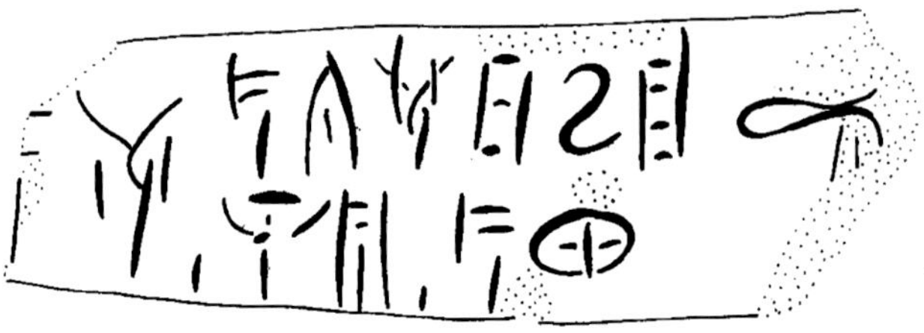

Fig. 17.4. Drawing of KN Dp 997. Drawing by L. Godart

KN Dp 997 + 7206 (I3/118)[10]
.a] po-ti-ni-ja-we-ja [
.b]to-sa , / ne-wa , po-ka OVIS^{f} [

"so many new fleeces of/belonging to Potnia, EWES xx (numbers broken)".

KN Dp 7742 (-)
.1 ?pe-ru-si-]nwa , po-ka [
.2]po-ti-ni-ja-we-i-jo [

line 1: "fleeces from last year (?)".

There may be, at the same site, a mention of "tufts of wool", *ti-ra /tilai/*:

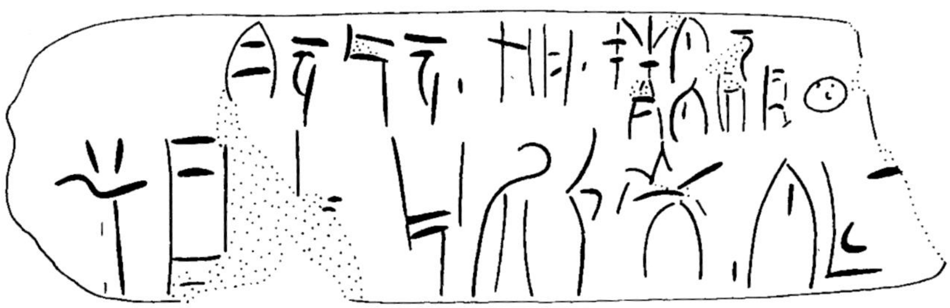

Fig. 17.5. Drawing of KN Od (1) 681. Drawing by L. Godart

KN Od (1) 681 (F18/103)
.a 'e-na-po-na , o-nu , pa-i-ti-jo' e-ti-wa-ja-qe ḶẠṆẠ[
.b qo-ja-ṭẹ ⌞ ⌟ a-pu-do-ke , ti-ra [
qo-ja-ṭẹ over ⟦ ⟧.

line .b: "*qo-ja-te* (personal name) has delivered fleeces ...".

[10] 118 is the indication of the scribe, I3 the findspot. For the findspots of Linear B tablets in the palace of Knossos, see Fig. 17.3.

There is no regular mention of lambswool as opposed to sheepswool, except in one Knossos tablet, Dk(2) 1066, where 200 lambs are recorded along with 19 units of wool, which have all been delivered to the administration (no deficit indication is attested):

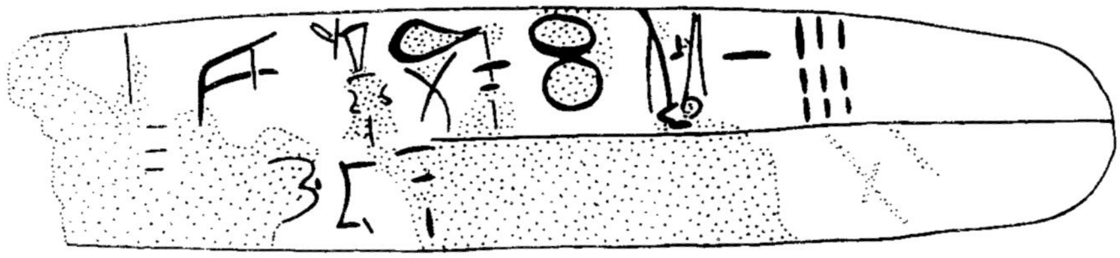

Fig. 17.6. Drawing of KN Dk (2) 1066. Drawing by L. Godart

```
KN Dk(2) 1066                                                        (J1/119)
      .A     ]          'ki' ne X OVIS^m 200    LANA 19
      .B     ]ṭẹ-u / ku-ta-to[                         ] vac.
```

"]*te-u* (shepherd's name) at *ku-ta-to* (place-name) new LAMBS 200, WOOL 19 units".

Since LANA 19 equals 57 kg , here there are ca. 285 g of wool per lamb, if the numbers are complete.

There is also one Mycenae record, MY Oe 111, which registers on lines 1 and 3 sheepswool, *o-u-ka /owika/* and on line 2 lambswool, *wo-ro-ne-ja /wroneia/* (cf. ὄϊς and ἀρήν).

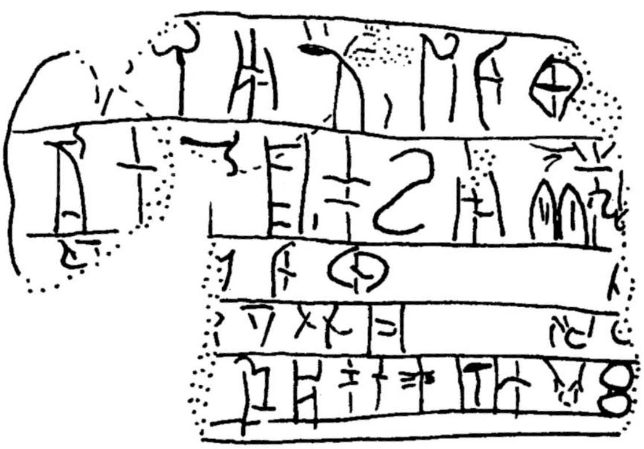

Fig. 17.7. Drawing of MY Oe 111+136. Drawing by L. Godart

```
MY Oe 111 + 136                          NMA 52–111 + 136               (51)
      .1       pe-ru-si-nwa , o-u-ka[
      .2      wo-ro-ne-ja , pa-we-si / [•]-me-'jo-i'  ḶẠṆẠ[
      .3      ne[-wa ]o-u-ka       ḶẠṆẠ[
      .4             ]-ḳị-ni-*5̣6̣ LANA 1̣0̣0̣[
      .5             ] o-ta-pa-ro-te-wa-ro       LANA 200[
      .6             ] vacat
```

Finally, textiles are described as *we-we-e-a* /*werwe*[h]*eha*/ "woollen" (cf. Gr. εἶρος) in two Knossos records: KN L 178 (cf. below *u-po-we/e-pi-ki-to-ni-ja?*, § 4.1.3) and L 870:

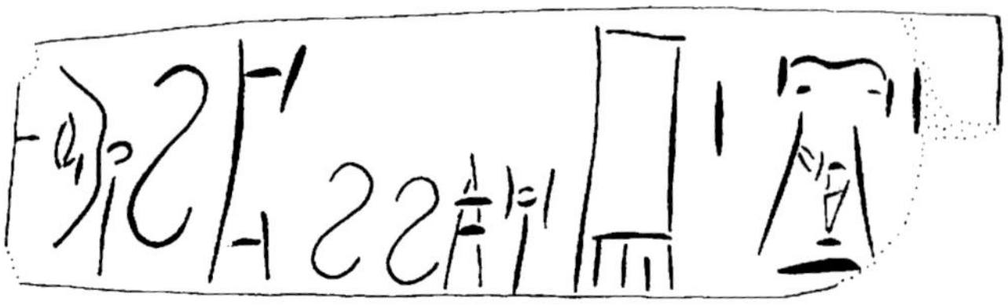

Fig. 17.8. Drawing of KN L 870. Drawing by L. Godart

```
KN L 870 + fr.                                          (I3/114?)
o-]ḍạ-ku-we-ta / we-we-e-a        TELA³ 1       TUN+KI 1
```

1.1.2. *Flax/linen*

Flax as a cultivated plant is designated by the syllabic sign **31* used logographically, and transcribed *SA*. It is attested at Pylos and Knossos in records related to fiscal matters. Since *SA* cannot be related to any Greek word indicating flax or linen, it is in general assumed that this is the abbreviation of an unknown foreign word.

The Mycenaean palace administration received a tax in kind on plots of land which were cultivated with flax (certainly identified as a cultivated plant by PY Na 520, with the formula *pu$_2$-te-re ki-ti-je-si*, "the planters cultivate".[11]

The adjective *ri-ta*, "linen (clothes, items)", or its abbreviation *RI* is applied to a variety of textile designations to indicate the fibre used for making them: **146*, **166+WE*, TUN+*KI*, TUN+*RI* as well as *ki-to* and *pa-we-a*. The adjective is attested at Knossos five or six times. Finally we have one attestation of *ri-no re-po-to* λίνον λεπτόν (KN L 693.1).

1.1.3. *Designations for thread?*

No obvious designations for thread have been identified in the Linear B records; the Classical Greek terms for warp (στήμων) and weft (πήνη) are not attested in the extant Linear B documentation. However, J. L. Melena has suggested[12] that the Mycenaean word *o-nu*, pl. *o-nu-ke*, might refer to the weft threads, whereas the mentions of *e-ne-ro* would correspond to the warp.[13] Lastly, he has hypothesized that another designation, *e-ta-wo-ne*, may be a kind of hyperonym (a word for any kind of thread).[14] Finally, there is one possible interpretation of the adjective *ko-ro-to* as /*klōston*/ "spun", applied to wool at Mycenae. Yet, the word could also be understood as /*khrōston*/ "dyed" (cf. below, § 4.2.1, *ki-ri-ta/ko-ro-to*). To sum up, although some words have been tentatively interpreted as designations for thread, these suggestions remain very fragile.

However, when it comes to flax *versus* linen, two different syllabograms are used logographically

[11] On this formula, see Del Freo 2001, 27–44.

[12] Melena 1975, 112–113.

[13] Melena 1975, 90.

[14] Melena 1975, 88–91. It is interesting to note that, according to Luján 1999, 129, *e-ta-wo-ne* does not exist in 1st millennium Greek, except perhaps through a personal name in Homer.

in the tablets, and, as argued elsewhere,[15] the coexistence of two notations might reflect two different stages in the processing of flax: on the one hand, *SA*, in the land/fiscal records, which is certainly identified as flax by the use of the word *ri-no /linon/* in the heading of the Pylos text Nn 228:

PY Nn 228 (S106–H1/Archives Room)
.1 o-o-pe-ro-si , ri-no , o-pe-ro
.2 u-ka-jo , *SA* 20 ro-o-wa , *SA* 35
.3 pu$_2$-ra$_2$-a-ke-re-u , *SA* 10 ke-i-ja-ka-ra-na
.4 *SA* 5 di-wi-ja-ta , *SA* 60
.5 a-pi-no-e-wi-jo *SA* 28
.6 po-ra-pi , *SA* 10 e-na-po-ro , *SA* 33
.7 te-tu-ru-we *SA* 38
.8–15 *vacant*
line 1: "they owe linen: still due...".

On the other hand, *RI*, attested in the Pylos Ma series as well as on PY Mm 11 and KN Nc 5100, which seems to be used in order to designate the processed flax fibre when ready to be spun, or even linen thread.

1.2. *Textile production through professional designations*

Various steps of the textile production are known in the Linear B documentation through nouns designating specialised workers.

1.2.1. *General professional designations, which can apply to any kind of fibre.*
They include mentions of "combers" (*pe-ki-ti-ra$_2$ /pektriai/*)[16] listed twice in records of personnel at Pylos (PY Ab 578, Ad 694); then of "spinners" (*a-ra-ka-te-ja* attested at Thebes, Knossos, and Pylos); there is also a professional designation related to the textile item *e-ne-ra*, *e-ne-re-ja*, and of which it has been suggested that it might be related to warp threads (?);[17] the designation is attested twice at Knossos (KN Ak [1] 638, X 522). The interpretation is by no means certain.

e-ta-wo-ne-we / e-ta-wo-ne-wo appears twice at Knossos. The question is whether there is a relationship with *e-ta-wo-ne*. Different interpretations have been suggested, but the designation remains obscure; J. T. Killen[18] has suggested "a finisher of cloth (fuller *vel sim.*)" which would imply a meaning close to that of *ka-na-pe-u /knapheus/* "fuller"; but then what would be its relationship with *e-ta-wo-ne*, which might supposedly be a kind/part of textile item?

The tablets mention "weavers" (*i-te-ja-o, i-te-we* and *pe-re-ke-u / pe-re-ke-we*),[19] "seamstresses" and/or "tailor" (*ra-pi-ti-ra$_2$* and *ra-pte*).

People called *o-nu-ke-ja/o-nu-ke-wi* are attested respectively at Pylos (PY Ab 194, Ad 675) and at Thebes (TH Oh 206.2). The word *o-nu-ke-ja* is a professional designation built on *o-nu-ke*, maybe "fringes", or "endings".[20]

---

[15] Rougemont 2007, 46–49. See also Perna 2004, 210.
[16] See also below, § 3. *Terminology and technical characteristics*, by M.-L. Nosch.
[17] Melena 1975, 91.
[18] Killen 1979, 159 ff.
[19] Cf. below, § 3.5. *Weaving terminology*, by M.-L. Nosch.
[20] Killen 1979. For an alternative interpretation, see Firth & Nosch 2002–2003.

Finally, the tablets also document specialists for the decoration or finishing (*a-ke-ti-ra$_2$/a-ke-ti-ri-ja*[21]*/a-ze-ti-ri-ja*); among the professional designations linked with textiles, Linear B texts also mention a more general designation: *a-pi-qo-ro*, "servants", known at Pylos and Thebes (PY Aa 804, Ad 690, TH Of 34).

1.2.2. *Professional designations strictly related to a specific fibre or technique*

First comes the problem of the "linen workers", *ri-ne-ja /lineiai/*: are they working only on the fabrication of linen textiles or do they also prepare the fibres? There is no way to decide, since among professional designations both possibilities are attested like *te-pe-ja*, "*te-pa* makers", or *me-re-ti-ri-ja/-ti-ra$_2$*, "corn grinders", i.e. women processing the "raw material".

At Pylos, a "linen collector", *ri-na-ko-ro /linagoros/* is recorded; this designation is unfortunately attested only outside the context of the professional activity of these people (PY An 129 is a record of personnel where people are counted and enumerated by name and sometimes also by professional designation). It is difficult to say more about the activity of the person designated by this word;[22] he could have been collecting the flax and/or linen (products) at various stages of the processing (from retting to the finished product). However, in the Linear B terminology, it seems that the closer one is to the 1st millennium Greek word for linen, the nearer one is to the finished product; if this observation is correct, it would make the idea of a collector of thread or finished textiles more plausible than a collector of retted flax fibres.

The word *ka-na-pe-u, /knapheus/* "fuller", is known at two sites, Pylos and Mycenae;[23] it is mentioned in this category since this technical process applies only to animal fibres, in Linear B texts sheepswool, and not vegetal fibres.

1.2.3. *Professional designations related to a particular cloth item*

a-pu-ko-wo-ko, meaning "headband makers", is a word attested twice in Pylos lists of personnel (PY Ab 210, Ad 671); this item (*a-pu-ke*, "headbands") may have existed in textile and in leather, as indicated, for the latter, by PY Ub 1315.

ko-u-re-ja, a word designating "(*pa-we-a*) *ko-u-ra* makers", has five attestations at Knossos.

Lastly, the documentation includes the designation *te-pe-ja*, "*te-pa* makers". It comes as no surprise that the word is attested at three sites (KN Le 641, TH Of 35, PY Ad 921), since it is built on one of the most common textile names.

Another designation in this category is more debated: *to-te-ja*, possibly "**to-ta* makers",[24] appears only once, at Knossos (KN Ak 611.1). The word *to-te-ja* has been interpreted by C. J. Ruijgh[25] as */storteia/*, feminine of */storteus/*, derived from **storton*, "objet qui peut être étendu, couverture, tapis". However, the case is considerably weaker than others, since it relies on a reconstructed garment/textile name (**to-ta*) which is, in fact, not attested in the documentation.

[21] Cf. Morpurgo Davies 1979, 91, 99 and n. 44.
[22] Rougemont 2007.
[23] PY Cn 1287, En 74/Eo 267, Eo 269, MY Oe 129, Oi 701.
[24] See Melena 1975, 116 ff.
[25] Ruijgh 1967, 252–253.

1.3. *Designations of textile items/garments*

1.3.1. *Textile items recorded by means of logograms*
The most frequent one, the TELA logogram, has the shape of a rectangle with small vertical strokes at the bottom. Numbers are conventionally added to the transcription of the logogram according to the number of strokes drawn by the scribe. The TELA logogram can be specified by a number of endograms /abbreviations, for example TELA+*TE*, for *te-pa*, a kind of cloth. However, there are also textile logograms with shapes which are less obviously similar to a textile, for example: **146*, its variation **160*, **164* or the ligatures **166+WE* and **168+SE.*

1.3.2. *Textiles designated by words written syllabically*
Some of these words, whose origin is either known or unknown, can be found in 1st millennium Greek:[26] for example *pe-ko-to /pekton/* from the verb πέκω, "to comb", *pa-we-a /pʰarweha/* (cf. φᾶρος, word of unknown origin), or *te-pa* (unknown origin, cf. perhaps Gr. τάπης?). The Linear B records also attest words which can be etymologized, but not in Greek, such as *ki-to /kʰitōn/* (cf. Gr. χιτών, a word which might be of Sumerian origin; it corresponds to Akk. *kitû*, Sum. GADA, "linen"),[27] but which have sometimes been integrated in words whose composition is Greek, like *e-pi-ki-to-ni-ja*, a designation for a piece of clothing worn "over the χιτών". Lastly, there are words which cannot be etymologized in Greek nor, so far, in any other language, like *tu-na-no.*

1.3.3. *Names of garments*
Linear B texts record a surprisingly small number of names of garments[28] (*e-pi-ki-to-ni-ja,* a garment worn "over the *khitôn*"; *e-pi-ro-pa-ja,* a garment worn "over the *lope*"; *ki-to, khitôn*; *u-po-we*, a garment "worn under"; and *we-a$_2$-no*, */wehanos/* "garment"). In some cases, it is difficult to say if the nouns designated a garment or a piece of textile.[29] The same can be said also for *pa-we-a*, */pʰarweha/*, since in Homer φᾶρος is either a garment or a piece of textile, according to the context.[30]

Furthermore, in Linear B, some garments are designated both by a noun written syllabically and by a logogram (e.g. **146*,[31] which contains the endogram *WE* / *we-a$_2$-no*).

1.3.4. *Coloured textiles*
In the Linear B texts, designations related to colours are apparently always associated with finished textile products, and never with fibres, nor with woollen or linen thread. This, however, does not imply that dyes were not applied to thread or fibres, only that the colour of woollen or

[26] The classification of words into the three categories described below goes back to Morpurgo Davies 1979, 90–91, who distinguishes three main categories of words, i.e. "words which are formally preserved in Greek", "words which are not attested as such in the Greek of the First Millennium but can be easily 'etymologized' from Greek material" and "words which have no definite equivalent in Greek and cannot be readily 'etymologized'".
[27] Cf. J. P. Vita, in this volume.
[28] For a linguistic analysis of names of garments, see Luján 1996–1997, 346–347.
[29] Luján 1996–1997, 346.
[30] On the attestations in Homer, see Luján 1996–1997, 337, with bibliography.
[31] On this logogram, see below § 2.1.

linen threads was not recorded by the Mycenaean scribes.[32] There seems to be a predominance of red hues.[33]

The colours attested are the following: *e-ru-ta-ra*, "red" (Gr. ἐρυθρός), *pa-ra-ku-ja*, perhaps "blue/green" (see also **56-ra-ku-ja*) (cf. Gr. σμάραγδος?), *po-ki-ro-nu-ka* "with multicoloured fringes/endings" (Gr. ποικίλος), *po-ni-ki-jo*, another shade of "red" (Gr. φοινίκιος), *po-pu-re-ja* / *po-pu-re-jọ[* / *po-pu-ro*$_2$, "purple" (Gr. πορφύρεος), *po-ri-wo*, "grey" (Gr. πολιός), *pu-ru-wa*, maybe "red-brown" (Gr. πυρρός), *re-u-ko/ka*, "white" (Gr. λευκός; see also *re-u-ko-nu-ka*, "with white fringes/endings").[34] Notations such as TELA+*PO* or *re* TELA/ *'re'* TELA+*TE* might also be understood as including abbreviations of the colours of the textiles (*PO*, *re*, see above, *po-pu-re-ja* / *po-ni-ki-jo*, and *re-u-ko*, etc.).

The tablets also contain words perhaps referring to the dyeing process (*ko-ro-ta*$_2$ and *ki-ri-ta*);[35] on these two words, see below § 4.2.1 *True or false synonyms?*

2. Logographic notations and textile terminology: a complex relationship

2.1. *Logograms in Linear B*

In order to record the economic data that were of interest to the palatial administration, the Linear B scribes used a mixture of logographic and syllabic notations; some items can be designated exclusively by means of logograms, or by means of logograms and of words written syllabically. Last but not least, logograms can be combined with abbreviations (adjuncts, endograms, ligatures), some of which can be related to words written syllabically.

This sometimes complex relationship can be traced back to Linear A, which will be treated below.[36]

For textiles, scribes have mainly used logograms, for example TELA, **146*, **158*, **160*, **166*, **189;* additional information is often provided by means of endograms, i.e. syllabic signs written inside the logograms; the best example is TELA, which can be associated with the following endograms (see Fig. 17.2): TELA+*KU,* perhaps to be compared with *ku* LANA at Thebes (?); TELA+*PA,* with *PA* standing for *pa-we-a* (cloth name); TELA+*PO,* where *PO* might be *po-ni-ki-jo, po-ki-ro-nu-ka,* or *po-pu-re-jo,* which are designations for colours or coloured items (in the case of *po-ki-ro-nu-ka,* which qualifies cloth items "with multicolored fringes or endings"); TELA+*PU,* with *PU* standing for *pu-ka-ta-ri-ja*; TELA+*TE,* with *TE* for *te-pa*; and lastly TELA+*ZO,* where *ZO* stands for a word so far unidentified.

2.2. *Logograms in Linear A*

2.2.1. *Reading the Linear A documents*

Studying the "Minoan" textile terminology through the Linear A documents is beyond our present capabilities for more than one reason.

---

[32] Although the word *ko-ro-to*, which might be understood, among other possibilities, as "dyed", is directly recorded with the LANA logogram on tablets of the Knossian Od series. See below § 4.2.1 *True or false synonyms*, and *DMic s. v. ko-ro-to.*

[33] On red coloured textiles, see Nosch 2004.

[34] For the linguistic analysis of these designations, see Luján 1996–1997, 347–352.

[35] From other texts (PY An 35, Un 443, maybe also TI X 6), we also know about *tu-ru-pte-ri-ja*, "alum", see Perna 2005 and Firth 2007.

[36] See § 2.2, *Logograms in Linear A*, by M. Del Freo.

More than 70% of the Linear B syllabograms are clearly derived from Linear A, but it is by no means certain that all these syllabograms had the same phonetic values in both scripts. Therefore, the transcriptions of Linear A documents with the phonetic values of the Linear B homomorphic syllabograms have to be considered as purely conventional.

It is generally agreed that the homomorphic syllabograms belonging to identical *groups* of signs in both scripts had the same phonetic values. However, identical groups of signs in Linear A and Linear B are very rare and a rigorous application of this rule makes it possible to "read" only a dozen of the syllabograms.[37]

Given these circumstances, it is clear that at present, even when for contextual reasons a syllabic sequence can be interpreted as a textile term, it can neither be read nor analysed with absolute confidence.

For these reasons, the study of "Minoan" textile terminology relies nearly exclusively on the evidence which is directly or indirectly provided by the logograms.

2.2.2. *LANA in Linear A (?)*

The ligature A 559, formed by syllabograms AB 80+26, closely resembles the Linear B logogram **145* LANA "wool".[38] The Linear A logogram is attested on four tablets respectively from Phaistos (PH 3a.3̣), Hagia Triada (HT <12>.4, HT 24a.1.2.3̣.4̣.5) and Khania (KH 43.1̣).[39]

Ligature AB 80+26 is recorded alone on HT 24a and, along with other commodities, on HT <12> (A 608, A 304, A 510, AB 38, A 626, AB 30 e A 511),[40] PH 3a (A 556, A557, A 563)[41] and KH 43 (A 527).[42]

The interpretation of AB 80+26 as "wool" seems to be confirmed by HT 24. On the side *b* of this tablet three different ligatures are followed by the sign AB 118 "talent" and by the figures 1+J+E, 1+J+E and 1 (where J and E are conventional

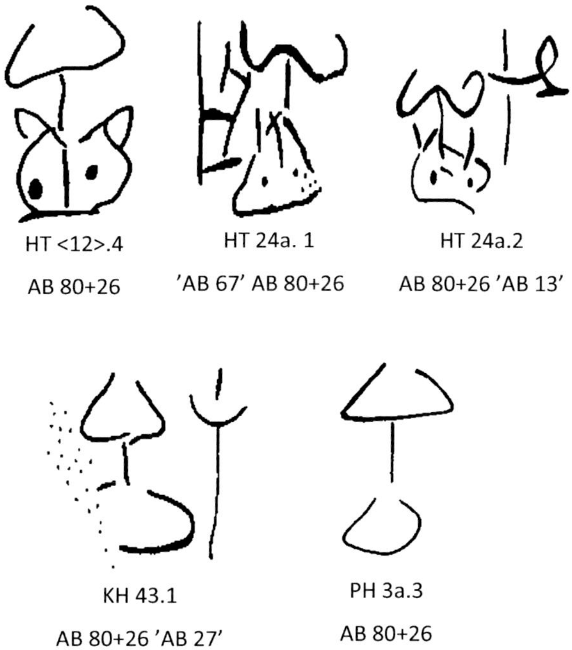

Fig. 17.9. Chart of possible Linear A logograms for wool. Drawing by L. Godart

[37] Cf. Olivier 1978; Godart 1984.

[38] Cf. *Docs*², 36, 52, 314; Palaima 1994, 317; Schoep 2002, 131–132.

[39] PH 3 was found in Room XLIV/38 and dates back to Middle Minoan (=MM) III (Militello 2002a, 67–69); HT <12>, which was perhaps unearthed in the area of Room 59 (Militello 2002b, 116), and HT 24, which was certainly found on the threshold between Corridor 9 and Room 26 (Militello 1988, 235; Militello 2002b, 113), are assignable to LM IB; KH 43, finally, has been brought to light in the Odos Katre excavation and dates back to LM IB (Hallager 1996a, 50–51).

[40] Some of these commodities can be interpreted as olive oil (?) (A 608 = 'A 303' 'AB 07'), cyperus / grain (?) (A 626 = A 303 'A 704') and figs (AB 30).

[41] All these ligatures contain the syllabogram AB 80: A 556 (AB 80+[•]), A 557 (]AB 8̣0̣+08), A 563 (]AB 8̣0̣+80).

[42] The ligature A 527 is formed by AB 40+7̣4̣[.

transcriptions of fraction signs). As it has been observed,[43] the sum of these figures, which probably amounts to 4.5 talents,[44] can be related to the 45 *noduli* found nearby.[45] Since the ratio 1: 10 between talents and *noduli* coincides with the ratio between the talent and the weight unit for wool in Linear B, it is likely that AB 80+26 was the Linear A sign for "wool".[46] In theory, the records on the two sides of HT 24 could have been unrelated. Yet, as it has been observed,[47] if one admits that the two missing figures on HT 24a.4–5 were close to the average of those preserved on lines 2–3 (]6, 10,]9 J, 6),[48] for side *a* of the tablet a total of 46–47 units of wool can be restored, i.e. a figure fully compatible with the 4.5 talents recorded on side *b*.[49] The relationship between AB 80+26 and AB 118 "talent", furthermore, is confirmed by HT <12>.4, where the ligature is followed by five units of AB 118.[50]

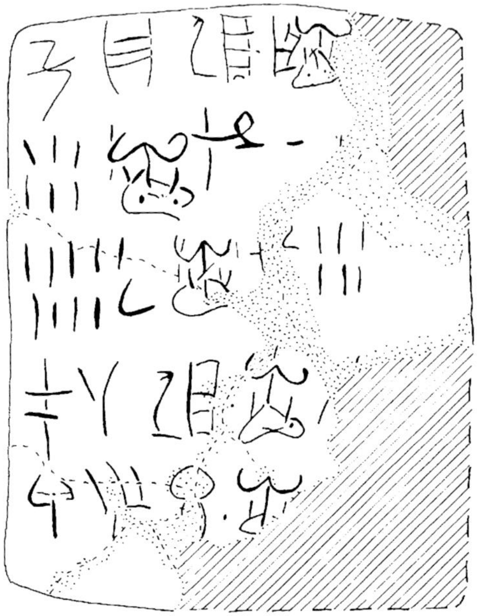

Fig. 17.10. HT 24. From GORILA I. Drawing by L. Godart

In Linear B, the phonetic values of AB 80 and AB 26 are respectively *ma* and *ru/lu*. Therefore, it has been proposed to read AB 80+26 as MA+RU and to relate it to Gr. μαλλός "tuft of wool" (cf. also Hsch. μάλλυκες· τρίχες). However, it is by no means certain that AB 80 and AB 26 had the same phonetic values in both scripts.

In some cases, ligature AB 80+26 is preceded or followed by isolated syllabograms, the aim of which was likely to specify acrophonically the quality or the destination of the commodity.

On HT 24a.1, ligature AB 80+26 is preceded by a syllabogram of uncertain reading, perhaps AB 6̣7̣, and immediately followed by a lacuna;[51] on lines 2–4, it is followed by the syllabogram AB 13;[52] on

[43] Hallager 2002.

[44] For J = ½ and E = ¼, see Bennett 1950, 204–222.

[45] The *noduli*, all of the same type and stamped by the same seal, were found on the windowsill between Corridor 9 and Room 27 at less than one meter from HT 24 (Militello 1988, 235; Hallager 1996a, 41).

[46] According to this hypothesis, each *nodulus* would represent a wool unit.

[47] Palaima 1994, 317–318.

[48] For the lacunae of lines 1 and 2, the following supplements can be suggested: ['AB 13'] and respectively [AB 80+26 'AB 13'] (see below).

[49] If this reconstruction is correct and each *nodulus* corresponded to one unit of wool, the fractional quantities on the tablet exclude that the figures were *directly* related to the 45 *noduli*. It is possible, therefore, that the *noduli* referred to a separate weighing of the wool recorded on HT 24. The uncertainty about the function of the *noduli* (Hallager 1996a, 130–133) makes it difficult to define the aim of such an operation (for a possible explanation, see Hallager 2002, 107–108).

[50] A quantity similar to that recorded on HT 24b. The quantities recorded on PH 3a and KH 43 are much lower: fraction A 717 (DD) (PH 3a) and two units (KH 43).

[51] In *GORILA* the juxtaposition 'AB 6̣7̣' AB 80+26 is conventionally classified as A 546.

[52] In *GORILA* the juxtaposition AB 80+26 'AB 13' is conventionally classified as A 561.

line 5, finally, it is immediately followed by a lacuna.[53] It can reasonably be assumed that AB 13 followed AB 80+26 also on lines 1 and 5.

In the Linear A tablets, AB 67–02 and its abbreviation AB 67, which may be read respectively as KI-RO and KI, convey the notion of "deficit" (like *o-pe-ro* and *o* in the Linear B documents, see above, § 1.1.3 *Designations for thread?*, with the tablet PY Nn 228). Thus it is possible that on HT 24a.1 a "deficit of wool" was recorded. AB 13 is not among the syllabograms which can be confidently read through the comparison with Linear B, so that the transcription ME remains hypothetical.

On KH 43.1 the ligature AB 8̣0̣+26 is preceded by a lacuna and followed by the syllabogram AB 27.[54] Like AB 13, AB 27 cannot be read with confidence. Thus the transcription RE is purely conventional and has to be regarded as a mere hypothesis.

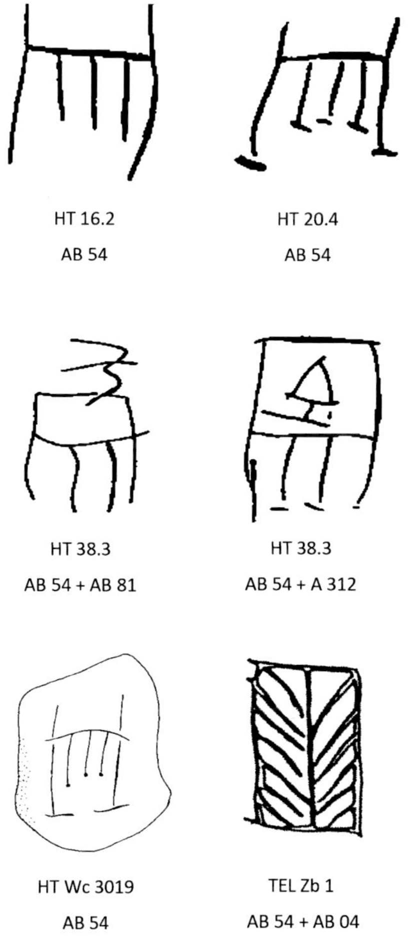

Fig. 17.11. Chart of possible Linear A logograms for cloth

2.2.3. *TELA in Linear A (?)*

The Linear A sign AB 54, when used logographically, is directly comparable to the Linear B logogram **159* TELA "cloth".[55] The logographic use of AB 54 is attested on six different documents: one roundel and three tablets from Hagia Triada (HT Wc 3019, HT 16.2, HT 20.4 and HT 38.3), a fragment of tablet from Akrotiri (Thera) (THE 8.2), and, apparently, a graffito from the Palestinian site of Tel Haror (TEL Zb 1).[56]

The logogram AB 54 is written alone on Wc 3019. On the rim of this roundel, three seal imprints are visible, one of which is superscribed by the fraction sign A 704 (E). On HT 16 and HT 20, AB 54 is listed with other logograms

[53] Since it could be followed by a syllabogram different from AB 13, AB 80+26[is conventionally classified as A 560 in *GORILA*. For similar reasons,]AB 8̣0̣+26 of PH 3a.3 is conventionally classified as A 558.

[54] In *GORILA* the juxtaposition]AB 8̣0̣+26 '27' is conventionally classified as A 562.

[55] Cf. *Docs*², 36, 49, 313; Palaima 1994, 317; Schoep 2002, 131. Apparently a logogram for "cloth" is attested also in Cretan Hieroglyphic: cf. the medallion *CHIC* #103 from the *Dépôt hiéroglyphique* at Malia and the observations on sign **163* in Oren & Olivier 1996, 101–102. The medallion, like the other texts from the hieroglyphic deposit, can be dated to the final phase of MM III (*CHIC*, 28).

[56] All the Hagia Triada documents come from the Villa: HT 16, 20 and 38 perhaps from Room 59 (Militello 2002b, 116); Wc 3019 probably from the area corresponding to Rooms 3, 11 and 13 (Hallager 1996b, 13, 31) or perhaps from Room 59 (Montecchi 2007, 15); THE 8 comes from Room Δ18α (Boulotis 1998, 407; Boulotis 2008, 72, Figs. 11–13); TEL Zb 1 has been found in a cult area (Oren & Olivier 1996, 92). These documents are datable respectively to LM IA (THE 8), LM IB (HT 16, 20, 38 and Wc 3019) and the 17th–16th cent. BC (TEL Zb 1). Theoretically, the graffito from Tel Haror could also be ascribed to Cretan Hieroglyphic (Oren & Olivier 1996, 109). According to the *Chronique des fouilles en ligne*, one new roundel with the AB54 logogram has been brought to light in Chania by M. Andreadaki Vlazaki in 2007 (see http://chronique.efa.gr/index.php/fiches/voir/273/).

and in both cases precedes the ligature A 604, which remains unidentified. As on the roundel, the commodity is measured by fraction signs, A 702 (B) and A 704 (E).[57] On HT 38, the logogram appears twice with other logograms and is followed by whole figures (2 and 1 respectively). Finally, on THE 8, AB 54 is apparently followed by the figure 200[.[58]

On HT 38.3, the logogram is ligatured with two different syllabograms: AB 54+81 and AB 54+A 312;[59] a third ligature, AB 54+04, is attested on TEL Zb 1,[60] while a fourth one, AB 54+09, can be read on THE 8.[61] In the first case, the syllabogram is placed on top of AB 54, while in the other three cases, it is placed inside the logogram, similar to the position of endograms of TELA in Linear B (see above, § 2.1, *Linear B*).[62]

The convention of specifying the meaning of a logogram by a ligatured syllabogram is common both to Linear A and Linear B administrations. In Linear A, many ligatures of this type are known, e.g. for grain (AB 120), wine (AB 131) and other commodities. In Linear B, it is clear that the function of these syllabograms was acrophonic. In the case of cloth, as already mentioned (see above, § 2.1, *Linear B*), *PA* of TELA+*PA* was the abbreviation of *pa-wo* /*pʰarwos*/, *TE* of TELA+*TE* was the abbreviation of *te-pa*, etc.

Fig. 17.12. HT 38. From GORILA I. Drawing by L. Godart

It is reasonable to think that in Linear A these syllabograms had the same function as in Linear B and that, consequently, AB 04, AB 09, AB 81 and A 312 were abbreviations of cloth names.

While the syllabograms AB 04 and AB 09 can be read as TE and SE, the phonetic value of the syllabogram AB 81 cannot be ascertained with confidence and consequently the transcription KU is simply a hypothesis.

It is interesting that ligatures AB 54+04 and AB 54+81 can be directly compared to ligatures TELA+*TE* and TELA+*KU* of Linear B. As it has been observed,[63] AB 54+A 312 might correspond to Linear B TELA+*ZO*. However, the endogram A 312 is slightly different from AB 20 *ZO*. For the ligature AB 54+09, finally, there are no direct parallels. However, it can be observed that in Linear

[57] It is likely that fractions referred to weight units as in the case of wool, but it cannot be excluded that cloths were fairly large and subdivided in fractional amounts (Palaima 1994, 317; Schoep 2002, 131). It is possible that the two other seal imprints on the roundel corresponded to one unit and that the roundel recorded a total of 2 E units (i.e. prob. 2¼ units: see above).

[58] On the Tel Haror graffito there are no figures.

[59] In *GORILA*, the two ligatures are conventionally classified as A 535 (AB 54+81) and A 536 (AB 54+A 312).

[60] See the discussion in Oren & Olivier 1996, 101–105.

[61] See the photograph in Boulotis 2008, 68, Fig. 2. The new ligature AB 54+09 should be classified as A 536*bis*.

[62] The logogram *163 of Cretan Hieroglyphic (see above) contains one sign, which does not match any of the signs attested so far for this script. Therefore, it is difficult to say if it was an endogram (cf. Oren & Olivier 1996, 102 n. 6).

[63] Melena 1975, 110.

B the syllabogram *SE* is attested in ligature with the logogram **168*,[64] which could be a product of the textile industry.[65]

Another Linear A logogram possibly related to cloths is AB 164.[66] This hypothesis is based on the fact that in Linear B **164* is attested in textile contexts (KN L 520, L 698). In Linear A, where it is also used as a syllabogram (HT 17.1 and 19.1), AB 164 is attested at Khania in four different variants (a, b, c, d) as an isolated logogram on eleven different roundels (KH Wc 2036–2045, Wc 2111).[67]

2.2.4. *Cloth names in Linear A*

The Linear A evidence for cloth names is limited to the above mentioned acrophonic abbreviations. In other words, there is evidence for different types of cloth whose names began with TE, SE, AB 81 (KU?) and A 312.

Comparing this evidence with that provided by the Linear B documents is not of much help. The correspondence between AB 54+04 and TELA+*TE* suggests that, as in Linear B, TE *could* be an abbreviation of *te-pa*, a cloth name with no clear Greek etymology.[68] As for AB 54+81, the possible comparison with Linear B TELA+*KU* (if AB 81 = KU) is unhelpful, as in this case it is not known which word *KU* abbreviates. In the case of SE, similarly, no comparison is possible, as in the Linear B documents there are no cloth names beginning with *se*.

If, as it has been argued above, the commodity recorded on HT 24b was wool, it is possible that the ligatures A 531 (AB 41+13 '67') and A 539 (AB 57+77) referred to cloth produced (or still to be produced) with that wool.[69] Read with the phonetic values of Linear B, AB 41+13 '67' and AB 57+77 would be respectively SI+NE 'KI' (AB 41+13 '67') and JA+KA (AB 57+77). Yet only JA and KI can be read with some confidence.

In conclusion, a brief comment needs to be made on the ostracon THE Zg 5 from Akrotiri. In her edition, A. Michailidou has suggested that, among the isolated signs that precede the figures, AB 80 MA was the acrophonic abbreviation of MA+RU "wool", and that the other signs were the abbreviations of at least

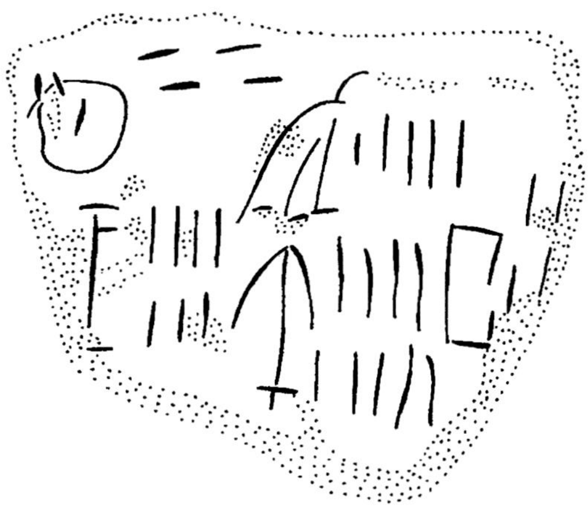

Fig. 17.13. THE Zg 5. From Michailidou 1992–1993. Drawing by L. Godart

[64] On all the Knossos Pp tablets, except on Pp 498.
[65] Cf. Melena 1975, 132.
[66] Cf. Schoep 2002, 132–133.
[67] The seal imprints on these roundels fluctuate from one to five.
[68] Cf. *DMic s.v.*
[69] The ligature AB 41+13 (A 530), followed by the figure 10, is attested on HT 23.5, a tablet with mixed commodities (wine, olive oil, cyperus, etc.).

three different cloth names.[70] In this particular case, one would have 40 units of AB 8̣0̣ ṂẠ (wool), as well as five PU (AB 50), seven ṬẸ (AB 0̣4̣) and nine ZO (AB 20) pieces of cloth.[71]

This hypothesis, which relies on the comparison with the Mycenaean words *pu-ka-ta-ri-ja*, *te-pa* and *zo-ta* as well as on the abundant archaeological evidence of textile activities in Akrotiri, is attractive, but regrettably still uncertain.

In fact, it can neither be proved that ṂẠ was a current abbreviation of MA+RU nor that ṬẸ is the correct reading for the third sign.[72] Moreover, it is not known if syllabograms AB 50 and AB 20 had the same phonetic values as in Linear B, i.e. PU and ZO. In any case, it is hardly conceivable that a Greek word like *pu-ka-ta-ri-ja* could have been attested in Minoan Akrotiri. Finally, as far as *zo-ta* is concerned, the contexts suggest that it was instead a personal name.

3. Terminology and technical characteristics

Textile tools are attested in the archaeological record, and iconography illustrates the Bronze Age loom. The types of textile tools used at the time of the Linear B records had already been in use for several millennia, and continued to be in use for a further two to three millennia.

The terminology for textile tools in Linear B is mainly attested indirectly through professional designations. We have thus far preserved evidence for four textile tools.

3.1. *Tools for spinning*

There is a professional designation *a-ra-ka-te-ja* "spinner", known from Thebes, Pylos and Knossos (TH Of 34, KN Ak 5009, Lc[1] 531, PY Aa 89, PY Aa 240). *a-ra-ka-te-ja* is built on the term for the spinning tool **a-ra-ka-ta* which is the Mycenaean predecessor for the 1st millennium Greek ἠλακάτη.

PY Aa 89 (S60–H4/Archives room)
a-ra-ka-te-ja MUL 37 ko-wa 26 ko-wo 16 TA 1
"37 women spinners, 26 girls, 16 boys, one supervisor"

PY Aa 240 (S240–H1/Archives room)
a-ra-ka-te-ja MUL 21 ko-wa 25 ko-wo 4 TA 1[
"21 women spinners, 25 girls, 4 boys, one supervisor"

PY Ad 677 (S290–H23/Archives room)
pu-ro a-ra-ka-te-ja-o ko-wo VIR 30 ko-wo 9
"At Pylos, the spinners' sons: 30 men, 9 boys"

70 Michailidou 1992–1993 [1995], 18.

71 The four TA (AB 59) are tentatively interpreted as oxhides, but a relationship with *TA* "supervisor" of textile workers in the Linear B texts cannot be excluded (Michailidou 1992–1993 [1995], 18).

72 L. Godart (cited by Michailidou 1992–1993 [1995], 12–13) reads the sign as **1̣7̣1̣*.

Fig. 17.14 . Drawing of KN Ak(1) 5009. Drawing by L. Godart

KN Ak(1) 5009 + 6037 + 8588 (103/F14[73])
.A] ka-pa-ra$_2$ [
.B] ko-wa , ṃẹ[
.C ạ-ra-ka-te-ja , / ḳọ[
.A Possibly ḌẠ[.

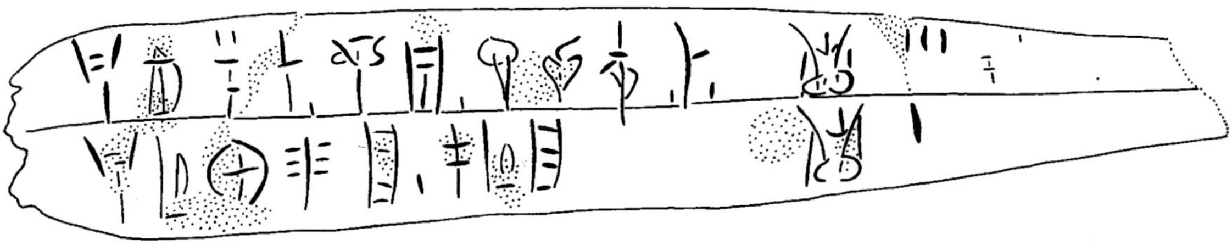

Fig. 17.15. Drawing of TH Of 34. Drawing by L. Godart

TH Of 34 (303/Epam. str.)
.1 a-pi-qo-ro , ne-wa , ko-tu-ro$_2$, *DA* , LANA 3 [[*PA* 1]] [
.2 a-ra-ka-te-ja , pa-ra-ja LANA 1 [

“Servants, young, the DA supervisor Κότυλος, 3 units of wool
Spinners, old, 1 unit of wool”

The word ἠλακάτη designates the distaff, not the spindle, according to the *Greek-English Lexicon* by Liddell-Scott-Jones.[74] In the 1st millennium, the spindle is termed ἄτρακτος. The spindle is the turning rod, whereas the distaff is the stick onto which the prepared wool fibres are fixed. For efficient and conventional spinning, a spindle is necessary, but a distaff is not.

Elizabeth Barber has reviewed the attestations of the term ἠλακάτη in 1st millennium Greek, and she demonstrates how the term has systematically been mistranslated as distaff.[75] Plato[76]

[73] Firth & Melena forthcoming.
[74] *DELG*, *s.v.* translates it as “quenouille” but adds : “p.-ê. aussi fuseau”.
[75] Barber 1991, 263–264.
[76] *Republic* 616c-617b.

describes a spindle as a shaft (ἠλακάτη) and a whorl (σφόνδυλος),[77] and it is reasonably clear that ἠλακάτη is not a distaff but the spinning device, the spindle. An older piece of evidence, but less strong, is the Odyssey,[78] in which a servant brings Helen her equipment: a silver wool basket and a golden ἠλακάτη. If ἠλακάτη is a distaff, then there is not much for Helen to do;[79] but if ἠλακάτη is instead a spindle, then she can immediately start to spin.[80]

The loss of this hand spinning technique in the medieval period has probably led to the present general confusion between spindles and distaffs, not only in dictionaries, but also in works on terminology and in archaeological classifications.[81] The mistranslation is especially prevalent in Anglo-Saxon scholarly literature but less in German publications, as demonstrated by W. P. Lehmann.[82]

Whatever the confusion between spindle and distaff, the translation of *a-ra-ka-te-ja* remains "female spinners". It is significant to notice that the *a-ra-ka-te-ja* women recorded in the Linear B documentation could never have spun all the wool produced in the villages. The average amount of wool available to be spun was about 500 kg per village and according to the calculations made by Andersson Strand, this would have produced approximately 1500 km of thread and required about 30,000 hours of spinning.[83] This suggests that the *a-ra-ka-te-ja* specialised in some types of spinning, for example producing warp yarn, or yarn of very fine quality. However, it must be emphasised that the *a-ra-ka-te-ja* women recorded on Lc(1) 531 by scribe 103 produce *pa-we-a* cloth, not thread.[84]

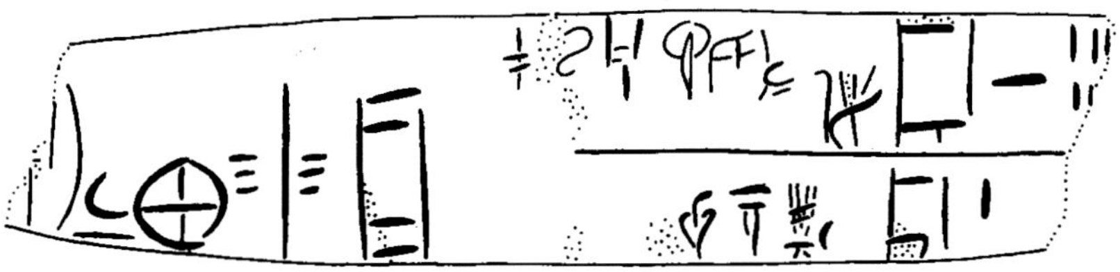

Fig. 17.16. drawing of KN Lc(1) 531 + 542. Drawing by L. Godart

```
KN Lc(1)     531 + 542                                        (103/F10)
      .A     ]               'pa-we-a ko-u-ra'    *161 TELA¹  15[
      .B     ]ạ-ra-ka-te-ja /           tu-na-no    TELA¹  1  [
```

"Spinners, 15 pieces of *pa-we-a ko-u-ra*, 1 piece of *tu-na-no*"

[77] No term is preserved for the spindle whorl in the Linear B records. Barber (1991, 263–264) notes that it is "curious that there is no common Indo-European word for the spindle whorl and that instead Greek and Latin (which uses *fusus*) seem to have picked up a loan word for this little utilitarian object."

[78] 4.125–135.

[79] *Pace* Due 1965, 8, note 22.

[80] Artemis' epithet χρυσηλάκατος is translated and understood as "with the golden spindle" by some scholars, "with the golden dart" by others. See discussion in Due 1965, 1–10.

[81] See also Gleba 2008.

[82] Lehmann 1995, 54.

[83] Andersson & Nosch 2003. For further data, see Andersson Strand, this volume.

[84] Nosch forthcoming B.

3.2. *Tools for wool preparation: combs*

Before spinning, the fibres need combing. The verb used in the 1st millennium is πέκω, "comb".[85] This tool whose 1st millennium Greek name is κτείς[86] is built on the same root (<*pkt-en-) and related to Latin *pecten*. The verb is again attested indirectly, through the formation of the occupational designation *pe-ki-ti-ra$_2$*, "female combers".

```
PY Ab 578                                          (Archives room/S186–H21)
.A GRA 2 T 4 TA
.B pu-ro , pe-ki-ti-ra2 MUL 7 ko-wa 4 ko-wo 4 NI 2 T 4
```

"At Pylos, 7 female combers, 4 girls, 4 boys, 1 TA supervisor, 230 litres of grain, 230 litres of figs"

Combing ensures that the fibres are free from foreign material and parallel. In several works, the term is translated as "carders"[87] but carding is an Iron Age technology in which the fibres are completely mixed and blended.

A second example of a term based on the same root for combing is the substantive or adjective *pe-ko-to/pektos* employed in connection with very heavy *te-pa* textiles (measured in terms of the 30 kilos of raw wool for one *te-pa pe-ko-to*). In this context, it may also be found in its abbreviated form *pe*. More technical knowledge and more documents are necessary to understand the meaning of the term in this context.

An example of *pe-ko-to* textiles is shown on tablet KN Lc(1) 526. It is written by the main textile scribe 103 at Knossos and comes from the same find-place as all other records of central Cretan production targets (F10 and F11). It registers a production target for the place *da-wo* located in southern Crete. The women at *da-wo* are commissioned to produce 10 *pe-ko-to* textiles, 14 *te-pa*, and 3 *tu-na-no*. After listing the three textile types, the scribe 103 calculated the necessary amount of wool for the production. The tablet is unfortunately broken, but John Killen has highlighted the ratios between wool and Mycenaean textiles.[88] Thus the production target can be calculated as follows:

10 *pe-ko-to* textiles = 100 units of wool = 300 kilos
14 *te-pa* textiles = 98 units = 294 kilos
3 *tu-na-no* = 9 units of wool = 27 kilos

The tablet records a total of 621 kilos of wool (= 207 units of wool) and this corresponds to wool from 828 sheep.[89] The production of the 10 *pe-ko-to* textiles alone consumes approx. half of the allocated wool. The *pe-ko-to* textiles are thus by far the heaviest textiles (30 kilos of wool per piece). How and whether this fact is technically related to combing is still an open issue.

---

85 *DELG*, s.v.
86 *DELG*, s.v.
87 *Docs*^{2}, 158, 570.
88 Killen 1966; Killen 1974.
1 *pa-wo* = 1.67 units of wool = 5 kilos of raw wool
1 *tu-na-no* = 3 units of wool =9 kilos of raw wool
1 *te-pa* = 7 units of wool = 21 kilos of raw wool
1 *te-pa pe-ko-to* = 10 units of wool = 30 kilos of raw wool
89 The calculation is based on wool yields from castrated wethers. See Killen 1964.

From the same root of πέκω "to comb" is derived the Mycenaean nominal form *po-ka* "fleece".[90] The meaning of πέκω in Mycenaean Greek therefore seems to cover both the treatment of wool and also a treatment of textiles.[91]

3.3. *Tools for weaving*

Textiles are made on a loom. There are several types of looms in prehistory, such as the horizontal loom, the back-strap loom, the two-beam-loom or the warp-weighted vertical loom. There is indirect evidence for the terminology for the vertical loom via the occupational designations for both male and female weavers:

- *i-te-ja-o*, feminine genitive plural form, "of the female weavers" (PY Ad 684)
- *i-te-we*, masculine dative singular (*histewei*) "for the male weaver", or nominative-accusative plural form (*histewes*) "male weavers" (PY Un 1322).

```
PY Ad 684                                        (S290–H23/Archives room)
(lat.sup.) a-pu-ne-we e-re-ta-o ko-wo
pu-ro ti-nwa-ti-ja-o i-te-ja-o ko-wo VIR 5 ko-wo 2
```

> "At Pylos, the women of Tinwato (toponym), the weavers' sons, 5 men, 2 boys,
> The Apunewe (toponym) rowers' sons."

Although the term for loom is not attested in Linear B, it seems plausible that the professional designation is built on the term for loom, and we can thus infer the existence of **i-to- /histos* "loom".[92]

A loom is a rather simple device with which a very complex technology is carried out and complex items result.[93] The loom in classical Greek is *histos,* from the verb *histamai.*[94] The root is related to the Indo-European root **sta-* for "standing", "upright". It indicates the upright concept of the vertical loom where both the weaver and the loom are placed in a vertical position. This idea of a standing device can also be found in the modern European languages, for example Ger. Webstuhl.[95] This contrasts to the horizontal loom in use in e.g. Mesopotamia.[96]

Most parts of the warp-weighted loom are not preserved archaeologically. The exception is the loom weights: the warp-weighted loom is attested through the thousands of loom weights found in Europe and in Anatolia. In 1st millennium Greek they are called ἀγνῦθες, a term without Indo-European etymology, or simply termed "stones", λᾶες.[97]

---

[90] Killen 1962.

[91] *DELG, s.v.*: "Cette famille de mots est surtout employée pour les brebis et pour la laine, d'une part au sens de 'tondre', de l'autre au sens de 'carder'. D'où l'emploi pour 'peigner' ".

[92] See also woman's personal name *i-ta-ja* in the formula *o-pi i-ta-ja* on tablet KN Xe 537, which is written by scribe 103?? Her name is possibly built on the same root.

[93] Harlizius-Klück 2004, 102: "Der Hauptanteil der Organisation liegt in der Hand und im Kopf der Weberin und ist nicht im Gerät externalisiert. Planung und Innovation beziehen sich eher aufs Gewebe und seine Bedeutungen, weniger auf die technischen Mittel. Der Webstuhl selbst besteht aus völlig unspezifischen Stangen, Stäben und Pfosten, die ihre Funktionalität und ihre Namen erst innerhalb des Webvorgangs erhalten."

[94] *DELG, s.v.* and supplement.

[95] Harlizius-Klück 2004.

[96] Breniquet 2008, and this volume.

[97] The term "stones" is attested in Linear B *ra-e-ja/ *lahaeia* PY Ta 642, 713.

Furthermore, as already seen by Barber, "the terms for the mechanisation of the weaving process are also without Indo-European etymologies".[98] This implies that terms for the devices, which are employed in operating the warp-weighted loom are of non-Greek origin: the term for the heddles (*mitos*) is of Semitic origin, and the words for the heddle bar (*kanon*) and shed bar (*kairos*) are also of non-Indo-European origin. These are integrated parts of the Bronze Age weaving technology but the words are only attested in 1st millennium Greek.

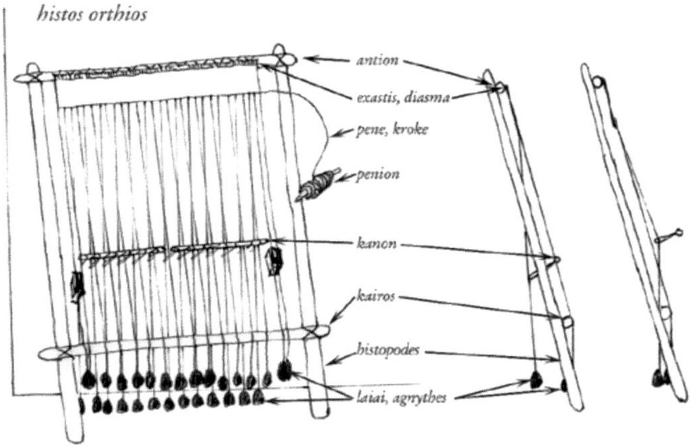

Fig. 17.17. The elements of the warp-weighted loom. From Harlizius-Klück 2004.

3.4. *Tools for sewing*

The term for needle is not preserved but needles are preserved archaeologically, and in the occupational designations in the feminine plural form *ra-pi-ti-ra$_2$/rhaptriai* "seamstresses", and in the masculine singular *ra-pte*/ῥαπτήρ "tailor". The perfect participle form *e-ra-pe-me-na, errhap(h)mena*, "sewn", and the adjectival form *ra-pte-ri-ja/rhapteriai* (PY Ub 1315), suggest that the Mycenaean needle was termed **ra-pi* /ῥαφίς.

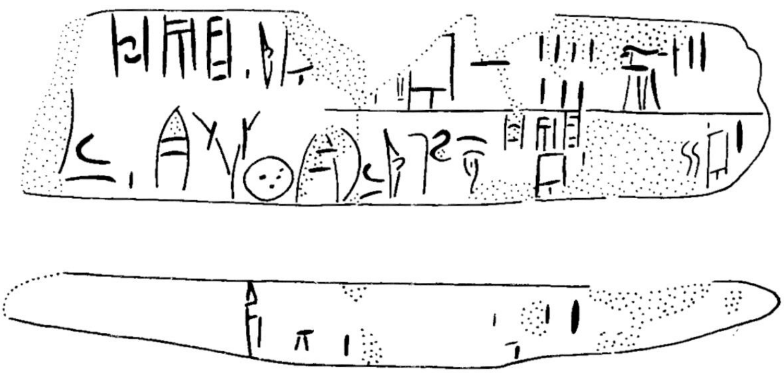

Fig. 17.18. Drawing of KN L(2) 647 + lower edge. Drawing by L. Godart

KN L (2) 647 + 2012 +5943+ 5974 (103/F14)
.A] 'nu-wa-ja , *pe* TELA^1['] TELA^1 17 TUN+*KI* 3
.B]ra , / e-ni-qe e-ra-pe-me-na 'nu-wa-ja' TELA^1 [ ]-ra$_2$ TELA^1
lat. inf.] [[ẹ-ṛị[ ] ṬẸḶẠ 1]]

[98] Barber 1991, 281.

PY Ub 1315 (H31/Room 99)
.1]-wo-ja a-ni-ja , te-u-ke-pi , 5 di-pte-ra_3 e-ru-ta-ra 1 ⌞ 6[
.2 ro-u-si-je-wi-ja 6 ra-pte-ri-ja a-ni-ja 3

> "5 sets of reins (fitted) with equipment; 16 red hides
> 6 (sets of reins) of Lousos type; 3 sets of reins with saddlers' work"[99]

However, tablets from Pylos (La 626, 630) record *pa-we-a* cloth, which is qualified as *a-ro-ta*, which could be ἄλωστα, meaning "not-sewn", and thus employing another root for the sewing technique. From the contextual analysis of the use of ῥάπτω in the Linear B inscriptions,[100] we learn that the same term for stitching or sewing is both used for textiles and for leather.[101] Barber notices that 1st millennium Greek preserves two terms for sewing: *kassuein*, a cognate of English "sew", and *raptein* which has no known etymology[102] but is the main form used in Mycenaean Greek.

3.5. *Weaving terminology*

A rather rich semantic field exists for the process of weaving. Mycenaean Greek, like other Indo-European languages, distinguishes between weaving (on a loom) *hyphainein*, and plaiting, *plekein*. To this comes the root yielding the terms for loom **histos*, and weaver **histeus.*

1. hist-. From the root **sta-*, "to stand", derives the terms *i-te-we*, "male weaver(s)" > **i-te-u* /*histeus*, here attested either in the dative singular *histewei*, or in the nominative-accusative plural *histewes* (PY Un 1322), and *i-te-ja-o/histeiaon*, genitive plural feminine form "of the (female) weavers" (PY Ad 684).

2. hyph-. Another Greek term for weaving is *hyphansis* "weaving" and the verb ὑφαίνω 'weave', and the weaver *hyphantès*. The root **webh-* is in use in other Indo-European languages as Engl. *weaving*, Toch. A. *wäp-*, German *weben.*[103] The Greek term *hyphainô* is a zero-vocalism verb (**ubh-*) with a specific suffix, which according to Pierre Chantraine[104] does not seem archaic.
In Linear B, this root is perhaps attested in the designation of a Goddess of Weaving (?), *u-po-jo po-ti-ni-ja* at Pylos (Fr 1225).

---

[99] Translation from *Docs*², 520.
[100] On the etymology, see Chadwick & Baumbach 1963, 241–242.
[101] According to Ventris & Chadwick (*Docs*², 578), the use of the same term for both leatherwork and textile-work suggests "that the masculine trade at least was concerned with sewing leather rather than cloth". This conclusion, however, is not necessarily true. It rather seems that both the *ra-pte* and the *ra-pi-ti-ra_2* are concerned with sewing, very much like the pair of *i-te-we* and *i-te-ja-o*, but that the term is also used to describe the technique of sewing in leather, and here there is no Linear B evidence for the gender of the person who stitched the leather.
[102] *DELG*, s.v.
[103] *DELG*, s.v.
[104] *DELG*, s.v.

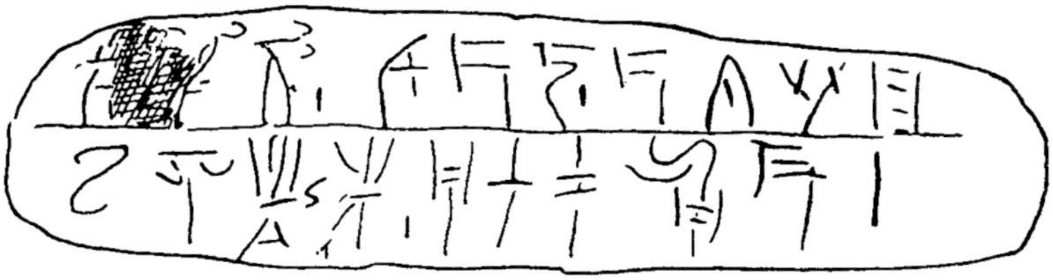

Fig. 17.19. Drawing of PY Fr 1225. Drawing by E. Bennett

```
PY Fr 1225                                              (Room 23/S1217-Cii)
.1     e-ra3-wo , u-po-jo , po-ti-ni-ja
.2     we-a2-no-i , a-ro-pa OLE+A S 1
```

"Oil for the u-po-jo Potnia
Ointment, for the garments, 9.6 litres of oil"

The designation is composed of the word Potnia, the Mistress, and *u-po-jo*. Some scholars interpret the goddess' designation as *Hypoíôn* < *hypo*, being the Goddess from below, chthonian, under ground.[105] Other scholars, however, interpret the designation as the goddess of weaving < **hyphoio* < ὑφαίνω.[106] Since the context on Fr 1225 concerns textiles, we would tend to prefer the interpretation as goddess of weaving.

3. *plek-*. Another term for weaving is related to the root of πλέκω found in the occupational designations *pe-re-ke-u* and *pe-re-ke-we/plekeus*. It is significant that this occupational designation occurs both at Mycenae (a record of wool MY Oe 130) and Pylos (a record of animals PY Cn 1287 and another record of men, PY Ae 574,) and has recently also been found at Thebes (TH Oh 208).[107] It is a designation for a male occupation related primarily to wool working.

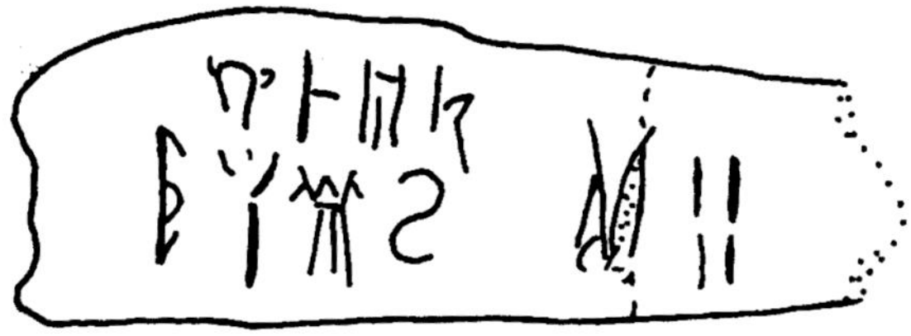

Fig. 17.20. Drawing of MY Oe 130. Drawing by L. Godart

```
MY Oe 130                                               (56/H. Oil Merch.)
          qa-da-wa-so
       pe-re-ke-we                    LANA 4

PY Ae 574                                               (Cii/Archives room)
       pe-re-ke-we VIR 13
```

[105] Van Leuven 1979.

[106] Boëlle 2004, 46 note 47. According to Boëlle, *hyphos* would be an unattested inverted thematic derivative from ὑφαίνω which is not attested. *Hyphos* is attested in Greek but it is a neutral sigmatic form (-es/-os) and its Mycenaean genitive form would thus be written **u-pe-o/huph-eh-os*.

[107] Nosch forthcoming A.

```
TH Oh 208                                   (-)
.1      ]1     pe-re-ke-we  ,  pe-re[
.2      ]-si    M 4                     [
```

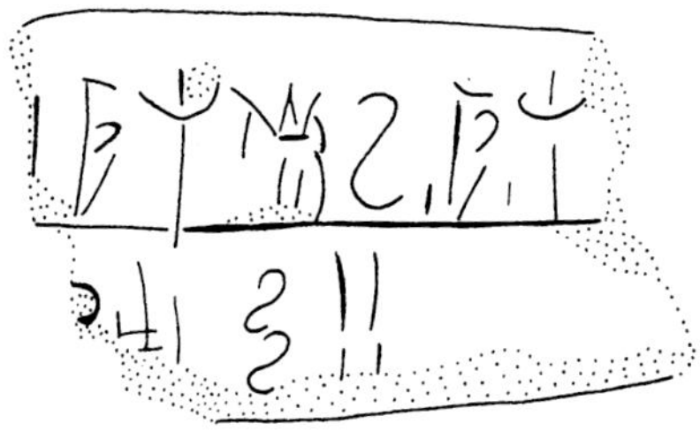

Fig. 17.21. Drawing of TH Oh 208. Drawing by L. Godart

On a tablet from Knossos, another term may be the verbal form of *plek-*, *pe-re-ke* (KN L 520).[108]

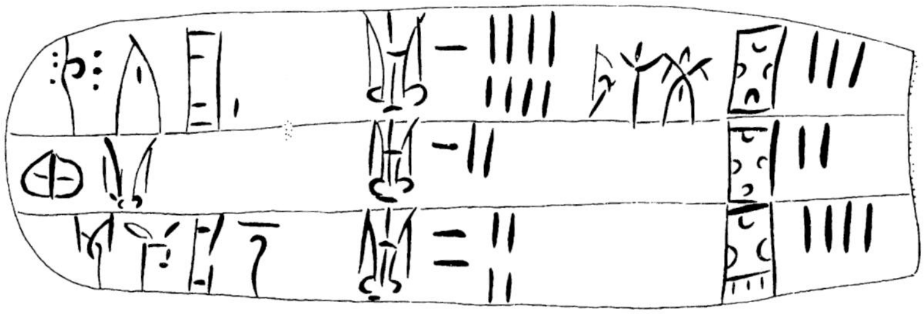

Fig. 17.22. Drawing of KN L 520. Drawing by L. Godart

```
KN  L  520                                                      (-/F8)
.1        do-ti-ja ,      LANA 18   pe-re-ke     *164 3
.2       ka-ma            LANA 12                *164 2
.3        sa-mu-ta-jo     LANA 24                *164 4
                     Cut at right.
```

The verb πλέκω[109] signifies braiding and is primarily used for basketry as well as for the braiding of hair and crowns, techniques which from a technological point of view are closely related to weaving. The term seems closely associated with wool and weaving in Mycenae and Thebes, and *pe-re-ke* is employed in connection with wool and textiles at Knossos. Thus, this term is strongly integrated within the textile manufacture and textile vocabulary at several palace sites, and seems to have a wide range of applications.

4. It is possible to see an e-vocalism in the development of **web-* in the form of the future participle *e-we-pe-so-me-na*, "to be woven", as was suggested by Chantraine[110] and Perpillou.[111]

[108] According to *Docs*², 488, *pe-re-ke* is rather not a verb but a man's name.

[109] *DELG*, *s.v.* Chantraine does not mention any Mycenaean evidence in his analysis of πλέκω. Today, in modern Greek, the verb has the sense of knitting. It should be noticed that in Althochdeutch the verb *flehtan* (flechten, "braid") is related to *flahs* (Germ. Flachs, Eng. flax). Barber 2001.

[110] *DELG*, *s.v.* 1123: "Enfin, on pourrait retrouver le vocalisme *e* en grec même, si l'on interprète le mycén. *ewepesesomena*, épithète de tissus, comme représentant un participe futur passif d'un verbe *ϝέφω, avec prothèse *ἐϝέφω, cf. Beekes, *Laryngeals* 67." Beekes (1969, 67) conjectures a verb *ἐϝέφω corresponding with prothetic *e*- to Old Hoch German weban, Eng. weave, later replaced by ὑφαίνω.

[111] Perpillou 1996.

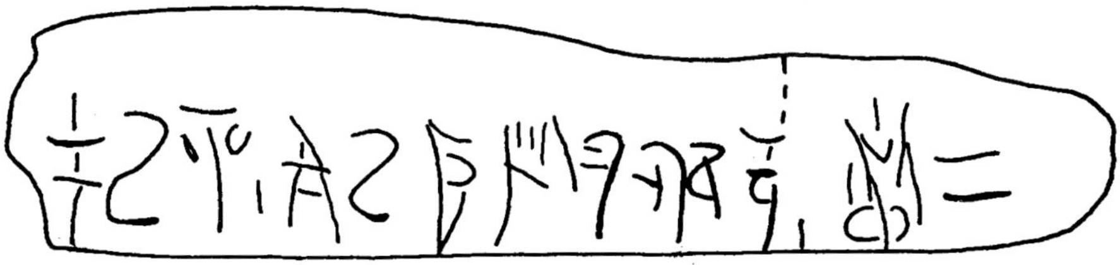

Fig. 17.23. Drawing of MY Oe 127. Drawing by L. Godart

MY Oe 127 (55/H. Oil Merch.)

pa-we-a_2 , e-we-pe-se-so-me-na , LANA 20

"*Paweha* cloth, to be woven, 20 units of wool (= 60 kilos)"

20 units of wool equals 60 kilos, and this is the necessary – and precise - quantity for weaving 12 pieces of *pa-we-a* cloth. If this etymology is followed, Oe 127 becomes equivalent to the Lc textile targets at Knossos in which the scribes allocate large quantities of wool to groups of workers for textile production.[112] However, the word *e-we-pe-se-so-me-na* is conventionally interpreted as future participle passive, "to be well boiled" < *hepsô*, "boil" (see below, § 4.2.1).

There are thus several terms available to express weaving in Mycenaean Greek.[113] The weavers and probably the loom take their name after the Indo-European root **sta-*, but the term for plaiting (*plekein)* is also well integrated in the textile vocabulary and does not reflect any clear technical differences from weaving.

4. Reflections on the use of pairs and groups of words in Linear B textile terminology

In the vocabulary linked with textile fibres and textile production in the Linear B corpus, some words seem to express opposite meanings (antonyms), whereas others apparently have very close, if not identical significations (synonyms). It is assumed here that by gathering and studying the pairs or groups of words concerned, it might be possible to obtain a better idea of technical differences, or technological details, which were particularly important to the scribes, and in general to the palatial administration, since they were carefully recorded.

4.1. *Pairs of antonyms expressing important technical differences?*

In the list of terms related to fibres, thread and textiles, some pairs of words functioning like antonyms can be identified; one hypothesis is that they correspond to the indication of (technical?) differences, which were particularly important to the scribes. One document, Knossos Ln 1568, seems to be especially revealing from that point of view, since it records repeated mentions of

[112] On the textile targets Lc(1) and Lc(2), see Killen 1966; Killen 1974.

[113] Finally, José Melena has suggested another term for weaving: he relates *ke-do-si-ja* with the Greek *gerdiós*, "weaver" (Melena 1975, 78–79). The term *ke-do-si-ja* is recorded after a list of male names on KN B 799, 804 and 8206. Gerdiós is only attested quite late, and according to Chantraine, Latin *gerdius* and Hebrew *girda'a* are loan words from Greek *gerdiós*. Chantraine, however, states that "de toute facon le mot grec risque fort d'être emprunté." (*DELG, s.v.*). Hesychius equals *gerdíos* with *hyphantès*. Thus, according to Melena, *ke-do-si-ja* is *gerdonsia*, with the meaning "textile workshop" or "loom team" (Melena 1975, 79). This interpretation, however, seems less plausible.

two words and their abbreviations. Other examples can be found, where the relevant words are attested not only in more than one document, but also in more than one corpus of texts.

4.1.1. *mi-ja-ro/pe-ko-to: [made of] "rough" [wool] versus "combed"*

The adjective *mi-ja-ro*[114] appears on KN Ln 1568,[115] once completely written, on line 1a; and in the form of the abreviation *mi*, written just before the logogram TELA+*TE* (on lines 1b, 2b, and 6):

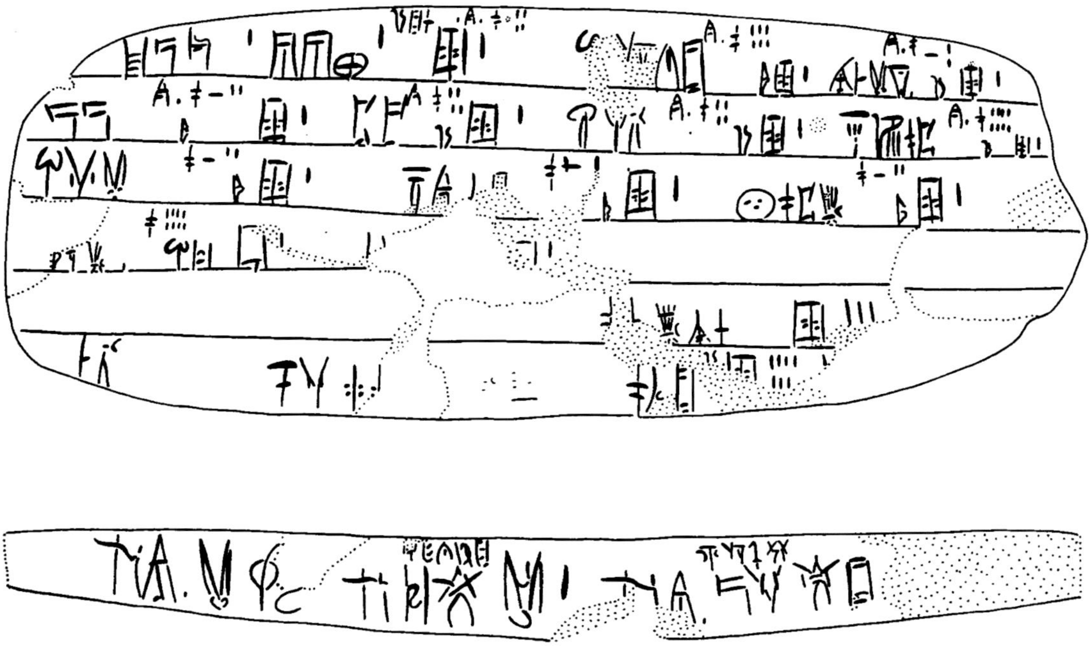

Fig. 17.24. Drawings of KN Ln 1568. Drawing by L. Godart

KN Ln 1568 (103/F14)

.1a mi-ja-ro , *e* , *pa* 4 *e* , *pa* 6 *e* , *pa* 12

.1b *56-po-so 1 wa-wa-ka 1 TELA^1+*TE* 1 ru-ki-ti-ja *pe* TELA+*TE* 1 wi-da-ma-ta$_2$, *mi* TELA^1+*TE* 1

.2a *e* , *pa* 12 *e* *pa* 4 *e* , *pa* 4 *e* , *pa* 8

.2b po-po *pe* TELA^1+*TE* 1 ta-su *mi* TELA+*TE* 1 ko-re-wo *mi* TELA+*TE* 1 di-*65-pa-ta *mi* TELA+*TE* 1

.3a *pa* 12 *pa* 1ḷ *pa* 12

.3b ru-sa-ma *pe* TELA^1+*TE* 1 na-e-ṛạ-jạ *pe* TELA+*TE* 1 qe-pa-ta-no *pe* TELA^1+*TE* 1

.4a *pa* 8

.4b]tu-na-no , ru-nu TELA^1 1 [ ]ṬẸḶẠ^x 1

.5 pạ-ṛọ no-si-ro TELA^x+*TE* 3

.6 da-wo to-sa te-[ ] ⟦*ṃị* ṬẸḶẠ+*ṬẸ*⟧ pa-ra-ja '*ṃị*' TELA^x+*TE* 7 [

lat. inf. .a a-ze-ti-ṛị-ja ne-ki-ri-de [

.b o-pi , ma-tu-wẹ o-nu-ke LANA 1 o-pi , po-ni-ke-ja [

114 On this adjective, see mainly the study by Killen 1974, esp. note 10, with bibliography, alternative interpretations proposed by other scholars, as well as a detailed analysis of the opposition between *pe* and *mi* on Ln 1568, where the author suggests that the sense of *mi-ja-ro* might be rather "rough or the like" and wonders if it did not "serve to designate cloth made from uncombed wool, as opposed to cloth made from combed (*pe-ko-to*: cf. πέκω??) wool (or, alternatively, 'rough' cloth as opposed to 'sheared' cloth [velvet *vel sim.*]) "

115 On this document see also Varias Garcia 1992.

The same document contains four instances of the abbreviation *pe* just in front of the same logogram TELA+*TE* (on lines 1b, 3b); *pe* could theoretically be the abbreviation of *pe-ru-si-nu-wo*, "from last year", or of *pe-ko-to*, "combed".

On the same document, the scribe wrote both *pa-ra-ja* "old" (line 6) and the abbreviation *pa* (lines 1a, 2a, 3a, 4a), which is an argument against the interpretation of *pe* as *pe-ru-si-nu-wo*, since, as demonstrated by Killen,[116] normally a scribe either writes *pe-ru-si-nwa/pe* or *pa-ra-ja/pa*.

The term which is believed to be the opposite of *mi-ja-ro*, namely *pe-ko-to*, appears in eight tablets (KN Lc[1] 526, 527, 535, 536, Lc 646, L 698, 5090, 8105), and six of them[117] are records of production targets and of cloth by the same scribe, 103, as Ln 1568.

On these tablets, the precision *pe-ko-to* is also systematically applied to the kind of textiles described by TELA+*TE*, just as *mi* and *pe* here, which reinforces the hypothesis of a pair of words functioning as antonyms (not merely for the meaning, but also in the use of the designations). Thus, it can be concluded that *pe-ko-to* is almost certainly the word behind the abbreviation *pe* on Ln 1568.

4.1.2. *]a-ro-ta / e-ra-pe-me-na "not sewn" versus "sewn"*

These two designations, attested at two different sites, may form a pair of antonyms; *a-ro-ta* (alph. Gr. ἄλωστα) is written on PY La 626 and La 630, whereas the second appears on a Knossos tablet (KN L[2] 647). These terms are both quite rare; *a-ro-ta*, at Pylos, is applied to the textile item designated by TELA+*PA*, where *PA* is an abbreviation of *pa-we-a*.

```
PY La 626                                   (S626–H13/Room 6, SW sector)
        supra mutila
.a            ]*1̣6̣6̣ 2 a-*3̣5̣-ṭọ 3̣[
.b            ]a-ro-ta TELA+PA 1   [
verso
              ]ku x 1 o 1          [

PY La 630                                   (S626–H13/Room 6, SE sector)
.a            ko-]u-ra
              ]ạ-ro-ta TELA+PA 1[
verso
              ]do P 7 PE 4̣[
```

[116] Killen 1972, 425–440.

[117] KN Lc(1) 526, 527, 535, 536, Lc 646; L 698.

Fig. 17.25. Drawing of KN L (2) 647. Drawing by L. Godart

KN L (2) 647 + 2012 +5943+ 5974 (103/F14)
.A] 'nu-wa-ja , *pe* TELA^1['] TELA^1 17 TUN+*KI* 3
.B]ra , / e-ni-qe e-ra-pe-me-na 'nu-wa-ja' TELA^1 [ ]-ra$_2$ TELA^1

lat. inf.] [[ẹ-ṛị[ ] ṬẸḶẠ 1]]

The references being exceptional, were the *pa-we-a* cloths normally sewn, but in some cases not? E. Luján,[118] referring to S. Marinatos,[119] has argued that *pa-we-a* were generally *not* sewn, which would confirm the interpretation of *a-ro-ta* as "not sewn". Yet, as a matter of fact, a number of different suggestions can be made: *pa-we-a* may have been given or stored without sewing at all, or before sewing; or it may in some cases have been sewn and then restored in its primary state.

The word *e-ra-pe-me-na*, at Knossos, is applied to plain TELA^1.

If the two words are indeed antonyms, it might be interesting to note that they are built on different roots, and that sewing thus must have been expressed by two terms, one which has yielded the occupational designations *ra-pte* and *ra-pi-ti-ra*$_2$, and another (less productive?) root for sewing (cf. Hesychius ἄλωστοι· ἄρραφοι ?).[120]

4.1.3. *u-po-we/e-pi-ki-to-ni-ja?*

The word *u-po-we* is attested once on KN L 178; *e-pi-ki-to-ni-ja* also appears at Knossos (KN L 693 and probably also on L 7514).

The mention *u-po-we /hupowes/* is interpreted as something "worn under", whereas *e-pi-ki-to-ni-ja /epik^{h}itōnia/* is probably a piece of clothing worn "over the χιτών"; a parallel may be suggested with *e-pi-ro-pa-ja /epilōpaia/* attested on KN Od 696, which might designate something worn "over the λώπη".

118 Luján 1996–1997, 336.
119 Marinatos 1967, A6.
120 Cf. *DMic s.v. a-ro-ta*. For the origin of the root, cf. *DELG s.v.* λῶμα.

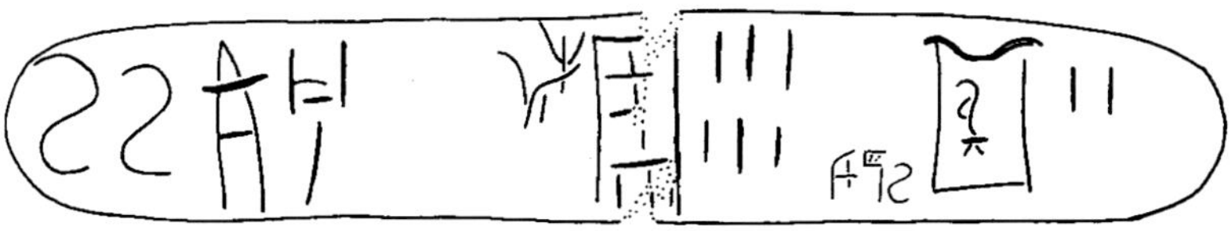

Fig. 17.26. Drawing of KN L 178. Drawing by L. Godart

KN L 178 + 281 ("124"/*RCT*)
we-we-e-a '**161*' TELA^{3}+*PA* 6 / u-po-we TUN+*RI* 2

On L 178, the scribe has recorded first a woollen cloth and then a linen one. The fibre used is coherent with the idea of something worn under (other clothes). The word *e-pi-ki-to-ni-ja* appears on the following Knossos tablet:

Fig. 17.27. Drawing of KN L 693. Drawing by L. Godart

KN L 693 (103/F19)
.1 ri-no , / re-po-to , 'qe-te-o' ki-to , AES M 1 [
.2 sa-pa P 2 Q 1 e-pi-ki-to-ni-ja AES M 1 [

".1 fine linen, to be paid, χιτών, 1 kg of BRONZE [
.2 *sa-pa* (ca.) 68 g *e-pi-ki-to-ni-ja* 1 kg of BRONZE["

This document, which mentions quantities of bronze after textiles, has been interpreted by specialists as a possible indication of the value or "price" of these textiles.[121]

4.2. *True or false synonyms?*

It has been suggested that some words/designations could be defined as antonyms. On the other hand, the scribes have used words the etymology of which points towards a very close meaning. The question is whether they were simply synonyms, or if they were used to indicate something different, and, if so, how different?

[121] Cf. Killen 1988, esp. 183: "The tablet may simply be a record of bronze which is due to be affixed to linen garments (...). On the other hand, it is perhaps a point in favour of the view that the tablet records the 'price' of linen expressed in terms of BRONZE that we find the phrase *ri-no re-po-to* written large as the heading of the tablet".

4.2.1. *ki-ri-ta / ko-ro-to*

E. Luján has argued[122] that the word *ki-ri-ta* /*k*^{h}*rista* /(cf. χρίω, "to rub", "to anoint") attested on KN Ld 785.1, may have designated a technical process in which the colour was applied onto the cloth, as opposed to *ko-ro-to* /*khrōston*/ (cf. χρώζω, "to dye") (KN Od 485, 486, 487, MY Oe 106), which would have implied the immersion of wool/cloth into a dye bath. However, he also indicates that it cannot be entirely excluded that the difference might be explained by a personal preference of the scribe, thus not necessarily corresponding to a technical difference.

It might be interesting to add to this pair of designations the participle form *e-we-pe-se-so-me-na*, which has been interpreted, *inter alia*, as "to be well boiled", since the meaning of the word could be related to dyeing techniques (by heating in a colour bath). It must be stressed that many other interpretations of this form have been suggested, and that P. Chantraine mentions the form under ἕπω "to take care of",[123] ἕψω, "to boil, to cook"[124] and ὑφαίνω "to weave".[125] It means that the possible interpretations range from "requiring attention, finishing"[126] to "to be well boiled"[127] and to "to be woven". In any case, the future form indicates a process *to be applied* to the wool (or to the *pa-we-a*$_2$, grammatically both solutions seem possible) recorded in the Mycenae tablet:

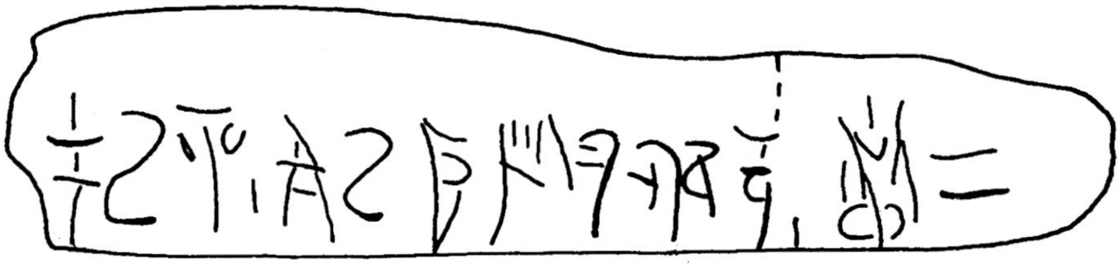

Fig. 17. 28. Drawing of MY Oe 127. Drawing by L. Godart

```
MY Oe 127                                    (55/House of Oil Merchant)
          pa-we-a2 , e-we-pe-se-so-me-na ,    LANA 20
```

The quantity is quite important (20 units = 60 kg of wool). With 60 kg of wool, it is possible to make 12 *pa-we-a* (3 *pa-we-a* = 5 units of wool).

4.2.2. *i-te-we/pe-re-ke-we*

Among textile related professional designations, two words seem to have described two types of activities related to weaving: *i-te-we*, /*histēwes*/ and *pe-re-ke-we* /*plekēwes*/.

On this subject, see above, § 3.5. *Weaving terminology.*

* * *

[122] Luján 1996–1997, 351.

[123] *DELG s.v.* ἕπω, Chantraine writes: "Doute également pour l'interprétation de *ewepesesomena* suggérée par Palmer, voir Chadwick-Baumbach 193".

[124] *DELG s.v.* ἕψω, Chantraine indicates: "Le mycénien a peut-être le futur passif *ewepesesomena* = εὖ ἑψησόμενα, dit d'étoffes, mais voir aussi sous ἕπω".

[125] *DELG s.v.* ὑφαίνω.

[126] See also Palmer 1963, 421, who explicitly rejects the interpretation "to be well boiled".

[127] *DELG s.v.* and *Documents*, 394.

One of the aims of this paper is to provide the reader with an overview, as complete as possible, of the variety of terms directly or indirectly related to textile fibres and fabrication in the Mycenaean Greek vocabulary, as attested by the extant documentation. The principle of this overview is the (attested) steps of the *chaîne opératoire* (§ 1. *From raw fibres to finished products*).

For the purpose of recording textile fibres and textile products, Mycenaean scribes used not only words, but also logographic notations; this practice has been inherited from the Linear A administration, as shown by the examples from Haghia Triada, Khania and Thera (§ 2.2 *Logograms in Linear A*). In particular, it is probable that Linear B logograms for LANA and TELA had their origin in Linear A, even if some doubts remain on the equation AB 80+26 = LANA. An interesting but sometimes problematic aspect of the notations used by the Linear B scribes is the joint use of endograms and adjuncts (syllabic signs abbreviating words, and placed inside or beside the logograms) on one hand, and of complete words written syllabically, on the other hand, to designate the same type of textile items. In the majority of cases, it has been possible to identify more or less securely the word referred to by the abbreviation, but some terms remain unclear. As this also is a practice inherited from Linear A, it has been attempted to study in parallel the same phenomenon in the Linear A documents (see § 2.2.4 *Cloth names in Linear A*), but the very limited number of preserved inscriptions, as well as the undeciphered state of the script prevents the drawing of any kind of definitive conclusion from this comparison. In any case, it seems reasonable to conclude that the Linear A signs AB 04, AB 09, AB 81 and A 312 (the first two readable as TE and SE) were abbreviations of Minoan cloth names, and that the ligatures AB 54+04 and AB 54+81 are comparable to the Linear B ligatures TELA+*TE* and TELA+*KU*.

When it comes to the use of words written syllabically, the working hypothesis has been to suggest that the vocabulary used by the Mycenaean scribes may reflect technical characteristics which were important and thus carefully recorded. The study of these words has shown that some of them had opposite meanings (e.g. *mi-ja-ro/pe-ko-to,]a-ro-ta / e-ra-pe-me-na, u-po-we / e-pi-ki-to-ni-ja*, see § 4.1. *Pairs of antonyms expressing important technical differences?*) whereas others seem to convey closely related concepts (e.g. *ki-ri-ta/ko-ro-to, i-te-we/pe-re-ke-we*, see § 4.2. *True or false synonyms?*); in the last case, the question is whether these words were indeed synonyms, or if the simple fact that the scribes chose different words implies that they signified different technical characteristics, and how different? This last question is far from being definitively answered. However it is suggested that an in-depth study of such lexical items might lead to a better understanding of some technical realities.

Within the long and continuous history of textile production, it is not surprising to find some similarities in the ways of recording wool and textiles from the Minoan to the Mycenaean administration.

E. Barber[128] notes as "interesting and highly significant the distribution in particular of the terms for looms and weaving" within the field of Greek textile terminology. She places these terms within the historical frame of the arrival of the Greek speaking populations in Greece, and the blend with other pre-Greek languages such as Minoan.

An important conclusion to be drawn is that Greek employs some terms for the action of weaving, such as the possible interpretation *u-po-jo*, which might be related to *hyphainein*, while they construct the term for the upright loom, *histos*, from yet another root. Finally, plaiting has

[128] Barber 1991, 279–280; 2001.

its own distinct terminology, although plaiting seems to have been integrated in the Late Bronze Age textile production and terminology. This suggests that plaiting, weaving and the loom could have come independently and perhaps at different periods into the Greek vocabulary.

It must be taken into consideration that the archaeology of Bronze Age Greece represents a particular situation regarding textile tools: spindle whorls are extremely abundant at Bronze Age sites in Mainland Greece and clay loom weights near to absent; in Crete, the archaeological excavations display a wealth of clay loom weights but only few spindle whorls are found. We know that thread was spun in abundance in the Bronze Age, and we know that the warp-weighted loom was in use. The use of the term λᾶας "stone" in the 1st millennium could thus be explained by a traditional use of stones instead of clay loom weights, in particular in the Bronze Age Mainland weaving tradition.

It is very important to integrate the research on terminology within the research of the development of textile technology. Plant fibres were processed some 4000 to 6000 years before animal fibres were introduced. This innovation must have enriched the vocabulary as well, because it required the appropriation of new techniques. Barber[129] points out that the textile terminology for wool fibre processing has partly been taken from non-Indo-European languages. Woollen textiles needed fulling (*ka-na-pe-u*), and woollen textile opened up for dye opportunities with a mordant (*tu-ru-pte-ri-ja*), two technical terms, which seem to have come from outside the Greek and Indo-European vocabulary.

Note

The authors of this chapter have divided the study of the various topics and categories of documents as follows below: section 2.2 *Logograms in Linear A* is the work of M. Del Freo; section 1 *From raw fibres to finished products*, as well as sections 2.1 *Logograms in Linear B* and 4 *Reflections on the use of words in textile terminology* are the work of Fr. Rougemont; section 3 *Terminology and technical characteristics* is the work of M.-L. Nosch.

Abbreviations

CHIC Olivier, J.-P. & L. Godart (eds.) (1996) *Corpus Hieroglyphicarum Inscriptionum Cretae*, Études crétoises 31, Paris.

CoMIK Chadwick, J. *et al.* (eds.) (1986–1998) *Corpus of Mycenaean Inscriptions from Knossos*, 4 vols. Rome & Cambridge.

DELG Chantraine, P. (2009) *Dictionnaire étymologique de la langue grecque: histoire des mots,* supplemented by A. Blanc, C. de Lamberterie, J.-L. Perpillou, Paris.

DMic. Aura Jorro, F. (1985–1993) *Diccionario micénico*, 2 vols. *Anejos I y II del Diccionario Griego-Español.* Madrid.

Docs² Ventris, M. & J. Chadwick (1973) *Documents in Mycenaean Greek*, 2nd ed., Cambridge.

GORILA Godart, L. & J.-P. Olivier (eds.) (1976–1985), *Recueil des inscriptions en linéaire A*, Études crétoises XXI-1/5, Paris.

KT⁵ Killen, J. T. & Olivier, J.-P. (eds.) (1989) *The Knossos Tablets. 5th Edition. Suplementos a Minos* 11. Salamanca.

[129] Barber 1991, 227.

Bibliography

Andersson, E., Mårtensson, L., Nosch, M.-L. B., and Rahmstorf, L. (2008) New Research on Bronze Age Textile Production, *Bulletin of the Institute of Classical Studies* 51, 171–174.

Andersson, E. & Nosch, M.-L. B. (2003) With a Little Help from my Friends: Investigating Mycenaean Textiles with the help from Scandinavian Experimental Archaeology. In Foster, K. P. & Laffineur, R. (eds.) *METRON. Measuring the Aegean Bronze Age. Proceedings of the 9th International Aegean Conference / 9e Rencontre égéenne internationale, Yale University, 18–21 April 2002*. 197–205 & table XLV. Aegaeum 24. Liège, Austin.

Barber, E. (1991) *Prehistoric Textiles. The Development of Cloth in the Neolithic and Bronze Ages with Special Reference to the Aegean*. Princeton.

Barber, E. J. W. (2001) The Clues in the Clothes. Some independent Evidence for the Movement of Families. In R. Drews (ed.) *Greater Anatolia and the Indo-Hittite Language Family. Papers presented at a Colloquium hosted by the University of Richmond, March 18–19, 2000*. Journal of Indo-European Studies, Monograph Series 38, 1–14.

Beekes, R. S. P. (1969) The Development of the Proto-Indo-European Laryngeals in Greek. Janua Linguarum, Series Practica 42. The Hague.

Bennett, E. L., Jr. (1950) Fractional Quantities in Minoan Bookkeeping. *AJA* 54, 204–222.

Boëlle, C. (2004) *PO-TI-NI-JA. L'élément féminin dans la religion mycénienne (d'après les archives en linéaire B)*. Études anciennes 26. Paris - Nancy.

Boulotis, C. (1998) Les nouveaux documents en linéaire A d'Akrotiri (Théra): remarques préliminaires. In F. Rougemont & J.-P. Olivier (eds.), *Recherches récentes en épigraphie créto-mycénienne*, 407–411. BCH 122. Paris.

Boulotis, C. (2008) Οι Πινακίδες Γραμμικής Α από το Ακρωθήρι (THE 7–12): Οψεις της οικονομικής Ζωής του Οικισμού. In *ΑΚΡΩΘΗΡΙ ΘΗΡΑΣ. Τριάντα χρόνια έρευνας (1967–1997). Επιστημονική συνάντηση, 19–20 Δεκεμβρίου 1997*, 67–94. Athens.

Breniquet, C. (2008) *Essai sur le tissage en Mésopotamie - Des premières communautés sédentaires au milieu du IIIe millénaire avant J.-C.* Travaux de la Maison René-Ginouvès 5. Paris.

Chadwick, J. & Baumbach, L. (1963) The Mycenaean Greek vocabulary, *Glotta* 41, 157–271.

Del Freo, M. (2001) Mycénien *pu-te-ri-ja* et le couple *ki-ti-me-na/ ke-ke-me-na*, *Revue de Philologie* LXXV-1, 27–44.

Del Freo, M. & Rougemont, F. (forthcoming) Observations sur la série Of de Thèbes. In *5th International Congress on Boeotian Studies, Thebes 10–13 September 2005*.

Due, O. S. (1965) The meaning of the Homeric formula χρυσηλάκατος κελαδεινή, *Classica et Medievalia* 26, 1–10.

Duhoux, Y. (1976) *Aspects du vocabulaire économique mycénien (cadastre - artisanat - fiscalité)*. Amsterdam.

Firth, R. (2007) Re-considering Alum on the linear B Tablets. In C. Gillis and M.-L. B. Nosch (eds.), *Ancient Textiles. Production, Craft and Society*, Ancient Textiles Series 1, Oxford, 130–138.

Firth, R. & Melena, J. L. (forthcoming) Findplaces of the Knossos tablets: The 5000 Series. In Th. G. Palaima & C. Shelmerdine (eds.), *11th International Mycenological Colloquium in Austin, Texas, 8–12 May 2000*.

Firth, R. & Nosch, M.-L. (2003) Scribe 103 and the Mycenaean Textile Industry at Knossos: the Lc(1) and Od(1) Sets, *Minos* 37–38, 121–142.

Frangipane, M., Andersson Strand, E., Laurito, R., Möller-Wiering, S., Nosch, M.-L., Rast-Eicher, A., Wisti Lassen, A. (2009) Arslantepe (Turkey): Textiles, Tools and Imprints of Fabrics from the 4th to the 2nd millennium BC, *Paléorient*. 35:1, 5–30.

Gleba, M. (2008) *Textile Production in Pre-Roman Italy*. Ancient Textiles Series 4. Oxford.

Godart, L. (1984) Du linéaire A au linéaire B. In C. G. Glotz (ed.) *Aux origines de l'Hellénisme. La Crète et la Grèce. Hommage à Henri van Effenterre*, 121–128. Paris.

Hallager, E. (1996a) *The Minoan Roundel and other Sealed Documents in the Neopalatial Linear A Administration*. Vol. I. Aegaeum 14. Liège–Austin.

Hallager, E. (1996b) *The Minoan Roundel and other Sealed Documents in the Neopalatial Linear A Administration*. Vol. II. Catalogue, Lists, Concordances. Aegaeum 14. Liège–Austin.

Hallager, E. (2002) One Linear A Tablet and 45 Nodules. *Creta Antica* 3, 105–109.

Harlizius-Klück, E. (2004) *Weberei als episteme und die Genese der deduktiven Mathematik*. Berlin.

Killen, J. T. (1962) Mycenaean po-ka: A Suggested Interpretation, *PP* 17, 26–31.

Killen, J. T. (1963) Mycenean po-ka: A Further Note, *PP* 18, 447–450.

Killen, J. T. (1964) The Wool Industry in Crete in the Late Bronze Age. *ABSA* 59, 1–15.
Killen, J. T. (1966) The Knossos Lc (Cloth) Tablets (Summary). *BICS* 13, 105–111.
Killen, J. T. (1968) The Knossos *o-pi* Tablets. In C. Gallavotti (ed.), *Atti e Memorie del 1° Congresso Internazionale di Micenologia (Roma 27 settembre-3 ottobre 1967)*. Incunabula Graeca 25:2, 636–643.
Killen, J. T. (1972) Two Notes on the Knossos Ak Tablets. In M. S. Ruipérez (ed.) *Acta Mycenaea. Actes du cinquième Colloque international des études mycéniennes, tenu à Salamanque, 30 mars–3 avril 1970.Vol.* II, 425–440. Salamanca.
Killen, J. T. (1974) A Problem in the Knossos Lc(1) (Cloth) Tablets. *Hermathena* 118, 82–90.
Killen, J. T. (1979) The Knossos Ld(1) Tablets. In E. Risch & H. Mühlestein (eds.), *Colloquium Mycenaeum. Actes du sixième colloque international sur les textes mycééniens et égéens tenu à Chaumont sur Neuchâtel du 7 au 13 septembre 1975*, 151–181. Neuchâtel – Genève.
Killen, J. T. (1982) TA and DA. In P. Oliva & A. Frolikova (eds), *Concilium Eirene XVI. Vol. III, Section IV: Mycenaeological Colloquium. Proceedings of the 16th International Eirene Conference*. Prague, 121–126.
Killen, J. T. (1984) The Textile Industries at Pylos and Knossos. In C. W. Shelmerdine & T. G. Palaima (eds.), *Pylos Comes Alive. Industry + Administration in a Mycenaean Palace*. New York, 49–63.
Killen, J. T. (1986) Two Mycenaean Words: ne-ki-ri-de, nu-wa(-i)-ja. In A. Etter (ed.), *O-o-pe-ro-si. Festschrift für Ernst Risch zum 75. Geburtstag*. Berlin – New York, 279–284.
Killen, J. T. (1988) Epigraphy and Interpretation in Knossos WOMAN and CLOTH Record. In J.-P. Olivier & Th. G. Palaima (eds.), *Studies in Mycenaean Epigraphy and Economy offered to Emmett L. Bennett, Jr.* Suplementos a Minos 10, 167–183. Salamanca.
Killen, J. T. (2001) Some thoughts on *ta-ra-si-ja*. In S. Voutsaki & J. T. Killen (eds.), *Economy and Politics in Mycenaean Palace States*. Cambridge Philological Society, supplementary volume 27, 161–180.
Lehmann, W. P. (1995) The Spindle or the Distaff. In B. B. Kachru & H. Kahane (eds.), *Cultures, Ideologies, and the Dictionary. Studies in Honor of Ladislav Zgusta*. Lexicographica series Maior 64, 53–60. Tübingen.
Liddell, H. G., Scott, R., Jones, H. S. (1940) A *Greek-English Lexicon,* 9th edition. Oxford.
Luján, E. R. (1996–1997) El léxico micénico de las telas. *Minos* 31–32, 335–370.
Luján, E. R. (1999) El léxico micénico de la lana. In Τῆς φιλίης τάδε δῶρα. *Miscelánea léxica en memoria de Conchita Serrano*. Manuales y Anejos de Emerita 41, 127–137. Madrid.
Marinatos, Sp. (1967) *Kleidung*. Archaeologia Homerica vol. 1 A. Goettingen.
Melena, J. L. (1975) *Studies on Some Mycenaean Inscriptions from Knossos Dealing with Textiles*. Minos Supl. 5. Salamanca.
Michailidou, A. (1992–1993 [1995]) 'Ostrakon' with Linear A Script from Akrotiri (Thera). A non-Bureaucratic Activity? *Minos* 27–28, 7–24.
Militello, P. (1988) Riconsiderazioni preliminari sulla documentazione in lineare A da Haghia Triada. *Sileno* 14, 233–261.
Militello, P. (2002a) Amministrazione e contabilità a Festòs. II. Il contesto archeologico dei documenti palatini. *Creta Antica* 3, 51–91.
Militello, P. (2002b) A Notebook by Halbherr and the Findspots of the Ayia Triada Tablets. *Creta Antica* 3, 111–120.
Montecchi, B. (2007) *Amministrazione e contabilità ad Haghia Triada nel TM IB*. Scuola Archeologica Italiana di Atene, Tesi di specializzazione. Athens.
Morpurgo Davies, A. (1979) Terminology of Power and Terminology of Work in Greek and Linear B. In E. Risch & H. Mühlestein (eds.) *Colloquium Mycenaeum. Actes du sixième colloque international sur les textes mycéniens et égéens tenu à Chaumont sur Neuchâtel du 7 au 13 septembre 1975*, 87–108. Neuchâtel – Genève.
Nosch, M.-L. (2000) *The Organization of the Mycenaean Textile Industry*. PhD. Salzburg.
Nosch, M.-L. (2004) Red coloured Mycenaean cloth. In L. Cleland, K. Stears & G. Davies (eds.) *Colours in Antiquity. Papers given at the Colour Conference, Dept. of Classics, University of Edinburgh, 10–13 September*. BAR IS 1267, Oxford, 32–39.
Nosch, M.-L. (2007) *The Knossos Od Series. An Epigraphical Study*, Veröffentlichungen der Mykenischen Kommission Band 25, Mykenische Studien 20, Österreichische Akademie der Wissenschaften, Philosophisch-historische Klasse, Denkschriften, 347. Band. Wien.
Nosch, M.-L. (forthcoming A) Les allocations de laine enregistrées dans les tablettes en Linéaire B de Thèbes, *Kadmos*.

Nosch, M.-L. (forthcoming B) Two Observations on the Knossos Lc(1) Targets. In T. G. Palaima & C. Shelmerdine (eds.), *11th International Mycenological Colloquium in Austin, Texas, 8–12 May 2000.*

Olivier, J.-P. (1978) 'Lire' le linéaire A? In J. Bingen *et al.* (eds.), *Hommages à Claire Préaux*, 441–449. Bruxelles.

Oren, E. D. & Olivier, J.-P. (1996) A Minoan Graffito from Tel Haror (Negev, Israel). *Cretan Studies* 5, 91–118.

Palaima, T. G. (1994) Seal-Users and Script-Users/Nodules and Tablets at LMIB Hagia Triada. In P. Ferioli *et al.* (eds.), *Archives Before Writing. Proceedings of the International Colloquium, Oriolo Romano, October 23–25, 1991*, 307–330. Roma.

Palmer, L. R. (1963) *The Interpretation of Mycenaean Greek Texts.* Oxford.

Perna, M. (2004) *Recherches sur la fiscalité mycénienne.* Études Anciennes 28. Nancy.

Perna, M. (2005) L'alun dans les documents en linéaire B. In P. Borgard, J.-P. Brun, M. Picon (eds.) *L'alun de Méditerranée, colloque international, Naples 4–6 juin, Lipari 7–8 juin 2003*, 39–42. Naples.

Perpillou, J. L. (1996) *Recherches lexicales en Grec Ancien. Etymologie, Analogie, Représentations.* Bibliothèque d'études classiques 7. Louvain

Rougemont, F. (2007) Flax and Linen Textiles in the Mycenaean palatial economy. In C. Gillis & M.-L. Nosch (eds.), *Ancient Textiles: Production, Craft and Society.* Ancient Textiles Series 1, 46–49. Oxford.

Rougemont, F. (forthcoming) Les phases de la production textile à l'époque mycénienne, de la fibre textile au produit fini: aperçu des données épigraphiques. In C. Breniquet (ed.), *Systèmes de production et de circulation. Le travail du fil. Journée organisée par la MAE, 16 juin 2008.* Cahiers des thèmes transversaux de la Maison René Ginouvès. Paris.

Ruijgh, C. J. (1967) *Études sur la grammaire et le vocabulaire du grec mycénien.* Amsterdam.

Schneider-Ludorff H. (1998) Filz in Nuzi? *SCCNH* 9, 163–168.

Schoep, I. (2002) *The Administration of Neopalatial Crete. A Critical Assessment of the Linear A Tablets and Their Role in the Administrative Process.* Minos Supl. 17. Salamanca.

Spantidaki, Y. & Moulherat, C. (2009) Cloth from Kastelli, Chania, *Arachne* 3, 8–15. [Available in French : *Rapport Identification des fibres* on http://ctr.hum.ku.dk/research/tools_and_textiles_/experimental_reports/]

Van Leuven, J. C. (1979) Mycenaean Goddesses Called 'Potnia'. *Kadmos* 18:2, 112–129.

Varias García, C. (1992) Un texto micénico singular sobre la industria textil de Cnoso: la tablilla KN Ln 1568. In J. Zaragoza & A. González Senmartí (eds.), *Actes del Xè Simposi de la seccio catalana de la SEEC (Tarragona, 28 a 30 de novembre de 1990)*, 441–446. Tarragona.

18. Mycenaean Textile Terminology at Work: The KN Lc(1)-tablets and the occupational nouns of the textile industry

Eugenio R. Luján

Textile production was one of the most important activities in the Mycenaean economy,[1] as evidenced by the great number of Linear B tablets which deal with the organization of textile production. We have information on textiles from various Mycenaean centres, specifically from Knossos, Pylos, Mycenae, and Thebes; the most extensive data comes from Knossos. On the Knossos tablets we can recover information about flocks, wool and the production of cloth.

Linear B tablets thus provide a rich textile terminology concerning types of cloths, garments and decoration. The complexity of the terminology employed by the scribes is indicative of the care and attention commanded by this industry inside the Mycenaean society.

Mycenaean textile terminology has been dealt with at length in the previous chapter, so in this chapter I will focus on a particular set of tablets which deal with textiles: the Lc(1)-set, written by the scribe 103 from Knossos. This is arguably the most informative set of tablets concerning the Mycenaean textile industry and contains extensive textile terminology. It is thus highly representative both of the kind of information that we can access from the Linear B tablets, and also of the problems that can arise from this type of information. From this perspective, special attention will be given to a particular terminological problem concerning the way in which groups of female workers are referred to by an occupational noun as it relates to the production of textiles. I will analyse the consistency of the use of those occupational nouns in relation to the production of the particular textiles for which these workers are responsible.

However, before we go into detail in the analysis of some of the terminology employed by this scribe, it is important to provide some background in order to contextualize this set of tablets. The so-called "hand 103" is one of the most active scribes from Knossos. He is responsible for various sets of tablets dealing with textiles, wool, and groups of female workers (sets Am(1), Lc(1), L(1), Od(1), M(1), etc.).[2]

The Lc(1)-tablets were found at the F10 find-spot in the area of the Western Magazines (XI) of the Knossos palace; tablet Lc(1) 561, though, was found nearby at find-spot F11 (Western

---

[1] An overview of Mycenaean economy can be found in Killen 2008. For the place of textile manufacturing in Mycenaean technology, see Bernabé & Luján 2008.

[2] See Landenius-Enegren 1995.

Magazine XII). In general, all the tablets written by scribe 103 were found in the area of Western Magazines of the palace of Knossos.[3]

1. The Lc(1)-set

In current editions,[4] the Lc(1)-set consists of 39 "regular" tablets plus two "tallying" tablets (535 and 536). Except for the latter, the information found on the tablets follows a regular pattern, typical of the pragmatic display of the information which is characteristic of the Linear B tablets.[5] The standardized structure of the entries of this set is shown on Table 18.1.

Table 18.1 Structure of the entries of the Lc(1)-set

<table>
<tr><td rowspan="2">Identification of group responsible for the production of cloths</td><td>type of cloth</td><td>cloth logogram + quantities</td><td>wool logogram + quantities</td></tr>
<tr><td>type of cloth (tu-na-no)</td><td>cloth logogram (TELA^{1}) + quantities</td><td>wool logogram + quantities</td></tr>
</table>

This structure is typified by tablet KN Lc(1) 525, which reads as follows (see Fig. 18.1):[6]

KN Lc(1) 525

.a		'wa-na-ka-te-ra'	TELA^{3}+*TE* 4̣0	LANA 100[
.b	se-to-i-ja , /	tu-na-no	TELA^{1}̣ 3	LANA [

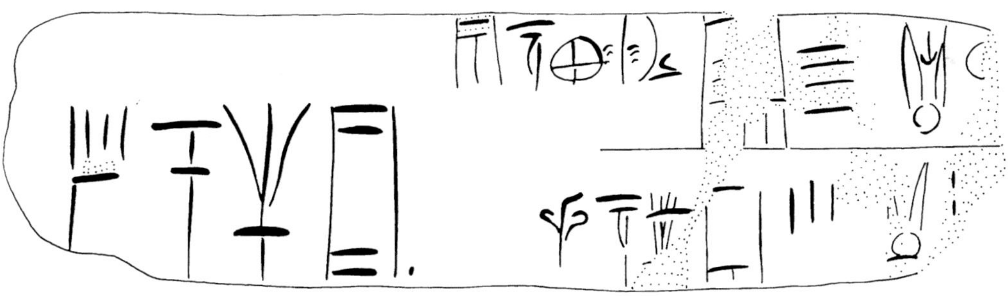

Fig. 18.1. Drawing of the tablet KN Lc(1) 525 [from CoMIK I]

[3] The find-places of the Knossos tablets have been extensively revised by Firth 1996–97. For the find-places of the set that will be analysed in this chapter, see Firth & Nosch 2002–2003, 121–125.
[4] *CoMIK*, *KT*[5]. See also Firth & Nosch 2002, 123–126.
[5] See Bernabé & Luján 2006, 205–210.
[6] For a comprehensive table of the data provided by the KN Lc(1)-set, see Appendix I.

However, we find a certain number of deviations from this scheme:[7]

- Tablets 526, 527, and 5746: on the first line there are two series (instead of just one) of cloth logogram plus quantity after *pe-ko-to*. However, only one quantity of wool is provided for both of them together.
- Tablet 551: on the second line we find an additional mention of *pe* (= *pe-ko-to*) plus two series of cloth logograms followed by quantities of wool.
- Tablets 530 and 532: on the second line there is an additional series of cloth logogram TELA+*TE*, followed by an additional quantity of wool.
- Tablet 558: *tu-na-no* appears on the first line. This tablet does not seem to follow the pattern of the other tables in this set.

As discussed in the previous chapter, the Mycenaean logogram TELA represents a woven piece of cloth, while the superindex denotes the number of small vertical traits (fringes) at the bottom of the cloth. In the case of this set, that number varies between 1 and 4.

The logogram TELA is qualified in this set by various terms, specifically:[8]

- *wa-na-ka-te-ra* /wanáktera/ 'royal' (cf. wa-na-ka /*wanaks*/ 'king', cf. Gk. ἄναξ);
- *pe-ko-to* /pektón/ 'combed' (cf. Gk. *pékō* 'to comb');
- *tu-na-no* (?), uncertain interpretation;
- *pa-we-a ko-u-ra* /pʰárweha koura(?)/ (cf. Gk. *phâros* 'cloak, mantle').

The identification of the group responsible for the production of the cloths listed on each tablet is made by one of these four possibilities – a personal name in the genitive case, a place name, the mention of a group of female workers or a combination thereof (a place name followed by a mention of a group of female workers). We find three personal names on this set: *e-me-si-jo-jo* (KN Lc(1) 551), *we-*]*we-si-jo-jo* (KN Lc(1) 7392), and]*ku-wo* (KN Lc(1) 532). *e-me-si-jo* is a well known character in the Knossian administration, one of the so-called "collectors",[9] involved also in the control of flocks, as shown by tablet KN De 1381. *we-we-si-jo* is another collector, who also plays a role in the control of flocks, wool and groups of workers, as shown by various Knossian tablets.[10] In fact, he might even have a "speaking" name related to his function if it can be rendered as *werwesios*, that is, a derivative by means of the suffix *-yos* from **werwos/werwes-* 'wool' (cf. Gk. εἶρος). However, the preservation of intervocalic -*s*- is problematic – one would expect that it had evolved into [h].[11] Of these three possibilities,]*ku-wo* can only be the final part of a personal name in the genitive case, but no name attested elsewhere on the Knossos tablets fits with it.

Two place names serve to identify the groups in charge of the productions listed on tablets KN Lc(1) 525 and 547. On the former we find *se-to-i-ja*, a place name that is mentioned on various tablets of the Knossos D-series dealing with flocks. The case of tablet KN Lc(1) 547 is slightly different: it is broken after *tu-ni-ja*, so that we cannot know for certain whether it was followed

[7] See Firth & Nosch 2002, 124–125, for a discussion of the classification of the tablets of this set.

[8] For further discussion of these terms see previous chapter.

[9] "Collectors" have been traditionally understood by scholars as certain palatial agents in charge of collecting the sheep, wool or cloth registered on the tablets. Recently, Rougemont 2001 has proposed to interpret them as inspectors sent by the palace to oversee production which was not proceeding to plan.

[10] References can be checked on the *DMic., s.u.*, as well as in F. Aura's *General Index of the Linear B*, available on-line: http://www.cervantesvirtual.com/portal/diccionariomicenico/.

[11] At Pylos there are also the *we-we-si-je-ja* workers attested on various tablets (PY Aa 762, Ab 217, Ad 318). This female noun poses similar problems; it is usually considered a derivative from the personal name *we-we-si-jo* rather than an occupational noun meaning 'wool workers'.

by something else. In any case, given that it appears frequently on the tablets of the D-series, it must have been an important centre for the raising of sheep, too.
Place names combine with the reference to a group of female workers on three tablets of this set:

- *a-mi-ni-so ko-u-re-ja* (KN Lc 550.B);
- *ko-no-so ko*[*-u-re-ja* (KN Lc 548);
- *ko-no-so te*[*-pe-ja* (KN Lc 549), cf. *ko-no-so te-pe-ja* (KN Le 641.4).

It should be noticed that this combination only happens with two of the biggest centres, *Amnisos* and *Knossos*, where it is expected that various groups of female workers were responsible for the production of textiles. Identifying them just by the place name, therefore, would have been impossible.

Mentions of groups of female workers alone can be found on various tablets of this set. They can be classified into two classes according to what the noun by which they are referred is based on:

- Feminine ethnic nouns: *da-wi-ja* (cf. *da-wo*), *da-*22-ti-ja* (cf. *da-*22-to*), *e-ki-si-ja* (cf. *e-ko-so*), *e-ra-ja* (cf. *e-ra*),]*ja-pu$_2$-wi-ja* (cf. *i-ja-pu$_2$-we*), *pa-i-ti-ja* (cf. *pa-i-to*), *qa-mi-ja* (cf. *qa-mo*), *ri-jo-ni-ja* (cf. *ri-jo-no*), and *tu-ri-si-ja* (cf. *tu-ri-so*). With the only exception of *i-ja-pu**, all place names appear on tablets of the D-series on which flocks are recorded. It is thus no wonder that specialized groups of female workers dealing with textiles are named after those place names.
- Feminine occupational nouns: *a-ra-ka-te-ja*, *e-ro-pa-ke-ja* (cf. *e-ro-pa-ke-u*), *ko-u-re-ja*, and *te-pe-ja*.

In a couple of instances there may be doubts concerning the classification of the noun. This is the case with *ne-we-wi-ja*, which is also attested on various tablets from Pylos (PY Aa 695, Ab 560, Ad 357) – this is probably an argument in favour of its interpretation as an occupational noun, but it cannot be ruled out *a priori* that two different places had the same name. Something similar happens with *da-te-we-ja*, which seems to be used as a place name on KN D 8174.

2. Groups of female workers named after their activities in the Mycenaean textile industry

Our focus now will be on the occupational nouns used for the identification of groups of female workers and their relation to Mycenaean textile terminology. Table 18.2 provides comprehensive information about occupational nouns of the textile industry and the products they are related to, including data both from the Lc(1)-set and other tablets.[12] Our goal will be to determine to what extent naming a group of textile workers after a textile product or activity is consistent with the data that we have about the production of that type of cloth and the distribution of the activities of the textile industry.

[12] Unless otherwise specified, the information is found on the Knossian tablets.

Table 18.2. Types of cloth and occupational nouns

Type of cloth or textile product	Groups of female workers (occupational nouns)	Masculine occupational nouns	Related nouns
pa-we-a ko-u-ra	ko-u-re-ja		
te-pa	te-pe-ja	te-pe-u (?)	
pe-ko-to	PY pe-ki-ti-ra$_2$		
	e-ro-pa-ke-ja	e-ro-pa-ke-u	
e-ne-ro	e-ne-re-ja		
o-nu	PY o-nu-ke-ja	TH o-nu-ke-wi (Dat.)	a-ke-ti-ri-ja/ a-ze-ti-ri-ja
	a-ra-ka-te-ja		
	to-te-ja	MY to-te-we- (Dat.)	
	ne-ki-ri-de		
	]ru-wo-we-ja		

2.1. *ko-u-re-ja*

The *ko-u-ra* type of cloth is attested at Knossos, Mycenae, and Pylos.[13] It is obviously a kind of *pa-we-a*, but no specific understanding of this adjective has been deduced at present. The *p^{h}arweha* (Gk. φᾶρος) were sizeable pieces of cloth that could be employed for various purposes, if we can rely on the evidence provided by the *Iliad* and *Odyssey*, in which they are used as cloaks for protection in cold weather, for making sails, and also as shrouds. It is precisely a *pharos* for her father-in-law that Penelope weaves and unweaves to deceive the pretenders in the *Odyssey*. According to the Mycenean data, the quantity of wool required to produce one *pharos* was LANA 1 M 2 (= 5 kg).[14]

The occupational noun *ko-u-re-ja* is only attested at Knossos and is always combined with a place name, as we have already seen – *ko-no-so ko-u-re-ja* can be read on KN Lc(1) 548 and *a-mi-ni-so ko-u-re-ja* on KN Lc(1) 550. However, the textile product called *pa-we-a ko-u-ra* was not only made by the specialized workers *ko-u-re-ja*, as one might think. We could speculate with the possibility that women working at other places were not so specialized in their tasks and produced a wider variety of cloths. However, according to the extant documentation as seen on Table 18.3, *pa-we-a ko-u-ra* were made by other groups that are referred to by other occupational nouns and not only by groups referred to by means of ethnic adjectives. The *pa-we-a ko-u-ra* were thus produced also by the *a-ra-ka-te-ja* (KN Lc(1) 531), the *e-ro-pa-ke-ja* (KN Lc(1) 534), and the *da-te-we-ja* (Lc(1) 540).

Unfortunately, the tablet Lc(1) 548, where the *ko-u-re-ja* women from Knossos are mentioned, is broken. If, as can be conjectured, it had the same structure as the rest of the set, we would expect it to record quantities of *tu-na-no* cloth on line 2. The *ko-u-re-ja* workers would thus not only produce *pa-we-a ko-u-ra*, but also *tu-na-no*. The same reasoning is valid for the *ko-u-re-ja* workers from Amnisos on KN Lc(1) 550.

[13] *pa-we-a$_2$ ko-u-ra* can be read on MY L 710.2. *ko-u-ra* appears on PY La 623 *v*., followed by TELA+*PA*, and is a likely restitution on PY La 630.a *ko-*]*u-ra*.

[14] Further information about all the types of cloth briefly discussed here can be found in the previous chapter.

Table 18.3. Types of cloths and groups responsible for their production

Type of cloth	Workers responsible for production		
pa-we-a ko-u-ra	a-ra-ka-te-ja	Lc(1) 531	**161* TELA^{1} 15[
	e-ro-pa-ke-ja	Lc(1) 534	'**161*' TELA^{1} 10[
	da-te-we-ja	Lc(1) 540	TELA^{1} 3[
	ko[-u-re-ja] ko-u-re-ja	Lc(1) 548 Lc(1) 550	? ?
	ne-we-wi-ja	Lc(1) 560	?
TELA+*TE*	ri-jo-ni-ja	Lc(1) 529	TELA^{2}+*TE* 30
	ri-jo-ni-ja	Le 642	TELA^{1}+*TE*[
	]ja	Lc(1) 530	TELA+*TE* 7
	]ja	Le 642	TELA^{1}+*TE* 2
	tu-ri-si-ja	Lc(1) 533	TELA^{1}+*TE* [
	]ja-pu$_2$-wi-ja	Lc(1) 541	TELA^{1}+*TE* 22
	qa-mi-ja	Lc(1) 543	TELA^{1}+*TE* 11[
	qa-mi-ja	Le 641	TELA^{1}+*TE* 1[
	e-ra-ja	Lc(1) 561	TELA^{1}+*TE* 1
	te[-pe-ja]	Lc(1) 549	?
	e-ki[-si-ja]	Le 5629	TELA^{1}+*TE* [
tu-na-no	a-ra-ka-te-ja	Lc(1) 531	TELA^{1} 1
	e-ro-pa-ke-ja	Lc(1) 534	TELA^{1} 1[
	da-te-we-ja	Lc(1) 540	?
	ko[-u-re-ja] ko-u-re-ja	Lc(1) 548 Lc(1) 550	? ?
	te[-pe-ja]	Lc(1) 549	?
pe-ko-to	da-wi-ja	Lc(1) 526	TELA^{1}+*TE* 10 TELA^{2}+*TE* 14[
	e-ki-si-ja	Lc(1) 527	TELA^{1}+*TE* 2 TELA^{2}+*TE* 19
	o-pi-si-ri-ja-we	L 8105	TELA^{X}[
pe	e-me-si-jo-jo	Lc(1) 551	TELA^{X}+*TE* 2 TELA^{1}+*TE* 10[
	pa-i-ti-ja	Le 641	TELA+*TE* 2
	da-wi-ja	Le 641	TELA^{X}+*TE* 1[
		Ln 1568	
mi	te-pe-ja	Le 641	TELA+*TE* 3
	pa-i-ti-ja	Le 641	TELA^{1}+TE 14
	do-ti-ja	Le 641	TELA+TE 6
mi-ja-ro		Ln 1568	
o-nu-ke	a-ze-ti-ri-ja	M 683, Ln 1568	

2.2. *te-pe-ja*

The word *te-pa* as such does not appear in the Lc(1)-set. However, it is found on other documents from Knossos (KN L 5090.2, Ws 8153.β) and also from Mycenae (MY Oe 107.1). Many of the tablets of the Lc(1)-set nonetheless record this kind of cloth. On Table 18.4 it can be seen that in many cases the logogram TELA (cloth) has a phonetic complement on it, the transcription of which is *TE*, that is, the abbreviation for *te-pa*. *te-pa* entries are thus found on tablets Lc(1) 525, 526, 529, etc., and they must have also appeared on other records which are now broken.

te-pa must have been heavy cloths; for the production of one *te-pa* seven units of wool were needed and that amounts approximately to 21 kg. Although there are some phonetic difficulties, it must be related to the Greek word τάπης. It is significant that we find the adjective οὖλος 'thick' qualifying some *tapetes* in the *Iliad* (XVI 224). In Homer *tapetes* are not used as garments, but to make a bed or a seat more comfortable.

We may thus assume that the *te*[*-pe-ja* workers mentioned on KN Lc(1) 549 were responsible for the production of *te-pa* cloths, just like the other groups of workers in this set. However, since that tablet is broken, we cannot know for certain. The word *te-pe-ja* is also attested at Thebes (TH Of 35.1) and Pylos (PY Ad 921). At Pylos the personal name *te-pe-u*, which may be simply the masculine form of the noun, also occurs (PY An 340.9).

In this case, according to the extant evidence, the *te-pa* type of cloth only combines with groups of workers referred to by an ethnic adjective or else identified by a place name or the name of a "collector". This is interesting when compared to what we just have seen in the case of the *pa-we-a ko-u-ra*, in the production of which some groups of female workers referred to by other occupational names were involved, too.

Table 18.4. Types of TELA+TE

'wa-na-ka-te-ra'	TELA^3+*TE* 40	LANA 100[	Lc(1) 525
'wa'			Le 654
'pe-ko-to'	TELA^1+*TE* 10 TELA^2+*TE* 14[		Lc(1) 526
'pe-ko-to'	TELA^1+*TE* 2 TELA^2+*TE* 19	LANA 153	Lc(1) 527
pe[-ko-to]	[TELA^1+*TE*] 18		Lc(1) 536
pe	TELA^X+*TE* 2 TELA^1+*TE* 10[	LANA 250[	Lc(1) 551
mi-ja-ro			Ln 1568
'mi'	TELA^1+*TE* 14		Le 641
	TELA+*TE* 3		Le 641
	TELA+*TE* 6		Le 641

The quality or type of these *te-pa* cloths can be further specified by means of various adjectives, as shown on Table 18.4. In the KN Lc(1)-set we find the adjectives *wa-na-ka-te-ra* and *pe-ko-to*. On tablets KN Le 641 and Ln 1568 (also written by scribe 103) we find the additional mention 'mi', which is an abbreviation for the adjective *mi-ja-ro*, attested on KN Ln 1568.

According to the extant evidence the *wa-na-ka-te-ra* kind of *te-pa* were only produced at one place, *se-to-i-ja*, which is mentioned on KN Lc(1) 525 and on Le 654 (both written by scribe 103).

These *te-pa* must have been of special quality given that they are qualified as 'royal' (*wanaktera*, cf. *wa-na-ka*, the Mycenaean equivalent of Gk. ἄναξ 'lord, king'). Since tablet KN Lc(1) 525 is broken we cannot know whether this involved using even more wool for the production of this cloth.

As for *pe-ko-to*, this word is a verbal adjective in *-to-* related to the Greek verb πέκω 'comb'. It has been proposed that it means 'shorn',[15] but this does not make sense given that all the wool must have been shorn before it can be used for producing cloth. J. L. Melena has argued that it refers instead to a special finishing labour, probably the taking of the nap of the garment so that it becomes especially soft.[16]

This *pe-ko-to* variety of *te-pa* was produced by various groups of workers, identified either by an ethnic adjective or by the name of the collector. Interestingly, there exists the occupational noun *pe-ki-ti-ra$_2$*, which occurs several times on the Pylos tablets (PY Aa 891, Ab 578.B, Ad 694, etc.) and which can also be safely supplied on two broken Knossos tablets: Ld(1) 656 and Xe 8537, the latter written by scribe 103.

The third possibility is *mi-ja-ro*, usually abbreviated into *mi*. It is the adjective μιαρός, which in first millennium Greek means 'stained'. In Mycenaean, however, it may mean that the cloth has been dyed. The use of various types of dyes is well attested on the Mycenaean tablets. No occupational noun related to this type of *te-pa* cloth seems to be attested on Linear B tablets, but the *te-pe-ja* are mentioned as responsible for the production of this type of cloth, too (see Table 18.5).

Table 18.5. Groups of workers producing various types of cloth

Groups pf workers	Type of cloth	
da-te-we-ja	pa-we-a ko-u-ra	Lc(1) 540
	[tu-na-no]	Lc(1) 540
	ri-ta pa-we-a	L(1) 594
te-pe-ja	[pa-we-a ko-u-ra]	Lc(1) 549
	[tu-na-no]	Lc(1) 549
	mi	Le 641

2.3. *e-ne-re-ja*, *o-nu-ke-ja*, and *a-ke-ti-ri-ja*

We can assume that the *e-ne-re-ja* workers, mentioned on KN Ak(1) 638.A,[17] were responsible for the production of the *e-ne-ro*, mentioned on several tablets dealing with textiles (KN L 695.4, Ai(2) 762). The interpretation of the word *e-ne-ro* is, however, problematic. Its occurrence on the tablet KN L 695.4 (*e-ne-ro re-u-ko* N 2), written by scribe 103, shows that this 'white (λευκόν) *e-ne-ro*' is measured as a quantity of wool (roughly equivalent to 500 g) and is not followed by the logogram TELA. This has been used as an argument to defend the theory that it is a kind of thread. The word *e-ne-ro* must be related in some way to the Greek word ἔνεροι (*éneroi*) 'those underneath, the dead', and Melena has argued that this must be the 'warp'.[18]

---

[15] See *DMic.*, s.u.

[16] Melena 1984, 37–38.

[17] It has been proposed that we have *pa-i-to* / *e*[-*ne-re-ja* on KN X 522 (Melena 1975, 89).

[18] Melena 1975, 88–91.

This brings us to the problem of how to interpret *o-nu*, a word which appears on various tablets (KN Od 681, Ln 1568, Ld 584, etc.) and is part of the compounds *re-u-ko-nu-ka* and *po-ki-ro-nu-ka*, meaning, respectively, 'with white *o-nu*' and 'with variegated *o-nu*', adjectives which are frequently found qualifying cloths.[19] J. L. Melena argued that *o-nu* was the weft,[20] but J. T. Killen has defended that it must have been some kind of ornament added to the cloth as part of the finishing process, probably some kind of wedge-shaped ornaments or else the fringes of the cloths.[21] We must acknowledge that it would be unexpected to find that there were workers specialized in producing either the thread of the weft or the thread of the warp and not both, even if their qualities may be different.

There is a feminine occupational noun related to *o-nu*, *o-nu-ke-ja*, but, interestingly, this noun does not seem to have been used at Knossos, though it appears on two Pylian tablets (PY Ab 194. B and Ad 675). Instead, at Knossos the noun *a-ke-ti-ri-ja/a-ze-ti-ri-ja* ἀσκήτριαι (*askḗtriai*) is found in contexts that suggest it was these workers who were responsible for the production of the *o-nu* (KN Ln 1568 and M 683, both written by the scribe 103).[22] On another tablet of this scribe (KN Xe 544.b) we even find the exceptional combination of ethnic name plus occupational name: *da-*22-ti-ja / a-ze-ti-ṛịị[-ja*.

Askḗtriai is related to the Greek verb ἀσκέω (*askéō*) 'work, form by art'. It is interesting to see that in the *Odyssey* the adjective ἀσκητός (*askētós*) is used with νῆμα (*nêma*) 'thread' (*Od.* IV 134), and in the *Iliad* and the *Odyssey* we find the formulaic phrase ἤσκειν εἴρια καλά (*ḗskein eíria kalá*) 'produce fair threads of wool' (*Il.* III 388). It would thus seem that these workers would be particularly concerned with the transformation of wool into thread.[23]

2.4. *a-ra-ka-te-ja*

There is yet another occupational noun that is clearly related with the transformation of the wool into thread to be used for the production of textiles. It is one of the nouns found in the Lc(1)-series: *a-ra-ka-te-ja* ἀλακατεῖαι 'spinners' (cf. Gk. ἠλακάτη [*ēlakátē*] 'distaff'). It also appears on KN Ak(1) 5009.C, as well as on some tablets from Pylos (PY Aa 89, 240, and Ad 380) and Thebes (TH Of 34.2). The Greek interpretation of this noun is clear and has not been questioned.[24] It would be expected, accordingly, that they produced spun wool. However, the information provided by the tablet KN Lc(1) 531 shows that this was not the case: just like other groups of female workers they too had to deliver cloths, specifically *pa-we-a ko-u-ra* and *tu-na-no*.

If we consider the descriptions of spinning and weaving in the *Iliad* and the *Odyssey*, we can gain some insight into why this is so. At *Odyssey* VII 103–106 we read:

> And fifty slave-women he had in the house, of whom some grind the yellow grain on the millstone, and others weave webs, or, as they sit, twirl the yarn, like unto the leaves of a tall poplar tree; and from the closely-woven linen the soft olive oil drips down.[25]

---

[19] See Luján 1996-1997, 347–349, with further references.
[20] Melena 1975, 112-113.
[21] Killen 1979, 164–167.
[22] The masculine noun *a-ke-te* ἀσκητήρ occurs twice on PY Jn 832, but in this case it refers to an activity related to bronze smiths.
[23] Luján 1996–1997, 347–349.
[24] See *DMic.*, *s.u.*
[25] Translation by Murray 1966.

This passage shows that the same female slaves were responsible for spinning and weaving. In fact, in the *Iliad* a formulaic way of telling a woman to mind her own business is to ask her to busy herself with the loom and the distaff, as shown, by the following passage from *Iliad* VI 490–492:

> Nay, go thou to the house and busy thyself with thine own tasks, the loom and the distaff, and bid thy handmaids ply their work.[26]

Even more than with other nouns, the case of the Mycenaean *a-ra-ka-te-ja* is a *caveat* against assuming, without further support from the sources, that the occupational noun by which a given group of workers was referred is sufficient proof that a univocal relation can be established between that noun and the actual activity that the group carried out.

2.5. Other occupational nouns

The noun *e-ro-pa-ke-ja* is attested on KN Lc(1) 534.B. It also appears on KN Ld(1) 595.1 and on a tablet from Mycenae (MY Fo 101.9). Its interpretation is uncertain, although it is generally agreed that it must refer to an activity inside the textile industry.[27] The masculine noun *e-ro-pa-ke-u* is attested at Knossos too (KN As 4493.2), showing thus that the activity referred to could be performed by both women and men. Like other groups in the Lc(1)-set the *e-ro-pa-ke-ja* women had to produce *pa-we-a ko-u-ra* and *tu-na-no* cloths; this cannot be used therefore as an indication of what their specific activity was.

Similar problems arise when deciphering the obscure nouns *ne-ki-ri-de*, and *ne-we-wi-ja*, mentioned on tablets written by scribe 103. For the former, if a relationship to Gk. νεκρός (*nekrós*) 'corpse' is possible, the interpretation 'shroud-makers' or the like would seem to follow. As for *ne-we-wi-ja*, as already stated, we are not even sure this is an occupational noun, just like *da-te-we-ja*.[28] The latter are responsible for the production of various types of cloth, as we saw on Table 18.5 above.

Finally, the noun *to-te-ja* should be mentioned for the sake of completion. It does not appear in the Lc(1)-set, but it is attested on a tablet written by the scribe 103 (KN Ak(1) 611.1) and can be possibly supplied on KN X 7846.b, where *to-te*[is to be read. The masculine form *to-te-we-* (Dat.) can be found on a tablet from Mycenae dealing with wool (MY Oe 106.1). J. L. Melena suggested that these *to-te-ja* produced rougher garments from which the nap has not been plucked, so that their product would be the opposite to the *pe-ko-to* type that we saw above (§ 2.2).[29] The most probable interpretation of this term seems to be *stórteiai*, thus makers of a cloth called **to-ta stortá*, which is not directly attested in Mycenaean (cp. Gk. στόρνυμι 'spread' and στρώματα 'bedcloths').

* * *

The analysis of the structure of the tablets of the Lc(1)-set from Knossos provides special insight into how Mycenaean scribes actually used the rich and developed terminology related to cloths and textiles available to them.

---

[26] Translation by Murray 1965.

[27] See *DMic.*, *s.u.* for the various proposals.

[28] See *DMic.*, *s.uu.*

[29] Melena 1975, 116–117.

Particularly important is the relationship between the products and the activities of the textile industry and the occupational nouns used to refer to certain groups of female workers responsible for the production of various types of cloth. Some of these occupational nouns are clearly derived from the nouns of those products or activities and provide additional key information in the field of textile terminology.

However, a careful study of the data provided by the Linear B tablets, especially those written by the scribe 103 from Knossos, clearly shows that we cannot take for granted that there is a univocal relation between the meaning of the occupational noun given to a specific group of workers and the actual activity performed by that group. We have found deviations in various ways: first, groups of workers who are not specialized in the production of the item after which they are named, such as the *te-pe-ja*, who produce *te-pa* but also *tu-na-no*; second, groups that carry out other activities than is to be expected according to their designation, such as the *a-ra-ka-te-ja*, who not only spin but also weave; third, textiles manufactured not only by the workers named after that product, but also by other groups, as is the case with the *pa-we-a ko-u-ra*, which were made not only by the *ko-u-re-ja*, but also by the *e-ro-pa-ke-ja*, *a-ra-ka-te-ja* and *da-te-we-ja*.

Linear B tablets are an invaluable source of textile terminology that can shed light on practices concerning textile production in the Bronze Age. However, the study of the relation between nouns and other data available, that is, the study of Mycenaean textile terminology *at work*, provides a cautionary lesson against arriving at quick conclusions.

Abbreviations

CoMIK Chadwick, J. *et al.* (1986–1998) *Corpus of Mycenaean Inscriptions from Knossos*, 4 vols. Roma & Cambridge.
DMic. Aura Jorro, F. (1985–1993) *Diccionario micénico*, 2 vols. *Anejos I y II del Diccionario Griego-Español*. Madrid.
KT[5] Killen, J. T. & Olivier, J.-P. (1989) *The Knossos Tablets. Fifth Edition. Suplementos a Minos* 11. Salamanca.

Bibliography

Bernabé, A. & Luján, E. R. (2006) *Introducción al griego micénico. Gramática, selección de textos y glosario*. Monografías de Filología Griega 18. Zaragoza.

Bernabé, A. & Luján, E. R. (2008) Technology. In Y. Duhoux & A. Morpurgo-Davies (eds.), *A Companion to Linear B. Mycenaean Greek Texts and their World*, vol. I, 201–233. Louvain-la-Neuve.

Firth, R. J. (1996–1997) [1999] The find-places of the tablets from the palace of Knossos, *Minos* 31–32, 7–122.

Firth, R. J. & Nosch, M.-L. B. (2002–2003) [2006] Scribe 103 and the Mycenaean textile industry at Knossos: the Lc(1) and Od(1)-sets, *Minos* 37–38, 121–141.

Killen, J. T. (1979) The Knossos Ld(1) tablets. In E. Risch & H. Mühlestein (eds.), *Colloquium Mycenaeum*, 151–182. Neuchâtel.

Killen, J. T. (1988) Epigraphy and interpretation in Knossos WOMEN and CLOTH records. In J.-P. Olivier & T. G. Palaima (eds.), *Texts, Tablets, and Scribes. Studies in Mycenaean Epigraphy and Economy offered to E. L. Bennett, Jr.*, 319–331. Suplementos a Minos 10. Salamanca.

Killen, J. T. (2008) Mycenaean economy. In Y. Duhoux & A. Morpurgo-Davies (eds.), *A Companion to Linear B. Mycenaean Greek Texts and their World*, vol. I, 159–200 Louvain-la-Neuve.

Landenius-Enegren, H. (1995) A prosopographical study of scribal hand 103. Methods, aims, and problems. In R. Laffineur & W.-D. Niemeier (eds.), *Politeia. Society and State in the Aegean Bronze Age*, 115–132. *Aegaeum* 12. Liège.

Luján, E. R. (1996–1997) [1999] El léxico micénico de las telas, *Minos* 31–32, 335–369.

Luján, E. R. (1999) El léxico micénico de la lana. In *Τῆς φιλίης τάδε δῶρα. Miscelánea léxica en memoria de Conchita Serrano*, 127–137. Madrid.

Melena, J. L. (1975) *Studies on some Mycenaean Inscriptions from Knossos dealing with Textiles*. Suplementos a Minos 5. Salamanca.
Melena, J. L. (1984) *Ex oriente lux*. Vitoria.
Murray, A. T. (1965) *Homer. The Iliad, with an English translation*, vol. I. London & Cambridge, Mass.
Murray, A. T. (1966) *Homer. The Odyssey, with an English translation*, vol. I. London & Cambridge, Mass.
Nosch, M.-L. (1998) L'administration des textiles en Crète centrale, hors des series Lc/Le/Ln, *Bulletin de Correspondance Hellénique* 122, 404–406.
Nosch, M.-L. (2007) *The Knossos Od Series. An Epigraphical Study*, Veröffentlichungen der Mykenische Kommission Band 25, Mykenische Studien 20, Österreichische Akademie der Wissenschaften, Philosophisch-historische Klasse. Wien.
Rougemont, F. (2001) Some thoughts on the identification of the 'collectors' in the Linear B tablets. In S. Voutsaki & J. T. Killen (eds.), *Economy and Politics in Mycenaean Palace States*, 129–138. Cambridge Philological Society, suppl. 27. Cambridge.

Appendix I: Structure and contents of KN Lc(1)-set

TABLET	IDENTIFICATION			TYPE OF CLOTH	CLOTH LOGOGRAM + QUANTITIES	WOOL LOGOGRAM + QUANTITIES
	PLACE NAME	PERSONAL NAME (GEN.)	GROUP OF FEMALE WORKERS			
525	se-to-i-ja			'wa-na-ka-te-ra'	TELA^{3}+*TE* 40	LANA 100[
				tu-na-no	TELA^{1} 3	LANA [
526			]da-wi-ja	'pe-ko-to'	TELA^{1}+*TE* 10 TELA^{2}+*TE* 14[	
				tu-na-no	TELA^{1} 3	LANA[
527			]e-ki-si-ja	'pe-ko-to'	TELA^{1}+*TE* 2 TELA^{2}+*TE* 19	LANA 153
				tu-na-no	TELA^{1} 2	LANA 6
528			e-ra-ja	'pa-we-a' ko-u-ra	TELA^{X}[	
				tu-na-no	TELA^{1} 1[	
529			ri-no-ni-ja		TELA^{2}+*TE* 30	
				tu-na-no	TELA^{X}[	
530			]ja	pa-we-a ko-u-ra	TELA^{1} 40	LANA 60[
				tu-na-no	TELA^{1} 3	LANA 12
					TELA+*TE* 7	LANA[
531			]a-ra-ka-te-ja	'pa-we-a ko-u-ra'	**161* TELA^{1} 15[	
				tu-na-no	TELA^{1} 1	

TABLET	IDENTIFICATION			TYPE OF CLOTH	CLOTH LOGOGRAM + QUANTITIES	WOOL LOGOGRAM + QUANTITIES
	PLACE NAME	PERSONAL NAME (GEN.)	GROUP OF FEMALE WORKERS			
532		]ku-wo		'pa-we-a' ko-u-ra	TELA^{1} 16	LANA 26 M 1 [
				tu-na-no	TELA^{1} 1	LANA 3
					TELA+*TE* 4	LANA 28[
533			tu-ri-si-ja		TELA^{1}+*TE* [	
534			e-ro-pa-ke-ja	pa-we-a ko-u-ra	'**161*' TELA^{1} 10[	
				tu-na-no	TELA^{1} 1[	
540			da-te-we-ja	'pa-we-a' ko-u-ra	TELA^{1} 3[	
541			]ja-pu$_{2}$-wi-ja		TELA^{1}+*TE* 22	LANA 154
543			qa-mi-ja		TELA^{1}+*TE* 11[	
				tu[-na-no		
546			pa-i-ti-ja			
547	tu-ni-ja[					
548	ko-no-so		ko[-u-re-ja			
549	ko-no-so		te[-pe-ja			
550	a-mi-ni-so		ko-u-re-ja[	'pa-we-a['		
551		e-me-si-jo-jo		'pa-we-a' [ko-u-ra]	TELA^{1} 110	LANA 140[
				tu-na-no	TELA^{1} 3	LANA 9
				pe	TELA^{x}+*TE* 2 TELA^{1}+*TE* 10[	LANA 250[
552						LANA 77
					TELA^{1} 1	LANA 3
553				pa-we-a ko-]u-ra	TELA^{1} 50	LANA 82[
				[tu-na-no]	TELA^{1}]2	LANA 6
					TELA^{1}+*TE* 10	LANA 70[
555						LANA 178[
					]2	LANA 6

TABLET	IDENTIFICATION			TYPE OF CLOTH	CLOTH LOGOGRAM + QUANTITIES	WOOL LOGOGRAM + QUANTITIES
	PLACE NAME	PERSONAL NAME (GEN.)	GROUP OF FEMALE WORKERS			
557					] TELA^{1} 80	LANA 130[
					TELA^{1} 7	LANA 11[
558				]tu-na-no	TELA^{2} 1	LANA 3
					TELA^{2} +*TE* 20[ ]	
560			ne-we-wi-ja	pa-we-a ko-u[-ra		
561			e-ra-ja	[[po-pu]] e-pi-qe-re-si		
					TELA^{1}+*TE* 1	
582					]TELA^{X} 30	LANA [
				tu-]na-no	TELA^{1} 6[	
1580						]LANA 56
						] LANA 3 [
5053	a-mi-ni-so[					
5746					TELA^{1}+*TE* 4 TELA^{2}+*TE* 10[	
				]tu-na-no	TELA^{1} 2	LANA[
7285					]30	LANA 100
7289					TELA^{1} 13	LANA 30 [
				tu-na-]no	TELA^{1} 1	LANA 3 [
7318						]LANA 14[
7321						] LANA[ ]100[
					TELA^{1} 2	LANA 1[
7376			]ja	'pa-we-a' ko-u-ra	'**161*' TELA^{1} 10[	
7392		we-]we-si-jo-jo		']pa-we[-a]' ko-u[-ra	]240[	
				[tu-na-no]	TELA^{1} 10[	
7549			]da-wi-ja	tu[-na-no]		
7901			da-]*22–ti-ja			
8572			]ne-ja			

19. Les textiles néo-assyriens et leurs couleurs

Pierre Villard

Une caractéristique de la documentation néo-assyrienne sur les textiles est l'existence d'une très riche iconographie, qu'il n'est cependant pas toujours facile de relier aux sources écrites. Les bas-reliefs, fresques ou ivoires qui offrent de nombreuses représentations de costumes, mais aussi de textiles faisant partie du harnachement des chevaux ou du revêtement des machines de guerre, ont déjà fait l'objet de nombreuses études, parmi lesquelles celles de Kristine Brown ou de Stephanie Dalley.[1]

Malgré le grand intérêt des représentations figurées, ces études soulignent aussi les limites de l'entreprise. S'il est possible d'analyser l'évolution des formes et des décorations, principalement pour les habits d'apparat ou les uniformes, il est plus difficile d'identifier les matières, les textures ou les techniques.[2] Il demeure surtout hasardeux, sauf dans quelques cas, d'identifier les vêtements représentés à des termes apparaissant dans la documentation écrite.

Les visées de la présente étude sont plus modestes. Elle se propose, à partir des sources écrites, d'une part de réfléchir sur l'organisation de la nomenclature des textiles à l'époque néo-assyrienne et d'autre part de présenter ce que l'on peut connaître de leurs couleurs.

1. Les listes de textiles

Étant donné que de nombreux termes ne peuvent encore recevoir d'équivalence précise, il n'est pas aisé de déterminer les catégories selon lesquelles les pièces d'habillement ou étoffes étaient classées. Une manière d'aborder la question consiste à envisager et à comparer les divers types de listes enregistrant des textiles, en commençant par celles dont la nature semble la plus claire.

1.1. *Listes de tribut*

Les listes de tribut incluses dans les inscriptions royales mentionnent fréquemment des textiles parmi les objets de valeur, mais sans y consacrer une très grande place. La formule standardisée, *lubulti birme u kitû*, « étoffes multicolores et étoffes de lin » revient le plus souvent,[3] avec

[1] Brown 1980 ; Dalley 1991.

[2] Voir cependant les importantes contributions d'Oppenheim 1949 et Barrelet 1977.

[3] Cf. e.g. Tadmor 1994, 56:1', kuš am-si zú am-si síg-za-gìn-sa_5 síg-za-gìn-gi_6 *lu-bul-ti bir-me* túg-gada-meš *lu-*⌜*bul*⌝*-[ti]*

quelquefois des mentions chiffrées, allant jusqu'à 3000 pièces reçues du royaume de Damas selon la stèle d'Adad-nērārī III et Nergal-ēreš de Tell Al Rimah.[4] Il est également parfois mentionné que ces étoffes figurent au nombre des produits de grande valeur[5]. Les évocations plus précises restent cependant rares.[6]

En dehors des inscriptions royales, les mentions d'étoffes provenant de tributs sont relativement rares. On peut néanmoins mentionner un document administratif du temple d'Aššur, enregistrant 338 textiles (*kuzippu*) originaires du pays de Hamath[7] ou la lettre de Sennacherib à Sargon II (SAA 1 34), dressant l'inventaire de divers produits, dont des étoffes, provenant de présents d'audience (*nāmurtu*) ou de tributs (*maddattu*) et distribués au palais et à de hauts dignitaires. Outre 15 tentes (^{túg}*zārutu*) haséennes destinées au palais, il est fait mention en assez grandes quantités d'étoffes de lin (*kitû*) et de *šaddīnu*.

Certains de ces termes ont manifestement une valeur générique. C'est le cas pour *kuzippu*, utilisé pour faire la somme d'habits de diverses sortes, et qui doit donc à cette époque signifier quelque chose comme « textile ».[8] D'autre part, *lubulti birme* est une expression qui ne se trouve que dans les inscriptions royales.[9] Les textes de la pratique utilisent simplement *birmu*, pour désigner des textiles multicolores,[10] *išpar birme*[11] étant le nom du tisserand capable de les produire. Le mot *birmu* a généralement été compris comme désignant des garnitures de type passementerie fixées aux vêtements de luxe,[12] mais M.-Th. Barrelet n'écartait pas l'hypothèse d'autres procédés pour la création d'étoffes décorées, tissage ou broderie.[13] *Lubulti birme* peut donc correspondre à une « pièce d'habillement pourvue de parures multicolores » mais aussi à toute sorte d'étoffes décorées. Dans les deux cas, l'expression semble fonctionner comme un générique pour divers types de vêtements ornés de motifs. Quant au terme *kitû*, qui correspond à l'étoffe de lin en général, il peut aussi renvoyer à une forme standardisée de vêtement, comme ce pourrait être le cas dans SAA 1, 34.[14] En revanche, le *šaddīnu* dont on peut préciser qu'il est de byssus[15] ou de lin,[16] ou grand,[17] doit correspondre à une forme particulière.[18]

---

kur-*šú-nu ma-'-at-tu*, « cuir d'éléphant, ivoire, laine de pourpre rouge, laine de pourpre bleue, étoffes multicolores et étoffes de lin, étoffes nombreuses de leurs pays » *et passim* dans les inscriptions royales néo-assyriennes (butin pris sur les États d'une coalition hittito-araméenne).

[4] Page 1968, 142:7, 3 *lim lu-bul-ti bir-me* ù gada-meš.

[5] E.g. Tadmor 1994, 170:12'-13', kù-gi kù-babbar an-na an-bar a-bár *lu-bul-ti bir-me* túg-gada *lu-bul-ti* kur-*šú-nu* síg-za-gìn-sa$_5$, *[mim-ma] aq-ru bi-nu-ut tam-tim na-ba-li*, « or, argent, étain, plomb, étoffes multicolores, étoffes de lin, étoffes de leur pays, laine de pourpre rouge, [toutes sortes de] produits coûteux originaires de la mer et du désert » (tribut des États syriens).

[6] Pour un exemple de description d'un vêtement dans une liste de butin, voir ci-dessous, note 78.

[7] StAT 3 1:35.

[8] Radner 1999, 117.

[9] CAD B, 258a, *s.v. birmu* ; CAD L, 235a, *s.v. lubuštu*.

[10] E.g. SAA 7, 70:i 2', ou SAA 16, 84:r.12.

[11] E.g. SAA 6, 42 : r.8. Cf Ak-Jadir 1971, 69.

[12] E.g.CAD B, 257b, *s.v. birmu* ; Dalley 1991, 126.

[13] Barrelet 1977, 59–83.

[14] La traduction « tunic » proposée dans Parpola 1957, 35–36 et 216, reste purement conventionnelle.

[15] SAA 1, 34 :11, 4 ^{túg}*šad-din bu-ṣi*. Cf SAA 7, 63:r. iii 3'.

[16] CAD S, 17, *s.v. saddinna*.

[17] SAA 1, 34:12, 10 ^{túg}*šad-din* kalag-meš.

[18] Parpola 1987, 35–36 et 227 propose la traduction « toga ».

1.2. *Les dots*

Parmi les documents juridiques, les dots,[19] bien qu'assez peu nombreuses, présentent l'intérêt de documenter à la fois des « étoffes d'ameublement » et des vêtements. Les informations les plus intéressantes à cet égard proviennent de deux textes de Kalhu.

CTN 2, 1 est un fragment de contrat de mariage dont il ne reste que l'inventaire de la dot. Parmi les items dont la lecture est sûre, sont enregistrés deux *dappastu*, [x] SI-LUH, [x] *kiṣiptu ša puškāyi*[20], [x] 1/2 *ša-hīli* de laine rouge, des *kiṣiptu* de laine pourpre accompagnés de textiles-[...],[21] 6 *nahlaptu* ordinaires, deux *ša-hīli*, 4 *gulēnu*, soit en tout 32 textiles (túg-pa-meš), suivis de divers ustensiles.

L'autre document, ND 2307 = *FNALD* n°14[22] est le contrat de mariage de la fille d'une gouvernante du harem (*šakintu*) du nouveau palais de Kalhu. L'inventaire de la dot comprend d'abord des métaux précieux et des bijoux, suivis de textiles de prix,[23] puis d'objets divers de moindre valeur, parmi lesquels d'autres textiles[24] (l32–33) et enfin de petits ustensiles domestiques.

Dans la liste des textiles de valeur, on trouve des *kuzippu*, dont deux de laine rouge « du port » (*kāri*), des *urnutu* dont deux de laine rouge « du port » et plusieurs de lin, des *huzūnu* et des túg-UŠ,[25] le tout étant estimé à la somme de 9 mines et 34 sicles d'argent.[26]

Dans la liste des objets de moindre valeur, on trouve, après la mention d'un lit de bronze et d'un marchepied de cuivre un rouleau de *qarrāru*, deux *dappastu*, deux *qirmu*, un *gulēnu*,[27] ainsi que deux *gammidu*, un *šaddīnu* de lin, dix É-SAG, des *kuzippu*, trois *urnutu* dont un de lin, des túg-UŠ et des *huzūnu*.

Il est clair qu'une partie des textiles mentionnés dans ces inventaires correspondait à des pièces d'étoffes plus qu'à des vêtements. Ce doit être le cas du *ša-hīli*, que l'on peut partager en deux selon CTN 2, 1, du *qarrāru*, qui peut se présenter enroulé, et aussi du *kiṣiptu*, pièce d'étoffe coupée selon la suggestion de J. N. Postgate.[28] Certaines étoffes apparaissent quant à elles comme des éléments de literie, comme on peut le voir en comparant ces inventaires de dots avec SAA 7, 117, un mémorandum concernant le mobilier d'un temple qui enregistre deux *dappastu*, un *qirmu*, un *gulēnu*, un *qarrāru*, un túg-SI-LUH « tout cela se rapportant au lit du temple de Šērū'a ».[29] On a manifestement ici un ensemble de literie, mais faute de savoir exactement comment les Assyriens se couchaient, il reste difficile de déterminer la fonction de chaque élément. « Couverture » pour *dappastu*, « dessus de lit » pour *qarrāru*, « oreiller$^?$ » pour SI-LUH sont davantage des propositions

[19] Radner 1997, 163–164.

[20] CTN 2, 1:5', [x t]úg *ki-ṣip-te ša pu-uš-ka-a-a*. Le CAD P, 543a, *s.v. pušku* A propose la traduction « larges chacun d'une paume. » Mais la même dénomination s'applique à 3 túg-bar-dul dans StAT 3, 1:22. On comprendrait mal que des robes-*kusītu* soient aussi étroites. Il pourrait alors s'agir d'un ethnique (variante de Hubuškia ?). Voir Faist 2007, 13.

[21] CTN 2, 1:7'-8', [x tú]g *ki-ṣip-[t]e ša* síg za-gìn-sa$_5$ *a-di* túg *ša* x[...], [x ^{túg}*k]i-ṣip-te ša* síg za-gìn-sa$_5$ *a-di* túg [k]imin.

[22] Parker 1954, 37–39 ; Postgate 1976, 103–107.

[23] ND 2307 = *FNALD* n° 14:14–21.

[24] ND 2307 = *FNALD* n° 14:22–33.

[25] Il s'agit probablement d'une autre réalité que la qualité ús attestée à Mari pour des vêtements de laine. Cf. Durand 2009, 14–16.

[26] ND 2307:20–21, pap 9 1/2 ma-na 4 gín kù-babbar *an-ni-ú* e$_{11}$-u, *ba-la-ta* kù-babbar *an-ni-e* « Cela ressort à 9 mines 34 sicles d'argent. » La formule qui suit sert à introduire la section suivante de l'inventaire : « Non inclus dans cet argent. »

[27] ND 207:22–25, giš-ná zabar ^{giš}*ki-tur-ru* urudu, *ki-ri-ka šá* ^{túg}*qa-ra-a-ru*, 2 ^{túg}*da-pa-sa-a-te* ^{túg}*qi-ri-mu*, ^{túg}*gu-igi*^{II}

[28] Postgate 1973, 37 n. 2. En faisant dériver le terme de *kaṣāpu*, qui semble une forme néo-assyrienne de *kasāpu*.

[29] SAA 7, 117 : 5–6 *an-nu-te ša* giš-ná, *ša* é ^{d}*še-ru-ú-a*.

plausibles[30] que des traductions assurées. D'autant que nous ne savons pas si ces dénominations se rapportaient à des fonctions précises ou simplement à des types d'étoffes.

Les autres items mentionnés dans les inventaires de dots pourraient être des vêtements, mais il n'est jamais précisé qu'il s'agisse de vêtements spécifiquement féminins. On notera enfin que l'ensemble des rubriques de CTN 2, 1 est repris dans le total par túg-pa-meš.[31] La lecture de cet idéogramme reste incertaine,[32] mais pourrait être *kuzippu*, terme utilisé dans l'inventaire du temple d'Aššur[33] pour faire la somme de textiles de diverses formes.

1.3. *Ensembles vestimentaires*

Pour ce qui concerne les textiles pouvant servir à l'habillement, les rares textes présentant des ensembles vestimentaires, soit des pièces d'habillement susceptibles d'être portées ensemble et destinées à une personne ou à un petit groupe, apportent des indications précieuses.

Une lettre adressée à Sargon II par Nabû-šumu-lēšir, responsable militaire en poste en Babylonie (SAA 17, 122), mentionne les riches habits remis à un membre de la famille royale nommé Abu-erība : une robe-*kusītu*, un *ša-hīli*, des *muṣīptu* en rouleau[?34] de Tukriš.[35] Le *kusītu* est connu pour être une robe, généralement de valeur, entrant en particulier dans l'habillement du roi, des hauts dignitaires et des dieux.[36] Elle est ici portée avec un *ša-hīli* et des *muṣīptu*,[37] qui ne doivent pas être des habits ayant une forme précise, mais plutôt des pièces d'étoffes susceptibles d'être drapées par dessus une robe.

Pour un ensemble plus complet, on peut considérer un document (SAA 11, 28) qui enregistre la contribution-*ilku* d'un individu, manifestement destinée à l'équipement d'un militaire. Sont enregistrés un *šupālītu halluptu*, un *gulēnu*, un couvre-chef (*kubšu*), un *ṣipirtu*, un *urnutu*, six mines de laine, un *sāgu*, des outres, des sandales neuves.[38]

Il doit s'agir de l'équipement standard d'un homme appelé à se déplacer avec un minimum de protection, puisque l'on retrouve plusieurs de ces items dans une liste de présents remis à des émissaires urartéens (SAA 7, 127) : des *šupalītu halluptu*, dont deux noirs, deux paires de jambières[?] (*šahartu*), deux paires de sandales, deux *elītu* noirs, deux *ša*-IŠ,[39] deux *ṣipirtu*.

Dans son très bel article, « Assyrian Uniforms », J. N. Postgate a étudié en détail les divers éléments de l'équipement militaire et a pu proposer une série d'équivalences.[40] Dans ces contextes, le *nahlaptu*, pourvu d'une encolure,[41] est une sorte de manteau, qui peut être éventuellement

[30] Fales & Postgate 1992, XXIX.
[31] CTN 2, 1:12'.
[32] Postgate 1973, 28.
[33] StAT 3, 1:35.
[34] L'expression est difficile à comprendre. Le CAD K, 217b, *s.v. karku*, traduit «*muṣiptu*-garment of twined (?) thread », Dietrich 2003, 108 proposant « a threaded work dress ». On pourrait aussi songer à un adjectif dérivé de *kirku*, « rouleau », terme qui peut s'appliquer à des textiles. Cf. CAD K, 408, s.v. *kirku* B, qui cite un texte néo-babylonien (*Nbk*. 369 :2), où le mot est associé à un *muṣiptu*.
[35] SAA 17, 122:7–9 túg-bar-dib, [túg]*šá-hi-il* [túg]*mu-ṣi-pe-ti, kar-ke-e-ti šá tuk-riš.*
[36] Oppenheim 1949, 179 ; Postgate 2001, 378–379.
[37] « A standard size piece of cloth » selon le CAD M II, 242, *s.v. muṣīptu*. Ces étoffes pouvaient être décorées par des ornements d'or. Cf. Oppenheim 1949, 173.
[38] SAA 11, 28:11–15.
[39] S'agit-il d'une étoffe destinée à se protéger de la poussière (sahar) de la route ? Voir Fales & Postgate 1992, XXIX.
[40] Postgate 2001, 387.
[41] Pour les divers qualificatifs du terme à Mari, cf. Durand 2009, 67–72.

renforcé par du métal, pour constituer une sorte d'armure.[42] *Elītu*, « vêtement de dessus » désignerait le châle à franges porté par dessus les autres habits, le *sāgu* serait une sorte de kilt, le *ṣipirtu* une ceinture tissée,[43] le *gulēnu* et l'*urnutu* des sortes de tuniques. Enfin, le *šupālītu*, « vêtement de dessous » serait une chemise, qui renforcée, formerait une sorte de cotte de mailles (*šupālītu halluptu*).

1.4. *Inventaires administratifs divers*

Les autres documents administratifs dans lesquels apparaissent des textiles sont assez variés.

A Ninive, des listes d'objets divers, généralement précieux, comportent quelques pièces d'étoffes,[44] mais on dispose aussi d'une série de textes qui ne concernent que des textiles (SAA 7, 93–116).

De Kalhu, provient un court billet (CTN 2, 152), probablement attaché à un sac contenant des étoffes : des traces de tissu sont en effet visibles sur la partie inférieure, anépigraphe, du verso.[45]

Deux inventaires ont enfin été retrouvés à Aššur. Le premier, assez court, appartenait à une archive privée (StAT 1, 39). Le second, provenant du temple d'Aššur (StAT 3, 1), est particulièrement intéressant par la variété des étoffes répertoriées et par les précisions concernant les couleurs.[46] Sont d'abord enregistrés 288 (étoffes) neuves, parmi lesquelles des *kusītu*, des *elītu*, des *šupālītu halluptu* dont certains pour conducteurs de chars, des *qarrāru*, des *niksu*, un *birmu*, des *gammidu*, des *maqāṭu*, des « vêtements blancs » ([túg]*pa-ṣu-tú*), des *dappastu* et un SI-LUH. La liste comprend ensuite 98 vieilles (étoffes), parmi lesquelles des *kusītu*, des *qirmu*, un SI-LUH, des *nēbettu*, un *nahlaptu*, un *šupālītu halluptu* avec ceinture, des *sasuppu*, des *pariktu*, une *étoffe pour couvrir le roi* (túg *ša* ugu lugal), un *niksu* du pays d'Akkad, une étoffe de lin, des *nahlaptu* courts.

Toutes ces étoffes, provenant du pays de Hamath et conservées en partie dans un *qabūtu*,[47] étaient placées sous l'autorité d'un économe-*lahhinu*.[48] Malgré la présence de « cottes de mailles » (*supālītu halluptu*), les types de textiles mentionnés et surtout la provenance du document montrent que l'on n'a sans doute pas affaire à une liste d'équipements militaires, mais plutôt à des étoffes de valeur pouvant être utilisées lors de cérémonies. Dans ce contexte, la « cotte de mailles » semble donc être un élément d'un costume d'apparat.

Or des remarques similaires peuvent être faites à propos des autres documents administratifs du corpus. Certains mentionnent des *šupālītu halluptu*,[49] mais il ne s'agit pas d'un élément discriminant,[50] car l'ensemble des étoffes qu'ils enregistrent ne diffère pas sensiblement de ce qui apparaît dans le reste de la documentation.

Par ailleurs, les fonctions précises de ces documents, parfois mal conservés et souvent de

[42] Cf. SAA 7, 89:r.8, gú-è urudu, *nahlaptu* de cuivre.
[43] Ces pièces étaient réalisées par des tisserands spécialisés, appelés *išpār ṣiprāti*. Cf. e.g. SAA 7, 115:r. i 7.
[44] E.g. SAA 7, 117, 119, 120, 124, 126.
[45] Postgate 1973, 165.
[46] Voir ci-dessous, section 2.4.
[47] On songe à un coffre (de grande contenance !), mais le terme désigne ordinairement un contenant de métal ou une sorte de bol ou cuve. Voir CAD Q, 43–44, *s.v. qabūtu*.
[48] StAT 3, 1:35–37.
[49] SAA 7, 97, 102, 104, 105, 108, 109, 119, 124, 126, StAT 1, 38:4, 9–10. StAT 1, 38:9–10 enregistre en particulier un *šupālītu halluptu* pour *gammidu*, ce qui suggère que le *gammidu* doit probablement être rangé dans la catégorie des habits.
[50] Pour un avis différent, Fales & Postgate 1992, XXVIII.

formulation elliptique, ne sont pas toujours faciles à déterminer. On peut cependant supposer que la tablette provenant d'une archive privée d'Aššur a été réalisée après l'inventaire d'un coffre,[51] puisqu'elle dresse la liste des étoffes manquantes.[52] Pour ce qui concerne des documents provenant du palais de Ninive, les quantités parfois très importantes des textiles[53] ou des fibres fournies[54] qui y sont mentionnés témoignent de l'activité des ateliers et de l'ampleur des magasins du palais. Il est également probable que certaines de ces listes furent établies lors de la fourniture à des invités ou de la récupération de textiles à l'occasion de diverses cérémonies organisées dans les palais. Selon SAA 7, 112, l'un des responsables chargé de fournir des vêtements était aussi chargé d'organiser des banquets.[55] Dans d'autres cas, les distributions devaient concerner des personnes résidant de façon permanente dans le palais. Un court billet portant la date du 2-ii-658 (SAA 7, 93) enregistre ainsi la livraison de deux habits-*maqāṭu bēte* (d'intérieur?), de la part d'un certain Ibbiya, dans le quartier domestique (*bētānu*). Et l'on peut aussi signaler la mention dans SAA 7, 107 de capes-*hullānu* d'intérieur? pour des femmes »,[56] qui est d'ailleurs pour ce corpus la seule attestation explicite d'habits féminins.

2. Qualifications, descriptions et couleurs

La liste importante, quoique non exhaustive des noms de textiles néo-assyriens mentionnés jusqu'ici, reste assez frustrante, car il n'existe dans la plupart des cas que peu d'indices pour proposer des traductions bien argumentées. Au moins, la documentation existante fournit-elle parfois des informations concernant les qualités de textiles, la façon de les décrire et les couleurs utilisées pour les teindre.

2.1. *Qualités de textiles*

2.1.1. *Les fibres*

Les fibres mentionnées dans la documentation écrite sont presque exclusivement la laine et le lin. Les documents administratifs en mentionnent des quantités parfois considérables : 1 talent 22 mines de laine dans SAA 7, 110, plus de 18 talents de lin dans SAA 7, 111, plus de 413 talents de lin dans SAA 7, 115,[57] d'importantes quantités de laine provenant de diverses provinces, dont 100 talents de laine de Karkemiš dans SAA 7, 116. Par ailleurs, les lettres adressées au roi font parfois allusion aux besoins des ateliers locaux. Un certain Marduk-šarru-uṣur se plaint à deux reprises à Assarhaddon de n'avoir pas été livré en laines teintes[58] et un autre message, de Nabû-šarru-uṣur, fait allusion aux stocks de fibres conservées dans le palais : « J'ai questionné Balasî

---

[51] Pour une mention de coffre servant à ranger des étoffes, cf. SAA 7, 119:ii 14', 1 giš-é–*ku-zip-pi* sud, « un coffre en bois pour textiles, vide. »

[52] StAT 1, 38:11, [pa]b *an-ni-u mu-ṭé-e* « Total de ce qui manque. »

[53] E.g. SAA 7, 101:r1 (total de plus de 300 étoffes) ; SAA 7, 104:r.5' (total de plus de 700 étoffes) ; SAA 7, 112:3' (400 textiles fournis par [NP] et r.1–3 (1500 *nahhaptu* fournies par Udini).

[54] E.g. SAA 7, 110, 111, 115, 116.

[55] SAA 7, 112:4'-9', [Id]pa-*š[e-zib-an-ni]*, en-*pi-qi-ti ša s[a* ...], túg-an-ta-meš túg-bar-d[ib-m]eš, *it-ti-din šu-ú* mí-*šú*, *qa-ri-i-a-ti*, *e-ta-ap-še*, « Nabû-š[ezibanni], le chargé de mission de NP, a donné des 'vêtements de dessus' et des robes. Lui et son épouse ont préparé les banquets. »

[56] SAA 7, 107:r.3', [0 0 0 0 gú]-lá *bé-te ša* mí-meš.

[57] SAA 7, 115:ii 4–5, pab 2 *me* 74 gú-un, síg-gada, r. i 9, pab 1 *me* 9 gú 10 ma, r. ii 2, pab 30 gú 21 ma-na.

[58] SAA 16, 82 et 83.

(et il m'a dit) : 'nous fournirons de la laine rouge provenant du palais et eux agiront selon leurs instructions.' J'ai aussi questionné Aplāya (et il a dit) : 'Ils nous donneront de la laine rouge. Les tisserands d'Ištar d'Arbèles viendront (la) travailler à Kurbail' ».[59]

Les autres fibres ne sont que très rarement mentionnées. Le byssus, tissu très fin et de grande valeur, réalisé à partir de filaments produits par des mollusques,[60] était réservé à quelques vêtements de luxe. Quant au coton, il demeurait une curiosité, même si Sennacherib se vante d'en avoir cultivé dans ses jardins : « on tondit les arbres porteurs de laine et on en tissa (le produit) pour faire des vêtements ».[61]

Par ailleurs, il est parfois précisé avec quelle fibre étaient réalisés certains types de textiles. Pour les étoffes de laine, sont ainsi attestés des *kiṣiptu* (CTN 2, 1:7'-8'), des *kuzippu* (*FNALD* 14:14) des *maqāṭu* (SAA 7, 111:2), des *ša-hīli* (CTN 2, 1:6') et des *urnutu* (*FNALD* 14:15).

Pour les étoffes de lin, sont attestés des *dappastu* (SAA 7, 115:i 11) des *maqāṭu* (SAA 7, 97:r.1 ; SAA 7, 109:iii 2' ; SAA 7, 112:10'), des *šaddīnu* (*FNALD* 14:26), des *šupālītu halluptu* (SAA 7, 108:r ii' 5') et des *urnutu* (*FNALD* 14:16, 18 ; SAA 7, 96:r3 ; SAA 7, 104:r2).

2.1.2. *Les origines*

Dans quelques cas, la mention d'un textile est accompagnée par un ethnique, qui constitue peut-être une indication des origines, mais pourrait aussi se référer à une forme particulière ou à un type de finition. On rencontre ainsi au fil des textes des *muṣiptu* de Tukriš (SAA 17, 122:8–9), des *urnutu* de Gubla (SAA 7, 108:r ii 4), des cottes de maille phrygiennes (SAA 7, 126:4), des *nahhaptu* guréennes (SAA 7, 112:r 1–2, SAA 7, 115:ii 18), un *niksu* d'Akkad (StAT 3, 1:31), un SI-LUH du Tabal (StAT 3, 1:24) ou des chaussures cimmériennes (SAA 7, 120:ii' 8–9).

On trouve d'autre part assez fréquemment les mentions kur, « du pays » ou kar, « du port », pour qualifier certains textiles. En certaines occurrences, ces mentions permettent de distinguer des habits ou pièces d'étoffes qui présentent par ailleurs les mêmes caractéristiques, par exemple dans SAA 7, 98 : « [x] *qirmu*, le « front » rouge, du port [...], [x] idem, le « front » rouge, du pays [...] ».[62] Il pourrait s'agir ici d'une distinction entre des productions locales et des étoffes de fabrication étrangère obtenues par le grand commerce. On remarque aussi que ces qualificatifs peuvent aussi s'appliquer à de la laine, par exemple dans SAA 16, 82[63] et 83.[64]

2.1.3. *L'état et la finition*

Les inventaires mentionnent enfin régulièrement le degré d'usure des textiles. La liste du temple d'Aššur (StAT 3, 1) oppose ainsi les pièces neuves (gibil) aux vieilles (sumun).[65] Dans StAT 1, 39, la distinction se fait entre les textiles neufs (gibil) ou élimés (*qalpu*),[66] terme que l'on retrouve dans CTN 2, 152.

---

[59] SAA 16, 54:12–r 11 *a-na* ᴵ*ba-la-si-i, as-sa-'a-al*, R. *ma-a* síg sa_5, ta šà é-gal *ni-dan, ma-a šú-nu a-na ṭè-mi-šú-nu, ep-pu-šú, ù a-na* ᴵ*a-ia, as-sa-'a-al ma-a* síg sa_5, *id-da-nu-na-ši, ma-a* lú-uš-bar-meš, *ša* ᵈ15 *ša* ᵘʳᵘ*arba-ìl, il-la-ku-u-ni, ina* ᵘʳᵘ*kur-ba-ìl ep-pu-šú*.

[60] Dalley 1991, 121.

[61] Luckenbill 1924, 116:viii 64, *iṣ-ṣu na-aš ši-pa-a-ti ib-qu-mu im-ha-ṣu ṣu-ba-ti-iš*.

[62] SAA 7, 98:9'-10', [x]+1 ᵗᵘ́ᵍ*qir-me* zag sa_5 k[ar 0 0 0], [x tú]g ditto zag sa_5 kur [0 0 0].

[63] Voir ci-dessus, n. 58.

[64] SAA 16, 83:r 1'-2', [... síg sa_5] kar, 20 [gú-u]n *li-di-nu-u-ni*, « Qu'on me donne de la laine rouge du port pour 20 talents » (à propos du travail des tisserands des Grands).

[65] StAT 3, 1:21 et 34.

[66] Postgate 1973, 165.

Par ailleurs, *biršu*, « de texture grossière » ou « feutré » peut qualifier toutes sortes de textiles de laine.[67] Il pourrait s'agir de l'indication d'un type de finition ordinaire. On peut aussi noter la mention isolée de 70 *maklulu* de troisième qualité.[68] Enfin, le qualificatif *zakiu* s'applique à un *qirmu* noir dans StAT 3, 1:12. Il ne s'agit probablement pas d'une étoffe nettoyée,[69] mais comme J.-M. Durand l'a mis en évidence à propos de la documentation de Mari, d'un type de finition donnant au tissu un aspect brillant.[70]

2.2. *Les éléments descriptifs*

Si les inventaires donnent rarement des descriptions détaillées, il est parfois précisé que certains textiles sont pourvus d'un « front » (zag, *pūtu*), qui peut correspondre au devant, mais plus probablement au petit côté. StAT 3, 1:10–11 enregistre en effet 38 *niksu* blancs, dont la longueur (ús) et la largeur (zag) sont rouges. D'autres étoffes comportent une bordure (nigín) et plus rarement, il est fait mention d'une « queue » (kun, *zibbutu*) qui s'oppose au front.[71]

Parmi les textiles comportant un « front », on trouve des *maqāṭu* (StAT 3, 1:15, SAA 7, 93:1, 94:4, 95:1, 97:r 1, 3, 98:4', 104:3', 108:r ii 3')[72] des *gulēnu* (SAA 7, 96:5', 98:8', 107:r 8'), des *maklulu* (SAA 7, 96:7', 102:5', 105:8'), des *naṣbutu* (SAA 7, 96:r 1), des *urnutu* (SAA 7, 96:r 2, 98:5', 109:iii 9'), des *qirmu* (SAA 7, 97:7', 10'), des *raddidu* (SAA 7, 105:3'), des *sasuppu* (SAA 7, 102:ii' 4–6), des *ša*-GIL (SAA 7, 96:r 4).

Les textiles attestés avec une bordure sont les *naṣbutu* (SAA 7, 96:11', 97:11', 102:1'), les *urnutu* (SAA 7, 102:2',109:ii 2') et les *šupālītu halluptu* (SAA 7, 105:9').

D'autres éléments descriptifs apparaissent ça et là. Certaines pièces d'étoffes (*maklulu, kuzippu*) sont décorées de pierres,[73] sans que plus de précisions ne soient données. D'autres sont qualifiés par le terme *sāiu*,[74] qui signifie peut-être « à point noué ».[75] Par ailleurs, les noms de végétaux ou d'animaux (grenade, bovins, caprins),[76] que l'on trouve parfois associés à des textiles correspondent sans doute à des motifs décoratifs, de même peut-être que les « dessins » (*uṣurtu*) associés à un *ša*-GIL.[77]

Les documents administratifs ne comportent pas de descriptions plus détaillées, mais on peut relever dans la célèbre lettre de Sargon II à Aššur, dans la liste du butin réalisé lors du pillage du temple du dieu Haldi à Muṣaṣir, une évocation précise quoique difficile à comprendre d'habits

---

[67] Cf. SAA 7, 95, 96, 97, 98, 99, 100, 102, 105, 107, 109, 119.

[68] SAA 7, 112:r 4 70, [túg]*muk-lál ša-la-šu-te*.

[69] Faist 2007, 13.

[70] Durand 2009, 137–138.

[71] SAA 7, 107:2' [...] x zag kun-*tú*. Les autres occurrences interviennent dans des passages encore plus lacuneux.

[72] Le *maqāṭu* est un textile qui comporte normalement un « front » puisqu'il est précisé dans SAAVII 107:r 9', nu zag, « sans front ».

[73] SAA 7, 96:7', 2 muk-lal bir zag sa_5 na_4–[meš 0], « deux *maklulu*, *feutrés*, le front rouge, avec des pierres [...] » ; SAA 7, 97:13', 1 ku-zip-pi na_4–me[š 0], « une pièce d'étoffe, avec des pierres. »

[74] Passim dans SAA 7, 108 et 109. Les contextes lacuneux ne permettent pas de préciser davantage, mais il est question d'*urnutu* dans SAA 7, 108:ii' 6' et SAA 7, 109:ii 2'-3'.

[75] Fales & Postgate 1992, 221.

[76] SAA 7, 109:i 2' [...] ditto ditto ditto nu-úr-ma ; SAA 7, 109:ii 2'-3', 1 *ur-nat* gi_6 nigin za[g 0], *sa-a* gud, « un *urnutu* noir, avec une bordure, le front [...], *noué*, (avec) un bovin. » Cf. SAA 7, 109:iv 4' ; SAA 7, 109:4' 1 ditto ditto nigin máš, « un (*urnutu* noir), avec une bordure, (avec) un chevreau. »

[77] SAA 7, 108:r ii' 6' [00 *š]a*-GIL giš-hur-meš.

destinés à revêtir des statues divines : « 9 vêtements de sa tenue divine, dont les *gaufrures* étaient garnies de disques d'or et de rosettes d'or *formant un liseré* ».[78]

2.3. *Teintures et fibres teintes*

Une grande partie des étoffes étaient teintes, mais la documentation écrite néo-assyrienne reste assez peu éclairante pour ce qui concerne les techniques et colorants utilisés.

Les listes de tribut des annales mentionnent assez souvent la laine teinte de pourpre rouge (*argamannu*) ou bleue (*takiltu*),[79] mais sans donner davantage de détails. Il s'agissait toutefois de produits très prisés et en une occasion, le rédacteur des annales de Tiglath-phalazar III mentionne même une curiosité sous la forme de « moutons vivants dont la laine est teinte de pourpre ».[80] On peut également citer la lettre de Sargon II à Aššur, qui mentionne parmi le tribut pris sur Muṣaṣir : « 130 étoffes multicolores et étoffes de lin, de la laine pourpre bleue et de la laine à tisser de couleur rouge des pays d'Urartu et Kilhu ».[81]

Quelques lettres font aussi allusion à des fournitures de fibres teintes ou de produits utilisés pour la teinture. La fin d'une missive de Sennacherib à Sargon, à propos du tribut de Commagène, montre que la laine teinte d'origine étrangère était un produit suffisamment prisé pour intéresser les marchands : « Ils ont aussi amené de la laine *rouge*. Les marchands m'ont dit : "nous avons choisi là-dessus 7 talents, mais les Commagéniens n'ont pas été d'accord et ont dit : 'qui pensez-vous donc être ? Vous n'avez pas à choisir (vous-mêmes). Qu'on apporte (la laine) pour que les tisseuses du roi fassent ici leur choix'" ».[82]

La production locale existait cependant. Dans une de ses lettres à Assarhaddon, Marduk-šarru-uṣur se plaint de ce que les livraisons attendues ne lui aient pas été fournies : « 31 talents de fil de lin, 80 talents de laine rouge du pays, 7 talents de (laine) noire du pays, 30 talents d'alun, 10 talents de natron, en tout 158 talents ».[83] La présence dans cette liste de natron et d'alun, servant de mordant, montre que les ateliers dont ce personnage était responsable s'occupaient à la fois de teinture et du traitement des fibres teintes. On trouve d'ailleurs un autre témoignage de cette activité dans un document administratif (SAA 7, 116), qui se présente sous forme de tableau. La première colonne enregistre des quantités de laine rouge, la seconde des quantités de garance[84] (*hūrutu*, *rubia tinctorum*), la troisième comporte une série de villes des provinces occidentales d'où ces produits étaient originaires. Enfin, un autre document administratif (SAA 7, 110) répertorie les quantités de laines rouges et noires nécessaires à la fabrication de diverses étoffes.

---

[78] Thureau-Dangin 1912, 60:386, 9 *lu-ba-ri-e* *<la>-al-bul-ti* dingir-*ti-šu ša ni-ip-hi* kù-gi *ia-ar* kù-gi *ši-biṭ-su-nu i-na mu-ur-di-e ṣu-ub-bu-tu*. Cf. Oppenheim 1949, 175.

[79] Dalley 1991, 123–124. Et voir ci-dessus les n. 3 et 5.

[80] Tadmor 1994, 68 Ann.14:3–4, udu-meš *bal-⌈ṭu⌉-[ti* ša síg-meš-*šú-nu*], *ar-ga-man-nu ṣar-pat* (Parmi le tribut de divers Etats syriens et anatoliens) ?

[81] Thureau-Dangin 1912, 56:366, [1] me 30 *lu-bul-ti bir-me* gada *ta-kil-tu* ù síg-meš *lu-bul-ti ta-bar-ri ša* ^{kur}*ur-ar-ṭi* ù ^{kur}*kil-hi*.

[82] SAA 1, 33:19–r1, síg-meš lum-lum-meš *is-se-niš na-ṣu-u-ni*, lú-dam-qar-meš *iq-ṭi-bu-u-ni*, *ma-a* 7 gú-un ta *lìb-bi ni-ib-ti-ar*, *ma-a* ^{kur}*ku-mu-ha-a-a la im-ma-gúr*, *ma-a ma-a'-at-tu-nu la ta-bi-ra*, *ma-a lu-bi-lu* mí-uš-bar-meš-*te*, *ša* lugal *am-ma-kam-ma li-bé-⌈e-ru⌉*.

[83] SAA 16, 82:r.5–10, 31 gú-un *ṭi-nu* gada, 80 gú-un síg sa$_5$ kur, 7 gú-un (síg) gi$_6$ kur, 30 gú-un na$_4$.*ga-bu-u*, 10 gú-un na$_4$*ni-ti-ru*, pab 1 *me* 58 gú-un.

[84] La nature des produits est explicitée en r 4', qui récapitule les rubriques précédentes [... g]ú hé-med 60 gú giš-hab x x, [x] talents de laine rouge, 60 talents de garance [...]. »

2.4. Les couleurs des textiles

Pour connaître l'utilisation de ces fibres teintes, les inventaires apportent des indications précieuses en précisant assez fréquemment les couleurs des textiles. Sont ainsi attestés le multicolore (*birmu*), le poupre avec les nuances rouge et bleu, le rouge (*sāmu*), le blanc (*peṣû*), et le noir (*ṣalmu*).

En considérant d'abord les textiles pour lesquels une seule couleur est mentionnée, on peut dresser la liste suivante.

2.4.1. *Textiles poupres*

Kiṣiptu de laine-*argamannu* (CTN 2 1:8') ; *pariktu* de laine-*takiltu* (StAT 3, 1:29)

2.4.2. *Textiles rouges*

Dappastu de laine rouge (*tabrību*) (StAT 3, 1:18) ; *elītu* (SAA 7, 105:11') ; *kusītu* (SAA 7, 105:7') ; *kuzippu* de laine rouge (*FNALD* n° 14:14) SI-LUH rouge sombre (sa gi_6) (StAT 3, 1:20) ; SI-LUH de laine rouge (SAA 7, 105:5') ; *šahartu* (SAA 7, 96:9') ; *ša-hīli* de laine rouge (CTN 2, 1:6') ; *ša*-IŠ (SAA 7, 105:7') ; *urnutu* de laine rouge (*FNALD* n° 14:15)

2.4.3. *Textiles blancs*

Kitû (StAT 3, 1:3 2) ; *kubšu* (SAA 7, 105:11') ; *nahlaptu* (StAT 3, 1:26) ; *niksu* (StAT 3, 1:10) ; *ṣubātu* (StAT 3, 1:16) ; *ṣubātu ša muhhi šarri* (StAT 3, 1:30) ; *šupālītu* (SAA 7, 94:1)

2.4.4. *Textiles noirs*

Dappastu (StAT 3, 1:19) ; *elītu* (SAA 7, 127:8') ; *šupālītu halluptu* (SAA 7, 127:9')

2.4.5. *Textiles multicolores*

Parmi les textiles comprenant plusieurs couleurs, les textiles dits multicolores, c'est-à-dire comportant probablement des motifs, étaient les plus estimés. Les inscriptions royales montrent qu'ils étaient réservés aux plus hauts personnages de l'État. Dans les textes administratifs, les attestations sont les suivantes : *kusītu* (SAA 7, 99:1, 105:6') ; ceinture-*nēbettu* (StAT 3, 1:25) ; *qarrāru* (StAT 3, 1:9) ; *qirmu* (StAT 3, 1:23) ; *sasuppu* (StAT 3, 1:28) ; *ša*-IŠ (SAA 7, 105:6').

2.4.6. *Textiles à plusieurs couleurs*

Si le terme *birmu* semble avoir été réservé aux étoffes à motifs, il arrivait assez fréquemment que des textiles soient ornés de bords d'une autre couleur, réalisés peut-être au moyen de bandes de tissu cousus à la pièce principale. Les inventaires précisent dans ce cas la couleur du « front » ou de la bordure : le rouge est alors de loin la couleur la mieux représentée, le noir venant ensuite.

Sur les couleurs des textiles, le témoignage des sources textuelles est en partie confirmé par les fresques subsistantes et les traces de couleurs encore visibles sur les bas-reliefs des palais[85]. Le vert et le jaune sont cependant totalement absents de la documentation écrite, alors même qu'ils figuraient sur certains textiles représentés dans le décor des palais.[86]

---

[85] Albenda 2005, 66–69.

[86] Cf. e.g. le génie ailé vêtu de jaune et vert sur une brique peinte du palais de Sargon II à Khorsabad : Albenda 1986, Pl. 150.

Il n'existe malheureusement presque pas de textiles anciens conservés, mais on a néanmoins retrouvé dans le bâtiment M de Sultantepe une pièce de tissu ornée de perles. Le décor présente des rangées de losanges, alternativement blancs avec une bordure rouge sombre et rouges avec une bordure jaune. Malgré leur intérêt, ces données sont insuffisantes pour permettre des comparaisons fructueuses.[87]

* * *

Au terme de cette étude, deux points méritent d'être soulignés. Pour ce qui concerne les principes organisant la nomenclature des textiles néo-assyriens, on constate que les termes renvoient plus souvent à une forme qu'à une fonction précise. Par exemple, *dappastu* est généralement traduit par « couverture ». Mais il s'agit d'une étoffe qui peut aussi bien être de laine que de lin.[88] Dans ce dernier cas, il s'agirait plutôt d'un drap. En outre, les scribes croient parfois utile d'indiquer « *dappastu* de lit »,[89] précision qui laisse entendre que l'étoffe pouvait servir à d'autres usages, par exemple de manteau. On remarquera aussi que des *maqāṭu*, des *hullānu* ou des *sasuppu* qui semblent être respectivement des sortes de toges, des capes et des serviettes, sont parfois dits « de maison », ou « d'intérieur » (*bēte*).[90] Là encore, ces termes doivent renvoyer à des formes d'étoffes dont on peut préciser la fonction particulière. Dans le même ordre d'idée, on peut signaler l'existence d'une « étoffe pa-[...] de salle de bain ».[91]

Pour ce qui concerne les couleurs, la documentation écrite, qui ne présente qu'une palette assez restreinte et probablement inférieure à ce qui existait dans la réalité, est sans doute biaisée par le prestige associé à certaines couleurs. Dans une longue lettre où il est question de malversations à Gūzāna, le compte-rendu d'un conseil tenu entre les hauts dignitaires de la province contient à cet égard une indication intéressante : « Šamaš-ēmuranni, le gouverneur, a tenu conseil avec Palṭī-Iaū et Nērī-Iaū disant : 'À qui devons-nous faire allégeance ?' Eux ont (répondu) au gouverneur : 'À celui qui t'a revêtu de laine rouge et t'a donné le bracelet d'or et la dague d'or' ».[92]

On dispose par ailleurs dans les inscriptions royales de multiples témoignages montrant que le don d'étoffes multicolores était pour les rois un moyen d'honorer leurs serviteurs. Les vassaux en recevaient parfois et le geste de Sennacherib en revêtant les travailleurs qui avaient creusé son canal dut être ressenti comme un honneur exceptionnel.[93]

Tout cela montre que les inventaires dont nous disposons, où la couleur rouge domine et où les étoffes à motifs sont bien représentés documentent principalement des textiles de valeur. Provenant du palais où de riches particuliers, ils ne donnent pas forcément une image fidèle des couleurs que portaient les Assyriens de l' époque, mais reflètent plutôt la mode du moment, avec sans doute des influences occidentales marquées, et les goûts d'une élite.

---

87 Barnett 1953, Pl. V/c. ; Brown 1980, 88–90.

88 Voir ci-desus, section 2.1.

89 SAA 7, 97:9', 6 ^{túg}*dáp-pa-sat* giš-ná ; SAA 7, 105:4'.

90 Pour le *maqāṭu*, SAA 7, 93:1, 94:4, 104:2'. Pour le *hullānu*, SAA 7, 107:r 3'. Pour le *sasuppu*, SAA 7, 120:ii' 4–5.

91 SAA 7, 120:ii' r 1–2, 1 ^{túg}*pa*-x-[0 0], *é-ra-ma-ki*.

92 SAA 16, 63:27–30 ^{Id}utu–igi-lal-*ni* lú-en-nam ta ^{I}*pal-ṭí-iá-u*, [ta] ^{I}*ni-ri-iá-u i-ta-ma-lik ma-a a-a-e-šá ni-ṣi-bat, ma-a šu-nu a-na* lú-en-nam *ma-a ša* síg sa_6 *ú-lab-bi-i[š]-ú-[ka-ni]*, [*ša* h]ar kù-gi gír kù-gi *i-di-nak-kan-ni*.

93 Luckenbill 1934, 82:33.

Abréviations

ARMT	Archives Royales de Mari. Traductions, Paris.
CAD	Chicago Assyrian Dictionary, Chicago.
CTN	Cuneiform Texts from Nimrud, Londres.
FNALD	J. N. Postgate, *Fifty Neo-Assyrian Legal Documents*, Warminster, 1976.
OIP	Oriental Institute Publications, Chicago.
Nbk.	J. N. Strassmaier, *Inschriften von Nabuchodonosor, König von Babylon*, Leipzig, 1889.
ND	Documents de Nimrud/Kalhu; Londres et Bagdad.
SAA	State Archives of Assyria, Helsinki.
SAAS	State Archives of Assyria, Studies, Helsinki.
StAT	Studien zu den *Assur*-Texten, Wiesbaden.

Bibliographie

Albenda, P. (1986) *The Palace of Sargon King of Assyria*, Paris.

Albenda, P. (2005) *Ornemental Wall Painting in the Art of the Assyrian Empire*, Leiden & Boston.

Al-Jadir, W. (1971) Le statut social des tisserands à l'époque des Sargonides, *Sumer* 27, 63–75.

Barnett, R. D. (1953) Assyrian Objects from Room M. 2, *Anatolian Studies* 3, 48–51.

Barrelet, M.-Th. (1977) Un inventaire de Kar-Tukulti-Ninurta : textiles décorés assyriens et autres, *Revue d'Assyriologie* 71, 51–92.

Brown, K. (1980) *The Question of Near Eastern Textile Decoration of the Early First Millenium B.C. as a Source for Greek Vase Painting of the Orientalizing Style*, University Microfilms International, Ann Arbor.

Dalley, S. (1991) Assyrian Textiles and the Origins of Carpet Design, *Iran* 29, 117–135.

Dietrich, M. (2003) *The Babylonian Correspondence of Sargon and Sennacherib*, SAA 17, Helsinki.

Durand, J.-M. (2009), *La nomenclature des habits et des textiles dans les textes de Mari*, ARMT 30, Paris.

Faist, B. (2007) *Alltagstexte aus neuassyrischen Archiven und Bibliotheken der Stadt Aššur*, StAT 3, Wiesbaden.

Fales, F. M. & Postgate, J. N. (1992) *Imperial Administrative Records, Part I*, SAA 7, Helsinki.

Fales, F. M. & Postgate, J. N. (1995) *Imperial Administrative Records, Part II*, SAA 11, Helsinki.

Kwasman, T. & Parpola, S. (1991) *Legal Transactions of the Royal Court of Nineveh*, Part I, SAA 6, Helsinki.

Luckenbill, D. D. (1924) *The Annals of Sennacherib*, OIP 2, Chicago.

Luukko, M. & van Buylaere, G. (2002) *The Political Correspondence of Esarhaddon*, SAA 16, Helsinki.

Oppenheim, A. L. (1949) The Golden Garments of the Gods, *Journal of Near Eastern Studies* 8, 172–193.

Page, S. (1968) A Stela of Adad-nirari III and Nergal-ereš from Tell al Rimah, *Iraq* 30, 139–153.

Parker B. (1954) The Nimrud Tablets, 1952– Business Documents, *Iraq* 16, 29–53.

Parpola, S. (1987) *The Correspondence of Sargon II, Part I*, SAA 1, Helsinki.

Postgate, J. N. (1973) *The Governor's Palace Archive*, CTN 2, Hertford.

Postgate, J. N. (1976) *Fifty Neo-Assyrian Legal Documents*, Warminster.

Postgate, J. N. (2001) Assyrian Uniforms. In W. H. van Soldt (éd.), *Veenhof Anniversary Volume*, 373–388, Leiden.

Radner, K. (1997) *Die neuassyrischen Privatrechtsurkunden*, SAAS 6, Helsinki.

Radner, K. (1999) *Ein neuassyrisches Privatsarchiv der Tempelgoldschmiede von Aššur* , StAT 1, Saarbrücken.

Tadmor, H. (1994) *The Inscriptions of Tiglath-Pileser III King of Assyria*, Jerusalem.

Thureau-Dangin (1912) *Une relation de la huitième campagne de Sargon*, Paris.

20. Textile Terminology in the Neo-Babylonian Documentation

Francis Joannès

The vocabulary related to the treatment of textiles is far from simple, since it is often dependent on and closely associated with a given moment in the history of techniques. For example, *mukabbu*, which refers to a Babylonian worker, can be translated into English as a "mender" or more exactly a "clothes mender" according to the CAD.[1] In French, however, it is translated according to the – quite poetic – activities of the professional designation, such as "rentrayeur", "stoppeur", "piqurier", "éplucheur", "épinceteur", "raccoutreur", "nopeur", and, when this person works with leather, "bichonneur". On the other hand, it is not easy to provide an exact translation, for instance, of the Akkadian term *lubāru*, which frequently appears in Neo-Babylonian textile terminology, as it is at the same time a fabric (French: "tissu"), a cloth (French: "étoffe"), a garment (French: "vêtement"), as well as a dress or a suit (French: "habit").

Nonetheless, in the current paper my intention is not to present the various aspects of the terminology of textile. Two major studies recently published, one by P.-A. Beaulieu,[2] the other by S. Zawadzki,[3] provide an exhaustive answer to many questions concerning the use of textiles in the Neo-Babylonian temples. Rather, my overall aim is to follow the production processes from the shearing of wool to the finishing of garments, mainly in the great Babylonian institutions of the 1st millennium BC. I will also examine the degree of documentation (i.e., recorded in detail, less documented, and no documentation at all), and the possible explanations for such differences in the levels of documentation. We will then examine the acquisition and storage of the raw textile material, followed by the first technical operations on the textile, especially the spinning and the weaving, and finally the manufacturing of fabrics.

1. Acquisition of the textile fibres

The concept of "wool circulation" has not been adequately illustrated or understood, despite the abundance of written attestations in connection with textile production. This state of affairs is due to the unevenly distributed documentation of the large institutions in 1st millennium BC

[1] CAD M_2, 181b.
[2] Beaulieu 2003.
[3] Zawadzki 2006.

Babylonia: we know when, who, and with which teams of workers the sheep shearing took place, i.e., generally during springtime. However, the following stage, the storage of the wool and how it was allocated to the spinners, remains scarcely documented.

1.1. *Storage of the wool*

What happened with the wool before it was spun? There were apparently three possibilities:

- direct allocation to the temple workers or to the royal palace workers, in relation to their system of maintenance: this refers to the "*lubuštu*/síg-ba" allocation.
- insertion into commercial networks. This suggests that most Babylonian families did not have sufficient wool resources for their personal use and that they had to buy it from peddlers. This would be most true for urban families; it is likely that the rural communities always kept a few cattle on hand.
- finally, wool was allocated to craftsmen to be processed, by means of an *iškaru*-contract: sometimes private families, mainly specialised workers in the service of the sanctuary, received wool to be transformed into a fixed number of pieces of cloths. It appears that most of the people in the weavers' workshops were working outside the temple, since there is no mention of a factory or a workshop in the inner area of the Babylonian sanctuaries.

It must be noted,[4] however, that in the temples, the wool was inspected and divided into several qualities and destinations: one qualification of the wool was termed "*sattukku*'s wool", which was probably reserved for the manufacture of the gods' clothes.

There are only scattered attestations of the places where the wool was stored. In Sippar, some wool came from the *bīt qāti*, i.e., a (or a group of) warehouse(s): *BIN* 1 26 is a letter, the sender of which asks for two talents of wool to be taken from the "northern building" (in the temple). He insists that the recipient of the letter should not take wool that is kept in store for the weavers.[5] Accordingly, the sanctuary of Šamaš in Sippar contained several buildings with that name, one of them located at the main entrance of the temple. In the same *bīt qāti*, garments made of linen were kept and piled up in wicker-baskets (*nakmaru*). Other storage buildings are known, sometimes without more detailed information. It is clear, however, that the finest wool reserved for weaving was kept in well-identified locations. Finally, the wool could have been stored in the royal warehouses, as shown by the text *Nbn* 788.[6]

Even if there is no specific vocabulary in connection with textile terminology, it is still possible to reconstruct a part of the wool circulation. For example, a group of cuneiform dockets concerning women in charge of spinning in the palace of Dūr-Yakīn, under the reign of Merodach-baladan II, the Chaldean king of Babylon, provides evidence for an interesting case study. It is a set of 24 clay dockets, each of them with a hole for attaching a cordon: these dockets consequently must have been tied to something. Each of them includes:

- A personal name (at least 16 of these are female names).
- An affiliation, probably administrative, following the formula "*ša* + (masculine) name"; most of these names, like Bāba-alsika-abluṭ, Hamkānu, Marnalu or Sîn-ereš, appear several times.
- A date, consisting of the month (always the eleventh) and the year of the reign of Merodach-baladan II (ninth, tenth, eleventh and twelfth years). For one docket, the date is unknown.

---

[4] Zawadzki 2006, 38.

[5] *ina* síg-há *šá ina* é-tu$_{15}$-si-sá 2 gun síg-há *kapdu šûbila ana muhhi* síg-há *ša išpari la taqarrub.*

[6] 1 gun síg-há *ultu šutummi šarri ana lubušti ša Šamaš.*

Spinner's (or Weaver's) Name	*Head of the Service*	*Month*	*Year*	*Reference*
^{f}Mannu-tammat	Bāba-alsika-abluṭ	11	9	JA 267 n°1
^{f}Iliti	Bāba-alsika-abluṭ	11	10	CTN 3 80
^{f}Nanaia-dayyānat	Bāba-alsika-abluṭ	11	10	JA 267 n°10
Šulundu	Hamkānu	11	10	JA 267 n°2
^{f}Šala'	Hamkānu	11	10	JA 267 n°4
^{f}Namirti	Hamkānu	11	10	JA 267 n°11
^{f}Uddu-harat	Hamkānu	11	10	JA 267 n°12
Bêl-ha'il	Marnalu	11	10	JA 267 n°3
^{f}Ma'adarhe	Ili-amar	11	10	JA 267 n°9
^{f}Kinnat	Sîn-ereš	11	10	JA 267 n°7
^{f}Rammenîti	Sîn-ereš	11	10	JA 267 n°8
^{f}Hepā	Sîn-ereš	11	10	K 3787
Nabû-išdī-ukīn *utullu ša* Dūr-Enlil	—	11	10	JA 267 n°16
Nabû-ušallim	Nabû-bêl-uṣur	11	10	UET 1 262
?	?	11	10	DŠ 32–11 (JCS 35 A.11)
^{f}Zudquti	Hamkānu	11	11	JA 267 n°14
^{f}Ili-šaptī-tazkur	Hamkānu	11	11	JA 267 n°15
[...]-sa-ru-rat	Marnalu	[o]	11	CTN 3 79
^{f}Halalat	Marnalu	11	11	JA 267 n°5
^{f}Hulumatu	Marnalu	11	11	JA 267 n°6
^{f}Mannu-kī-Bāba	Hussanni-Bêl	11	11	JA 267 n°13
Šamaš-lumur *re'û*	Emūq-Nabû ša Kār-Nabû	26–11	11	JA 267 n°17
Nabû-ipuš	Nabu-ukinanni	11	12	CTN 3 81
unknown	*unknown*	*unknown*	12	"texte d'Orléans" (JA 267, 247 note 2)

One distinctive feature of these dockets is that they were found in different places: primarily in Khorsabad, but also in Kalhu, and even in Ur. The presence of the tags has been associated with the capture and plunder in 709 BC of the southern capital of Merodach-baladan II, the city of Dūr-Yakīn, which was totally destroyed in 707 by Sargon II of Assyria.[7] The explanation proposed by J.-M. Durand for the *raison d'être* of these dockets is rather convincing: the dockets were tied to bundles of wool ready to be spun and which were already assigned to workers. After the capture of Dūr-Yakīn, these bundles would have been carried to Assyria, with their dockets, and subsequently distributed between several Assyrian royal palaces. According to G. van Driel, however,[8] the shearing operation took place in southern Babylonia mainly during the spring. The dating of the dockets, i.e., referring to month eleven, is therefore not in accordance with the shearing time. We thus have to imagine an intermediate period of several months, between the time of shearing and the time of the distribution of wool to the workers in the palace of Dūr-Yakīn and of the preparation of the wool for spinning. It seems equally possible here that this was not combed wool for spinning, but wool already spun and prepared as bundles of yarn ready for weaving.

[7] Durand 1979.
[8] Van Driel 1998, 228–229.

This wool must have been the yearly quota of raw material assigned to female weavers either working in a royal workshop or working at home according to the system of an *iškaru*-contract; that is, a contract of employment by which a major economic institution supplies the raw material (and sometimes the equipment itself) to some individuals, provided they deliver back the finished product after a certain period of time.

There are some other scattered references concerning an allocation to spinners of the wool collected by the large Neo-Babylonian religious or royal institutions. For example, in the text *BRM* 1 7, we find such an allocation for spinning or for combing of eight talents (210 kg) of wool with the aim of producing combed wool.[9]

The process of combing the wool was referred to in Akkadian by the term *ṣuppu* or *ṣuppātu* according to the CAD, which considers *ṣuppu* an Aramaic loan, a term denoting "a strip (or layer) of combed wool".[10] This product is thus mentioned in the text *Nbk* 286 next to *ishunu*, another kind of strip.[11] In *Nbn* 222, we find the transfer of strips of red combed wool (*ṣuppātu ša tabāri*) by an *išpar birmi* ("weaver of multicoloured cloth") to a *mukabbu* ("mender"): this transfer seems to concern mainly wool used for repairs.

For the time being, there has been only one reference to the process of spinning: in *CT* 56 454: rev. 8, we find the purchase of spindles (*pilaqqu*) by the temple of Šamaš, for [x] shekels of silver given to an individual.[12]

Regarding linen, the finest quality of this textile was often imported as a fabric already woven.[13] Furthermore the mention of "byssus" (Akkadian *būṣu*) as a fabric of extra-fine linen is only attested in texts from the neo-Assyrian period or, for Babylonia, in texts dated to the beginning of the 1st millennium. Thereafter *būṣu* is no longer recorded.

However, in addition to imported linen, there was also linen of local origin. The discussion by S. Zawadzki[14] concerning the linen fabrics referred to by the Akkadian words *salhu/šalhu* and *qatān* indicates that at least *qatān* denotes that the fibres were in a state in which they were ready to be woven. A possible translation could be "bundle of linen", which could well correspond to the modern expression "bundle of yarn" ready for weaving, that is to say, retted, beaten, combed and spun.

1.2. *From spinning to weaving*

Spinning yarn (called *ṭīmu*, *ṭimītu* or *ṭimūtu*) was either of wool or linen. The references to yarn might concern not only the raw material, but also yarn already dyed (sometimes imported) and allocated directly to specialised craftsmen: actually it was chiefly for embroidery, or for sewing together pieces of complex sets of clothing.

As noted by S. Zawadzki,[15] the mixture of wool and linen, which is prohibited in the Bible (Lev.

---

9 síg-há zi-ga mu-ni, 8 gun mí-uš-bar-meš, *a-na* síg *pu-šik-ki*, iti izi u$_4$ 20–kam, mu 7–kam ^{d}nà-pap lugal.

10 CAD Ṣ p. 249 b.

11 14 ma-na *dul-lu gam-ru*, ki-lá 10 síg *is-hu-nu*, ù 3–ta síg *ṣu-up-pa-a-ta*, ^{l}dub-numun *a-na* é-babbar-ra, *it-ta-din* iti še, u$_4$ 19-kam mu 35–kam, ^{d}nà-níg-du-uri$_3$, lugal tin-tir^{ki}.

12 *CT* 56 454:rev. 8 [x] gín kù-babbar *šá a-na pi-la-qu a-na* ^{Id}utu-sig$_5$-KAL sì-[*na*], [......] x ½ gín kù-babbar *ina pi-la-qí* 1 gín *re-hi*.

13 cf. *CT* 2 40, for instance.

14 Zawadzki 2006, 107–109.

15 p. 32, note 70.

19, 19; Deut. 22, 11), is not *a priori* excluded in the Mesopotamian texts. However, except for a single example of the manufacture of a curtain (*tunšu*) mixing wool and linen – perhaps only for embroidery or for strengthening the edges – there is no example in the administrative records of the textile documentation of such a mixture.

1.3. *Manufacturing the fabrics*

References to the next stage, i.e., to the process of weaving strictly speaking, are almost totally absent in the textual documentation. This activity is in fact strongly attached to a home-made production. It seems logical to suppose that the *išparu* weavers carried out their activity for the Ebabbar temple either at home or in specialised workshops for those fabrics demanding more elaborate work. Women are also mentioned as producing fabrics. The text *Dar.* 43, for example, makes clear that widows taken care of by the Ebabbar temple had to weave three coats (*gulēnu*) a year for the temple.[16] The yearly *gulēnu* coat, which is practically never attested in a religious context, but is considered private clothing, seems to be the standard production of one person working in a private environment.[17] Finally, a text from Uruk, published by M. Jursa, dated to the year 31 of Nebukadnezzar II, reveals the functioning of the system in temples.[18] We can deduce from this reference that an individual weaver at home could produce at least 2.5 kg of wool fabric in 6 months. But it cannot be concluded that 5 kg of wool would yield only 2.5 kg in terms of fabric. Without doubt, the remuneration of the weaver has to be included in the initial allocation.

2. Technical operations on the textiles

The management of fabrics and garments is especially well documented by the administrative documentation of the Neo-Babylonian temples. However, the management appears complex due to the fact that two spheres of economic and artisanal activity are linked: on the one hand, a group of highly specialised craftsmen, authorised to work within the sanctuary and to use expensive dyeing materials, especially the different kinds of purple, and, on the other hand, another group of craftsmen working outside, in their urban family workshops. As already noted by H. Bongenaar,[19] however, some of these outside specialists were oblates, i.e., slaves of the temple, and it is often mentioned that they live in houses rented to them by the temple of Šamaš. The analysis by S. Zawadzki of the Sippar documentation[20] is particularly enlightening from this point of view: he shows without doubt that there was a *vertical* division of labour, with a succession of stages like shearing / spinning / weaving / bleaching /dyeing / and final ornamentation. Each of these stages required the intervention of a specialised craftsman. Simultaneously, however, there was

[16] *Dar.* 43: "Among them, Idintu, Mistaia and Bazîtu will have to supply, from their own production, 3 *gulēnu*-coats a year, following the rules of the *iškaru*, to the god Šamaš" (*ina lìb-bi* ^{f}*i-DIN-tu*$_4$ ^{f}*mi-is-ta-a u* ^{f}*ba-zi-tu*$_4$ (6') *ina* mu-an-na 3 túg *gu-le-en iš-ka-ri a-na* ^{d}utu (7') ta *ra-man-ši-na i-nam-din-na-'a*).

[17] Cf. the texts (*NBDMC*)-*Michigan* 47, where a woman has to weave a *gulēnu*-coat a year, and *VS* 5 24, where in the yearly obligations of the gardener of Ṭābiya appears the production of one *gulēnu*-coat for his landowner, from 5 *mana* of wool supplied to him by Ṭābiya.

[18] Jursa 1997, n°13: "5 mana of clothes, equivalence of 10 mana of wool, property of the Treasury of Ištar and Nanaia are charged on Tuqnaia. She will have to deliver it in the month of Du'ūzu" (5 ma-na túg *mi-ih-ṣi* šám 10 ma-na síg-há níg-ga ^{d}gašan *šá* unug^{ki} *u* ^{d}*na-na-a ina* ugu ^{f}*tuq-na-a* dumu-mí-*su šá* ^{Id}en-mu-gar-*un ina* iti šu *ta-nam-din*).

[19] Bongenaar 1997, 302.

[20] Zawadzki 2006, 57–67.

a *horizontal* division of labour depending on whether wool or linen was used, or whether simple wool or purple wool was used, and so on. This system produced specialisations (for instance *ašlāku* and *puṣṣayu*) which took place together at a given stage of the fabrication process or at several levels, such as the *išparu*, whose job was to weave and to finish the garments. From this we can see a kind of grid, which can seem very complex and not rational, but which in fact has its own logic.

Moreover, in the administrative documentation of the neo-Babylonian temples, at least two systems of circulation of the textiles can be found: on the one hand, a circuit of manufacture from the raw material until the final decoration; on the other hand, a circuit of maintenance, implying cleaning and restoration for the re-using of the gods' clothes. All these operations correspond to terminology which can be summarised in a table, although of course not exhaustive:

shearing	*gazāzu*	*gāzizu*
spinning	*ṭamû*	*ṭāmītu*
weaving	*kaṣāru, šapû, maḫāṣu*	*išparu*
fulling	[*ašlākūtu epēšu* (?)]	*ašlāku*
bleaching	*pūṣu*	*pūṣāya, ašlāku*
cleaning	*zikûtu*	*ašlāku*
dyeing	*ṣab/pu, ṣubītu*	*išpar birme, ṣābû*
finishing	*šapû(?)*	*išparu*
repair	*batqa ṣabātu*	*mukabbu*

It has also been noted that the jobs of the *ašlāku* and the *pūṣāya* are still quite unclear: it seems that when a new garment was given to them for work, they had to bleach it; on the contrary, when the garment had already been worn, they mostly had to clean it. But many other questions remain, for instance, concerning the nature and format of the products used as dyes: dyeing products were mostly of vegetable origin, with the exception of the mineral alum used as mordant. The use of yarn already dyed is also well attested, as shown by some allocations of blue or red purple wool, of which is said they are given for repair (*ana batqa*): certainly, this concerns dyed yarn.

As demonstrated in the written inventories of clothing taken from storage for the divine processions (*ana tabê*), the management of the different sets of clothing was a rather complicated task: clearly, entirely new clothes (*eššu*) manufactured especially for these occasions are combined with "old" clothes which had already been used for a while (*labīru*). Thus, we can conclude that the garments of the gods were not kept in the Treasury for a very long time, nor that such clothes were cleaned frequently. As a consequence of the continual search for a completely pure environment for the divine statues, the ceremonies of clothing (*lubuštu*) mostly implied a presentation of new clothes.

At the same time, however, we may assume that most of the really valuable clothes (especially those made of fine linen) were kept safely in coffers or some kind of chests in the Treasury of the temple: a text dated to the Hellenistic period reports, for instance, the fact that the Seleucid king Antiochos III visiting Babylon had the opportunity to see the ceremonial coat of Nebukadnezzar II, preserved for three centuries in the Treasury of the Esagil.

Finally, it should be noted that there is no attestation of simple manufacture of garments in the records of allocations to specialised craftsmen; on the contrary, there are numerous records of transfers of precious or rare products from the temple's store rooms: chemicals for bleaching, and also products for dyeing as well as bundles of purple wool of various shades. Such distributions to craftsmen resulted in many administrative documents in the form of receipts and inventories.

Besides the operations of laundering, dyeing, and repair, the craftsmen certainly also intervened to decorate the edges or the main surfaces of cloth. Techniques included binding, fringes, hems, embroidery or braiding. There are only few indications of these activities, but P.-A. Beaulieu, resuming a study by L. Oppenheim,[21] collected data concerning the application of ornaments made from precious metal on the *kusītu* dress, the *mēṭu* garment, and the *nēbeḫu* belt of the statues of Ištar and others goddesses in Uruk. He terms these ornaments of precious metal "*sequins*" in the form of lions, stars, crosses and rosettes. The manufacture of these pieces of gold or silver introduced a new type of craftsmen, specialised in the processing of precious metals, into the circuit of garments.

Moreover, besides wearing single-coloured garments and multi-coloured garments decorated with ornaments or combining elements of different colours, the divine statues were decorated with great assemblages of jewellery, particularly on the headgear, which combined complex headdress with ornaments in precious metals and stones.

3. Neo-Babylonian clothing

As has been noted before, there is not always a clear distinction between "fabric" and "garment" in the neo-Babylonian terminology. P.-A. Beaulieu considers, for instance, *mihṣu* a generic term which can designate either a coat or a fabric. S. Zawadzki establishes the basic signification of the Akkadian *ṣibtu* as a fabric that was probably rectangular and draped to form a garment. On the other hand, some inventories mentioned rolls of a fabric called *kirku*, a term of Aramaic origin.[22]

It is well known that during the Neo-Babylonian period, some very general designations were used for clothing such as *muṣiptu* or túg-kur-ra, for which the best translation would probably be "dress" (in French: "habit"). According to M. Roth, in her article concerning the Neo-Babylonian dowry,[23] the word *muṣiptu* is "a generic term for garments" and may refer to small or large, simple or complex garments. The same generic function is fulfilled by the Sumerian sequence túg-kur-ra, even if it is not the ideographic equivalent of Akkadian *muṣiptu*.

The same use of generic designation may be applied to the word *lubāru* as a kind of standard garment; for instance, in some Neo-Babylonian contracts of sale, a clause is added concerning a gift for the seller's wife, with the expression: "... the dress for the owner's wife".[24] In some cases, *lubāru* is even substituted by túg-há.

Finally, the frequent designation *šir'am/sariyam* of Aramaic origin applies to a common outdoor garment, a sort of coat or military jacket. It seems to be the secular equivalent of *nahlaptu*, which during the Neo-Babylonian period applied only to religious or royal ceremonial-dress.

---

[21] Beaulieu 2003, 21. Cf. Oppenheim 1949.

[22] Cf. in *TBER* 93 a *kirku* "woven at home" (*ša ina bīti mahṣu*).

[23] Roth 1989/90.

[24] *lubāri ša bēlet bīti.*

As a matter of fact, the entry *sir'iam* in the CAD[25] shows that during the Neo-Babylonian period, a great variety of *šir'am* existed, found mainly in the text *Nrg* 28 from the Ebabbar-archive, combined with designations such as:

šir'am ša zikāri	"men's š."
šir'am ša kitī (ša) amilti	"linen š. for a lady"
šir'am ša kitī	"simple linen š."
šir'am ša tabāri	"red š."
šir'am ša inzahurēti	"blue š."
šir'am ša síg hé-me-da	"š. made of red purple wool"
šir'am šupālītu eššetu babbanītu	"new fine š. to wear as under-garment"
šir'am elēnītu murruqītu babbanītu	"very fine š. to wear as upper-garment"
šir'am ša qallatu	"small š."

and so on ...

To conclude, two main facts must be considered concerning textiles and clothing during the Neo-Babylonian period: we have, on the one hand, some very precise and functional terms for the garments created for and adapted to specific occasions and mainly used in the luxurious environment of the temples; on the other hand, we have many quite vague cloth and clothing designations, often of Aramaic origin, and commonly used for the ordinary people. To identify the exact meaning and translation of the name of garments remains almost impossible, but we do have the possibility of perceiving some general features or combinations in costumes like headdress/coat/belt. The investigation of these features can lead to a better understanding of the series of garments mentioned in the Neo-Babylonian administrative documentation.

Abbreviations

BIN	Babylonian Inscriptions in the Collection of J. B. Nies. New Haven.
BRM	Babylonian Records in the Library of J. Pierpont Morgan, I–IV. New Haven.
CAD	*The Assyrian Dictionary of the Oriental Institute of the University of Chicago.* Chicago (1956–).
CT	Cuneiform Texts from Babylonian Tablets in the British Museum. London.
CTN	Cuneiform Texts from Nimrud. London.
Dar	Strassmaier J. N. (1897) *Inschriften von Darius, König von Babylon (521–485. Chr.).* Leipzig.
JA	*Journal Asiatique.*
NBDMC	S. Michigan coll.
Nbk	Strassmaier J. N. (1889) *Inschriften von Nabuchodonosor, König von Babylon (604–561 v. Chr).* Leipzig.
Nbn	Strassmaier J. N. (1889) *Inschriften von Nabonidus, König von Babylon (555–538 v. Chr.).* Leipzig.
TBER	F. Joannès, Textes économiques de la Babylonie récente: étude des textes. Paris, 1982.
UET	Ur Excavations: Texts. London.

Bibliography

Beaulieu, P.-A. 2003 *The Pantheon of Uruk During the Neo-Babylonian Period*, Cuneiform Monographs 23.

Bongenaar A. C. V. M. 1997 *The Neo-Babylonian Ebabbar Temple at Sippar: Its Administration and its Prosopography*, Publications de l'Institut historique et archéologique néerlandais de Stamboul 80, Nederlands Historisch-Archaeologisch Instituut te Istanbul, Leiden.

Van Driel, G. 1998 Neo-Babylonian sheep and goats. In J. N. Postgate & M. Powell (eds.), *Domestic Animals in Mesopotamia*, Part I, Bulletin on Sumerian Agriculture 7, 219–258.

[25] CAD S, 314b.

Durand, J.-M. 1979. Les "slave documents" de Merodach-Baladan, *Journal Asiatique* 267, 245–260.

Jursa, M. 1997 Neu- und spätbabylonische Texte aus den Sammlungen der Birmingham Museums and Art Gallery, Iraq 59, 97–174.

Oppenheim, A. L. 1949 The golden garments of the Gods, *Journal of the Near Eastern Studies* 8, 172–193.

Roth, M. 1989–1990 The Material Composition of the Neo-Babylonian Dowry, *Archiv für Orientforschung* 36/37, 1–55.

Zawadzki, S. 2006 *Garments of the Gods, Studies on the Textile Industry and the Pantheon of Sippar according to the Texts from the Ebabbar Archive*, Orbis biblicus et orientalis 218.

21. Garments in Non-Cultic Context (Neo-Babylonian Period)

Stefan Zawadzki

Research on the terminology of garments, clothes and fabrics in non-cultic contexts encounters a fundamental obstacle caused mainly by the character of the documents on which they are based. Whilst for studies of garments of the gods great importance is given to the so-called tabulated lists enumerating garments taken from a temple's wardrobe to be repaired or cleaned and returned before the celebration,[1] no such category of documents exists for the study of the textile terminology in non-cultic context. The production of garments in private houses did not need written documentation.

1. Sources

The abundance of documents concerning the textile industry, which comes mostly from temple archives, should be interpreted with great caution to prevent the drawing of false conclusions. Let us take an example of the prosopographical data. Among weavers in the Neo-Babylonian temple household only a few names of women are known, which might be interpreted as proof that the manufacture of garments there was almost exclusively in the hands of men. Such an opinion might be based on the data from the Neo-Babylonian Ebabbar temple at Sippar. Except for the woman called Muranātu who appears quite often,[2] the names of only two other women are known: Ri'du[3] as a member of the group of five weavers, and Ubartu as manufacturer of garments for the *lubuštu* ceremony for the month Arahsamnu;[4] additionally Dar 43 mentions three women manufacturing garments for the Ebabbar temple as their labour assignment (*iškaru*). The poverty of data is striking and it is in sharp contrast with the data from Old-Babylonian Mari,[5] where the

[1] The texts, which are discussed in Zawadzki 2006, will be published in his *Garments of the Gods, Part II: Texts* (abbreviated: GG II). I would like to thank John MacGinnis for his comments and improvement of my English.

[2] Bongenaar 1997, 331–332 and Zawadzki 2006, 219.

[3] Unpublished BM 66172: 4'.

[4] See BM 69225 (= GG II 494). She was supplied with blue-purple wool and made her job in the workshop called *bīt asî*. For the meaning of the term, see the comments to GG II 494.

[5] A few years ago I had the chance to familiarize myself with the Mari documents concerning the manufacture of the textiles when Jarosław Maniaczyk, Ph.D., studied them while preparing his dissertation written under supervision of prof. J-M. Durand and myself.

number of women manufacturing cloth is quite high.[6] The difference can easily be explained, i.e. in the Neo-Babylonian period the temple textile industry was organized in the prebendary system, owned by men, while in Mari the textile industry was an important part of the palace economy, located partly in the women's interior quarter of the palace. At least some of the ladies mentioned in the documents from Neo-Babylonian Sippar appear probably because they inherited the prebends of their fathers, who might have had no male successors.[7]

2. Basic terminology

In the Neo-Babylonian period the textile terminology in non-cultic context appears mostly in documents dealing with the following categories:

- dowries[8]
- issues of rations (as an obligation to a benefactor)[9]
- payment for wet-nurses[10]
- as part of the soldier's uniform[11]
- worker's clothing[12]

where garments are only one component among many others such as real estate, slaves, silver, food, furniture, etc. Some texts used the terms in their precise meaning, in other the terms are used in their generic meaning, simply to stress that the person is obliged to provide a regular supply of clothes to another person (wet-nurse, former owner who gave his property for a care, etc.). It means that the same term can describe a specific garment or only denote "clothing" in general, regardless of their specific function. In such different meanings the terms *lubuštu*, *lubāru*,[13] *muṣiptu*, but also túg-kur-ra were used.

[6] It seems that similar situation was also in the Neo-Babylonian palace economy, see Kleber 2008, 246 ff. (5.3. Textilherstellung im Palast).

[7] Concerning the possibility of inheritance of the prebends by women, see Jursa 1999, 61 and Wunsch 2003, No. 17 and No. 43, cf. Jursa 2005, 32.

[8] For analysis of garments terminology in such a context, see Roth 1989/1990, 29–33.

[9] Cf. for example Nbn 697 (Wunsch 1993, No. 211), where Iqīša freed his slave under the condition he would supply him with *kurummatu* and *lubuštu*. However, the freed slave escaped and Esagila-ramat, the wife of Iddin-Marduk/ Iqīša took over the care of Iqīša supplying him with *ipri piššatu u lubuštu*. In such a situation Iqīša cancelled the previous agreement and established the woman as the owner of the disloyal slave. Cf. also VS 5, 21 (= NRV 12) from Dilbat where Uraš-šum-iddin who was ill, asked his daughter Ṭabatu for care because his brother disappeared and his son escaped. According to an agreement a father established his daughter the heir of his measurement (*mandidūtu*)-prebend for the ration (*kurummatu*) comprising *ipri piššatu u lubuštu*, obliging himself not to sell the prebend, not to grant it to anybody else, and not give it as a pledge or deduct any (other) expenses. For other such texts, see CAD under *ipru*, *piššatu* and *lubuštu*. Cf. also VS 5 47 (Baker 2004, No. 31) establishing the supply of one *gulēnu* garments yearly by the adopted son to his adoptive father.

[10] Wunsch 2003/2004, 173f.

[11] Concerning soldiers' uniform, see MacGinnis, in press.

[12] No special studies exist, although the data appears in many documents. The most popular were túg-kur-ra and *širam* garments.

[13] The *lubuštu* appears in a document establishing a type of remuneration for a person who needed special care, for example because of old age or illness, see examples cited in Note 9. It is interesting to note that in the contract concerning the sale of real estate in which additional payment (*atru*) and a garment are sometimes added to the full price, the garment is always described with the term *lubāru*, cf. for example Cyr 345:26 (sale of a built plot in the district Šuanna in Babylon); Cam 423:15 (sale of a plot with a derelict house and a hut in the district Tê in Babylon); Nbk 4 (sale of a small plot with a derelict house in the same district as in the previous document); Wunsch 1993, No. 292,

Let us consider first the term *lubāru*, which usually describes clothing in general. The term does however have a more specific meaning in Uruk in comparison with Sippar. In Sippar the *lubāru* is an outer dress of three gods: Šamaš, Bunene and Adad, and only one goddess: Anunītu, while the outer dress of goddesses (except Anunītu)[14] was called *kusītu*. In Uruk[15] the situation is more complicated as some goddesses (but not all, an exception is for example Nanaya) could be dressed in a *lubāru*, made of white wool, the heaviest item in their dress, but their wardrobe (again with four notable exceptions: Bēlti-ša-Rēš, Uṣur-amassu, Urkayītu and Ladies) also contained a *kusītu*, comparable with the *kusītu* of goddesses in Sippar. Might the *lubāru* be dressed over the *kusītu*? A positive answer can be suggested, as the weight of *lubāru* of Lady-of-Uruk was more than double that of the *kusītu*.[16] This suggests that *lubāru* was a very large loose poncho laid over the richly adorned *kusītu*. The tradition of Sippar was different, as *lubāru* did not belong to the attire of goddesses there.

How to explain such a difference between these two cultic centers? At least one of a few possibilities is that there existed in society an association of *lubāru* with the highest or dominant position of a deity, held in all important cities by gods, except for Uruk, where the dominant positions were held by goddesses. By dressing the goddesses in Uruk in the *lubāru*s their pre-eminent position, comparable to the position of gods, was stressed. It should be noted, however, that known texts from Sippar and Uruk never mention adornments of the *lubāru* with precious metals or precious stones; such adornments concern exclusively the *kusītu* and *muṣīptu* garments. One can say that some goddesses in Uruk had two outer garments – the *lubāru*, and when this was taken off, a *kusītu*.

Evidently the term *muṣiptu* was also used in such a broad manner. This is clear from BM 76968/72,[17] where *muṣiptu* describes 13 items enumerated earlier as a part of a dowry. A similar situation appears in Stigers, No. 42 where the first line, to be treated as a heading, reads "garments and household goods of the *abarakku*" (^{túg}*mu-ṣip-tu*$_4$ *ù ú-di-e šá* ^{lú}agrig), followed by a listing of túg-kur-ra ^{túg}*širam*, *lubāru ša qabli*,[18] and one additional garment which cannot be identified due to the condition of the tablet. Basing herself only on first text, M. Roth recognized that *muṣiptu* "is a generic term which may be used to refer to several types of garments".[19] It seems true, and one can add additional observations, based on a few details present in some other texts. For example, according to Dar 530 *muṣipētu* were placed rolled in the chest (*arannu*); rolled *muṣipētu*, in addition to túg-ḫum-ḫum, are mentioned also in Nbk 369. This means that, after having been taken of, off, they could have been easily rolled up. If so, the *muṣiptu* must be a rather large piece of fabric or cloth, which could be shaped around the person. Specific threaded *muṣipētu*, typical of the region around Tukriš are mentioned in the Neo-Assyrian letter.[20] It is striking

Rev. 13' (sale of a plot with date garden). For more such texts in which *lubāru* is added, see CAD A II 502. Although it seems that in both categories of the documents *lubuštu* and *lubāru* were used in their generic meaning, the consequent use of a different word should be noted.

[14] It is treated in detail in Zawadzki 2006.

[15] For the list of items which are present in the attire of particular goddesses in Uruk, see Beaulieu 2003, 15–16.

[16] Cf. Zawadzki 2006, 89 and 118.

[17] Roth 1989, No. 42.

[18] To be added to the few known references. Probably the same garment written g a d a *šá* murub$_4$ appears in the unpublished tablet BM 60071: 4.

[19] Roth 1989/1990, 28.

[20] ABL 511: 8–9 ($^{\text{túg!}}$*mu-ṣi-pe-ti/kar-ke-e-ti*); see the last edition in Dietrich 2003, No. 122.

that the term is almost always used in non-military contexts. I know only one text BM 70342,[21] where *muṣiptu* were given to soldiers who went on a military expedition, but in all other cases *muṣiptu* was a term describing civilians' clothes. This is supported by BM 103452: 16,[22] where the garments stolen from a private house are designated *muṣiptu*.[23] In one text *muṣiptu* was part of remunerations paid to a wet-nurse.[24]

The túg-kur-ra usually refers to a specific garment, but sometimes, as demonstrated by BM 63956, the term can have a generic meaning:

BM 63956 (= Bertin 1441); collated:

1. 3 túg-kur-ra *la ḫi-ri*
2. 6 ^{túg}*il-ta-'-pi ḫi-ru-*[tu_4]
3. 5 túg-kur-ra *ḫi-ru-t*[u_4]
4. pap 14 túg-kur-ra-m[eš ...

In this text *iltapi* is recognized as a special type of garment included in the general category of túg-kur-ra garments.[25] However, the same term could also be used in a highly imprecise way as is demonstrated by the comparison of two texts, where in the first one the name of the garment is written ideographically túg$^{\text{hi-a}}$, for which the syllabic writing *lubāru* is well established,[26] while in the second, where evidently the same garment is meant, the ideogram túg-kur-ra is written. In fact túg$^{\text{hi-a}}$ = *lubāru* cannot be the same as túg-kur-ra, for which the reading *lubāru* is excluded.

CT 55, 859

1. ½ *ma-na* ⌜6 gín⌝ síg-za-gìn-kur-ra
2. *a-na* túg$^{\text{hi-a}}$ *šá* ^{d}utu *ù*
3. ^{d}*Bu-ne-ne*

BM 59621[27]

3. ½ *ma-na* síg-za-gìn-<kur>-ra
4. *a-na* túg-kur-ra *šá* [^{d}]utu
5. 6 gín *a-na* túg-kur-ra
6. *šá* ^{d}*Bu-ne-ne*

There are many texts similar to BM 59621 where 30 shekels are given for *lubāru* (túg$^{\text{hi.a}}$) of Šamaš and 6 shekels for *lubāru* of Bunene, so it is evident that túg-kur-ra in BM 59621 stands for túg$^{\text{hi.a}}$.[28] Although two different ideograms are used, obviously the same garment, *lubāru* is meant in both texts.

A similar imprecise use of a term concerns *taḫapšu* denoting a type of blanket used as cover and made quite often of old used items, such as *salḫu* or *kibsu*.[29] However, in BM 65145,[30] instead of the typical data that the garments listed below are destined for the *lubuštu* ceremony, we have the entry [*šá*] *ta-ḫap-š*[*ú* ..../ [*šá* u]d-15-kam "for the *taḫapšu* of the 15th day" what is unparalleled. A similar situation appears in BM 70592,[31] where Ardiya, the well-known weaver producing multi-coloured cloths, received wool "for (manufacturing) the blanket for the month Ayaru" ([*a*]-*na*

[21] Edited by Zawadzki 2003, 280*.
[22] Jursa, Paszkowiak, Waerzeggers 2003/2004, 265.
[23] Cf. also BM 64880 (MacGinnis 1998, No. 4: 9) where *muṣiptu* denotes also garment in general.
[24] Wunsch 2003/2004, 214.
[25] Cf. also BM 73306: 4'-7' (Appendix No. 3).
[26] In the many parallel texts published in GG II the writing túg$^{\text{hi.a}}$ or *lu-ba-ri*, ^{túg}*lu-ba-ra*, ^{túg}*lu-ba-ri*, ^{túg}*lu-ba-ru* is used.
[27] See GG II 455.
[28] Additionally it should be noted that túg-kur-ra is not noted as a garment for the gods.
[29] See Zawadzki 2006, 134–135.
[30] See GG II 275.
[31] See GG II 423.

ta-ḫap-šú šá iti-gu$_4$), again instead of expected *a-na lubuštu šá* iti-gu$_4$ "for the *lubuštu* ceremony in the month Ayaru", known from many other parallel texts.

The problem, which in my opinion is not definitely resolved concerns the Akkadian equivalent of the above discussed term túg-kur-ra, for which different readings have been suggested. The oldest proposal – to read túg-kur-ra as *muṣiptu*[32] – was already refuted by Oppenheim because *muṣiptu* is feminine while túg-kur-ra is followed by masculine adjectives,[33] and is now definitely excluded by the above quoted BM 76968/72, where túg-kur-ra is one of additional 12 garments described together as 13 *muṣiptus*. The next proposed reading was *sadru*,[34] refuted, however, by Borger.[35] The latest reading, *suḫattu* or *supātu,* is suggested by CAD S, 346 based on a comparison of two texts, UCP 9, 271 and Dar 253. The first of these enumerates the equipment for one soldier, the second for 12 soldiers:

UCP 9, 271	**Dar 253**
one *suhattu*	12 túg-kur-ra (i.e. one for each soldier)
one *sir'annu* (= *sir'am*) *parzilli*	12 *sir'am*
one *kūrapānu ša suḫattu*	–
one *karballatu ša sir'annu*	12 *karballatu*
12 *nūtus*	
24 *šēnus* (= 12 pairs)	

The parallelism between both texts is striking, and the probability that túg-kur-ra should be read *suḫattu* or *supātu* is high, though some doubt still exists. The problem is that by using the same method one could propose reading túg-kur-ra as *muṣiptu*, which, as already mentioned, can be excluded. Such an obviously wrong equation might also be formulated by comparison of the texts concerning payments to wet-nurses or to the adoptive parents for raising a child.[36] The last entry in these texts is usually túg-kur-ra, though in a similar text (GC I 14, concerning the repayment of the costs of nutrition and raising), in the case that the sister demands to take her son back from his adoptive father (his uncle) she has to pay all costs including, among other garments, the *muṣiptu*.

An interchangeable use of ideogram and syllabic writing as a proof that a syllabic writing is the equivalent of an ideogram should therefore be used with great precaution, though equally it cannot be discarded. I used this method when establishing that *patinnu* is the syllabic writing of túg-murub$_4$.íb-lá. Despite the reservation to this equation expressed by the authors of CAD P, 277,[37] who describe my proposal as a "suggestion", I still maintain my view. When searching for convincing arguments for identifying an ideogram and the corresponding Akkadian word at least four criteria should be observed:

1. The texts should belong to the same category of documents and be laid according to the same format.
2. A generic meaning for both the ideographic and the syllabic writing should be excluded.
3. The term discussed should denote an element of a specific costume.
4. Data from texts outside the group being compared should not be in conflict with the proposed equation.

[32] Ungnad 1937, 96.
[33] Oppenheim 1950, 188–189; cf. Borger 1957, 7.
[34] CAD Ṣ, 225b.
[35] Borger 1981, 187, No. 536.
[36] For the list of components, see Wunsch 2003/2004, 214.
[37] The CAD authors knew only my short text in NABU 1997, no. 16, now enlarged a little in Zawadzki 2006, 121–122.

There are some significant differences in the processes by which the equations túg-murub$_4$.íb-lá = *patinnu* and túg-kur-ra = *suḫattu/supātu* were established. Firstly, for túg-murub$_4$.íb-lá a generic meaning is excluded and all items in the texts being compared comprised the attire of just a single god or goddess. Only in such a context, where the entries occupy the same position in a stereotyped list, can an interchangeable use of idiomatic and syllabic writing be supported. This gives a good base for the recognition of the equivalence. Because in the text quoted above *ilta'pi* was included in the general category of túg-kur-ra we cannot exclude that also *suḫattu/supātu* is used with a similar sense, i.e. that it is one of the small number of terms covered by the generic term túg-kur-ra. To be sure that *suḫattu/supātu* is a real equivalent of túg-kur-ra more examples are needed.

An interesting observation relating to túg-kur-ra can be made on the basis of BM 79658,[38] which concerns the dispatch of equipment for 30 soldiers fighting against Tyre. Apart from one túg-kur-ra for each soldier, 9 túg-kur-ra were given "for their tents" (*ana maškanatušunu*), which evidently suggests that it was a loose long coat to be used as a cover by soldiers in their tents. Such an opinion is also supported by the fact that instead of 8 *šir'am* the soldiers received an additional 4 túg-kur-ra, which suggests that túg-kur-ra were longer/larger than *šir'am*. The important thing is that two texts mention "half" (*mišil*) túg-kur-ra,[39] which suggests that two types of túg-kur-ra, long and short, were in use.[40] Of all garments, túg-kur-ra was evidently the most popular garment used by people working in every branch of human activity. It seems to me that soldiers used it as a daily garment, while *šir'am* was a piece of military equipment, put on before a battle against enemies.

Although *šir'am* was an important element of the uniform of soldiers and of the archers employed by the temple as a guard for temple flocks, slaves and workers (such as for example millers),[41] it also appears as a component of dowries.[42] If the same term was used to describe the clothing of persons active in such different fields and of different social positions (free and non-free), it is obvious that there must have been some elements in common, making it possible to define it as *šir'am*. The use by members of such different groups as soldiers, archers or millers rules out the possibility that it was a long suit. Rather it must be a close-fitting garment, otherwise it would hinder workers from performing their work. A similar meaning is proposed by John MacGinnis in his forthcoming book, where the translation "jerkin" is given. Depending on the use *šir'am* might be manufactured from linen, wool or even hides. It is well known that *šir'am* for soldiers in earlier periods were produced with leather, reinforced with metal pieces or scales; in the Neo-Babylonian period it was manufactured mostly of wool because the term is regularly preceded by the determinative túg, not kuš as in the earlier periods. There were probably differences between the kinds of *šir'am* depending on the wearer's profession and position in the society. Both men and women used the garment, there was an outer (*elēnītu*) and an inner (*šupalītu*) type, and it could be cheap (non-dyed) or expensive, dyed with red and blue-purple dye (see CAD S, 314).[43]

[38] Zawadzki 2003, 279*.

[39] Nbn 662: 12–13 and BM 75889 = Bertin 3059: 1, 10–11; cf. Appendix No. 7.

[40] Basing himself on this and a few additional texts Oppenheim 1950, p. 189 suggests that túg-kur-ra was simply a blanket.

[41] Discussed by Bongenaar 1997, 13, and Kessler 1999, 251–252.

[42] Roth 1989/1990, 31.

[43] Cf. the full list given by F. Joannès in this volume.

In the following I wish to present and discuss an unpublished fragment of a text belonging to the collections of in the British Museum, as it includes a number of words which are either new or have new attributes. Although the tablet reached the British Museum together with the vast collection of tablets uncovered in Sippar, it does not look like a typical text written in the Ebabbar temple. The signs are well written, precise and elegant; the tablet was prepared in advance and before use it was covered with a cloth impression of which are preserved in a few places. Only part of 12 lines of the obverse is preserved; the reverse is totally damaged; based on l. 11' which mentions "this silver" it is certain that a significant part of the beginning of tablet is missing.

3. New evidence: BM 76136

BM 76136 (83–1–18, 1501)[44]
7.5 × 5.8 cm

1'. 2 ⸢gada(?)$^{\text{hi.a(?)}}$ 1^{en} túg-kur-ra *eš-šú {eš-šú}*⸣ 1^{en} túg-kur-ra *iš-ḫi eš-šú*
2 '. 2 gada$^{\text{hi.a}}$ 2–*ta* ^{túg}*mu-ṣip-tu*$_4$ *ra-ki-is-tu*$_4$ 2 ^{túg}*gu-nak-ku*
3'. 1$^{\text{it túg}}$*mu-ṣip-tu*$_4$ *šá šu-mu-un-du eš-še-tu*$_4$2–*ta gam-mi-da-*⸢*tu*$_4$⸣
4'. 1^{it} *eš-še-tu*$_4$ 1^{it} *ga-di-it-tu*$_4$ 1^{en} *gu-li-en-nu*
5'. 1^{it} túg-lam-lam 2–*ta šá ri-še-e-tu*$_4$ *eš-šu-tu*$_4$ 1^{en} *kab-ṭu*(?) *šá ṣi-*⸢*in*⸣$^?$-[x]
6'. 2 gada *a*-mur-*sak-ku* 1^{en} *šam-tu*$_4$ *pa-ri-it-ta-nu* 2 ⸢*šap*(?)- x⸣ -[x]
7'. 20 *dan-nu gu-ra-bi* 2–*ta nam-zi-tu*$_4$ 2 *nam-ḫa-r*[*i*
8'. 1^{it} *šid-da-tu šá nam-zi-tu*$_4$ 1^{en}*gan-gan-*[*na*
9'. 1^{it} *e-si-it-tu*$_4$ na$_4$ na$_4$ *ma uš /dug t*[*u*$_4$-
10'. ^{md}ag-su ^{f}geme-^{d}gašan-*iá* ^{f}[....
11'. [pap x *á*]*š-ta-pi-ri* kù-babbar-*a*$_4$ 1 [.....
12'. [.....] ⸢x x⸣ GAL-⸢*tu*$_4$⸣ [...

.... one new túg-kur-ra garment, one new tied túg-kur-ra garment, two linen garments, two tied *muṣiptu* garments, two *gunakku* garments, one new *muṣiptu*-garment dyed with *šumuttu*-plant, two *gamidātu* garments, (of which) one is new (and) one *gadittu*,[45] one *gulēnu* garment, one túg-lum-lum, two new bolsters, one *kabṭu*-garments with *ṣin*[..], two linen *a-mur-sakku* garments, one *šamtu* garment of bright colour, two [...], 20 vats with *gurābu*-wrappings, two fermenting vats, 2 vats [....], one stand for fermenting vat, one pot stand [......], one stone pestle, ..., Nabû-erība, Amat-Bētiya, P[Nf, total three?] slaves, the above mentioned silver- 1+ [x minas] .... large (?) [....]

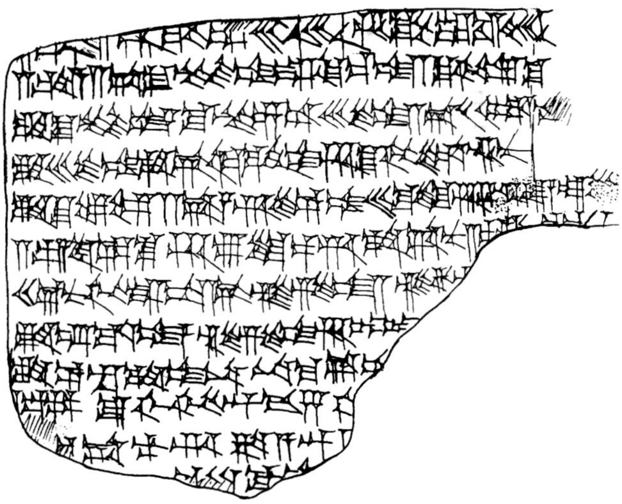

BM 76136

[44] Published with the kind permission of the Trustees of the British Museum.
[45] Or: "two *gamidātu* garments, one new garment, one *gadittu* garment".

The interpretation of the text is not easy, as we do not know its provenience or the reason for its composition. Three possibilities can be suggested; that it is a marriage agreement with partly preserved dowry list; the division of an inheritance; or a part of a business inventory. It is possible to interpret it as a list of dowry components, because all these goods – garments, vessels for producing and storing of beer, stone tools, slaves and money – are known from such lists.[46] However, in the dowry lists, real estate is listed first, followed by money, slaves, and further on garments, sometimes preceded by other "household goods" (furniture, jars, etc.). In the text under discussion, silver was indeed mentioned before the garments, but the slave only after the garments. The number of garments is high (20 or a little more), but similar quota are known from other published texts: 13 garments in BM 76968/72,[47] 20 *muṣibtu rabītu u qallat* in BM 47492[48] and even 50 *muṣibtus* in BM 76029.[49]

The second possibility is that the text concerns the division of an inheritance, but in such documents also the sequence is totally different: first real estate (fields, orchards, houses), then money and slaves, and other less expensive goods are enumerated at the end. It might be part of a business inventory, but this is less probable because larger numbers would be expected, whereas here only one or two items of each category are enumerated. It seems to me that even despite the abnormal order, an interpretation as a dowry is the most probable. If this is correct, a major part of the text, which would most likely have taken the form of a dialogue contract, is missing. This would have included stipulations typical for marriage agreements; the list of witnesses, place of composition and date are also missing.[50]

The first item cannot be identified; the preserved parts of the wedges make the reading ^{túg}*šir-a-am* (garments which are not mentioned further) improbable. Next there is one new túg-kur-ra garment,[51] and one new túg-kur-ra described by the attributive *išḫi*. The word is known from a Neo-Babylonian text from Ur as the name of a leather object followed by a quiver: ^{kuš}*iš-ḫi* and ^{kuš}*šal-ṭu*.[52] It seems to me that the translation "túg-kur-ra garment with a quiver" has no sense and another possibility is to recognize *išhi* as a separate garment, as ^{túg}*iš-ḫi* is known from one Neo-Assyrian tablet.[53] The third and in my opinion most probable idea is based on reading *íz-ḫi*, "tied". Such a description is known from the list of garments for Šamaš of Sippar (BBSt 36): 6, 10 and 11. The reading was established by Christopher E. Woods:[54]

> 1^{en} túg-íb-lá *iz-hi*, "1 tied *nēbeḫu*-belt" (l. 6), supported now with two additional attestations: BM 50209+: 8 (⸢1⸣ *ni-bi-ḫu šá iz*-hi**),[55] and BM 50963, rev. 5': ^{túg}*ni-bi-ḫu šá iz-*⸢*ḫi*⸣[56]

---

[46] See Roth 1989/90.

[47] Roth 1989, No. 42.

[48] Wunsch 2003, No. 7:13.

[49] Roth 1989, No. 33.

[50] For the structure of such texts, see Roth 1989.

[51] The first two signs following túg-kur-ra are badly preserved; the second is almost certainly read *šú*, but the reading of the first one (*eš*?) is not (over *eš* part of *Winkelhaken* is seen, but it is not certain if it is accidental or it belong to the previous line). The preliminary interpretation is that *eš-šú* was repeated mistakenly by the scribe.

[52] UET 4 117: 8.

[53] Fales & Postgate 1992, No115 II 15 (3 gú *a-na* túg *iš-ḫi*).

[54] Woods 2004, 100f.

[55] Read wrongly in Zawadzki 2006, 120, note 364 as *ni-bi-ḫu šá* giš-ban; the text was subsequently cleaned and the reading *iz-ḫi* is now certain, see Zawadzki, GG II, No. 171.

[56] Zawadzki, GG II, No. 383.

> 1^{en} *ku-lu-lu* za-gìn-<kur>-ra *ku-lu-lu šá* / kù-gi *iz-ḫi*, "1 blue-purple *kulūlu*-headband with golden (ornament) fastened on it (ll. 10–11)
> 1^{en} túg-ud-a za-gìn-<kur>-ra / *iz-ḫi* ká, "1 blue-purple *lubāru*-garment with golden (ornament in shape of the) gate fastened to it" (ll. 10–11).[57]

If the suggested reading is correct, the difference between the first and the second túg-kur-ra lies in that the first one was loose while the cut of the second was more closely fitted.

There are two types of *muṣiptu*s, described with the word *rakistu*, or with the word *šumundu*, which, according to my knowledge appears here for the first time. The first description is not new, as it appears in a text dated to the time of Artaxerxes published by Strassmaier as early as in 1893[58] concerning a dowry (*nudunnu*) consisting of silver, households utensils (kù-babbar *ú-du-ú*) and ^{túg}*mu-ṣip-tu*$_4$ *ri-kis-tu*$_4$, obviously derived from the verb *rakāsu* "to bind" with the possible translation "tied, attached, joined". As no details are given in our text, it is difficult to say whether the adorned garments were described in such a way, or rather the fashion was meant. The first possibility should be taken into account by comparison with CT 56 388, where, as one may think from the preserved fragment, *anṣabtu* (ring-shaped adornments) were attached to a *pišannu* bag: *an*]-*ṣab-tu*$_4$ *šá* gada *pi-šá-an-ni* [.....] 4/*šá ra-ki-su-tu*$_4$. Such an interpretation is also suggested by the authors of CDA[59] who tried tentatively to derive *muṣiptu* from the verb *ṣuppû* II "to decorate, inlay". This seems, however, doubtful, as according to my knowledge, there is no basis for thinking that *muṣiptu*s were decorated (with the exception of those that formed part of divine attire).[60] A rather more ordinary, simple item of clothing must be taken into account as it was used by ordinary people: slaves, shepherds, workmen, apprentices, wet-nurses. The clear suggestion that *muṣiptu* was an ordinary garment appears in GC 2 349 where these garments were given to people working for the temple under *rab ḫanše*, chiefs of 50 workers (though in fact the group there usually consisted of only ten persons, and only once of 40 and once of 30 workers). The named supervisors are responsible for these garments described as their *u'iltu ša* Eanna "the debit of the Eanna temple" and for their return to the temple by the month of Ulūlu, i.e. at the time when it was expected that the workers would finish the work. As we see, the garments were lent to the workers for the period when they were employed by the temple. This is the best evidence that *muṣiptu*s cannot be expensive, but rather inexpensive garments made of cheap (or lower quality material). Coming back to the *muṣiptu rakistu*, in our text no attachments are mentioned, which makes a second possibility, that it was more tied than the other widely used *muṣiptu*s, more probable. BM 74577 from Sippar, 18th year of Darius reveals the garments were also subject of loans.[61] A very well known prebendary Luṣi-ana-nūr//šangû Ištar Bābilî obliged himself to deliver in the next 20 days one *uzāru*-cover of good quality (*babbanitu*) to Šamaš-iddin/Šamaš-rē'ušunu; if he fails he should pay Šamaš-iddin 10 shekels of silver. We do not know why Šamaš-iddin needed the garment, but it is obvious that it was destined for a

[57] My translation follows the traditional translation of ká as "gate", while Woods translates "itemization". It would be justifiable if it had been used at the beginning of the counting, i.e. in l. 2.
[58] Strassmaier 1893, No. 27.
[59] Black, George, Postgate 2000 (further: CDA).
[60] Decorated *muṣiptu*, as main item of the goddesses' attire, used there probably as a synonym for *kusītu* are known only from Uruk, cf. Beaulieu 2003, 204 ff.
[61] Waerzeggers 2000–2001, No. 100. Cf. also Oppenheim 1950, 189.

special occasion, as he demanded an additional warranty giving him the possibility of buying it in case the contractor did not fulfil his obligation.

The second *muṣiptu* is described as *šumundu*, which might be derived from *šamādu* "to apply ornaments, inlay" except for the fact that no elements which should be attached are mentioned. I would suggest connecting it with the name of the plant [ú]*šumuttu*, known already from the Sumerian period, written sumun-dar or su-an-dar. Mixed with different plants it was used widely in pharmacology for producing poultices, salves, lotions, etc. According to a Middle Babylonian literary text it grows in meadows together with *arsuppu* barley, "bitter barley" (*šigūšu*), emmer (*kunāšu*), *inninnu*-plant and *kakkû*-lentils (?) (CAD S, 383). Because in two texts the plant is compared to blood (CAD Š/3, 301) CAD tentatively suggests identifying it with beetroot. This is possible, as the use of local plants for the dyeing of clothes is well established. If *šumundu* stands for *šumuttu*, this would be the first attested use of *šumuttu* as a dyeing substance and we can suggest recognizing here a *muṣiptu* of red beetroot or of beetroot-like colour.[62]

L. 2' enumerates 2 *gunakkus*. Except for our text the term is known from two additional documents, one dated to the time of Alexander,[63] where two sons give their mother 1/3 of their share of the house, the *gunakku* and a small bronze pot, and from BM 76968/72: 13[64] from Borsippa, dated to the 8th year of the Seleucid's reign, i.e. from the time of Antiochus III (203 BC), where one *gunakku* is given as a part of the dowry, summed as 13 *muṣiptus*. Our text might also belong to the late Babylonian or Hellenistic period. CAD G, 134 suggests identifying it with the Greek *kaunakes* and Aramaic *gonakkā*, however a different possibility was suggested by CDA, 96, where an Old Persian etymology is suggested.[65] The term appears also in the Targum where it is a counterpart of Hebrew *machber* "covering or mat" and *smicha* blanket in contemporary Hebrew.[66]

The text enumerates next two *gamidātus*, followed by one *eššetu* and one *gadittu*. We can translate each one separately, or recognize in the second and third more precise descriptions of the first one. If this is correct, one can say that the first one was new, while the second one differed from the former; the lack of any acceptable etymology makes a translation impossible. As the first described quality ("new"), the second might be a counterpart of the first, i.e. "threadbare, worn", but this is not so obvious as others factors may have had to be taken into account, such as the material used or the method of weaving. Such a possibility is suggested by GC 2, 361 from Uruk where it is stated that [x] *gamidāti* were made of *šundu*-wool and *šupêtu* wool, according to CAD Š/3, 328 "thick wool". According to one text, a *gamidatu* garment was probably destined for a female singer as the request was made by a *rabītu ša nārê*.[67] If our list is part of a marriage agreement this would be the first attestation of *gamidātu* as a component of a dowry. It is interesting to note that *gamidatu* could be borrowed together with money.[68] BM 64798[69] counts 20 shekels of red wool, 50 shekels of blue-purple wool, 1 mina of alum and [1 mi]na(?) of apple-colour-dye ([giš]ḫašḫur) given

[62] The last possibility is the kind suggestion of Marie-Louise Nosch.

[63] BRM 2, no. 50 and dupl. RIAA^2 295 (Boiy 2003, 28ff with copy on p. 58).

[64] Roth 1989, No. 42.

[65] According to the kind information of Miguel Ángel Andrés-Toledo it exists in OIr. the word **gaunaka*, an adjective from **gauna* "hair", Av. *gaona*. "hair, colour" and Middle Pers. *gœnag*, "colour, kind".

[66] Jastrow 1903, 224 (courtesy M. Sandowicz).

[67] BM 80711, Jursa 2002, 107–108.

[68] VS 4, 157.

[69] See below, Appendix No. 4.

to Šāpik-zēri, son of Šamaš-ah-iddin, a well-known weaver producing multi-coloured cloths and dyer (*ṣāpû*) for dyed fabric (*a-na* [*ṣi*]-*bu-tu*)[70] for two *gamidātu*-garments. At least some part of a *gamidatu* or its applications could be made of different kinds of coloured wool.

gulēnu garments appears three times as part of a dowry, so it must be a garment for young women. In TMH 2/2, No. 2 two *gulēnus* precedes four túg-kur-ra *šam-tu* and four *šir'am*, i.e. it was recognized as more precious than four túg-kur-ra and four *šir'am*. However, from VS 5, 47[71] we know that *gulēnu* was also worn by men, as the adoptive father obliged his adoptive son to deliver him one *gulēnu* yearly. In a new agreement made only three month later[72] between the adoptive son and the wife of the adoptive father (because her husband was at that time seriously ill), she relinquished her right to the *gulēnu*, i.e. it suggests that in her opinion her husband did not need it anymore and she also did not wish to take it for herself. Perhaps because of its size and fashion it would be difficult to adapt it for woman. An interesting situation appears in VS 6, 168[73] where the person acting as the performer of the prebendary owner is paid with one *gulēnu*, one *lubāru* and one *hullānu* yearly. Again, *gulēnu* was recognized here as the most expensive of three items. According to a promissory note[74] in the space of five days one person must deliver a *gulēnu* or pay 8 shekels of silver.[75] Probably this represents the formalisation of a pre-existing obligation. According to Dar 43 *gulēnu* garments were manufactured for the Ebabbar temple by three women as their labour assignment (*iškaru*), one garment each per year. From BE 8/1, no 138 where 32 *gulēnus* are destined for the *ḫarranu* enterprise we know that the garment was the subject of commerce.[76]

Although *gulēnu* and *gamidatu* are known already from the Neo-Assyrian period, the earliest data from Babylonia comes from the end of the Neo-Babylonian period. It cannot be excluded that this type of garment came to Babylonia with Assyrian prisoners after the fall of Assyria and found wider acceptance in Babylonia only after a few decades.

Here it is the third attestation of túg-lam-lam, the first in TBER 94,[77] where a multi-coloured one is mentioned,[78] the next, chronologically probably the earliest attestation (Borsippa, 10th year of Nabonidus, 546 BC) appears in L 1639,[79] where 1 túg-lam-lam was sold for 12 kur 2 pi 4 *sūtu* of dates, i.e. it was a quite expensive garment.

The 2 *šá re-še-e-tu*$_4$ can be compared with the following attestations:

2 túg *šá re-eš-tu*$_4$	VS 6 275: 3
1^{en} ^{túg}é *re-še-e-tú*	BM 61494: 3[80]

[70] *ṣibûtu*, see CAD Ṣ, 171b "dyed fabric".
[71] Baker 2004, No. 31.
[72] BM 114728 (Baker 2004, No. 35).
[73] Baker 2004, No. 67.
[74] Baker 2004, No. 252 and her comments in p. 81.
[75] Cf. unp. BM 60585, badly preserved text concerning *gulēnu*-garments, where in te first entry (ll.1–2) the price of 4 *gulēnus* is 53 shekels of silver.
[76] We can mention here the performer, who was paid partly with *uzāru* garment by prebendaries for the work done by him on their order (VS 5, 87/88 = Baker 2004, no. 39).
[77] Durand 1981.
[78] For the correct reading, based on collation of the text, see Joannès 1984, 72: ⌜1^{en}⌝ túg-lam-lam* *šá** *bir-mu*. The reading suggested by Roth 1989, No. 34: 17 (3 túg$^{hi.a?}$ 1^{en} *šá bir-mu*) did not take into account Joannès' collation of the tablet.
[79] Such number is given in Joannès 1989, 239, while in Joannès 1984, 74 as L. 1659.
[80] See Appendix, No. 1.

1[it] *re-eš-tu*$_4$	CT 56 298: 18
1 é *re-še-tu*$_4$	BM 60282: 5[81]

CAD R, 378 suggests, based on Ner 28: 13 (1 túg é *re-še-tu*$_4$) plural form *rēšētu*, supported now by our text.

There are two different proposals concerning the meaning of the term. Roth, based on the word *rīštu* "rejoicing, jubilation", gave a tentative translation as "festival garment",[82] accepted by CAD R. A new meaning was suggested by Jursa and accepted by Kessler. According to Jursa, who compared the word with Aramaic *by s'dy'* it is a bolster: "Behältnis des Kopfendes", also "Polster", (Kessler used the German word "Kopfstütze"). In Kessler's opinion the lack of determinative túg suggest that this object was made of a wooden box ["Kasten aus Holz"] wrapped around with textiles.[83] The question is if there is any significance that texts mention two or only one *ša rēšētu*. One can postulate one long bolster for two people in bed or each person had his/her own.

I refrain from any comments on the next garment as the reading of the sign following *kab* is uncertain (it looks as if it was written over an erasure); the *ṣi* is sure but the following sign is not.

The linen *a*-mur-*sak-ku* appears also as a component of a dowry in a text (with duplicate) published by M. Roth[84] and dated to Seleucid period (203 BC). It must accordingly be a female garment. Our text might belong to the same category of documents and might be dated also to the late period, i.e. this garment might have appeared in Babylonia in that late period of Mesopotamian history.

The next entry is *šamtu parittanu*. Two things need to be commented on. Up until now *šamtu*, written š*am-tu* or *ša-an-ṭu* was attested as an attributive of túg-kur-ra (CAD Š I, 339 and BM 67367).[85] Here however it is evidently a noun, not an adjective, qualified with the attributive *parittannu*. This means that *šamtu* was used as an additional element of a túg-kur-ra garment or as a separate garment. My suggestion is that it might be an "apron", a small one added to túg-kur-ra and a large one functioning as a separate piece of clothing. The word *parittannu* is known till now only from documents from the Nuzi archives as a description of the colour of horses.[86] It is obvious that it reached Mesopotamia with the Hurrians, famous for their great experience in the training of horses and their use in army. The word is based on Indic *bharita* or *palitá*[87] and, by comparison with the colour of other horses mentioned in the Nuzi texts, it was suggested by the authors of CAD that it denotes a light colour, while the CDA 39 gave a tentative translation "grey" (with question mark). The text discussed here is important because it demonstrates that despite

81 Jursa 2003, 229 with copy on p. 238; Ner 28: 13; PTS 3853: 8; Kessler 1999, 246, with copy on p. 257
82 Roth 1989/1990, 31.
83 Kessler 1999, 253.
84 Roth 1989, No. 42.
85 See Appendix, No. 5.
86 Cf. CAD B, 112a.
87 Miguel Ángel Andrés-Toledo drew my attention to OIr. **pari-tanū*, with the meaning "around the body".

the passage of a few centuries, the word known previously only in northern Mesopotamia later appears in the south of the country and no longer restricted to describing the colour of horses, but also other objects such as garments.

A few additional words known from the Neo-Babylonian period deserve our attention:

bultu. The *bultu* garment appears in the administrative text BM 61528[88] belonging to the Ebabbar archives at Sippar, written in the 3rd year of Cambyses. The only parallel can be found in Amarna and Nuzi texts, where the writing *bul-da* (EA) or *bu-ul-du* (Nuzi) is standardized by dictionaries to *bultu*. The tentative translation "quilt or blanket" has been suggested (CAD B 311, CDA 48). This is, according to my knowledge, the first attested use of the word in Mesopotamia after a gap of almost 800 years, and, just like the word *parittannu*, it spread from north to south. Despite the fact that Babylonia witnessed the intrusion or resettlement of new people of different origin, the *bultu* was still worn by inhabitants of Babylonia at the end of VI century BC.

il-ḫu. It appears in BM 29378: 16, 20,[89] known before only from the synonym list Malku and the lexical series An= *Anum* (CAD I/J 71).

il-ta -a'-pi qalpi. "Threadbare *iltapi*" appears in the above-cited BM 29378: 16, 20. This is known also from Nbn 703: 5 with the qualification *la ḫi-ri*, while in two consecutive lines it appears túg-kur-ra *la ḫi-ri* and túg-kur-ra *ḫi-ri* . Scholars have connected it with *iltepītu* and translated "loincloth".[90] However, as in Malku list VI 123 *iltepitum* is equated with *naḫlaptu* with the qualification *ú-ri,* and *naḫlaptu* is certainly not a loincloth but a type of shirt or short garment, such a meaning should be taken into account. Instructive is the above-cited BM 63956 where *ilta'pi* is included into the broad type of garments called túg-kur-ra. Obviously it is not "loincloth" but a separate garment comparable in some way to túg-kur-ra.

^{túg}***kišku.*** Known earlier from three texts,[91] ^{túg}*kišku* appears also in FLP 628: 4 in a clearer context: 1$^{\text{en giš}}$ná *šá me-suk-kan-na a-di-i* ^{túg}*ki-iš-ki-šú* making it possible for Waerzeggers to recognize it as "a kind of bedcloth, possibly suspended above the bed".[92] The word is based on the verb *karāku*, "to roll up", so it was presumably a rectangular blanket, which can be easily rolled up and might also be used for other purposes.

tarpuštu. It appears again in the above cited FLP 628:5–7 as a garment for women and men 1$^{\text{en túg}}$*tar-pu-uš-tu*$_4$ / ^{lú}nita 2–*ta* ^{túg}*tar-pu-uš-tu*$_4$ / ^{lú}*a-mil-tu*$_4$, derived according to Waerzeggers from *rapāšu*. That however does not really help us in identifying it further beyond the fact that it was a garment for both men and women.

mu-ri-si **and**

(*lubāri*) *ša qātê* (TÚG.ŠU^{ii}), appears only in BM 49913: 6, 10,[93] all subsumed in the *muṣipētu*, which included additionally a linen *salḫu* tunic, a garment called *pān muṣê*, *ṣibtu*, *naḫlaptu birmu* and *kusītu*. The etymology of the last item suggests that it was a type of cover for hands, maybe comparable to our gloves.

^{túg}***ra-šu-ti-ik-ka.*** This is known from BM 25630 /25653 (Borsippa, 23.1.Dar 20), transliterated and discussed in her unpublished dissertation concerning Šaddinnu of the Belija'u family by J. Wojciechowska, who kindly

[88] See Appendix, No. 2.

[89] Jursa, Paszkowiak & Waerzeggers 2003/2004, 260.

[90] CAD I/J 89.

[91] Roth 1989/1990, 30f.

[92] Waerzeggers 2002, 333, n. 5.

[93] Cf. the copy of the text in Lorenz 2009, No. 3. My copy (to be published in GC II 356) differs in a few places from Lorenz copy and readings, although I accept his reading of line 3 (in Lorenz, l. 2).

gave me permission to quote the appropriate fragment.[94] This is an agreement establishing Šaddinnu as heir of a date palm orchard in the *ḫanšû* (50^{e}) in Bīt-Habaṣirtu on the following conditions:

> ^{12}*u$_4$-mu* 2 ½ *qû* še-bar *ù* [x qû] zú-lum-ma mun$^{\text{hi.a}}$
> ^{13}*saḫ-li-e* 1 bán še-giš-ì [(x) *i*]-*na* mu-an-na
> $^{14\ \text{m}}$*Šad-din-nu* ⸢*a-na* ^{md} en⸣-[mu] *i-nam-din*
> ^{15}1^{en} túg-kur-ra 1$^{en\ \text{túg}}$*š*[*ir*]-⸢*a-am*⸣ 1$^{en\ \text{túg}}$*ra-šu-ti-ik-ka*
> ^{16}*eš-šu-tu i-na* mu-an-na [^{m}*Šad-din-nu*] ⸢*a-na*^{md} en⸣-[mu] *i-nam-din*
> (the lines according to BM 25630).
>
> "Šaddinnu will supply Bēl-iddin with 2½ *qû* of barley and [x *qû*] of dates daily (plus) salt, cress (and) 1 *sūtu* of sesame yearly. (Additionally) each year [Šaddinnu] will deliver Bēl-[iddin] one túg-kur-ra, one *šir'am* and one new *rašutikku*-garment."

It seems to me that the same garment appears also in Stigers 1976, No. 42: 5 *r*[*a-šu*]-*ti-ik-*[*ka*]. To make any suggestion on the function of these garments additional attestations are needed.

1^{en} GADA *maš-šá-nu* known only from BM 61494: 6[95]

bīt panē (part of the bowmen equipment).
BM 61345 (to be edited by John McGinnis,[96] quoted with his kind permission)

> 2. (...) 1^{en} *šir-'a-am*
> 3. 1^{en} é *pa-ni-e* 1 ^{kuš}*ṣal-li*
> 4. 2 *šib-ib-ba-nu*

"one jerkin, one ..., one shield? (and) 2 belts."

BM 61425 (to be edited by John McGinnis, quoted with his kind permission)

> 4. (...) 2–*ta* ^{túg}*šir-'a-a*[*m*]
> 5. [x túg *šá*] murub$_4$ *šá* 1^{en} ⸢^{túg}⸣é *pi*(?)-⸢*ni*$^{?}$⸣-*e*$^{?}$
> 6. [x x x (x)-*zu*^{meš} 1$^{\text{en túg}}$ *kar-ra-*⸢*bal*⸣-*tu*

L. 5. The filling of the gap is mine.
"two jerkins, [x] loincloth belt(s) .... one *karballatu* cap."
By comparison with *bīt rešētu* "bolster" one might suggest that *bīt panē* also might be an item made partly of wood and covered with fabric.

* * *

To sum up, just as current terminology is unable to keep pace with the thousands of new patterns and production techniques of clothing and textiles, a similar phenomenon held in the ancient world despite the incomparably smaller scale of production, technologies and materials. It is not surprising that quite frequently the same term was used for widely different clothing. Given the relative poverty of available iconography, combined with the relatively imprecise information contained in these texts, working out the functions and differences between individual pieces of clothing and textiles will be a long and probable unending process.

---

[94] See Paszkowiak 2005, No. 186/186a.
[95] See Appendix, No. 1.
[96] MacGinnis (in press)

Appendix

1. BM 61494 (82-9-18, 1468)
4.2 × 3.2 cm
3.9.Nbn 12

1.1 túg-kur-ra
2.1^{en} *na-ṣi-ba-tu4*
3.1$^{en\ túg}$é *re-še-e-tú*
4.*ina* igi ^{md}amar-utu-mu-ùru
5.^{lú}kù-dím *pa-qi-id*
6.⸢1^{en}⸣ ^{gada}*maš-šá-nu* [(x)]
Rev. 7.iti-gan ⸢u$_4$⸣-3-kam
8.mu-12-kam ⸢^{md}⸣ag-i
9.lugal tin-tir^{ki}

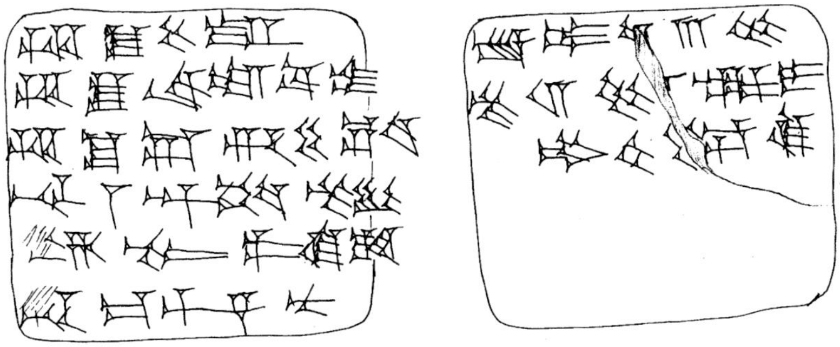

BM 61494

L. 2. Concerning *naṣbatu*, see Kessler 1999, 255, according to him it was a long coat with a hood.
L. 4. Marduk-šum-uṣr, the *kutimmu* was active in the period from Nbn 12 to the beginning of Cambyses' reign, cf. the list of attestation in Bongenaar 1997, 389, to which our text should be added.

Because the garments were delivered to a goldsmith it seems that they belonged to the attire of cultic statues and the best candidate is Anunītu-ša-Sippar-Anunītu, a warrior goddess worshipped in Sippar. As they were entrusted to a goldsmith it is logical to think that all of them were adorned with gold or silver ornaments.

One túg-kur-ra garment, one *naṣbatu*-coat, one bolster handed over to Marduk-šum-uṣur, the goldsmith. (Additionally) one linen *maššanu* garment (was given to him).
Month Kislimu, 3rd day, 12th year of Nabonidus, king of Babylon.

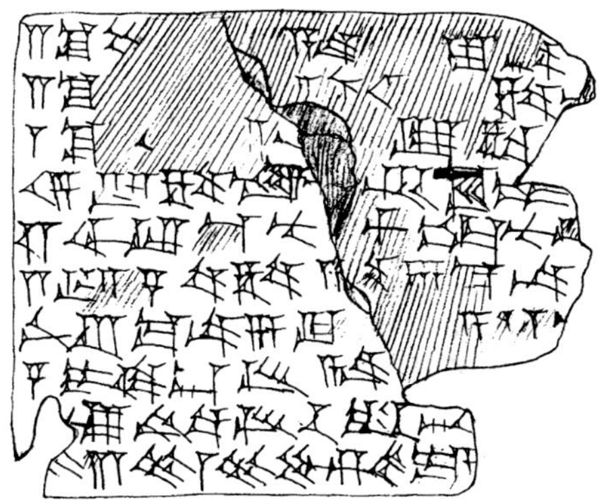

BM 61528

2. BM 61528 (82-9-18, 1501)
7.1 × 6.0 cm
Cam 3

1. 2 túg-kur-[ra-meš *šá*] *a-na* [x] gín ⸢kù-babbar⸣ [(x)]
2. 2 túg-[x x x x] ⸢*a-na*⸣ 11 [gín] kù-babbar ⸢*ina*⸣ [(x x)]
3. 1 túg ⸢x x x⸣ [x x] gín ⸢kù-babbar⸣ [....
4. 15 gín kù-babbar šá[m] 1^{en} *bul-tu$_4$* [(x x)]
5. 2! (written 1/3) *ši-ib-*⸢*ba*⸣*-nu* [x] *maš-ka-nu*⸢$^{meš?}$⸣[(x x)]
6. 2 gín 4-*tú* kù-babar *a-na* ⸢6 *ma*⸣*-na* [x x]
7. pap 2/3 *ma-na* 6 gín [kù-babbar] *a-na* ⸢x⸣[x x]
8. *šá i-di* é-meš *a-na* [PN ^{lú}x x x x]
9. *ù* ^{lú}erín-meš-*šú* ⸢*it-ta*⸣-[*din* iti.x u$_4$-kam]
10. mu-3-kam ^{m}*Kam-bu-zi-ja* [lugal e-ki lugal kur-kur]

Rev. 11. ⸢4⸣ túg-kur-ra-meš [x x x] gín (?) [kù-babbar (?)....]
12. ^{md}en-kád *it-tan-nu* túg?-[x]-⸢ra?⸣.[me]š(?) .... [PN]
13. *ù* ^{lú}erín-meš-*šú* sum^{in}

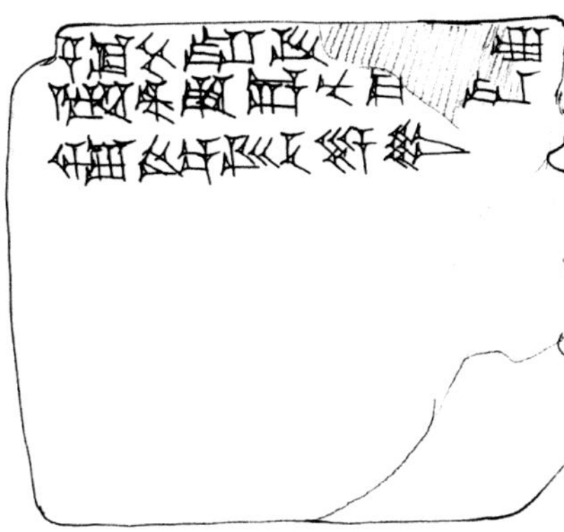

2 túg-kur-ra garments for [x] shekels [of silver]
2 ... –garments for 11 [shekels] of silver [....
1 ... –garments for [x] shekels of silver [....
15 shekels, the price of one *bultu*-garment
2 ¼ shekels for 6 minas of ....,
total 46 shekels [of silver] for ....from the rent of houses for [PN,(the ...)] and his workers he has given.
[Month x, nth day], 3rd year of Cambyses,[king of Babylon, king of Lands].
5 túg-kur-ra garments [....for x] shekels [of silver (?)] Bēl-kāsir has given .... and his workers he has given.

3. BM 73306 (82-9-18, 13317)
5.1 × 5.9 cm

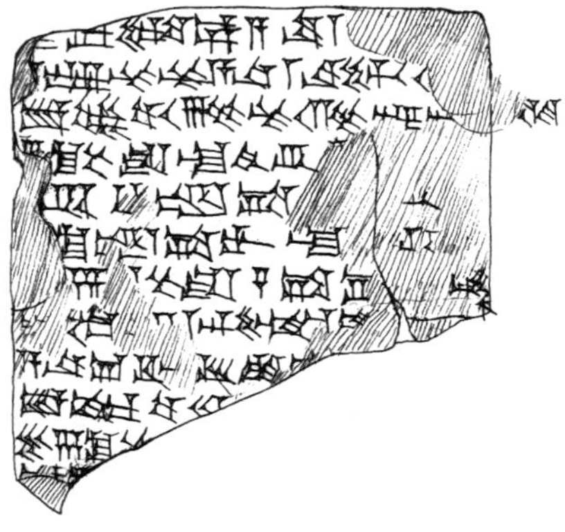

BM 73306

1’. [x bá]n (?) *saḫ-li-e a-na* 1/^{m} [....
2’. ^{md}ag-mu-mu *a-na* ^{m}*Na-ṣir u* [^{m}*Šu-la-a* sum]
3’. iti-sig$_4$ u$_4$-16-kam mu-12-kam ^{d}ag-⌜I⌝ [lugal tin-tir]^{ki}
4’. ⌜3⌝ túg-kur-ra *la ḫi-ri* ⌜*a*?⌝ [...
5’. 1^{en} ^{túg}*il-ta-[pi ḫi]-*⌜*ru*⌝*-[tu*$_4$
6’. [2] ^{túg}*il-ta-pi la [ḫi]-*⌜*ri*⌝
7’. [pap] 6 [tú]g-kur-ra *šá* ta ⌜é⌝ [x x] *zi*
8’. ⌜x x x⌝ ^{md}*Bu-ne-n[e-* ...
9’. *a-na* é-gu[r$_7$]-meš *it-t[a-din*
10’. iti-ne u$_4$-21-[kam ...
11’. 56 túg-ku[r-ra ...
12’. iti.⌜x⌝ [...

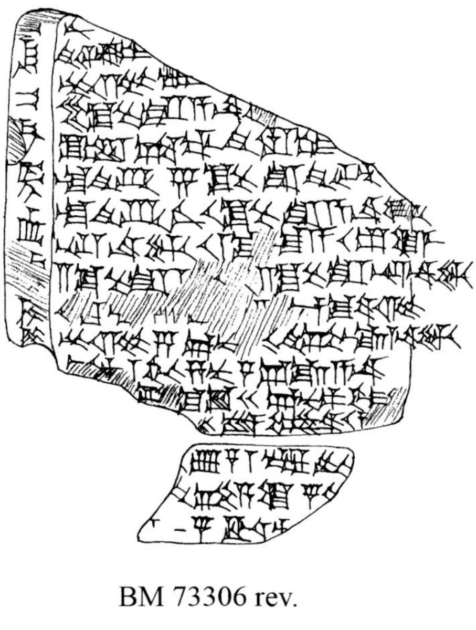

BM 73306 rev.

Rev. 1’. pap [...
2’. *ḫi-ši-iḫ-tu*$_4$ [....
3’.mu-12-kam ^{d}ag-[i lugal tin-tir^{ki} ..
4’. 50 túg-kur-ra *a-na* ⌜x⌝[...
5’. ^{túg}*il-ta-pi ḫi-ru-tu*$_4$ [...
6’. *la ḫi-ri* 5 túg-kur-ra *ḫi-ši-i[ḫ-tu*$_4$]
7’. *la ḫi-ri* pap 10 túg-kur-ra *a-na* ⌜x x⌝
8’. *ina* igi ^{m}*Na-ṣir u* ^{m}*Šu-la-a* 10 kùr *qé-me*
9’. 2 túg-kur-ra ⌜x x⌝ 2 túg-kur-ra *ina* igi ^{m}*Na-ṣir*
10’. *u* [^{m}]⌜*Šu-la-a*⌝ ⌜x x x iti⌝-šu u$_4$-12-kam
11’. mu-12-kam 6 *dan-nu* ⌜kaš⌝ dùg-ga *ina* šu^{ii} ^{m}*Na-ṣir*
12’. ⌜*ina* igi⌝ ^{m}*Gu-za-nu šá* é-šu^{ii} *a-na*
13’. [x x] ⌜ga(?)⌝ *šu-bul* 20 *dan-nu pa-ṣu-*[*ú*]
14’. [x x x x]-⌜*din*⌝(?) iti sig$_4$ u$_4$-12 +[x-kam]
L.e. 15’. [x x x x]-⌜x⌝-*ú šá* ^{md}ag-šeš-[..
16’. [x x x ^{uru}]⌜gú⌝-du$_8$-a^{ki} 4–*t*[*ú*?....
17’. [x x x] ⌜x⌝ *šá*/4 *kab* ⌜x⌝ [....
L.h e. 18’. [x x x] *ri* 2 ni-gál-la ⌜x⌝ [x]

Obv. 2’ and rev. 8’. Naṣir and Šula. They should be identified with traders acting usually together in the time of Nabonidus (see Bongenaar 1997, 285), however the state of preservation of the text makes it impossible to explain their activity, though it is at least clear that they also issued garments.
Obv. 10’. Or u$_4$-20-[1–lal-kam], i.e. 19th of Simanu.

Only about half of the text is preserved, which causes a serious problem with its interpretation. The text might be a yearly settlement account from objects issued in the 12th year of Nabonidus. In rev. 16' the city of Kutha is mentioned, but the content is broken. Quite an important number of garments is enumerated, i.e. in total about 120 túg-kur-ra including *iltapi* garments.
Rev. 11'. Guzānu *ša bīt qāti* is a well-known person active in the period from Nbn 4 till Nbn 12 (see Bongenaar 1997, 216), where, as in this text, he is responsible for providing beer.

[x *sūt*]*u* (?) of cress for 1/^{m}? [...] Nabû-šum-iddin [has given] to Naṣir and [Šulā]. Month Simanu, 16th day, 12th year of Nabonid[us, king of Baby]lon.
⸢3⸣ túg-kur-ra garments without *hīru* f[or$^?$
1 *iltapi* garment [......]
[2] *iltapi* garments without [*hī*]*ru*, [total] 6 túg-kur-ra garments, which were [taken from x storehouse ....] Bunen[e-x] has delive[red] to *bīt karê*–storehouse. Month Abu, 21st [day, 12th year of Nabonidus, king of Babylon].
65 túg-ku[r-ra] garments [....] month Ul[ūlu ?
Rest lost

Rev.
....] total [....], demand [...] for 12th year of Nabon[idus, king of Babylon]
50 túg-kur-ra garments for ⸢x⸣ [...
[x] *iltapi* garments with *hīru* [x] without *hīru,* 5 túg-kur-ra garments, the dema[nd] without *hīru*, total 10 túg-kur-ra garments for [x] at disposal of Naṣir and Šulā (was given). 10 kur of flour, 2 túg-kur-ra garments ... 2 túg-kur-ra garments at disposal of Naṣir and ⸢Šula⸣ (?)... month⸣ Du'uzu, 12th day, 12th year; 5 *dannu* vats of good quality beer by the hand of Naṣir at disposal of Guzānu, of the rear house are brought to .... ; 20 *dannu* vats of strong beer [..... has been delive]red(?). Month Simānu, day 12+[4?] ..... of Nabû-aḫ-[x] ... of Kutha ... ¼? ... sicles ...

4. BM 64798 (82-9-18, 4779)
4.3 × 3.3 cm

1. ⸢1/3⸣ *ma-na* ^{síg}*ta-bar-ri*
2. 5/6 *ma-na* ^{síg}*ta-kil-tu*$_4$
3. ⸢1⸣ *ma-na* ^{na_4}*gab-bu-ú*
4. [1$^?$ *ma*]-*na* giš-ḫašḫur *a-na*
5. [*ṣi*]-*bu-tu*$_4$ *šá* 2–[*ta*]
6. [^{túg}]*gam-mi-da-*⸢*tu*$_4$⸣

Edge 7. [*a-na*] ^{m}dub-numun

Rev. 8. [A-*šú šá* ^{md}]utu-šeš-mu
9. [sum^{in}] iti-apin
10. [u$_4$-x]-kam mu-3-kam
11. [^{md}U+GUR-lugal]-ùru
12. [lugal] tin-tir^{ki}

L. 4. The narrow damaged space excludes any significantly different reconstruction of the numeral.
L. 7. Šāpik-zēri, son of Šamaš-ah-iddin, head of *išparātu birmi*/*ṣapû* was active between Nbk 35 and Cyr (Bongenaar 1997, 349–351 and Zawadzki 2006, 230–231), which gives the argument for dating the text to the 3rd year of Neriglissar.

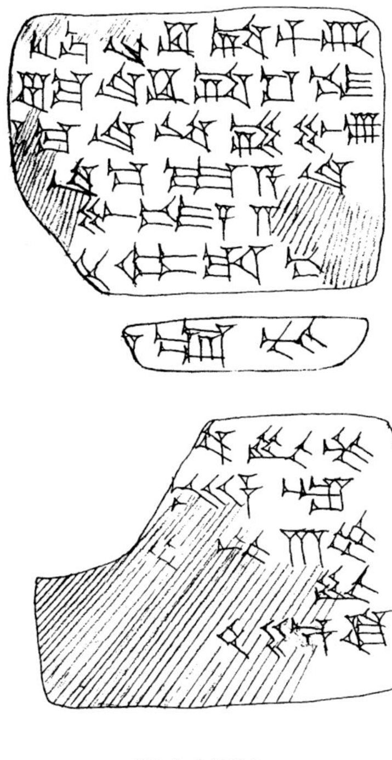

BM 64798

20 shekels of red wool, 50 shekels of blue-purple wool, one mina of alum, [1? mi]na of apple-coloured dye for dying for two *gamidatu* garments [for] Šāpik-zēri, [son of] Šamaš-aḫ-iddin [was given]. Month of Arahsamna, 3rd year of [Neriglis]sar, [king of Baby]lon.

5. BM 67367 (82-9-18, 7363)
4.3 × 3.4 cm

1. 2 túg-kur-ra *šam-tu*
2. *šá* ^{md}utu-mu *a-na*
3. ⸢8⸣ gín kù-babbar kù-babbar *úl* sur
4. [x tú]g-kur-ra *šam-tu*
5. [*šá* ^{m}]^{d}utu-su *a-na*
6. [x] gín kù-babbar kù-babbar *úl* sur

Edge 7. [it]i-sig$_{4}$ u$_{4}$-9-kam
Rev. 8. [m]u-6-kam

2 túg-kur-ra garments with apron(?) Šamaš-iddin sold for 8 shekels of silver (but) silver has not been paid (to him).
[x] túg-kur-ra garments with apron(?) Šamaš-erība sold for [x] shekels of silver (but) silver has not been paid (to him).
Month Simānu, 6th year.

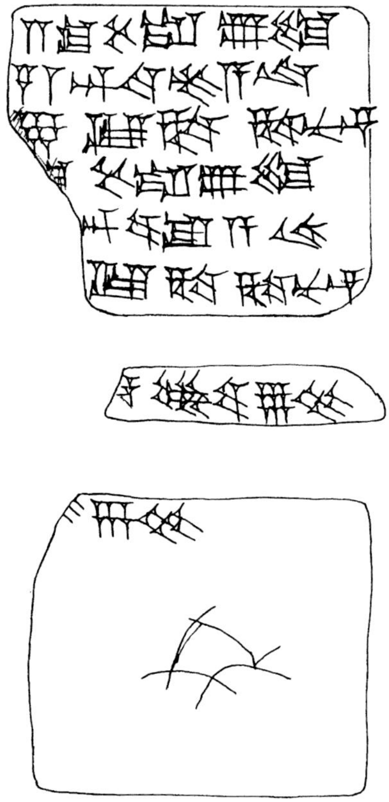

BM 67367

6. BM 62740 (82-9-18,2709)
5.0 × 2.9 cm

1. 14 túg-kur-ra-meš
2. ^{md}en-*ka-ṣir* a-*šú*
3. *šá* ^{md}en-kar^{ir}
4. sum^{na} iti-gu$_{4}$

Rev. 5. u$_{4}$-25-kam
6. mu-31-kam

14 túg-kur-ra garments were given to/by (?) Bēl-kāṣir, son of Bēl-eṭir.
Month Du'uzu, 25th day, 31st year.

7. BM 75889 (83-1-18, 1244). Copy: Bertin 3059; Str. II 315/4
5.0 × 4.2 cm

1. ⸢6⸣ túg-kur-ra-meš *šá* ^{m}*Ina*-giš-mi-^{d}U+GUR
2. ⸢*a*⸣-*na É-babbar-ra id-din-nu*
3. *ina lìb-bi* 1^{en} túg-kur-ra *a-na*
4. ^{md}utu-šeš-meš-*bul-liṭ* sum^{na}
5. 1^{en} túg-kur-ra <*a-na*> ^{m}*Dan-nu*-^{d}U+GUR sum^{na}
6. *mi-šil* túg-kur-ra <*a-na*> ^{md}AG-⸢x⸣
7. (erased) [(x x)]-⸢x-*tu*$_{4}$⸣ [(x)] ⸢x⸣

Edge 8. *mi-šil* t[úg-kur-r]a *a-na*
Rev. 9. ^{md}DI.KUD-⸢x⸣-[x]-KAM SUM[^{na}]

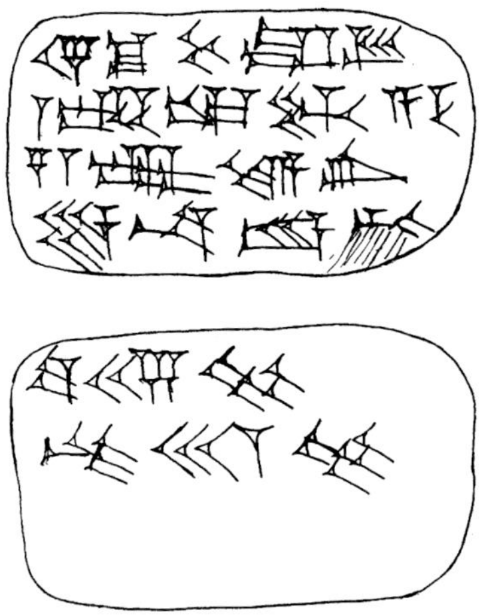

BM 62740

10. *mi-šil* túg-kur-ra ^{md}AG-*na-din**-[x]
11. *mi-šil* túg-kur-ra ^{md}AG-KIŠ-*di-nu*
12. *šá* KÁ *me-e*

L. 1. At the beginning of the line we see now 2+2, which suggests the reconstruction ⌜6⌝, not ⌜5⌝ as in the Strassmaier's copy. Note, however, that in Nbn 662 two "half túg-kur-ra" are recognized in total as one túg-kur-ra.
L. 6. *mi-šil* is written over erasure.
L. 7. The idea that whole line was erased is supported by the fact that 6 túg-kur-ra- garments for 6 persons are enumerated.
L. 10. *din,* contra Strassmaier ⌜*ṣir*⌝ and Bertin *ṣir.*

⌜6⌝ túg-kur-ra garments, which were delivered to the Ebabbar temple by Ina-ṣilli-Nergal, including:
one túg-kur-ra garment which was given for Šamaš-aḫḫē-bulliṭ.
one túg-kur-ra garment (for) Dannu-Nergal was given
half túg-kur-ra garment (for) Nabû-⌜x⌝[(-x) was given]
half t[úg-kur-r]a garment for Māndanu-[x]-ereš was given
half túg-kur-ra garment (for) Nabû-nādin-[x was given]
half túg-kur-ra garment (for) Nabû-kiššat-dīnu of the water gate.

Abbreviations

ABL	R. F. Harper, *Assyrian and Babylonian Letters.* Chicago, 1892–1914.
AfO	*Archiv für Orientforschung.*
AOAT	Alter Orient und Altes Testament. Neukirchen-Vluyn.
Akkadica	*Akkadica.* Périodique bimestriel de la Fondation Assyriologique Georges Dossin. Brussels.
BBSt	BBSt- King L. W. (1912) *Babylonian Boundary-Stones and Memorial-Tablets in the British Museum.* London.
BE	The Babylonian Expedition of the University of Pennsylvania. Philadelphia.
Bertin	Unpublished tablet copy by G. Bertin housed in the British Museum.
BM	British Museum. London. Tablet inventory number.
BRM	Babylonian Records in the Library of J. Pierpont Morgan, I–IV. New Haven.
CAD	*The Assyrian Dictionary of the Oriental Institute of the University of Chicago.* Chicago (1956–).
Cam	Strassmaier J. N. (1890) *Inschriften von Cambyses, König von Babylon (529–521 v. Chr.).* Leipzig.
CDA	J. Black, A. George, N. Postgate (eds.), *A Concise Dictionary of Akkadian.* SANTAG Arbeiten und Untersuchungen zur Keilschriftkunde 5. Wiesbaden, 1999–2000.
CT	Cuneiform Texts from Babylonian Tablets in the British Museum. London.
Cyr	Strassmaier J. N. (1890) *Inschriften von, König von Babylon (538–529 v. Chr.).* Leipzig.
Dar	Strassmaier J. N. (1897) *Inschriften von Darius, König von Babylon (521–485. Chr.).* Leipzig.
FLP	Free Library of Philadelphia. Philadelphia. Tablet inventory number.
GC	R. P. Dougherty, *Goucher College Cuneiform Inscriptions I/II.* New Haven, 1923/1933.
GG II	S. Zawadzki, *Garments of the Gods, Part II: Texts* (to be published).
JCS	*Journal of Cuneiform Studies.* New Haven – Boston – Ann Arbor.
NABU	*Nouvelles Assyriologiques Brèves et Utilitaires.*
Nbk	Strassmaier J. N. (1889) *Inschriften von Nabuchodonosor, König von Babylon (604–561 v. Chr).* Leipzig.
Nbn	Strassmaier J. N. (1889) *Inschriften von Nabonidus, König von Babylon (555–538 v. Chr.).* Leipzig.
Ner	Evets, B. T. A (1892) *Inscriptions of the Reigns of Evil-Merodach, Neriglissar and Laborosoarchod.* Leipzig.
NRV	M. San Nicolò & A. Ungnad, *Neubabylonische Rechts- und Verwaltungsurkunden I.* Leipzig, 1935.
OIr.	Old Iranian.
PTS	Princeton Theological Seminary. Princeton. Tablet inventory number.
RA	*Revue d'assyriologie et d'archéologie orientale.* Paris.

RIAA	L. Speleers, *Recueil des inscriptions de l'Asie antérieure des Musées Royaux du Cinquantenaire à Bruxelles: Textes sumériens, babyloniens et assyriens*. Brussels, 1925.
SAA	State Archives of Assyria. Helsinki.
Str.	Unpublished tablet copy by J. N. Strassmaier housed in the British Museum.
TBER	F. Joannès, Textes économiques de la Babylonie récente: étude des textes. Paris, 1982.
TMH	Texte und Materialien der Frau Professor Hilprecht Collection of Babylonian Antiquities. Universität Jena. Leipzig.
UET	Ur Excavations: Texts. London.
VS	Vorderasiatische Schriftdenkmäler der (Königl.) Museen zu Berlin. Berlin.

Bibliography

Baker, H. D. (2004) *The Archive of the Nappāḫu Family.* AfO Beiheft 30. Wien.

Beaulieu, P.-A. (2003) *The Pantheon of Uruk During the Neo-Babylonian Period.* Cuneiform Monographs 23. Leiden, Boston.

Black, J., George, A., Postgate, N. (2000) *A Concise Dictionary of Akkadian, 2nd (corrected) printing.* Wiesbaden.

Boiy, T. (2003) RIAA[2] 293+300. Hellenistic Legal Documents from Uruk in the Royal Museums of Art and History (Brussels), *Akkadica* 124, 19–64.

Bongenaar, A. C. V. M. (1997) *The Neo-Babylonian Ebabbar Temple at Sippar: Its Administration and Its Prosopography.* Istanbul.

Borger, R. (1957) Assyriologische und altarabische Miszellen, *Orientalia N.S.* 26, 1–11.

Borger, R. (1981, *Assyrisch-babylonische Zeichenliste*, Neukirchen-Vluyn.

Dietrich, M. (2003, *The Babylonian Correspondence of Sargon and Sennacherib* (SAA 17), Helsinki.

Durand J.-M. (1981), *Textes babyloniens d'epoque récente.* Paris.

Fales, F. M. & Postgate, J. N. (1992) *Imperial Administrative Records, Part I. Palace and Temple Administration,* SAA 7, Helsinki.

Jastrow, M. (1903) *A Dictionary of the Targum, the Tamudic Babli and Yerushalmi, and the Midrashic Literature*, vol. I, Jerusalem.

Joannès, F. (1984) Contracts de Marriage d' époque Récente, *RA* 78, 71–81.

Joannès, F. (1989) *Archives de Borsippa. La famille Ea-ilûta-bâni. Étude d'un lot archives familiales en Babylonie du VIIIe au Ve siècle av. J.-C.* Geneva.

Jursa, M. (1999) *Das Archiv des Bēl-rēmanni.* Istanbul.

Jursa, M. (2002) Florilegium babyloniacum: Neue Texte aus hellenistischer und spätbabylonischer Zeit. In C. Wunsch (ed.), *Mining the Archives. Festschrift for Christopher Walker on the Occasion of His 60th Birthday.* Dresden, 107–127.

Jursa, M. (2003) Betten, Kopfpölster und 'Kühlschränke. *Festschrift für Burkhart Kienast zu seinem 70. Geburtstage dargebracht von Freuden, Schülern und Kollegen versammelt by G. J. Selz.* AOAT Bd 274. Münster, 229–238.

Jursa, M. (2005) *Neo-Babylonian Legal and Administrative Documents. Typology, Contents and Archives.* Guides to the Mesopotamian Textual Records vol. 1. Münster.

Jursa, M., Paszkowiak, J. & Waerzeggers, C. (2003/2004) Three Court Records, *AfO* 50, 255–268.

Kessler, K.-H. (1999) Ein Einbruch in ein *Bīt Šutummu* eines Tempelbäckers. In B. Böck, E. Cancik-Kirschbum, Th. Richter (eds.) *Minuscula Mesopotamica. Festschrift für Johannes Renger.* AOAT Bd 267. Münster, 245–257.

Kleber, K. (2008) *Tempel und Palast. Die Beziehungen zwischen dem König und dem Eanna-Tempel im spätbabylonischen Uruk* (Veröffentlichungen zur Wirtschaftsgeschichte Babyloniens im 1. Jahrtausend v. Chr.,Bd 3). Münster.

Lorenz, J. *Nebukadnezar III/IV. Die politischen Wirren nach dem Tod des Kambyses im Spiegel der Keilschrifttexte.* Dresden.

MacGinnis, J. (1998) Ordering the House of Šamaš: texts from the Mangement of the Neo-Babylonian Ebabbara, *Iraq* 60, 207–217.

MacGinnis, J. (in press), *The Arrows of the Sun: Armed Forces in Sippar in the First Millennium B.C.*

Oppenheim, A. L. (1950) Review of H. H. Figulla, Business Documents of the New-Babyloninan Period, Philadelphia – London 1949, *JCS* 4, 188–195.

Paszkowiak (Wojciechowska), J. (2005) *Archiwum Šaddinnu z rodu Bēlija'u. Działalność nowobabilońskiego przedsiębiorcy w świetle tekstów z Borsippy*, Poznań (unpubl. diss.).
Roth, M. (1989) *Babylonian Marriage Agreements 7th-3rd Centuries B.C.* AOAT Bd 222. Neukirchen-Vluyn.
Roth, M. (1989/1990) The Material Composition of the Neo-Babylonian Dowry, *AfO* 63, 1–55.
Stigers, H. G. (1976) Neo- and Late Babylonian Business Documents from the John Frederick Lewis Collection, *JCS* 28, 3–59.
Strassmaier, J. N. (1893) Eine kleine babylonische Keilschrifttexte aus dem Britischen Museum. In *Actes du 8e Congrès International des Orientalistes*, Leiden.
Ungnad, A. (1937) *Neubabylonische Rechts - und Verwaltungsurkunden. Glossar.* Leipzig.
Waerzeggers, C. (2000–2001) *Het archiefs van Marduk-rēmanni. Studie van de familie Ṣāḫit-ginê in Nieuwbabylonisch Sippar* (548–484 v. Chr.), vol. 1–2. Ghent (unp. diss.)
Waerzeggers, C. (2002)Endogamy in the Neo-Babylonian Period. In C. Wunsch (ed.), *Mining the Archives. Festschrift for Christopher Walker on the Occasion of His 60th Birthday.* Dresden, 319–342.
Wunsch, C. (1993) *Die Urkunden des babylonischen Geschäftsmannes Iddin-Marduk. Zum Handel mit Naturalien im 6. Jahrhundert v. Chr, II,* Gröningen.
Wunsch, C. (2003) *Urkuneden zum Ehe- Vermögens- und Erbrecht aus verschiedenen neubabylonischen Archiven.* Dresden.
Wunsch, C. (2003/2004) Findelkinder und Adoption in der neubabylonischen Quellen, *AfO* 50, 174–244.
Woods, Ch. E. (2003) The Sub-God Tablet of Nabû-apla-iddina Revisited, *JCS* 56, 23–103.
Zawadzki, S. (2003) Nebuchadnezzar and Tyre in the Light of New Texts from the Ebabbar Archives in Sippar. In *Eretz-Israel. Archeological, Historical and Geographical Studies. Hayim and Miriam Tadmor Volume,* vol. 27, 276*–181*.
Zawadzki, S. (2006) *Garments of the Gods. Studies in the textile Industry and the Pantheon of Sippar according to the Texts from the Ebabbar Archive.* Orbis Biblicus et Orientalis vol. 218, Fribourg-Göttingen.
Zawadzki, S. (forthcoming) *Garments of the Gods, Part II: Texts.*

22. Some Considerations about Vedic, Avestan and Indoiranian Textile Terminology

Miguel Ángel Andrés-Toledo

The method of the historical comparative linguistics allows us to reconstruct the language of the Indoeuropeans. As Watkins[1] states, "a language necessarily implies a society, a speech community, and a culture, and a proto-language equally necessarily implies a proto-culture". So we can also reconstruct part of the Indoeuropean culture by means of this method and observe that there was a common vocabulary related to textile.

Because of their close relationship the Indian and the Iranian languages offer a good chance to approach this common vocabulary, which was inherited from Indoiranian and sometimes even from Indoeuropean times. As a matter of fact, as Puhvel[2] states, "a firm Indo-Iranian prototype is also a doubly secure foundation for further cross-Indo-European comparison".

A word can surely be traced back to Indoiranian when it is preserved in both Old Indian and Old Iranian. In the Old Indian group Vedic must be preferred with regard to Sanskrit, while in the Old Iranian group Avestan offers a broader material than Old Persian. With less degree of certainty, languages of a middle or even modern period can also be used in one group for the reconstruction.

According to Schlerath,[3] when at least two etymologically related words appear together in the same syntagm in both Old Indian and Old Iranian, and provided that they are used in the same or at least similar context, they can be reconstructed as an Indoiranian formula. If we find parallels of this formula in other Indoeuropean languages, it can also be traced back to Indoeuropean. However, as Campanile and Watkins[4] have observed, sometimes a single word of this formula was substituted by a synonym in one language, so that this formula could have been semantically, though not morphologically nor lexically, preserved.

I will focus on how the Indoiranian and Indoeuropean linguistic reconstruction can contribute to identify which textile terminology existed among the Indoiranians before they became separated in the Indian and the Iranian groups, that is, at a certain period before c. 1500 BC. Some Old Indian and Old Iranian textile terminology was inherited from Indoiranian, but lacks

[1] Watkins 1995, 7.
[2] Puhvel 1987.
[3] Schlerath 1968, viii–xv.
[4] Campanile 1993; Watkins 1995, 15.

in other Indoeuropean languages. Conversely, other Indoiranian words can be traced back to an Indoeuropean textile terminology. As we will see, some of them belong to Indoiranian and even to Indoeuropean poetic formulas.

In my analysis of this terminology I have followed Adams & Barber's[5] classification according to the different processes of the elaboration of the textile. To it I have added some roots lacking in their list and especially other words of Indoiranian languages not quoted by them.

1. Wool

Concerning the material of the textile, the only Indoiranian root which can be traced back to Indoeuropean is that for "wool", IE. **h₂u̯l̥h₁neh₂-* > IIr. **Hu̯l̥HnaH-* "wool", which is attested in the following Indoeuropean languages:

a) Old Indian: Ved. *ū́rṇā-* "wool, woolen thread".[6]
b) Iranian:
   - Old Iranian: YAv. *varənā-* "wool".[7]
   - Middle and New Iranian: Khwar. *w'n*, Šuγnī *wůn* "wool".[8]
c) Other IE. languages: Hit. */hulana-/*, */huli̯a-/*, Dor. Gr. *lā̃nos*, Lat. *lāna*, OHG. *wolla*, Lith. *vìlna*, Serbocr. *vȕna* "wool".[9]

Besides this, another root existed in Indoiranian which mainly designated the goat's or camel's wool: IIr. **dr̥Hćá-*.

a) Old Indian Ved. *dūrśá-* "rough clothing (maybe from goat's skin or wool)".[10]
b) Middle and New Iranian: Khot. *dairśa-* "of goat's wool"; Waxī *δIrs, δürs*, Yidγa *lirs, lurs*, Šuγnī *δoxi* "goat's wool"; NP. *dirs* "camel's hair; threadbare garment".[11]

These words for wool are the only ones for a material of textile which can be reconstructed to Indoiranian, since words for flax, cotton or any other material do not stem from a common source in the Old Indian and Old Iranian languages.

2. Thread

From that wool single threads were drawn. The animal origin of this material is manifest through the terminology of the only word for "thread" that can be traced back to Indoiranian as well as to Indoeuropean: IE. root **deḱ-* "thread, hair"[12] > IIr. **dać-* "thread". This root yielded some Iranian words referred to goat's wool and other words meaning "hair" in some Indoeuropean languages:

[5] Adams & Barber 1997.
[6] Mayrhofer 1992–2001, 1.243.
[7] Bartholomae 1904, 1372.
[8] Morgenstierne 1974, 90.
[9] Pokorny 1959–1969, 1139–1140.
[10] Mayrhofer 1992–2001, 1.740.
[11] Bailey 1979, 165a.
[12] Adams & Barber 1997, 569.

a) Old Indian: Ved. *daśā-* "fringe".[13]
b) Middle and New Iranian: Khot. *dasa* "thread"; Bal. *dasag* "thread"; NP. *dasah*, Munǰī *lâsa* "rope of goat's wool"; Yidya *loso, lāsa* "neck-rope".[14]
c) Other IE. languages: OIcel. *tāg* "thread, fibre"; MHG. *zāch* "wick"; OIr. *dūal* "lock of hair"; Goth. *tagl* "a single hair"; OHG. *zagel*, OE. *tæg(e)l* "tail".[15]

3. Braid, plait

Concerning the processes of preparation of textiles, at least four Indoiranian roots referred to braid and plait can be traced back to Indoeuropean. The first, IE. **derbʰ-* "to braid" > IIr. *darbʰ-* "to braid", gave at least one Indian word and some Iranian words related to sewing and needles:

a) Old Indian: Ved. *darbh-* "to knot, to tie", Ved. *darbhá-* "bundle, tuft", Ved. *sáṁdr̥bdha-* "tied together".[16]
b) Iranian:
   - Old Iranian: YAv. *dərəβδa-* "bundle of muscles".[17]
   - New Iranian: Kurd. *dirūn* "to sew"; Sariq. *δef* "needle-hole"; Waxī *drəv-/drəvd-*, Parāčī *andarf-* "to sew", etc.[18]
c) Other IE. languages: Lith. *dirbti* "to work"; Rus. *dórob* "basket".[19]

This root seems to belong properly to the vocabulary of basketry. However, it has further implications in the Indoiranian context. On one hand, the New Iranian words indicate that the meaning "to braid" of this root was very close to the semantic field of "sewing" and "needle" in Iranian, as Kurd. *dirūn* "to sew", Sariq. *δef* "needle-hole", Waxī *drəv-/drəvd-* and Parāčī *andarf-* (< **ham-darf-*) "to sew"[20] demonstrate. On the other hand, while Ved. *darbhá-* "bundle, tuft" usually refers to a bundle of a particular kind of grass, and therefore it has no relation with "sewing" or "needle", Ved. *sáṁdr̥bdha-* could belong to the textile terminology. Indeed, it is accompanied by Ved. *sū́tra-* "thread" in ŚBM 14.6.7.6b:[21] *sū́treṇa sáṁdr̥bdhāni bhavanti* "(the creatures, Ved. *bhūtā́ni*) are sewn by a thread".

This Vedic passage is very important for the Comparative Mythology, because it reveals that the Indoeuropean image of the thread of life, from which the creatures are made, is also found in India. But let us return to its implications for the Indoiranian textile terminology.

Ved. *sáṁdr̥bdha-* means literally "tied together". However, the Vedic verb *sáṁ-darbh-*, from which this participle stems, is etymologically identical with Parāčī *andarf-* (< **ham-darf-*) "to sew". As this Vedic participle is accompanied by Ved. *sū́tra-* "thread" in the above mentioned passage, I think that it must not be understood as "tied together", but as "sewn", like its Parāčī verbal cognate.

Therefore, the Indoeuropean root meaning "to braid" designated in Indoiranian a type of sewing. We could think of some techniques, like the Nålebinding or sprang technique, which is still used in Iran, or the ply-splitting, which is used in Rajasthan with camel's hair. However,

[13] Mayrhofer 1992–2001, 1.710.
[14] Bailey 1979, 154a.
[15] Pokorny 1959–1969, 191.
[16] Mayrhofer 1992–2001, 1.703–704.
[17] Bartholomae 1904, 742.
[18] Cheung 2007, 60.
[19] Pokorny 1959–1969, 211.
[20] Cheung 2007, 60.
[21] Weber 1849.

this information cannot be inferred by the only comparison between Ved. *sáṁdr̥bdha-* and Parāčī *andarf-*.

The second Indoeuropean root inherited by Indoiranian which refers to braid and plait is IE. **kert-* "to plait"[22] > IIr. **kart-* "to twist".

a) Old Indian: Ved. *kart-* "to spin", Ved. *káṭa-* "mat".[23]
b) Middle Iranian: Khwar. *kncy-* "to turn (one's head)"; Khwar. *'nknc* "needle".[24]
c) Other IE. languages: Hit. *kar-za, kar-za-na-aš* "weaver's tool to make yarn" (?);[25] Gr. *kártallos* "basket"; Lat. *crātis* "wickerwork"; Goth. *haúrds* "door"; OIcel. *hurð*, OHG. *hurt*, OE. *hyrdel* "hurdle"; OPrus. *corto* "hedge".[26]

Although Hit. *kar-za, kar-za-na-aš* and Ved. *kart-* are more closely related to the meaning "to spin", it is unlikely that they reflect the oldest Indoeuropean meaning. In fact, the meaning "to plait" is preserved not only in the words of western Indoeuropean languages, but also in Ved. *káṭa-* "mat" < **kŕ̥ta-*. Since the meaning of the western languages is confirmed in the eastern part by this Vedic word, a meaning "to plait" matches better the Indoeuropean root **kert-*. According to the words stemming from this root, it should have designated some kind of wickerwork, from which the meaning "to spin" developed in some eastern Indoeuropean languages.

The third Indoeuropean root related to braid and plait, IE. **pleḱ-* "to plait",[27] does not actually belong to the textile terminology. Moreover, it is not present in Indoiranian. As a matter of fact, the only Vedic word stemming from it, Ved. *praśna-*, which is translated as "turban, headband", must rather be understood as some kind of wickerwork used to protect the head.

a) Old Indian: Ved. *praśna-* "turban, headband (?)".[28]
b) Other IE. languages: Gr. *plékō*, Lat. *plectō* "I plait", OHG. *flehtan*, OE. *fleohtan* "to plait", OCS. *pletǫ* "plait"; Gr. *plókos, plókamos* "braid, plait"; OHG. *flahs*, OE. *fleax* "flax".[29]

The fourth Indoeuropean root preserved in Indoiranian which belongs to the semantic field of braid and plait, IE. **resg-* "to wattle"[30] > IIr. **rasg-* "to wattle" must be excluded from the textile terminology too.

a) Old Indian. Ved. *rájju-* "rope, cord",[31] Ved. *rajjusarja-* "rope-maker".
b) Middle and New Iranian: Sogd. *rγzy* "cloth"; Yidγa *roγz* "woman's cloak"; NP. *raγza* "woolen cloth".
c) Other IE. languages: Lat. *restis* "rope, cord"; OCS. *rozga* "root, branch"; Latv. *rezg(i)ù* "I knit".[32]

As the Indoeuropean words stemming from this root demonstrate, IE. **resg-* designated the process of wattling fibres in order to make a rope or cord. Since a meaning related to textile is only present in the Iranian group, it must be excluded from the Indoeuropean and even from the Indoiranian textile terminology.

22 Adams & Barber 1997, 571.
23 Mayrhofer 1992–2001, 2.316, 1.290.
24 Cheung 2007, 244–245.
25 Eichner 1973, 98, n.78.
26 Pokorny 1959–1969, 584–585.
27 Adams & Barber 1997, 570–571.
28 Mayrhofer 1992–2001, 2.185.
29 Pokorny 1959–1969, 834.
30 Adams & Barber 1997, 571.
31 Mayrhofer 1992–2001, 2.427.
32 Pokorny 1959–1969, 874.

4. Twist > Spin

As Barber[33] already observed, roots meaning "to twist, to turn" were used for "spinning" in some Indoeuropean languages, because the process of spinning implies the twisting of fibres together. Likewise these roots yielded words for "sinew", where fibres are naturally twisted. This is the case of the first root of this semantic field: IE. *sneh*$_1$- > IIr. **snaH*-:

a) Old Indian: Ved. *snā́yu*-, Ved. *snā́van*- "sinew".[34]
b) Old Iranian: YAv. *snāuuar*- "sinew", YAv. *snāuuarə.bāzura*- "having sinews like arms".[35]
c) Other IE. languages: Hit. *išḫunau̯ar*, Gr. *neurá́*, Lat. *nervus*, Arm. *neard*, Toch. B *ṣñaura* "sinew"; Gr. *néō*, Lat. *neō* 'I spin'; Goth. *nēþla*, OIcel. *nāðl*, OE. *nǣdl* (< IE. **sneh*$_1$-*tleh*$_2$-) "needle"; OHG. *nā(w)en* "sew"; OHG. *sena(wa)*, OSax. *sinewa*, OE. *seonu* "sinew"; Latv. *snāju* "to twist together, to spin".[36]

Bader[37] considered that these words stem from a deverbative variant of a present stem with nasal infix of the Indoeuropean root **seh*$_2$- "to bind". However, I agree with Adams & Barber's[38] reconstruction as IE. *sneh*$_1$- "to twist fibres together".

Although the connection of this root with the textile terminology is corroborated by some Indoeuropean languages, it is not shared by the Indoiranian group, which only inherited the Indoeuropean heteroclite noun **sneh*$_1$*u̯r*-/**sneh*$_1$*u̯n*- "sinew".

The second Indoeuropean root with the meaning "to twist" which could have been applied to the textile terminology since Indoeuropean times is IE. **terk*(^{u})- "to twist"[39] > IIr. **tark*- "to twist":

a) Old Indian: Ved. *tark*- "to twist", Ved. *tarku*- "spindle".[40]
b) Middle Iranian: OKhot. *hataljs*- "to flutter" < **fra-tark*- (?).[41]
c) Other IE. languages: Hit. *tarku*- "to turn oneself, to dance"; Gr. *átraktos* "spindle"; Lat. *torqueō* "I twist, I torment"; OE. *þrǣstan* "to turn, to twist"; Toch. B *tärk*- "to twist around"; OCS. *trakŭ* "band, belt"; Alb. *tjerr* "to spin".[42]

As we observe, the meaning "to spin" and its derivate "spindle" are attested in Alb. *tjerr* "to spin" as well as in Ved. *tarku*- and Gr. *átraktos* "spindle". Hence we can conclude that it is very likely that the Indoeuropean root **terk*(^{u})- "to twist" was already used with the meaning "to spin" since Indoeuropean times. However, it was not inherited by the Iranian group, but only by the Indian, so that it cannot be traced back to Indoiranian.

There is a third Indoeuropean root which means "to twist", IE. **u̯rei̯ḱ*- "to twist", and was continued by IIr. **u̯rai̯ć*- "to twist", but it must be excluded from the textile terminology.

a) Old Indian: Ved. *vréśī*- "whirling (epithet of the waters)".[43]
b) Iranian:

---

33 Barber 1975, 297.
34 Mayrhofer 1992–2001, 2.770–771.
35 Bartholomae 1904, 1629.
36 Pokorny 1959–1969, 977.
37 Bader 1990, 25.
38 Adams & Barber 1997, 571.
39 Adams & Barber 1997, 572.
40 Mayrhofer 1992–2001, 1.633.
41 Bailey 1979, 488a.
42 Pokorny 1959–1969, 1077.
43 Mayrhofer 1992–2001, 2.598.

- Old Iranian: YAv. *uruuaēs-* "to twist, to turn", YAv. *uruuaēsa-* "whirl".[44]
- Middle and New Iranian: MP. *ristag* "rope, string, thread"; CSogd. *rwyšt* "spun"; Khwar. *rys* "thread"; NP. *rēštan, rēs- / rēsīdan, rēs-*, Paṣ̌tō *rēšəl* "to spin", etc.; Bal. *rēštan, rēs-* "to twist threads".[45]

c) Other IE. languages: Gr. *riknós* "bent"; OE. *wrīgian* "to tend, to incline"; Latv. *rìšti* "to bind, to tie up".[46]

The only connection of this root with the textile terminology is found in Middle and New Iranian languages, so that it can neither be reconstructed as Indoiranian nor as Indoeuropean. Therefore, it seems that the meanings "thread" and "to spin" of the Iranian group developed from a general meaning "to twist", which was applied in Middle Iranian and hence in New Iranian to the twisting of threads.

5. Stretch

When spinning the thread, and in particular after the spinning is finished, the thread must be more or less stretched, according to the technique and the type of textile to be made. When it is put on the loom it must be stretched and tightened too. This action is expressed in some Indoeuropean languages by the Indoeuropean root **ten-* "to extend, to stretch", which was inherited by IIr. **tan-* "to extend, to stretch" and yielded some words related to the textile in Indian and Iranian languages.

a) Old Indian: Ved. *tan-* "to extend, to tighten",[47] Ved. *tantí-* "rope", Ved. *tántu-* "thread", Ved. *tántra-* "loom"; Ved. *prācīna-tāná-* "warp" (< *prācī́na-* "turned towards the front, eastward" + *tan-* "to stretch"); Ved. *tásara-* "shuttle".[48]

b) Iranian:
- Old Iranian: YAv. *tan-* "to extend, to stretch".[49]
- Middle and New Iranian: MP. *tadan, tan-* "to spin a thread"; NP. *tanīdan, tan-* "to turn, to spin"; Kurd. *tan-* "to weave", etc.[50]

c) Other IE. languages: Gr. *teínō* "I draw, I tighten"; Lat. *ten-d-ō* "I stretch"; OHG. *den(n)en* "to stretch"; OPrus. *sasin-tinclo* "rabbit-snare"; Lith. *tiñklas* "snare".[51]

Indoeuropean languages other than Indian and Iranian did not develop words specifically related to the textile from this root. Several words from IIr. **tan-* developed in Vedic, mainly related to the weaving on a loom. With regards to Iranian, the connection with weaving is only found in Kurd. *tan-*, while the rest of Middle and New Iranian words point to the spinning of threads. Therefore, it is possible that this root was related to the textile terminology in Indoiranian, but it cannot be assured that it designated the same process.

---

[44] Bartholomae 1904, 1533–1535.

[45] Cheung 2007, 437.

[46] Pokorny 1959–1969, 1158–1159.

[47] Mayrhofer 1992–2001, 1.618–619.

[48] From IIr. **tan-s-* < IE. **ten-s-* "to stretch". vid. Ved. *tam̐s-* "to shake, to tear"; Goth. *at-þinsa* "I draw"; Lith. *tęsti* "to stretch" in Mayrhofer 1992–2001, 1.609–610.

[49] Bartholomae 1904, 633.

[50] Cheung 2007, 377.

[51] Pokorny 1959–1969, 1065–1066.

Furthermore, it is noteworthy that this root was applied to an Indoeuropean image connected with textile: that of the Night as a woman who spreads her dress through the sky. As Christol[52] already observed, this action is expressed by the same Indoeuropean root **ten-* in Vedic and Greek, as we see in RV 1.115.4:[53]

tát sū́ryasya devatvám̐ tán mahitvám madhyā́ kártor vítatam̐ sám̐ jabhāra
yadéd áyukta harítaḥ sadhásthād ā́d rā́trī vā́sas tanute simásmai

This is the divinity of the Sun, this is his greatness. In the midst of the work she (i.e. the Night) has withdrawn the spread (dress). / When he has harnessed his fallow (horses) from their abode, then the Night spreads her dress for herself.

And in Homer, *Od.* 11.19:[54]

all' epì nùx oloḕ tétatai deiloĩsi brotoĩsi
but baneful Night is spread over wretched mortals.

Therefore, the image of the spread dress of the Night is attested in an Indoeuropean formula: "Night" (in Nom. Sing.: Ved. *rā́trī-* and Gr. *nùx*) + IE. **ten-* (Ved. *tanute* and Gr. *tétatai*, both in 3rd. Sing. Middle).

With regards to Indoiranian, the root **tan-* appears in an Indoiranian formula referred to the same image, that of a woven sky. As Skjærvø[55] has noticed, Vedic and Avestan attest the same image of the universe as a textile. However, he has not observed that this image is expressed by the Indoiranian formula "the thread of Truth". Ved. *tántu-*, a noun from the Indoiranian root **tan-* "to extend", is used together with the genitive singular *ṛtásya* "of Truth" in AVŚ 2.1.5:[56]

pári víśvā bhúvanāny āyam ṛtásya tántum̐ vítatam̐ dṛśé kám
yátra devā́ amṛ́tam ānaśānā́ḥ samāné yónāv ádhy aírayanta

Around all beings I went in order to see the stretched thread of Truth,[57] in the same womb where the gods, having attained immortality, were begotten.

In Old Avestan, Ved. *tántu-* was substituted by the synonym OAv. *hiθā-* "thread" in Y 34.10,[58] but the exact Old Avestan cognate of the Vedic genitive singular *ṛtásya*, namely OAv. *aṣ̌ahiiā* (< **ṛ́tasi̯a*), remained:

ahiiā. vaŋhə̄uš. manaŋhō. š́iiaoθanā. vaocat̰. gərəbąm. huxratuš.
spəṇtąmcā. ārmaitīm. dąmīm. vīduuā̊. hiθąm. aṣ̌ahiiā.
tācā. vīspā. ahurā. θβahmī. mazdā. xšaθrōi. ā. vōiiaθrā

By means of the action of this Good Thought, the well-minded proclaims his understanding, he who knows the Beneficent Ārmaiti, the creator, the thread of Truth, and all the tools of the loom in her power, o Ahura Mazdā.

---

[52] Christol 1986, 191, 193.
[53] Geldner 1951, 1.152; Christol 1986, 191, 193.
[54] Von der Mühll 1962, 193.
[55] Skjærvø 2005.
[56] vid. Whitney & Lanman 1905, 39; Lüders 1959, 472; YV,VS 32.10; TĀ 10.1.4 and MnārUp 2.5. AVŚ 2.1.5a=AVP 2.6.4a; AVŚ 2.1.5b=AVP 2.6.5b.
[57] cf. Ved. *ṛtásya tántur vítataḥ* in RV 9.73.9 and Ved. *ṛtásya tántum̐* in AVŚ 13.3.19.
[58] Geldner 1886; Skjærvø 2005, 272.

Therefore, in both Vedic and Old Avestan there was a mere variation of a common Indoiranian formula "thread" (in Acc. Sing.: Ved. *tántum* and OAv. *hiϑąm*) + IIr. **r̥tási̯a / ŕ̥tasi̯a* "of Truth" (Gen. Sing.).

To summarise, it is obvious that the Indoeuropean root **ten-* "to extend" was associated to the textile in Indoiranian and possibly in Indoeuropean times, as the Indoiranian formula for the "thread of Truth" and the Indoeuropean formula for the spread dress of the Night demonstrate.

6. Weave

One of the most famous Indoeuropean roots of the textile terminology is IE. **h₂eu̯-* "to weave" and its extension IE. **(h₂)u̯ebʰ-* "to weave"[59] > IIr. **(H)u̯abʰ-* "to weave".

The first yielded some technical terms of the process of weaving in Vedic, but it seems not to have been inherited by the Iranian group, so that it cannot be traced back to Indoiranian.

a) Old Indian: Ved. *o-* "to weave",[60] Ved. *ótu-* "woof", Ved. *vayitrī-* "weaver", Ved. *vāya-* "weaver", Ved. *veman-* "loom".
b) Old Iranian: YAv. *[u]zgərəsnāuuaiiō* (*°uuaiiō* ?).[61]
c) Other IE. languages: OIcel. *vāð* "weaving", OHG. *wāt* "clothes"; Lith. *áusti* "to weave", Rus. *usló* "weaving".[62]

Conversely, its extension IE. **(h₂)u̯ebʰ-* "to weave",[63] through IIr. **(H)u̯abʰ-* "to weave", is well attested in both the Indian and the Iranian groups as well as in other Indoeuropean languages.

a) Ved. *vabh-* "to bind"; Ved. *°vábhi-* "binding, weaving".[64]
b) Iranian:
   - Old Iranian: OAv. / YAv. *vaf-* "to sing, to praise" < "to weave songs"; YAv. *ubdaēna-* "made from textile".[65]
   - Middle and New Iranian: MP., Parth. *waf-*, Sogd., Khwar. *w'f-*, NP. *bāftan, bāf-*, Oss. I. *wafyn/wæft*, Oss. D. *wafun/uft* "to weave, to intertwine", etc.[66]
c) Other IE. languages: Hit. *ḫuppai-* "to ensnare, to interlace"; Hit. *ḫuppala-* "net"; Hit. *ḫup(p)ar(a)-* "name of a textile";[67] Gr. *huphaínō* "I weave"; Toch. A *wäp-*, Toch. B *wāp-* "to weave"; OHG. *weban* "to weave".[68]

It is noteworthy that Hit. *ḫuppai-* and Ved. *vabh-* did not mean "to weave", but "to interlace" and "to bind" respectively and that the Iranian words for "to weave" also meant "to intertwine, to braid". In my opinion, it is very likely that IIr. **(H)u̯abʰ-*, and even IE. **(h₂)u̯ebʰ-*, developed a specialized meaning regarding IE. **h₂eu̯-* "to weave". Provided that its extension designated a kind of weaving by means of a technique of binding or knotting, it could have been applied to the manufacture of knotted rugs. Accordingly, the relation between the knotted rugs and the poetic

59 Adams & Barber 1997, 572.
60 Mayrhofer 1992–2001, 1.275–276.
61 Bartholomae 1904, 411.
62 Pokorny 1959–1969, 75–76.
63 Adams & Barber, 1997 572.
64 Mayrhofer 1992–2001, 2.506.
65 Bartholomae 1904, 1346 401.
66 Bailey 1979, 305; Cheung 2007, 402.
67 Kronasser 1967, 47.
68 Pokorny 1959–1969, 1114.

composition, widespread in the Indoeuropean literatures and expressed by this Indoeuropean root *(h_2)u̯eb^h-* in both cases,[69] becomes clearer. As the same verb designated the composition of poetry and the knotting of rugs, Tuck's[70] theory about the common mnemonic fundaments of the patterns of Indoeuropean textile and those of the Indoeuropean poetry seems even more likely.

7. Dye

The process of dying is also represented in the Indoeuropean textile terminology by IE. **(s)reg-* "to redden, to colour"[71] and was inherited by IIr. **rag-* "to redden, to colour".

a) Old Indian: Ved. *raj-* "to redden, to colour", Ved. *rajayitrī́-* "dyer (Fem.)", Ved. *rāga-* "colour".[72]
b) Middle and New Iranian: Khwar. *rxtk* "red"; Khot. *rrāṣa-* "dark-coloured" < **raxša-*; Kurd. *raš* "black"; Waxī *rakṣ̌* "grey, brown".[73]
c) Other IE. languages: Gr. *rézō* "I colour", Gr. *rogeús* "dyer"; Gr. *rē̃gos* "rug, blanket".[74]

As Adams & Barber[75] observed, "to colour" was the same as "to redden" in the Indoeuropean terminology of dyes, because red was the first colour used as a dye in Europe and the Near East.

Concerning Indoiranian, Vedic agrees with Greek in the use of this root as a technical term for dying, but the Iranian group offers a non-specialized broader spectrum of terms related to colour, among which only Khwar. *rxtk* shares the meaning "red".

8. Sew

Another widespread Indoeuropean root related to the textile terminology is IE. **si̯eu̯h_1-/*si̯uh$_1$-* (> **suh$_1$-*) "to sew"[76], which was continued by IIr. **si̯au̯H-/*si̯uH-*.

a) Old Indian: Ved. *sī́v-* "to sew", Ved. *syū́-* "seam", Ved. *syū́man-* "bond, lace", Ved. *sū́tra-* "thread".[77]
b) Middle Iranian: Khot. *hīya-* "sewn stuffs".[78]
c) Other IE. languages: Hit. *sum(m)anza(n)* "thread"; Gr. *humḗn* "membrane", OPrus. *schumeno* "shoemaker's thread"; Lat. *suere*, Goth. *siujan*, OIsl. *sȳja*, OHG. *siuwen*, OE. *sēowian*, OCS. *šiję*, Toch. A *su-*, Lith. *siū́ti*, Latv. *šũt* "to sew".[79]

Besides this, another root was used in Indoiranian for the meaning "to sew" or a similar one: IIr. **ćai̯H-* "to sew".[80]

69 West 2007, 36–38.
70 Tuck 2006.
71 Adams & Barber 1997, 572–573.
72 Mayrhofer 1992–2001, 2.424.
73 Bailey 1979 362a–b; Cheung 2007, 313–314.
74 Pokorny 1959–1969, 854.
75 Adams & Barber 1997, 572–573.
76 Adams & Barber 1997, 573. According to Bader 1990, IE. **sh$_2$-*, which is enlarged as IE. **sh$_2$-i̯-* and **sh$_2$-u̯-*.
77 Mayrhofer 1992–2001, 2.734.
78 Bailey 1979, 483.
79 Pokorny 1959–1969, 891.
80 Cheung 2007, 29.

a) Old Indian: Ved. *cī́ra-* "strip, rag", Ved. *cela-* "dress".[81]
b) Middle and New Iranian: MMP. *'bzyn-* (+ **upa-*) "to sew"; Parth. *'bjyn'gr* "tailor"; Sogd. *šwm-* "to sew"; Khwar. *bcn-* (+ **upa-*) "to mend, to patch up (clothing)"; Yayn. *šiy-*, *šíta* "to sew";[82] OKhot. *cile* (Pl.) "dress", Rōš. *cīl* "large woman's head-dress", Yazγ. *cil* "head-dress".[83]

No Indoeuropean word inherited by Indoiranian attests the meaning "needle". Nevertheless, an Indoeuropean root with the meaning "sharp" was already used in Indoiranian to designate it: IE. **ḱuH-* "sharp" > IIr. **ćuH-* "sharp > needle".

a) Old Indian: Ved. *śū́la-* "pike"; Ved. *sūcí-* "needle" < **śūcí-*, probably due to contamination with the verbal root Ved. *sī́v-* "to sew".[84]
b) Iranian:
   - Old Iranian: YAv. *sūkā-* "needle".[85]
   - Middle and New Iranian: Khot. *suṃjsañu*, MP. *sōzan*, NP. *sōzan*, Bal. *sūčīn*, Parāčī *sīčīn*, Oss. *sūʒīn/soʒīnæ*, Šuγnī *sij* "needle".[86]
c) Other IE. languages: Lat. *culex*, OIr. *cuil* "mosquito"; Arm. *slak'* "pike, dagger".[87]

9. Manufactured textiles

Only two words designating manufactured textiles can be traced back to Indoiranian. The first stems from the Indoeuropean root **drep-* "to cut", which gave the word for "rag" in some Indoeuropean languages and was inherited by IIr. **drap-* "rag". In the Indian and the Iranian groups it yielded the word for "banner", but also other words related to manufactured textiles, such as Ved. *drāpí-* "mantle, garment".

a) Old Indian: Ved. *drāpí-* "mantle, garment", Ved. *drapsá-* "banner".[88]
b) Iranian:
   - Old Iranian: YAv. *drafša-* "banner", YAv. *drafšakauuaṇt-* "having banners".[89]
   - Middle and New Iranian: MP. *drab-* "to wear"; Parth. *drb-* "to put (clothes)"; MP. *drafš*, NP. *drafš* "banner".
c) Other IE. languages: Gr. *drépō* "I cut off"; Latv. *drãpana* "clothing"; Cz. *zdraby* "rags"; Illyr. **drap(p)-* > Gallo-Romance *drappus* "piece of cloth" > Ital. *drappo*, Cat. *drap* "rag".[90]

The second word related to manufactured textiles is that for "mantle": IIr. **atka-*.

a) Old Indian: Ved. *átka-* "mantle".[91]
b) Old Iranian: YAv. *aδka- / at̰ka-* "mantle".[92]

---

[81] Mayrhofer 1992–2001, 1.545.
[82] Cheung 2007, 29.
[83] Bailey 1979 ,103.
[84] Mayrhofer 1992–2001, 2.651, 2.739.
[85] Bartholomae 1904, 1582.
[86] Bailey 1979, 427a.
[87] Pokorny 1959–1969, 626–627.
[88] Mayrhofer 1992–2001, 1.758, 1.754–755.
[89] Bartholomae 1904, 771–772.
[90] Cheung 2007, 77.
[91] Mayrhofer 1992–2001, 1.58.
[92] Bartholomae 1904, 61.

10. Dress, wear

Although it does not belong to the terminology of the materials and processes of preparation of textiles, there is a widespread Indoeuropean root with the meaning "to dress, to wear clothes" which yielded many words for "dress" and "cloth": IE. **u̯es-* "to dress" > IIr. **u̯as-* "to dress".

a) Old Indian: Ved. *vas-* "to dress";[93] Ved. *vásana-* "dress"; Ved. *vástra-* "dress", Ved. *vastravant-* "having a dress"; Ved. *vásman-* "dress", Ved. *vā́sas-* "clothing", Ved. *vāsaḥpalpūlī-* "washer of clothes"; Ved. *upa-vā́sana-* "covering".
b) Old Iranian: Av. *vah-* "to dress";[94] YAv. *°vaŋha-* "dressing", YAv. *vaŋhana-* "dress", YAv. *vastra-* "dress", YAv. *vastran-* "cammer for dresses", YAv. *vastrauuaṇt-* "having a dress".[95]
c) Other IE. languages: Hit. *u̯a-aš-še-iz-zi* "he wears"; Gr. *(F)eîmai* "I dress"; Lat. *vestis* "cloth, garment"; Goth. *wasjan* "to wear".[96]

As Schlerath[97] observed, this root was used as a *figura etymologica* in the Indoiranian formula **u̯astra- / u̯asana- u̯as-* "to dress a dress", as the comparison between Vedic and Avestan demonstrates.[98] However, he did not observed that this formula could be traced back to Indoeuropean, because it is found in many passages in Homer[99] and it is applied to the Moon's dress in *h. Hom.* 32.8,[100] where Gr. *eímata essaménē* is attested. Therefore, we can state that an Indoeuropean formula **u̯es° u̯es-* "to dress a dress" existed and that it referred to gods that dressed the sky.

In other non-Indoeuropean cultures gods dress the sky.[101] As this seems to be a widespread image, the most striking in the Indoeuropean cultures is not that they also attest it, but that they expressed it by means of the same poetic formula.

Notwithstanding, this is not the only formula that this root attests. It was used in a further Indoeuropean formula, IE. **u̯es-* "to dress" + **pei̯ḱ-* "to adorn",[102] which reveals an important fact for the Indoeuropean textile terminology: the existence of textiles with precious stones set, maybe pearls. Indeed, the gods spreading their dress adorned with stars through the nocturnal sky were assimilated to mortals who wear a dress made with pearls. This Indoeuropean image is manifest in the Indoeuropean formula **u̯es-* "to dress" + **pei̯ḱ-* "to adorn", as we see in Vedic, Avestan and Greek:

- Vedic: RV 10.1.6[103] *sá tú vástrāṇy ádha péśanāni vásāno agnír nā́bhā pr̥thivyā́ḥ* "He, Agni, who is wearing now adorned dresses, (is) in the navel of the earth."
- Avestan: Yt 13.3[104] *yim. mazdā̊. vaste. vaŋhanəm. stəhrpaēsaŋhəm. mainiiu.tāštəm* "(the sky), which Mazdā

[93] Mayrhofer 1992–2001, 2.529.
[94] Bartholomae 1904, 1394.
[95] Bartholomae 1904, 1385–1386.
[96] Cheung 2007, 405.
[97] Schlerath 1990.
[98] See for instance RV 10.5.4b *adhīvāsáṁ ... vāvasāné* and Yt 13.3 *vaste. vaŋhanəm. stəhrpaēsaŋhəm.*
[99] Homer, *Il.* 5.905, 16.670, 16.680, 18.517, 23.67; *Od.* 2.3, 4.253, 4.308, 6.228, 7.265, 8.366, 10.542, 11.191, 14.320, 14.516, 15.338, 16.457, 17.203, 17.338, 19.72, 19.218, 20.125, 23.115, 24.59, 24.158; and also *h. Hom.* 6.6.
[100] Allen 1912, 90.
[101] cf. Sum. AN.MA and Akk. *nalbaš šamê* "dressed with the sky", referred to the god Nabû. According to Parpola 1985, 35, this image could have influenced the Harappan representation of gods dressed with the sky.
[102] Jackson 2001; West 2007, 353.
[103] Geldner 1951, 3.123; Schlerath 1968, 84.
[104] Geldner 1886; Kellens 1975, 12; Ichaporia 1999, 26; Malandra 2003, 266–267.

dresses as a dress adorned with stars (/ or with set stars), fashioned by the Spirit"

- Greek: Aischylos, *Pr.* 24[105] *hē poikileímōn nùx* "the Night dressing an adorned dress"; Nonus, *Dion.* 40.416–417 *poikílon eĩma phérōn, túpon aithéros, eikóna kósmou* "dressing an adorned dress, image of the sky, similar to the universe".

As we observe, this Indoeuropean formula makes manifest that, among the Indoeuropeans, the gods of the nocturnal sky were thought as humans who wear a dark dress adorned with some kind of bright or precious stones. As this image can be linguistically reconstructed to a common Indoeuropean period, it must necessarily reflect the existence of this type of textile with precious stones set.

* * *

To summarise, the linguistic reconstruction of the Indoiranian and the Indoeuropean yields many details about the textile terminology. Concerning the materials, it seems that the Indoiranians mainly used wool in their textiles. It is unlikely that they did not use plant fibres. However, no word for them can be traced back to Indoiranian. Conversely two roots for "wool" can be traced back to Indoiranian: a) IIr. **Hu̯l̥HnaH-* "wool", which stems from IE. **h₂u̯l̥h₁neh₂-* "wool" and possibly is applied to the sheep's wool; and b) IIr. **dr̥Hćá-* "wool", which probably designates the goat's wool, but is used for the camel's wool in New Persian. The connection between the wool and the name for "thread" in Indoiranian and Indoeuropean (IE. **dek̂-* "thread, hair" > IIr. **dać-* "thread") remained in some Indoeuropean languages, where this word means "hair".

With regards to the processes and techniques of the textile and other terms related to textile, words for "to stretch (threads / manufactured textile)", "to weave, to knot rugs", "to redden, to colour", "to sew", "needle", "rag", "mantle" and "to dress" can be traced back to Indoiranian.

At least two Indoeuropean roots inherited by Indoiranian were applied to the technique of braiding thread. The first, IE. **derbʰ-* "to braid", became specialized in Indoiranian and designated a type of sewing, maybe the Nålebinding, sprang, or the ply-splitting. The second, IE. **kert-*, designated some kind of wickerwork, but developed the meaning "to spin" in some Vedic words. However, an Indoiranian meaning "to spin" cannot be reconstructed only by means of Ved. *kart-* "to spin" and Khwar. *'nknc* "needle". Moreover, Ved. *káṭa-* "mat" is closer to the words related to the wickerwork of the rest of Indoeuropean languages than to Ved. *kart-* "to spin", and demonstrates that this meaning developed secondarily. Those Indoeuropean roots with the meaning "to twist" also developed secondarily and independently the meaning "to spin" in Indian and Iranian, so that this meaning did not belong to the Indoiranian textile terminology.

Conversely, the Indoeuropean root **ten-* "to stretch" was related to the textile terminology in Indoiranian, although it did not designate the same process in the Indian and the Iranian groups. Nevertheless, it was part of an Indoeuropean formula "Night" + IE. **ten-*, referring to the dress of the Night, spread through the sky.

Of the two roots with the meaning "to weave", IE. **h₂eu̯-* and its extension IE. **(h₂)u̯ebʰ-*, only the second can be traced back to Indoiranian. It could have been applied to the manufacture of knotted rugs and is also connected with the poetic composition in some Indoeuropean languages.

Concerning dyes, the Indoeuropean root **(s)reg-* "to redden, to colour", which is continued by

---

[105] Jackson 2001, 123.

IIr. *rag- "to redden, to colour", demonstrates that red was the first colour used as a dye among the Indoeuropeans. This is confirmed by the archaeological evidence in Europe and the Near East and is connected with the selection of sheep with pale to white wool.[106]

Two common words for "sewing" and one for "needle" can be traced back to Indoiranian. On one hand, we can reconstruct IIr. *si̯au̯H-/*si̯uH-* "to sew", which stems from IE. *si̯eu̯h$_1$-/*si̯uh$_1$-* (> *suh$_1$-*) "to sew", and IIr. *ćai̯H-* "to sew". There is apparently no difference in meaning between both. On the other hand, the word for "needle", IIr. *ćuH-*, surely stems from the Indoeuropean root *ḱuH-* "sharp".

Two common words for "rag" and "mantle", IIr. *drap-* "rag" and IIr. *atka-* "mantle", also stem from Indoiranian. The first stems from the Indoeuropean root *drep-* "to cut", while the second is only found in the Indoiranian group.

Finally, there was an Indoeuropean root *u̯es-* "to dress", inherited by IIr. *u̯as-* "to dress", which yielded some terms for "dress" and "cloth" in the Indian and the Iranian languages, but was also used in two Indoeuropean poetic formulas. The first, IE. *u̯es° u̯es-* "to dress a dress", was applied to gods who dressed the sky. The second, IE. *u̯es-* "to dress" + *pei̯ḱ-* "to adorn", was applied to gods dressing the nocturnal sky and reveals the existence of textiles adorned with some kind of bright or precious stones among the Indoeuropeans.

Abbreviations

Acc.	Accusative
Akk.	Akkadian
Alb.	Albanian
Arm.	Armenian
Av.	Avestan
AVP	Atharvaveda Saṁhitā Paippalāda
AVŚ	Atharvaveda Saṁhitā Śaunakīya
Bal.	Baločī
Cat.	Catalonian
CSogd.	Christian Sogdian
Cz.	Czech
D.	Digorian
Dor.	Doric
Fem.	Feminine
Gen.	Genitive
Goth.	Gothic
Gr.	Greek
h. Hom.	*Hymni Homerici*
Hit.	Hittite
I.	Ironian
IE.	Indoeuropean
IIr.	Indoiranian
Il.	*Ilias*
Illyr.	Illyrian
Ital.	Italian
Khot.	Khotanese
Khwar.	Khwaresmian

[106] Barber 1975, 296.

Kurd.	Kurdish
Lat.	Latin
Latv.	Latvian
Lith.	Lithuanian
MHG.	Middle High German
MnārUp	Mahānārāyaṇa Upaniṣad
MMP.	Manichaean Middle Persian
MP.	Middle Persian
Nom.	Nominative
NP.	New Persian
OAv.	Old Avestan
OCS.	Old Church Slavonic
Od.	*Odysseia*
OE.	Old English
OHG.	Old High German
OIr.	Old Irish
OIcel.	Old Icelandic
OKhot.	Old Khotanese
OP.	Old Persian
OPrus.	Old Prusian
OSax.	Old Saxon
Oss.	Ossetian
Parth.	Parthian
Phl.	Pahlavi
Pl.	Plural
Pr.	*Prometeus*
Rōš.	Rōšānī
Rus.	Russian
RV	R̥gveda
Sariq.	Sariqōlī
Serbocr.	Serbocroatian
Sing.	Singular
Sogd.	Sogdian
Sum.	Sumerian
ŚBM	Śatapatha Brāhmaṇa Mādhyandina
TĀ	Taittirīya Āraṇyaka
Toch.	Tocharian
Ved.	Vedic
Y	Yasna
Yayn.	Yaynōbī
YAv.	Young Avestan
Yazy.	Yazyulāmī
Yt	Yašt
YV,TS	Yajurveda Taittirīya Saṁhitā
YV,VS	Yajurveda Vājasaneyi Saṁhitā

Bibliography

Adams, D. Q. & Barber, E. J. W. (1997) "Textile" and "Textile Preparation", in Mallory, J. P. & Adams, D. Q. (eds.). *Encyclopedia of Indo-european culture*. London, Fitzroy Dearborn, 569–574.

Allen, Th. W. (1912) *Homeri Opera. Tomus V. Hymnos, Cyclum, Fragmenta, Margiten, Batrachomyomachiam, Vitas continens.* Oxford – New York, Oxford University Press.

Bader, F. (1990) Le liage, la peausserie et les poètes-chanteurs Homère et Hésiode: la racine *seh_2- 'lier'. *Bulletin de la Société de Linguistique* 85, 1–59.

Bailey, H. W. (1979) *Dictionary of Khotan Saka*. Cambridge, Cambridge University Press.

Barber, E. J. W. (1975) The PIE notion of cloth and clothing. *Journal of Indo-European Studies* 3–4, 294–320.

Bartholomae, C. (1904) *Altiranisches Wörterbuch*. Strassburg, K. J. Trübner.

Campanile, E. (1993) Reflexions sur la réconstruction de la phraséologie poétique indo-européenne. *Diachronica* 10, 1–12.

Cheung, J. (2007) *Etymological Dictionary of the Iranian verb*. Leiden – Boston, Brill.

Christol, A. (1986) De φόως ἐρέων à *ā dyām tanoṣi*. Note de phraséologie. *Bulletin de la Société de Linguistique* 81, 181–204.

Eichner, H. (1973) Die Etymologie von heth. *mehur. Münchener Studien zur Sprachwissenschaft* 31, 53–107.

Geldner, K. F. (1886) *Avesta. The sacred books of the Parsis*. Stuttgart, Kohlhammer.

Geldner, K. F. (1951) *Der Rig-veda aus dem Sanskrit ins Deutsche übersetzt und mit einem laufendem Kommentar versehen*. Cambridge, Massachusetts, Harvard University Press.

Ichaporia, P. R. (1999) Fravardin Yasht – Translation of the Text. *Journal of the K. R. Cama Oriental Institute* 63, 25–50.

Jackson, P. (2001) A Graeco-Avestan parallel regarding the adornment of the sky with stars. *Indogermanische Forschungen* 106, 122–125.

Kellens, J. (1975) *Fravardīn Yašt (1–70). Introduction, édition et glossaire*. Wiesbaden, Reichert.

Kronasser, H. (1967) Heth. *we-/wa- : ḫu-*. In Meid, W. (ed.), *Beiträge zur Indogermanistik und Keltologie, Julius Pokorny zum 80. Geburtstag gewidmet*. Innsbruck.

Lüders, H. (1959) *Varuna*. Göttingen, Vandenhoeck & Ruprecht.

Malandra, W. W. (2003) *Gōhr ī āsman*: A Problem in Avestan Cosmology. In Adhami, S. (ed.). *Paitimāna, Essays in Iranian, Indo-European and Indian Studies in Honor of Hanns-Peter Schmidt*. Costa Mesa, California, Mazda, 266–274.

Mayrhofer, M. (1992–2001) *Etymologisches Wörterbuch des Altindoarischen*. Heidelberg, Carl Winter-Universitätsverlag.

Morgenstierne, G. (1974) *Etymological Vocabulary of the Shughni group*. Wiesbaden, Harrassowitz.

Pokorny, J. (1959–1969) *Indogermanisches etymologisches Wörterbuch*. Bern, Francke.

Puhvel, J. (1987) *Comparative Mythologie*. Baltimore – London, The Johns Hopkins University Press.

Schlerath, B. (1968) *Awesta-Wörterbuch. Vorarbeiten: I. Index locorum zur Sekundärliteratur des Awesta. II. Konkordanz*. Wiesbaden, Harrassowitz.

Schlerath, B. (1990) Indoiranisch *vástra-* ntr. "Gewand, Kleidung". *Orientalia-NS* 59, 251–254.

Skjærvø, P. O. (2005) Poetic and Cosmic Weaving in Ancient Iran. Reflections on Avestan *vahma* and Yasna 34.2. In Haug, D. & Welo, E. (eds.). *Haptačahaptāitiš. Festschrift for Fridrik Thordarson*. Oslo, Novus, 267–279.

Tuck, A. (2006) Singing the rug: patterned textiles and the origins of Indo-European metrical poetry. *American Journal of Archaeology* 110, 539–550.

Von der Mühll, P. (1962) *Homeri Odyssea*. Stutgard, Teubner.

Watkins, C. (1995) *How to kill a dragon: aspects of Indo-European poetics*. Oxford – New York, Oxford University Press.

Weber, A. (1849). *The Çatapatha-Brāhmaṇa in the Mādhyandina-Çākhā with extracts from the commentaries of Sāyaṇa, Harisvāmin and Dvivedānga*. Berlin.

West, M. L. (2007) *Indo-European poetry and myth*. Oxford – New York, Oxford University Press.

Whitney, W. D. & Lanman, C. R. (1905) *Atharvaveda Saṁhitā*. Cambridge, Massachussets, Harvard University Press.